T0180722

Lecture Notes in Computer Science 12453

More information about this series at http://www.springer.com/series/7407

Meikang Qiu (Ed.)

Algorithms and Architectures for Parallel Processing

20th International Conference, ICA3PP 2020
New York City, NY, USA, October 2–4, 2020
Proceedings, Part II

 Springer

Editor
Meikang Qiu 🆔
Columbia University
New York, NY, USA

ISSN 0302-9743   ISSN 1611-3349 (electronic)
Lecture Notes in Computer Science
ISBN 978-3-030-60238-3   ISBN 978-3-030-60239-0 (eBook)
https://doi.org/10.1007/978-3-030-60239-0

LNCS Sublibrary: SL1 – Theoretical Computer Science and General Issues

This Springer imprint is published by the registered company Springer Nature Switzerland AG
The registered company address is: Gewerbestrasse 11, 6330 Cham, Switzerland

Preface

This three-volume set contains the papers presented at the 20th International Conference on Algorithms and Architectures for Parallel Processing (ICA3PP 2020), held during October 2–4, 2020, in New York, USA.

There were 495 submissions. Each submission was reviewed by at least 3 reviewers, and on the average 3.5 Program Committee members. The committee decided to accept 147 papers. We will separate the proceeding into three volumes: LNCS 12452, 12453, and 12454. Yielding an acceptance rate of 29%.

ICA3PP 2020 was the 20th in this series of conferences started in 1995 that are devoted to algorithms and architectures for parallel processing. ICA3PP is now recognized as the main regular event of the world that is covering the many dimensions of parallel algorithms and architectures, encompassing fundamental theoretical approaches, practical experimental projects, and commercial components and systems. As applications of computing systems have permeated in every aspect of daily life, the power of computing systems has become increasingly critical. This conference provides a forum for academics and practitioners from countries around the world to exchange ideas for improving the efficiency, performance, reliability, security, and interoperability of computing systems and applications.

Following the traditions of the previous successful ICA3PP conferences held in Hangzhou, Brisbane, Singapore, Melbourne, Hong Kong, Beijing, Cyprus, Taipei, Busan, Melbourne, Fukuoka, Vietri sul Mare, Dalian, Japan, Zhangjiajie, Granada, Helsinki, Guangzhou, and Melbourne, ICA3PP 2020 was held in New York, USA. The objective of ICA3PP 2020 is to bring together researchers and practitioners from academia, industry, and governments to advance the theories and technologies in parallel and distributed computing. ICA3PP 2020 will focus on three broad areas of parallel and distributed computing, i.e., Parallel Architectures and Algorithms (PAA), Parallel computing with AI and Big Data (PAB), and Parallel computing with Cyberseucrity and Blockchain (PCB).

We would like to thank the conference sponsors: Springer LNCS, Columbia University, North America Chinese Talents Association, and Longxiang High Tech Group Inc.

October 2020   Meikang Qiu

Organization

Honorary Chairs

Sun-Yuan Kung   Princeton University, USA
Gerard Memmi   Télécom Paris, France

General Chair

Meikang Qiu   Columbia University, USA

Program Chairs

Yongxin Zhu   Shanghai Advanced Research Institute, China
Bhavani Thuraisingham   The University of Texas at Dallas, USA
Zhongming Fei   University of Kentucky, USA
Linghe Kong   Shanghai Jiao Tong University, China

Local Chair

Xiangyu Gao   New York University, USA

Workshop Chairs

Laizhong Cui   Shenzhen University, China
Xuyun Zhang   The University of Auckland, New Zealand

Publicity Chair

Peng Zhang   Stony Brook SUNY, USA

Finance Chair

Hui Zhao   Henan University, China

Web Chair

Han Qiu   Télécom-ParisTech, France

Steering Committee

Yang Xiang (Chair)   Swinburne University of Technology, Australia
Weijia Jia   Shanghai Jiao Tong University, China

Yi Pan	Georgia State University, USA
Laurence T. Yang	St. Francis Xavier University, Canada
Wanlei Zhou	University of Technology Sydney, Australia

Technical Committee

Dean Anderso	Bank of America Merrill Lynch, USA
Prem Chhetri	RMIT, Australia
Angus Macaulay	The University of Melbourne, Australia
Paul Rad	Rackspace, USA
Syed Rizvi	Penn State University, USA
Wei Cai	Chinese University of Hong Kong, Hong Kong, China
Abdul Razaque	University of Bridgeport, USA
Katie Cover	Penn State University, USA
Yongxin Zhao	East China Normal University, China
Sanket Desai	San Jose State University, USA
Weipeng Cao	Shenzhen University, China
Suman Kumar	Troy University, USA
Qiang Wang	Southern University of Science and Technology, China
Wenhui Hu	Peking University, China
Kan Zhang	Tsinghua University, China
Mohan Muppidi	UTSA, USA
Wenting Wei	Xidian University, China
Younghee Park	San Jose State University, USA
Sang-Yoon Chang	Advanced Digital Science Center, Singapore
Jin Cheol Kim	KEPCO KDN, South Korea
William de Souza	University of London, UK
Malik Awan	Cardiff University, UK
Mehdi Javanmard	Rutgers University, USA
Allan Tomlinson	University of London, UK
Weiwei Shan	Southeast University, China
Tianzhu Zhang	Télécom Paris, France
Chao Feng	National University of Defense Technology, China
Zhong Ming	Shenzhen University, China
Hiroyuki Sato	The University of Tokyo, Japan
Shuangyin Ren	Chinese Academy of Military Science, China
Thomas Austin	San Jose State University, USA
Zehua Guo	Beijing Institute of Technology, China
Wei Yu	Towson University, USA
Yulin He	Shenzhen University, China
Zhiqiang Lin	The University of Texas at Dallas, USA
Xingfu Wu	Texas A&M University, USA
Wenbo Zhu	Google Inc., USA
Weidong Zou	Beijing Institute of Technology, China
Hwajung Lee	Radford University, USA

Yuxuan Jiang	The Hong Kong University of Science and Technology, Hong Kong, China
Yong Guan	Iowa State University, USA
Chao-Tung Yang	Tunghai University Taiwan, China
Zonghua Gu	Zhejiang University, China
Gang Zeng	Nagoya University, Japan
Hui Zhao	Henan University, China
Yong Zhang	The University of Hong Kong, Hong Kong, China
Hanpin Wang	Peking University, China
Yu Hua	Huazhong University of Science and Technology, China
Yan Zhang	University of Oslo, Norway
Haibo Zhang	University of Otago, New Zealand
Hao Hu	Nanjing University, China
Zhihui Du	Tsinghua University, China
Jiahai Yang	Tsinghua University, China
Fuji Ren	Tokushima University, Japan
Long Fei	Google Inc., USA
Tianwei Zhang	Nanyang Technological University, Singapore
Ming Xu	Hangzhou Dianzi University, China
Golden Richard	Louisiana State University, USA
Virginia Franqueira	University of Derby, UK
Haoxiang Wang	Cornell University, USA
Jun Zhang	Shenzhen University, China
Xinyi Huang	Fujian Normal University, China
Debiao He	Wuhan University, China
Vijayan Sugumaran	Oakland University, USA
Ximeng Liu	Singapore Management University, Singapore
Zhan Qin	The University of Texas at San Antonio, USA
Dalei Wu	The University of Tennessee at Chattanooga, USA
Kathryn Seigfried-Spellar	Purdue University, USA
Jun Zheng	New Mexico Tech, USA
Paolo Trunfio	University of Calabria, Italy
Kewei Sha	University of Houston - Clear Lake, USA
David Dampier	The University of Texas at San Antonio, USA
Richard Hill	University of Huddersfield, UK
William Glisson	University of South Alabama, USA
Petr Matousek	Brno University of Technology, Czech Republic
Javier Lopez	University of Malaga, Spain
Dong Dai	Texas Tech University, USA
Ben Martini	University of South Australia, Australia
Ding Wang	Peking University, China
Xu Zheng	Shanghai University, China
Nhien An Le Khac	University College Dublin, Ireland
Shadi Ibrahim	Inria Rennes – Bretagne Atlantique, France
Neetesh Saxena	Bournemouth University, UK

Contents – Part II

Poster Paper

Parallel Computing with AI and Big Data (PAB)

Distributing Data in Real Time Spatial Data Warehouse

Wael Hamdi[1(✉)] and Sami Faiz[2]

[1] Manouba University, Campus Universitaire de la Manouba, 2010 Manouba, Tunisia
hamdiwael2009@gmail.com
[2] LTSIRS Laboratory, BP 37 Le Belvedere, 1002 Tunis, Tunisia
sami.faiz@insat.rnu.tn

Abstract. Nowadays, there are many real-time spatial applications like location-aware services and traffic monitoring and the need for real time spatial data processing becomes more and more important. As a result, there is a tremendous amount of real-time spatial data in real-time spatial data warehouse. The continuous growth in the amount of data seems to outspeed the advance of the traditional centralized real-time spatial data warehouse. As a solution, many organizations use distributed real-time spatial data warehouse (DRTSDW) as a powerful technique to achieve OLAP (On Line Analytical Processing) analysis and business intelligence (BI). Distributing data in real time data warehouse is divided into two steps: partitioning data and their allocation into sites. Several works have proposed many algorithms for partitioning and allocation data. But with a huge amount of real-time spatial data generated, the system performance degrades rapidly, especially in overload situations. In order to deal with this volumetry and to increase query efficiency, we propose a novel approach for partitioning data in real-time spatial data warehouse to find the right number of clusters and to divides the RTSDW into partitions using the horizontal partitioning. Secondly, we suggest our allocation strategy to place the partitions on the sites where they are most used, to minimize data transfers between sites. We have evaluated those proposed approaches using the new TPC-DS (Transaction processing performance council, http://www.tpc.org, 2014) benchmark. The preliminary results show that the approach is quite interesting.

Keywords: Distributed real-time spatial data warehouse · Data partitioning approach · Data allocation · Real time transactions

1 Introduction

These recent years, there is a huge increase in the use of spatio-temporal applications. Such applications result dynamic environments. Several works dedicated to spatio-temporal query processing (e.g., see [1,4,6,8,9,11,16], [20]) rely mainly on the ability of indexing and storing spatio-temporal data.

© Springer Nature Switzerland AG 2020
M. Qiu (Ed.): ICA3PP 2020, LNCS 12453, pp. 3–13, 2020.
https://doi.org/10.1007/978-3-030-60239-0_1

Today, the issue of real-time seems increasingly clear. Researching on the real-time spatial analysis becomes a hot topic. This calls for new real-time spatio-temporal query processing algorithms. Indeed, numerous research efforts are devoted to real-time spatial data warehouses management. In fact, the real time spatial data warehouses (RTSDW) is benefiting from a combination of mechanisms developed in Real Time Systems (RTS) and classic spatial data warehouse. In fact, RTSDW allows the management of huge amounts of spatial information with respect to time constraints, including their goal is to minimize the number of transactions that miss their deadline.

But, the gradual increase in complex applications requires the use of real-time data warehouse for handling data as well as treatments over multiple nodes which appears clearly insufficient. As a solution, the distributed real time data warehouse was entered to meet these requirements and complex distributed applications. Several works are proposed for distributing data in data warehouse.

In our work, we propose two steps for distributing the traditional centralized RTSDWH: data partitioning and their allocation on sites. Data partitioning is a fragmentation of a logical database into distinct independent units [14]. Whereas, data allocation is a physical fragmentation of different partitions over different geographically distant sites.

This paper is organized as follows: In Sect. 2, we present the system model that we consider. Then, in Sect. 3, we discuss the related works. Our contribution for distributing data in RTSDW is introduced in detail in Sect. 4 where we present our partitioning approach and our data allocation strategy. We conclude this paper by a discussion of this work and then we present our future works.

2   Real Time Spatial Data Warehouse Model

In this section, we give an overview of real-time spatial data model and transaction model.

2.1   Real-Time Spatial Data Model

Stored data in real-time spatial data warehouse are from heterogeneous sources. We propose the use of ETL (Extract-Transform-Load) process as follows:

- Data extraction: extracts data from heterogeneous data sources.
- Data transformation: transforms the data for storing it in the proper format or structure for the purposes of querying and analysis.
- Data loading: loads it into the final target (data warehouse).

A real-time spatial data have the ability to change their locations continuously. Thus, the arrival of a new location information about the data, say p, at some time t2 ($t2 > t1$) may result in expiring the previous location information of p at time t1. This is in contrast to traditional data where data are expired only after its deadline as it becomes in the system [10]. The validity of a spatial data depends on localization and the time [5].

$$validity(d_i) = validS(d_i) * validT(d_i) \tag{1}$$

where $validS(d_i) = 1$ if d_i does not change its location else $validS(d_i) = 0$ and $validS(d_i) = 1$ if d_i doesn't pass its deadline $validT(d_i) = 1$ if $t_{current} < t_{AV}$ else $validT(d_i) = 0$) where $t_{current}$ denotes the current time of our system and t_{AVI} is defined as follows:

$$t_{AVI} = timestamp + validity \tag{2}$$

d_i, as a real time spatial data, is valid when $validity(d_i) = 1$ ($validS(d_i) = 1$ and $validT(d_i) = 1$).

2.2 Transaction Model

In RTSDWH, there are two types of transactions: global and local. Global transactions are distributed real-time transactions executed at more than one site, whereas the local transactions are executed at the originating site (parent site) only.

Each local site receives update transactions (write-only) and query transactions (read-only). Also it supports its own base tables, and derived tables (views).

Update Transaction: Update transactions, like their name indicates, update the values of real-time data in order to reflect the real world status. They are executed periodically and provided new values to real-time data. For each update transactions Ti, the deadline of Ti is estimated to be $(r_i + P_i)$ where r_i is its release time and P_i is its period.

Query Transaction: Query transactions, representing user requests, arrive aperiodically and may read data. In order to guarantee QoS, in each query transaction (user transaction) user should provide two parameters, acceptable response time delay and acceptable result staleness.

3 Related Works

In distributed data warehouse (DDWH), data are physically distributed across multiple DWs in multiple processors at different sites. Distributing data in data warehouse provides a coherent and single view of data. As a result, it facilitates the policy and decision makers. Indeed, several works proposed new approaches for distributing data warehouse and uses two design approaches: Top-Down or Bottom-Up. The Top-Down approach is used for a very large system where the centralized data warehouse is built first. Then, the data marts are then created.

Whereas, the Bottom-Up approach aims to integrate local data warehouse into a global data warehouse.

In this context, Inmon, in [7], proposed three types of distributed data warehouses as follow:

- Local and Global Data Warehouses: Data are distributed geographically or over multiple sites. In this case, there is a local data warehouse and a global data warehouse. The local data warehouse represents data at a remote site, and the global data warehouse represents the global volume of data that is integrated across the business.
- The technologically distributed data warehouse: Data in the data warehouse environment become more and more important. As a solution, the volume of data is distributed over multiple processors. Logically there is a single data warehouse, but physically there are many data warehouses.
- The independently evolving distributed data warehouse: The data warehouse environment grows up in an uncoordinated manner-first one data warehouse appears, then another.

In [17], authors proposed a new approach, named as "Two Tier Data Warehouse", which is a combination of a centralized data warehouse and decentralized data marts.

Noaman et al. in [12] proposed an architecture for DDW. It uses a Top-Down approach and presents two fundamental issues: fragmentation and allocation of the fragment to various sites. An extension of this work was proposed by author in [12] by describing the functionality of DDW architecture components, by giving definition of relational data model for DW and by giving a horizontal fragmentation algorithm.

Two other approaches are proposed by Zhou et al. in [18] and by Bernardino et al. in [2] for distributing data in data warehouse. The first approach designed a Hierarchical distributed data warehouse (HDDW) which integrates the local data marts into a hierarchy. It used the Bottom-Up approach for building the global data warehouse. The second approach designed a new technique called data warehouse stripping (DWS) which is a round robin data partitioning approach for relational data warehouse. The limitation of these two approaches (HDDW and DWS) is that they are not suitable in data warehouse with big dimensions.

Therefore, as a solution, the authors proposed a new method called selective loading to deal with data warehouses with big dimensions in DWS systems. But this method faces a problem when the data size will be bigger over time and the round-robin partitioning approach must be applied again.

All these approaches are unable to manage spatial data. Thus, in [3], Gorawski et al. proposed a distributed spatial data warehouse system designed for storing and analyzing a wide range of spatial data. This system is based on a new model called the cascaded star model and use data stream which is not suitable for distributed data warehouse.

Table 1 compares the different approaches cited in related works. This comparison highlights their weak sides. As we show in Table 1, all these approaches

Table 1. Related works.

	Big dimensions	Real time processing	Spatial data	Freshness of data
Local and global DW	No	No	No	No
Technologically distributed DW	No	No	No	No
Independently evolving DDW	No	No	No	No
Two Tier DW	No	No	No	No
Noaman et al. [12]	No	No	No	No
HDDW	No	No	No	No
DWS	No	No	No	No
Selective loading	Yes	No	No	No
Gorawski et al. [3]	No	No	Yes	No

are not suitable for distributing data in real-time spatial data warehouse. In the next section, we present our approach for distributing data in RTSDWH.

4 Data Distributing in RTSDWH

Our approach uses a Top-Down design and presents three fundamental steps: partitioning data, allocation of the partition to various sites and variation computing as shown in Fig. 1.

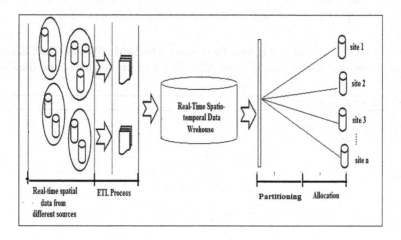

Fig. 1. Data distributing in RTSDWH.

4.1   Partitioning Data in RTSBD

Data partitioning is a fragmentation of a logical database into distinct independent units. It divides the RTSDWH into partitions equally using the horizontal partitioning. This phase is divided into seven steps that are detailed in the following sections:

- Predicate Usage Matrix (PUM) construction.
- Query Usage Matrix (QUM) construction.
- Predicate Interrelationship Matrix (PIM) construction.
- Predicate clustering with the G-means method.
- Frequency Usage Matrix (FUM) construction.
- Construction of the Matrix of the Extraction Costs (MEC).
- Construction of the Matrix of the Access Costs (MAC).

Predicate Usage Matrix (PUM) Construction. Similar to authors in [5], we use a predicate usage matrix, PUM, that defines predicate usage of each query q_i in the workload W. PUM lines represent workload queries and columns represent predicates. PUM (i, j) is set to one if q_i includes predicate p and to zero otherwise.

Query Usage Matrix (QUM) Construction. QUM include the uses of the queries by the sites. QUM lines represent workload queries and columns represent sites. QUM (i, j) is set to one if q_i is used by a site and to zero otherwise.

Predicate Interrelationship Matrix (PIM) Construction. For the generation of this matrix, we use the two matrixes PUM and QUM. PIM lines represent workload queries and columns present predicates and sites.

Predicate Clustering with the G-means Method. We use the Gaussian-means (G-means) algorithm [1] to classify the set of predicates of the dimension tables.

Algorithm 1 G-means(PIM, α) algorithm

Inputs
PIM : Predicate Interrelationship Matrix
A confidence level α which is used in the statistical
test
Begin
1: Let C be the initial set of centers (usually C ← {x̄}).
2: C ← kmeans(C, PIM).
3: Let {xi|class(xi) = j} be the set of data points assigned to center cj
4: Use a statistical test to detect if each {xi|class(xi) = j} follow a Gaussian distribution (at confidence level α).
5: If the data look Gaussian, keep cj. Otherwise replace cj with two centers split from cj.
6: Repeat from step 2 until no more centers are added.
End.

This algorithm takes as input the predicate interrelationship matrix PIM. Then, it divides the set of predicates into clusters. It starts with a small number of k-means [4] centers (k = 1), and grows the number of centers. Each iteration of this algorithm splits into two those centers whose data appear not to come from a Gaussian distribution. A statistical test decides whether to split a k-means center into two centers.

Frequency Usage Matrix (FUM) Construction. FUM include the frequencies of use of the fragments on all different sites. FUM lines present different fragments and columns present sites. FUM(i,j) includes the frequency of use of a fragment i by the users of a site j.

Construction of the Matrix of the Extraction Costs (MEC). MEC include the costs of loading of the different fragments in every site. MEC lines present different fragments and columns present sites. MEC(i, j) includes the extraction cost of the fragment i in the site j.

Construction of the Matrix of the Access Costs (MAC). MAC include the access costs of the different fragments in every site. MAC lines present different fragments and columns present sites. MAC(i, j) includes the access cost of the fragment i in the site j.

4.2 Data Allocation for Distributed Real-Time Spatial Data Warehouse (DRTSDWH)

In this section, we present our data allocation strategy for RRTSDWH processing. Indeed, fragment assignment on sites is decided based on the expected origin of the queries that served for fragmentation. The goal is to place the fragments on the sites where they are most used, to minimize data transfers between sites. In this context, two thresholds are proposed as follow:

- *Threshold_AC*: Threshold of the access costs. Below this threshold, remote access are acceptable.
- *Threshold_EC*: Threshold of the extraction costs (*Threshold_EC* > *Threshold_AC*). Below this threshold, we propose the use of full replication where all data in the fragment are replicated on the site. Otherwise, we choose a partial replication where only the most accessed data can be replicated on the site.

Our allocation strategy is performed off-line as shown in Algorithm 2. During this level, we identify, for each site, the most accessed partitions from a given most frequent OLAP queries.

Algorithm 2 Data allocation algorithm (Thresold_AC, Threshold_EC, Q$_j$,)

```
Inputs:
Let P₁, P₂... Pₙ be the initial set of partitions
nbSites
i ∈ [1, n]
Begin
    For j from 1 to nbSites do
            If MAC(i,j)< Threshold_AC then
            GetAccessedPartition(Qⱼ)
            Else if MEC(i,j)< Threshold_SC then
                FullReplication(Pᵢ)
        Else
                PartialReplication(Pᵢ)
        End if
        End If
    End
End.
```

4.3   Variation Computing

This step is made for the control and the analysis of the variations of the distribution parameters.

- Case 1: If a new transaction is introduced on the system, it is necessary to verify its impact on the diagram of fragmentation of the RTSDWH.
- Case 2: When a new site is added, we must take into account the its needs.
- Case 3: When a site is not more operational, a reorganization of the RTS-DWH's distribution is necessary.
- Case 4: If the costs of loading of a fragment increase, it risks to damage the performances aimed by the retained distribution diagram.
- Case 5: When the costs of access increase, it is necessary to study the impact of this increase on the global performance of the system.

5   Experiments

To assess the performance of our novel approach during query execution, we implemented a simulator in JAVA language; the queries from the TPC-DS Benchmark were run.

Experiment 1: Execution time Evaluating.

We compare execution time in DRTSDWH with RTSDWH.

Results show that the execution time using our proposed approach outperform the standard centralized RTSDW. Therefore, the applicability of our approach led to the average diminution of the execution time by 27%. The simulation results shown in Fig. 2 confirm this conclusion.

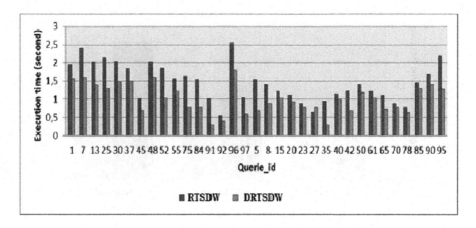

Fig. 2. Execution time comparison between RTSDW and DRTDW.

Experiment 2: Average Retention Time Evaluating. The average retention time (ART), for a given workload W, is the average time between the arrival and the execution of all queries $q_i \in W$ processed by the system; ART is given by the following formula:

$$ART = \frac{\sum_{qi \in W} et_{qi} - t_{qi}}{|W_q|} \tag{3}$$

where :

- et_{qi} is the execution time of the query qi
- t_{qi} is the arrival time of the query qi.

In our system, there are three loads: high, medium and low. Simulation parameters are given in Table 2.

Table 2. Simulation parameters

	High	Medium	Low
Arrival numbers of queries/second	2	5	10
Arrival numbers of updates/second	5	8	12

Simulation results show that ART is better when using our proposed approach under various loads than when using RTSDW, the enhancement is about 10%. (cf., Fig. 3).

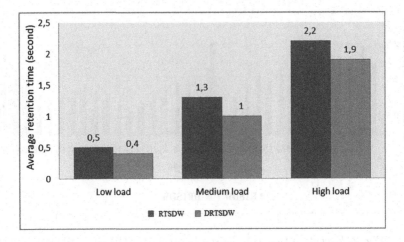

Fig. 3. Execution time comparison between RTSDW and DRTDW.

6   Conclusion

The main objective of our work that we have presented in this paper is to improve the system performances in RTSDW. Therefore, we enhance the quality of service (QoS) in RTSDW. So, we presented an efficient horizontal partitioning approach in the context of RTSDW in which we uses the G-means-based fragmentation approach to define automatically the initial partitions. Then, we focus on how to place those fragments on the sites where they are most used.

We are interested in the following two directions for future work. First, we are considering to integrate our materialized view management system in our proposed architecture in [3] called DDFCSA-DRTSDWH (Distributed Double Feedback Control Scheduling Architecture in Distributed Real Time Spatio Temporal Data warehouse)and to evaluate its performance. Second, we intend to focus on others techniques for query optimization to ensure a high QoS for DRTSDW such as materialized views; an effective method to respect such a time constraint and to minimize the overall execution time in a DRTSDW. We expect to address these issues in future papers.

References

1. Bai, L., Lin, Z., Xu, C.: Spatiotemporal operations on spatiotemporal XML data using XQuery. In: 2016 12th International Conference on Natural Computation, Fuzzy Systems and Knowledge Discovery (ICNC-FSKD), pp. 1278–1282. IEEE (August 2016)
2. Bernardino, J., Madeira, H.: Experimental evaluation of a new distributed partitioning technique for data warehouses. In: International Database Engineering and Applications Symposium, pp. 312–321 (2001)
3. Gorawski, M., Malczok, R.: Distributed spatial data warehouse indexed with virtual memory aggregation tree. In: STDBM, pp. 25–32 (2004)

4. Hadjieleftheriou, M., Kollios, G., Gunopulos, D., Tsotras, V.J.: On-line discovery of dense areas in spatio-temporal databases. In: Hadzilacos, T., Manolopoulos, Y., Roddick, J., Theodoridis, Y. (eds.) SSTD 2003. LNCS, vol. 2750, pp. 306–324. Springer, Heidelberg (2003). https://doi.org/10.1007/978-3-540-45072-6_18
5. Hamdi, S., Bouazizi, E., Faiz, S.: A speculative concurrency control in real-time spatial big data using real-time nested spatial transactions and imprecise computation. In: 2017 IEEE/ACS 14th International Conference on Computer Systems and Applications (AICCSA), pp. 534–540. IEEE (October 2017)
6. Huang, C.Y., Liang, S.H.: LOST-Tree: a spatio-temporal structure for efficient sensor data loading in a sensor web browser. Int. J. Geogr. Inf. Sci. **27**(6), 1190–1209 (2013)
7. Inmon, W.H.: Building the Data Warehouse. Wiley, Hoboken (2005)
8. Iwerks, G.S., Samet, H., Smith, K.: Continuous k-nearest neighbor queries for continuously moving points with updates. In: Proceedings of the 29th International Conference on Very Large Data Bases, vol. 29, pp. 512–523. VLDB Endowment (September 2003)
9. Lee, M.L., Hsu, W., Jensen, C.S., Cui, B., Teo, K.L.: Supporting frequent updates in R-trees: a bottom-up approach. In: Proceedings of the 29th International Conference on Very Large Data Bases, vol. 29, pp. 608–619. VLDB Endowment (September 2003)
10. Mokbel, M.F., Xiong, X., Aref, W.G.: SINA: scalable incremental processing of continuous queries in spatio-temporal databases. In: Proceedings of the 2004 ACM SIGMOD International Conference on Management of Data, pp. 623–634. ACM (June 2004)
11. Mokbel, M.F., Xiong, X., Aref, W.G., Hambrusch, S.E., Prabhakar, S., Hammad, M.A.: PALACE: a query processor for handling real-time spatio-temporal data streams. In: Proceedings of the 13th International Conference on Very Large Data Bases, vol. 30, pp. 1377–1380. VLDB Endowment (August 2004)
12. Noaman, A.Y., Barker, K.: A horizontal fragmentation algorithm for fact relation in a distributed data warehouse. In: Proceedings of the 8th International Conference on Information and Knowledge Management, CIKM 1999, pp. 154–161 (November 1999)
13. Noaman, A.Y., Barker, K.: Distributed data warehouse architectures. J. Data Warehouse. **2**(2), 37–50 (1997)
14. Phansalkar, S., Ahirrao, S.: Survey of data partitioning algorithms for big data stores. In: 2016 4th International Conference on Parallel, Distributed and Grid Computing (PDGC), pp. 163–168. IEEE (December 2016)
15. Saltenis, S., Jensen, C.S., Leutenegger, S.T., Lopez, M.A.: Indexing the positions of continuously moving objects. ACM SIGMOD Rec. **29**(2), 331–342 (2000)
16. Tao, Y., Papadias, D., Shen, Q.: Continuous nearest neighbor search. In: Proceedings of the 28th International Conference on Very Large Data Bases, pp. 287–298. VLDB Endowment (August 2002)
17. White, C.: A technical architecture for data warehousing. InfoDB J. **9**(1), 5–11 (1995)
18. Zhou, S., Zhou, A., Tao, X., Hu, Y.: Hierarchically distributed data warehouse. In: Proceedings of the 4th International Conference/Exhibition on High Performance Computing in the Asia-Pacific Region, Beijing, China, pp. 848–53 (2000)

Accelerating Sparse Convolutional Neural Networks Based on Dataflow Architecture

Xinxin Wu[1,2(✉)], Yi Li[1,2], Yan Ou[1,2], Wenming Li[1], Shibo Sun[3],
Wenxing Xu[3], and Dongrui Fan[1,2]

[1] State Key Laboratory of Computer Architecture,
Institute of Computing Technology, Chinese Academy of Sciences,
Beijing 100190, China
{wuxinxin,liyi,ouyan,liwenming,fandr}@ict.ac.cn
[2] School of Computer and Control Engineering, University of Chinese Academy
of Sciences, Beijing 100190, China
[3] Beijing Institute of Petrochemical Technology, College of Information Engineering,
Beijing 100190, China
771586376@qq.com, xuwenxing@bipt.edu.cn

Abstract. Convolutional Neural Networks (CNNs) achieve state-of-the art performance in a wide range of applications including image recognition, speech recognition, and natural language processing. Large-scale CNNs generally have encountered limitations in computing and storage resources, but sparse CNNs have emerged as an effective solution to reduce the amount of computation and memory required. Though existing neural networks accelerators are able to efficiently process sparse networks, the strong coupling of algorithms and structures makes them inflexible. Dataflow architecture can implement different neural network applications through flexible instruction scheduling. The dataflow architecture needs to be initialized at execution time to load instructions into the computing array. Running a dense convolutional layer only needs to be initialized once due to regular calculations. However, running a sparse convolutional layer requires multiple initializations, which takes a long time to fetch instructions from memory, resulting in the computing array being idle and degrading performance. In this paper, we propose an instruction sharing strategy based on the field content in the instruction, which can reduce initialization time and improve performance. Moreover, we use an extended instruction sharing strategy based on the static nature of filters to remove filters-related instructions, further reducing initialization time. Experiments show that our strategies achieve 1.69x (Alexnet), 1.45x (VGG-16) speedup and 37.2% (Alexnet), 34.26% (VGG-16) energy reduction compared with dense networks. Also, they achieve on average 2.34x (Alexnet), 2.12x (VGG-16) and 1.75x (Alexnet), 1.49x (VGG-16) speedup over Titan Xp GPU and Cambricon-X for our benchmarks.

Keywords: Dataflow architecture · Sparse convolutional neural networks · Instruction scheduling · Instruction sharing

© Springer Nature Switzerland AG 2020
M. Qiu (Ed.): ICA3PP 2020, LNCS 12453, pp. 14–31, 2020.
https://doi.org/10.1007/978-3-030-60239-0_2

1   Introduction

Application-specific accelerators [4,16] have proposed as a high performance and low power alternative to traditional CPUs/GPUs due to stringent energy constraints. However, CNNs continue to evolve towards larger and deeper architectures as applications diversify and complicate, which leads to a heavy burden on processing computation, memory capacity, and memory access of accelerators [6]. To address these challenges, many methods for reducing model parameters are proposed to turn dense networks into sparse networks using the redundant characteristics of model parameters [7], such as pruning [13], low rank [15]. In order to make full use of the advantages of sparse network computing and storage, many accelerators have emerged to accelerate sparse networks [1,11,12,29,30] due to the low efficiency of CPU/GPU for sparse network acceleration.

However, these application-specific accelerators suffer from inflexible characteristics due to the tightly coupled nature of algorithms and structures. For example, DianNao family [4], a series of customized accelerators without sparsity support can not benefit from sparsity. Cnvlution [1] has completely modified DaDianNao's [5] microstructure to support sparse networks. EIE [12] and ESE [11] accelerators supporting sparse networks are no longer suitable for dense networks.

Dataflow architecture has advantage in good flexibility and high data parallelism and power efficiency for today's emerging applications such as high performance computing [2], scientific computing [21,27] and neural networks [25,28]. It implement different applications through flexible instruction scheduling. Compared with the traditional control flow (controlled by PC), it consists of a simple control circuit composed of processing unit array (PEs) that can communicate directly avoiding frequent memory access. Based on codelet model (a dataflow-inspired parallel execution model) [9], an application is represented as a Codelet Graph (CDG) which is a directed graph consisting of codelets, which are composed of instructions, and arcs, which represent data dependencies among codelets. At the same time, a codelet is fired once all data is available and all resource requirements are met, which maximizes instruction codelet-level and data-level parallelism. Its natural parallel characteristics fit perfectly with the inherent parallelism of neural network algorithms. Based on this architecture, we study the data characteristics and instruction characteristics of the neural network to optimize the mapping and execution of the network to maximize the structural benefits.

In the dataflow architecture, codelets instructions in the CDG need to be loaded from the memory into the instruction buffer in the PE through initialization process when the PE array is running. A codelet in the buffer waits for the condition to be satisfied and then is emitted. For dense convolutional layer of CNN, convolution operation of each channel is mapped on the PE array in the form of a CDG diagram. The codelets instructions formed by different channels are the same due to the regular calculation mode. In this case, codelets instructions need only be loaded once to implement convolution operations in all channels.

However, for a sparse convolutional layer of CNN obtained using pruning method such as [13], the following problem will occur when the PE array is running. The codelet instructions in the CDG of different channels are no longer the same, due to the pruning operation makes the convolution calculation irregular. It must load the codelets instructions of different channels from different memory address when running, which causes frequently memory access and the idleness of the PE array. Eventually results in a significant performance degradation.

Based on the above problem, in this paper we use two strategies to accelerate sparse CNNs based on dataflow architecture. Our contributions are as follows:

- By analyzing the data and instruction characteristics of the sparse convolution layer, we share same instructions in the convolutional layer based on the field contents in instructions, and proposed an instruction sharing strategy (ISS), which can reduce initialization time and computing array idle time.
- Based on the ISS strategy, we use the extended instruction sharing Strategy (EISS) to delete filter-related instructions by using the static nature of filters, further reducing the initialization time.
- These two strategies achieve on average 1.45x (Alexnet), 1.24x (VGG-16) and 1.69x (Alexnet), 1.45x (VGG-16) speedup and 36.57% (Alexnet), 33.05% (VGG-16) and 37.2% (Alexnet),34.26% (VGG-16) energy reduction respectively compared with dense convolutional layers. In addition, they achieve on average 2.34x (Alexnet), 2.12x (VGG-16) and 1.75x (Alexnet), 1.49x (VGG-16) speedup over Titan Xp GPU and current state-of-the-art Cambricon-X accelerator.

This paper is organized as follows. In Sect. 2, we introduce the background of dataflow architecture and CNN. In Sect. 3 we analyze the existing problems in implementing sparse CNN based on a dataflow architecture. In Sect. 4, we describe two strategies in detail to accelerate sparse convolution. In Sects. 5 and 6, we present our evaluation methodology and experimental results respectively. In Sects. 7 and 8, we provide related work and a conclusion to this work.

2   Background

2.1   Dataflow Architecture

This section explains microarchitecture, execution model, and instruction format of a dataflow process unit (DPU), that resembles coarse-grained instruction level dataflow architecture, such as Runnemede [2], TERAFLUX [10].

Microarchitecture. Figure 1 is an instantiated dataflow accelerator, which includes a micro controller (Micc), processing element (PE) array and a network on chip (NoC). The structure of the DPU is similar to the traditional many-core architecture [8], and each PE inside the array is a control-flow core. The micro controller manages the overall execution of the PE array and is responsible for communicating with the host. Each PE contains a buffer for storing instructions,

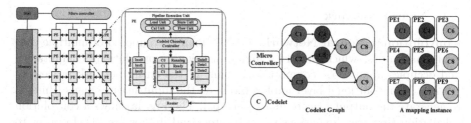

Fig. 1. DPU: An instantiated dataflow accelerator.

Fig. 2. Codelet execution model.

a buffer for storing data, a codelet choosing controller unit, a codelet status register unit and a pipeline execution unit.

Before running each PE, it needs to go through an initialization phase to load all required codelets in the memory into the instruction buffer by routers. Then the codelet choosing controller selects ready codelet instructions and sends them to the pipeline execution unit according to the codelet status register. The pipeline execution unit contains load, store, calculation and flow unit to execute the corresponding instructions.

Opcode	F0	F1	F2	Fuction
Load	Operand index0	Address offset	Base address index0	Value(F0)=DRAM(Base address(F2) + F1)
Store	Operand index0	Address offset	Base address index0	DRAM(Base address(F2) + F1)=Value(F0)
Madd	Operand index0	Operand index1	Operand index2	Value(F2)=Value(F0)*Value(F1)+Value(F2)
Mul/Add/Sub	Operand index0	Operand index1	Operand index2	Value(F2)=Value(F0)*/+/-Value(F1)
Flow	Operand index0	Operand index1	PE index1	Value(F1,F2)=Value(F0)

Fig. 3. Instruction format.

Execution Mode. In order to achieve high utilization of computing elements in each PE, an application is written for a codelet model, which is a dataflow-inspired parallel execution model [9]. All code are partitioned into codelets containing a sequence of instructions, which are linked together based on codelets dependencies to form a Codelet Graph (CDG) and then are mapped in the PE array, as shown in the Fig. 2. A codelet will only fire when all data is available and all resource requirements are met, which maximizes instruction codelet-level and data-level parallelism.

Instruction Format. The instruction format of DPU is shown in Fig. 3. And the instruction set is fixed-length. Each instruction is composed of instruction code, source operand index and destination operand index. DPU contains basic arithmetic instructions (Add, Sub, Mul, Madd) and memory access instructions (Load, Store) and direct communication instructions (Flow) between PEs.

2.2 Neural Network

Convolutional Neural Network. CNN is mainly composed of multiple convolutional layers, which performs high-dimensional convolutions computation and occupies about 85% computational time of the entire network processing [19,24]. A convolutional layer applies filters on the input feature maps (ifmaps) to generate the output feature maps (ofmaps). The Fig. 4a shows a convolution calculation that the dimensions of both filters and ifmaps are 3D. In order to reduce memory access and save data movement energy cost, data needs to be reused [3]. In the PE array, convolution can be reused in each PE, and ifmap and filter are reused between PEs through flow operations. Figure 4b shows the process of convolution reuse in PE. To generate first row of output, three rows ($row1, row2, row3$) of the ifmap and the filter are mapped in PE. The PE implements convolution reuse by sliding window. Figure 4c shows ifmap and filter reuse between PEs. Filter weights are reused across PEs vertically. Rows of ifmap values are reused across PEs vertically and horizontally. In Fig. 4c, through flow operation, PE3 reuses filter1 of PE1 and $row2$–$row3$ of PE1 ifmap. PE2 reuses $row1$–$row3$ of PE1 ifmap. PE4 reuses filter2 of PE2, and also reuses $row2$–$row3$ of PE2 ifmap, and $row4$ of ifmap of PE3.

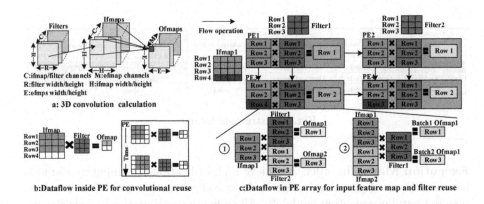

Fig. 4. Opportunities for convolution reuse.

Sparse Neural Network. Due to the challenge of large-size CNN models on hardware resources, researchers have proposed many methods to compress CNN models (e.g. pruning [13], low rank [15], short bit-width [14]) that reduce models size without loss of accuracy or slight loss. Among them, using the pruning method to generate a sparse network is one of the effective methods. The state-of-art pruning method [13] using three-step method (First, network is trained to learn which connections are important. Second, unimportant connections are pruned based on a pre-set threshold. Third, network with remaining connections is retrained to get final weights) achieves a sparsity of 11% for Alexnet [17] and 7% for VGG-16 [23].

3   Motivation

Dense networks can easily achieve high performance and energy efficiency on application-specific accelerator platforms due to dedicated hardware support [4]. However, these accelerators lack dedicated hardware support for sparse networks due to strong coupling of algorithms and structures. Even though the sparse network greatly reduces the amount of calculation and memory access, its performance and energy efficiency gains are small. For data flow architecture, flexible instruction scheduling provides the possibility for different algorithm implementations.

Based on the execution model of the dataflow, the convolution operations of different channels of the dense network have the same instruction codelets due to the regular calculation of features. Therefore, it only needs one initialization operation, that is, loading instructions of codelets from memory, and then all channels convolution operations can be implemented. The Fig. 5 shows the instructions required to calculate the partsum value of two channels. The operand indexes of ifmap, filter, and ofmap are respectively represented by $IF0$–$IF3$, $F0$–$F3$, and $OF0$. The base address indexes of ifmap, filter, and ofmap are 0, 1, and 2, respectively.

In Fig. 5, channel 1 and 2 perform the same convolution operation by using different data, which is obtained by channel offset, and address offset and base address index (points to base address) in the instruction. For the data in the same position of each channel, their channel offsets are different (Ifmap are 0×0, 0×400. Filter are 0×0, 0×100 respectively), but the instructions are the same due to the operand index, the address offset and the base address index are the same. For example, the base address, address offset, operand index0 of the $inst1$ instruction of two channels are respectively 0×0, 0, and $IF0$. Therefore, convolution can be done without interruption by one initialization, which ensures the full utilization of the computing resources of the PE array.

However, compared to dense CNN, the execution way of sparse CNN has changed. As shown in the Fig. 5, the pruning operation removes the instructions required for the zero weight. The load instructions ($inst6, inst7$), madd instructions ($inst10, inst11$), flow instructions ($inst19, inst20$) of channel 1 are removed. For channel 2, the load instructions ($inst5, inst8$), madd instructions ($inst9, inst12$), flow instructions ($inst18, inst21$) are removed. It can be seen that the instructions of different channels are no longer exactly the same, which makes it necessary to reinitialize the PE array when performing the convolution operation. That is, new instructions of codelet are loaded from memory into the instruction buffer. When instructions are loaded, the PE array is in an idle state, making the computing resources underutilized, and eventually causing a serious performance degradation of sparse convolution.

The Fig. 6 shows the time taken by the DPU to execute several sparse convolutional layers of Alexnet and VGG-16. Since dense convolution requires only one initialization operation, the initialization time is almost negligible compared to the execution time (accounting for 1.04% of the total time). However, for sparse convolution, it can be seen that the execution time of the PE array is significantly reduced, but multiple initialization operations take a long time, and they account for an average of 50.2% of the total time. Obviously, the total time of sparse networks does not decrease much compared with dense networks, and some are even longer than dense networks, which makes the sparse network unable to accelerate at all. The multiple initialization seriously hindered the benefits of sparse networks.

The reason for sparse convolution initialization multiple times is that the pruning operation removes filter-related instructions and data, which makes the instructions of different channels no longer the same. In order to reduce the initialization time of sparse networks, an intuitive idea is to use one load instruction to load multiple values from memory. This requires adding multiple operand indexes in the instruction and increasing the data transmission bandwidth to support the operation. However, all instructions are fixed-length, and the data transmission bandwidth is also fixed, so this method is undesirable.

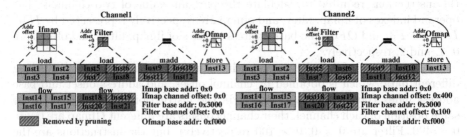

Fig. 5. The instructions required for the two channels to perform a convolution operation in PE1.

By analyzing these instructions and data, it is found that the pruning operation only removes the instructions related to the zero weight in filters, these instructions include load, flow, and madd. However, the relevant instructions of ifmap and partsum are not affected. These instructions are still the same between different channels. For example, in the Fig. 5, the instructions $inst1$–$inst4, inst13$, and $inst14$–$inst17$ of channel 1 and channel 2 are still the same. These same instructions can be shared between different channels. Multiple accesses of these instruction addresses increase the cache hit rate during each initialization, which reduces instruction load time, also reduces the idle time of the PE array, and increases the full use of computing resources.

The above observation and analysis motivate us using highly efficient strategies to take advantage of the sparsity of neural networks.

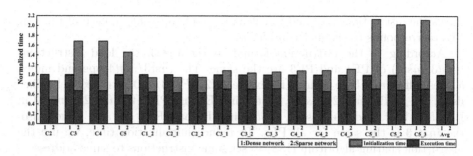

Fig. 6. Execution time breakdown of convolutional layers normalized to dense networks.

4 Acceleration Strategy for Sparse Networks

In this section, we present the detailed strategies of our proposed, including the instruction sharing strategy (ISS) and extended instruction sharing strategy (EISS). We use the advanced pruning method [13] to obtain sparse networks. This method resets the weights below the threshold to 0 and retains the weights above the threshold based on a preset threshold.

Using Eyeriss [3] analysis framework can achieves energy saving mapping of convolutional layers statically based on the reuse opportunities of the above dataflow in the Fig. 4. It folds multiple logical PE arrays, which are required for full convolutional layer operations, into the physical PE array for data reuse and ofmaps accumulation. For example, it folds logical PE from different channels at the same position onto a single physical PE to achieve ofmaps accumulation. It also folds multiple ifmaps and multiple filters to a same physical PE for ifmaps and filters reuse, as shown in the Fig. 4c. In this paper, the ISS and EISS strategies are implemented on the logical PE array to realize the instruction sharing of the convolutional layer.

The reason for instruction sharing is that the convolution calculates characteristics regularly, that is, the same multiply-accumulate operation is performed using different data. Pruning method destroys part of the characteristic of regular calculation. Therefore, our instruction sharing idea is still applicable to other dataflow methods, such as WS, OS, NLR, RS [3], whose purpose is to maximize data reuse to achieve excellent performance and energy efficiency.

4.1 Instruction Sharing Strategy

Multiple initializations of the sparse convolution cause the PE array to be idle, which severely hinders performance improvement. By analyzing these instructions, we have found that the same instructions exist in the convolution operation between different channels, which include instructions related to ifmaps and partsums. More aggressively, the same instruction also exists in the convolution operation in a channel, because the PE array reuses filters in the vertical direction and reuses ifmaps in the vertical and horizontal directions. The Fig. 7

shows the instructions required to calculate an ofmap value in each of the four PEs (corresponds to channel 1 in Fig. 5).

According to the instruction format in the Fig. 3, the load instructions ($inst3, inst4$) of PE1 and PE3 are the same. The $inst3/inst4$ operand index, address offset and base address index are $IF2/IF3$, 3/4, 0/0 respectively. Similarly, the $inst9$, $inst12$, $inst16$, $inst17$ in PE1 and PE3, the $inst22$, $inst23$ in PE1 and PE2, and the $inst11$, $inst12$ in PE2 and PE4 are also same. Also, the $inst13$ is the same for all PEs. To reduce initialization time, we use the instruction sharing algorithm (ISS) to set same instructions to same address.

We divide these instructions into three categories, ifmaps instructions (IF_type), which include load instructions and flow instructions with ifmaps. Filters instructions (F_type), these include load instructions, flow instructions, and madd instructions with filters. Partsum/Ofmap (P_type) instructions, including store instructions with partsums/ofmaps. For the same instruction, we use Algorithm 1 and Algorithm 2 to achieve the instruction sharing of different channels and the same channel respectively.

For convolution instructions of different channels, as previously analyzed, the pruning operation makes the filters instructions of different channels different, but has no effect on ifmaps instructions and partsums instructions. These instructions are still the same in different channels. Algorithm 1 achieves the sharing of the same instructions in different channels. Based on the instruction of channel 1 ($channel = 1$), for each instruction in other channels ($channel > 1$), the instruction address is updated to the instruction address in channel 1 if the type of this instruction is an ifmap (IF_type) or a partsum type (P_type), and it is the same as an instruction in channel 1. In the Fig. 5, the load instructions $inst1$–$inst4$, flow instructions $inst14$–$inst17$ in channel 2 belong to the ifmap type (IF_type), and store instruction $inst13$ belongs to partsum type (P_type). These instructions are the same as those in channel 1. Using algorithm 1, their addresses are updated to the instructions in channel 1.

Algorithm 2 implements instruction sharing between different PEs in a channel. According to the convolution reuse rules above, we divide the PE array into groups by columns because PEs in the vertical direction are mapped to the same filters. For each PE in the group, the filters type instructions are partially the

---

Algorithm 1: Instruction sharing algorithm for different channels.

Input: Codelet instruction, $Inst$; Convolution layer channels, C; Instruction type, F_type; IF_type; P_type;
Output: Address of instruction: $Addr$;

1 **for** *each channel* $\in [2, C]$ **do**
2   **for** *each Inst* $\in$ *channel* **do**
3   **if** *(Inst type belongs to IF_type or P_type)* && *(Inst is the same as Inst' of channel1)* **then**
4              $Addr(Inst) = Addr(Inst')$

Algorithm 2: Instruction sharing algorithm for the same channel.

Input: Codelet instruction, *Inst*; First PE, *PE*1; PE array group number, *G*;
           PE number in group, *P*; First PE in the group, *P*1; Instruction type,
           F_type; *IF_type*; *P_type*;
Output: Address of instruction: *Addr*;

1 **for** *each* $g \in [1, G]$ **do**
2   **for** *each* $p \in [1, P]$ **do**
3   **for** *each* $Inst \in p$ **do**
4   **if** *(p > 1) && (Inst type belongs to F_type) && (Inst is the same*
              as Inst′ of P1) **then**
5             $Addr(Inst) = Addr(Inst')$
6   **if** *(inst type belongs to IF_type or P_type) && (Inst is the same*
              as Inst′ of PE1) **then**
7             $Addr(Inst) = Addr(Inst')$

---

same. Based on the first PE in the group ($p = 1$), for each instruction in the other PEs in the group ($p > 1$), the address is updated to the instruction address in $P1$ if the instruction type is a filter type (F_type) and is the same as one in $P1$. In the Fig. 7, Algorithm 2 updates the madd instructions $inst9$ and $inst12$ addresses in PE3 to the $inst9$ and $inst12$ addresses in PE1, and also updates the $inst11$ and $ins12$ addresses in PE4 to the $inst11$ and $inst12$ addresses in PE2. For instructions of ifmaps (IF_type) and partsums type (P_type), using PE1 as the benchmark, an instruction address in another PE is updated to the instruction address in PE1 if it is the same as an instruction in PE1. In the Fig. 7, the $inst13$, $inst22$, $inst23$ of PE2, $inst3$, $inst4$, $inst13$, $inst16$, $inst17$ of PE3 and $inst13$ of PE4 are all updated to the addresses in PE1.

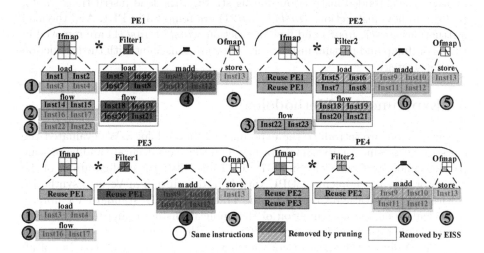

Fig. 7. Instructions required to perform a convolution operation in one channel. The PE array, filter, ifmap and ofmap size are 2 * 2, 2 * 2, 4 * 4 and 2 * 2, respectively.

4.2   Extended Instruction Sharing Strategy

To further reduce the initialization time, we use the static sparsity of weights to extend the instruction sharing strategy. For static sparsity, zero value weights are permanently removed from the network. And the sparseness of weights does not change with the input data. Once they are mapped on the PE, their related instructions will not change.

As mentioned above, the same instructions come from the ifmap and partsum related instructions, and the instructions where PEs are mapped to the same filters in the same channel. Due to the pruning operation, the related instructions of different filters are different, and they cannot benefit from the instruction sharing strategy. Based on its static characteristics, we use immediate multiply-accumulate instructions instead of multiply-accumulate instructions, which are not needed for dense instructions.

Opcode	F0	F1	F2	Fuction
Imm-madd	Operand index 0	Imm	Operand index 2	Value(F2)=Value(F0)*Imm)+Value(F2)

Fig. 8. Imm-madd instruction format.

The Fig. 8 shows the format of the imm-madd instruction, which directly carries a value. For all filters, it eliminates the need to load non-zero weights from memory using load instructions. At the same time, flow instructions are no longer required to pass non-zero weights to other PEs. Therefor, it reduces the loading of filter-related instructions and non-zero weights, which is beneficial to memory access and performance. In structure, we added the control logic of immediate multiply-accumulate instruction decoding and transmission to achieve this operation. It only adds a little hardware overhead. In the Fig. 7, by using the extended instruction sharing strategy, the load instructions ($inst5$, $inst8$) and flow instructions ($inst18$, $inst21$) are removed in PE1. And the load instructions ($inst7$, $inst8$) and flow instructions ($inst20$, $inst21$) are removed in PE2. For ifmap and partum instructions, they are consistent with the instruction sharing strategy.

5   Experimental Methodology

In this section, we introduce the experimental methodology. We evaluated the our strategies using a DPU simulator based on the cycle-accurate and large-scale parallel simulator framework SimICT [26]. Table 1 lists the configurations of the simulator. We also implement DPU with RTL description in Verilog, synthesize it with Synopsys Design Compiler using TSMC 12 nm GP standard VT library. We calculate energy consumption of the applications according to circuit-level of atomic operations with Synopsys VCS using PrimeTime PX.

Figure 1 shows the structure of the simulation system which consists of all components of the DPU. The DPU in the simulator consists of 8×8 PE array, and PE nodes are connected by 2D mesh networks. Each PE contains one 16-bits MAC (fix point multiply accumulate) in the SIMD8 model. There are 8 KB instruction buffer and 32 KB data buffer in each PE. To provide high network bandwidth, the on-chip networks consist of multiple independent physical 2D mesh networks. And to provide fast memory access, a 1 MB cache is added to the structure.

Table 1. Configuration of the instantiated dataflow accelerator.

Module	Configuration	Module	Configuration
Host	ARM core	Micc	Control messages, data transmission
PE	8 * 8, SIMD-8, 16-bits MAC, 8 KB Instruction Buffer, 2 KB Data Buffer	NoC	2D mesh, 1 control network, 1 data/memory-access network, 1 inter-PE network
Memory subsystem	DDR3 1333 MHz with 1 MB cache	Peak performance	1.887 TOPS

Benchmarks. To evaluate our strategies, we use representative neural networks AlexNet [17] and VGG-16 [23] convolution layers as benchmarks which have different sizes and parameter scales. Table 2 lists the corresponding sparsity of different kinds of layers. Each layer is translated into a CDG through our designed compiler based on LLVM [18] platform. With the limited space, the details of the compiler are not included in this paper.

Table 2. Configuration of convolutional layer parameters and sparsity for Alexnet and VGG-16.

	Layer	Ifmap	Filter	Sparsity	Layer	Ifmap	Filter	Sparsity
Alexnet	C2	$31^2 * 96$	$5^2 * 256$	0.62	C3	$15^2 * 256$	$3^2 * 384$	0.65
	C4	$15^2 * 384$	$3^2 * 384$	0.63	C5	$15^2 * 384$	$3^2 * 256$	0.63
VGG-16	C1_2	$224^2 * 128$	$3^2 * 128$	0.78	C2_1	$112^2 * 64$	$3^2 * 128$	0.66
	C2_2	$112^2 * 128$	$3^2 * 128$	0.64	C3_1	$56^2 * 128$	$3^2 * 256$	0.47
	C3_2	$56^2 * 256$	$3^2 * 256$	0.76	C3_3	$56^2 * 256$	$3^2 * 256$	0.58
	C4_1	$28^2 * 256$	$3^2 * 512$	0.68	C4_2	$28^2 * 512$	$3^2 * 512$	0.73
	C4_3	$28^2 * 512$	$3^2 * 512$	0.66	C5_1	$14^2 * 512$	$3^2 * 512$	0.65
	C5_2	$14^2 * 512$	$3^2 * 512$	0.71	C5_3	$14^2 * 512$	$3^2 * 512$	0.64

Evaluation Metric. In this paper, we refer to the instruction sharing strategy, and extended instruction sharing strategy as ISS, EISS, respectively. By applying these benchmarks, we evaluate our strategies in different ways. For the ISS and EISS, we compare with dense networks and verify the effectiveness of our method in terms of instruction execution times, execution time (performance) and energy. At the same time, we report the speedup of our DPU and Cambricon-X (peak sparse performance 544 GOP/s) over NVIDIA Titan Xp GPU (12 TFLOP/s peak performance, 12 GB GDDR5x, 547.7 GB/s peak memory bandwidth), a state-of-the-art GPU for deep learning. To run the benchmark, We use cuSparse library based on CSR indexing implement sparse network (GPU-cuSparse) [20]. We use nvidia-smi utility to report the power.

6   Experimental Results

6.1   Instruction Execution Times

Compared to dense networks, pruning operation removes redundant filter weights, thereby removing related filters instructions, which reduces the number of instruction executions. As shown in the Fig. 9, the load, calculation, and flow instructions execution times of Alexnet (VGG-16) sparse network are reduced by 10.38% (5%), 63.38% (66.95%), and 61.69% (48.06%), respectively, and the total instructions execution times are reduced by an average of 55.03% (54.24%). The EISS uses imm-madd instructions to further remove load instructions and flow instructions of non-zero weights and does not affect the execution times of other instructions, which reduces the load and flow instructions execution times by an average of 0.4% (1.33%) and 12.18% (8.77%) based on the pruning operation. Compared with dense networks, the EISS reduces load, calculation, and flow instructions execution times of Alexnet (VGG-16) by 10.78% (6.33%), 63.38% (66.95%), and 73.87% (56.83%), respectively, and reduces the total instructions execution times by an average of 55.8% (55.8%).

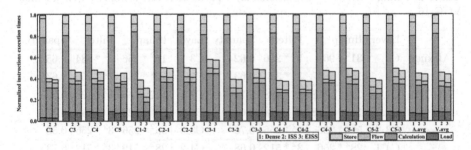

Fig. 9. Instruction execution times breakdown of convolutional layers normalized to dense layers.

6.2   Execution Time and Performance

The pruning method reduces the number of instruction executions and memory access to redundant data. It also reduce the execution time of the sparse network. By implementing ISS and EISS strategies, the initialization time of the sparse network is reduced, which reduces the total time and improves performance. As can be seen from the Fig. 10, compared with the dense network, the total execution time of ISS was reduced by 30.84% (Alexnet) and 19.65% (VGG-16) on average, respectively. In terms of performance, the average performance of the sparse network under the ISS is 1.45x (Alexnet) and 1.24x (VGG-16) that of the dense network and the maximum is 1.84x (Alexnet C2 layer). The EISS uses imm-madd instructions to replace the madd instructions of the dense network, eliminating non-zero weights instructions and data memory access, which further reduces the initialization time, execution time and improves performance. Compared with the dense network, the total execution time of EISS has been reduced by an average of 40.74% (Alexnet) and 31.35% (VGG-16), respectively. For performance, the EISS improves average performance by 14% on the basis of the ISS, which is 1.69x (Alexnet) and 1.45x (VGG-16) that of the dense network.

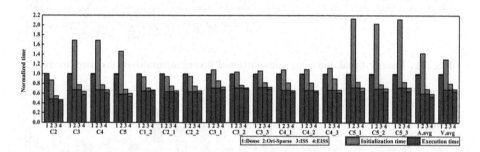

Fig. 10. Execution time breakdown of convolutional layers normalized to dense networks.

6.3   Energy

The ISS method improve the performance of sparse convolution, and also reduce energy consumption due to the reduction in instruction execution times, instruction and data memory access. We show the energy breakdown for convolutional layers in Fig. 11. Compared to dense convolution, the ISS reduces the total energy on average by 36.57% (Alexnet) and 33.05% (VGG-16). The EISS reduces the total energy by an average of 37.20% (Alexnet) and 34.26% (VGG-16). The EISS only slightly reduces the energy consumed by the data buffer, instruction buffer, and transmission due to the removal of non-zero weight instructions (load and flow) and data loading. Although it reduces the memory access of instructions and data, the energy proportion of memory access is small. Based on the ISS, the total energy is reduced by less than 1% on average.

6.4 Compare with Other Accelerators

In addition to comparing with dense networks, we also compare the performance and energy efficiency with GPU and Cambricon-X accelerator. In Fig. 12, we report the speedup comparison of DPU and Cambricon-X over GPU baseline across all the benchmarks. Compared to the GPU platform, on average, DPU achieves 2.34x (Alexnet), 2.12x (VGG-16) speedup for the sparse convolution. Cambricon-X achieves 1.34x (Alexnet), 1.42x (VGG-16) speedup. Therefore, DPU speedup 1.75x (Alexnet) and 1.49x (VGG-16) over Cambricon-X. Table 3 reports the energy efficiency comparison of the GPU, Cambricon-X and our DPU. Our energy efficiency is 12.56x of GPU, while Cambricon-X is 18.9x

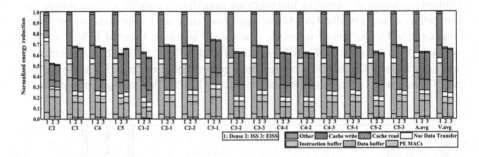

Fig. 11. Energy breakdown for convolutional layers normalized to dense layers.

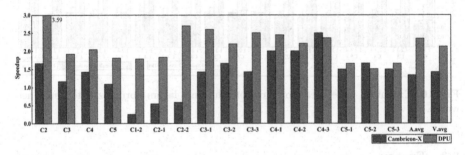

Fig. 12. Speedup of DPU and Cambricon-X sparse convolution over GPU baseline.

Table 3. GPU and Cambricon-X compared to DPU.

	GPU (Titan Xp)	Cambricon-X	This work
Technology	16 nm	65 nm	12 nm
Area	-	6.38 mm^2	13.82 mm^2
Power	243 W	954 mW	1.93 W
Throughput	7320 GFLOPS	544 GOPS	730 GOPS
Energy efficiency	30.1 GFLOPS/W	570.2 GOPS/W	378.2 GOPS/W

of GPU. It can be seen that the energy efficiency of DPU is not as good as Cambricon-X, but the performance is higher than it.

7   Related Work

Sparse neural networks have emerged as an effective solution to reduce the amount of computation and memory required. Many application-specific accelerators have been proposed to accelerate sparse networks.

Although DaDianNao [5] uses wide SIMD unit with hundreds of multiplication channels to achieve efficient processing of the neural network, it cannot take advantage of the sparsity of modern neural networks. CNV [1] decouples these lanes into finer-grain groups, and by using activation sparseness, only non-zero activation values are passed to the arithmetic unit, which accelerates the convolutional layer. However, it does not take advantage of the sparsity of weights. Eyeriss [3] implements a data gating logic to exploit zeros in the ifmap and saves processing power. However, it only has energy saving effect without acceleration effect. At the same time, it cannot accelerate the network with sparse weights. EIE [12], ESE [11] were proposed for leveraging the sparsity of full-connected layers in neural networks with CSC sparse representation scheme. However, EIE and ESE are no longer suitable for dense networks and sparse convolutional layer. Compared with these accelerators, we use the sparsity of the weights to accelerate the convolutional layer of the neural network.

Cambricon-X [29] designs Indexing Module (IM) efficiently selects and transfers no-zero neurons to connected PEs with reduced bandwidth requirement. Finally, it accelerates the convolutional layer and the fully connected layer, including dense networks and sparse networks. Cambricon-S [30] uses a coarse-grained pruning method which reduces the irregularity drastically sparse network. And designs a hardware structure to leverage the benefits of pruning method. It also achieves acceleration of convolutional layers and fully connected layers. SCNN [22] uses Cartesian product-based computation architecture which eliminates invalid calculations. Moreover, it designs the hardware structure to efficiently deliver weights and activations to a multiplier array. However, those application-specific accelerators sacrifice the flexibility of the hardware architecture to achieve the highest performance and energy efficiency, and cannot adapt to new applications. In this paper, applications can be implemented through flexible instruction scheduling based on the dataflow architecture.

8   Conclusion

In this paper, we propose two strategies to accelerate sparse CNN in the dataflow architecture. These strategies improve the performance of sparse networks and reduces energy consumption. We are the first to propose instruction sharing method to accelerate sparse networks based on a dataflow architecture. Our method is only applicable to networks with sparse weights, not to activate sparse networks because it is dynamically generated by using Relu function after each operation. In the future, we will continue to study sparse acceleration at the hardware level.

Acknowledgement. This work was supported by the National Key Research and Development Program (2018YFB1003501), the National Natural Science Foundation of China (61732018, 61872335, 61802367, 61672499), Austrian-Chinese Cooperative R&D Project (FFG and CAS) Grant No. 171111KYSB20170032, the Strategic Priority Research Program of Chinese Academy of Sciences, Grant No. XDC05000000, and the Open Project Program of the State Key Laboratory of Mathematical Engineering and Advanced Computing (2019A07).

References

1. Albericio, J., Judd, P., Hetherington, T., Aamodt, T., Jerger, N.E., Moshovos, A.: Cnvlutin: ineffectual-neuron-free deep neural network computing. ACM SIGARCH Comput. Arch. News **44**(3), 1–13 (2016)
2. Carter, N.P., et al.: Runnemede: an architecture for ubiquitous high-performance computing. In: 2013 IEEE 19th International Symposium on High Performance Computer Architecture (HPCA), pp. 198–209. IEEE (2013)
3. Chen, Y.H., Krishna, T., Emer, J.S., Sze, V.: Eyeriss: an energy-efficient reconfigurable accelerator for deep convolutional neural networks. IEEE J. Solid-State Circuits **52**(1), 127–138 (2016)
4. Chen, Y., Chen, T., Xu, Z., Sun, N., Temam, O.: Diannao family: energy-efficient hardware accelerators for machine learning. Commun. ACM **59**(11), 105–112 (2016)
5. Chen, Y., et al.: DaDianNao: a machine-learning supercomputer. In: Proceedings of the 47th Annual IEEE/ACM International Symposium on Microarchitecture, pp. 609–622. IEEE Computer Society (2014)
6. Dean, J., et al.: Large scale distributed deep networks. In: Advances in Neural Information Processing Systems, pp. 1223–1231 (2012)
7. Denil, M., Shakibi, B., Dinh, L., Ranzato, M., De Freitas, N.: Predicting parameters in deep learning. In: Advances in Neural Information Processing Systems, pp. 2148–2156 (2013)
8. Fan, D., et al.: SmarCo: an efficient many-core processor for high-throughput applications in datacenters. In: 2018 IEEE International Symposium on High Performance Computer Architecture (HPCA), pp. 596–607. IEEE (2018)
9. Gao, G.R., Suetterlein, J., Zuckerman, S.: Toward an execution model for extreme-scale systems-runnemede and beyond. CAPSL Tecnhical Memo 104, Department of Electrical and Computer Engineering, University of Delaware (2011)
10. Giorgi, R.: TERAFLUX: harnessing dataflow in next generation teradevices. Microprocess. Microsyst. **38**(8), 976–990 (2014)
11. Han, S., et al.: ESE: efficient speech recognition engine with sparse LSTM on FPGA. In: Proceedings of the 2017 ACM/SIGDA International Symposium on Field-Programmable Gate Arrays, pp. 75–84. ACM (2017)
12. Han, S., et al.: EIE: efficient inference engine on compressed deep neural network. In: 2016 ACM/IEEE 43rd Annual International Symposium on Computer Architecture (ISCA), pp. 243–254. IEEE (2016)
13. Han, S., Pool, J., Tran, J., Dally, W.: Learning both weights and connections for efficient neural network. In: Advances in Neural Information Processing Systems, pp. 1135–1143 (2015)

14. Holi, J.L., Hwang, J.N.: Finite precision error analysis of neural network hardware implementations. IEEE Trans. Comput. **42**(3), 281–290 (1993)
15. Jaderberg, M., Vedaldi, A., Zisserman, A.: Speeding up convolutional neural networks with low rank expansions. arXiv preprint arXiv:1405.3866 (2014)
16. Jouppi, N.P., et al.: In-datacenter performance analysis of a tensor processing unit. In: 2017 ACM/IEEE 44th Annual International Symposium on Computer Architecture (ISCA), pp. 1–12. IEEE (2017)
17. Krizhevsky, A., Sutskever, I., Hinton, G.E.: Imagenet classification with deep convolutional neural networks. In: Advances in Neural Information Processing Systems, pp. 1097–1105 (2012)
18. Lattner, C., Adve, V.: LLVM: a compilation framework for lifelong program analysis & transformation. In: Proceedings of the International Symposium on Code Generation and Optimization: Feedback-Directed and Runtime Optimization, p. 75. IEEE Computer Society (2004)
19. Long, J., Shelhamer, E., Darrell, T.: Fully convolutional networks for semantic segmentation. In: Proceedings of the IEEE Conference on Computer Vision and Pattern Recognition, pp. 3431–3440 (2015)
20. Naumov, M., Chien, L., Vandermersch, P., Kapasi, U.: Cusparse library. In: GPU Technology Conference (2010)
21. Oriato, D., Tilbury, S., Marrocu, M., Pusceddu, G.: Acceleration of a meteorological limited area model with dataflow engines. In: 2012 Symposium on Application Accelerators in High Performance Computing, pp. 129–132. IEEE (2012)
22. Parashar, A., et al.: SCNN: an accelerator for compressed-sparse convolutional neural networks. In: 2017 ACM/IEEE 44th Annual International Symposium on Computer Architecture (ISCA), pp. 27–40. IEEE (2017)
23. Simonyan, K., Zisserman, A.: Very deep convolutional networks for large-scale image recognition. arXiv preprint arXiv:1409.1556 (2014)
24. Szegedy, C., et al.: Going deeper with convolutions. In: Proceedings of the IEEE Conference on Computer Vision and Pattern Recognition, pp. 1–9 (2015)
25. Xiang, T., et al.: Accelerating CNN algorithm with fine-grained dataflow architectures. In: 2018 IEEE 20th International Conference on High Performance Computing and Communications; IEEE 16th International Conference on Smart City; IEEE 4th International Conference on Data Science and Systems (HPCC/SmartCity/DSS), pp. 243–251. IEEE (2018)
26. Ye, X., Fan, D., Sun, N., Tang, S., Zhang, M., Zhang, H.: SimICT: a fast and flexible framework for performance and power evaluation of large-scale architecture. In: Proceedings of the 2013 International Symposium on Low Power Electronics and Design, pp. 273–278. IEEE Press (2013)
27. Ye, X., et al.: An efficient dataflow accelerator for scientific applications. Fut. Gener. Comput. Syst. **112**, 580–588 (2020)
28. Ye, X.: Applying cnn on a scientific application accelerator based on dataflow architecture. CCF Trans. High Perform. Comput. **1**(3–4), 177–195 (2019)
29. Zhang, S., et al.: Cambricon-X: an accelerator for sparse neural networks. In: The 49th Annual IEEE/ACM International Symposium on Microarchitecture, p. 20. IEEE Press (2016)
30. Zhou, X., et al.: Cambricon-S: addressing irregularity in sparse neural networks through a cooperative software/hardware approach. In: 2018 51st Annual IEEE/ACM International Symposium on Microarchitecture (MICRO), pp. 15–28. IEEE (2018)

DAFEE: A Scalable Distributed Automatic Feature Engineering Algorithm for Relational Datasets

Wenqian Zhao[1] , Xiangxiang Li[1] , Guoping Rong[2] , Mufeng Lin[1] ,
Chen Lin[1] , and Yifan Yang[1]([⊠])

[1] Transwarp Technology (Shanghai) Co., Ltd., Shanghai, China
{wenqian.zhao,xiangxiang.li,mufeng.lin,chen.lin,yifan.yang}@transwarp.io
[2] The Joint Laboratory of Nanjing University and Transwarp on Data Technology,
Nanjing University, Nanjing, China
ronggp@nju.edu.cn

Abstract. Automatic feature engineering aims to construct informative features automatically and reduce manual labor for machine learning applications. The majority of existing approaches are designed to handle tasks with only one data source, which are less applicable to real scenarios. In this paper, we present a distributed automatic feature engineering algorithm, DAFEE, to generate features among multiple large-scale relational datasets. Starting from the target table, the algorithm uses a Breadth-First-Search type algorithm to find its related tables and constructs advanced high-order features that are remarkably effective in practical applications. Moreover, DAFEE implements a feature selection method to reduce the computational cost and improve predictive performance. Furthermore, it is highly optimized to process a massive volume of data. Experimental results demonstrate that it can significantly improve the predictive performance by 7% compared to SOTA algorithms.

Keywords: AutoML · Automatic feature engineering · Relational dataset · Big data · Feature selection · Machine learning

1 Introduction

Nowadays, as most businesses in the world embrace the opportunity of data science, people are dramatically impressed by the magic of data mining. An experienced data scientist is capable of generating and transforming useful features based on his or her high degrees of skill. However, this procedure, which is called *feature engineering*, is highly manual and typically requires domain knowledge. In addition, it is usually unrepeatable and non-scalable and occupies the majority of time in a machine learning task.

W. Zhao and X. Li—These authors contributed equally to this work.

M. Qiu (Ed.): ICA3PP 2020, LNCS 12453, pp. 32–46, 2020.
https://doi.org/10.1007/978-3-030-60239-0_3

As a result, automatically extracting informative features from data has become the forefront of both academia and industry. Although priory knowledge is essential, there are still some regular routines of generating informative features from the original attributes. However, since the actual structured data is typically residing in the relational database management systems (RDMS), being represented as a set of tables with relational links, the combinations of original features' mathematical transformations among relational tables may also be important, which makes the number of potential features grows exponentially.

Recently, researchers try to automatically extract the information among the relational datasets in two ways. One way is to use deep neural networks to implicitly complete the feature generation work, such as R2N [18]. However, since the created features are uninterpretable, this approach is hardly applicable in scenarios requiring strict justifications, such as fraud detection in banks. The other is to generate a tree representation for the connected datasets and then constructs features automatically when searching through the tree. Deep Feature Synthesis (DFS) [12] uses the Depth-First-Search algorithm. After finding out the farthest table from the target table, DFS sequentially applies different kinds of mathematical functions and generates features according to the relationship along the search path. To prevent generating too many features, DFS introduces a hyper-parameter called max-depth to control the farthest table it can reach. However, some information may be lost since not all tables are used to generate features. Furthermore, as DFS only supports limited mathematical functions, it can not extract complex features in practice.

Based on these observations, we propose the Distributed Automatic FEature Engineering (DAFEE) algorithm to improve the performance of automatic feature engineering. DAFEE uses a Breadth-First-Search type algorithm to search through the datasets. Each time the features are joined back to the target table, feature selection strategies would be applied to remove useless features. Additionally, DAFEE generates high-order features by applying sophisticated transformations upon multiple keys, which are both informative and interpretable. The major contributions of this paper are summarized as follows:

- DAFEE improves the DFS algorithm and can produce more useful features by increasing interaction among entities efficiently.
- DAFEE generates advanced features to improve model performance in ways that SOTA algorithms cannot, such as generating features based on the interest values, by connecting multiple keys simultaneously and so on.
- DAFEE is implemented with the Spark [31] distributed framework and utilizes its advantage to be better applied to large datasets.
- DAFEE reduces the complexity of feature explosion and improves the robustness by applying pruning and feature selection strategy to trim trivial features before the feature expansion process.

The remaining part of this paper is organized as follows: Sect. 2 reviews some related work; Sect. 3 introduces our research motivation, DAFEE algorithm and its implementation in detail; Sects. 4 and 5 present the datasets and experiment results; Sect. 6 summarizes the work.

2 Related Work

Most automatic feature engineering works contain both automatic feature generation and feature selection. Some of them explicitly expand the dataset with all transformed features and apply a feature selection afterward. ExploreKit [13] is such a framework that generates a large number of candidate features with identified common operators and conducts meta-feature based ranking and the greedy-search evaluating approaches. AutoLearn [14] proposes a regression-based feature learning algorithm which constructs regression models to study the feature correlations. These correlation scores are deemed as new features to conduct feature selections after all features are generated.

On the other hand, FEAture DIScovery (FEADIS) [5] employs a wide range of feature combinations and includes constructed features greedily. Cognito [17] expresses feature generation as a directed acyclic graph, which is called a Transformation Tree. Each non-root node in the tree represents a transformed dataset with newly generated features and is associated with a score that measures the importance. Several heuristics tree traversal strategies such as depth-first and balanced traversal could be done with it. [29] introduces Genetic Programming (GP) into feature generation and selection of high-dimensional classification problems, where feature selection selects the leaf nodes used to construct the GP tree. [30] presents AutoCross to automatically generate cross features for categorical features. AutoCross also ensures efficiency, effectiveness, and simplicity with mini-batch gradient descent and multi-granularity discretization.

Some novel techniques such as meta-learning and reinforcement learning are also applied on automatic feature generation. Learning Feature Engineering (LFE) [26] proposes a meta-learning method to evaluate the features' score on the Transformation Tree. It trains a classifier to recommend a set of useful transformations from historical feature engineering experiences on other datasets. [16] derives a strategy for efficiently searching the Transformation Tree by reinforcement learning (RL). Under a budget constraint, RL could be used for performance-driven exploration of the tree.

All researches above only consider single table, Data Science Machine (DSM) [12] is the first end-to-end system that automates feature engineering for relational tables. The core of DSM is Deep Feature Synthesis (DFS), which automatically generates features from interconnected tables. In addition to the DFS-based feature generation method, DSM autotunes an entire machine learning pipeline, including data preprocessing, Truncated SVD based feature selection, and automatic model selection with Bayesian Copula Process for Bayesian hyperparameter optimization. One Button Machine [19] extends DSM to handle large datasets while compensating for the disadvantage of DSM to handle unstructured data. Neural networks are also taken into account as [18] proposes a novel Relational Recurrent Neural Network (R2N) that maps the relational tree to a target value. However, the generated features are not interpretable.

3   Method

Let $\mathcal{D}$ be a set of relational tables, where each entry is related to at least another table in $\mathcal{D}$. Assume any relationship is treated as an edge, and one or more connected edges started from target table constitute a path p. The path set $\mathcal{P}$ includes all possible paths. Suppose $g_p^{(t)}(.)$ is an arbitrary mapping function that takes any features X as input and uses the transformation t along the path p. Then the result feature set F could be written as $\mathcal{F} = \{F : F = g_p^{(t)}(X), \forall \text{ path } p, \text{ transformation } t\}$. To simplify the setting, we assume each table could be split into the training data set $\mathcal{D}_{train}$ and validation data set $\mathcal{D}_{validate}$ by a universal criterion. Assume $\ell(.)$ is the loss function (such as RMSE or F1-score) and $L(.)$ is a learning algorithm. Our target is to derive:

$$F^\star = \underset{F \subseteq \mathcal{F}}{\arg\min}\, \ell(L(\mathcal{D}_{train}, F), \mathcal{D}_{validate}, F) \tag{1}$$

In this manuscript we propose DAFEE, an automatic scalable feature engineering algorithm that is implemented on the distributed computation engine Spark [31]. Based on the relationship of tables, it connects all the tables and convey the information of each table to the target table through joining any two related tables in an ordered manner. Finally, DAFEE applies a feature selection after each time joining other tables to the target table, in order to reduce the number of features. In this section, we explain the motivation for DAFEE and introduce the algorithm in detail.

3.1   Motivation

Most of the researches on feature engineering [7,24,25,28], especially automatic feature engineering [13,17,20,30], only consider single table. However, real-world data is usually much more complicated. In most cases, features are derived from multiple tables. Consider the customer anomaly detection scenario in banking industry as an example. In order to predict whether a customer is abnormal, we need to fully utilize different aspects of information, such as customer profile and transactions. However, these data are usually collected in separate tables, making it impossible for automatic feature engineering algorithms mentioned above to handle. Besides, generating useful features, especially interaction features extracted from multiple tables requires domain knowledge and is time-consuming. In this manuscript, we propose DAFEE to tackle these problems.

3.2   Combination Strategy

To search and join related tables, a revised version of deep feature synthesis (DFS) [12] is used. In DFS, a table is called an *entity*, which is capable to handle numeric, categories, timestamps and free text features. We use the same concepts and symbols introduced in DFS.

A *forward* relationship could be simply regarded as a one-to-many or one-to-one relationship, while a *backward* relationship is just the opposite.

Direct features (DFEAT) are features directly delivered through *forward* relationships. Relational features (RFEAT) are generated by applying aggregation functions like MAX, MIN and SUM on a *backward* relationship. Different from direct features and relational features, Entity features (EFEAT) are created by taking transformations on the entity's current features, regardless of any *forward* or *backward* relationship.

DFS uses a search strategy similar to depth-first-search, which starts from the target entity, traverses through the relation paths and stops searching when finding any leaf. On each leaf entity, entity features would be generated at first. Then direct features or relational features are generated on the leaf and joined to its parent according to the relationship between them. These operations would be taken sequentially until all features are joined back to the target entity.

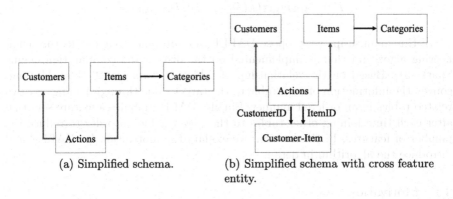

(a) Simplified schema.

(b) Simplified schema with cross feature entity.

Fig. 1. Simplified Schema for e-commerce scenario.

However, this strategy is unable to generate some complicated but useful features. Figure 1(a) shows a simplified schema with four entities targeting for predicting actions of customers in an e-commerce scene. Customers entity contains personal information like age, gender, registration date, and so on. Items and Categories entities hold features describing items and categories. Actions entity is the entity with labels that records customers' actions, such as browse and purchase. The direction of arrows represents the direction of relationships illustrated above. Complicated features, such as *"the number of products whose sales ranks 1st last month in its category that customer A bought in a week"*, are often used to describe the customer but could not be generated by DFS.

To solve this problem, we introduce Combine functions to join *forward* entity back to its father and create new features after generating relational features. With Combine functions, we can first construct relational features like *"whether the sales of this product ranks 1st last month in its category"* and then construct the feature illustrated in the last paragraph based on that. Besides, since a Combine function merges the visited *forward* entity into current entity, the entity tree is simplified as its depth reduces. In addition, we introduce more types of relational features, such as Rank and Proportion features. Conditional features filtered by Time Windows and Interesting Values are also revised to make them more applicable.

Algorithm 1: Naive features generation

---

1 **Function** DAFEE(E^i, E_M, E_V):
2      $E_V \leftarrow E_V \cup E^i$;
3      $F^i \leftarrow F^i \cup$ EFEAT(E^i);
4      $E_F \leftarrow Forward(E^i, E_M)$;
5   **while** $E_F \neq null$ **do**
6   **for** $E^j \in E_F$ **do**
7              $F^j \leftarrow F^j \cup$ RFEAT(E^j, E^i) $\cup$ EFEAT(E^j);
8              $F^i \leftarrow F^i \cup$ DFEAT(E^i, E^j);
9              $Combine(E^i, E^j)$;
10         $E_F \leftarrow Forward(E^i, E_M)$;
11     $E_B \leftarrow Backward(E^i, E_M)$;
12   **for** $E^j \in E_B$ **do**
13   **if** $E^j \cup E_V$ **then**
14   continue;
15   **DAFEE**(E^j, E_M, E_V);
16         $F^i \leftarrow F^i \cup$ RFEAT(E^i, E^j);

Here E and F are entity and feature set corresponding. A^i indicates the i-th element of set A. A_V represents the subset of A annotated by V. For example, E^j presents the j-th entity while E_V stands for the set of visited entities.

3.3   Cross Features Generation

Another common scenario is the multi-key connection scenario, such as two entities connected by two separated keys. In the sample case illustrated in Fig. 1(b), the relationship between Actions entity and Customer-Item entity is a typical multi-key connection relationship since the two entities are connected both by CustomerID and ItemID. As DFS only supports one foreign key in each connection of entities, Customer-Item can not be connected to Actions directly. This causes information lost of the Actions entity in this case.

To support multi-key connection, we introduce a new relationship called *cross* relationship in DAFEE. Operations that are used to create relational features, such as MAX and MIN, are also used to create cross features upon a cross relationship. Besides, some more complicated operations that utilizes both backward and cross relationships, e.g. *"the proportion of milk A purchased by customer B among all milk brands that B purchases in a week"*, are also supplied. With cross features, the procedure of automated feature engineering is more close to manual feature engineering.

In general, we can create more meaningful features using DAFEE. However, at the same time, we also generate more useless features. To deal with such side effect, we introduce a trim algorithm to be explained in Sect. 3.4 and a feature selection module to be illustrated in Sect. 3.5.

3.4 Trim Strategy

A heuristic algorithm with two principle rules is used to trim the useless features. Both rules are introduced according to experts' experiences, some useful features may be dropped, but with a negligible chance:

- The first rule discards those numeric transformations such as SUM or AVG on a date related feature. For example, "sum of the months of transaction dates" is a useless feature, which is an encryption over the month of transaction dates without further useful information.
- The second rule aims to trim cross features. If one entity has a *forward* entity and a *cross* entity that share one common foreign key, then the features generated through DFEAT from the *forward* entity will be passed directly to the *cross* entity. For example, in Fig. 1(b) the features in the Actions entity that are inherited from the Customer entity will be joined to Customer-Item entity directly without any RFEAT operations.

3.5 Feature Selection

Feature selection has been proven to be an effective and efficient way to reduce dimension and achieve a better result to solve many machine learning problems [22]. In general, feature selection algorithms could be categorized into three types: filter, wrapper and embedded [27]. Filter methods select features according to the intrinsic property of dataset, without using any classifier. Typical filter methods include similarity based methods [6,9], information theoretical based methods [2,21] and statistical based methods [4,23]. Wrapper methods iteratively learn and predict feature scores by using certain machine learning algorithm and select part of them as outputs. One such commonly used wrapper method is recursive feature elimination (RFE) [8], which recursively ranks features by the importance score. Embedded methods integrate the feature selection process with the learner training process, and complete both processes in the same optimization stage.

Since the label information is included, and embedded methods usually take less time than wrapper methods, an embedded methods based on XGBoost [3] is used. XGBoost is an end-to-end tree boosting system that is widely used in different machine learning tasks. It also designs mechanisms to achieve high scalability, which is essential in handling the large dataset scenario. Moreover, its feature importance method using boosting trees has already been proven to be able to get good performance in industry [10]. Such feature importance is used to rank and output the top K (or p percent) of the features.

In order to control the number of features, feature selection is applied when the current entity is the target entity. Breadth-First-Search (BFS) allows more control on feature selections, which is also one of the reasons why a BFS type strategy is preferred to search forward entities to a Depth-First-Search alike strategy that are used in DFS. The final feature generation algorithm with trim strategy and feature selection is presented in Algorithm 2.

Algorithm 2: Feature generating

---

1 **Function** DAFEE($E^i, E_M, E_V, target$):
2      $E_V \leftarrow E_V \cup E^i$;
3      $F^i \leftarrow F^i \cup \mathsf{EFEAT}(E^i)$;
4      $E_F \leftarrow Forward(E^i, E_M)$;
5   **while** $E_F \neq null$ **do**
6   **for** $E^j \in E_F$ **do**
7              $F^j \leftarrow F^j \cup \mathsf{RFEAT}(E^j, E^i) \cup \mathsf{EFEAT}(E^j)$;
8              $F^i \leftarrow F^i \cup \mathsf{DFEAT}(E^i, E^j)$;
9              $Combine(E^i, E^j)$;
10   **if** $target == i$ **then**
11                 $F^i \leftarrow FeatureSelect(F^i)$;

12         $E_F \leftarrow Forward(E^i, E_M)$;

13     $E_B \leftarrow Backward(E^i, E_M)$;
14   **for** $E^j \in E_B$ **do**
15   **if** $E^j \cup E_V$ **then**
16   **continue**;
17   **DAFEE**($E^j, E_M, E_V, target$);
18         $F^i \leftarrow F^i \cup \mathsf{RFEAT}(E^i, E^j)$;
19   **if** $target == i$ **then**
20             $F^i = FeatureSelect(F^i)$;

21     $E_C \leftarrow CrossRelation(E^i, E_M)$;
22   **for** $E^j \in E_C$ **do**
23   **if** $E^j \cup E_V$ **then**
24   **continue**;
25   **DAFEE**(E^j, E_M, E_V);
26         $F^i \leftarrow F^i \cup \mathsf{CFEAT}(E^i, E^j)$;
27   **if** $target == i$ **then**
28             $F^i = FeatureSelect(F^i)$;

29   **if** $target == i$ **then**
30         $F^i = FeatureSelect(F^i)$;

3.6 Implementation

Since the amount of data grows rapidly recently, it is necessary to implement algorithms on a scalable framework. Our system is based on Spark for the capability of distributed computation. Beside the high performance, Spark provides a module called Spark SQL [1], which offers sufficient commonly used operations, e.g. SUM, MIN, MAX, SELECT and JOIN, to process relational data. On top of Spark, we have developed additional operations like MODE and MEDIAN to fulfill the needs. All codes are optimized following the Spark code optimization skills, such as caching the intermediate results. Otherwise, the whole DAG may be computed multiple times.

Moreover, the mechanism to filter data based on some specified condition is implemented. Time Window and Interesting Value are two typical filters. A Time Window is a period of time, such as "*1 day*" and "*3 months*". After specifying the time column and cutoff time, one can use the Time Window to filter data by time. For example, if one wants to generate "the number of transactions of each customer in 1 week", one can filter "the number of transactions of each customer" by Time Window "1 week". Since data scientists usually generate the same features with different Time Windows, some commonly used Time Windows are implemented into a TimeWindows object in a convenient manner.

Interesting Value is another widely used filter. Data scientists may be interest in studying the conditions that one certain column's value is equal to a specific value. One example is "the number of actions that a user put ABC items into shopping cart within 1 week". If there is a column called "action type" with value "1" indicating "put into shopping cart", then the interesting Value is the "1" value of column "action type". One side effect is that the number of newly generated features grows dramatically as the number of interesting value increases. For example, if we have two Interesting Values, then the total number of generated features is nearly doubled. To solve this problem, we introduce a hyper-parameter called target columns for Interesting Values to restrict the columns that interesting values filters applied on.

4   Evaluation

We conducted experiments to demonstrate the performance and scalability of our automatic feature engineering approach. We mainly compared DAFEE with the DFS algorithm and its Python implementation Featuretools. The latter is the SOTA algorithm implementation and widely used by data scientists. To evaluate fairly, a four-node Spark cluster (2×8 cores E5-2620 V4 with 8×16 GB ram memory) is set up, where DAFEE and DFS are both ran in the yarn mode. The software environment is based on Linux 3.10.0-957 with Java (1.8.0-131), Hadoop (2.7.2) and Spark (2.4.3).

To justify the performance of our method, we test our implementation on four datasets with different sizes and configurations: HI GUIDES[1], KDD Cup 2014[2], Coupon Purchase (See footnote 2) and IJCAI 2015[3]. The completed machine learning pipeline is introduced in the following sections. For scalability, we compare the speedup on four different datasets mentioned above between our framework and Featuretools on Spark, which is the implementation of DFS based on Spark. All the experiments described in the following paragraphs are repeatedly conducted for three times to reduce the randomness. Average scores are taken as the final results.

4.1   Dataset Overview

Four datasets used in experiments will be described briefly in the chapter.

HI GUIDES predicts whether users will purchase quality travel services in the short term. There are five related tables: UserProfile describes users' profile information; Action contains user behavioral information; OrderHistory includes

---

[1] http://www.dcjingsai.com.
[2] http://www.kaggle.com.
[3] http://ijcai15.org.

users' historical order information; OrderFuture contains user id and order type for training dataset; UserComment includes users' evaluation and comments.

KDD 2014 Cup helps donorschoose.org to find the exciting projects that are more likely to be funded by website users. There are five related tables: Projects includes information regarding project itself and it's corresponding teacher and school; Essays includes a short description and full essay of each project; Donations: contains information about donations on each project; Resources: contains requested resources of each project; Outcomes: contains outcome of each project with the label column "*is_exciting*". Table projects, essays and resources provide both training and testing data, while donations and outcomes only give information regarding projects in training set.

Coupon Purchase predicts the coupons that a customer will buy in a certain period of time. The dataset contains a whole year's information. There are five tables: User_list includes user profile and registration information; Coupon_list includes coupon information; Coupon_visit includes browsing log of coupons; Coupon_detail includes purchase log and more details of coupons; Coupon_area includes area information of coupons.

IJCAI 2015 predicts the repeated buyers for merchants in Tmall. The data could be extracted into four different table: Users includes user profile, including age range and gender; Merchant includes merchant ID; Log includes merchant id, and action information between users and items; User_merchant includes indicates whether a user repeatedly buy a merchant.

4.2 Preparation

To be comparable to DFS, we applied the same machine learning pipeline on DFS and DAFEE, except for the feature generation part. Figure 2 shows the full procedure. As DFS's effectiveness highly relies on its parameter settings such as the choice of aggregation functions and interesting values, we carefully follow Featuretools' guidance to ensure that the result reasonably reflects the true performance of DFS.

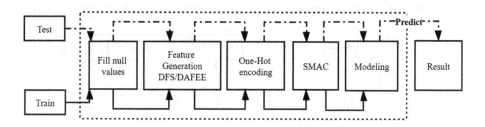

Fig. 2. Machine learning pipeline for effectiveness experiment.

In the preprocessing part, null values are filled with zeros at the beginning. One-hot encoding is put after feature generation to avoid the curse of dimensionality. In modeling section, we choose different model for different task. We use random forest classifier for IJCAI 15 task while applying LightGBM [15] for other three tasks. In order to get the best models, we tuned their hyper-parameters automatically using SMAC [11], which is a widely used hyper-parameter optimization algorithm based on Bayesian optimization. The hyper-parameters we choose to tune are the same as defined in paper of DFS. Specifically, we set $n \in [50, 500]$ as the number of decision trees with default value 50; $m_d \in [1, 20]$ as the maximum depth of the decision trees without any default value; $\beta \in [1, 100]$ as the maximum percentage of features used in decision trees with default value 50; $rr \in [1, 10]$ as the ratio to re-weight underrepresented classes with default value 1.

5   Experimental Results and Analysis

In order to prove that our method is superior to DFS, we compared them on both the aspect of predictive performance and scalability. In this testing phase, DAFEE and DFS treat the four datasets in the same view (Fig. 3). The sample size and configuration information is shown in Table 1. The entity relationship of HI GUIDES and KDD Cup 2014 remain unchanged as they were illustrated originally, while those of the other two's are revised slightly for a better performance. In Coupon Purchase, all features of userCoupon entity are discarded except for user_list, coupon_list, and coupon_visit. userCouponPurchase and user-Purchase entities are generated from the coupon_detail table. By the way, cross features are only generated upon IJCAI15.

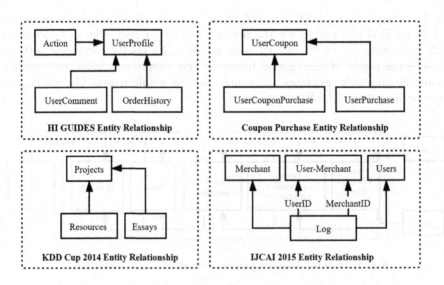

Fig. 3. The entity relationship of four datasets.

Table 1. The number of observations and features per entity in each experiment.

HI GUIDES			IJCAI 2015		
Entity	Rows	Features	Entity	Rows	Features
Action	1,666,060	3	Log	38,743,689	7
OrderHistory	27,512	7	Merchant	1,993	1
UserComment	12,337	5	User-Merchant	260,864	2
UserProfile	50,383	4	Users	212,062	3
KDD Cup 2014			Coupon Purchase		
Entity	Rows	Features	Entity	Rows	Features
Essays	664,098	2	UserCoupon	7,776,447	24
Projects	664,098	35	UserCouponPurchase	241,823	20
Resources	3,667,217	7	UserPurchase	168,996	13

5.1 The Performance of DAFEE and DFS

In this section, we compare the performance of DAFEE and DFS on four datasets in Table 1. Since IJCAI 2015 does not accept submissions any more, we use AUC score with 3-fold cross-validation as the performance metric. The results are illustrated in Table 2.

Table 2. The experiment performances. Online AUC for HI GUIDES and KDD Cup 2014, top 10 online MAP for Coupon Purchase and local AUC for IJCAI 2015.

Algorithm	HI GUIDES	KDD Cup 2014	Coupon Purchase	IJCAI 2015
DFS	0.84**0[a]	0.61485	0.00434	0.647241
DAFEE	**0.87150**	**0.6166**	**0.0053**	**0.66835**

[a]The website shows the exact score as 0.84**0, which omits two decimal values.

It can be observed that DAFEE outperforms DFS on all of the four datasets. Specifically, the performance of DAFEE exceeds which of DFS **3.6%** on HI GUIDES, **0.3%** on KDD Cup 2014, **21.43%** on Coupon Purchase and **3.3%** on IJCAI 2015. The most significant improvement achieved on Coupon Purchase is over 20%. Obviously, DAFEE is able to generate more useful features than DFS, especially in the prediction of transaction types with time related variables.

5.2 The Scalability of DAFEE and DFS

In this experiment, the scalability of the DAFEE and DFS algorithm is measured by the speedup rate when the computation resource increases. Data preparation is applied beforehand to enable FeatureTools on Spark to work properly. Specifically, we divide each dataset into several partitions so that all worker nodes can apply DFS in parallel. However. This method has two obvious shortcomings. First, if one non-target entity has a one-to-many relationship with the target entity, the samples in the non-target entity may be copied several times. Second, the data may be skewed if it is not uniformly distributed.

During the experiments, the number of driver cores and its memory is fixed to 1 and 16 GB. Furthermore, we set the total executor memory to 96 GB and the number of cores in each executor to 4 to ensure that the memory for feature expansion is sufficient. The number of executors are set to 2, 4, 6 and 8 to get the speedup curve. The number of data partitions is set to 96. Furthermore, two executors are treated as one node to make the graph more clear.

Since DFS segments the dataset before feature generation, the time of such segmentation should also be counted. Hence we provide both DFS Main time and DFS Total time, which stands for the time of main DFS process and the time including segmentation time. Because in the IJCAI 2015 experiment DFS failed to complete due to OOM exception, we only draw the speedup curve of DAFEE in Fig. 4.

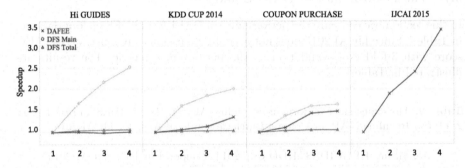

Fig. 4. The speedup performances (in times) of DAFEE and DFS main/total. 1, 2, 3, and 4 on the horizontal axis correspond to the cases where the number of executors is 2, 4, 6, and 8, respectively. The vertical axis is Speedup (run time/run time when executors equal to 2).

Since we manually assign the required data to each partition beforehand, we can find that in Fig. 4 the speedup of DFS Main is always the highest. However, after considering the overhead of data preparation, the speedup of DFS Total drops dramatically, which indicates that the data preparation takes the majority of execution time. On the DAFEE side, we find that it reaches its highest speedup, which is 3.388, on the large IJCAI 2015 dataset, while achieves a low speedup when the size of dataset is small. The reason may be that the communication overhead overwhelm that of computation when the dataset is small. Since the real data are usually huge, DAFEE may achieve a potential high speedup on them. To sum up, it can be inferred from Fig. 4 that DAFEE has a higher Speedup than DFS Total in most scenarios, especially when the amount of data is huge.

6 Discussions and Conclusions

In this paper we introduce DAFEE, a scalable distributed automatic feature engineering algorithm for multiple tables in real-world applications. By providing a novel search method with flexible time windows and new cross relationships, DAFEE can generate more useful features like proportional feature and high-order features than DFS to improve the prediction performance. Besides, DAFEE includes feature selection strategies to remove useless features and reduce the memory consumption. Additionally, experiment results demonstrate DAFEE is more scalable than DFS, which is valuable in the era of big data.

For the future work, we plan to dig deeper into the feature selection part. Since XGBoost used in DAFEE is a supervised learning algorithm, we can only apply feature selection on the target entity. To generalize the feature selection strategies, unsupervised feature selection methods other than XGBoost are needed. In addition, although the scalability of DAFEE is better than DFS, it is still not optimal. In the future, we intend to improve our algorithm and its Spark implementation to get better scalability. Finally, we plan to revise the framework to generate more useful high-order features.

References

1. Armbrust, M., et al.: Spark SQL: relational data processing in spark. In: Proceedings of the 2015 ACM SIGMOD International Conference on Management of Data, pp. 1383–1394 (2015)
2. Battiti, R.: Using mutual information for selecting features in supervised neural net learning. IEEE Trans. Neural Netw. **5**(4), 537–550 (1994)
3. Chen, T., Guestrin, C.: XGBoost: a scalable tree boosting system. In: Proceedings of the 22nd ACM SIGKDD International Conference on Knowledge Discovery and Data Mining, pp. 785–794 (2016)
4. Davis, J.C., Sampson, R.J.: Statistics and Data Analysis in Geology, vol. 646. Wiley, New York (1986)
5. Dor, O., Reich, Y.: Strengthening learning algorithms by feature discovery. Inf. Sci. **189**, 176–190 (2012)
6. Duda, R.O., Hart, P.E., Stork, D.G.: Pattern Classification. Wiley, Hoboken (2012)
7. Guo, H., Jack, L.B., Nandi, A.K.: Feature generation using genetic programming with application to fault classification. IEEE Trans. Syst. Man Cybern. Part B (Cybern.) **35**(1), 89–99 (2005)
8. Guyon, I., Weston, J., Barnhill, S., Vapnik, V.: Gene selection for cancer classification using support vector machines. Mach. Learn. **46**(1–3), 389–422 (2002)
9. He, X., Cai, D., Niyogi, P.: Laplacian score for feature selection. In: Advances in Neural Information Processing Systems, pp. 507–514 (2006)
10. He, X., et al.: Practical lessons from predicting clicks on ads at Facebook. In: Proceedings of the Eighth International Workshop on Data Mining for Online Advertising, pp. 1–9 (2014)
11. Hutter, F., Hoos, H.H., Leyton-Brown, K.: Sequential model-based optimization for general algorithm configuration (extended version). Technical report TR-2010-10. Computer Science, University of British Columbia (2010)

12. Kanter, J.M., Veeramachaneni, K.: Deep feature synthesis: towards automating data science endeavors. In: 2015 IEEE International Conference on Data Science and Advanced Analytics (DSAA), pp. 1–10. IEEE (2015)

13. Katz, G., Shin, E.C.R., Song, D.: ExploreKit: automatic feature generation and selection. In: 2016 IEEE 16th International Conference on Data Mining (ICDM), pp. 979–984. IEEE (2016)

14. Kaul, A., Maheshwary, S., Pudi, V.: AutoLearn—automated feature generation and selection. In: 2017 IEEE International Conference on Data Mining (ICDM), pp. 217–226. IEEE (2017)

15. Ke, G., et al.: LightGBM: a highly efficient gradient boosting decision tree. In: Advances in Neural Information Processing Systems, pp. 3146–3154 (2017)

16. Khurana, U., Samulowitz, H., Turaga, D.: Feature engineering for predictive modeling using reinforcement learning. In: Thirty-Second AAAI Conference on Artificial Intelligence (2018)

17. Khurana, U., Turaga, D., Samulowitz, H., Parthasrathy, S.: Cognito: automated feature engineering for supervised learning. In: 2016 IEEE 16th International Conference on Data Mining Workshops (ICDMW), pp. 1304–1307. IEEE (2016)

18. Lam, H.T., Minh, T.N., Sinn, M., Buesser, B., Wistuba, M.: Neural feature learning from relational database. arXiv preprint arXiv:1801.05372 (2018)

19. Lam, H.T., Thiebaut, J.M., Sinn, M., Chen, B., Mai, T., Alkan, O.: One button machine for automating feature engineering in relational databases. arXiv preprint arXiv:1706.00327 (2017)

20. Leather, H., Bonilla, E., O'Boyle, M.: Automatic feature generation for machine learning based optimizing compilation. In: 2009 International Symposium on Code Generation and Optimization, pp. 81–91. IEEE (2009)

21. Lewis, D.D.: Feature selection and feature extraction for text categorization. In: Proceedings of the Workshop on Speech and Natural Language, pp. 212–217. Association for Computational Linguistics (1992)

22. Li, J., et al.: Feature selection: a data perspective. ACM Comput. Surv. (CSUR) **50**(6), 1–45 (2017)

23. Liu, H., Setiono, R.: Chi2: feature selection and discretization of numeric attributes. In: Proceedings of 7th IEEE International Conference on Tools with Artificial Intelligence, pp. 388–391. IEEE (1995)

24. Markovitch, S., Rosenstein, D.: Feature generation using general constructor functions. Mach. Learn. **49**(1), 59–98 (2002)

25. Mitra, P., Murthy, C., Pal, S.K.: Unsupervised feature selection using feature similarity. IEEE Trans. Pattern Anal. Mach. Intell. **24**(3), 301–312 (2002)

26. Nargesian, F., Samulowitz, H., Khurana, U., Khalil, E.B., Turaga, D.S.: Learning feature engineering for classification. In: IJCAI, pp. 2529–2535 (2017)

27. Sheikhpour, R., Sarram, M.A., Gharaghani, S., Chahooki, M.A.Z.: A survey on semi-supervised feature selection methods. Pattern Recogn. **64**, 141–158 (2017)

28. Tang, J., Alelyani, S., Liu, H.: Feature selection for classification: a review. In: Data Classification: Algorithms and Applications, p. 37 (2014)

29. Tran, B., Xue, B., Zhang, M.: Genetic programming for feature construction and selection in classification on high-dimensional data. Memetic Comput. **8**(1), 3–15 (2015). https://doi.org/10.1007/s12293-015-0173-y

30. Yuanfei, L., et al.: AutoCross: automatic feature crossing for tabular data in real-world applications. arXiv preprint arXiv:1904.12857 (2019)

31. Zaharia, M., Chowdhury, M., Franklin, M.J., Shenker, S., Stoica, I., et al.: Spark: cluster computing with working sets. HotCloud **10**(10–10), 95 (2010)

Embedding Augmented Cubes into Grid Networks for Minimum Wirelength

Jingjing Xia[1], Yan Wang[1(✉)], Jianxi Fan[1], Weibei Fan[2], and Yuejuan Han[1]

[1] School of Computer Science and Technology,
Soochow University, Suzhou 215006, China
20185427013@stu.suda.edu.cn, {wangyanme,jxfan,hyj}@suda.edu.cn
[2] School of Computer, Nanjing University of Posts and Telecommunications,
Nanjing 210003, China
wbfan@njupt.edu.cn

Abstract. Deriving an effective VLSI layout for interconnected network is important, since it increases the cost-effectiveness of parallel architectures. Graph embedding is the key to solving the problems of parallel structure simulation and layout design of VLSI. Wirelength is a criterion measuring the quality for graph embedding. And it is extensively used for VLSI design. Owing to the limitation of the chip area, the total wirelength of embedded network becomes a key issue affecting the network-on-chip communication performance. AQ_n, the n-dimensional augmented cube, is an important interconnection network topology proposed for parallel computers. In this paper, we first study the minimum wirelength of embedding augmented cube into a linear array based on the maximum induced subgraph problem. Furthermore, we obtain the exact wirelength of embedding augmented cubes into grids and propose a linear embedding algorithm to prepare for further study of efficient layout areas.

Keywords: Graph embedding · Wirelength · Augmented cube · Linear array · Grid

1 Introduction

The tremendous engineering advances made in Very Large Scale Integration (VLSI) manufacturing technology has aroused great theoretical interest in VLSI circuit layout issues. Through the effective VLSI layout of the interconnection network, the cost-effectiveness of the parallel architecture can be improved. These efforts have focused on minimizing the layout area of the circuits on the chip. This is partly due to the fact that the layout, which consumes a large amount of chip area, is more expensive to manufacture, less reliable, and more difficult to test, than the VLSI layout which consumes less chip area [2].

In order to meet the requirements of scalability, energy consumption, size, clock asynchronization, reusability etc. in large-scale integrated circuits, the new

© Springer Nature Switzerland AG 2020
M. Qiu (Ed.): ICA3PP 2020, LNCS 12453, pp. 47–61, 2020.
https://doi.org/10.1007/978-3-030-60239-0_4

design method Network-on-Chip (NoC) came into being, which is a new innovation compared to the original design patterns [25]. Grid is one of the most mainstream NoC interconnection structures. It is to connect components together in the form of a matrix. The topology of grid is simple, and it has good scalability and low power consumption. Many researchers early focused on the embedding of simple graphs into complex graphs. They studied the embedding of grids into exchanged crossed cube, crossed cubes, locally twisted cubes, faulty crossed cubes and twisted-cubes [9,20,21,24]. Then, another kind of embedding is to study the embedding of complex graphs into simple graphs. They studied embedding hypercubes, locally twisted cubes, exchanged hypercube and 3-ary n-cubes into grids [1,5,6,18,19]. There is no research on the embedding of augmented cubes into grid networks. Thus, in this paper, we mainly study the embedding of augmented cubes into grids.

Then, we mainly introduce related work and the contribution of this paper in the following subsections.

1.1 Related Work

Augmented cube, an enhancement to the hypercube, proposed by Choudum and Sunitha [4], not only retains some excellent properties of hypercube but also contains some embedding properties that hypercube does not have. For example, n-dimensional augmented cube AQ_n contains cycles of all lengths from 3 to 2^n, but Q_n contains only even cycles [14]. Since its introduction, AQ_n has attracted the interest of many researchers because of its favorable properties. Hsu et al. studied the fault hamiltonicity and the fault hamiltonian connectivity of the augmented cubes [12]. Ma et al. mainly studied panconnectivity and edge-fault-tolerant pancyclicity of augmented cubes [14]. They also studied the super connectivity of augmented cubes [15]. Edge-independent spanning trees have important applications in networks. Thus, Wang et al. studied the edge-independent spanning trees in augmented cubes and proposed an $O(N \log N)$ algorithm that constructs $2n - 1$ edge-independent spanning trees in AQ_n [22]. Mane et al. studied the construction of spanning trees in augmented cubes, and constructed $n-1$ edge-disjoint spanning trees of the augmented cubes [16]. With the development of optical networks, Li et al. studied the routing and wavelength assignment for the augmented cube communication pattern in a linear array wavelength division multiplexing optical network [13].

Graph embedding is the operation of mapping a guest graph into a host graph. Embedding the graph into a linear array is also called linear layout (or linear arrangement) problem. The minimum linear layout problem was first proposed by Harper in 1964 and proved to be NP-Complete [7]. The grid embedding is not only related to the ability of grid to simulate other structures, but also the layout of different structures on the chip. Owing to the limitation of the chip area, the total wirelength of the embedding network has become a key issue affecting the communication performance of the on-chip network. In [1], Bezrukov et al. obtained an approximate result of embedding the hypercube to the grid and lower bound estimate of the wirelength. Rajasingh et al. proposed a minimum wirelength for embedding hypercubes into grid networks [18]. In [19], Shalini et al. proposed a linear algorithm for embedding locally twisted cube into a grid network and obtained the minimum wirelength. In [5], Fan et al. studied embedding exchanged hypercube layout into a grid and obtained an exact formula of minimum wirelength. They also studied the exact wirelength for embedding 3-ary n-cubes into grids [6].

1.2 Contribution

The graph embedding problem is a very worthwhile topic in the field of parallel computing. Regarding the layout of the chip, most researchers let grid be the guest graph because of its simple structure, good scalability and easy to implement on the chip. Augmented cube is an enhancement to the hypercube by adding the complement edge, which makes it more complex than other variants of hypercube. To the best of our knowledge, there are no research results on embedding augmented cubes into grids. In this paper, we study the embedding of n-dimensional augmented cubes into grids with minimum wirelength. The major contributions of the paper are as follows:

(1) By studying embedding AQ_n into linear array L_N, where $N = 2^n$, the minimum wirelength of embedding can be obtained.
(2) We first study embedding AQ_n into grid $M(2^{\lfloor \frac{n}{2} \rfloor}, 2^{\lceil \frac{n}{2} \rceil})$ and calculate the exact wirelength. Then we propose a linear algorithm for the embedding.
(3) We compare the embedding method mentioned in this paper with the random embedding through simulation experiments.

The rest of this paper is organized as follows: In Sect. 2, some preliminaries are described. In Sect. 3, the wirelength of embedding an augmented cube into a linear array is obtained. Then we study the minimum wirelength of embedding AQ_n into a grid. Section 4 gives simulation and experimental results. The last part is the conclusion of this paper.

2 Preliminaries

In this section, we will introduce some definitions and notations used in this paper. Let $G = (V, E)$ be a graph, where $V(G)$ and $E(G)$ denote vertex set and edge set of graph G, respectively. Let (u, v) be an edge with end vertices u and v. And we call u, v neighbors for each other. Given a simple graph G, if $V' \subseteq V$, the subgraph of G induced by the vertex subset V' is denoted by $G[V']$.

Let $G = (V(G), E(G))$ and $H = (V(H), E(H))$ be two connected graphs. G is isomorphic to H (represented by $G \cong H$) if and only if there exists a bijection ψ from $V(G)$ to $V(H)$, such that if $(u, v) \in E(G)$ then $(\psi(u), \psi(v)) \in E(H)$. For a subset $S \subseteq V(G)$, let $T = \{x \in V(H) | there\ is\ y \in S,\ such\ that\ y = \psi(x)\}$. Then, we write $T = \psi(S)$ and $S = \psi^{-1}(T)$.

For two connected graphs G and H, an embedding $\pi = (\psi, P_\psi)$ of G into H is defined as follows [11]:

(i) ψ is a bijective map from $V(G) \rightarrow V(H)$.

(ii) P_ψ is a one-to-one map from $E(G)$ to $\{P_\psi((u,v)) \; : \; P_\psi((u,v))$ *is a path in* H *between* $\psi(u)$ *and* $\psi(v)$ *for* $(u,v) \in E(G)\}$.

Definition 1 [17]. The congestion of an embedding π of G into H is the maximum number of edges of the graph G that are embedded on any single edge of H. Then the congestion of an edge e in H is the number of paths in $\{P_\psi((u,v))\}_{(u,v)\in E(G)}$ such that e is in the path $P_\psi((u,v))$ and denoted by $EC_\pi(e)$. In the other words:

$$EC_\pi(e) = |\{(u,v) \in E(G) : e \in P_\psi((u,v))\}|. \tag{1}$$

Thus, The edge congestion of an embedding π of G into H is given by,

$$EC_\pi(G,H) = max\{EC_\pi(e)|e \in E(H)\}. \tag{2}$$

$$EC(G,H) = min\{EC_\pi(G,H)|\pi \text{ is an embedding from } G \text{ to } H\}. \tag{3}$$

Edge congestion is one of the important factors of embedding problem. The wirelength we mainly study in this paper is another important factor. And there is a significant relationship between edge congestion and wirelength.

Definition 2 [17]. The wirelength of an embedding π of G into H is given by

$$WL_\pi(G,H) = \sum_{(u,v)\in G} d(\psi(u),\psi(v)), \tag{4}$$

where $d(\psi(u),\psi(v))$ denotes the shortest length of the paths $P_\psi((u,v))$ in H.

Lemma 1 [17]. Under the embedding $\pi = (\psi, P_\psi)$, the graph H is divided into two subgraph H_1 and H_2 if one edge cut S is removed. Let $G_1 = G[\psi^{-1}(V(H_1))]$ and $G_2 = G[\psi^{-1}(V(H_2))]$. If S satisfies the following conditions:

(i) For every edge $(a,b) \in E(G_i)$, $i = 1,2$, $P_\psi(a,b)$ has no edges in S.

(ii) For every edge $(a,b) \in E(G)$ with $a \in V(G_1)$ and $b \in V(G_2)$, $P_\psi((a,b))$ has exactly one edge in S.

(iii) G_1 or G_2 is a maximum subgraph.

Then $EC_\pi(S)$ is minimum and $EC_\pi(S) \leq EC_g(S)$ for any other embedding g of G into H.

Since $EC_\pi(S)$ is minimum based on Lemma 1, and according to the definitions of edge congestion and wirelength, the relationship between edge congestion and wirelength is as below.

Lemma 2 [17]. Let $\pi : G \rightarrow H$ be an embedding, and $S_1, S_2, \ldots, S_p$ be p edge cuts of H such that $S_i \cap S_j = \varnothing, i \neq j, 1 \leq i, j \leq p$. Then

$$WL_\pi(G,H) = \sum_{i=1}^{p} EC_\pi(S_i). \tag{5}$$

Then, we would introduce the definition of augmented cube in the following. Let AQ_n denote the n-dimensional augmented cube [4]. And it has 2^n vertices, each of which corresponds to an n-bit binary string. AQ_n can be defined recursively as below [13]:

(1) For $n = 1$, AQ_1 is a complete graph with two vertices labeled 0 and 1, respectively. The edge $(0, 1)$ is called 0-dimensional hypercube edge.
(2) For $n \geq 2$, an AQ_n can be recursively constructed by two copies of AQ_{n-1}. We denote the two copies as AQ_{n-1}^0 and AQ_{n-1}^1, where $V(AQ_{n-1}^0) = \{0u_{n-2}u_{n-3}\ldots u_0|u_i \in \{0,1\} \text{ for } 0 \leq i \leq n-2\}$ and $V(AQ_{n-1}^1) = \{1u_{n-2}u_{n-3}\ldots u_0|u_i \in \{0,1\} \text{ for } 0 \leq i \leq n-2\}$. Then, we add 2^n edges between AQ_{n-1}^0 and AQ_{n-1}^1, and these edges can be divided into the following two sets:
(a) $\{(0u_{n-2}u_{n-3}\ldots u_0, 1u_{n-2}u_{n-3}\ldots u_0)|u_i \in \{0,1\} \text{ for } 0 \leq i \leq n-2\}$, where the edges in this set are called $(n-1)$-dimensional hypercube edges, denoted by HE_{n-1}.
(b) $\{(0u_{n-2}u_{n-3}\ldots u_0, 1\overline{u}_{n-2}\overline{u}_{n-3}\ldots \overline{u}_0)|u_i \in \{0,1\} \text{ for } 0 \leq i \leq n-2\}$, where the edges in this set are called $(n-1)$-dimensional complement edges, denoted by CE_{n-1}.

For example, AQ_1, AQ_2, AQ_3, and AQ_4 are shown in Fig. 1. We can see that Q_n is the subgraph of AQ_n, thus AQ_n retains all favorable properties of Q_n. It is proved in that AQ_n is $(2n-1)$-regular, and $(2n-1)$-connected graph with 2^n vertices for any positive integer n.

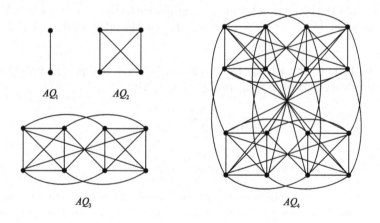

AQ₁   AQ₂

AQ₃   AQ₄

Fig. 1. AQ_n *for* $n = 1, 2, 3, 4$.

3   Embedding Augmented Cubes into Grids

In this section, we first study embedding AQ_n into a linear array and obtain the minimum wirelength of embedding. Then, we further study embedding AQ_n into grid to obtain the minimum wirelength of embedding.

3.1   Embedding Augmented Cubes into Linear Arrays

In this section, we will study the minimum wirelength of embedding AQ_n into a linear array. Before discussing the issue, we first study maximum induced subgraph of AQ_n. It is the key to our research on the wirelength problem.

In [3], Chien et al. solved the maximum induced graph for augmented cube. For any positive integer m, it can be uniquely expressed as $m = \sum_{i=0}^{r} 2^{p_i}$, where $p_0 > p_1 > \ldots > p_r$. Chien et al. proposed a useful function [13]:

$$f(m) = \begin{cases} 0, & m \le 1 \\ \sum_{i=0}^{r}(p_i + 2i - \frac{1}{2})2^{p_i}, & m \text{ is even and } m \ge 2 \\ \sum_{i=0}^{r-1}(p_i + 2i - \frac{1}{2})2^{p_i} + 2r, & m \text{ is odd and } m \ge 2 \end{cases} \tag{6}$$

Property 1 [13]. $f(2^k + m) = f(2^k) + f(m) + 2min\{2^k, m\}$ if $k \ge \lfloor \log_2 m \rfloor$.

Let $\xi_{AQ_n}(m)$ be the number of edges among induced subgraphs with m vertices. The following lemmas can be proved by the function $f(m)$:

Lemma 3 [13]. For any $n \ge 1$ and $0 < m \le 2^n$, then $\max \xi_{AQ_n}(m) = f(m)$.

Lemma 4 [23]. Let $\{U_0, U_1, \ldots, U_k\}$ be a partition of U, where $U \subseteq V(G)$. Let $\xi(U)$ denote the number of edges of the graph $G[U]$, and $\xi(U_i, U_j) = |\{(u, v)|u \in U_i, v \in U_j, \text{ where } 0 \le i < j \le k\}|$. Then $\xi(U) = \sum_{i=0}^{k} \xi(U_i) + \sum_{0 \le i < j \le k} \xi(U_i, U_j)$.

We use L_N to represent a linear array graph with the size of N, where $V(L_N)$ is the vertex set $\{l|0 \le l \le N-1\}$ and $E(L_N)$ is the edge set $\{(l-1, l)|1 \le l \le N-1\}$.

Notation 1. Let $lex : V(AQ_n) \to \{1, 2, \ldots, 2^n\}$ be a mapping, where $N = 2^n$ and for arbitrary vertex $u = u_{n-1}u_{n-2}\ldots u_0$ in AQ_n,

$$lex(u) = \sum_{i=0}^{n-1} u_i * 2^i + 1. \tag{7}$$

which is actually the decimal number of u.

In [13], Chien et al. studied the wavelengths of embedding augmented cube into linear array by considering the congestion. They proved that the natural embedding is an optimal scheme in embedding augmented cube into linear array. There is a significant relationship between edge congestion and wirelength. So, we can use the similar way to prove the following lemma.

Lemma 5. For each edge $e \in E(L_N)$, $EC_{lex}(e) = l(2n-1) - 2f(l)$. And the embedding lex is an optimal scheme which has minimized the congestion of each edge.

Proof. Let $U_{n,l}$ denote a vertex set of l vertices in AQ_n defined by $U_{n,l} = \{u| \sum_{i=0}^{n-1} u_i 2^i < l\}$. Hence $U_{n,l} \subseteq V(AQ_n)$ consists of l vertices and maps to the first l vertices in L_N by the embedding lex. Then $EC_{lex} = l(2n-1) - 2\xi(U_{n,l})$, where $e = (l-1,l)$. Then we would prove that $\xi(U_{n,l}) = f(l)$ by induction n. Clearly, the statement holds for $n = 1$. Suppose that the claim is true for $n \leq k$, i.e., $\xi(U_{k,l}) = f(l)$ for $0 \leq l \leq 2^k$. Then consider that $n = k+1$, that is $0 \leq l \leq 2^{k+1}$. The cases are as below.

Case 1: $l = 0$. Obviously, $U_{k+1,l} = \varnothing$ and $f(0) = 0$. Hence $\xi(U_{k+1,0}) = f(0)$.

Case 2: $l \neq 0$. We consider the following subcases.

Case 2.1: $1 \leq l \leq 2^k$. It implies that $\sum_{i=0}^{n-1} u_i 2^i < 2^k$ for $u \in U_{k+1,l}$. So $u_k = 0$ for $u \in U_{k+1,l}$, i.e., $U_{k+1,l}$ is a subset of $V(AQ_k^0)$. $U_{k,l}^1 = \varnothing$ and $U_{k+1,l} = U_{k,l}^0$. Since AQ_{k+1}^0 is isomorphic to AQ_k and by the induction hypothesis, $\xi(U_{k,l}^0) = f(l)$ which implies $\xi(U_{k+1,l}) = f(l)$.

Case 2.2: $2^k < l \leq 2^{k+1}$. This implies that $|U_{k+1,l}^0| = |V(AQ_k^0)|$ and let $l' = |U_{k,l}^1|$ where $l' = l - 2^k$. Thus for any vertex $u \in U_{k,l}^1$, there are exactly two vertices in $U_{k,l}^0$ adjacent to u. This implies that $\xi(U_{k,l}^0, U_{k,l}^1) = 2|U_{k,l}^1| = 2l'$. Since $\{U_{k,l}^0, U_{k+1,l}^1\}$ is a partition of $U_{k+1,l}$, by Lemma 4 we have $\xi(U_{k+1,l}) = \xi(U_{k,l}^0) + \xi(U_{k,l}^1) + \xi(U_{k,l}^0, U_{k,l}^1)$. By the induction hypothesis, we have

$$\begin{aligned}
\xi(U_{k+1,l}) &= \xi(U_{k,l}^0) + \xi(U_{k,l}^1) + \xi(U_{k,l}^0, U_{k,l}^1) \\
&= f(|U_{k,l}^0|) + f(|U_{k,l}^1|) + \xi(U_{k,l}^0, U_{k,l}^1) \\
&= f(2^k) + f(l') + 2l'
\end{aligned}$$

Therefore, by Property 1, we have $\xi(U_{k+1}^l) = f(l)$.

It is obvious that $EC_{lex} = l(2n-1) - 2\xi(U_{n,l}) = l(2n-1) - 2f(l)$. Thus we can prove that the embedding lex is an optimal scheme which has minimized the congestion of each edge. □

Lemma 6. Under the embedding lex of AQ_n into L_N, where $N = 2^n$, we have

$$WL_{lex}(AQ_n, L_{2^n}) = 2WL_{lex}(AQ_{n-1}, L_{2^{n-1}}) + 2^{2n-1}. \qquad (8)$$

Proof. Let $\pi = lex$. For $n \geq 2$, AQ_n can be partitioned into two disjoint subgraphs AQ_{n-1}^0 and AQ_{n-1}^1 by the definition of AQ_n. Let edge cut $e = (2^{n-1}, 2^{n-1}+1) \in E(L_N)$. Let $(u,v) \in E(AQ_n)$, where $u \in V(AQ_{n-1}^0)$ and $v \in V(AQ_{n-1}^1)$, then we consider the hypercube edges and the complement edges between AQ_{n-1}^0 and AQ_{n-1}^1, respectively.

Case 1. $(u,v) \in HE(n-1)$. For each vertex $u \in V(AQ_{n-1}^0)$, there is a vertex v in AQ_{n-1}^1 adjacent to u. Then the distance of $\psi(u)$ and $\psi(v)$ in linear array is 2^{n-1}. See Fig. 2. There are 2^{n-1} vertices in AQ_{n-1}^1, so $\sum d(\psi(u), \psi(v)) = 2^{n-1} \times 2^{n-1}$, where $u \in V(AQ_{n-1}^0)$ and $v \in V(AQ_{n-1}^1)$.

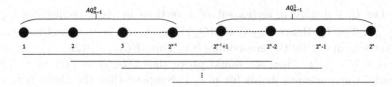

Fig. 2. The paths of embedding $(n-1)$-dimensional hypercube edges into linear array.

Case 2. $(u,v) \in CE(n-1)$. For any vertex u in AQ_{n-1}^0, there always is a $(n-1)$-complement edge (u,v), where $v \in V(AQ_{n-1}^1)$. See Fig. 3. Besides, the distance of $\psi(u)$ and $\psi(v)$ in linear array forms an arithmetic sequence with a tolerance of 2. There are 2^{n-1} vertices in AQ_{n-1}^0, so we have $\sum d(\psi(u), \psi(v)) = 1 + 3 + \ldots + 2^n - 1 = 2^{n-1} \times 2^{n-1}$, where $u \in V(AQ_{n-1}^0)$ and $v \in V(AQ_{n-1}^1)$. Therefore

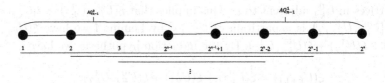

Fig. 3. The paths of embedding $(n-1)$-dimensional complement edges into linear array.

$$WL_\pi(AQ_n, L_{2^n}) = 2WL_\pi(AQ_{n-1}, L_{2^{n-1}}) + 2^{2n-1}$$

$\square$

Theorem 1. The minimum wirelength of AQ_n into L_N under embedding *lex* is:

$$WL_{lex}(AQ_n, L_{2^n}) = 2^{2n} - 3 \times 2^{n-1} \tag{9}$$

Proof. We derive this theorem from Lemma 6, then we would prove the result by induction on n. For $n = 1$, $WL_{lex}(AQ_1, L_2) = 2^2 - 3 \times 2^0 = 1$. Thus, assume that the result is true for $n = k - 1$. Then we prove the result for $n = k$.

$$WL_{lex}(AQ_k, L_{2^k}) = 2WL_{lex}(AQ_{k-1}, L_{2^{k-1}}) + 2^{2(k-1)-1}$$
$$= 2(2^{2(k-1)} - 3 \times 2^{(k-1)-1}) + 2^{2k-1}$$
$$= 2^{2k} - 3 \times 2^{k-1}$$

Then the theorem is proved. $\square$

3.2 Embedding Augmented Cubes into Grids

In this section, we study the minimum wirelength of embedding AQ_n into a grid $M[2^a, 2^b]$, where $a = \lfloor \frac{n}{2} \rfloor$, $b = \lceil \frac{n}{2} \rceil$ and $a + b = n$. Firstly, the definition of grid is given as below:

Notation 2 [5]. An $m \times n$ grid $M(m,n)$ is denoted by an $m \times n$ matrix

$$\begin{pmatrix} \alpha_{11} & \alpha_{12} & \cdots & \alpha_{1n} \\ \alpha_{21} & \alpha_{22} & \cdots & \alpha_{2n} \\ \cdots & \cdots & \cdots & \cdots \\ \alpha_{m1} & \alpha_{m2} & \cdots & \alpha_{mn} \end{pmatrix}$$

Where $V(M) = \{\alpha_{ij} | 1 \le i \le m,\ and\ 1 \le j \le n\}$, $(\alpha_{i,j}, \alpha_{i,j+1}) \in E(M)$ for $1 \le i \le m$, $and\ 1 \le j \le n-1$, and $(\alpha_{k,l}, \alpha_{k=1,l}) \in E(M)$ for $1 \le k \le m-1$, $and\ 1 \le l \le n$. $\langle \alpha_{11}, \alpha_{12}, \cdots, \alpha_{1n} \rangle$ and $\langle \alpha_{m1}, \alpha_{m2}, \cdots, \alpha_{mn} \rangle$ are called the row-borders, while $\langle \alpha_{11}, \alpha_{21}, \cdots, \alpha_{m1} \rangle$ and $\langle \alpha_{1n}, \alpha_{2n}, \cdots, \alpha_{mn} \rangle$ are called the column-borders.

Definition 3. Let $lex\ :\ AQ_n \to M(2^a, 2^b)$ be an embedding, where $a = \lfloor \frac{n}{2} \rfloor$, $b = \lceil \frac{n}{2} \rceil$ and $a + b = n$. Embedding lex can be defined as follows: The first row is labeled from 1 to 2^b from top to bottom, and the ith row is labeled as $(i-1)2^a + 1$, $(i-1)2^a + 2, \ldots, i2^a$ from left to right where $i = 1, 2, \ldots, 2^b$.

The embedding lex of hypercube into grid has been proved in [18]. And augmented cube is an enhancement on hypercube. Then, we first introduce some lemmas about embedding of hypercube.

Lemma 7 [10]. For $i = 1, 2, 3, \ldots, 2^n$, $P_i = \{0, 1, \ldots, i-1\}$ is an optimal set in Q_n.

Let A_i be a horizontal edge cut of the grid $M[2^a, 2^b]$, where $i = 1, 2, \ldots, 2^a - 1$, $a = \lfloor \frac{n}{2} \rfloor$, $b = \lceil \frac{n}{2} \rceil$ and $a + b = n$, such that A_i disconnects $M[2^a, 2^b]$ into two components X_i and $X_{i'}$. For X_i, there are $i2^b$ vertices. Then we let $R_i^{lex} = \{1, \ldots, i2^b\}$ denote the vertices of X_i.

By Lemma 7, the following lemmas can be easily proved.

Lemma 8. $R_i^{lex} = \{1, \ldots, i2^b\}$ is an optimal set in AQ_n for $i = 1, 2, \ldots, 2^a$.

Similarly, let B_j be a column edge cut of the grid $M[2^a, 2^b]$ where $j = 1, 2, \ldots, 2^b - 1$, $a = \lfloor \frac{n}{2} \rfloor$, $b = \lceil \frac{n}{2} \rceil$ and $a + b = n$, such that B_j disconnects $M[2^a, 2^b]$ into two components Y_j and $Y_{j'}$. For Y_i, let C_j^{lex} be the vertex set of Y_j.

Lemma 9. For $j = 1, 2, \ldots, 2^a$

$$C_j^{lex} = \left\{ \begin{array}{llll} 1 & 1 \times 2^b + 1 & \cdots & (2^a - 1) \times 2^b + 1 \\ 2 & 1 \times 2^b + 2 & \cdots & (2^a - 1) \times 2^b + 2 \\ \cdots & \cdots & \cdots & \cdots \\ j & 1 \times 2^b + j & \cdots & (2^a - 1) \times 2^b + j \end{array} \right\}$$

is an optimal set in AQ_n, where $a = \lfloor \frac{n}{2} \rfloor$, $b = \lceil \frac{n}{2} \rceil$ and $a + b = n$.

Lemma 10. The embedding lex of AQ_n into $M(2^a, 2^b)$ induces a minimum wirelength $WL_{lex}(AQ_n, M[2^a, 2^b])$, where $a = \lfloor \frac{n}{2} \rfloor$, $b = \lceil \frac{n}{2} \rceil$ and $a + b = n$.

Proof. Horizontal edge cut A_i disconnects $M[2^a, 2^b]$ into two components X_i and $X_{i'}$. Similarly, column edge cut B_j disconnects $M[2^a, 2^b]$ into two components Y_j and $Y_{j'}$. Let G_i and $G_{i'}$ be the inverse images of X_i and $X_{i'}$ under *lex* respectively. The edge cuts A_i and B_i of the partition, satisfy conditions (i) and (ii) of Lemma 1. In order to show that $EC_{lex}(A_i)$ is minimum, we just need to prove that $|E(G_i)|$ is maximum by Lemma 1.

Since G_i is a subcube derived from the vertices of R_i^{lex} by Lemma 8, it is true that G_i is a maximum induced subgraph of augmented cube. Thus by Lemma 1, $EC_{lex}(A_i)$ is minimum for $i = 1, 2, \ldots, 2^a - 1$.

Similarly, let G_j and $G_{j'}$ be inverse images of Y_j and $Y_{j'}$ under *lex*, respectively. By Lemma 9, it is true that G_j is a maximum induced subgraph of augmented cube derived from the vertices of C_j^{lex}. Thus by Lemma 1, $EC_{lex}(B_j)$ is minimum for $j = 1, 2, \ldots, 2^b - 1$.

Thus by Lemma 2, $WL_{lex}(AQ_n, M(2^a, 2^b))$ is minimum.   □

Lemma 11 [17]. $WL_{lex}(Q_n, P_{2^n}) = 2^{2n-1} - 2^{n-1}$, where P_{2^n} is a path with 2^n vertices.

Firstly, we study the paths of embedding hypercube edges into grid, and the ends of each path are in different columns. The problem can be transformed into calculating the wirelength of embedding a hypercube into a linear array. By Lemma 11 Manuel *et al.* solved the problem of embedding Q_n into $M[2^a, 2^b]$, and calculated the wirelength of a path. Then the exact wirelength of embedding hypercube edges into grid where the ends of each path are in different columns is $2^b(2^{2a-1} - 2^{a-1})$.

Then, we study the paths of embedding complement edges into grid, and the ends of each path are in different columns. Let the embedding $\pi = lex$. Then for the embedding $\pi(\psi, P_\psi)$ of AQ_n into $M(2^a, 2^b)$, where $a = \lfloor \frac{n}{2} \rfloor$, $b = \lceil \frac{n}{2} \rceil$ and

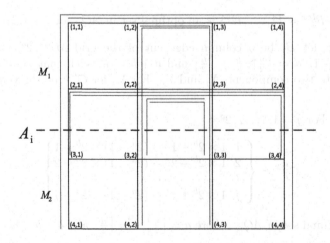

Fig. 4. The paths of embedding 3-dimensional complement edges of AQ_4 into $M[4 \times 4]$.

$a + b = n$, horizontal edge cut A_i disconnects $M[2^a, 2^b]$ into two components X_i and $X_{i'}$. Let G_i and $G_{i'}$ be the inverse images of X_i and $X_{i'}$ under π, respectively. Let $e = (u, v)$ be a complement edge of AQ_n, where $u \in V(G_i)$ and $v \in V(G_{i'})$, thus $\psi(u) = \alpha_{ij} \in V(X_i)$ and $\psi(v) = \alpha_{i'j'} \in V(X_{i'})$. So $d(\psi(u), \psi(v))) = |i' - i| + |j' - j|$. Then, we use W_a to denote the sum of $d(\psi(u), \psi(v)))$, where 2^a is the number fo rows in grid.

Lemma 12. For embedding m-dimensional complement edges into $M[2^a, 2^b]$, where $a < m \le n - 1$ and $a = 2^{\lfloor \frac{n}{2} \rfloor}$, the wirelength of embedding is

$$W_a = 2W_{a-1} + 2^{a+b-2}(2^a + 2^b). \tag{10}$$

Proof. Let $i = 2^{a-1}$, so the horizontal edge cut A_i disconnects $M[2^a, 2^b]$ into two $M[2^{a-1}, 2^b]$, M_1 and M_2, as depicted in Fig. 4. Let $e = (u, v)$ be a complement edge of AQ_n, where $\psi(u) = \alpha_{i,j} \in V(M_1[2^{a-1}, 2^b])$ and $\psi(v) = \alpha_{i'j'} \in V(M_2[2^{a-1}, 2^b])$. Then the distance of e in grid is $d(\psi(u), \psi(v)) = |i - i'| + |j - j'|$. There are 2^{n-1} vertices in $M[2^{a-1}, 2^b]$, so there are 2^{n-1} complement edges that one end vertex is in M_1 and the other end is in M_2. Let y_i be a vertex of the ith row. Then for the first row R_1, there are 2^b vertices, so

$$\sum_{\psi(y_1) \in V(R_1), \psi(v) \in V(M_2)} d(\psi(y_1), \psi(v)) = 2 \times (2^a + (2^a + 2) + (2^a + 4) + \dots$$

$$+ (2^a + 2^b - 2))$$
$$= 2^b[2^{a+1} + 2(2^{b-1} - 1)]$$

It is similar for the rest rows from 2 to 2^{a-1}, then the distance of all the paths of embedding complement edges into $M[2^a, 2^b]$, where for each path one end is in M_1 and the other end is in M_2.

$$\sum_{R_1} d(\psi(y_1), \psi(v)) + \sum_{R_2} d(\psi(y_2), \psi(v)) + \dots + \sum_{R_{2^{a-1}}} d(\psi(y_{2^{a-1}}), \psi(v))$$
$$= 2^{a+b-2}(2^{a-1} + 2^b)$$

So

$$W_a = 2W_{a-1} + 2^{a+b-2}(2^a + 2^b).$$

By Lemma 12, we obtain the recursion formula of the wirelength about embedding complement edges into grids. In the following, we will calculate the exact formula of minimum wirelength about embedding augmented cubes into grids.

Theorem 2. The minimum wirelength of embedding AQ_n into $M[2^a, 2^b]$ under *lex* is:

$$WL_{lex}(AQ_n, M) = 2^a(2^{2b} - 3 \times 2^{b-1}) + 2^b(2^{2a-1} - 2^{a-1})$$
$$+ 2^{a+b-2}(a2^b + 2^{a+1} - 2) \tag{11}$$

where $a = \lfloor \frac{n}{2} \rfloor$, $b = \lceil \frac{n}{2} \rceil$, $n \ge 2$ and $a + b = n$.

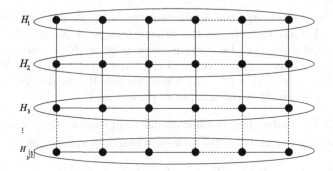

Fig. 5. Mapping vertices of AQ_n to the ith row from left to right.

Proof. There are 2^a rows in $M[2^a, 2^b]$, where $a = \lfloor \frac{n}{2} \rfloor$, $b = \lceil \frac{n}{2} \rceil$ and $a + b = n$. Let H_i denote the ith row, where $1 \leq i \leq 2^a$, as depicted in Fig. 5. Then the inverse images of $H_1, H_2, \ldots, H_{2^a}$ are disjoint sets in AQ_n, and $|(H_i)| = 2^b$. Obviously, vertices in H_i are mapped to the ith row from left to right. There are 2^b vertices in each row. Each row can be considered as a linear array. And the embedding for each row can be seen as the embedding of AQ_b into L_{2^b}. So for each row, $WL_{lex}(H_i, L_{2^b}) = WL(AQ_b, L_{2^b}) = 2^{2b} - 3 \times 2^{b-1}$.

Then, we study the paths of hypercube edges and complement edges in grids. And the ends of each path are in different rows. For the hypercube edges, by Theorem 11, we derived that the wirelength along the columns is $2^b(2^{2a-1} - 2^{a-1})$.

For complement edges, we derived the result by the recursion formula in Lemma 12. The wirelength of complement edges in different rows is $2^{a+b-2}(a2^b + 2^{a+1} - 2)$. We prove the result by induction on a. The base case is trivial. Assume that the result is true for $a = k - 1$. Then we prove the result for $a = k$.

$$
\begin{aligned}
W_k &= 2W_{k-1} + 2^{k+b-2}(2^k + 2^b) \\
&= 2(2^{k+b-3}((k-1)2^b + 2^k - 2)) + 2^{k+b-2}(2^k + 2^b) \\
&= 2^{k+b-2}(k2^b + 2^{k+1} - 2)
\end{aligned}
$$

Thus the minimum wirelength of AQ_n into $M[2^a, 2^b]$ under embedding *lex* is:

$$WL_{lex}(AQ_n, M) = 2^a(2^{2b} - 3 \times 2^{b-1}) + 2^b(2^{2a-1} - 2^{a-1}) + 2^{a+b-2}(a2^b + 2^{a+1} - 2)$$

where $a = \lfloor \frac{n}{2} \rfloor$, $b = \lceil \frac{n}{2} \rceil$, $n \geq 2$ and $a + b = n$.
Then the theorem is proved.   □

For each vertex $u = u_{n-1}u_{n-2}u_{n-3} \ldots u_0 \in V(AQ_n)$, let j $(0 \leq j \leq 2^n - 1)$ be a decimal representation of u. Thus we can use u_n^j to represent each vertex in AQ_n. We present Algorithm 1 for embedding AQ_n into $M(2^a, 2^b)$, where $a = \lfloor \frac{n}{2} \rfloor$, $b = \lceil \frac{n}{2} \rceil$ and $a + b = n$.

Algorithm 1. Embedding AQ_n into $M(2^a, 2^b)$

Input: The augmented cube AQ_n and grid $M(2^a, 2^b)$.
Output: Embedding *lex* of AQ_n into $M(2^a, 2^b)$ with minimum wirelength.

1: /*Label the vertices of AQ_n */
2: Set count = 1;
3: For each vertex in $u \in V(AQ_n)$, let the decimal value of $u = null$;
4: **for** $j = 0$ to $2^n - 1$ **do**
5:    $num(u_n^j) = count$;
6:    $count = count + 1$;
7: **end for**
8: /*Label the vertices of $M(2^a, 2^b)$ */
9: **for** $j = 0$ to $2^n - 1$ **do**
10:   Label the *ith* row of $M(2^a, 2^b)$ as $(i - 1)2^b + 1, (i - 1)2^b + 2, \cdots, i2^b$ from left to right where $i = 1, 2, \cdots, 2^a$.
11: **end for**
12: **return** *lex*.

---

4   Simulation and Experiments

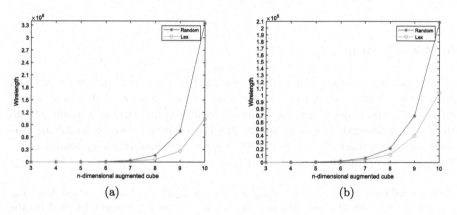

Fig. 6. (a) Wirelength of embedding augmented cubes into linear arrays. (b) Wirelength of embedding augmented cubes into grids

In this section, we compare the result with the other embedding scheme to verify that the proposed embedding is superior to the random embedding [8]. The random embedding (short for Random)is the bijection $f : \{1, .., n\} \rightarrow \{1, \ldots, n\}$ is random.

Firstly, we consider the wirelength of embedding augmented cubes into linear arrays. As seen in Fig. 6(a), compared with the random embedding, the proposed embedding has lower wirelength. With the increasing of the dimension, it also has better performance than random embedding. Besides, the wirelength increases rapidly when we rise the dimension of the augmented cubes. In Fig. 6(a), when n is less than 7, the difference is not obvious. We list the exact wirelength in Table 1. Linear array is a special grid $M[1, n]$, then the comparison of two embedding schemes in gird would be similar to the linear array. In Fig. 6(b), compared with the random embedding, the proposed embedding has lower wirelength. And the exact wirelength is shown in Table 1.

Table 1. Wirelength of embedding augmented cubes into linear arrays and grids in different dimensions

Dimension	Linear array		Grid	
	Lex	Random	Lex	Random
3	52	60	36	42
4	232	300	120	156
5	976	1568	432	606
6	4000	7604	1248	1864
7	16192	36018	4288	6474
8	65152	162366	11648	20782
9	261376	745656	39680	69460
10	1047040	3351960	103936	208526

5   Conclusions

In this paper, we study embedding augmented cubes into grid networks and obtain the exact wirelength of embedding. Firstly, we prove that augmented cubes can be embedded into linear arrays with minimum wirelength and obtain the exact wirelength based on the maximum induced subgraph problem. Furthermore, we obtain the minimum wirelength of embedding augmented cubes into grids and propose a linear algorithm.

Acknowledgment. This work is Supported by the Joint Fund of the National Natural Science Foundation of China (Grant No. U1905211) and A Project Funded by the Priority Academic Program Development of Jiangsu Higher Education Institutions (PAPD).

References

1. Bezrukov, S.L., Chavez, J.D., Harper, L.H., Röttger, M., Schroeder, U.-P.: Embedding of hypercubes into grids. In: Brim, L., Gruska, J., Zlatuška, J. (eds.) MFCS 1998. LNCS, vol. 1450, pp. 693–701. Springer, Heidelberg (1998). https://doi.org/10.1007/BFb0055820
2. Bhatt, S.N., Leighton, F.T.: A framework for solving VLSI graph layout problems. J. Comput. Syst. Sci. **28**(2), 300–343 (1984)

3. Chien, M., Chen, J., Tsai, C.: Maximum induced subgraph of an augmented cube. Int. J. Comput. Electr. Autom. Control Inf. Eng. **8**(5), 807–810 (2014)
4. Choudum, S.A., Sunitha, V.: Augmented cubes. Networks **40**(2), 71–84 (2002)
5. Fan, W., Fan, J., Lin, C.-K., Wang, G., Cheng, B., Wang, R.: An efficient algorithm for embedding exchanged hypercubes into grids. J. Supercomput. **75**(2), 783–807 (2018). https://doi.org/10.1007/s11227-018-2612-2
6. Fan, W., Fan, J., Lin, C., Wang, Y., Han, Y., Wang, R.: Optimally embedding 3-ary n-cubes into grids. J. Comput. Sci. Technol. **34**(2), 372–387 (2019)
7. Garey, M.R., Johnson, D.S.: Computers and Intractability: A Guide to the Theory of NP-Completeness. W.H. Freeman and Company, New York (1979)
8. Glantz, R., Meyerhenke, H., Noe, A.: Algorithms for mapping parallel processes onto grid and torus architectures. Computer Research Repository, abs/1411.0921 (2014)
9. Han, Y., Fan, J., Zhang, S., Yang, J., Qian, P.: Embedding meshes into locally twisted cubes. Inf. Sci. **180**(19), 3794–3805 (2010)
10. Harper, L.H.: Global Methods for Combinatorial Isoperimetric Problems. Cambridge University Press, Cambridge (2004)
11. Heydemann, M., Opatrny, J., Sotteau, D.: Embeddings of complete binary trees into star graphs with congestion 1. In: Hawaii International Conference on System Sciences, pp. 546–554 (1995)
12. Hung, C., Hsu, H., Liang, K., Hsu, L.: Ring embedding in faulty pancake graphs. Inf. Process. Lett. **86**(5), 271–275 (2003)
13. Li, T.K., Chien, M.J., Chen, J.C., Tsai, C.H.: Routing and wavelength assignment for augmented cubes in array-based wavelength-division-multiplexing optical networks. J. Parallel Distrib. Comput. **70**(1), 59–68 (2010)
14. Ma, M., Liu, G., Xu, J.: Panconnectivity and edge-fault-tolerant pancyclicity of augmented cubes. Parallel Comput. **33**(1), 36–42 (2007)
15. Ma, M., Zhu, L.: The super connectivity of exchanged hypercubes. Inf. Process. Lett. **111**(8), 360–364 (2011)
16. Mane, S.A., Kandekar, S.A., Waphare, B.N.: Constructing spanning trees in augmented cubes. J. Parallel Distrib. Comput. **122**, 188–194 (2018)
17. Manuel, P.D., Rajasingh, I., Rajan, B., Mercy, H.: Exact wirelength of hypercubes on a grid. Discrete Appl. Math. **157**(7), 1486–1495 (2009)
18. Rajasingh, I., Arockiaraj, M., Rajan, B., Manuel, P.D.: Minimum wirelength of hypercubes into n-dimensional grid networks. Inf. Process. Lett. **112**(14–15), 583–586 (2012)
19. Shalini, A., Abraham, J., Arockiaraj, M.: A linear time algorithm for embedding locally twisted cube into grid network to optimize the layout. Discrete Appl. Math. **286**, 10–18 (2018)
20. Wang, X., Fan, J., Jia, X., Zhang, S., Yu, J.: Embedding meshes into twisted-cubes. Inf. Sci. **181**(14), 3085–3099 (2011)
21. Wang, X., Qi, D., Zhou, N.: Embedding strategies of mesh network into the exchanged crossed cube. J. Interconnect. Netw. **18**(02–03), 1850011 (2018)
22. Wang, Y., Shen, H., Fan, J.: Edge-independent spanning trees in augmented cubes. Theoret. Comput. Sci. **670**, 23–32 (2017)
23. Yang, X., Evans, D., Megson, G.: Maximum induced subgraph of a recursive circulant. Inf. Process. Lett. **95**(1), 293–298 (2005)
24. Yang, X., Dong, Q., Tang, Y.Y.: Embedding meshes/tori in faulty crossed cubes. Inf. Process. Lett. **110**(14), 559–564 (2010)
25. Zhang, H., Gu, H., Wang, C.: Research of network on chip topology. China Integr. Circ. **16**(11), 42–46 (2007)

ELVMC: A Predictive Energy-Aware Algorithm for Virtual Machine Consolidation in Cloud Computing

Da-ming Zhao[1], Jian-tao Zhou[1(✉)], and Shucheng Yu[2]

[1] College of Computer Science, Inner Mongolia University, Hohhot, China
1477166037@qq.com, cszhoujiantao@qq.com
[2] Department of Electrical and Computer Engineering, Stevens Institute
of Technology, Hoboken, NJ, USA
shucheng.yu@stevens.edu

Abstract. Virtual machine consolidation (VMC) is a technology that aggregates virtual machines distributed on multiple physical machines into a small number of physical machines to improve resource utilization and energy efficiency of data center. However, excessive virtual machine aggregation and migration can also have a significant negative impact on performance. In this paper, an algorithm named ELVMC with multiple resource prediction is proposed for optimal virtual machine consolidation. It applies a modified Best-Fit Decreasing (BFD) algorithm for resource optimization at both overloaded hosts and underloaded hosts with consideration of load balancing. Different from current research, ELVMC aims to obtain an optimal virtual machine (VM) placement during each consolidation process by simultaneously optimizing multiple system performance metrics in terms of energy consumption, VM migrations and QoS guarantees while keeping the load balanced. Simulation results show that ELVMC is superior to the state of the arts, including the traditional BFD and SABFD-HS algorithms as well as recent research VMCUP-M and MUC-MBFD.

Keywords: Energy consumption · Load balance · Resource prediction · Virtual machine consolidation · Virtual machine migration

Supported by the National Natural Science Foundation of China under Grant No. 61662054, 61262082, Inner Mongolia Colleges and Universities of Young Technology Talent Support Program under Grant No. NJYT-19-A02, the Major Project of Inner Mongolia Natural Science Foundation: Research on Key Technologies of Cloud Support for Big Data Intelligent Analysis under Grant No. 2019ZD15, Natural Science Foundation of Inner Mongolia under Grand No. 2015MS0608, Inner Mongolia Science and Technology Innovation Team of Cloud Computing and Software Engineering, and Inner Mongolia Application Technology Research and Development Funding Project "Mutual Creation Service Platform Research and Development Based on Service Optimizing and Operation Integrating".

1   Introduction

Cloud computing is a popular computing model that provides scalable, on-demand services worldwide via the Internet [1]. It is leased to users in a pay-as-you-go way, allowing them to customize the right amount of resources according to their requirements. However, the widespread use of cloud computing increases the energy consumption of data centers [2], which has become a significant challenge to the resource management of cloud data centers. According to the U.S. Department of Energy statistics, the data center accounts for 1.5% of all U.S. energy consumption [3]. The high energy consumption of a cloud data center is not only due to a large number of servers but also because of its inefficient resource usage, which results in significant energy waste. The latter is mainly attributed to the low CPU utilization in cloud systems – a recent IBM report shows that the average CPU utilization of the physical machine (PM) in the data center is only 15% to 20% [4]. This is mainly due to the energy consumed by other sub-systems such as memory, mainboard and hard drive, even when the CPU is idle. Previous research reveals that an idle physical host usually consumes as much as 50% of its peak energy consumption [5]. To reduce the overall energy consumption of a cloud system, one strategy is, therefore, to aggregate VMs to as few PMs as possible so as to reduce the number of idle or underloaded PMs. For this purpose, virtual machine consolidation (VMC) has recently been widely applied, wherein VMs are live migrated to appropriate PMs for resource aggregation. PMs without VMs running are then switched to low-energy mode (e.g., sleep mode) if not completely shut down.

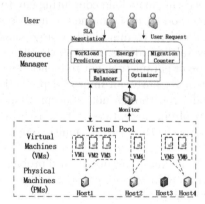

Fig. 1. Energy-efficient resource management framework

Although promising, achieving optimal VMC in cloud systems is by no means trivial. This is mainly caused by the uncertainty and dynamics of real-time load in a cloud center. To accommodate uncertainty and dynamics, oversubscription is usually adopted in real-world cloud computing systems, that is, the resources provided typically exceed the maximum actual resource required. But this method, in turn, results in massive waste in PMs and energy consumption. Therefore, there shall be a prediction algorithm in place to actually foresee the future load of the cloud in a timely fashion. What impacts a VMC strategy also includes load balancing and Quality of Service (QoS). Load balancing distributes the resource utilization equally across the nodes in the cloud to avoid the situation where few nodes are overloaded while others are idle. It helps improve the performance and resource utilization of the data center to achieve required QoS, an important factor for user satisfaction. QoS requirements are commonly formalized in the form of Service Level Agreements (SLAs) that usually defines critical business requirements including the minimum throughput or maximum response time characteristics of the system. Last but not least, VM migration itself also introduces extra energy consumption and adversely impacts QoS because of additional delays. Therefore, an optimal VMC strategy shall minimize the number of VM migrations.

To address the aforementioned problems, an optimal VMC scheme shall consider the strategies for source/destination PM selection as well as which VMs to be migrated. To this end, predictive VMC has been adopted in the literature, which is compared to one of the traditional methods which allocate resources in advance to meet the peak requirements by estimating users' needs to guarantee the QoS. Specifically, time series analysis [6] is a common approach to be utilized for VMC, which predicts a load of next time slot based on the historical information and reallocates resources in advance according to the expected value. Traditional forecasting methods, such as exponential smoothing [7] and moving average [8], do not fit the load trend well with low prediction accuracy and are not suitable for load forecasting in a cloud computing environment. The artificial intelligence prediction method [9] based on neural networks has high prediction accuracy and can be realized by training with a large number of data samples. However, it exhibits several limitations, such as the slow convergence rate, its sensitivity to the selection of parameters, not to mention its tendency to fall into a locally optimal solution. Recently, an improved prediction method for traditional prediction methods [10], though not as accurate as artificial intelligence prediction methods, can obtain acceptable solutions in a faster time and has become a promising approach. However, its multi-step prediction performance is unknown.

In general, VMC has been known as an NP-hard problem, for which optimization algorithms can find an approximate optimal solution. Heuristic algorithms such as First-Fit Decreasing (FFD), Best-Fit Decreasing (BFD) [11] and their variants such as MBFD, MFFD, space aware best fit decreasing (SABFD) have been proven effective in solving the VMC problem [12] which is essentially as a bin-packing problem. However, these methods usually consider the sole objective of minimizing the number of PMs and inevitably produce a significant number of unnecessary migrations and increase the risk of SLA violations. Nguyen et. al. [13] proposed a VMC method with multiple usage prediction to overcome the drawbacks mentioned above. Recently, Moghaddam et. al. [14] designed a fine-tuned prediction model to recorrect the error of prediction algorithm to improve energy efficiency, reduce SLA violations, and minimize VM migrations. But these two methods do not take load balance into consideration and hence the response time will increase because of the resource competition in high-loaded PM. In contrast to most of the existing works that only minimize the energy consumption or migrations, this paper considers the VMC problem as a two-dimensional bin packing problem to simultaneously optimize multiple performance objectives of minimizing energy consumption and VM migration while maintaining the load balance. To solve this multi-objective optimization problem, a predictive energy-aware algorithm with an optimal load balance named ELVMC, which improves the VM selection process and the consolidation process based on BFD, is designed to determine the strategy of VMC.

The main contributions of this paper are as follows:

- A new framework is proposed for predicting and analyzing resource requirements for VM consolidation implements to simultaneously minimize energy consumption, migrations, and improve load balance of resources among servers.
- A novel VM selection method is designed, which considers the variation of VMs' resource utilization in the future, and selects the VM with the largest migration efficiency to reduce VM migrations.
- The VMC problem is formulated as a two-dimensional bin-packing problem considering the CPU and memory utilization and a heuristic solution is designed. The solution adopts four metrics to evaluate the energy consumption, VM migrations, and keep the load balance of the data center at the desired service level.
- Experimental evaluation is performed on real-world workloads of the Google Cluster. Simulation results show that the proposed algorithm exhibits better performance compared to the traditional BFD algorithm, VMC algorithm with multiple usage prediction (VMCUP-M), and SABFD under typical evaluation criteria.

The rest of this paper is organized as follows. Section 2 discusses the related work. The problem definition and model formulation are presented in Sect. 3. Section 4 describes the proposed algorithm. Section 5 shows the experimental results and compares them with other methods. Section 6 concludes the paper.

2   Related Work

VMC has been proved as an NP-hard problem, which is usually formulated as a bin packing problem. Previous research usually address it as a one-dimensional bin packing problem or multi-dimensional bin packing problem depending on the resources considered. Additionally, this problem can be divided into a single-objective or multi-objective problem according to the objective to be optimized. An early [15] presents a scheme called PADD, which utilizes single-objective optimization for this problem. It saves energy by dynamically migrating VMs and packing them into fewer physical machines as much as possible. As one of the multi-objective optimization solutions in literature, Raju et al. [16] proposed an algorithm called EAMOCA in hybrid cloud to minimize execution time and energy consumption while maximizing the resource utilization.

Recent research on this problem started to consider multiple factors pertaining to VMC and formulate the problem as multi-objective optimization. Aziz et al. [17] proposed a server consolidation algorithm called Sercon, which assigns scores to VMs to be migrated according to their CPU and memory utilization. The purpose of this algorithm is to reduce the number of servers and the number of migrations. However, it does not consider energy consumption. As Sercon is non-predictive, unnecessary migrations will occur. Alsadie et al. [18] designed an algorithm called LIFE to decide a target PM for a VM based on the correlation coefficient and predicted future requirements of computing resources. With the help of prediction, it reduces energy consumption and the number of VM migrations. Wu et al. [19] presented an algorithm called IGGA to obtain the consolidation policy with the highest overall score of energy consumption and migration overhead by comparing the consolidation scores. Meanwhile, CPU, memory, and network bandwidth are considered in this algorithm. However, the algorithm only focuses on the current resource usage of each physical host without predictive analysis.

Cloud computing resource requirements are highly time-varying, nonlinear, chaotic and indeterminacy [20]. In the previous studies on cloud computing load forecasting, researchers applied multiple methods and models to predict which host would be overloaded or underloaded. Nguyen et al. [21] proposed a VMC algorithm called VMCUP, which uses linear regression to estimate the short-term future CPU utilization and ultimately reduce total migrations. However, such traditional forecasting methods do not fit the load trend well with low prediction accuracy when multi-step prediction is required. Pacheco-Sanchez et al. [20] innovatively designed the MAP/MAP/1 queuing model for performance prediction of applications deployed in the cloud and presented a theory for parameterizing the model. Farahnakian et al. [11] utilized two regression prediction models, linear regression and K-Nearest Neighbor (KNN), while the experimental results show that the prediction error increases with the number of prediction steps. Calheiros et al. [22] obtained a high prediction accuracy by applying the autoregressive integrated moving average (ARIMA) model and improved the resource utilization rate. But the energy consumption of the data center is not considered in this work. Alsadie et al. [10] also took ARIMA as their predictive method to reduce performance degradation because of the VM migration.

Although meta-heuristic algorithms can usually yield better results, researchers prefer greedy heuristic algorithms to meta-heuristic algorithms because of their high execution efficiency in large data centers. Moreover, some of them pay attention to the VM selection instead of utilizing traditional methods, such as minimum migration time or randomly select. This is because multiple resources are considered, and their relationship to each other affects the number of migration. Beloglazov et al. [23] proposed an improved BFD algorithm named

PABFD to consolidate servers with the least increase in energy consumption. In order to avoid excessive migration while reducing energy consumption, Hui et al. [12] designed an algorithm called SABFD to decide the placement policy for the VMs with high CPU utilization. However, it ignores the migration time affected by memory. Nguyen et al. [13] also applied modified PABFD with utilization prediction. And they considered the minimum resource temperature to find out the reason for overloaded caused by VM resource utilization to reduce the number of migrations. Sayadnavard et al. [24] presented an algorithm based on each PM score to improve energy efficiency. VMs are assigned a weight according to the results of the two sorts about the CPU usage and memory usage. Then the heuristic VMC algorithm is processed. Thiam et al. [25] considered credits and migration costs to select the PM with the minimum weighed by the migration cost between the current position and the target PM. But it is non-predictive, some unnecessary migrations will be avoided. Liu et al. [26] studied an enhancing energy-efficient and QoS dynamic VMC (EQVC) method. It determines whether to migrate VMs from the overloaded host by comparing the loss of CPU capacity of the VM migration and the PM overload. The relevant research is summarized in Table 1. The proposed method is at the bottom of the table.

In general, there has yet few research in literature that focuses on load balancing between resources while considering energy consumption and migration. Given that the improvement of each sub-objective may cause other performance degradation, it is necessary to design a new predictive method to dynamically and quantitatively deploy the VMs according to the workload changes over some time while considering these constraints.

Table 1. Different aspects of the notable recent VMC algorithms

Reference	Resource	Predictive	Objective			
			Energy	Migration Times	QoS	Load Balance
[15]	CPU	✗	✓	✗	✗	✗
[16]	CPU	✗	✓	✓	✓	✗
[17]	CPU, Memory	✗	✗	✓	✗	✗
[18]	CPU	✓	✓	✓	✗	✗
[19]	CPU, Memory, Bandwidth	✗	✓	✓	✗	✗
[21]	CPU	✓	✓	✓	✗	✗
[20]	HTTP Requests	✓	✗	✗	✓	✗
[13]	CPU, Memory	✓	✓	✓	✗	✗
[22]	HTTP Requests	✓	✗	✗	✓	✗
[12]	CPU, Memory	✓	✓	✗	✗	✗
[23]	CPU	✗	✓	✓	✗	✗
[14]	CPU	✗	✓	✗	✓	✗
[13]	CPU, Memory, Bandwidth	✓	✓	✓	✓	✗
[25]	CPU, Memory	✗	✓	✓	✓	✗
[26]	CPU	✗	✓	✓	✗	✗
[27]	File System	✗	✓	✗	✗	✗
ELVMC	CPU, Memory	✓	✓	✓	✓	✓

3   Problem Statement and Model Formulation

Figure 1 shows the energy-efficient resource management framework considered in this paper. Monitor module is for collecting CPU and memory utilization, and send these data to other modules. At the same time, overload and underload hosts are determined. The optimizer module receives this information. Then the workload predictor module predicts the resource utilization in the future according to the data from the monitor module. The energy consumption module is employed to calculate the energy consumption of each server according to the CPU utilization in the monitor module. The migration counter module calculates migrations of VMs on overloaded and underloaded hosts based on the information from the monitor module and the optimizer. The current load of each server will be calculated by the workload balancer module and sent it to the optimizer module. Next, the optimizer module will migrate VMs on underloaded or overloaded hosts to obtain the most effective consolidation scheme based on the information of each module. The virtualization module performs VM migration and actual resource allocation according to the consolidation scheme of the optimizer module. Then, it will change the servers' state. The above optimization process is periodically performed.

3.1   Energy Consumption Modeling

It is assumed that the data center consists of m heterogeneous PMs, $H = \{H_1, H_2, ..., H_m\}$. At any given time, users submit their requests for provisioning n VMs, $V = \{V_1, V_2, ..., V_n\}$. The energy consumption of a physical host consists of the energy consumption of multiple system subcomponents. Like most work, the energy consumption of CPU is demonstrated as the most significant contributor to the energy consumption of a system, and the relationship between the energy consumption of a physical host and the utilization of CPU can be considered linear [2,27]. Here, shutting down the idle host is replaced with switching it to sleep mode as the cost of restarting the new host would be prohibitive. Therefore, each physical host can be classified into two states, which can be presented as E_j^{busy} and E_j^{idle}. And the energy consumption model proposed in [2] is also applicable to the experimental environment of this paper. The energy consumption of jth physical host at time t can be calculated as:

$$E_j\left(U_j\left(t\right)\right) = \begin{cases} (E_j^{busy} - E_j^{idle}) \times U_j\left(t\right) + E_j^{idle} & U_j > 0 \\ E_j^{idle} & otherwise \end{cases} \tag{1}$$

where E_j^{idle} and E_j^{busy} are the average measured energy consumption of the jth host when it is in sleep mode and full workloads, respectively. $U_j\left(t\right)$ represents the CPU utilization rate of the jth physical host at time t. It can be formalized as:

$$U_j\left(t\right) = \frac{\sum_{i \in H_j} vc_i\left(t\right)}{C_j} \tag{2}$$

where C_j represents the CPU capacity of the jth PM. $vc_i(t)$ is the used CPU capacity of the ith VM on the jth PM at time t.

The energy consumption that the jth PM consumes in period $[t_0, t_1]$ is defined as below:

$$E_{H_j} = \int_{t_0}^{t_1} E_j(U_j(t)) \, dt \tag{3}$$

The value of the t_0 and t_1 will be explained in Sect. 5.

3.2 Resource Usage Prediction

The purpose of the resource usage prediction is to adjust the supply of resources gradually in advance by predicting the future requirements from the cloud consumer. The resource utilization time series is taken as the input of the load prediction model, and evaluate the demand value in the next cycle through the process of data preprocessing, model identification, and future demand prediction. The resource utilization prediction model adopted in this paper is based on the ARIMA model, which is characterized by high prediction accuracy and is capable of one-step or multi-step prediction based on loads matching multiple patterns. Most importantly, it has been proven to be simple to use for forecasting resource utilization in a cloud environment [10].

The experimental data used in this paper represent the CPU and memory utilization of each node as a set of time series values. This historical data is obtained through the monitor module. The collected historical data is selected as the input of the prediction model, and a model suitable to match the time series is trained through this historical data. Finally, a one-step or multi-step prediction can be made for future demand. The predicted utilization for the CPU and memory resource for PM P_j at time t is computed as:

$$PU_j(t) = ARIMA(U_j(t)) \tag{4}$$

$$PM_j(t) = ARIMA(M_j(t)) \tag{5}$$

where $PU_j(t)$ and $PM_j(t)$ denote the predicted CPU and memory utilization of P_j respectively. $M_j(t)$ is the memory utilization rate of the jth physical host at time t. It can be calculated as:

$$M_j(t) = \frac{\sum_{i \in H_j} vm_i(t)}{MEM_j} \tag{6}$$

where MEM_j represents the memory capacity of the jth PM. $vm_i(t)$ is the used memory capacity of the ith VM on the jth PM at time t.

3.3 The Number of Migrations

Because excessive resource utilization will increase resource competition, it is necessary to reserve some physical resources for each physical host to guarantee its normal operations. The upper bound of CPU and memory usage for each host are denoted as T_C and T_R. These two values are usually set to 80% and 100% of their resource utilization respectively, according to the research in [26]. Overloaded hosts and underloaded hosts are distinguished by CPU and memory thresholds.

The number of virtual machine migrations $N_{total}(t)$ in the data center consists of the number of migrations of virtual machines on overloaded hosts $N_{ol}(t)$ and virtual machines on underloaded hosts $N_{ul}(t)$ at time t. It can be calculated as:

$$N_{total}(t) = N_{ol}(t) + N_{ul}(t) \tag{7}$$

$$N_{ol}(t) = \sum_{j=1}^{m} N_{ol\_j} \tag{8}$$

$$N_{ul}(t) = \sum_{j=1}^{m} N_{ul\_j} \tag{9}$$

For an overloaded host OL_j at time t, some VMs on it are migrated out so that the host will not exceed the maximum capacity. Its migrations N_{ol_j} is an uncertain value that depends on the VM selection strategy and it will be discussed in the next section. For an underloaded host UL_j at time t, all VMs on it will be migrated out so that the host can enter sleep mode to save energy. Obviously, for an underloaded host, its migrations N_{ul_j} is a fixed value that is equal to the number of VMs on it. It can be calculated as:

$$N_{ul\_j} = \sum_{i=1}^{n} D_{iUL_j} \tag{10}$$

D_{iUL_j} is a binary number, and it is set to 1 if the ith VM can be migrated from the underloaded host UL_j. Otherwise, it is set to 0. In particular, if there are no hosts that can accommodate all VMs of an underloaded host at time t, no migrations will happen and N_{ul_j} will be set to 0. In particular, for overloaded hosts, it is assumed that there will always be a destination host to accommodate the VMs that are going to be migrated.

3.4 Load Balance Modeling

The purpose of load balance is to ensure that the resources of each physical host are effectively allocated and equitably used to guarantee the cloud user's QoS requirements. In order to avoid excessive resource competition among virtual machines on the same host, $L_j(t)$ is used to evaluate the residual space between the resource demand of the jth PM and the resource threshold of the jth PM at time t. It can be calculated as:

$$L_j(t) = \sqrt{\left((U_j(t) - T_C)^2 + (M_j(t) - T_R)^2\right)} \tag{11}$$

The smaller of $L_j(t)$, the more reasonable allocation of resources on jth PM.

4   Algorithm Design

In this section, a VM consolidation algorithm called ELVMC is proposed. VMC is conventionally considered as a bin packing problem. The bin represents a physical host with multi-dimensional resources such as CPU, memory and network bandwidth. In this paper, VMC is considered as a two-dimensional bin packing problem with considering CPU and memory of PMs. And the items are the VMs to be deployed. VMC problems are typically decomposed into three subproblems, namely, source PM selection strategy, VM selection criteria and destination PM selection strategy [28], which are proved to be NP-hard. The solution to each of them will be discussed in this section.

4.1   Problem Definition

Suppose that there are n VMs to be consolidated into m PMs. The objective is to simultaneously minimize energy consumption, migrations and load balance, so as to obtain a new mapping between VMs and PMs by migrating VMs. The consolidation problem can therefore be formulated as:

Minimize:

$$E = \sum_{j=1}^{m} E_{H_j} \tag{12}$$

$$N = \int_{t_0}^{t_1} N_{total}\left(t\right) \tag{13}$$

$$L = \int_{t_0}^{t_1} \sum_{j=1}^{m} L_j\left(t\right) \tag{14}$$

In addition, this paper focuses on short-term load prediction to avoid unnecessary migrations, so the predictive utilization of resources also need to be considered as constraints. Therefore, the constraints of the optimization problem should be subjected to:

$$U_j\left(t\right) \times C_j + vc_i^{'}\left(t\right) \le C_j \times T_C \tag{15}$$

$$M_j\left(t\right) \times MEM_j + vm_i^{'}\left(t\right) \le MEM_j \times T_R \tag{16}$$

$$PU_j\left(t + \eta\right) \times C_j + vc_i^{'}\left(t + \eta\right) \le C_j \times T_C \tag{17}$$

$$PM_j\left(t + \eta\right) \times MEM_j + vm_i^{'}\left(t + \eta\right) \le MEM_j \times T_R \tag{18}$$

where the ranges of i and t are limited as: $i \in V$, $t \epsilon [t_0, t_1]$. $vc_i^{'}\left(t\right)$ and $vm_i^{'}\left(t\right)$ are the CPU and memory capacities of the VMs that are going to be migrated at time t respectively. Similarly, $vc_i^{'}\left(t + \eta\right)$ and $vm_i^{'}\left(t + \eta\right)$ are the CPU and memory capacities of the VMs that are going to be migrated at time $t + \eta$ respectively. η is the time interval between consecutive monitoring events. $PU_j(t + \eta)$ and $PM_j(t + \eta)$ are the predicted CPU and memory utilization of the destination host j at time $t + \eta$ respectively.

Algorithm 1: ELVMC

Input: Initial placement scheme: *population*
Output: The optimal placement scheme of VMs to satisfy multiple objectives

1 /*Overloaded server migration with ELVMC*/
2 $Migrations = 0$
3 $overMark = 0$
4 **for** H_j in H **do**
5      $VMList \leftarrow sortVM(H_j)$
6   **while** $U_j(t) > T_C$ or $M_j(t) > T_R$ **do**
7          $population, Migrations \leftarrow ELBFD(VMList, H_j, overMark)$
8   Update $U_j(t), M_j(t)$
9   **if** $Migrations == 0$ **then**
10   Select a sleep host h_{sleep} from H_{sleep} and switch it to an active mode
11             $population, Migrations \leftarrow ELBFD(VMList, H_j, overMark = 1)$
12   Update $U_j(t), M_j(t), U_{h_{sleep}}(t)$ and $M_{h_{sleep}}(t)$
13   **if** $Migrations == 0$ **then**
14   Buy one or more new hosts h_{new} from the cloud provider H_{new} so that the H_j is not overloaded.
15   Update $U_j(t), M_j(t), U_{h_{new}}(t)$ and $M_{h_{new}}(t)$
16                 $H \leftarrow H \cup \{h_{new}\}$
17   **end**
18   **end**
19   **end**
20 **end**
21 /*Underloaded server migration with ELVMC*/
22 **for** H_j in H **do**
23   /*Sort hosts in ascending order by E_{H_j}/L_j */
24     $H_{underloadedHost} \leftarrow sortHost(H)$
25     $VMList \leftarrow sortVM(H_{underloadedHost})$
26     $population, Migrations, VMList \leftarrow$ $ELBFDUnderloaded(VMList, H_{underloadedHost})$
27   Update $U_{underloadedHost}(t), M_{underloadedHost}(t)$ and $VMList$
28   **if** $len(VMList) == 0$ **then**
29   Switch $H_{underloadedHost}$ to sleep mode
30         $H_{sleep} \leftarrow H_{sleep} \cup \{H_{underloadedHost}\}$
31   **end**
32   **else**
33   The VMs on the underloaded PM $H_{underloadedHost}$ can not be migrated out
34   **end**
35 **end**

4.2 Source PM Selection Strategy

The source PM is the host from which the VMs will migrate out. By setting the upper threshold, which has been defined in the previous section, the host is regarded as overloaded if its CPU utilization or memory utilization is higher than its thresholds. If there is no host overload, then the host with the smallest E_{H_j}/L_j value is selected as the underloaded host. This is because a smaller E_{H_j} means that the resource utilization of the request received by the host at this time is small, which leads to lower energy consumption. Furthermore, a larger L_j means that there is a high probability that the CPU and memory resources on the host have a large probability of unbalanced distribution. It is easy to cause waste of specific resources.

4.3   VM Selection Criteria

Once the source PM is determined, the following step of VMC is to select one or more VM(s) from the source PM. In order to avoid the excessive competition of resources, once a physical host is overloaded, some VMs on it should be migrated out to mitigate the overload risk. In this case, the selected VMs should follow the principles listed below:

– The resource utilization of the destination PM should not exceed the upper threshold after migration to avoid the possibility of the destination PM being overloaded in the future due to the migration. Therefore, the VM on the overloaded PM with the largest increase in resource requirements is considered as a candidate VM to be migrated. The resource utilization change of a VM can be defined as:

$$incVM = (vc_i^{'}(t + \eta) - vc_i^{'}(t)) \times (vm_i^{'}(t + \eta) - vm_i^{'}(t)) \qquad (19)$$

The value of it may be negative, which means that the VM will not increase its resource requirements so that it tends to be kept in the source PM.

– Maximize migration efficiency (ME), that is, minimize migration times and minimize migration time so that the source host is not overloaded. The CPU utilization of the host directly affects its energy consumption, and in most hosts, excessive CPU utilization is often the main factor causing its overload [12]. At the same time, in order to minimize the migration time, virtual machines with small memory should be migrated first [12]. Taking the above two points into consideration, this paper defines the priority of ith VM migration on the jth PM at time t is calculated as:

$$priority_{ij} = \frac{vc_i(t)}{vm_i(t)} \times incVM \qquad (20)$$

The VMs' priority will be sorted in descending order to achieve the optimization goal. It can be easily found that the VM with increased resource requirements, which is also with more CPU and less memory, will obtain the highest migration priority.

Then those VMs are added into the migration queue until the resource requirement of the source PM is lower than its thresholds. On the contrary, once a physical node is underloaded, all VMs on it is added into the migration queue. The purpose of it is to accommodate VMs with as few PMs as possible to reduce energy consumption.

4.4   Destination PM Selection Strategy

The destination PM selection strategy plays a significant role in increasing the energy-efficiency of the data center. It usually occurs after the above two steps to avoid the overload of the PM or to change the mode of source PM to sleep mode for energy saving. Therefore, selecting an appropriate destination PM for the VMs on the source PM is another problem to be considered in this paper. However, if the destination PM is not suitable, it may cause two situations:

- The load of the destination PM may increase rapidly, which may even lead to a system crash in severe cases.
- The destination PM has the ability to accommodate the migrated VM currently but cannot satisfy resource requirements at subsequent time slot in the future, which results in unnecessary migrations.

ELVMC algorithm addresses these problems based on energy consumption combined with resource utilization prediction. To be specific, the traditional heuristic algorithm BFD is modified for energy-aware and load balance with the ME virtual machine selection policy. Algorithm 1 describes the ELVMC algorithm in detail.

Algorithm 2: $ELBFD(VMList, H_{source}, overMark)$

---

1 $migration_{count} = len(VMList)$
2 **for** v_i in $VMList$ **do**
3 $\quad$ Set $minEL \leftarrow MAX$
4 $\quad$ Set $flag \leftarrow false$
5 $\quad$ **if** $overMark == 0$ **then**
6 $\quad\quad$ **for** H_j in H **do**
7 $\quad\quad\quad$ **if** H_j satisfies the constraints (17)-(20) **then**
8 $\quad\quad\quad\quad$ $oldEnergy \leftarrow getEnergy(H_j)$
9 $\quad\quad\quad\quad$ $oldLoadBalance \leftarrow getLoadBalance(H_j)$
10 $\quad\quad\quad\quad$ Place v_i on H_j, Update $U_j(t), M_j(t), U_{source}(t), M_{source}(t)$
11 $\quad\quad\quad\quad$ $newEnergy \leftarrow getEnergy(H_j)$
12 $\quad\quad\quad\quad$ $newLoadBalance \leftarrow getLoadBalance(H_j)$
13 $\quad\quad\quad\quad$ $incEL =$
   $\quad\quad\quad\quad$ $(newEnergy - oldEnergy) * (newLoadBalance - oldLoadBalance)$
14 $\quad\quad\quad\quad$ **if** $incEL < minEL$ **then**
15 $\quad\quad\quad\quad\quad$ $minEL \leftarrow incEL$
16 $\quad\quad\quad\quad\quad$ $flag \leftarrow true$
17 $\quad\quad\quad\quad$ **end**
18 $\quad\quad\quad$ **end**
19 $\quad\quad$ **end**
20 $\quad\quad$ **if** $flag == true$ **then**
21 $\quad\quad\quad$ Update $VMList$
22 $\quad\quad\quad$ Update $population$
23 $\quad\quad$ **end**
24 $\quad$ **end**
25 $\quad$ **else**
26 $\quad\quad$ **for** H_j in H_{sleep} **do**
27 $\quad\quad\quad$ same as line7-line22
28 $\quad\quad$ **end**
29 $\quad$ **end**
30 $\quad$ $migration_{count} \leftarrow migration_{count} - len(VMList)$
31 **end**
32 **return** $population, migration_{count}$

Firstly, for overloaded hosts, sorting VMs according to the value of $priority_{ij}$ in descending order by $sortVM(H_j)$. Then, set $overMark = 1$. The ELBFD is used to find appropriate destination PMs for VMs on them, and the resource utilization of the source host is updated. If no host is available, select a host from the sleeping host set H_{sleep} and switch it to active mode, then use ELBFD to find an appropriate destination PM until the overloaded host's resources utilization below the threshold value. Otherwise, the cloud provider will buy a new host

---

Algorithm 3: $ELBFDUnderloaded(VMList, H_{source})$

---

1 $migration_{count} = 0$
2 $VMList_{copy} \leftarrow VMList.copy()$
3 **for** v_i in $VMList$ **do**
4 | Set $minEL \leftarrow MAX$
5 | Set $flag \leftarrow false$
6 | **for** H_j in H **do**
7 | | same as line7-line22 in Algorithm 2
8 | **end**
9 | **if** $flag == true$ **then**
10 | | Update $VMList$
11 | | Update $population$
12 | **end**
13 **end**
14 **if** $flag == true$ **then**
15 | $migration_{count} \leftarrow migration_{count} + len(VMList_{copy})$
16 **end**
17 **else**
18 | restore population and all VMs to their source PM
19 **end**
20 **return** $population, migration_{count}, VMList$

---

until there are no overloaded hosts in the data center to guarantee the SLA with the cloud consumer. When a feasible solution is obtained, update the current placement scheme and add the new host to the host set H. After that, the host with the smallest E_{H_j}/L_j in the data center is regarded as underloaded. The ELBFDUnderloaded algorithm is adopted to find destination PMs to accommodate all VMs on the underloaded host, which improves the overall resource utilization of the data center at the same time. Every time, if a feasible placement scheme is found, the resource utilization of each host is updated, and the new placement scheme is returned. Then the host will turn to sleep model to reduce energy consumption. Otherwise, no migration occurs.

4.5   Destination PM Selection Strategy for Overloaded Host

The ELBFD algorithm proposed for overloaded hosts is presented in Algorithm 2. A modified BFD named ELBFD is used to select destination PMs for the VMs in the overloaded host which is denoted as H_{source}. It takes increased energy consumption and load balance into consideration at the same time. If a host's CPU and memory utilization are less than their thresholds, the VM can be temporarily placed on this host and update the host's resource utilization. Finally, the placement scheme can be considered feasible and update. For the case that there is no feasible solution, a sleeping host from H_{sleep} host will be activated until there is no overloaded host in the data center.

4.6   Destination PM Selection Strategy for Underloaded Host

For underloaded hosts, the objective for the underloaded host is to consolidate VMs with the least PMs so that the energy consumption can be reduced and improve the resource utilization. The BFDUnderloaded is presented in Algorithm 3. At the beginning of this algorithm, the original placement of VMs on the underloaded PM should be saved in case the migration fails. The destination PM does not contain sleep hosts, and the search process is the same as Algorithm 2. Finally, if there is a VM on the underloaded host which can not be accommodated by any PM, then the host should keep on. Otherwise, the VM placement should be updated.

5   Experimental Evaluation

In this section, the simulation environment and performance metrics are introduced to analyze the performance of the proposed approach. Experimental results are presented to evaluate the effectiveness of the ELVMC algorithm from the aspects of energy consumption, the load balance, and the migrations of the consolidation policy. Then the ELVMC is compared with the existing VMC algorithms which are classified as with or without utilization prediction.

5.1   Experiment Setup

For the experiments described in this section, the Google Cluster Data (GCD) dataset [29] is adopted to reflect a real environment using a total of 800 heterogeneous PMs as servers [10,14]. The energy consumption E_j^{idle} and E_j^{busy} mentioned in Sect. 3.1 are assumed as 162 and 215 W. The ratio of VMs to PMs is assumed to be 2:1 [13]. The fact that the simulation environment is similar to an IaaS cloud environment such as Amazon EC2, where several independent users create and manage VMs, with the exception that all VMs are single-core, explains the fact that the workload data is used to simulate from the single-core VM. In this paper, the utilization of the VMs follows the GCD dataset which is also used as a benchmark to evaluate the proposed work. Such traces data comprises 672,003 jobs of the real data over about a month in May 2011, and are known as the second version of Google cloud traces. In the proposed experiments, each PM was initially allocated the CPU capacity from 2 to 32 randomly, and the unit is GHz. The memory capacity of each PM was randomly chosen from 256 to 2048 in the GB. VMs were also heterogeneous, with different sizes and resource requirements. The implementations of all algorithms were coded in Python and were carried out on a laptop with an Intel Core i7-8750H processor with 2.2 GHz CPU and 8 GB of RAM usable. The operating system running on the computer was Windows 10.

To create the CPU and memory utilization of VMs, the tasks of each job were aggregated by summing their CPU and memory consumption every five minutes in 24 h. Thus, the consolidation process was executed for 288 intervals, which corresponded to a change in resource demand over the day. In addition, the values of t_0 and t_1 mentioned in Eq. 3 were 0 and 288, respectively. Then, 1600 VMs were extracted with the utilization of CPU and memory under T_C and T_R from the GCD. The time interval between consecutive monitoring events

η was set to 6, which was detailed in Sect. 4. It meant that the ELVMC would decide whether VMs should be migrated according to their resource utilization in the next half an hour to avoid unnecessary migrations. Moreover, the ELVMC started after one hour, and it would be applied to PMs in the data center every five minutes as in [13].

5.2 Performance Metrics

The VMC scheme is determined under different configuration scenarios. Then the mean value of each evaluation metric is adopted for statistics to obtain the energy consumption, the number of virtual machine migrations, and the load balance of the data center. The metrics taken in [24] are used to evaluate the proposed ELVMC approach, which comprise energy consumption (EC), virtual machine migrations (VMM), load balance and service level agreement violation (SLAV).

5.3 Experimental Results

Table 2. Simulation results of VM selection strategies

VMC	VMs migration($\times 1000$)	EC(KWh)	Load Balance($\times 1000$)	SLAV($\times 0.00001$)
ARIMA-MMT-ELVMC	4.984	237.512	7.642	3.32
ARIMA-RS-ELVMC	5.212	246.103	7.984	4.05
ARIMA-HS-ELVMC	4.853	232.067	7.405	2.73
ARIMA-MC-ELVMC	4.512	223.211	7.139	2.31
ARIMA-LBO-ELVMC	4.474	219.254	7.112	2.27
ARIMA-ME-ELVMC	4.461	215.053	7.093	2.25

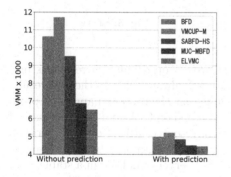

Fig. 2. Number of migrations

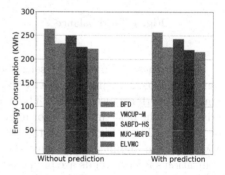

Fig. 3. Energy consumption

For the metrics mentioned above, four existing heuristic VMC algorithms, including classical BFD, SABFD, VMCUP-M [13], and MUC-MBFD [14], are compared with ELVMC. Each of them consists of a VM selection policy and a destination host selection policy. Among them, the first two algorithms do not include the utilization prediction, while the latter two algorithms and the ELVMC do. In order to illustrate the need for prediction, the performance metrics with or without utilization prediction will be considered. Moreover, the impacts of five VM selection strategies are compared against the proposed VM selection method ME which is based on the priority. These strategies include:

- Minimum migration time (MMT): selecting the VM with the minimum migration time for migration.
- Random select (RS): selecting the VM from the source host randomly.
- High CPU utilization (HS): selecting the VM that makes the highest CPU utilization from the source host firstly.
- Maximize correlation (MC): selecting the VM that has the largest correlation with the load change of the source host.
- Lexicographic bi-criterion objective (LBO): selecting the VM that maximizes the predicted CPU along with memory size and network bandwidth usage overall VMs while minimizing the migration time.

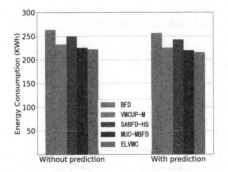

Fig. 4. Load balance

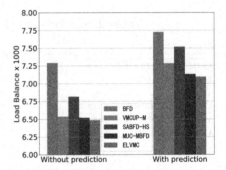

Fig. 5. SLAV

The impact of various VM selection strategies on simulation results is presented in Table 2. It shows that ELVMC with ME always achieves the best results among all performance metrics. The reason is that the proposed approach properly selects the best suited VM for migration based on CPU and memory instead of just one. Then, Fig. 2 shows the impact of resource utilization prediction on VMM. It can significantly reduce the number of migrations because resource utilization's future trends are predicted so that servers can keep in a stable state without aggressive reconfigurations, and unnecessary migrations can be avoided.

Moreover, the impacts of different VMC algorithms on energy consumption and load balance are discussed. The results are shown in Fig. 3 and 4. As can be seen, ELVMC is better than other algorithms in terms of these two metrics. The reason is that ELVMC takes into account the load balance of the new destination host to be selected, avoiding the excessive consolidation caused by other BFD algorithms due to the pursuit of low energy consumption. However, the result of the load balance without prediction is less than that with a prediction, as shown in Fig. 4. Compared with Fig. 2, it can be found that this is because more migrations are required so that more VMC algorithms are scheduled, resulting in lower load balance values without prediction. But in the long run, finding the more reliable destination hosts for VMs will save more energy and resources can be more reasonably allocated to cope with abrupt load changes during the consolidation process. This is shown in Fig. 3.

Finally, Fig. 5 shows the simulation results and comparison of algorithms in terms of SLAV. ELVMC's performance outperforms the other methods in both cases, which means that the proposed algorithm can deal with the change of resource utilization in time, and the destination servers do not become overloaded while migrating VMs. This is due to the fact that the prediction ensures the destination PM will not be overloaded for the current and future periods. The resource utilization of each host is more balanced compared to other methods with ELVMC, which also reduces PMs' overload risk due to load fluctuation. Although the difference in SLAV between the methods is small without prediction, the overload risk can be reduced with the help of prediction. As a consequence, the results demonstrate that the proposed approach is performing well while consolidating VMs.

6 Conclusion

Energy efficiency has become one of the major concerns for today's cloud data centers. VMC has been proved as an effective approach for reducing energy consumption. In this paper, an energy-aware algorithm named ELVMC with resource prediction is proposed for solving the VMC problem in cloud computing. In particular, a modified heuristic algorithm based on BFD is adopted for the consolidation scheme. It provides energy-efficiency improvement compared to three popular VMC algorithms. Predictive resource utilization and reasonable VM selection policy based on maximizing migration efficiency adopted in ELVMC can reduce migrations significantly. Moreover, ELVMC consolidates VMs into a reduced PM set while balancing the workload of each PM with a modified BFD algorithm. The real workload of Google is adopted in experiments, and it shows that the ELVMC can obtain a more energy-efficient and reasonable consolidation policy compared with other heuristic algorithms.

References

1. Liu, Z., Tong, W., Gong, Z.X., Liu, J., Yue, H., Guo, S.: Cloud computing model without resource management center. In: International Conference on Cyber-Enabled Distributed Computing and Knowledge Discovery (2011)
2. Gao, Y., Guan, H., Qi, Z., Yang, H., Liang, L.: A multi-objective ant colony system algorithm for virtual machine placement in cloud computing. J. Comput. Syst. Sci. **79**(8), 1230–1242 (2013)
3. Chiu, D., Stewart, C., Mcmanus, B.: Electric grid balancing through lowcost workload migration. ACM SIGMETRICS Perform. Eval. Rev. **40**(3), 48–52 (2012)

4. Birke, R., Chen, L.Y., Smirni, E.: Data centers in the cloud: a large scale performance study. In: IEEE International Conference on Cloud Computing (2012)
5. Hameed, A., et al.: A survey and taxonomy on energy efficient resource allocation techniques for cloud computing systems. Computing **98**(7), 751–774 (2016). https://doi.org/10.1007/s00607-014-0407-8
6. Amekraz, Z., Hadi, M.Y.: An adaptive workload prediction strategy for non-gaussian cloud service using ARMA model with higher order statistics. In: 2018 IEEE 11th International Conference on Cloud Computing (CLOUD) (2018)
7. Zhang, Z., Xiao, L., Li, Y., Li, R.: A VM-based resource management method using statistics. In: IEEE International Conference on Parallel and Distributed Systems (2012)
8. Ishak, S., Al-Deek, H.: Performance evaluation of short-term time-series traffic prediction model. J. Transp. Eng. **128**(6), 490–498 (2002)
9. Prevost, J.J., Nagothu, K.M., Kelley, B., Mo, J.: Prediction of cloud data center networks loads using stochastic and neural models. In: International Conference on System of Systems Engineering (2011)
10. Alsadie, D., Tari, Z., Alzahrani, E.J.: Online VM consolidation in cloud environments. In: 2019 IEEE 12th International Conference on Cloud Computing (CLOUD), pp. 137–145. IEEE (2019)
11. Farahnakian, F., Pahikkala, T., Liljeberg, P., Plosila, J., Hieu, N.T., Tenhunen, H.: Energy-aware VM consolidation in cloud data centers using utilization prediction model. IEEE Trans. Cloud Comput. **PP**(99), 1 (2016)
12. Hui, W., Tianfield, H.: Energy-aware dynamic virtual machine consolidation for cloud datacenters. IEEE Access **6**(99), 15259–15273 (2018)
13. Nguyen, T.H., Di Francesco, M., Yla-Jaaski, A.: Virtual machine consolidation with multiple usage prediction for energy-efficient cloud data centers. IEEE Trans. Serv. Comput. **13**, 186–199 (2017)
14. Moghaddam, S.M., OâSullivan, M., Walker, C., Piraghaj, S.F., Unsworth, C.P.: Embedding individualized machine learning prediction models for energy efficient VM consolidation within cloud data centers. Future Gener. Comput. Syst. **106**, 221–233 (2020)
15. Min, Y.L., Rawson, F., Bletsch, T., Freeh, V.W.: PADD: Power aware domain distribution. In: IEEE International Conference on Distributed Computing Systems (2009)
16. Raju, R., Amudhavel, J., Kannan, N., Monisha, M.: A bio inspired energy-aware multi objective chiropteran algorithm (EAMOCA) for hybrid cloud computing environment. In: International Conference on Green Computing Communication and Electrical Engineering (2014)
17. Murtazaev, A., Oh, S.: Sercon: Server consolidation algorithm using live migration of virtual machines for green computing. IETE Tech. Rev. **28**(3), 212 (2011)
18. Alsadie, D., Tari, Z., Alzahrani, E.J., Zomaya, A.Y.: Life: a predictive approach for VM placement in cloud environments. In: IEEE International Symposium on Network Computing and Applications (2017)
19. Wu, Q., Ishikawa, F., Zhu, Q., Xia, Y.: Energy and migration cost-aware dynamic virtual machine consolidation in heterogeneous cloud datacenters. IEEE Trans. Serv. Comput. **PP**(99), 1 (2016)
20. Pacheco-Sanchez, S., Casale, G., Scotney, B., Mcclean, S., Parr, G., Dawson, S.: Markovian workload characterization for QoS prediction in the cloud. In: IEEE International Conference on Cloud Computing (2011)

21. Hieu, N.T., Francesco, M.D., Yla-Jaaski, A.: Virtual machine consolidation with usage prediction for energy-efficient cloud data centers. In: IEEE International Conference on Cloud Computing (2015)
22. Calheiros, R.N., Masoumi, E., Ranjan, R., Buyya, R.: Workload prediction using ARIMA model and its impact on cloud application' QoS. IEEE Trans. Cloud Comput. **3**(4), 449–458 (2015). 2
23. Beloglazov, A., Buyya, R.: Optimal online deterministic algorithms and adaptive heuristics for energy and performance efficient dynamic consolidation of virtual machines in cloud data centers. Concurr. Comput. Pract. Exper. **24**(13), 1397–1420 (2012)
24. Sayadnavard, M.H., Haghighat, A.T., Rahmani, A.M.: A reliable energy-aware approach for dynamic virtual machine consolidation in cloud data centers. J. Supercomput. **75**(4), 2126–2147 (2019)
25. Thiam, C., Thiam, F.: Energy efficient cloud data center using dynamic virtual machine consolidation algorithm. In: Abramowicz, W., Corchuelo, R. (eds.) BIS 2019. LNBIP, vol. 353, pp. 514–525. Springer, Cham (2019). https://doi.org/10.1007/978-3-030-20485-3_40
26. Liu, Y., Sun, X., Wei, W., Jing, W.: Enhancing energy-efficient and QoS dynamic virtual machine consolidation method in cloud environment. IEEE Access **6**(99), 1 (2018)
27. Tian, W., et al.: On minimizing total energy consumption in the scheduling of virtual machine reservations. J. Netw. Comput. Appl. **113**, 64–74 (2018)
28. Beloglazov, A.: Energy-efficient management of virtual machines in data centers for cloud computing. Department of Computing and Information Systems (2013)
29. Liu, Z., Cho, S.: Characterizing machines and workloads on a google cluster. In: International Conference on Parallel Processing Workshops (2012)

Design of a Convolutional Neural Network Instruction Set Based on RISC-V and Its Microarchitecture Implementation

Qiang Jiao[1(✉)], Wei Hu[1], Yuan Wen[2], Yong Dong[1], Zhenhao Li[1], and Yu Gan[1]

[1] Wuhan University of Science and Technology, Wuhan, China
qjtxjq@gmail.com
[2] Trinity College Dublin, Dublin, Ireland

Abstract. The success of Convolution Neural Network (CNN) in computer vision presents a continuing challenge on performance requirement in both training and inference processes. Various software optimization has been examined towards existing hardware devices such as CPU and GPU to meet the computation needs; however, the performance gap between ideal and reality will keep going if there is short of hardware support. In this paper, we propose a customized CNN processor by extending the RISC-V instruction set. We have added six primary instructions by analyzing and abstracting the characteristics of conventional CNN models. The target micro-architecture has been upgraded accordingly to exploit the parallelism in the massive data access. We evaluated our work on the broadly used CNN model, LeNet-5, on Field Programmable Gate Arrays (FPGA) for the correctness validation. Comparing to traditional x86 and MIPS ISAs, our design provides a higher code density and performance efficiency.

Keywords: CNN · RISC-V architecture · RISC-V processor · Custom instruction · FPGA

1 Introduction

Convolution Neural Network (CNN) has been one of the most attractive technique in the past decade due to its outstanding prediction capability in a wide range of computer vision applications [5,10,11,20,22]. Such a high precision comes from extensive training over numerous training samples by well-structured models which usually have a large number of trainable weights to capture the characteristics of the model inputs. An example is a model with a larger size generally outperforms its smaller counterparts in terms of prediction accuracy [17]. But, at the same time, it also introduces a significantly greater number of operations which requires more time and resources to perform the calculation [6]. The computation burden can prohibit a comprehensive CNN model's deploying in practice, particularly for those applications with performance requirement.

© Springer Nature Switzerland AG 2020
M. Qiu (Ed.): ICA3PP 2020, LNCS 12453, pp. 82–96, 2020.
https://doi.org/10.1007/978-3-030-60239-0_6

The mainstream machine framework selects the most appropriate form of convolution among the options allowed by the runtime for the given platform. For CPU processors, a general matrix multiplication (GEMM) routine is a popular replacement of direct convolution [8,23], as most machines already have a fast implementation of the Basic Linear Algebra Subprograms (BLAS), while for the GPU counterpart, it often relies on the runtime library provided by the device vendors, such as Nvidia CuDNN and AMD ROCm. Because neither CPU nor GPU is initially designated for performing machine learning applications, it requires a significant effort for software to optimize code towards those processors. As a result, the generated code is often cumbersome, and its performance falls into suboptimal if there is a short of customized hardware support.

GPU (Graphics Processing Unit) is the broadest used hardware in training machine learning models because of its high throughput in parallel computing together with the massive bandwidth [9,13,21]. The GPU programming model allows a single instruction to perform on a group of independent data on separate PEs (Computing Elements) in a synchronized manner, which called SIMD (Single Instruction Multiple Data), to upgrade the computation throughput. The programmability of GPU makes it an easy-to-use platform that is popular among data scientists and CNN designers. However, the GPU also comes with obvious limitations that constrain its usage in a narrower range of scenarios. First, workload performance is not portable. The program has to be tuned before migrating from one GPU to another if the two processors have different architectures or various hardware resources. Second, GPU is not transparent to the programmer, and it requires explicit management of data and the functionality of processing the data. Transforming legacy code and library to the GPU is not a trivial effort. Finally, GPU requires the code to be compiled on-the-fly by a Just-in-Time compiler which shipped by the device vendor. Such a process introduces a nonnegligible runtime overhead.

In this paper, we propose hardware optimization for Convolutional Neural Networks. Instead of implementing a discrete accelerator, we extended the RISC-V instruction set by characterizing and abstracting the working pattern of mainstream CNN models. The proposed instruction set is transparent to the programmer and can be easily supported by compilers. It enhances the readability of the assembly program by decreasing the size of the code. In our experiment, we validated our design by implementing in on FPGA (Field Programmable Gate Arrays) fabrics. We have also examined and optimized the corresponding micro-architecture to improve the data parallelism.

The paper makes the following contributions:

- Design an instruction set based on RISC-V to optimize classic CNN operations
- Validate the extended instruction set on the FPGA fabric
- Optimize the micro-architecture to upgrade data access parallelism.

2 Background

2.1 Convolution Neural Network

Convolution Neural Network (CNN) has won considerable attention due to its outstanding performance in many sub-domains in computer vision, including image classification, object detection, motion estimation, video tracking, and so on. It outperforms the human being's capability in many of the above aspects via extensive training over a vast amount of samples for well-designed model structures. A typical CNN model consists of an input layer, an output layer and many hidden layers in between. Each layer performs a distinct operation on its input which is the output from its leading layer and outputs the processing outcome to its following layer. The most remarkable operation is the convolution that applies the multiply-accumulate operation for each element of the input (usually a 4-dimensional array called tensor) by a group of trainable variables, called weights. Behind each convolution, a CNN model usually presents an activation and pooling layer to perform the non-linear transformation and downsampling. The typical sub-structure of CNN shows in Fig. 1. Assisted by the activation layer, the convolution captures the feature hidden in the input by adjusting its weights and passes the abstracted information to a deeper layer for further analysis. Due to the intensive operation, the convolution layers dominant the overall execution time of a CNN model. Most of the optimization is targetting this type of layer because of its overwhelming runtime cost.

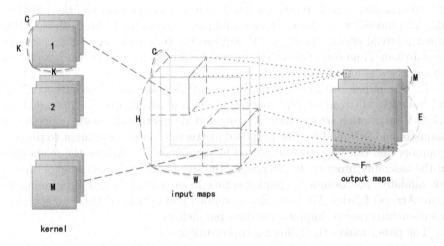

Fig. 1. Typical substructure of convolution neural network.

2.2 RISC-V

RISC-V is an open standard instruction set architecture (ISA) under open source licenses that allow hardware researchers to study and develop on top of it without having to pay fees to use it [12, 24]. The design of RISC-V follows reduced instruction set computer principles that made up by a small but highly-optimized set of instructions. To the contrary of x86 and ARM, the modular nature of RISC-V promises an incremental ISA design. The core of RISC-V is called RV32I which is the primary ISA that never changes; therefore, it provides a stable hardware target for the compiler and operating system designers. The principle specifies the basic ISA in terms of instruction code, the number of registers, memory addressing mode and so on. Following the principle, architecture researchers are able to extend ISA by adding new instructions to perform customized functionalities. Through incrementally modifying the compiler to support the extended instruction, such enhanced service can be delivered to the software. Besides the upgrade brought by the customization, the whole process is transparent to the programmer and no explicit program changing is required.

3   RISC-V CNN Processor Design

Developing a dedicated ISA and implementing the corresponding processor with Verilog is a complex process that requires a proper abstracting of the computation and efficient implementation. However, for CNN, such a process can be simplified due to its regular building structures. A group of operations, such as convolution, pooling, and activation, are the building block for standard CNN networks and are often the type of compute-intensive. Hence in this paper, we designate to expend RISC-V standard towards these operations. We also optimize the storage structure because of the massive data access required by CNNs. In this section, we present the details of accelerating CNN model by combining scalar, logic, jump, control and other instructions provided in rv32i with CNN specialized instructions. In order to simplify instruction decoding, all instructions are 32 bits. A data buffer memory is implemented, and peripheral devices such as flash, SRAM and SDRAM are added, including 32 32-bit registers.

3.1   CNN Oriented Instruction Set

We design our customized instruction set based on top of RV32I. The instructions include matrix loading and storing, together with other operations such as activation, pooling and so on. The customized instructions are designed to share common opcode bits as many as possible for those ones that perform on the same data path to simplify the control logic implementation. The combination of standard RV32I instructions and CNN oriented customization fulfills the functionality of CNN computation. Thereby, CNN can be implemented in assemblies with the extended ISA.

We describe the customized commands in this section. The inst[6:0] is the opcode part of the instruction. We use 0001011 of this part to identify the instruction is customized for a special operation. It used together with the function code (funt), which is inst[14:12], to decode the functionality of a specific instruction. The bits in inst[11:7], inst[19:15], and inst[24:20], contains the register address respectively.

Matrix Load and Store Instructions

The instruction format is shown in Fig. 2. The function code 0000 indicates matrix load, MLOAD. It loads matrix, such as feature map and convolution kernel from the main memory to the on-chip cache. The m_addr, m_size and dest_addr specifies the memory address, size of the fetching data and the on-chip memory address, respectively. The bits in ker_str indicates the rs3 register with its upper 16 bits describe the size of convolution window while the lower 16 bits represent the stride of the convolution. The loaded matrix is kept in the on-chip scratchpad. For the computation efficiency reason, we reformat the matrix while storing it. The layout rearrangement changes with the size of convolution, and we will show the details in the following section.

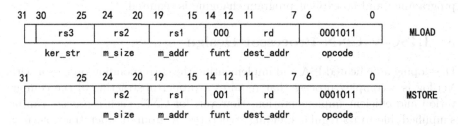

Fig. 2. Instruction for matrix load and store.

The function code 001 stands for matrix store command, MSTORE. It writes a matrix from the on-chip scratchpad to the off-chip memory. The decoding parts are functionally similar to MLOAD. The mnemonic of load and store is described as:

```
MLOAD rd, rs1, rs2, rs3
MSTORE rd, rs1, rs2
```

Matrix Operation Instruction

The instruction format is shown in Fig. 3: Function codes 010 and 011 define the operation of matrix multiplication (MCONV) and addition (MADD). Intrinsically, both instructions perform multiply and accumulations. For MCONV and MADD, dest_addr is the address to store the computation result, while m_addr1 and m_addr2 keep the address to the two operands separately. Different from MLOAD and MSTORE, the inst[30:25] indicates register rs3. In MCONV, the upper 16 bits of rs3 register keep the number of rows for the first matrix while the lower 16 bits store the number of columns for the second one. This mechanism is related to the matrix rearrangement and details is shown in the following section. In MADD, the value of the rs3 is the size of the matrix.

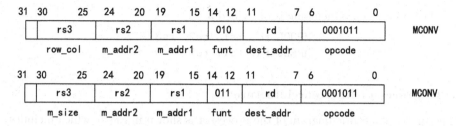

Fig. 3. Instruction of Matrix operation.

The instruction mnemonic is described as:

```
MCONV rd, rs1, rs2, rs3
MADD  rd, rs1, rs2, rs3
```

Pooling and Activation Instructions

The instruction format is shown in Fig. 4. The function code of 100 means the maximum pooling instruction MPOOL. The maximum pooling is selected here because, in CNN, the effects of maximum pooling, minimum pooling and average pooling will not be much different. But the minimum and maximum pooling have advantages in implementation simplicity. The dest_addr represents the starting address of the result after pooling. The m_addr and m_size respectively represent the starting address and size of the matrix to be pooled. The ker_str identifies the rs3 register. The upper 16 bits of the register store the size of the pooling window, and the lower 16 bits stores the size of the stride.

The function code 101 indicates the activation command MRELU. The ReLU activation has been broadly used since the appearance of AlexNet. Comparing to the function of Sigmod and tanh, ReLU has a much simpler structure and easy to implement, particularly in hardware. The dest_addr specifies the starting address to the result after activation. The m_addr and m_size specify the address and size of the matrix feeding to the activation function.

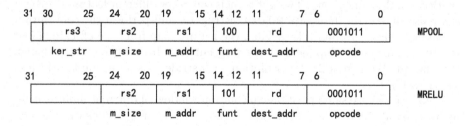

Fig. 4. Instruction of pooling and activation.

The instruction mnemonic is described as:

```
MPOOL rd, rs1, rs2, rs3
MRELU rd, rs1, rs2
```

3.2 Processor Micro-architecture

The simple structure diagram of the processor is shown in Fig. 5, which includes five stages of operations: fetch, decode, execute, access to memory and write back. The execution unit is the main part of the design, which includes a general-purpose computing unit (ALU unit) and a CNN (M_ALU unit) computing unit, in which the data processed by the convolution computing operation interacts indirectly with the off-chip memory. Apart from the execution unit, there is no significant difference between the other four levels and the classic five-level pipeline. Therefore, we focus on introducing the CNN computing unit and data access optimization method in detail.

After the basic instructions are decoded, the logic, arithmetic, shift and other instructions are executed in the general computing unit. Our CNN instructions, from matrix loading, convolution, to pool activation, and the final matrix storage will be completed in CNN computing unit. All phases of CNN processing are taken place in the chip, including matrix loading, convolution, pooling, activation and writing the results. Such a process does not involve the main memory accessing. Data exchange and management are performed within the in-chip buffer (rerang_buffer).

Memory Access Optimization
Given the fact that the speed of computation comes faster than data loading for convolution, we have to optimize data access.

As we can see from Fig. 1, for classic convolution layer, multiple multi-channel convolution kernels convolve input feature map by stride. The result of one layer is the input of the next layer. Based on this truth, we optimize data access in two directions.

A: Because the same group of weight kernels convolves different feature maps, we can reuse the weight data with M_LOAD instruction and perform the convolution in parallel. The convolution kernel and the feature map from the off-chip SDRAM will be sent to the on-chip memory (rerang_buffer), from where we can change their layout. We take a 2 * 2 convolution with stride 2 as an example, which is shown in Fig. 6.

Here, we use the same position convolution window of different input feature maps in parallel with the calculation of the corresponding convolution kernel. Using the M_LOAD instruction requires multiple cycles to load the convolution kernel matrix, and each convolution kernel only needs to be accessed once. The convolution kernel is fully reused, and the input feature map may need to be loaded multiple times or completed in a single time depending on its size. The inst [31:25] field of M_LOAD specifies the size of the convolution window.

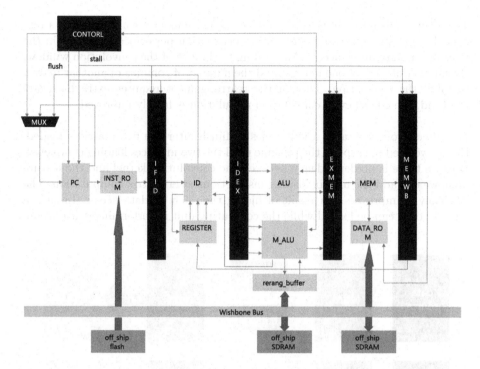

Fig. 5. Structure diagram of the processor.

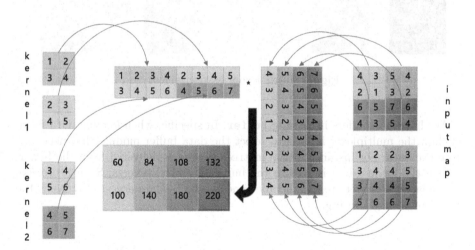

Fig. 6. Matrix layout rearrangement.

This value is also used in the convolution calculation, so it will be kept in a register. In this way, when the convolution instruction performed, the data in the rerang_buffer can be accessed according to the size of the convolution window, the number of rows of one matrix, and the number of columns of another matrix to fulfill the calculation. Though the rearrangement requires extra hardware overhead, the efficiency of convolution calculation is hugely improved.

B: The convolution takes a few cycles to finish after the two matrices loaded. Hence, we need to suspend the pipeline until the two matrices loading completed. Because there is no dependency between the convolution kernel and the feature map, we can add the matrix loading instructions after matrix computation. The efficiency is improved as a result of hiding the delay of data access. In such a scheme, data can be loaded while the computing unit is performing convolution.

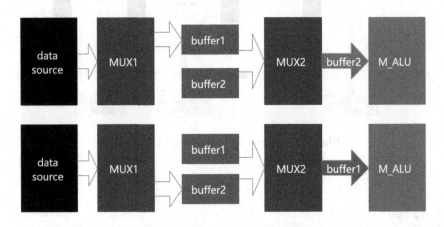

Fig. 7. Matrix layout rearrangement.

We have two buffers in rerang_buffer. In specific, when loading data from off-chip, the multiplexer is used to select the data buffer module that needs to store the data. For instance, if the matrix operand is delivered from the buffer 1, then the loaded data is sent to the buffer 2, and vice versa. Such a process takes place alternately to use the idle time caused by data access effectively. Its structure is shown in Fig. 7.

Calculation Unit

In our M_ALU, there are 32 tree-shaped multipliers and adders. Hence, it can support eight 2 * 2 convolution kernels, three 3 * 3, two 4 * 4 or one 5 * 5 kernel's parallel processing. We need compiler supports for generating the correct code. Figure 8 presents the structure.

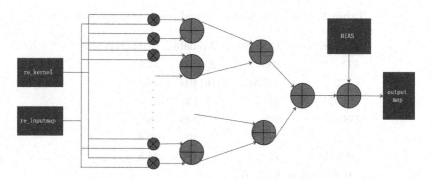

Fig. 8. Diagram of the calculation unit.

To give a clearer example, we present the pseudocode convolutional layer as follows:

```
LI $1,#imm1        //size of kernel ($1 high 16 bits)
                   //sliding step ($1 low 16 bits)
MLOAD $3,$4,$5,$0  //LOAD kernel map
                   //$3 dest M1 addr
                   //$4 kernel addr
                   //$5 kernel size
...
MLOAD $6,$7,$8,$1  //LOAD feature map
                   //$6 dest M2 addr
                   //$7 input map addr
                   //$8 input map size
...
LI $2,#imm2        //row of M1 ($2 high 16 bits)
                   //cow of M2 ($2 low 16 bits)

MCONV $9,$3,$4,$2  //$9 temporary output addr1
...
MADD $10,$9,$11,$2 //$10 temporary output addr2
MSTORE $13,$10,$2  //$13 store output addr
```

4 Experiment

We have implemented the hardware micro-architecture and customized instructions by Verilog. The design is synthesized by Xilinx toolset on a Artix-7 FPGA. To validate the hardware implementation, we have tested and simulated it by Vivado2019.1.

Vivado report regarding occupancy and power consumption is shown in Table 1.

Table 1. Resources occupancy and power consumption

Resource	Utilization	Available	Utilization (%)
FF	29015	202800	14.3
LUT	19720	101400	19.4
BRAM	31	135	22.9
DSP48	158	840	18.8
I/O	83	400	20.8
BUFG	11	32	34.8
Multi-column	0.362 W		

According to the synthesis report, the implementation is hardware friendly. It requires 19.4% LUT and 14.3% Flip-Flop of the FPGA chip. The memory consumption dominants the area usage of the design. As we can see from the result, the RISC-V extension uses 22.9% of the RAM and 34.8% of the BUFG. This high memory occupancy is determined by the nature of the CNN network, which introduces a large amount of data processing. The power consumption is 0.362 W. Combining the result, we can expect our implementation accommodates three CNNs on a single chip under a 1-W power budget.

We use LeNet-5 on MNIST dataset to validate our design for the reason of simplicity. LeNet-5 is a well-known machine learning model that has been reported with outstanding image classification accuracy. It comes with a simple structure that contains three sets of convolutional and pooling layers, and two sets of fully connected layers. As a CNN model, LeNet-5 has all classic structural features while requires minimum effort to implement, which is particularly critical for hardware design and validation. For the same reason, we select MNIST dataset, which consists of a group of handwriting images of a consistent size by 28 * 28 pixels. All these selections aim at testing the correctness of hardware with the simplest settings.

We use risc32-unknown-linux-gun-series cross-compilation tool chain to generate the code. The customized instructions are implemented with embedded instruction codes.

We used the MNIST data set to test our design. Limited by the length of the paper, here we present the simulation result of classifying digit 8 and 9 in this section. We observe the similar result while doing all the rest handwriting classifications.

Figure 9 shows the simulation of predicting handwriting digit 8 in MNIST. The result buffer num_result [0:9] carries the confidence of the prediction and it is a vector of real number in hexadecimal that indicate the probability of digit 0 to 9. In Fig. 9, the values in num_result [0:9] are: 3dcf81e8, 3c14f69c, 3dab3793, 3dbee02a, 3baa9717, 3c592b7f, 3dbcbab6, 3a85d313, 3f761144, 3de57108. Once transformed in to the decimal, the corresponding values are: 0.101322, 0.009092, 0.0836021, 0.093201, 0.005206,0.013255, 0.092153, 0.001021, 0.961201, 0.112032.

As num_result [8] has the highest value (0.961201), we know that the digit 8 has been correctly classified.

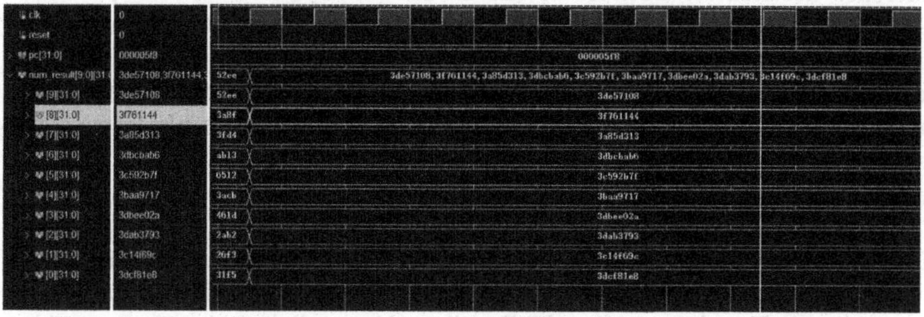

Fig. 9. Simulation of classifying MNIST digit 8

Similarly, Fig. 10 provides the result in predicting digit 9. The hexadecimal values in num_result [0:9] are 3d004b7f, 3ba6d267, 3da8255b, 3da85a0b, 3a9e12a5, 3ca5e99e, 3e3a78f2, 3b46a3bd, 3e0157ee, 3f78a0b1, which are 0.031 322, 0.005091, 0.0821025, 0.082203, 0.001206, 0.020253, 0.182102, 0.003031, 0.126312, 0.971202 in decimal. The num_result [9] which has the largest probability 0.971202, is the correct prediction. Hence, we can conclude that our RISC-V extension are performing correctly.

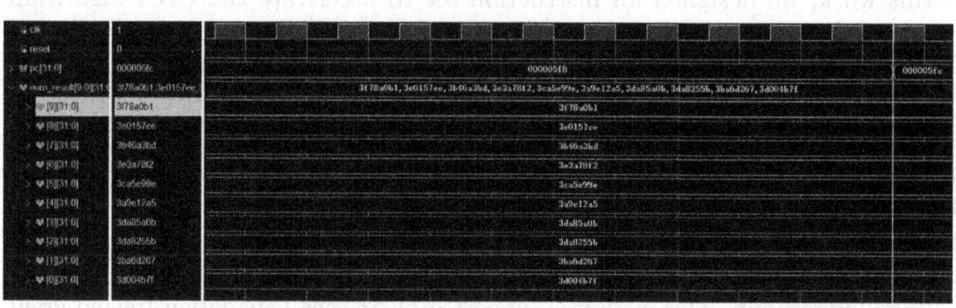

Fig. 10. Simulation of classifying MNIST digit 9

5   Related Work

Convolution Neural Networks is a typical compute-intensive application that involves massive tensor operations. The performance of CNN impacts significantly on its deployment, particularly for those models used in inference. Various methods have been examined to boost computation efficiencies, such as using matrix multiplication, Fast Fourier Transform (FFT) [1,3], or Winograd [7,25] to replace the direction convolution. Such type of algorithm generally have a better data locality and therefore capable of shortening the execution by strengthened data reuse.

Mainstream CNN frameworks intrinsically support GPU as a training and inference accelerator. Because of the SIMD model, the GPU can process a GEMM-based (General Matrix Multiplication) convolution in massive parallelism [14]. Though GPU has high performance in computation, it is tricky to release its full horsepower because of the architectural difference within processors by various vendors. Hand tuning code towards different platforms is not feasible. Hence, automatic tools, like TVM [2,18], has been developed to perform auto-optimization for diverse hardware platforms.

The automatic tools also open the window to explore the territory of software define hardware (SDH). Accelerators implemented on FPGA fabrics have been widely studied because of the wide bandwidth, high performance and low energy consumption. For the moment, SDH still requires collaboration from machine learning experts and hardware designers, because of the complexity.

RISC-V is an open standard instruction set architecture (ISA) that follows the reduced instruction set computer (RISC) principles [24]. It provides a framework to facilitate microprocessor and accelerator design. For CNN accelerating, solutions such as custom processor [15,16,19], integrated/discrete accelerator, SoC for edge IoT computing [4] have been proposed built on top of RISC-V. The design works correctly and efficiently according to our FPGA simulation.

6   Conclusion

In this work, we designed an instruction set to accelerate the CNN algorithm based on the computational characteristics of the CNN network algorithm. We propose a micro-architecture and two schemes in reformatting data layout to optimize data access. The design has been implemented in FPGA fabric by Verilog. In our experiment, we validate our hardware design by implementing LeNet-5 on MNIST dataset. The simulation result by Vivado shows that our RISC-V extension for CNN performs efficiently with a small hardware consumption.

Acknowledgement. This work was supported by Science Foundation Ireland grant 13/RC/2094 to Lero - The Irish Software Research Centre.

References

1. Abtahi, T., Kulkarni, A., Mohsenin, T.: Accelerating convolutional neural network with FFT on tiny cores, pp. 1–4 (May 2017). https://doi.org/10.1109/ISCAS.2017.8050588
2. Chen, T., et al.: TVM: end-to-end optimization stack for deep learning. CoRR abs/1802.04799 (2018)

3. Chitsaz, K., Hajabdollahi, M., Karimi, N., Samavi, S., Shirani, S.: Acceleration of convolutional neural network using FFT-based split convolutions. CoRR abs/2003.12621 (2020)
4. Flamand, E., et al.: GAP-8: a RISC-V SoC for AI at the Edge of the IoT. In: 2018 IEEE 29th International Conference on Application-Specific Systems, Architectures and Processors (ASAP), pp. 1–4 (2018)
5. He, K., Zhang, X., Ren, S., Sun, J.: Deep residual learning for image recognition. In: CVPR, pp. 770–778. IEEE Computer Society (2016)
6. Iandola, F.N., Moskewicz, M.W., Ashraf, K., Han, S., Dally, W.J., Keutzer, K.: SqueezeNet: AlexNet-level accuracy with 50x fewer parameters and <1 mb model size. CoRR abs/1602.07360 (2016)
7. Kala, S., Jose, B.R., Mathew, J., Nalesh, S.: High-performance CNN accelerator on FPGA using unified Winograd-GEMM architecture. IEEE Trans. Very Large Scale Integr. (VLSI) Syst. 27(12), 2816–2828 (2019)
8. Kala, S., Mathew, J., Jose, B.R., Nalesh, S.: UniWiG: unified Winograd-GEMM architecture for accelerating CNN on FPGAs. In: 2019 32nd International Conference on VLSI Design and 2019 18th International Conference on Embedded Systems (VLSID), pp. 209–214 (2019)
9. Kim, H., Nam, H., Jung, W., Lee, J.: Performance analysis of CNN frameworks for GPUs. In: 2017 IEEE International Symposium on Performance Analysis of Systems and Software (ISPASS), pp. 55–64 (2017)
10. Krizhevsky, A., Sutskever, I., Hinton, G.E.: ImageNet classification with deep convolutional neural networks. In: NIPS, pp. 1106–1114 (2012)
11. Lecun, Y., Bottou, L., Bengio, Y., Haffner, P.: Gradient-based learning applied to document recognition. Proc. IEEE 86(11), 2278–2324 (1998)
12. Lee, Y.: An agile approach to building RISC-V microprocessors. IEEE Micro 36(2), 8–20 (2016)
13. Li, C., Yang, Y., Feng, M., Chakradhar, S., Zhou, H.: Optimizing memory efficiency for deep convolutional neural networks on GPUs. In: Proceedings of the International Conference for High Performance Computing, Networking, Storage and Analysis, SC 2016, pp. 633–644 (2016)
14. Li, X., Liang, Y., Yan, S., Jia, L., Li, Y.: A coordinated tiling and batching framework for efficient GEMM on GPUs. In: PPoPP, pp. 229–241. ACM (2019)
15. Li, Z., Hu, W., Chen, S.: Design and implementation of CNN custom processor based on RISC-V architecture. In: 2019 IEEE 21st International Conference on High Performance Computing and Communications; IEEE 17th International Conference on Smart City; IEEE 5th International Conference on Data Science and Systems (HPCC/SmartCity/DSS), pp. 1945–1950 (2019)
16. Lou, W., Wang, C., Gong, L., Zhou, X.: RV-CNN: flexible and efficient instruction set for CNNs based on RISC-V processors. In: Yew, P.-C., Stenström, P., Wu, J., Gong, X., Li, T. (eds.) APPT 2019. LNCS, vol. 11719, pp. 3–14. Springer, Cham (2019). https://doi.org/10.1007/978-3-030-29611-7_1
17. Luo, J., Zhang, H., Zhou, H., Xie, C., Wu, J., Lin, W.: ThiNet: pruning CNN filters for a thinner net. IEEE Trans. Pattern Anal. Mach. Intell. 41(10), 2525–2538 (2019)
18. Moreau, T., Chen, T., Jiang, Z., Ceze, L., Guestrin, C., Krishnamurthy, A.: VTA: an open hardware-software stack for deep learning. CoRR abs/1807.04188 (2018)
19. Porter, R., Morgan, S., Biglari-Abhari, M.: Extending a soft-core RISC-V processor to accelerate CNN inference. In: 2019 International Conference on Computational Science and Computational Intelligence (CSCI), pp. 694–697 (2019)

20. Simonyan, K., Zisserman, A.: Very deep convolutional networks for large-scale image recognition. In: ICLR (2015)
21. Strigl, D., Kofler, K., Podlipnig, S.: Performance and scalability of GPU-based convolutional neural networks. In: 2010 18th Euromicro Conference on Parallel, Distributed and Network-Based Processing, pp. 317–324 (2010)
22. Szegedy, C., et al.: Going deeper with convolutions. In: CVPR, pp. 1–9. IEEE Computer Society (2015)
23. Vasudevan, A., Anderson, A., Gregg, D.: Parallel multi-channel convolution using general matrix multiplication. In: 2017 IEEE 28th International Conference on Application-Specific Systems, Architectures and Processors (ASAP), pp. 19–24 (2017)
24. Waterman, A., Lee, Y., Avizienis, R., Cook, H., Patterson, D.A., Asanovic, K.: The RISC-V instruction set. In: Hot Chips Symposium, p. 1. IEEE (2013)
25. Yu, J., et al.: Instruction driven cross-layer CNN accelerator with Winograd transformation on FPGA. In: 2017 International Conference on Field Programmable Technology (ICFPT), pp. 227–230 (2017)

Optimizing Accelerator on FPGA for Deep Convolutional Neural Networks

Yong Dong[1,2(✉)], Wei Hu[1,2], Yonghao Wang[3], Qiang Jiao[1,2], and Shuang Chen[1,2]

[1] College of Computer Science and Technology, Wuhan University of Science and Technology, Wuhan, China
salonmoner@gmail.com
[2] Hubei Province Key Laboratory of Intelligent Information Processing and Real-time Industrial System, Wuhan, China
[3] Digital Media Technology Lab, Birmingham City University, Birmingham, UK

Abstract. With the development of deep learning, the traditional neural networks architecture has been gradually met the bottleneck of the performance. Convolutional neural networks (CNNs) has been widely concerned because of its high precision advantage. However, CNNs are usually computationally large. And in addition to the widely used GPUs, but which has higher energy. And FPGA is gradually used to achieve CNNs acceleration due to its high performance, high concurrency, fast development cycle and re-configurability characteristics. Although previous works have made considerable progress, few researches have been used to address data dependence in data structure of CNN. Data dependence greatly affects the performance of accelerators. In this paper, we present a way to greatly improve the read efficiency of the accelerated hardware by reconstructing the original digital set and using the preload mechanism. This effectively reduces the problem of data dependence. In this way, pipeline technology can speed up the CNN computing process more effectively. We implemented the accelerator architecture on the XC 7z045 board, and the proposed accelerator has a clear advantage over previous studies to improve the efficiency and processing speed of CNNs.

Keywords: Convolutional neural networks · FPGA · Hardware acceleration

1 Introduction

In recent years, due to the high accuracy of CNN, CNN has been widely used by scholars and enterprises in the fields, such as image classification [1, 2], speech recognition [3], target detection [4, 5], and so on. However, CNN improves prediction accuracy by consuming a lot of resources. In general, GPUs are the preferred platform because they are able to compute large-scale data in parallel. However, GPUs have high energy consumption and they have fixed logic in hardware. On the other hand, ASIC and FPGAs have received extra attention for CNN acceleration, with the high throughput and energy efficiency. But ASIC requires a high time-consuming and complex design process. So, FPGA becomes a promising solution for image recognition because of FPGA's highly flexible and highly parallel bit-oriented architecture.

© Springer Nature Switzerland AG 2020
M. Qiu (Ed.): ICA3PP 2020, LNCS 12453, pp. 97–110, 2020.
https://doi.org/10.1007/978-3-030-60239-0_7

Previous works have proposed FPGA-based hardware acceleration designs for CNNs or related structures [6–8, 10, 13, 14, 16]. But most of them improve computing performance by refactoring FPGAs' hardware architecture. And it will reduce the accuracy of the number of operations. There are still some problems when FPGA is used in CNN acceleration. It is very important to consider the logical resource utilization or bandwidth utilization optimization to improve the performance. CNNs are computation-intensive, which requires a lot of data and data weights during the transferring. Such processing will result in huge bandwidth requirements. Thus, during the data processing of CNNs in FPGA, data management for FPGA memory has significant impact on the processing speed. Although some studies have proposed improvements to the memory management methods, such as the memory-centric accelerator proposed in [9], the effect is not been good in practice, but poor in the application.

In this paper, we propose a method to improve accelerator performance by optimizing data management, and improve the computing efficiency and parallelism of accelerators through three optimization strategies. The main contributions of this paper are described as follows. (1). The original data set is resized into hardware-friendly data. (2). The three key methods are optimized including data transferring, data processing and data computing respectively. (3). And the performance of our CNN accelerator is better than the previously proposed accelerator through the experimental results.

The rest of the paper is arranged as follows. Section 2 briefly introduces the research background and some related works. Section 3 provides an exploration of accelerator optimization. Section 4 provides the details the FPGA accelerator architecture proposed in this paper. Section 5 gives the results of our experiments. Finally, conclusions are shown in Sect. 6.

2 Background

2.1 CNN Structure

Since each neuron of the traditional fully connected neural network is connected to the neuron of the ring layer during image processing, a large number of parameters will be generated during back propagation, which will cause huge consumption of resources and calculations, and may cause overfitting. Figure 1 shows a typical fully connected neural networks. Therefore, it is necessary to use a new neural network structure, which is a convolutional neural network.

Convolutional neural networks (CNN) is a multi-level feedforward neural networks inspired by biological sciences. Because CNN can reduce the number of parameters through the sharing of local perception fields and parameters, the local perception field means that the neuron only perceives the local image, and then synthesizes the obtained local perception to get all the features. And the weight sharing is to use multiple filters to convolve the image to obtain multiple feature maps. CNNs usually consist of three parts: convolutional layer, pooling layer, and fully connected layer.

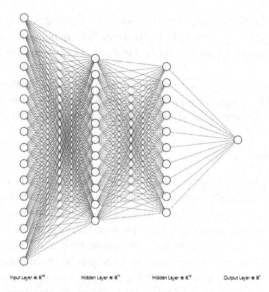

Fig. 1. Fully connected neural network

Figure 2 shows a typical LNet-5 neural network. The entire network of LNet-5 has 7 layers. The input two-dimensional image will be processed by two convolution operations and passed through two pooling layers. After passing through the fully connected layer, the input image will finally be classified by the softmax function [17].

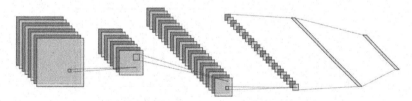

Fig. 2. LeNet-5 architecture

Typical CNN has the following three important parts including convolutional layer, pooling layer and fully connected layer. The convolutional layer is the core part of CNN. After receiving multiple input feature maps, the convolutional layer extracts the feature information in the feature map through a filter. Even after the multi-layer convolution operation, the model is still a linear process, and the output feature mapping needs to be non-linearly mapped. The excitation function generally used by CNN is the Rectified Linear Unit (ReLu), as shown in Formula 1.

$$Relu(x) = max(0, x) \tag{1}$$

After the processing of the convolutional layer, the dimension of the data is still very large, and it is difficult to be processed by the fully connected layer for the high complexity. It is necessary to reduce the dimensions through the pooling layer and prevent the model from overfitting. Common methods include maximum pooling, minimum pooling and average pooling, so that the data scale after convolution and pooling will be greatly reduced.

The fully connected layer as the last layer is usually used as the final prediction. The fully connected layer integrates highly abstracted features after multiple convolutions and pooling. The purpose of classification is achieved by using the softmax function.

2.2   Related Works

There has been a large number of FPGA-based accelerated CNN researches before. Specifically, Zhang has deeply explored the digital dependencies in convolutional layers and used a uniform loop expansion factor for each layer to avoid complex hardware structures [10]. Li proposed an optimized accelerator architecture based on a binary neural networks, which achieved a large increase in computation efficiency [11]. Zhang proposed a technique that uses deep pipelines and uses quantitative models to map multiple CNNs to FPGA clusters, increasing parallel speed [12]. Lu implemented the Winograd algorithm on FPGA, which reduced the complexity of the image processing, and the convolution speed is greatly improved, but it is only suitable for smaller convolution kernels [13].

Obviously, the previous researches on FPGA-based CNN acceleration had achieved good results. Figure 3 showed the design architecture of traditional accelerator. But

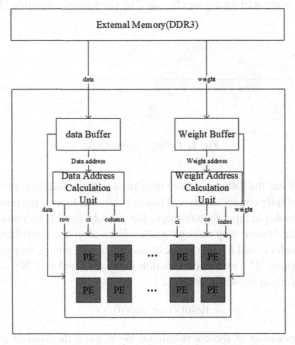

Fig. 3. Traditional accelerator design architecture

due to the limited on-chip resources, some data needs to be stored in off-chip memory. During the processing of data reading and moving, it needs to be specifically mapped to hardware, which will bring a lot of time consumption. This paper proposed a way to remove the address computing unit through the optimization of data store, thereby improving the efficiency of CNN accelerated data access.

3 Architecture Overview

Figure 4 shows the overall architecture of the accelerator. The architecture proposed mainly includes two parts in this paper. One part is composed of CPU and off-chip ram, which is mainly responsible for data storage and data rearrangement, and FPGA provides computing power for CNNs. After the CPU reorganizes the data, it is provided to FPGA. And the corresponding data compute on FPGA and results will transmit to CPU until the entire convolution processing is completed. FPGA is responsible for the most complicated part of the calculation in the convolution operation, and CPU provides control assistance.

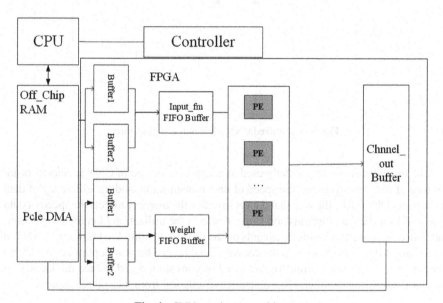

Fig. 4. CNN accelerator architecture.

The data transmission module is mainly realized by ping-pang buffer and FIFO buffer. The ping-pang buffer read the data as needed from the off-chip memory based on the preloading mechanism. And the quantitative data is processed in blocks in the on-chip FIFO buffer, so that the PE arrays can perform calculations more quickly.

4 CNN Accelerator Design Exploration

According to the analysis in Sect. 2.2, it is clear that the current FPGA-based CNN accelerator architecture cannot adapt to the current acceleration system. Some problems described in the above sections should be processed. The optimized architecture is shown in Fig. 5. The on-chip computing units obtains the data and return the data to the off-chip ram after processing.

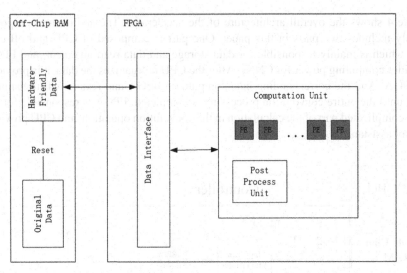

Fig. 5. Optimized CNN accelerator architecture

The three directions that usually used to address the design of CNN accelerators are the way of data management, the speed of data transmission, and the efficiency of data processing. Obviously, the way that data is stored in the memory affects the speed of data access. As for the data transmission stage, it affects the utilization of computing power, and data processing places high demands on the processing unit. We know that CNN is a computation-intensive process in the convolutional layer. Our work optimizes the above three processes to greatly improve the speed of convolution and reduce the latency of the algorithm. So, we will conduct a detailed discussion from these three aspects.

4.1 Data Set Resize

The data normally required by convolutional layers includes input feature maps, deviations, and weights. And we know that the order of reading the data in the convolutional layer is certain. Similarly, if the data stored off-chip is reorganized into a form that is suitable for the hardware, the processing unit on the FPGA can read data from off-chip memory in the order. As shown in Fig. 6, the off-chip data is sequentially restored in the on-chip buffer in this way, and the processing unit can read the data on the on-chip buffer one by one by reading data.

But in fact, there are other problems to deal with in this way. Because the resources of the on-chip memory are limited, it is impossible to put all the off-chip data into the on-chip memory. Therefore, we use the FIFO buffer to cache only part of the data. Thereby it can effectively avoid the huge resource consumption caused by data reading and writing in the on-chip memory. The whole process is shown in Fig. 6. By using the optimization of the data method, the FPGA does not need to compute the required data address each time when reading data, so we can delete the address computation unit from the traditional FPGA architecture in Fig. 3, then the processing unit can be updated and it can speed up the data read operations and reduce the latency.

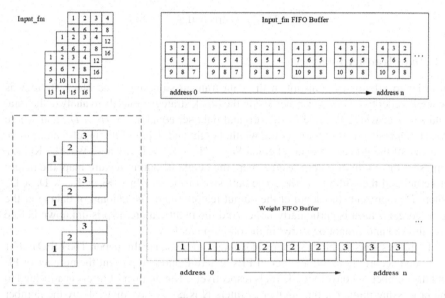

Fig. 6. Data resize

In addition, although our method can effectively speed up data read operations, it will also cause certain resource consumption when performing convolution operations. As shown in Fig. 7, when the striders is 1 and the size of convolution kernel is 3 * 3, a single data is read twice. In order to settle this problem, a different approach is proposed. It also uses a 3 * 3 matrix. During the data reading process, it only needs to read one column of data, and obtain a new matrix based on the previous data, and store a matrix again when changing rows. In this way, multiple readings of the same data from off-chip memory are avoided. The feature map shown in Fig. 7 is used to illustrate. First, it reads a 3 * 3 matrix buffer and compute with the filter. Then it only needs to read data 4, 8, 12, and fills the newly obtained data to the matrix buffer, re-assign the buffer during line break. Such operations are shown in Fig. 7.

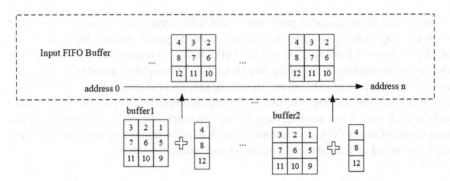

Fig. 7. Matrix buffer

4.2 Data Transfer Interface

In order to effectively transmit in the data transmission stage, the block strategy is generally adopted. However, when using the block strategy, it needs to analyze the total data set size, where the size of the original data set equal $W_i \times H_i \times D_i$, and where W_i, H_i, D_i respectively represent the width, height and depth of the input feature map. The size of the processed data set equal $W_o \times H_o \times K_h \times K_w$, where W_o, H_o, K_h, K_w represent the width of output feature map, the height of output feature map, the height of kernel and the width of kernel. The data size of the weight is $K_h \times K_w \times D_i \times D_o$ where D_o represents the depth of the output feature map. The amount of data after the data reorganization is particularly large. And the input feature map is unknown before it is allowed and cannot be stored in the on-chip memory.

Figure 8 shows the block strategy of storage mode used in this paper. The FIFO buffer is divided according to the number of PEs, where P and N represent the number of PE and the number of convolution kernels respectively. The number of filters is allocated to the processing units, but the number of filters N is not always divisible by the number of PEs P. Therefore, it is necessary to fill the filter by adding 0, but the final result will not be changed. Input feature maps are assigned to the processor in order. In this way, each processor can operate independently and keep parallel efficiency.

And the off-chip memory can store enough data. There are bandwidth limitations between off-chip memory and on-chip memory. Therefore, a pre-loading mechanism is adopted through the ping-pang structure to separate the data of read and write operations in the buffers. First, buffer 1 reads the required data from the off-chip memory. When the data is transferred to the next buffer, buffer 2 begins to read data. When data is transferred in buffer 2, buffer 1 begins to read data. By using this preloading method, the read and write of a single buffer can be separated, which can effectively improve the speed of data access. The same scheme is used for the weight buffer. The bandwidth pressure of off-chip memory can be effectively reduced via this method, while the parallelism of data processing is increased. The detail is shown in Fig. 9.

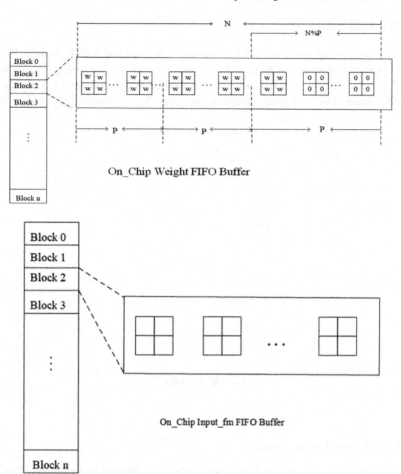

Fig. 8. The data storage in buffer

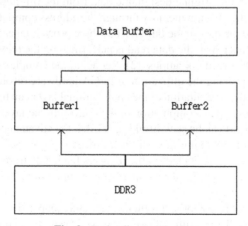

Fig. 9. Preloading mechanism

4.3 Data Processing

In the data processing module, each PE contains a register and a DSP module. So, the PEs will have the storage and processing abilities. During the processing, the PE reads $K_h \times K_w \times P$ elements from the FIFO buffer in order, and distributes $K_h \times K_p$ data to each PE in order. The similar operations are designed for the weight buffer (Fig. 10).

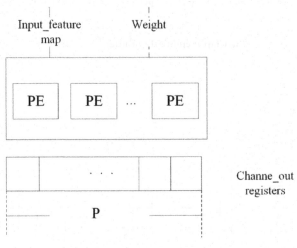

Fig. 10. PE array

4.4 Deep-Level Pipeline Process

In order to improve the throughput of the accelerator, we adopted the Deep-level Pipeline technology.

The adoption of these technologies has the following advantages. (1). Improving the data read efficiency. In the original design, the data from the off-chip memory needs to obtain the address of the off-chip memory through the address computation module, and then retrieve and send the data to the buffer. The whole process takes a long time. In the scheme proposed in this paper, the data read is performed on the register, which greatly reduces the time of data read operations. (2). Reducing the complexity of FPGA based development. (3). Improving the utilization of FPGA hardware resources.

After using DLP, the computation of the convolutional layer can be divided into multiple unrelated processes. The computation of the convolutional layer will be processed in the form of pipeline. It can be seen from Fig. 11 that the advantages of using pipeline technology to reduce the waiting time of each processing process. At the same time, CPU gets the data from the FIFO buffer, the registers fetch data from the buffer, and the PE unit reads the data in the register and performs computation. These three steps can be performed simultaneously.

At the same time, there is another problem, the feature map and weight in the off-chip memory will be accessed many times. On-chip memory usually cannot store all the data,

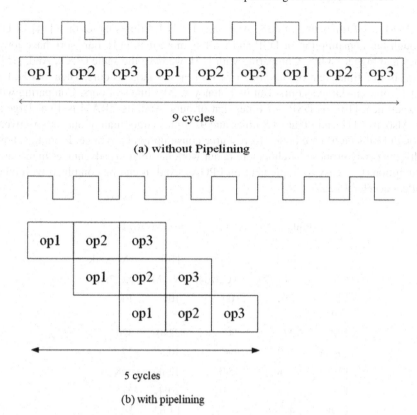

(a) without Pipelining

(b) with pipelining

Fig. 11. Pipeline technology

and this can avoid this situation by using FIFO buffer. CPU reads the data in the buffer at a certain rate, and then stores the data in the FIFO buffer. When the data in the FIFO is read, CPU will fill the data into the FIFO buffer again.

5 Experimental Results

5.1 Experimental Setup

The work proposed in this paper was implemented as an accelerator by using Vivado Hign-level Synthesis (HLS) v2019. And the Vivido HLS tool can directly program Xilinx's FPGA board using C language. XC 7z045 development board was used for the experiments. The model SqueezeNext was verified on the development board with the operating clock of 100 MHz and the XC 7z045 board.

5.2 Experimental Results

Table 1 shows the comparison of state-of-art deep CNN implementations on the other FPGA boards. Compared with [14], The [14]'s DSP is approximately one tenth of ours.

Its BRAM is 7.7 times that of ours. And our LUT is 1.4 times larger than [14]'s LUT. Although our consumption of DSP and LUT is more than [14], our work have lower latency and bigger GOPS. Comparing with [15], we have the same Clock. And, [15]'s BRAM is 45.2 times that of ours. Its FF is 1.4 times as large as ours. Then, our DSP is about 13 times as large as ours. But its Latency is 20.3 times of ours. Comparing with the experiment [10], its DSP is 7 times that of ours. And, its BRAM is 46 as large as ours. Also, its LUT and FF are 4.8 times and 9.7 times larger than of ours, respectively. Although [10]'s GOPS are better, its resource consumption is so huge. From the above results, the newly proposed architecture in our work has obvious advantages in resource consumption. It is easy to implement on FPGAs. And, it can be combined with other excellent accelerators.

Table 1. Comparison with previous works

	[14]	[15]	[10]	Our work
Platform	ZU3EG	XC 7z045	VC707	XC 7z045
Clock (MHz)	250	100	100	100
DSP	37	24	2240	320
BRAM	170	996	1024	22
LUT	24K	154K	186K	38K
FF	30K	137K	205K	21K
GOPS	47.09	–	61.62	59
Latency	104.3 ms	1955 ms	–	96 ms

According to the experiments, the accelerator architecture proposed in this paper has lower resource consumption, and it has a better performance in reducing latency. It is found that the acceleration architecture proposed in our work has obvious advantages and is more suitable to accelerate CNN.

6 Conclusions

CNNs have the features of computation-intensive and memory-intensive. However, the features are rarely considered as key factors when designing FPGA based optimization. In this paper, the data features of CNNs are analyzed to optimize the data access operations during CNN processing on FPGA. The optimization of data access and data store in this paper can provide better data operations with higher parallelism for CNN processing. According to the analysis of data dependency of CNN, a novel reconfigurable architecture was proposed and described in this paper. It provided the data access mechanism in the architecture design, which greatly improved the parallel efficiency and computation speed of CNN. This work was implemented and verified on the XC

7z045 board. The experimental results showed that our design can achieve better performance in memory bandwidth utilization and computational efficiency. According our design and experiments, there are still more features which can be used to optimize the performance based on FPGA. They are our future work.

References

1. LeCun, Y., Bottou, L., Bengio, Y., Haffner, P.: Gradient-based learning applied to document recognition. Proc. IEEE **86**(11), 2278–2324 (1998)
2. Krizhevsky, A., Sutskever, I., Hinton, G.E.: Imagenet classification with deep convolutional neural networks. In: Advances in Neural Information Processing Systems, pp. 1097–1105 (2012)
3. Zhang, Y., Pezeshki, M., Brakel, P., Zhang, S., Bengio, C.L.Y., Courville, A.: Towards end-to-end speech recognition with deep convolutional neural networks. arXiv preprint arXiv:1701.02720 (2017)
4. Liu, Wei., et al.: SSD: single shot multibox detector. In: Leibe, Bastian, Matas, Jiri, Sebe, Nicu, Welling, Max (eds.) ECCV 2016. LNCS, vol. 9905, pp. 21–37. Springer, Cham (2016). https://doi.org/10.1007/978-3-319-46448-0_2
5. Szegedy, C., et al.: Going deeper with convolutions. In: Proceedings of the IEEE Conference on Computer Vision and Pattern Recognition, pp. 1–9 (2015)
6. Li, H., Fan, X., Jiao, L., Cao, W., Zhou, X., Wang, L.: A high performance FPGA-based accelerator for large-scale convolutional neural networks. In: 2016 26th International Conference on Field Programmable Logic and Applications (FPL), pp. 1–9. IEEE (2016)
7. Suda, N., et al.: Throughput-optimized OpenCL-based FPGA accelerator for large-scale convolutional neural networks. In: Proceedings of the 2016 ACM/SIGDA International Symposium on Field-Programmable Gate Arrays, pp. 16–25 (2016)
8. Maki, A., Miyashita, D., Nakata, K., Tachibana, F., Suzuki, T., Deguchi, J.: Fpga-based cnn processor with filter-wise-optimized bit precision. In: 2018 IEEE Asian Solid-State Circuits Conference (A-SSCC), pp. 47–50. IEEE (2018)
9. Peemen, M., Setio, A.A.A., Mesman, B., Corporaal, H.: Memory-centric accelerator design for convolutional neural networks. In: 2013 IEEE 31st International Conference on Computer Design (ICCD), pp. 13–19. IEEE (2013)
10. Zhang, C., Li, P., Sun, G., Guan, Y., Xiao, B., Cong, J.: Optimizing fpga-based accelerator design for deep convolutional neural networks. In: Proceedings of the 2015 ACM/SIGDA International Symposium on Field-Programmable Gate Arrays, pp. 161–170 (2015)
11. Li, Y., Liu, Z., Xu, K., Yu, H., Ren, F.: A 7.663-TOPS 8.2-W energy-efficient FPGA accelerator for binary convolutional neural networks. In: FPGA, pp. 290–291 (2017)
12. Zhang, C., Wu, D., Sun, J., Sun, G., Luo, G., Cong, J.: Energy-efficient CNN implementation on a deeply pipelined FPGA cluster. In: Proceedings of the 2016 International Symposium on Low Power Electronics and Design, pp. 326–331 (2016)
13. Lu, L., Liang, Y., Xiao, Q., Yan, S.: Evaluating fast algorithms for convolutional neural networks on FPGAs. In: 2017 IEEE 25th Annual International Symposium on Field-Programmable Custom Computing Machines (FCCM), pp 101–108. IEEE (2017)
14. Yang, Y., et al.: Synetgy: algorithm-hardware co-design for convnet accelerators on embedded fpgas. In: Proceedings of the 2019 ACM/SIGDA International Symposium on Field-Programmable Gate Arrays, pp. 23–32 (2019)
15. Gschwend, D.: ZynqNet: an FPGA-accelerated embedded convolutional neural network. ETH Zurich (2016)

16. Hu. W, Chen. S, Li, Z.-H., Liu, T.-Y., Li, Y.-N.: Data optimization CNN accelerator design on FPGA. In: 2019 IEEE International Conference on Parallel & Distributed Processing with Applications, Big Data & Cloud Computing, Sustainable Computing & Communications, Social Computing & Networking (ISPA/BDCloud/SocialCom/SustainCom), pp. 294–299. IEEE (2019)
17. LeCun, Y., et al.: Gradient-based learning applied to document recognition. Proc. IEEE **86**(11), 2278–2324 (1998)

HpQC: A New Efficient Quantum Computing Simulator

Haodong Bian[1]iD, Jianqiang Huang[1,2(✉)]iD, Runting Dong[1], Yuluo Guo[1],
and Xiaoying Wang[1]

[1] Department of Computer Technology and Application,
Qinghai University, Xining, China
HPC_BHD@163.com, hjqxaly@163.com, eleanordrt@163.com, qhugyl@163.com,
Wangxiaofu163@163.com
[2] Department of Computer Science and Technology,
Tsinghua University, Beijing, China

Abstract. With the continuous popularization of quantum computing, high-efficiency quantum computing simulators have attracted
researchers' attention because the running time and memory overhead
of quantum computing is increased exponentially, which means that it
is challenging to be simulated on a traditional computer. The current
mainstream work solves this problem by using multi-node clusters, and
we find that its single-node performance has not been effectively exerted.
This paper proposes HpQC (High-performance Quantum Computing),
a simulator that can efficiently parallel quantum computing on a single-
node multi-core processor. First, HpQC used AVX2 and FMA instruction
sets to maximize the advantages of SIMD (Single Instruction Multiple
Data) vectorizations; second, it reduced the CPU calculation cycle by
using faster and more efficient bit operations; and finally, we designed
innovation data structure to utilize spatial locality of cache effectively.
Besides, this article selects the state-of-the-art quantum computing simulator, QuEST (the Quantum exact simulation toolkit), as the benchmark
for performance evaluation. For the quantum fourier transform, experimental results show that HpQC can achieve an average acceleration
of 2.20x (GNU compiler) and 1.91x (Intel compiler), respectively, compared to QuEST. As for the random quantum circuit program, HpQC
can achieve an average speedup of 1.74x (GNU compiler) and 1.51x (Intel
compiler), respectively, compared to QuEST.

Keywords: HpQC · QuEST · SIMD · Memory access optimization ·
Quantum fourier transform · Random quantum circuits

1 Introduction

As we all know, quantum computing [1] is a new type of computing model. Due to
the characteristics of quantum mechanics, it has higher computing efficiency than

© Springer Nature Switzerland AG 2020
M. Qiu (Ed.): ICA3PP 2020, LNCS 12453, pp. 111–125, 2020.
https://doi.org/10.1007/978-3-030-60239-0_8

traditional computing models. At least in the currently known quantum computing algorithms, such as quantum random walk [2], quantum fourier transform [3], and other algorithms, its processing speed is much faster than traditional computing. Therefore, quantum computing is usually applied to large-scale and sophisticated data processing and calculation, such as cryptography [4–6], artificial intelligence [7–9], and other fields. With the development of science and technology in various fields, quantum computing with inherent time complexity has been unable to meet the needs of related applications. For this reason, many researchers in related fields try to obtain higher performance through parallel quantum computing.

The realization of quantum computing must depend on the corresponding computing equipment. Admittedly, many scholars have currently used quantum computers to complete some quantum simulations or achieve more efficient optimization [10–13], due to their expensive cost and immature technology, we need to find cheap and practical alternatives to simulate quantum computing. In this regard, many research scholars have developed efficient simulators for simulating quantum computing based on traditional computers, such as the simulation of random quantum circuit implemented on Sunway TaihuLight [14], the realization of efficiently implementing quantum computing simulations by using CUDA programming on GPUs [15], and the implementation of quantum circuits based on FPGA [16].

Due to the exponential growth of quantum computing runtime and memory overhead in traditional computers [17], it is challenging to perform quantum computing simulations on traditional computers. To further improve the performance of the simulator, some research scholars have made corresponding optimizations based on the characteristics of the processor architecture. Simulation may be difficult because of memory limitations, which follows that the simulation of quantum algorithms instead of single gate operations can give the simulator significant performance improvement [18]. It is also possible to use more efficient data compression to obtain the simulation of more qubits [19]. However, the ideal simulator needs to balance memory space and computing time. ProjectQ [20] is a quantum computing emulator developed based on a single-machine multi-core processor. It uses OpenMP, AVX instruction set, and cache blocking to obtain efficient parallelization. On distributed-memory machines, qHiPSTER (the Quantum High Performance Software Testing Environment) [21] improves performance by using multi-threading, vectorization, and cache partitioning on a single machine; it also efficiently performs quantum calculations on multiple machines by dividing the load equally, while reducing communication time as much as possible. Some researchers have also realized automatic code generation and calculation optimization on multi-machine multi-core processors with excellent portability [22].

Good vectorization and cache blocking operation can bring excellent performance to the simulator, and balanced load processing capacity is the key to large-scale parallel computing. At present, QuEST (High-performance Quantum Computing) [23] is the first open-source quantum computing simulation simulator that supports three different architectures of single-machine multi-threading, multi-machine multi-threading, and GPU. Due to its balanced load processing capability, QuEST has comparable and better computing performance than ProjectQ on single-node multi-core processors, although it does not use SIMD (Single Instruction Multiple Data) vectorization and cache blocking optimization.

As the quantum simulators developed quantum simulators, it is becoming more and more common to use multi-node clusters to solve simulation problems. However, its single-node performance has not been effectively exerted. First, the simulator fails to utilize the advantages of SIMD vectorization fully; second, a large number of redundant calculations lead to reduced performance; third, the traditional data structure can not give full play to the advantages of spatial locality of cache.

This paper proposes HpQC (High-performance Quantum Computing), which is a simulator that can provide efficient parallel quantum computing on a multi-core processor. First, HpQC has better vectorization performance; second, it replaces the time-consuming multiplication and division operations by introducing low-overhead bit operations; third, it uses innovative data structures to maximize the spatial advantages locality of the cache. This paper selects the state-of-the-art QuEST quantum computing simulator as the benchmark for performance evaluation. Experimental results show that HpQC has better performance than QuEST. When executing the program of QFT, HpQC can achieve an average speedup of 2.20x (GNU compiler) and 1.91x (Intel compiler), respectively, compared to QuEST. As for the implementation of random quantum circuits, HpQC can achieve an average speedup of 1.74x (GNU compiler) and 1.51x (Intel compiler), respectively, compared to QuEST.

Specifically, our contributions:

(1) A parallel computing method that efficiently uses AVX2 and FMA instruction sets is designed to optimize its vectorization operation.
(2) An innovative and fast shift operation method is designed to optimize the calculation of qubits.
(3) An creative data structure is proposed so that the advantages of spatial locality of cache can be maximized and the memory access optimization can be achieved.

2   Background

2.1   Basic Operations of Quantum Computing

The basis of quantum computing is qubit. Not only can qubits realize the classic $|0\rangle$ and $|1\rangle$ states, but they can also express the superposition state through linear combination. For an m-qubit, it can have 2^m states, which is also the key to quantum parallelism [12].

Large-scale quantum computing programs are usually composed of many quantum logic gates. CNOT gate, Hadamard gate, Unitary Transformation often appear in many programs, because they all play the essential role in state transition.

CNOT Gate. CNOT gate can usually control the controlled quantum object and is generally used to entangle the quantum. The conversion results of $|00\rangle$, $|01\rangle$, $|10\rangle$, $|11\rangle$ by using the CNOT gate are demonstrated in Fig. 1 (a). If the control bit is 0, the target bit remains unchanged; otherwise, the control bit is 1 and the target bit is inverted.

Hadamard Gate. In quantum computing, it is usually possible to perform a Hadamard gate operation on $|0\rangle$ or $|1\rangle$ to make it in a superimposed state. Figure 1 (b) shows the Hadamard gate and demonstrates the conversion process of $|0\rangle$ and $|1\rangle$.

Unitary Transformation. Quantum states are not immutable. One quantum state can also evolve into another quantum state through the unitary matrix, and the process of transformation is called unitary transformation. In Fig. 1 (c), the results of $|0\rangle$ obtained after Unitary Transformation are demonstrated.

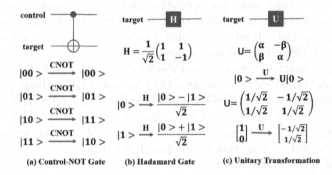

(a) Control-NOT Gate   (b) Hadamard Gate   (c) Unitary Transformation

Fig. 1. CNOT gate, Hadamard gate and Unitary transformation.

2.2 Classic Quantum Computing Program

Quantum Fourier Transform. The QFT (quantum fourier transform) [24] is the foundation of quantum logic gates, and it is also the core component of various quantum algorithms. Because of its extremely low time complexity [3], it is often used to solve complex problems such as large prime factorization. The calculation process of QFT is periodic, and the program is commonly composed of basic quantum gates such as CNOT and Hadamard, which made it easier to master and understand. Figure 2 shows the QFT operation of 3 qubits.

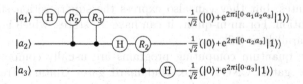

Fig. 2. The QFT operation of 3 qubits.

Random Quantum Circuits. Random quantum circuits are outstanding candidates for proving quantum supremacy [25]. Normally, the simulation of quantum computing is random and irregular, which lead that the user cannot accurately determine the final result of the calculation in advance, so we cannot perform special optimization operations by searching for regularity or other similar methods. It can be seen that many logic gates present in the random quantum circuit must be executed in order. Therefore, it is vital to improving the performance of the single-door operation because this is the key to the overall performance improvement.

3   Optimization Method

3.1   SIMD Vectorization

SIMD vectorization has always existed in many calculation optimizations, and its existence can make parallel operations better perform on a single calculation of multiple data. For the QuEST quantum computing simulator with load balancing advantages, it still has comparable or even better performance than ProjectQ although without the SIMD optimization. However, a large amount of data in the QuEST for a single calculation operation does not perform well. After analyzing the specific program by using the Perf tool, we found that although the single calculation operation inthe program has been automatically vectorized by the compiler, it still occupied a sizeable overall overhead. To avoid this situation, we need to add the SIMD inline function for vectorization optimization manually in HpQC simulator.

Algorithm 1. QuEST compactUnitary

Input: sizeBlock, sizeHalfBlock, thisBlock, indexUp, indexLo, numTasks, RU, RL, IU, IL, aI, aR, bI, bR, numTasks, stateVecReal[], stateVecImag[]

1: $numTasks \leftarrow (qureg.numAmpsPerChunk >> 1)$
2: $sizeHalfBlock \leftarrow (1 << targetQubit)$
3: $sizeBlock \leftarrow 2 * sizeHalfBlock$
4: **for** $i \leftarrow 0$ to $numTasks$ **in parallel do**
5:    $thisBlock \leftarrow i/sizeHalfBlock$
6:    $indexUp \leftarrow thisBlock * sizeBlock + i\%sizeHalfBlock$
7:    $indexLo \leftarrow indexUp + sizeHalfBlock$
8:    $RL \leftarrow stateVecReal[indexUp]$, $IL \leftarrow stateVecImag[indexUp]$
9:    $RU \leftarrow stateVecReal[indexLo]$, $IU \leftarrow stateVecImag[indexLo]$
10:    $stateVecReal[indexUp] \leftarrow aR * RU - aI * IU - bR * RL - bI * IL$
11:    $stateVecImag[indexUp] \leftarrow aR * IU + aI * RU - bR * IL + bI * RL$
12:    $stateVecReal[indexLo] \leftarrow bR * RU - bI * IU + aR * RL + aI * IL$
13:    $stateVecImag[indexLo] \leftarrow bR * IU + bI * RU + aR * IL - aI * RL$
14: **end for**

To embody the specific method of SIMD optimization in more detail, in Algorithm 1, we give the algorithm of the CompactUnitary function in the QuEST simulator. It is used to describe the evolution of a single qubit from one state to another under the action of the unitary matrix U. As can be seen from Algorithm 1, there is a large amount of data in the CompactUnitary function perform the operations of a single multiplication, addition, and subtraction operation. Where SIMD vectorization, as a single-instruction multiple-data stream, is more suitable. Therefore, adding SIMD vectorization operations to these computationally intensive function that has similar operations to the CompactUnitary function can achieve higher calculation effect to get overall performance improvement.

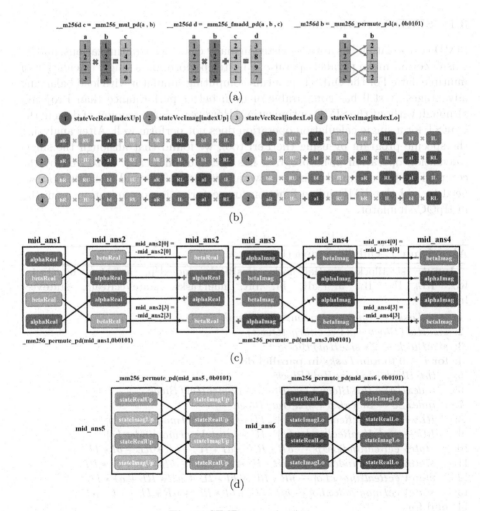

Fig. 3. SIMD vectorization.

In selecting the SIMD instruction set, we used INTEL's AVX2 and FMA instruction sets and gave detailed steps for **CompactUnitary** function optimization. As shown in Fig. 3, Fig. 3 (a) shows the three inline functions appearing in the vectorization operation, which are used to deal with multiplication, multiplication and addition, and data exchange. The left side of Fig. 3 (b) is the calculation operation that needs to be optimized. The right side of Fig. 3 (b) demonstrates the swap operation of positions 2 and 4. This operation needs to be performed because it is very time-consuming to update the four variables, such as **stateRealUp** multiple times in the loop, to perform the _mm_set_pd operation. Besides, It is also essential to use data exchange for multiple updated variables and get different operations, just as shown in Fig. 3 (d). The values of four constants, such as **alphaReal** in the loop are always fixed, so it is very cost-effective to perform a simple data inversion operation before the start of the loop, and Fig. 3 (c) shows the constant data exchange and inversion operations.

3.2   Bit Computing Optimization

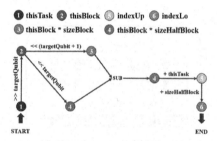

Fig. 4. Bit computing optimization.

QuEST has a very balanced load processing capability and has good scalability. However, After analyzing using the **Perf** tool of the Linux system, we found that most functions will perform **idiv** operation after being compiled into assembly language occupies nearly 50% of the total function overhead. The result of the analysis shows that the time-consuming **idiv** operations are concentrated in lines 5 to 7 of Algorithm 1 (the rest of the functions with the same operations are also applicable). It can be seen in Algorithm 1 that this part of the operation contains division and modular remainder operations. As we all know, performing a division operation in the CPU often requires more than ten clock cycles, and performing a modular remainder operation requires tens or even hundreds of clock cycles, which has a significant impact on the overall performance of the program. In contrast, bitwise operations of the integers in the CPU takes at most one clock cycle. In the HpQC, to better improve the program's performance, it is essential to convert the division and modular remainder operations into bitwise operations as much as possible. Since the compiler does not automatically complete the conversion of this part of the operation, we need to optimize it manually.

For **sizeHalfBlock** inline 2, it is equal to the result of the number 1 shifted **targetQubit** to the left. Then **thisBlock** is equal to the value of **thisTask** shifted **targetQubit** to the right. For **sizeBlock**, it is equal to the value of number 1 shifted left by **targetQubit** + 1 bit. Therefore, multiplying **thisBlock** by **sizeBlock** equals the value of **thisBlock** shifted left by **targetQubit** + 1 bit. For **thisTask** modulus residual **sizeHalfBlock** operation. First, divide **thisTask** by **sizeHalfBlock** and record the result as **a**. Then, multiply a by **sizeHalfBlock** and record the result as **b**. Finally, use **thisTask** to subtract **b** and get the final result. The operation of calculating the values of **a** and **b** can be performed by bitwise operations. After the above simplification steps, the **idiv** operation can be successfully eliminated. The specific operation conversion is shown in Fig. 4, starting from **START** following the direction of the arrow to **END** to complete the entire calculation process.

3.3 Memory Access Optimization

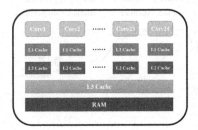

Fig. 5. A architecture of CPU processor with 24 cores.

With the continuous acceleration of CPU computing, slow memory bandwidth has become the main bottleneck limiting the system's overall speed. In CPU architectures, increasing the cache between the CPU and memory to obtain higher data access speed is the key to improving system performance. As shown in Fig. 5, this is a CPU processor with 24 cores, each of which has its L1 cache and L2 cache, and all cores share L3 cache and memory. In this CPU architecture, the closer to the CPU, the less storage space the cache has, and the higher the speed of the CPU accessing memory. However, under normal circumstances, most programs cannot improve the system performance due to inadequate cache statuses such as insufficient cache data reuse and frequent cache line replacement. Therefore, solving the problem of the efficient and reasonable use of limited cache space is the key to improving system performance.

Innovative Data Structure. In the QuEST, the state vector is divided into the real part vector and imaginary part vector, and continuous address space is allocated according to specific needs, facilitating the sequential access to data. However, in some quantum computing operations, a relatively long step-by-step data access operation is usually required, which is very unfriendly for a limited size cache. As shown in the left of Fig. 6, for the real part vector and the imaginary part vector, we need to obtain the values at the **indexup** and **indexlo** positions in the real part and the imaginary part vector, and then perform the calculation. Suppose the step between **indexup** and **indexlo** is **X**, and the length of **X** is much larger than the length of the cache line in the current CPU. It can be seen from the left of Fig. 6 that the cache line replacement operation needs to be performed four times to obtain the values at four different positions. As we all know, the cache in CPU accesses data in units of the length of a cache line, as for the data exceeding the length of the cache line, it has to perform replacement operations, which is very unfavorable for the performance of the cache.

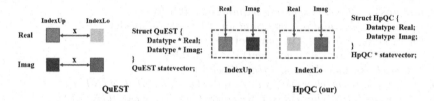

Fig. 6. Data structure of QuEST and HpQC.

To solve this problem, in the developed HpQC, we introduced a simple and efficient data structure. As shown in the right of Fig. 6, the real part vector and the imaginary part vector, which were stored in two consecutive stages, are compressed into a structured vector containing the real part and the imaginary part. In this way, initial values stored respectively in the indexup position of the real vector and the imaginary vector can be merged into the indexup position of the new structure vector. Now, the indexup position only needs to be accessed once, and the real and imaginary values can be continuously obtained. Compared with QuEST, the merged data storage structure can better achieve cache data reuse. This is because a cache line access operation can simultaneously obtain the real and imaginary values at the same position, reducing replacements of the cache line.

Store Byte Alignment. In modern CPU architecture, the cache plays a significant role between CPU and memory, which can speedup data transmission and minimize CPU waiting time. When accessing data, the cache is usually carried out with a cache line as the unit length. At present, the size of most cache lines is 64 bytes. Sequentially stored data can better utilize the advantages of cache data reuse when acquiring data in the cache line unit length. However, the situation shown in Fig. 7 (a) will sometimes happen when the data that is not byte-aligned is acquired with the cache line. The figure shows that the **Double**

data that was originally acquired only once needs to be operated twice without byte alignment. This is because a single cache line fails to obtain the required data fully, so the cache has to obtain the remaining data through the cache line replacement operation. Figure 7 (b) shows the data alignment operation, and we can find that the data under byte alignment no longer has access across the cache line. This dramatically improves the efficiency of data access.

Algorithm 2. HpQC compactUnitary

Input: sizeHalfBlock, indexUp, indexLo, RU, RL, IU, IL, numTasks, stateVec[], mid_ans1, mid_ans2, mid_ans3, mid_ans4, mid_ans5, mid_ans6, mid_ans0, aI, aR, bI, bR

1: $mid\_ans1 \leftarrow \_mm256\_set\_pd(aR, bR, bR, aR)$
2: $mid\_ans2 \leftarrow \_mm256\_permute\_pd(mid\_ans1, 0b0101)$
3: $mid\_ans2[0] \leftarrow -mid\_ans2[0]$, $mid\_ans2[3] \leftarrow -mid\_ans2[3]$
4: $mid\_ans3 \leftarrow \_mm256\_set\_pd(aI, -bI, bI, -aI)$
5: $mid\_ans4 \leftarrow \_mm256\_permute\_pd(mid\_ans3, 0b0101)$
6: $mid\_ans4[0] \leftarrow -mid\_ans4[0]$, $mid\_ans4[3] \leftarrow -mid\_ans4[3]$
7: **for** $i \leftarrow 0$ to $numTasks$ **in parallel do**
8:     $indexUp \leftarrow ((i >> targetQubit) << targetQubit) + i$
9:     $indexLo \leftarrow indexUp + sizeHalfBlock$
10:     $\_mm\_prefetch((char*)\&stateVec[indexLo + 64], \_MM\_HINT\_T1)$
11:     $\_mm\_prefetch((char*)\&stateVec[indexUp + 64], \_MM\_HINT\_T1)$
12:     $RL \leftarrow stateVec[indexLo].real$, $IL \leftarrow stateVec[indexLo].imag$
13:     $RU \leftarrow stateVec[indexUp].real$, $IU \leftarrow stateVec[indexUp].imag$
14:     $mid\_ans5 \leftarrow \_mm256\_set\_pd(IU, RU, IU, RU)$
15:     $mid\_ans6 \leftarrow \_mm256\_set\_pd(IL, RL, IL, RL)$
16:     $mid\_ans0 \leftarrow \_mm256\_mul\_pd(mid\_ans1, mid\_ans5)$
17:     $mid\_ans0 \leftarrow \_mm256\_fmadd\_pd(mid\_ans3, \_mm256\_permute\_pd(mid\_ans5, 0b0101), mid\_ans0)$
18:     $mid\_ans0 \leftarrow \_mm256\_fmadd\_pd(mid\_ans2, mid\_ans6, mid\_ans0)$
19:     $mid\_ans0 \leftarrow \_mm256\_fmadd\_pd(mid\_ans4, \_mm256\_permute\_pd(mid\_ans6, 0b0101), mid\_ans0)$
20:     $stateVec[indexUp].real \leftarrow mid\_ans0[0], stateVec[indexLo].imag \leftarrow mid\_ans0[1]$
21:     $stateVec[indexLo].real \leftarrow mid\_ans0[2]$, $stateVec[indexUp].imag \leftarrow mid\_ans0[3]$
22: **end for**

Data Software Prefetch. In actual program processing, data transmission is also quite time-consuming. If the program can achieve in advance to the data to be used in the cache in the future, the data transmission will be reduced inevitably. In reality, CPU can prefetch part of the data into the cache at the same time as the calculation to mask the time-consuming data transmission. However, this requires the program itself to have high local accessibility. By modifying the state vector structure, our program is more suitable for using data prefetching to get better performance. The innovative structure puts the elements that need to be calculated in adjacent positions, which is very friendly to cache with the advantages of spatial locality.

In the HpQC, although the hardware prefetching method cannot achieve the expected effect, we use the software prefetching method for data prefetching. In terms of technology, we uses the **_mm_prefetch** function of the **SSE** instruction set to prefetch the data. According to the experimental results, we found that prefetching the data into the cache other than the L1 cache has a better effect.

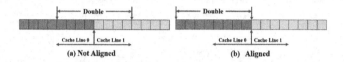

Fig. 7. Data alignment operation.

To further reflect the optimized program details, we have given an optimization algorithm, as shown in Algorithm 2.

4 Performance Evaluation

4.1 Device Information

The experimental platform was intel's fourth-generation Haswell architecture CPU. The frequency of this CPU is 2.30 GHz, 24 cores, and the L1 cache size is 32 KB, the L2 cache size is 256 KB, the L3 cache size is 30720 KB, DRAM memory size is 128 GB, Single max GFlops is 1766.4, Double max GFlops is 883.2. As for the SIMD instruction set, we selected the **AVX2** instruction set and **FMA** instruction set supported by the CPU. Second, we chose two different compilers, **Intel** and **GNU**, as the source code compilation environment, because the QuEST simulator also supports these two compilers. This can reflect the optimization effect of the HpQC simulator in different compilation environments in many aspects. Finally, the fast and efficient **OpenMP** parallel language is used for multi-thread implementation.

4.2 Test Program Selection

QFT is usually used as a benchmark program for the optimization of high-performance quantum computing. This is because it not only has an excellent structural law but also has a reasonable calculation regularity. Quantum random circuit simulation is usually more random and has no rules to find, so it is more suitable for testing the universality of a quantum computing simulator. In the experiment, the benchmark test programs consist of a QFT of 21 to 30 qubits and a random quantum circuit of 30 qubits. Besides, the experiment thoroughly considered the versatility of the test program. In the simulator's optimization process, we did not make any particular optimization for any specific program, so the optimization operation is more versatile in quantum computing.

4.3   Performance Comparison and Analysis

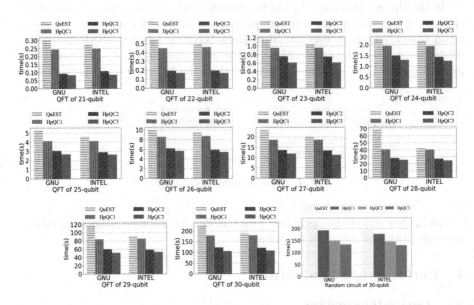

Fig. 8. QFT performance of 21 to 30 qubits and random circuit performance of 30 qubits.

As shown in Fig. 8, we demonstrate the QFT performance of 21 to 30 qubits and compares the performance in four different states. In Fig. 8, HpQC1 represents HpQC that only realizes SIMD vectorization; HpQC2 represents HpQC that realizes SIMD vectorization and calculation optimization; HpQC3 represents HpQC that implement all optimization methods.

In the same compilation environment, we can find that HpQC1 has a lower time overhead than QuEST. This is because sufficient vectorization can execute SIMD faster to achieve higher performance. Secondly, HpQC2, which achieves calculation optimization, has a significant performance improvement, because compared to the time-consuming multiplication and division operations, the use of fast and efficient bitwise operations can significantly reduce the CPU calculation time. Finally, HpQC3, which implements memory access optimization, has a higher calculation speed than HpQC2, because the innovative data structure can better utilize the spatial locality of the cache. For different compilation environments, the performance effects of HpQC vary since different compilers automatically optimize the code in different ways. What is exciting is that both of them have the same performance improvement trend. We can also find that the QFT performance of 21 and 22 qubits has excellent performance because its space overhead approximates the size of the cache space, and smaller memory consumption is more suitable for taking advantage of spatial locality of the cache.

As shown in Fig. 8 lower right corner, we demonstrated a 30-qubit random quantum circuit by using the GNU and INTEL compilers. First of all, HpQC1 has a pronounced speed increase compared to QuEST, which shows that SIMD vectorization plays a better role in stochastic quantum circuit optimization. Secondly, HpQC2 has a higher speedup than HpQC1, which means that computational optimization is also applicable to random quantum circuits. Finally, the performance brought by memory access optimization is also very significant, which means that the innovative data structure plays a better effect on random circuits.

As shown in Table 1, the QFT of 21 and 22 qubits with different compilers have achieved performance improvements that approach or even exceed 3x acceleration ratios. Besides, QFT has achieved performance improvements of 2.20x and 1.91x harmonic average speedup with the GNU and INTEL compilers, respectively. With the INTEL compiler, we can also see that QFTs of more than 22 qubits are less affected by the number of qubits and have a stable performance improvement, which means that our optimization will not cause performance degradation with the increase of qubits. In a 30-qubit random quantum circuit, HpQC achieves performance improvements of 1.74x and 1.51x acceleration ratios with GNU and INTEL compilers compared to QuEST, respectively. This shows that the HpQC simulator is suitable for the simulation of random quantum circuits and maintains the best state.

Table 1. Speedup of HpQC compared with QuEST.

	SpeedUp (GNU)	SpeedUp (INTEL)
QFT-21	3.64x	3.22x
QFT-22	3.18x	2.93x
QFT-23	1.91x	1.71x
QFT-24	1.75x	1.75x
QFT-25	1.97x	1.77x
QFT-26	1.76x	1.75x
QFT-27	1.96x	1.78x
QFT-28	2.67x	1.73x
QFT-29	2.32x	1.72x
QFT-30	2.16x	1.78x
QFT Harmonic mean	2.20x	1.91x
Random-30	1.74x	1.51x

Note: SpeedUp = (QuEST runing times)/(HpQC runing times)

5   Conclusion

This paper presents an HpQC simulator. On a single node, the methods of SIMD vectorization, calculation, and memory access optimization are used to achieve a more efficient performance. Moreover, it has more stable performance and will not cause performance loss with the increase of qubits. Besides, we also found that qubit operations close to the cache storage space can achieve higher performance. In the future, we will continue to conduct multi-node tests further to verify the scalability of HpQC.

Acknowledgment. This paper is partially supported by the National Natural Science Foundation of China (No.61762074, No.61962051), National Natural Science Foundation of Qinghai Province (No. 2019-ZJ-7034). "Qinghai Province High-end Innovative Thousand Talents Program - Leading Talents" Project Support. The Open Project of State Key Laboratory of Plateau Ecology and Agriculture, Qinghai University (2020-ZZ-03).

References

1. Bennett, C.H., Divincenzo, D.P.: Quantum information and computation. Nature **404**(6775), 247–255 (2000)
2. Douglas, B.L., Wang, J.: Efficient quantum circuit implementation of quantum walks. Phys. Rev. A **79**(5), 052335 (2009)
3. Cleve, R., Watrous, J.: Fast parallel circuits for the quantum Fourier transform. In: Foundations of Computer Science (2000)
4. Nejatollahi, H., Dutt, N., Ray, S., Regazzoni, F., Banerjee, I., Cammarota, R.: Post-quantum lattice-based cryptography implementations: a survey. ACM Comput. Surv. **51**(6), 1–41 (2019)
5. Bruss, D., Erdelyi, G., Meyer, T., Riege, T., Rothe, J.: Quantum cryptography: a survey. ACM Comput. Surv. **39**(2), 6-es (2007)
6. Elliott, C., Pearson, D., Troxel, G.D.: Quantum cryptography in practice. In: ACM Special Interest Group on Data Communication (2003)
7. Pudenz, K., Lidar, D.A.: Quantum adiabatic machine learning. Quantum Inf. Process. **12**(5), 2027–2070 (2013)
8. Biamonte, J., Wittek, P., Pancotti, N., Rebentrost, P., Wiebe, N., Lloyd, S.: Quantum machine learning. Nature **549**(7671), 195–202 (2017)
9. Ying, M.: Quantum computation, quantum theory and AI. Artif. Intell. **174**(2), 162–176 (2010)
10. Ash-Saki, A., Alam, M., Ghosh, S.: QURE: Qubit re-allocation in noisy intermediate-scale quantum computers. In: Proceedings of the 56th Annual Design Automation Conference 2019 (DAC 2019). Association for Computing Machinery, New York, NY, USA, Article 141, pp. 1–6 (2019)
11. Murali, P., Linke, N.M., Martonosi, M., Abhari, A.J., Nguyen, N.H., Alderete, C.H.: Full-stack, real-system quantum computer studies: architectural comparisons and design insights. In: Proceedings of the 46th International Symposium on Computer Architecture (ISCA 2019). Association for Computing Machinery, New York, NY, USA, pp. 527–540 (2019)

12. Liu, J., Byrd, G.T., Zhou, H.: Quantum circuits for dynamic runtime assertions in quantum computation. In: Proceedings of the Twenty-Fifth International Conference on Architectural Support for Programming Languages and Operating Systems (ASPLOS 2020). Association for Computing Machinery, New York, NY, USA, pp. 1017–1030 (2020)
13. Das, P., Tannu, S.S., Nair, P.J., Qureshi, M.: A case for multi-programming quantum computers. In: Proceedings of the 52nd Annual IEEE/ACM International Symposium on Microarchitecture (MICRO 52). Association for Computing Machinery, New York, NY, USA, pp. 291–303 (2019)
14. Li, R., Wu, B., Ying, M., Sun, X., Yang, G.: Quantum supremacy circuit simulation on sunway taihulight. IEEE Trans. Parallel Distrib. Syst. **31**(4), 805–816 (2020)
15. Gutierrez, E., Romero, S., Trenas, M.A., Zapata, E.L.: Quantum computer simulation using the CUDA programming model. Comput. Phys. Commun. **181**(2), 283–300 (2010)
16. Aminian, M., Saeedi, M., Zamani, M.S., Sedighi, M.: FPGA-based circuit model emulation of quantum algorithms. In: Proceedings of the 2008 IEEE Computer Society Annual Symposium on VLSI (ISVLSI 2008). IEEE Computer Society, USA, pp. 399–404 (2008)
17. Feynman, R.P.: Simulating physics with computers. Int. J. Theor. Phys. **21**(6), 133–153 (1999)
18. Häner, T., Steiger, D.S., Smelyanskiy, M., Troyer, M.: High performance emulation of quantum circuits. In: Proceedings of the International Conference for High Performance Computing, Networking, Storage and Analysis (SC 2016). IEEE Press, Article 74, pp. 1–9 (2016)
19. Wu, X.-C., et al.: Full-state quantum circuit simulation by using data compression. In: Proceedings of the International Conference for High Performance Computing, Networking, Storage and Analysis (SC 2019). Association for Computing Machinery, New York, NY, USA, Article 80, pp. 1–24 (2019)
20. Steiger, D.S., Häner, T., Troyer, M.: ProjectQ: an open source software framework for quantum computing. Quantum **2**, 49 (2018). Crossref. Web
21. Smelyanskiy, M., Sawaya, N.P., Aspuruguzik, A.: qHiPSTER: the quantum high performance software testing environment. arXiv: Quantum Physics (2016)
22. Häner, T., Steiger, D.S.: 0.5 petabyte simulation of a 45-qubit quantum circuit. In: Proceedings of the International Conference for High Performance Computing, Networking, Storage and Analysis (SC 2017). Association for Computing Machinery, New York, NY, USA, Article 33, pp. 1–10 (2016)
23. Jones, T., Brown, A., Bush, I., Benjamin, S.C.: QuEST and high performance simulation of quantum computers. Sci. Rep. **9**(1), 1–11 (2019)
24. Weinstein, Y.S., Pravia, M.A., Fortunato, E.M., Lloyd, S., Cory, D.G.: Implementation of the quantum Fourier Transform. Phys. Rev. Lett. **86**(9), 1889–1891 (2001)
25. Guo, C., et al.: General-purpose quantum circuit simulator with projected entangled-pair states and the quantum supremacy frontier. Phys. Rev. Lett. **123**(19), 190501 (2019)

Outsourced Privacy-Preserving Reduced SVM Among Multiple Institutions

Jun Zhang[1]($\boxtimes$), Siu Ming Yiu[2], and Zoe L. Jiang[3]

[1] Shenzhen University, Shenzhen, China
jzhang3@cs.hku.hk
[2] The University of Hong Kong, Hong Kong, China
smyiu@cs.hku.hk
[3] Harbin Institute of Technology (Shenzhen), Shenzhen, China
zoeljiang@gmail.com

Abstract. Executing data mining algorithms locally is usually computationally intensive. A promising solution is to outsource the heavy data mining tasks and datasets. On the other hand, combining data from multiple institutions for a big and varied training set helps enhance the performance of data mining. Due to privacy concerns, different institutions should encrypt their datasets with different keys. Support Vector Machine (SVM) is a popular classifier. It is challenging to train SVM on encrypted datasets in the cloud. Existing schemes use either the multikey fully homomorphic encryption on one server, or partially homomorphic encryption on two non-colluding servers. The former is inefficient and the institutions have to remain online, while the latter relies too heavily on the assumption of two non-colluding servers. To remove these limitations, we demonstrate how to train SVM for both horizontally and vertically partitioned datasets. To reduce training complexity and enhance security, we focus on reduced SVM with a secure kernel matrix. We proved the security of our scheme and the experimental results validated its efficiency.

Keywords: Privacy-preserving SVM · Reduced SVM · Outsourced dataset · Multiple keys

1 Introduction

With the popularity of cloud computing, an increasing number of institutions are motivated to upload their data to a cloud system. On the other hand, data mining techniques are widely used in many areas. Data mining algorithms are usually computationally intensive, especially for large datasets. The cloud service providers have powerful and elastic computing abilities. A promising solution is to outsource both the datasets and data mining tasks to the cloud. However, the privacy of the outsourced data is a major concern. Leaking personal information such as medical or financial records may even violate the laws (i.e., HIPAA).

© Springer Nature Switzerland AG 2020
M. Qiu (Ed.): ICA3PP 2020, LNCS 12453, pp. 126–141, 2020.
https://doi.org/10.1007/978-3-030-60239-0_9

To make sure the legal control of the data, it is common for the institutions to encrypt their data before outsourcing to the cloud. Different institutions encrypt their datasets under *different keys* to avoid leaking information to other institutions. Otherwise, if different institutions encrypt their datasets with a common key, each institution has the ability to decrypt the encrypted data of other institutions. Moreover, combining data from different institutions for a big and varied training set helps enhance the performance of data mining. Data might be horizontally (different records with same attributes) or vertically (same records with different attributes) partitioned among the institutions. For example, each branch of a company owns part of a customer database (horizontally partitioned). According to the privacy agreement, they are not allowed to disclose any information of individual customers to other branches except using the information for statistical analysis.

Support vector machine (SVM) is a popular classifier, which is a typical example of data mining algorithms. In this paper, we aim to implement the SVM classifier on encrypted datasets (horizontally or vertically partitioned) under different keys among multiple institutions. Our design goals are listed as follows.

- **Data Outsourcing (DO):** The institutions outsource their datasets to a third-party cloud system.
- **Multiple Keys (MK):** Each institution is responsible for maintaining the confidentiality and privacy of the outsourced dataset. Therefore, different institutions should encrypt their datasets by different keys.
- **Multiple Institutions (MI):** Collaboration between multiple institutions should be made possible without privacy breach.
- **Single Server (SS):** A scheme based on only one single server is preferred.
- **Horizontally and Vertically Partitioning (HVP):** A scheme can handle with horizontally or vertically partitioned datasets.

Table 1. Comparison between Existing Works

	DO	MK	MI	SS	HVP
[15]	✔	✗	✗	✔	✗
[9]	✔	✗	✗	✔	✔
[5]	✔	✗	✔	✗	✔
[25]	✔	✗	✔	✗	✗
[2]	✗	✗	✔	✗	✔
[19]	✔	✗	✔	✔	✔
[27,30]	✔	✔	✔	✗	✔
our scheme	✔	✔	✔	✔	✔

Comparison between existing schemes is shown in Table 1. The approaches in [2,5,9,15,19,25] cannot support multiple encryption keys. The assumption of two non-colluding servers is a must in [27,30]. Our scheme aims to perform the training and testing of SVM on one cloud server and allow the usage of multiple keys. We focus on homomorphic encryption based approach in this paper. Zhang et al. [32] propose a secure dot product protocol, which uses an integer vector encryption that supports addition, linear transformation and weighted dot product. But their scheme has a security weakness when applied to SVM that the cloud knows the gram matrix in clear. To be specific, we assume that the size of the training dataset is n and the dimension of each training sample is m (n is much larger than m in most cases). If an attacker in the cloud gets m encrypted training samples that are linearly independent and succeeds in obtaining their plaintexts, this attacker can recover any unknown training sample through solving linear equations with the known training samples and the gram matrix. To remove the security weakness in [32], we replace the gram matrix with secure kernel matrix. The size of secure kernel matrix is much smaller, reducing the training complexity of SVM in terms of memory and time. We also conduct extensive experiments to evaluate the efficiency and accuracy of our scheme. The results show that we can achieve similar accuracy as the traditional SVM. The advantage of our scheme becomes significant when considering large-scale datasets. We construct a well-designed protocol for the training and classification of reduced SVM. Our contributions are listed as follows.

- We only require one cloud server and thus our system model is simple.
- We use reduced SVM with secure kernel matrix, which enhances security and reduces computational complexity.
- Our scheme works for horizontally or vertically partitioned dataset.
- The institutions in our system can stay offline during the protocol execution.

2   Related Work

A privacy preserving dot product protocol on horizontally partitioned dataset was proposed in [2]. Secure dot protocol acts as an important building block in association rule mining [28], Naïve Bayes classifier [13] and decision tree classifier [20] on vertically partitioned dataset, k-means clustering [33] on arbitrarily partitioned dataset, as well as neural network [8,22] and SVM [9,14,23,29].

Roughly speaking, there are two ways to achieve privacy preserving data mining in the cloud. One is perturbation based approach [15]. Data sent to the cloud is perturbed by a random transformation, which considers only one institution. The other is cryptography based approach [2,19,27,30], such as secret sharing [9], Oblivious Transfer (OT) and Fully Homomorphic Encryption (FHE). The cryptography based approach provides a higher level of privacy compared to the perturbation based approach, but incurs higher computation/communication overhead.

With one cloud server, Liu et al. [19] propose to use BGN cryptosystem [4] which supports homomorphic additions and one homomorphic multiplication under single key to compute the dot product. However, the scheme in [19] is in essence a single key protocol. There indeed exists a multikey FHE primitive that allows computation on data encrypted under multiple keys [21]. However, its efficiency is still far from practice and it requires interactions among all the institutions during the decryption phase.

With the existence of two non-colluding servers, each institution can secret-share its data among two servers [5]. Then two servers compute on the shares of the input interactively and send the shares of the result to the institutions to reconstruct the final output. Although the secret sharing based approach is better in terms of computation cost, it incurs higher communication cost [26] and cannot deal with data encrypted under multiple keys. Homomorphic encryption and Yao's garbled circuits are combined to perform privacy preserving ridge regression in [25], which focuses on the single key setting and is not suitable for the case of multiple keys. Peter et al. put forward a scheme that transforms the ciphertexts under different keys into those under the same key [27], incurring a huge amount of interactions between the servers. To reduce communication overhead, proxy re-encryption [3] can be utilized to transform ciphertexts [30]. However, the amount of interactions is still heavy. If the underlying cryptosystem is additively homomorphic, they rely on interactions between two servers to compute multiplication and vice versa. These schemes only work on linear mean classifier, a much simpler classifier than SVM.

3   Model Description

3.1   System Model

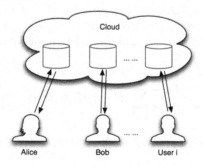

Fig. 1. The outsourced model.

We focus on the outsourced model shown in Fig. 1. The involved parties are multiple institutions and the cloud.

(1) Multiple Institutions (MIs): there are multiple institutions in our model. We list several institutions in Fig. 1, such as Alice, Bob and User i. They encrypt their datasets using different keys, then upload the encrypted records to the cloud.

(2) Cloud (C): the cloud acts as a storage and computing platform. It runs data mining algorithms on the encrypted datasets uploaded by MIs. For example, if the involved parties are interested in SVM, the cloud will train SVM on the uploaded datasets and obtain an encrypted classifier for future classification.

3.2   Threat Model

We assume that both the cloud and the institutions are honest-but-curious. And there is no collusion between the institutions and the cloud. All will execute the protocol honestly, but are curious to infer sensitive information. To be specific, we handle with two types of attacks in terms of an attacker's background information.

- Type 1 attacker in the cloud observes only the encrypted records without knowing anything about the unencrypted records.
- Type 2 attacker in the cloud collects m encrypted records which are linearly independent and obtains the corresponding m plaintexts.

The capability of type 2 attacker is stronger than type 1 attacker. Type 2 attacker can get m pairs of ciphertext and plaintext, while type 1 attacker knows nothing about the plaintexts of the training dataset.

4   Preliminaries

4.1   Support Vector Machine

We are given a training dataset $\{(\mathbf{x_i}, y_i)\}_{i=1}^{n}$, SVM finds the best separating hyperplane to partition data into two classes according to the label. The hyperplane is denoted as $\mathbf{w}^T\mathbf{x} - b = 0$. The primal form of SVM is often solved in its dual form (Equation (1)).

$$\min_{\alpha} \frac{1}{2} \sum_{i=1}^{n} \sum_{j=1}^{n} \alpha^T y_i y_j \mathbf{K}(\mathbf{x_i}, \mathbf{x_j})\alpha - \sum_{i=1}^{n} \alpha_i$$

$$s.\ t.\quad 0 \le \alpha_i \le C \ and \ \sum_{i=1}^{n} \alpha_i y_i = 0, \ i = 1, \cdots, n \tag{1}$$

where $\alpha \in R^n$ and $\mathbf{K}(\mathbf{x_i}, \mathbf{x_j})$ is called a kernel. The **kernel matrix** $\mathbf{K} \in R^{n \times n}$ contains the kernel values for every pair of training samples. For linear SVM, $\mathbf{K}$ is called **gram matrix** computed as $\mathbf{K}(\mathbf{x_i}, \mathbf{x_j}) = \mathbf{x_i}^T\mathbf{x_j}$. After we get α by solving Equation (1), we can further compute the weight vector $\mathbf{w} = \sum_{i=1}^{n} \alpha_i \mathbf{x_i} y_i$ and the bias item $b = \mathbf{w}^T\mathbf{x_i} - y_i$. Given the weight vector $\mathbf{w}$ and bias item b, the separating plane is $\mathbf{w}^T\mathbf{x} - b = 0$.

4.2   Integer Vector Encryption

Encryption: Given an integer vector $\mathbf{x} \in Z_p^m$ and secret key $\mathbf{S} \in Z^{m \times n}$, where m is the dimension and p is the alphabet size, the ciphertext $\mathbf{c} \in Z^n$ is a vector that satisfies $\mathbf{Sc} = t\mathbf{x} + \mathbf{e}$, where t is a large integer and e is an error term with elements smaller than $t/2$. The secret key $\mathbf{S}$ is restricted to a simple form $\mathbf{S} = [\mathbf{I}, \mathbf{T}]$, where $\mathbf{I}$ is an identity matrix concatenated with matrix $\mathbf{T}$.

Decryption: With secret key S, decryption can be done as $\mathbf{x} = \lceil \frac{\mathbf{Sc}}{t} \rfloor$.

Dot Product: Given two plaintext-ciphertext pairs under different keys - $(\mathbf{x_1}, \mathbf{c_1})$ under key $\mathbf{S_1}$, $(\mathbf{x_2}, \mathbf{c_2})$ under key $\mathbf{S_2}$, the encrypted dot product of $\mathbf{x_1}^T \mathbf{x_2}$ is computed as Equation (2), where the secret key for decrypting $\mathcal{E}(\mathbf{x_1}^T \mathbf{x_2})$ is $vec(\mathbf{S_1}^T \mathbf{S_2})^T$.

$$\mathcal{E}(\mathbf{x_1}^T \mathbf{x_2}) = \lceil \frac{vec(\mathbf{c_1}\mathbf{c_2}^T)}{t} \rfloor \tag{2}$$

4.3   Dot Product of Vectors from the Same Institution

Based on the integer vector encryption [31], secure dot product protocol is designed in [32]. They assume that Alice owns vector $\mathbf{x_a}$ under key $\mathbf{S_a}$ and Bob has vector $\mathbf{x_b}$ under key $\mathbf{S_b}$. The cloud needs $vec(\mathbf{S_a}^T \mathbf{S_a})^T$ to decrypt $\mathcal{E}(\mathbf{x_a}^T \mathbf{x_a})$, where vec represents a column vector by concatenating all the columns of $\mathbf{S_a}^T \mathbf{S_a}$ and $\mathcal{E}(\ )$ denotes the integer vector encryption. They use the key-switching technique to transform $\mathcal{E}(\mathbf{x_a}^T \mathbf{x_a})$ from key $vec(\mathbf{S_a}^T \mathbf{S_a})^T$ to secret key $[\mathbf{I}, \mathbf{T_c}]$ of the cloud. They compute key-switching matrix $\mathbf{M_a}$ as Equation (3), where $*$ means binary representation and $\mathbf{A_a}$ is a random matrix, and $\mathbf{E_a}$ is a random noise matrix.

$$\mathbf{M_a} = \begin{bmatrix} (vec(\mathbf{S_a}^T \mathbf{S_a})^T)^* - \mathbf{T_c}\mathbf{A_a} + \mathbf{E_a} \\ \mathbf{A_a} \end{bmatrix} \tag{3}$$

4.4   Dot Product of Vectors from Different Institutions

Alice owns vector $\mathbf{x_a}$ under key $\mathbf{S_a}$ and Bob has vector $\mathbf{x_b}$ under key $\mathbf{S_b}$, $\mathbf{x_a}$ and $\mathbf{x_b}$ are two vectors from different institutions. The cloud needs $\mathbf{S_a}^T \mathbf{S_b}$ to decrypt $\mathcal{E}(\mathbf{x_a}^T \mathbf{x_b})$. They divide $\mathbf{S_a}^T \mathbf{S_b}$ into three parts and compute key-switching matrix $\mathbf{M_{ab}}$ through breaking it into the addition of three key-switching sub-matrices.

$$\mathbf{S_a}^T \mathbf{S_b} = \begin{bmatrix} \mathbf{I} & \mathbf{0} \\ \mathbf{T_a}^T & \mathbf{0} \end{bmatrix} + \begin{bmatrix} \mathbf{0} & \mathbf{T_b} \\ \mathbf{0} & \mathbf{0} \end{bmatrix} + \begin{bmatrix} \mathbf{0} & \mathbf{0} \\ \mathbf{0} & \mathbf{T_a}^T \mathbf{T_b} \end{bmatrix} \tag{4}$$
$$= \mathbf{S_{ab_1}} + \mathbf{S_{ab_2}} + \mathbf{S_{ab_3}}$$

Alice owns $\mathbf{T_a}$, Bob owns $\mathbf{T_b}$, and the cloud can get $\mathbf{T_a}^T \mathbf{T_b}$ through randomization. They compute the key-switching sub-matrices $\mathbf{M_{ab_1}}$, $\mathbf{M_{ab_2}}$ and $\mathbf{M_{ab_3}}$ in Equation (5), where $\mathbf{A_{ab_i}}$ is random matrix and $\mathbf{E_{ab_i}}$ is random error matrix for $i \in \{1, 2, 3\}$. They obtain the final key-switching matrix $\mathbf{M_{ab}}$ through adding up all the sub-matrices $\mathbf{M_{ab}} = \sum_{i=1}^{3} \mathbf{M_{ab_i}}$.

$$\mathbf{M_{ab_i}} = \begin{bmatrix} (vec(\mathbf{S_{ab_i}})^T)^* - \mathbf{T_c A_{ab_i}} + \mathbf{E_{ab_i}} \\ \mathbf{A_{ab_i}} \end{bmatrix} \tag{5}$$

5   Our Scheme

The privacy-preserving SVM scheme in [32] takes a *full kernel matrix* as input, of which the size is n^2 where n denotes the number of training samples. Considering the existing large-scale training datasets, it is impractical to store such a huge matrix in the main memory. As a result, we resort to reduced SVM which makes use of a *reduced kernel matrix* instead of full kernel matrix. Each element of the reduced kernel matrix is a dot product between a sample in the original training dataset and a sample in the reduced training dataset [10].

The reduced kernel matrix is much smaller than the full kernel matrix. For type 1 attacker (see Sect. 3.2), it is infeasible to recover training samples from the full kernel matrix or the reduced kernel matrix. However, for type 2 attacker, both the full kernel matrix and the reduced kernel matrix have a weakness in security. Provided that the dimension of each training sample is m, if an attacker succeeds to obtain at least m linearly independent samples, the attacker can recover all the training samples from the kernel matrix. To remove this security weakness, we focus on *secure kernel matrix* so that an attacker cannot deduce anything from this secure kernel matrix.

5.1   Reduced SVM with Secure Kernel Matrix

Given a training dataset $\{(\mathbf{x_i}, y_i)\}_{i=1}^{n}$, where the training sample $\mathbf{x_i} \in R^m$ and the class label $y_i \in \{+1, -1\}$, we use $\mathcal{R} = \{\mathbf{r_j} | \mathbf{r_j} \in R^m, j = 1, \cdots, \bar{n}\}$ where $\bar{n} \ll n$ to denote the *reduced set* (typically less than 10% of the original training dataset). The full kernel matrix $\mathbf{K}$ is a $n \times n$ matrix, consisting of dot products among all the training samples, while the reduced kernel matrix $\mathbf{K}^\mathcal{R}$ is a $n \times \bar{n}$ matrix computed as

$$\mathbf{K}_{i,j}^\mathcal{R} = k(x_i, r_j), \ i = 1, \cdots, n, \ j = 1, \cdots, \bar{n} \tag{6}$$

It is shown that the reduced set is not necessarily to be a subset of the training set. The reduced set consists of random vectors in [16,24]. In this case, each element of the reduced kernel matrix is a dot product between a training sample and a random vector. As long as the random vectors are kept secret, the attacker cannot learn more from the kernel matrix than the leaked training samples (see Theorem 1). We call such a kernel matrix *secure kernel matrix* (reduced kernel matrix with random vectors as the reduced set).

The reduced SVM (RSVM) is defined as

$$\min_{\mathbf{v}, b, \xi} \frac{1}{2}(||\mathbf{v}||^2 + b^2) + C \sum_{i=1}^{n} \xi_i^2 \tag{7}$$

$$s.\, t. \quad y_i(\mathbf{K}^\mathcal{R}\mathbf{v} + b) \geq 1 - \xi_i, \ i = 1, \cdots, n$$

where $\mathbf{v}$ is a vector of dimension $\bar{n}$, b is a constant and C is the cost parameter of slack variables ξ_i $(i = 1, \cdots, n)$. The decision function $f(x)$ is composed of the solutions $\mathbf{v}$, b, and the vectors $\mathbf{r_j}$ of the reduced set.

$$f(x) = \sum_{j=1}^{\bar{n}} v_j k(\mathbf{x}, \mathbf{r_j}) + b \qquad (8)$$

To solve the optimization problem of the RSVM, we adopt the smooth SVM method [12]. The idea is to first convert Eq. (7) into an unconstrained optimization problem, then use smoothing techniques to achieve twice differentiability. Finally, a quadratically convergent Newton algorithm with an Armijo stepsize [6] is used to solve the unconstrained smooth reformulation of SVM.

5.2 Privacy-Preserving SVM on Horizontally Partitioned Dataset

We show the steps of our Privacy-preserving SVM scheme on horizontally partitioned dataset in Protocol 1. Alice and Bob first agree on the random vectors of the reduced set. Once the random reduced set is ready, Alice will encrypt it using her key and upload to the cloud. The encrypted secure kernel matrix is computed in step 1. And the ciphertext transformation is performed in step 2 with the key-switching matrices. The cloud gets the plaintext of secure kernel matrix in step 3 and solve the optimization problem using smooth SVM method in step 4.

Protocol 1: Training RSVM with Secure Kernel Matrix

Input: For the cloud: Encrypted training samples $\{\mathcal{E}(\mathbf{x_i})\}_{i=1}^{n}$ outsourced by Alice and Bob, encrypted random reduced set $\mathcal{E}(\mathcal{R})$, class labels $\{y_i\}_{i=1}^{n}$, secret key $\mathbf{S_c}$, key-switching matrices $\mathbf{M_a}$, $\mathbf{M_b}$ and $\mathbf{M_{ab}}$.
Output: The cloud gets parameters of the decision function $\mathbf{v}$ and b.

1. Encrypted Secure Kernel Matrix: The cloud computes $\mathcal{E}(\mathbf{K}^{\mathcal{R}})$ as $\mathcal{E}(\mathbf{K}_{ij}^{\mathcal{R}}) = \mathcal{E}(\mathbf{x_i}^T \mathbf{r_j})$.
2. Ciphertext Transformation: Transform $\mathcal{E}(\mathbf{K}_{ij}^{\mathcal{R}})$ to what the cloud can decrypt using its own secret key $\mathbf{S_c}$. To be specific, the cloud will calculate $\mathbf{M_a}\mathcal{E}(\mathbf{K}_{ij}^{\mathcal{R}})^*$ if $\mathbf{K}_{ij}^{\mathcal{R}}$ is the dot product of two samples encrypted by Alice's Key. In a similar way, if two samples are encrypted by Alice's and Bob's key separately, then the cloud will compute $\mathbf{M_{ab}}\mathcal{E}(\mathbf{K}_{ij}^{\mathcal{R}})^*$.
3. Secure Kernel Matrix $\mathbf{K}^{\mathcal{R}}$: The cloud decrypts the transformed $\mathcal{E}(\mathbf{K}^{\mathcal{R}})$ using $\mathbf{S_c}$ to get $\mathbf{K}^{\mathcal{R}}$ in plaintext.
4. Compute v and b: Incorporate $\mathbf{K}^{\mathcal{R}}$ into Equation (7) and solve the optimization problem using smooth SVM method to obtain $\{\mathbf{v}\}_{i=1}^{\bar{n}}$ and b.

We demonstrate how to classify a new testing sample in Protocol 2. As the random reduced set is known by all the institutions, each institution can conduct classification tasks locally. After running Protocol 1, the cloud obtains the parameters of the decision function. In Protocol 2, we require the cloud sends these parameters to the involved institutions. In step 2, the institution will first compute the dot products of the testing sample and all the elements of the reduced set. Then the institution finishes computing the decision function (see Eq. (8)) and get the class label.

Protocol 2: RSVM Classification

Input: For the institution: test sample $\mathbf{x_t}$, and the random reduced set $\mathcal{R}$.
Output: Class label.

1. The cloud sends the parameters of the decision function $\mathbf{v}$, b to the institution.
2. The institution computes $\sum_{j=1}^{\tilde{n}} v_j k(x_t, r_j)$ locally and get the class label.

5.3   Extension to Vertically Partitioned Dataset

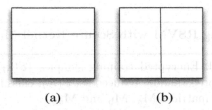

(a)   (b)

Fig. 2. Two data partitioning strategies - (a) Horizontally partitioned; (b) Vertically partitioned.

As shown in Fig. 2, the horizontally partitioned dataset contains different records with same attributes and vertically partitioned dataset includes same records with different attributes. Different institutions should generate different parts of the random vectors for the reduced set according to the pattern of vertical partitioning. Then the random vectors are encrypted by the keys of the institutions and uploaded to the cloud. The cloud computes secure kernel sub-matrix for each vertical partition. The final secure kernel matrix is derived by adding up all the secure kernel sub-matrices.

6   Security Analysis

We first prove the security of secure kernel matrix in Theorem 1. Then we analyze the security of Protocol 1 in Theorem 2 by evaluating the security of each step separately according to the Composition Theorem defined in [7]. If every step in Protocol 1 is secure, we can prove that the entire protocol is secure. Then we show the security of Protocol 2 in Theorem 3.

Theorem 1. *Type 1 or type 2 attacker cannot recover any data record from the secure kernel matrix.*

Proof. Type 1 attacker observes only the encrypted records, while type 2 attacker collects a set of plaintexts. It is obvious that the attack ability of type 2 attacker is stronger than type 1 attacker. Type 1 attacker cannot recover any record from the secure kernel matrix based on the semantic security of the underlying encryption algorithm. As for the security against type 2 attacker, we first show type 2 attacker cannot know the content of random vectors of the reduced set, based on which we then prove it cannot recover any data record from the secure kernel matrix.

Without loss of generality, we assume the type 2 attacker successfully collects the plaintexts of $n-1$ training samples from external information sources, among which at least m samples are linearly independent. If the attacker succeeds to figure out the correspondences between the released training samples and the elements of the secure kernel matrix, then it can set up linear equations to recover the content of random vectors of the reduced set. However, it contradicts with the fact that the random vectors are completely random and kept secret. As a result, it is infeasible for the attacker to know the correspondences between the released training samples and the elements of the secure kernel matrix. Therefore, type 2 attacker cannot know the content of random vectors of the reduced set.

According to Eq. (6), the elements of secure kernel matrix is computed as $\mathbf{K}_{i,j}^{\mathcal{R}} = k(x_i, r_j)$. Therefore, to recover any data record $\mathbf{x}_i$ from $\mathbf{K}^{\mathcal{R}}$, we need to know $\mathbf{r_j}$ in advance. Given that we proved type 2 attacker cannot know the content of random vectors of the reduced set above, type 2 attacker cannot recover any data record from the secure kernel matrix.

Theorem 2. *The cloud C learns nothing more than the result of secure kernel matrix. It is computationally infeasible for C to infer any information about the institutions' original sample $\mathbf{x_i}$ as long as the Integer Vector Encryption is semantically secure.*

Proof. We discuss Protocol 1 step by step. We analyze it with the Real and Ideal paradigm. We need to prove that the dot product computation of any two cipertexts is secure against a honest-but-curious ($\mathcal{HBC}$) adversary $\mathcal{A}_C^{\mathcal{HBC}}$ corrupting C in the real world. Besides, we build a simulator $\mathcal{F}^{\mathcal{HBC}}$ in the ideal world to simulate the view of the honest-but-curious adversary $\mathcal{A}_C^{\mathcal{HBC}}$.

Step 1: The view of $\mathcal{A}_C^{\mathcal{HBC}}$ includes its input $\{\mathcal{E}(\mathbf{x_i}), \mathcal{E}(\mathbf{r_j})\}$ and the output $\mathcal{E}(\mathbf{x_i}^T \mathbf{r_j})$. The simulator $\mathcal{F}^{\mathcal{HBC}}$ computes $\mathcal{E}(I), \mathcal{E}(2I)$, where I is an identity vector and $2I$ is also a vector of which each element is 2. Then $\mathcal{F}^{\mathcal{HBC}}$ computes

$\mathcal{E}(I^T 2I)$ and returns $\{\mathcal{E}(I), \mathcal{E}(2I), \mathcal{E}(I^T 2I)\}$ to $\mathcal{A}_C^{\mathcal{HBC}}$. Since the view of $\mathcal{A}_C^{\mathcal{HBC}}$ are ciphertexts generated under Integer Vector Encryption and $\mathcal{A}_C^{\mathcal{HBC}}$ has no knowledge of the private key $\mathbf{S}$. If $\mathcal{A}_C^{\mathcal{HBC}}$ can distinguish the view from the real world and the one from the ideal world, then it indicates $\mathcal{A}_C^{\mathcal{HBC}}$ could succeed to distinguish ciphertexts generated by Integer Vector Encryption, which contradicts to the assumption that Integer Vector Encryption is semantically secure. Therefore, $\mathcal{A}_C^{\mathcal{HBC}}$ is computationally infeasible to distinguish from the real world and the ideal world.

Step 2: We can prove the security of this step in the same way as Step 1. The view of $\mathcal{A}_C^{\mathcal{HBC}}$ is $\{\mathbf{M_a}\mathcal{E}(\mathbf{K}_{ij}^{\mathcal{R}})^*\}$, $\{\mathbf{M_b}\mathcal{E}(\mathbf{K}_{ij}^{\mathcal{R}})^*\}$ and $\{\mathbf{M_{ab}}\mathcal{E}(\mathbf{K}_{ij}^{\mathcal{R}})^*\}$. The simulator $\mathcal{F}^{\mathcal{HBC}}$ computes $\{\mathbf{M_a}\mathcal{E}(I^T 2I)^*\}$, $\{\mathbf{M_b}\mathcal{E}(I^T 2I)^*\}$, $\{\mathbf{M_{ab}}\mathcal{E}(I^T 2I)^*\}$. The ciphertexts in the view of $\mathcal{A}_C^{\mathcal{HBC}}$ cannot be distinguished.

Step 3: The secure kernel matrix $\mathcal{E}(\mathbf{K}^{\mathcal{R}})$ is decrypted in this step. We already proved its security in Theorem 1.

Step 4: The cloud obtains $\{\mathbf{v}\}_{i=1}^{\bar{n}}$ and b through solving the optimization problem with smooth SVM method. No sensitive training samples are involved in this process. Hence this step is secure.

Theorem 3. *The cloud cannot deduce the class label of the testing sample, and thus Protocol 2 is secure.*

Proof. The parameters of the decision function $\mathbf{v}$, b are sent to the institution. Each institution knows about the reduced set. Protocol 2 runs $\sum_{j=1}^{\bar{n}} v_j k(x_t, r_j)$ locally, which implies its security.

7   Experimental Evaluation

The configuration of our PC is 64-bit Windows 7 operating system with Intel(R) Core(TM) i5-3570 CPU, 3.4 GHz, and 16 GB memory. We use Math Kernel Library (MKL) to accelerate our implementation.

As shown in [10,11,17], the reduced SVM can achieve similar classification performance to a traditional SVM and the reduced set might contain random vectors [16,24]. To validate the classification performance of the reduced SVM with random vectors as the reduced set, we downloaded six datasets of different sizes from the UCI machine learning repository and LIBSVM website. Four datasets include sensitive medical records or personal information - Pima Indian diabetes (Pima), German credit numeric version (German), Taiwan credit card clients data (CreditCard) and census bureau database (Adult). The other two datasets are less sensitive - Ionosphere and Svmguide3. The statistics of the datasets are listed in Table 2. The datasets from the LIBSVM website are already scaled to $[-1, 1]$. We also pre-process the datasets from the UCI machine learning repository to the range $[-1, 1]$ and use 1 and -1 as class labels. We use the smooth SVM method to solve the reduced SVM problem in this paper. We compare the classification performance between the **R**educed **SVM** with **R**andom vectors as the reduced set (**RRSVM**), Reduced **SVM** with training data subset as the reduced set (**RSVM**) and **LIBSVM** [1]. For RRSVM, we generate random vectors whose elements are between -1 and 1. For RSVM, we randomly select 10% of each dataset as the reduced set. The large-scale Adult dataset has separate training and testing set, we can compute classification accuracy directly on the testing set. For the remaining datasets, we use 10-fold cross validation to measure the average classification accuracy.

Table 2. Dataset statistics

Dataset	Ionosphere	Pima	German
Instances	351	768	1000
Attributes	8	34	24
Dataset	Svmguide3	CreditCard	Adult
Instances	1243	30000	32562
Attributes	21	24	123

We conduct experiments with linear kernel and the results are shown in Fig. 3. It is shown that RRSVM, RSVM and LIBSVM achieve similar classification accuracy. RRSVM and RSVM performs even better than LIBSVM on the Svmguide3 dataset, which might be contributed to the reduced risk of overfitting. One may argue that using nonlinear kernel might improve the accuracy of LIBSVM. We also test on the Gaussian kernel. The classification performance of LIBSVM only increases a little from 75.97% to 76.37% for the Svmguide3 dataset.

We record the training and testing time for each dataset in Table 3. Apart from CreditCard and Adult, we use a reduced set of size 10% for both RRSVM and RSVM. It is shown that RRSVM and RSVM run faster than LIBSVM due to the reduced set (except for the training of Svmguide3). The advantage of

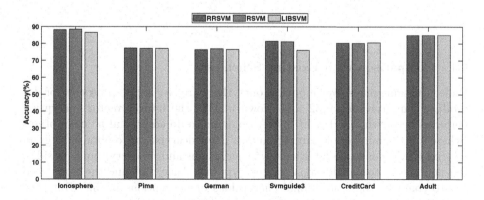

Fig. 3. Comparison of classification performance

RRSVM or RSVM becomes significant when it comes to large-scale datasets, for example, CreditCard and Adult. We test on a much smaller reduced set of size 1%. The accuracy rates of CreditCard and Adult are 80.16% (compared to 80.13% for 10% reduced set) and 84.94% (compared to 84.93% for 10% reduced set), respectively. Therefore, we achieve similar classification performance with a 1% reduced set but run much faster, which validates that our enhanced scheme with secure kernel matrix (random vectors as the reduced set) is effective.

Table 3. Accuracy and Running Time

Dataset		RRSVM	RSVM	LIBSVM
Ionosphere	accuracy	88.03%	**88.31%**	86.58%
	training	2.6×10^{-3}	$\mathbf{2.5 \times 10^{-3}}$	3.5×10^{-3}
	testing	3.93×10^{-5}	$\mathbf{3.92 \times 10^{-5}}$	3.09×10^{-4}
Pima	accuracy	**77.21%**	77.08%	77.07%
	training	8.8×10^{-3}	$\mathbf{6.0 \times 10^{-3}}$	8.4×10^{-3}
	testing	$\mathbf{6.69 \times 10^{-5}}$	6.87×10^{-5}	6.71×10^{-4}
German	accuracy	76.30%	**76.90%**	76.50%
	training	$\mathbf{9.50 \times 10^{-3}}$	4.70×10^{-2}	3.80×10^{-2}
	testing	$\mathbf{2.14 \times 10^{-4}}$	2.76×10^{-4}	3.3×10^{-3}
Svmguide3	accuracy	**81.26%**	81.02%	75.97%
	training	6.2×10^{-2}	9.4×10^{-2}	$\mathbf{3.6 \times 10^{-2}}$
	testing	$\mathbf{2.85 \times 10^{-4}}$	4.29×10^{-4}	2.80×10^{-3}
CreditCard	accuracy	80.21%	80.16%	**80.50%**
	training	1.78	**1.45**	41.99
	testing	$\mathbf{4.10 \times 10^{-3}}$	4.20×10^{-3}	2.07
Adult	accuracy	84.93%	84.94%	**84.97%**
	training	**3.37**	47.8	76.02
	testing	**0.09**	0.28	12.25

7.1 Comparison with Existing Schemes

Under the same one server model as ours, there are two secure protocols for dot product computation [19,32]. As shown in [32], the integer vector encryption performs better than FHE. In [18], Alice needs to do additional n^2 encryptions (where n is the number of training samples), $2n^2$ decryptions to obtain the gram matrix. Bob needs to do additional n^2 encryptions and n^2 decryptions. Besides, the cloud needs to do additional $2n^2$ encryptions. Alice, Bob and the cloud need not to compute these additional encryptions or decryptions in [32]. But they use full kernel matrix of which the size is n^2. However, the size of our secure kernel matrix is $n \times \bar{n}$, where $\bar{n} \ll n$ is the size of the reduced set (typically less than 10% of the original training dataset). Therefore, we successfully reduce the computational complexity of our scheme. The communication overhead of our scheme is mainly caused by the upload of key-switching matrices. No matter how many training samples we have, the communication overhead is constant and relies on the dimension m of the training sample and dimension T of the secret key $[\mathbf{I}, \mathbf{T}]$.

8   Conclusions

In this paper, we proposed a privacy preserving scheme for SVM on horizontally and vertically partitioned dataset among multiple institutions. Instead of using the full kernel matrix, we adopt the reduced SVM with a secure kernel matrix. We proved the security of our protocol. We evaluated the performance of our reduced SVM with secure kernel matrix and concluded that it can not only achieve similar classification accuracy to the traditional SVM but also has a great advantage over large-scale datasets. One limitation of our scheme is that it only works for the integer domain (or fixed-point number with scaling) although it covers a wide range of applications already. Designing a scheme which supports operations on real domain is one of our future work.

References

1. LIBSVM: A library for support vector machines. https://www.csie.ntu.edu.tw/~cjlin/libsvm/. Accessed 15 Apr 2020
2. Amirbekyan, A., Estivill-Castro, V.: A new efficient privacy-preserving scalar product protocol. In: Proceedings of the Sixth Australasian Conference on Data Mining and Analytics, vol. 70, pp. 209–214 (2007)
3. Blaze, M., Bleumer, G., Strauss, M.: Divertible protocols and atomic proxy cryptography. In: Advances in Cryptology EUROCRYPT 1998, pp. 127–144 (1998)
4. Boneh, D., Goh, E.J., Nissim, K.: Evaluating 2-DNF formulas on ciphertexts. In: Theory of Cryptography Conference, pp. 325–341 (2005)
5. Demmler, D., Schneider, T., Zohner, M.: ABY-A framework for efficient mixed-protocol secure two-party computation. In: NDSS (2015)
6. Dennis Jr., J.E., Schnabel, R.B.: Numerical Methods for Unconstrained Optimization and Nonlinear Equations. SIAM, Philadelphia (1996)
7. Goldreich, O.: Foundations of Cryptography: Volume 2, Basic Applications. Cambridge University Press, New York (2004)
8. Hesamifard, E., Takabi, H., Ghasemi, M.: Deep neural networks classification over encrypted data. In: Proceedings of the Ninth ACM Conference on Data and Application Security and Privacy (CODASPY 2019), pp. 97–108 (2019)
9. Laur, S., Lipmaa, H., Mielikäinen, T.: Cryptographically private support vector machines. In: Proceedings of the 12th ACM SIGKDD International Conference on Knowledge Discovery and Data Mining, pp. 618–624 (2006)
10. Lee, Y.J., Huang, S.Y.: Reduced support vector machines: A statistical theory. IEEE Trans. Neural Netw. 18(1), 1–13 (2007)
11. Lee, Y.J., Mangasarian, O.L.: RSVM: Reduced support vector machines. In: Proceedings of the 2001 SIAM International Conference on Data Mining, pp. 1–17 (2001)

12. Lee, Y.J., Mangasarian, O.L.: Ssvm: A smooth support vector machine for classification. Comput. Optim. Appl. **20**(1), 5–22 (2001)
13. Li, X., Zhu, Y., Wang, J.: Secure Naïve Bayesian classification over encrypted data in cloud. In: Chen, L., Han, J. (eds.) ProvSec 2016. LNCS, vol. 10005, pp. 130–150. Springer, Cham (2016). https://doi.org/10.1007/978-3-319-47422-9_8
14. Li, X., Zhu, Y., Wang, J., Liu, Z., Liu, Y., Zhang, M.: On the soundness and security of privacy-preserving SVM for outsourcing data classification. IEEE Trans. Dependable Secure Comput. **15**(5), 906–912 (2018)
15. Lin, K.-P., Chang, Y.-W., Chen, M.-S.: Secure support vector machines outsourcing with random linear transformation. Knowl. Inf. Syst. **44**(1), 147–176 (2014). https://doi.org/10.1007/s10115-014-0751-1
16. Lin, K.P., Chen, M.S.: Privacy-preserving outsourcing support vector machines with random transformation. In: Proceedings of the 16th ACM SIGKDD International Conference on Knowledge Discovery and Data Mining, pp. 363–372 (2010)
17. Lin, K.M., Lin, C.J.: A study on reduced support vector machines. IEEE Trans. Neural Netw. **14**(6), 1449–1459 (2003)
18. Liu, F., Ng, W.K., Zhang, W.: Encrypted SVM for outsourced data mining. In: EEE 8th International Conference on Cloud Computing, pp. 1085–1092 (2015)
19. Liu, F., Ng, W.K., Zhang, W.: Secure scalar product for big-data in mapreduce. In: IEEE International Conference on Big Data Computing Service and Applications (BigDataService), pp. 120–129 (2015)
20. Liu, L., Chen, R., Liu, X., Su, J., Qiao, L.: Towards practical privacy-preserving decision tree training and evaluation in the cloud. IEEE Trans. Inf. Forensics Secur. **15**, 2914–2929 (2020)
21. López-Alt, A., Tromer, E., Vaikuntanathan, V.: On-the-fly multiparty computation on the cloud via multikey fully homomorphic encryption. In: Proceedings of the Forty-Fourth Annual ACM Symposium on Theory of Computing, pp. 1219–1234 (2012)
22. Ma, X., Zhang, F., Chen, X., Shen, J.: Privacy preserving multi-party computation delegation for deep learning in cloud computing. Inf. Sci. **459**, 103–116 (2018)
23. Maekawa, T., Kawamura, A., Kinoshita, Y., Kiya, H.: Privacy-preserving SVM computing in the encrypted domain. In: Asia-Pacific Signal and Information Processing Association Annual Summit and Conference (APSIPA ASC), pp. 897–902 (2018)
24. Mangasarian, O.L., Wild, E.W., Fung, G.M.: Privacy-preserving classification of vertically partitioned data via random kernels. ACM Trans. Knowl. Discovery Data (TKDD) **2**(3), 12 (2008)
25. Nikolaenko, V., Weinsberg, U., Ioannidis, S., Joye, M., Boneh, D., Taft, N.: Privacy-preserving ridge regression on hundreds of millions of records. In: IEEE Symposium on Security and Privacy (SP), pp. 334–348 (2013)
26. Pedersen, T.B., Saygın, Y., Savaş, E.: Secret sharing vs. encryption-based techniques for privacy preserving data mining (2007)
27. Peter, A., Tews, E., Katzenbeisser, S.: Efficiently outsourcing multiparty computation under multiple keys. IEEE Trans. Inf. Forensics Secur. **8**(12), 2046–2058 (2013)
28. Qiu, S., Wang, B., Li, M., Liu, J., Shi, Y.: Toward practical privacy-preserving frequent itemset mining on encrypted cloud data. IEEE Trans. Cloud Comput. **8**(1), 312–323 (2020)
29. Vaidya, J., Yu, H., Jiang, X.: Privacy-preserving SVM classification. Knowl. Inf. Syst. **14**(2), 161–178 (2008)

30. Wang, B., Li, M., Chow, S.S., Li, H.: A tale of two clouds: Computing on data encrypted under multiple keys. In: 2014 IEEE Conference on Communications and Network Security (CNS), pp. 337–345 (2014)
31. Yu, A., Lai, W.L., Payor, J.: Efficient integer vector homomorphic encryption (2015). https://courses.csail.mit.edu/6.857/2015/files/yu-lai-payor.pdf
32. Zhang, J., Wang, X., Yiu, S.M., Jiang, Z.L., Li, J.: Secure dot product of outsourced encrypted vectors and its application to SVM. In: Proceedings of the Fifth ACM International Workshop on Security in Cloud Computing, pp. 75–82 (2017)
33. Zou, Y., Zhao, Z., Shi, S., Wang, L., Peng, Y., Ping, Y., Wang, B.: Highly secure privacy-preserving outsourced k-means clustering under multiple keys in cloud computing. Secur. Commun. Netw. **2020**, 1–11 (2020)

A Distributed Business-Aware Storage Execution Environment Towards Large-Scale Applications

Feng Jiang[1], Yongyang Cheng[1(✉)], Changkun Dong[1], Zhao Hui[1], and Ruibo Yan[2]

[1] China Telecom Cloud Computing Corporation, Beijing, China
chengyy2@chinatelecom.cn
[2] State Key Laboratory of Networking and Switching Technology, Beijing University of Posts
and Telecommunications, Beijing, China

Abstract. The rapid development of distributed storage technology has attracted growing attention from both industry and academia. Although distributed system has been studied for many years, however, it is widely applied to engineering practice since the rise of cloud computing recently. Cloud storage constructs a storage resource pool with massive common storage devices through networks, which could be allocated to authorized users on demand. In this paper, we present a distributed business-aware storage execution environment towards large-scale applications. Initially, we present the overall of our proposed storage execution environment and different data delivery models according to corresponding business needs. Furthermore, we design a globally shared and cross-regional deployed metadata service to minimize the access delay of metadata. Finally, the approach presented in this paper has been validated to be effective through a series of evaluation experiments and actual use cases.

Keywords: Cloud computing · Object-oriented storage · Distributed system · Large scale applications · High-performance architecture

1 Introduction

Cloud computing is a technology that integrates distributed computing, parallel computing and grid computing, which divides complex computing programs into innumerable smaller subroutines and submits them to a huge system composed of multiple servers for calculation and analysis [1–3]. The ultimate goal of cloud computing is to provide common users with essential computing power, just like water, electricity and coal. Cloud storage is a concept that extends and derives from cloud computing, which not only saves the hardware cost, but also has characteristics of scalability, transparency and load balancing. In general, cloud storage is the underlying support of cloud computing. It virtualizes the storage clusters composed of multiple common PC servers into a scalable, flexible and transparent storage resource pool. Then, authorized users could arbitrarily access and manage the storage resource pool through a unified interface and pay for it [4, 5].

F. Jiang and Y. Cheng are co-first authors of the article.

© Springer Nature Switzerland AG 2020
M. Qiu (Ed.): ICA3PP 2020, LNCS 12453, pp. 142–156, 2020.
https://doi.org/10.1007/978-3-030-60239-0_10

Cloud storage centralizes all kinds of storage resources and automatically manages them through specialized software without any human involvement. Thus, users could dynamically use storage resources without considering the technical details of complex distributed storage system (e.g. data distribution, automatic expansion, load balancing, etc.). In this case, users could pay more attention on business processes [6], which is conductive to improving efficiency, reducing cost and innovating technology. Although many companies and research institutes have offered a variety of storage services for users [7–12], challenges still remain to be resolved. First of all, the mainstream cloud storage services are usually deployed within a Metropolitan Area Network (MAN). Service providers might add new available areas to meet the needs of development. In order to ensure continuous reading and writing performance, users have to constantly switch and migrate data among storage services in different regions. Furthermore, with the rapid development of data backup, medical imaging and other industries, the demands for reducing writing latency and improving upstream speed are particularly prominent. However, the existing cloud services lack flexibility in trade-offs between consistency and availability models. Finally, the traditional metadata service in object-oriented cloud storage systems is regionally isolated, which means that logically adjacent metadata and data might have a large distance in physical location, leading to the access delay of metadata becomes a bottleneck in the bandwidth-limited network environment.

The main contributions of our work in this paper contain the following:

- We propose a wide execution environment interconnected through dedicated high-speed communication networks, which provides different data delivery models for users according to business needs.
- We design a globally shared metadata storage service to minimize the access delay of metadata in a bandwidth-limited network environment.
- We take full account of feasibility and performances in our approach and test it through a series of evaluation experiments and actual use cases.

The rest of this paper is organized as follows. We discuss the related work in Sect. 2. The overview of our proposed approach is illustrated in Sect. 3. In Sect. 4, we introduce the execution mechanism of metadata storage service. In Sect. 5, we evaluate the performances of our proposed approach through quantitative experiments. We give a series of actual case studies based on our proposed execution environment in Sect. 6. Concluding remarks are made in Sect. 7.

2 Related Works

In this section, we will compare our proposed large-scale high performance object-oriented cloud storage architecture with other existing approaches and clearly point out the similarities and differences with them. Kanthavar et al. [13] propose an architecture to build the infrastructure of secure and scalable cloud storage service, which is designed using open source technologies and tools. This low-cost approach provides secure storage, upload and download of data. Maksutov et al. [14] describe an approach of interaction with the cloud storage where it is possible to conceal information stored in files. However, the cloud services in these approaches are deployed within a MAN. Service providers might add new available areas to meet the needs of development. In order to ensure continuous reading and writing performance, users have to constantly switch and migrate data between storage services in different regions. Dhanya et al. [15] propose a two level auditing model to experiment to verify the veracity of the consistency and reliability of the data between a data cloud and an audit cloud as promised by the data cloud provider. SureshPatil et al. [16] present a system that merges the schemes replication and erasure code for data availability to avoid redundancy in the dat. However, with the rapid development of data backup, medical imaging and other industries, the demands for reducing writing latency and improving upstream speed are particularly prominent. These two approaches lack flexibility in trade-offs between consistency and availability. Smit et al. [17] present a methodology for a service-oriented application that provides relevant metadata information describing offered cloud services via a uniform RESTful web service. The data provided by this service is automatically acquired and mapped to a standard ontology. Verginadis et al. [18] discuss a metadata schema for data-aware multi-cloud computing which aspires to form the appropriate background vocabulary that will aid the big data-aware application deployments. However, the metadata service in the two approaches is regionally isolated, which means that logically adjacent metadata and data might have a large distance in physical location, leading to the access delay of metadata becomes a bottleneck in the bandwidth-limited network environment.

3   Our Proposed Object Oriented Storage (OOS) Architecture

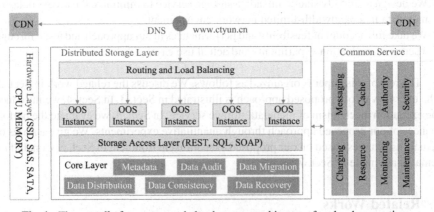

Fig. 1. The overall of our proposed cloud storage architecture for cloud computing.

In this section, we will present our proposed distributed business-aware storage execution environment towards large-scale applications. The overall of our proposed execution environment is hierarchically illustrated in Fig. 1. Distributed storage layer is the core and the most difficult part of our proposed architecture. A high performance cloud storage system requires the ability to evenly distribute data across multiple storage nodes. Furthermore, in order to ensure the reliability and availability, it is necessary to realize the redundant storage of data on the premise of consistency. When a running storage node fails, its services could be automatically migrated to other normal storage nodes in the cluster. To access these issues, we construct a computational instance group from multiple object-oriented storage instances. When an instance fails, the system could automatically migrate its load to other instances of the same computational instance group. In addition, our proposed architecture also supports dynamically adding or reducing the instances to make the processing power of computational instance group scalable. At the top of Fig. 1, CDN components perform routing and load balancing, which would cache hot data in the cloud storage system to the nearest location to users to reduce the access latency and bandwidth. Unlike the traditional ones, we adopt a hierarchically hybrid storage strategy to implement the CDN components, including SSD, SAS and SATA. We combine the high performance of SSD and the cost advantage of SAS and SATA. Thus, data would migrate with heat. The hottest data would be storage in SSD, the moderate data in SAS and the lightest data in SATA. In this way, after being processed by computational instances, data could be returned to the client as soon as possible.

To improve the service availability, the data recovery service could be divided into fragment, object and disk levels. In the phase of writing, if some fragments fail to write, fragment-level data recovery is required. The data recovery service restores the failed fragments to the corresponding physical disk using other successful fragments in the same fragment group. In the phase of data auditing, if some objects lose data, object-level data recovery is required. The data recovery service compares the object hash value provided by the metadata service to find the fault fragment and restores it. In the phase of disk monitoring, if some physical disks are damaged, disk-level data recovery is required. In this case, the corresponding relationship between virtual disk and physical disk recorded in the routing table needs to be adjusted. We take Table 1 as an instance, if the physical disk Pd1 is damaged, then all locations of the physical disk Pd1 in the routing table would be replaced by other available physical disks.

Table 1. The disk-level data recovery service in our proposed approach.

Virtual disk	Physical disk					
Vd0	**Pd1 [Pd7]**	Pd2	Pd3	Pd4	Pd5	Pd6
Vd1	Pd2	Pd3	Pd4	Pd5	Pd6	**Pd1[Pd8]**
Vd2	Pd3	Pd4	Pd5	Pd6	**Pd1[Pd9]** Pd2	
...	...	...	...	...	...	...
Vd_{65535}	Pd5	Pd6	**Pd1[Pd7]** Pd2	Pd3	Pd4	

With the rapid development of data backup, medical imaging and other industries, the demands for reducing writing latency and improving upstream speed are particularly prominent. In this case, a flexible data delivery that allows users to make further trade-offs between consistency and availability according to specific business scenarios is needed. From the perspective of clients, it could be divided into strong, weak and final consistencies. If a strong consistency data delivery model is adopted, the data consistency could be guaranteed. However, when network or other faults occur among master and backup replicas, the writing operation would be clocked, leading to the service availability could not be satisfied. As mentioned above, some latency-sensitive application scenarios could tolerate a certain amount of data inconsistency, but they have higher requirement for service availability. Thus, we provide different data delivery models for users to make further trade-offs between service consistency and availability according to their specific business application needs.

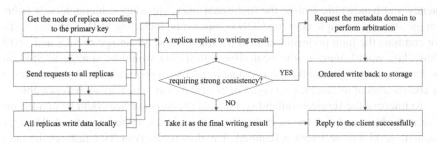

Fig. 2. The Writing process in our proposed data delivery model.

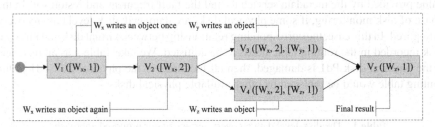

Fig. 3. The mechanism of vector clock in our proposed data delivery model.

When writing data in our proposed data delivery model, initially, it calculates the storage node of each data replica according to the primary key. Then, writing requests are sent to all data replicas, which would perform the writing data operation locally. Once a data replica replies to the writing result, users could choose whether to require strong consistency model. If the strong consistency model is chosen, it would request the metadata domain to perform arbitration until all data replicas reply to writing results. Finally, the data delivery model writes back to the storage and replies to the client successfully. Figure 2 illustrates the process of writing data in our proposed data delivery model. Due to the differences among computational instance, it might occur that the same data is updated by multiple writing operations, which means that the orders of different writing operations could not be strictly guaranteed. To access this issue, we design a mechanism of vector clock to resolve the conflict. Vector clock could be expressed as [instances, counters], where counters represent the writing update number of instances. Initially, W_x writes an object once to produce an object version V1 ([W_x, 1]). Then,

Wx writes this object again and produces a new object version V2 ([Wx, 2]). Next, Wy and Wz perform writing operations to this object and produce two new object versions V3 ([Wx, 2], [Wy, 1]) and V4 ([Wx, 2], [Wz, 1]). In this case, there are two versions of writing conflicts. We add timestamps to the metadata of each object, which would be introduced in the next section. Based on this, we could choose the data version whose timestamps is latest as the final writing result V5([Wx, 2], [Wy, 1], [Wz, 1]), which is illustrated in Fig. 3.

Then, we would take a practical application instance mentioned in Sect. 3 to describe the data writing process in two different data delivery models. We assume that the metadata domain of object A is configured in North China and the data domain requested by the client is written in a computational instance of Shanghai. Thus, the data writing process under the strong consistency model is as follows: 1. The client sends data writing request to Shanghai; 2. The interface layer in Shanghai writes data to its corresponding storage layer; 3. The storage layer returns a successful result; 4. The computational instance in Shanghai requires writing arbitration to the metadata domain in North China; 5. After writing arbitration, the metadata domain in North China records a successful result and returns it to the computational instance in Shanghai; 6. Finally, the computational instance returns a successful result to the client. Differ to the strong consistency model, if users choose the strong availability model, the interface layer in Shanghai would insert a metadata into the messaging queen after step 3, which contains a monotonically increasing ID. If this step fails, the computational instance in Shanghai returns a failed result to the client. Otherwise, it returns a successful result to the client intermediately. Meanwhile, the metadata in the messaging queue should be asynchronously submitted to the metadata domain in North China to perform writing arbitration. The right part of Fig. 4 illustrates the data writing process, where the blue arrow represents the strong consistency model and the red one represents the strong availability model.

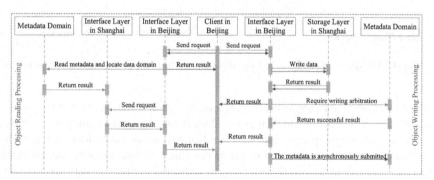

Fig. 4. The mechanism of vector clock in our proposed data delivery model. (Color figure online)

When reading data in our proposed data delivery model, the first step is the same as writing data. It needs to calculate the storage node of each data replica according to the primary key. Then, reading requests are sent to all data replicas, which would perform the reading operation locally. Once a data replica replies to the reading result, users could choose whether to require strong consistency model. If the strong consistency model is chosen, it needs to wait for all replicas to complete the reading operation before performing reading arbitration. If all results are consistent, the data delivery model replies to the client successfully. Otherwise, it would merge the results of multiple replicas according to specific principles. Finally, the data delivery model updates the result of inconsistent replicas and returns it to the client successfully. Then, we would take a practical application instance mentioned in section to describe the data reading process. We assume that the metadata domain of object B is configured in North China, the data domain is written in a computational instance of Shanghai and the data reading request is sent by a client in Beijing. Thus, the data reading process under the strong consistency model is illustrated as left part of Fig. 4.

4   Globally Shared Metadata Service

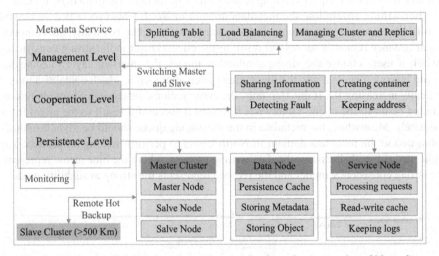

Fig. 5. The levels of global shared metadata service from the perspective of hierarchy.

Based on our proposed execution environment, the whole China is divided into several metadata domains according to their physical locations. A metadata domain is usually a large area covering multiple provinces, such as North China metadata domain, East China metadata domain and South China metadata domain. Different metadata domain communicates and exchange information through dedicated high-speed networks. Besides cross-province deploying, multiple metadata domains could also be deployed in one province, depending on the specific business demand for latency. The global shared metadata service maintains multiple globally shared tables, which are used to save records of objects, attributes of containers and account of clients. The row keys in each table are ordered and divided into several sub-tables according to a specified range. These sub-tables would be used to persist object data to the disk. Although each metadata domain is discrete deployed, no record with the same row key exists in any two metadata domain, ensuring that all metadata could be serviced as a whole. From the perspective of hierarchy, the global shared metadata service could be divided into three levels, which is illustrated in Fig. 5.

The management level is mainly responsible for the division and distribution of sub-tables. All creation, querying and deletion operations of tables are handled by the management level. There are two application processes in the management level, one is the master and the other is the slave. At the same time, only the master process is in the working state, while the slave process is in the ready state. The switching between them depends on services of the cooperation level. The object data of these two management processes are persisted in the whole metadata service and could be started on any computational instance, avoiding the single point failure. When the global shared metadata service is initialized, the management processes would scan all sub-tables and assign each sub-table to a service node in the persistence layer. In addition, the management processes would monitor the status of each computational instance through services of the cooperation level. If a computational instance is found to be faulty, the sub-tables run on it would be reassigned to other available computational instances. Finally, the management level maintains a mapping from the sub-table to computational instance. When accessing a sub-table for the first time, the client of metadata service needs to query the computational instance that is currently responsible for processing the running sub-tables, and then initiate a request for the queried instance. In the subsequent accesses, the client would cache the information until its status changes, instead of querying the management processes frequently

The cooperation layer provides a distributed collaboration service, which is used to maintain a shared state of the global metadata service. By establishing a heartbeat connection with the metadata service node, the cooperation level could sense the failure in time. Through the notification mechanism provided by the cooperation level, the management level could discover the failed computational instances and conduct further operations, such as sharing information, assigning sub-tables and switching nodes. Furthermore, the cooperation level also obtains and caches the address information of the management processes to avoid frequent queries on the cooperation level. Finally, the main function of global shared metadata service is to maintain a table that records all object metadata. The metadata belonging to a same container in the table would be stored in a same metadata domain. Thus, the metadata service allocates and manages metadata domain and fault domain based on the container. The specific processes of assigning metadata domain to containers are as follows:

1. Client sends requests to create a container and designates the metadata;
2. Data domain sends requests to a cooperation level to create a container;
3. The cooperation level assigns containers to specified metadata domains.

The persistence level contains multiple data nodes, service nodes and index nodes, which would be deployed to each metadata domain. A computational instance could be used as multiple types of nodes or only plays one role. Each sub-table corresponds to a file, which would be divided into multiple data blocks according to a fixed size when writing to the persistence level. Then, these data blocks would be created three replicas and stored in discrete computational instances. In addition, the corresponding relationship of files, data blocks and computational instances is stored in the index cluster composed of index nodes. Each index node is a single high performance server and provides the data storage service with a persistence of no less 99.99999999999%. Among these index nodes, the persistence level would designate one index node as the master node and the other two as the slave nodes, which are interconnected with the master node through a dedicated line with low latency and high availability. According to the quorom mechanism, as long as the condition W + R > N is met, the client could read at least one valid data when there is no more than one computation instance failure, where N refers to the total number of replicas, R refers to the minimum number of computational instances in a successful reading operation and W refers to the minimum number of computational instances in a successful writing operation. Thus, the persistence level could ensure the high availability of services and strong consistency of data based on quorom and vector clock mechanisms. In this case, the slave service nodes could immediately take over the data and keep it consistent when the master service node breaks down.

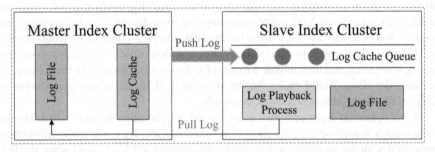

Fig. 6. The synchronization of master and slave index clusters in our proposed approach.

In order to ensure the security of index data, we deploy a remote slave index cluster for hot backup, which is located in another high-level data center and no less than 500 km away from the mater index cluster. The slave index cluster continuously obtains an ordered set of update operations from the master index cluster and commits them to the local storage node as an atomic transaction. Thus, the next set of operations would not be accepted until the transaction is committed successfully. Based on this, it ensures that the data in slave index cluster is a snapshot of the master index cluster at a certain time in the past. Figure 6 illustrates the mechanism of synchronization between master and slave index clusters. The master index cluster pushes operation logs to the slave index cluster. The receiving thread in the slave cluster gathers the operation logs and writes them to a global shared log cache queue. Once the slave index cluster receives operation logs, it would return a successful result of synchronization to the master index cluster. Then, the master index cluster updates its local memory and writes log files to the disks.

$$New\ activate\ cache\ =\ Frozen\ cache\ +\ Writing\ cache \tag{1}$$

$$Result\ =\ New\ activate\ cache\ +\ Disk\ file \tag{2}$$

$$Result = Frozen\ cache + Writing\ cache + Disk\ file \qquad (3)$$

The service node maintains a cache area for reading in the memory, which caches the most frequently accessed row keys. When processing a reading request, the service node would first search in the reading cache. If found, the service node would return a successful result to the client. Else it would further search in the writing cache to see if the row key is newly written. However, if neither of the reading and writing cache areas caches the target row key, the service node has to read the file persisted on the disk. In order to avoid the invalid operation of reading files, we take advantage of the bloom filter mechanism. We define the reading cache area as the frozen cache area. Thus, the new active cache area is the sum of the frozen cache area and the writing cache area, which could be expressed as Eq. (1). The querying result for the client could be expressed as Eqs. (2) and (3).

In addition, the global shared metadata service divides multiple metadata failure domains, each of which is composed of several data service nodes. These data service nodes are usually distributed in a converged area, such as a single data center or multiple data centers that are interconnected with low latency, large bandwidth and high availability networks. For the data blocks in the persistence level, all of their replicas are persisted in the same failure domain, which should not be stored across multiple failure domains. The purpose of this design is to reduce the number of affected data blocks when the metadata service nodes break down suddenly.

5   Experiment Evaluation

We have introduced the main features of our proposed cloud storage architecture in the previous sections. In order to further illustrate the effectiveness and rationality, we would conduct a series of simulation experiments for performance comparisons in this Section. The simulation experiments were deployed on three same PCs, which had 16 GB of RAM, i7-8550U of Core, 3.80 GHz of CPU and 1 TB of disk space. PC A was simulated as a master to access the stored data object and PC B was simulated as a slave to implement highly available services. PC C was simulated as a storage server to store designed data objects, which had been deployed three different file systems: HDFS-based, GFS-based and our proposed approaches. A data object was randomly assigned a size that ranged from 1 KB to 1 MB. To get closer to the real scene, we designed a threshold timeout. All data objects whose response time exceeded the threshold would be abandoned. In addition, during the sampling process, we shut down server A and let server B continue to provide services. We repeatedly conducted the experiments in the condition that the values of timeout were set to 15 ms and 30 ms. The sampling time was 1000 ms and the total stored data were 20 thousand.

Response time was the total amount of time that a storage system took to response to a request for execution. Ignoring transmission time for a moment, the response time was the sum of waiting time and execution time. Figure 7 (a) illustrated the average response times based on different storage systems. The average response times of HDFS-based, GFS-based and our proposed approaches were 16.23 ms, 15.18 ms and 10.34 ms, which in our proposed approach was 36.29% and 31.88% less than the other two approaches. In our proposed approach, we design a globally shared metadata service to break the data barrier and minimize the access delay of metadata. However, the logically adjacent metadata might be regionally isolated in HDFS-based and GFS-based approaches. Furthermore, the file data structure in our proposed approach was more flat than the other two ones. Thus, the average response time was the shortest.

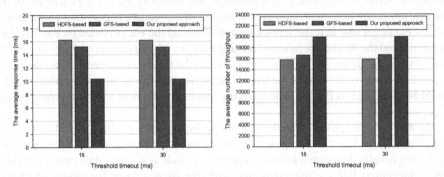

Fig. 7. The response times and throughputs based on different approaches (timeout = 15 and 30).

Throughput was the data objects that had been accessed by Server A within a unit time. Figure 7 (b) showed the throughputs based on different approaches. The average throughputs of HDFS-based, GFS-based and our proposed approaches were 15798, 16632 and 19890, which in our approach was 25.90% and 19.59% more than the other two approaches. The value of throughput was directly affected by the average response time. As mentioned above, the average response time in our approach was shortest. Thus, it could respond to more data objects during the sampling time.

Input/Output Per Second (IOPS) referred to the number of I/O requests that could be processed by the storage system within a unit time. For applications with frequent large file transferring (size over 1 MB), bandwidth was the key aspect to restrict the system performance, while for applications with frequent small file reading, sufficient I/O capability was required. Figure 8 (a) illustrated the average IOPS, including sequential and random reading, based on different approaches. Due to the flexible data delivery and caching models in our proposed approach, it could respond to more I/O requests during the sampling time than the other two approaches.

In order to improve the system utilization, we should try our best to keep the I/O busy and reduce the time consuming of CPU computing. Figure 8 (b) showed the disk utilization of server C based on different approaches. Due to the switching mechanism between the master and the slave in our proposed approach, it ensured that the data in server B was a snapshot of server A at a certain time in the past. We intercepted the results of the period from 500 ms to 580 ms. The final results showed that the thrashing in our proposed approach was the slightest compared with the HDFS-based and GFS-based approaches when server A suddenly became unavailable.

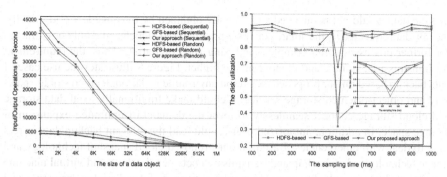

Fig. 8. The average IOPS and disk utilization based on different approaches.

6 Use Cases

Rapid evolution of the cloud storage technology brings a number of use cases that could benefit from the integration of cloud computing and object-oriented storage. In this section, we would choose some typical application cases to illustrate the potential of our proposed distributed business-aware storage execution environment.

6.1 Data Backup

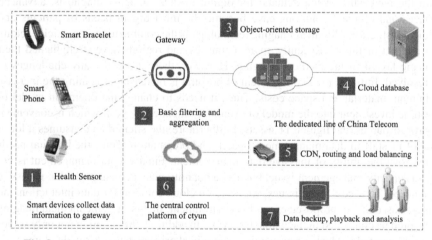

Fig. 9. The use case of data backup based on our proposed execution environment.

With the development of society and technology, the requirement for the backup of massive data is increasing. However, a lot of challenges remain to be issued. Initially, the data backup time is long and the efficiency is low, which seriously affect the core business system. Thus, the storage architecture for data backup should have good scalability to meet the development of business needs. Furthermore, the backup data has the characteristics of high privacy and strong consistency. The storage architecture needs to ensure that the data would not be invaded and illegally obtained by the outside world. Finally, the raw data collected by the smart sensors is unstructured, which could only be extracted for further processing after complex analyzing and computing. Thus, the requirement for transmission, storage and computing are higher. Figure 9 illustrates the use case of data backup based on our proposed execution environment. Through cloud storage gateway and object-oriented storage service, it could make traditionally local data storage and cloud storage connect seamlessly to meet the demand of unlimited expansion of storage space. In addition, authorized users could read the relevant backup data in real time to deal with emergencies. Based on our proposed cloud snapshot technology and image technology, data backup and transmission between database and virtual machine could be realized quickly. Integrated hierarchical storage of standard, low frequency and archive could effectively cope with backup data of different business values. In this way, the business system could be guaranteed and the backup data could be transferred by peak load shifting, which greatly reduces the network consumption. In addition, it supports remote host-standby, three centers and multiple backup mechanisms in two places, and the cloud storage space is not capped. To make the data backup more smooth and stable, we use two China Telecom's dedicated lines for remote debugging of servers, so that the data could not only achieve the stable transmission, but also could be backed up safely.

6.2 Medical Imaging

With the increasing of the demand for digital medical imaging diagnosis, advanced services and efficient solutions have become the most urgent needs of patients and medical institutions. Medical imaging technology is the combination of medicine, cloud and internet of things. Recently, many countries and regions have strengthened their strategic layout in the medical imaging. However, there are still many challenges to be resolved. Initially, the IT investment of hospital is increasing, meaning the increase of human, material and space costs. Thus, it needs to change the operation model of one-time investment into the model of even investment every year, which is convenient to save resources and lighten IT assets. Furthermore, the silent data consumes a lot of time of IT staff in data management, search and preparation. Thus, the hospital needs to improve the ability of massive data management. Finally, data management is not only a capacity management issue, but also a quantity management issue that is easy to be ignored. To access these issues, we design a cloud solution for data interaction and storage based on our proposed execution environment. In this case, the medical imaging data could be controlled in the whole life cycle. In addition, this solution not only solves the problem of cloud leakage prevention on medical privacy data (e.g. medical imaging, record, diagnosis, etc.), but also solves the problems of data sharing between hospitals and patients, hospitals and hospitals, hospitals and governments. Figure 10 illustrates the use case of medical imaging based on our proposed execution environment, which could easily support a unified management of 10 billion level small files and 100 GB level single large file, which solves the problem of massive small file management. Due to the frequency of requiring, data could be hierarchically stored to reduce the storage cost. Meanwhile, a data object is fragmented and distributed stored in the cloud to realize minute-level retrieval of billion-level small files. Finally, we support multiple-link transmission, scheduled download and local cache to improve the transmission speed and break the component barrier.

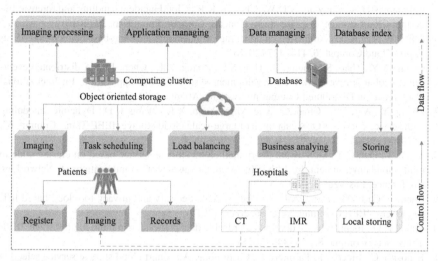

Fig. 10. The use case of medical imaging based on our proposed execution environment.

7 Conclusions

In this paper, we have proposed a distributed business-aware storage execution environment towards large-scale applications. To achieve this goal, we initially present the overall of our proposed execution environment and different data delivery models according to corresponding business needs. Then, we design a globally shared and cross-regional deployed metadata service to minimize the access delay of metadata. Finally, the approach presented in this paper has been validated to be effective through a series of evaluation experiments and actual use cases. The system data durability is an important index of performance evaluation. However, the existing approaches are based on retrogressive models, which are not applicable when all data replicas lost. Plan for future work is to model the data durability of this execution environment and quantitatively evaluate the corresponding performances.

Acknowledgment. The authors would like to thank Lilian Huang, the vice general manager of China Telecom Cloud Computing Corporation at that time, Markus Fischer, the senior director of Apple corporation, Daniel Christianto, the senior manager of Apple corporation at that time, for helpful discussions during the initial phase of this work.

References

1. Yang, C.W., Huang, Q.Y., Li, Z.L., Liu, K., Hu, F.: Big Data and cloud computing: Innovation opportunities and challenges. Int. J. Digital Earth **10**, 13–53 (2017)

2. Cao, Z.J., Lin, J., Wan, C., Song, Y.H.: Optimal cloud computing resource allocation for demand side management. IEEE Trans. Smart Grid 8(4), 1943–1955 (2017)
3. Al-Dhuraibi, Y., Paraiso, F., Merle, P.: Elasticity in cloud computing: State of the art and research challenges. IEEE Trans. Serv. Comput. 99 (2018)
4. Rahman, N.H.A., Cahyani, N.D.W., Choo, K.K.R.: Cloud incident handling and forensic-by-design: Cloud storage as a case study. Concurr. Comput. Pract. Exp. 29 (2016)
5. Chi, P.W., Lei, C.L.: Audit-free cloud storage via deniable attribute-based encryption. IEEE Trans. Cloud Comput. 6, 414–427 (2018)
6. Cheng, Y.Y., Zhao, S., Cheng, B., Hou, S.L., Zhang, X.L., Chen, J.L.: A distributed event-centric collaborative workflows development system for IoT application. In: International Conference on Distributed Computing Systems, Atlanta, USA (2017)
7. Zhang, Y., Wei, Q.S., Chen, C., Xue, M.D., Yuan, X.K., Wang, C.D.: Dynamic scheduling with service curve for QoS guarantee of large-scale cloud storage. IEEE Trans. Comput. 67, 457–468 (2017)
8. Su, W.T., Dai, C.Y.: QoS-aware distributed cloud storage service based on erasure code in multi-cloud environment. In: 14th IEEE Annual Consumer Communications \& Networking Conference, Las Vegas, NV, USA (2017)
9. Zhou, J., Cao, Z.F., Dong, X.L., Vasilakos, A.V.: Security and privacy for cloud-based IoT: challenges. IEEE Commun. Mag. 55, 26–33 (2017)
10. Wang, H.Q.: Identity-based distributed provable data possession in multicloud storage. IEEE Trans. Serv. Comput. 8, 328–340 (2015)
11. Esposito, C., Ficco, M., Palmieri, F., Castiglione, A.: Smart cloud storage service selection based on fuzzy logic, theory of evidence and game theory. IEEE Trans. Comput. 65, 2348–2362 (2016)
12. Dou, W.C., Zhang, X.Y., Liu, J.X., Chen, J.J.: HireSome-II: Towards privacy-aware cross-cloud service composition for big data applications. IEEE Trans. Parallel Distrib. Syst. 26, 455–466 (2015)
13. Kanthavar, S., Simeen, S., Sarayu, S.: Design of an architecture for cloud storage to provide infrastructure as a service. In: India Council Conference, India (2018)
14. Maksutov, A.A., Kutepov, S.V., Hrapov, A.S.: Efficient processing and storage of data on untrusted cloud storage services. In: IEEE Conference of Russian Young Researchers in Electrical and Electronic Engineering, St. Petersburg, Russia (2017)
15. Dhanya, D., Vidhya, R., Narendra, M., Davamani, A., Anbarasi, L.J.: Consistency and reliability of services in cloud environment. In: International Conference on Current Trends in Computer, Electrical, Electronics and Communication, Mysore, India (2017)
16. SureshPatil, D., Mane, R.V., Ghorpade, V.R.: Improving the availability and reducing redundancy using deduplication of cloud storage system. In: International Conference on Computing, Communication, Control and Automation, Pune, India (2017)
17. Smit, M., Pawluk, P., Simmons, B., Litoiu, M.: A web service for cloud metadata. In: IEEE Eighth World Congress on Services, Honolulu, HI, USA, (2012)
18. Verginadis, Y., Patiniotakis, I., Mentzas, G.: A metadata schema for data-aware multi-cloud computing. In: Innovations in Intelligent Systems and Applications, Thessaloniki, Greece (2018)

QoS-Aware and Fault-Tolerant Replica Placement

Jingkun Hu[1], Zhihui Du[2(✉)], Sen Zhang[3], and David A. Bader[2(✉)]

[1] Worldmoney Blockchain Management Limited, Hong Kong, Hong Kong
kun.hu@worldmoney.org
[2] New Jersey Institute of Technology, Newark, NJ, USA
{zhihui.du,bader}@njit.edu
[3] State University of New York, College at Oneonta, Oneonta, USA
zhangs@oneonta.edu

Abstract. As emerging applications become more and more distributed
and decentralized, it has become a more challenging problem to design
and build fault-tolerant network systems with high Quality of Service
(*QoS*) guarantee. In this paper, an optimal replica placement problem
is formulated in terms of minimizing the replica placement cost subject
to both *QoS* and fault-tolerant constraints. Based on the generalized
graph model, the optimal replica placement problem is proved to be
NP-hard. The essential properties of the proposed problem are investigated and two criteria, which can quantitatively measure the replica
demand of each node and the contribution of one replica to other nodes
in the graph, are proposed. The criteria are employed to develop efficient
heuristic algorithms. Finally the proposed algorithms are evaluated with
extensive network configurations and the experimental results show that
the proposed heuristic algorithms can generate solutions very close to
the optimal results.

Keywords: Replica placement · Quality of service · Fault tolerance ·
Heuristic algorithm · Distributed system

1 Introduction

Replica placement [10] is a critical technology that can be used for many different purposes, such as reducing latency, enhancing fault tolerance, optimizing
bandwidth use, or improving scalability of network. So it has been employed in
extensive applications such as Content Delivery/Distribution Network (CDN)
[15], data grid [10], cloud [9] and edge computing environment [1]. The theoretical model of replica placement is named as facility location [4], K-median
[2], minimum K-center [13] and so on under different application scenarios. The
great variety and constant changes in application requirements are essentially
part of the reasons why so many different replica placement policies have been
developed and why replica placement problems remain to be an active research
field. Emerging applications, such as smart city [11] and autonomous unmanned

© Springer Nature Switzerland AG 2020
M. Qiu (Ed.): ICA3PP 2020, LNCS 12453, pp. 157–172, 2020.
https://doi.org/10.1007/978-3-030-60239-0_11

cars [19], entail critical requirements in both Quality of Service (QoS) and fault tolerance for supporting networks behind these applications. How to provide guaranteed service in real time with optimal cost yet simultaneously meeting constraints with both fault tolerance and QoS is a challenging problem.

In this paper, we investigate a novel optimal replica placement problem where both QoS and fault-tolerant constraints must be met simultaneously, and propose heuristic algorithms to tackle the problem. The major contributions of our work are as follows.

1. A novel optimal replica placement problem with both QoS and fault-tolerant constraints has been formulated and proved to be NP-hard.
2. Two critical criteria (metrics) *Replica Demanding Priority* and *Replica Profit Ratio* have been developed to quantitatively estimate the replica demand degree and the contribution of given replica to other nodes.
3. Efficient heuristic replica placement algorithms have been developed and extensive experimental results show that the proposed algorithms can achieve results very close to optimal solutions.

2   Related Work

Sahoo et al. [15] provides a comprehensive survey of data replica in CDN systems. Aral et al. [1] summarizes the replica placement in even broader environments and applications, which could be centralized or decentralized, with complete or partial information, and static or dynamic. These surveys show that replica placement problems vary from each other to meet the varying underneath constraints. Most replica placement algorithms take QoS as their constraint or optimization object due to the importance of QoS under different scenarios. Tang et al. [16] proposed an early and typical replica placement research that presents their algorithm solution based on general problem formulation with QoS constraint. At the same time, there are many QoS related replica placement researches focus on algorithm efficiency under practical and specific application scenarios [5] instead of analyzing the problem's hardness under general cases.

There are also replica placement problems subject to constraints related to fault tolerance. Khan et al. [12] developed the replica placement techniques to guarantee a fault-tolerant performance under the uncertainties of arbitrary server failures. However, this work did not consider the QoS requirement. There are some further research [14,18] on fault-tolerant requirement.

In principal, the QoS requirement will reshape each node's local topology and fault-tolerant requirement will entail redundant resources. So the QoS and fault-tolerant constrains together will make the problem very different from existing ones. Up to date, little research has been found on optimal replica problem subject to both *QoS* and fault-tolerant constraints with locality freedom. Du et al. [6] attacks the proposed problem by proposing an approximation algorithm. Their basic idea and contribution is that based on a linear programming (LP) relaxation, the replica problem will be constructed and solved by LP solvers to achieve a real number solution in range [0,1] first. According to the LP solution, the integer programming (IP) solution with a strict upper bound or approximation ratio will be achieved by developing the corresponding rounding algorithm. This work is different from that paper since the major contribution of this paper is developing efficient heuristic algorithms based on the proposed two criteria to obtain good performance on typical data instances but without guaranteeing that a theoretical approximation ratio will stand for any inputs.

3   Problem Description

The replicated infrastructure considered in this work consists of multiple geographically distributed servers that can be classified into two categories: the servers with a replica placement and the servers serving as proxies only. A proxy server doesn't have a replica on itself, but it can delegate requests to other replica servers in its close vicinity. Each node is associated with a certain cost for hosting a replica.

QoS requirement can be distance, response time, latency or throughput, depending on the concerned environment and application. We allow QoS to be individualized (localized) and the situation where a uniformed QoS constraint across all nodes is a naive case of our problem.

A fault-tolerant network means that it can continue to provide service in the presence of failures. Failures may happen at different levels of a network. In our work we consider failures at the replica level rather than at the node level or any other lower levels. The fault-tolerant requirement is also allowed to vary from node to node and the situation where a uniformed fault tolerance configuration on all nodes is thus also a naive case of our problem.

3.1   System Model

We model the replicated network topology using a connected undirected graph $G = (V, E)$, where $V = \{v_0, v_1, \ldots, v_{N-1}\}$ is the set of nodes whose cardinality is N, which can also be simply represented by the unique values from 0 to $N - 1$. And $E = \{(u, v) | u, v \in V\}$ is a set of edges over V. Each node $v \in V$ is an abstract representation of the site that can be placed with replica. An edge $e = (u, v) \in E$ represents that there is *direct* link between the two sites identified by u and v.

For a pair of nodes v and u belonging to V, we use $l(u, v)$ or $l(v, u)$ to represent the distance of the direct link between them. If there is no direct link from v to u, the value of $l(v, u)$ would be infinite and we define $l(v, u) = \infty$. Let L denote the set of weights of all these direct links. We use $d(u, v)$ to represent the shortest distance among all paths between v and u. Since the graph is connected, $d(u, v)$ always exists.

For each node v, we introduce a parameter $s(v)$ to represent the total cost for placing a replica on the node v.

Now let us consider QoS requirement for each node. We use $q(v)$ to quantitatively describe the QoS requirement of a node v. Without loss of generality, here $q(v)$ is defined as the largest distance that is allowed for the node v to possibly get replica service from other nodes. If a node v has a replica, it can meet the request immediately; if a node v does not have a replica, it should send requests to a nearby node with a replica placed, denoted by u, that satisfies $d(v, u) \leq q(v)$. We use Q to represent all the QoS requirements of all the nodes.

Because a network is composed of N nodes, the entire replication scheme can be represented by a vector $X = <x_0, x_1, \ldots, x_{N-1}>$ where x_i gets 1 if a replica is stored at node v_i, and 0 otherwise. Let R represent the set of nodes with replicas placed. Then the total cost C of the replicated network is simply calculated by the following objective function: $C(R) = \sum_{i=0}^{N-1} x_i \times s_i$.

Furthermore, we use another parameter $m(v)$ to stand for the fault-tolerant level of node v and all nodes are represented by M. More specifically, let $mrft(v)$ represent the maximum number of replica failures the node v can tolerate, then we define $m(v) = mrft(v) + 1$. It means that in order to tolerate $m(v) - 1$ arbitrary replica failures, at least $m(v)$ replicas should be accessible for the node v. When $M = <1, \ldots, 1>$ for all the nodes, the network has zero failure tolerance, which can only happen in a replica-fault-free system.

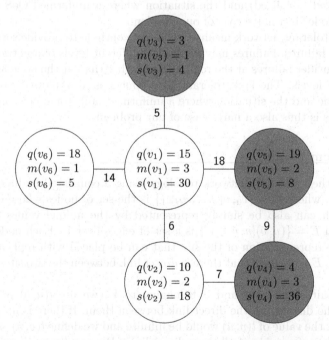

Fig. 1. An example of replica network with QoS and fault-tolerant requirements where each node v has a QoS requirement $q(v)$, fault-tolerant requirement $m(v)$ and replica storing cost $s(v)$. The numbers associated with the edges represent the distances between the incident nodes. The nodes with replica placed are colored in gray and those without white.

Figure 1 shows an example of a replicated network with QoS and fault-tolerant requirements defined for each node. We can see only node v_3 and node v_5's requirements can be met at the same time.

In summary, we have established the following concepts and notations to describe the network in the rest of the discussion. Let $G = (V, E)$ be a graph representing a replicated network, where $V = \{v_0, v_1, \ldots, v_{N-1}\}$ represents the set of nodes of the graph, and E is the set of the edges of the graph built over V. Let $L = <l_0, l_1, \ldots, l_{|E|-1}>$ be the weights associated with each $e \in E$. Let $S = <s_0, s_1, \ldots, s_{N-1}> = <s(v_0), s(v_1), \ldots, s(v_{N-1})>$ be the cost vector associated with each $v \in V$. Let $Q = <q_0, q_1, \ldots, q_{N-1}> = <q(v_0), q(v_1), \ldots, q(v_{N-1})>$ stand for the QoS requirements on V with q_i corresponding to the QoS requirement of v_i. Let $M = <m_0, m_1, \ldots, m_{N-1}> = <m(v_0), m(v_1), \ldots, m(v_{N-1})>$ stand for the fault-tolerant requirements on V with m_i corresponding to the fault-tolerant level of node v_i. R refers to a set of nodes which are placed replicas. M refers to the whole vector of fault-tolerant levels. The same convention also applies to S, Q, L and other notations. In the following discussion, all these notations will be used whenever they don't cause any ambiguity.

3.2 Optimal QoS-Aware and Fault-Tolerant Replica Placement

Given a graph $G = (V, E)$ with each node annotated by s, q and m values, and each edge annotated by l, if any replica placement solutions exist for the network, then the objective of the problem is to find a replica placement solution R whose total cost of replicas is minimized subject to the constraints of QoS and fault tolerance of every node. Here we name it as OQFRP problem. Alternatively, the proposed problem can also be modeled as an optimization problem whose objective is to minimize the total cost of the network subject to the Q and M constraints:

$$Min : C(R) = \sum_{i=0}^{N-1} x_i \times s_i, \text{subject to} \tag{1}$$

$$\sum_{d(i,j) \leq q(i), \forall j \in \{0, \ldots, N-1\}} x_j \geq m_i, \forall i \in \{0, \ldots, N-1\}, \text{and} \tag{2}$$

$$x_i \in \{0, 1\} \tag{3}$$

The problem turns out to be an integer linear programming problem, which, generally being NP-hard, suggests that our problem is likely also NP-hard. In next subsection, we will provide a proof the NP-completeness of the decision version of the OQFRP problem.

3.3 Intractability of the Problem

The decision form of the OQFRP, denoted by $DOQFRP$, asks whether or not a network has a replica solution that has the cost less or equal to a specified value

k:

$$\{<G' = (V', E'), L, Q, M, k>|\exists R(V') \wedge \sum_{v \in R(V')} s(v) \le k\}$$

It basically asks that given a k and a graph G that satisfies the preconditions L, Q, and M [1], whether there exists a replica placement solution set $R(V')$ which has a total cost less than or equal to k.

We now show that DOQFRP is NP-complete. First, it is easy to see that DOQFRP is in NP, since given a solution R and k, we know these input can always be encoded in polynomial length and whether or not the total cost of R is less than k can always be verified in polynomial time. Then, we show that DOQFRP can be reduced from the dominating set problem (DS), a well-known NP-complete problem, i.e., $DS \le_p DOMQRP$.

Dominating Set(DS): For a graph $G = (V, E)$, a Dominating Set $DS(V)$ of G is defined as a subset of V such that $\forall v \in (V - DS(V))$ is adjacent to $\exists v' \in DS(V)$. It can be formally described using the following language.

$$\{< G = (V, E), k > |\exists DS(V) \wedge |DS(V)| \le k\}$$

First, we show that a polynomial time reduction [7] can be constructed from any instance G of DS into an instance G' of DOMQRP problem using the following straightforward method. For $\forall v \in V$, we create a node $v' \in V'$, and assign $q(v') = 1$, $m(v') = 1$ and $s(v') = 1$ to the node. For every edge in E, we create an edge for E' and associate the edge with $l(u, v) = 1$, Obviously this reduction can be carried out in polynomial-time. We then need to show that a positive solution to one problem implies a positive solution to the other problem under the above described polynomial time transformation. More specifically, we need to show the instance $<G' = (V', E'), L, Q, M, k> \in DOMQRP$ has a replica set of whose total cost is no more than k if and only if the instance $<G = (V, E), k> \in DS$ has a dominating set of which the cardinality is no greater than k.

Sufficient Condition $\rightarrow$ If there exists a dominating set $DS(V)$ for G with cardinality less than k for G, for each $v \in DS(V)$, following the above construction mapping scheme, we can always mark the mapped node $v' \in V'$. Since the mapping is a one-to-one relationship and the costs, links, QoS, and fault-tolerant level in G's are all unit values, we can claim that all the marked v's constitute the $R(V')$ that is a replica solution set whose total cost is less than k for G'.

Necessary Condition $\leftarrow$ If $R(V')$ exists for G', following the above one-to-one mapping construction, for each node $v' \in R(V')$, we can mark a node v in V on the graph G. Then we claim all the marked nodes form the $DS(V)$ for G. Since every node $v' \in V' - R(V')$ can be directly linked to at least a node in $R(V')$ due to unit M, QoS, links and costs, a node not in $DS(V)$ must have a neighbor in $DS(V)$. Clearly, $|DS(V)| \le k$ due to the one-to-one mapping of the

---

[1] For those graphs that inherently cannot meet the preconditions of Q, L M, it is trivial to conclude that no feasible solutions can be found.

transformation. So we can find a solution $DS(V)$ mapped from $R(V')$ for the reduced dominating set instance.

Hence we have proved that $DOQFRP$ is NP-complete, which immediately follows that $OQFRP$ is NP-hard.

3.4   Influencing Set and Influenced Set

Two critical data structures, which will be used in the subsequent sections to facilitate the analysis of the problem, are defined here.

Influencing Set: $\forall v \in V$, the influencing set of v is the set of nodes u whose distances to v are within the QoS requirement of v.

$$IG(v) = \{u|d(v,u) \leq q(v)\}.$$

Influenced Set: $\forall v \in V$, the influenced set of v is the set of nodes u whose distances to v are within the QoS requirement of u.

$$ID(v) = \{u|d(v,u) \leq q(u)\}.$$

From the perspective of a given node v, $IG(v)$ denotes a set of nodes that can meet the QoS requirement of v, which means if any member in $IG(v)$ is placed replica, it can serve the request of node v. $ID(v)$ denotes a set of nodes whose QoS requirements can be met by v, which means if v is placed a replica, those nodes in $ID(v)$ will be served. From fault tolerance perspective, $IG(v)$ denotes a set of nodes whose failures may influence node v; while $ID(v)$ denotes a set of nodes whose service may be influenced by failure of replica on node v.

3.5   The Essential Properties of the Problem

The following properties of the problem and the interplay between parameters are very important for us to design effective heuristics.

1. *Existence of a solution.* Without considering cost constraints, QoS can always be guaranteed. Given fault-tolerant levels $M = <m_0, m_1, \ldots, m_{N-1}>$, the necessary and sufficient conditions for the existence of a replica placement solution R is as follows:

$$m(v) \leq |IG(v)|, \forall v \in V.$$

This property reveals the essential relationship between M and Q.
2. *Relationship among M, Q and R.* Furthermore, if a solution R exists, then for every node v the following inequality holds.

$$m(v) \leq |IG(v) \cap R|, \forall v \in V.$$

It means that given node v, $m(v)$ must not exceed the cardinality of the replica placed nodes belonging to the influencing set of v, since the fault-tolerant level of a node v simply refers to the number of replicas being placed on a subset of $IG(v)$. The size of the replica solution for the whole network should be no smaller than $m(v)$ of any node v and the replicas that meet m requirements for node v should be the intersection of $IG(v)$ and R.

3. *Lower bound and upper bound of M.* Lower bound (or the lowest level) and upper bound (or the highest level) of fault tolerance are the limitations inherent in the graph subject to QoS constraints. The lowest level of fault tolerance occurs when there is no redundant replica. It means that the fault-tolerant value of every node is assigned with 1. Thus, the lower bound of fault-tolerant value is $M = <1, 1, \ldots, 1>$. The highest fault-tolerant value a network can achieve occurs at the most aggressive case where all the nodes of the influencing set of any v are placed replicas. As a result, the upper bound for the fault-tolerant value of the whole network system is $M = <|IG(v_0)|, |IG(v_1)|, \ldots, |IG(v_{N-1})|>$. The two bounds show the range of M values. This analysis also shows that in our problem the QoS and fault-tolerant requirements are not independent, they will interact with each other.

4   Heuristic Algorithms

Based on the two ranking criteria proposed in last section, we will propose two heuristic algorithms for generating efficient solutions to the proposed problem.

4.1   Two-Level Selection Criteria

For any given node, we would like ask the following questions. How necessary is it to put replicas on all nodes of its influencing set to support the node? How much one replica place on one specific node can contribute to other nodes? We propose two criteria named **Replica Demanding Priority** and **Replica Profit Ratio** respectively to quantitatively answer the two questions and then employ the metrics to develop our heuristic algorithms.

Replica Demanding Priority(RDP): For each node v, $RDP(v)$ is defined as follows.

$$RDP(v) = \frac{1}{C_{|IG(v)-IG\_R(v)|}^{m(v)-|IG\_R(v)|}} \qquad (4)$$

where $IG\_R(v)$ represents the set of nodes that are already placed replicas inside $IG(v)$. The higher demand priority is, the less flexibility or possibility for a node with various configurations is. For example, if $RDP(v) = 1$, it means that there is only one replica placement configuration can meet the requirements of v, so we can safely put replicas in $IG(v)$ to meet v without considering other potential better solutions. This measurement therefore enables us to rank all the nodes so that our algorithms can systematically choose candidate nodes to place replicas in an ordered way. According to this criterion, we would like to select the nodes to place replicas inside their influencing sets according to the decreasing order of their RDP values.

Replica Profit Ratio (RPR): For each node v, $RPR(v)$ is defined as follows.

$$RPR(v) = \frac{\sum_{u \in (ID(v) - ID\_R(v)) - \{v\}} AvgCost(u)}{s(v)} \qquad (5)$$

where $AvgCost(u) = \frac{\sum_{p \in (IG(u) - IG\_R(u))} s(p)}{|IG(u) - IG\_R(u)|}$ is the average cost to meet one replica failure of given node u. Here, $ID\_R(u)$ means the nodes belonging to the influenced set of u have been placed replicas. So $RPR(v)$ means the profit ratio can be obtained if put one replica on node v. As some nodes in its influenced set may have been placed replicas, we exclude them from the cost calculation. We would like to pick those nodes that can bring the most profit to their individual influenced set first, because the more profit to influenced nodes, the less total cost may be needed to meet the overall M requirement. For example, to achieve the same effect as we put a replica on v, $RPR(v) = 10$ means that the total average cost is 10 times of $s(v)$. We would also like pick those nodes with the most profit ratio node first to place replicas for the objective of minimizing the cost.

These two selection criteria are used in tandem in our algorithm design because we would like to know which node we need to meet its fault tolerance first, then choose where the replicas should be placed within its influencing set based on their contribution to other nodes.

4.2   Framework of the Two Heuristics

The general ideas of the two heuristics are outlined in the following steps, and their more detailed pseudo-codes are given in Algorithms 1 and 2.

1. Calculate the shortest distance of all pairs.
2. Build the influencing set and the influenced set for every node.
3. Calculate the initial value of RDP for every node (the value will be iteratively updated as our algorithms increasingly place replicas).
4. For node v with the largest RDP, select nodes without replica from its influencing set and generate the candidate node set CNS.
   (a) Sort the nodes in CNS according to their RPR in decreasing order.
   (b) **Algorithm 1**: Select $m(v) - |IG\_R(v)|$ nodes in descending RPR order from CNS. Update the corresponding $IG\_R(v)$ and $ID\_R(v)$;
       Algorithm 2: Select *one* node with the largest RPR value from CNS. Update the corresponding $IG\_R(v)$ and $ID\_R(v)$.
5. If there is unsatisfied node left, update their RDP, RPR value and repeat step 4.

Algorithm 1: The pseudo code of batched replica placement

/* **Procedure:** Main Routine   */
input : $G = (V, E), Q, M$
output: a replica placement solution X

1 *Find all-pairs shortest paths distance;*
2 *Build influencing sets and influenced sets;*
3 **for** $\forall v \in V$ **do**
4 | $IG\_R(v) = ID\_R(v) = \phi, r(v) = 0;$
5 **end**
6 *mark all nodes as unsatisfied;*
7 *calculate RDP and RPR for each node;*
8 **while** *there exists unsatisfied node* **do**
9 | *Find one unsatisfied node p with the highest RDP value;*
10 | *Select $L = m(p) - |IG\_R(p)|$ nodes from $IG(p) - IG\_R(p)$ according to the*
   | *descending order of their RPR values as $u_1, ..., u_L$ with $r(u_i) == 0$;*
11 | **for** $j \leftarrow 1$ **to** L **do**
12 | | *Let $r(u_j) = 1$ and update all IG_R and ID_R sets related with u_j;*
13 | | **for** $\forall v \in V \&\&v$ *is unsatisfied* **do**
14 | | | **if** $|IG\_R(v)| \geq m(v)$ **then**
15 | | | | *mark node v as satisfied;*
16 | | | **end**
17 | | **end**
18 | **end**
19 **end**
20 **return** $X = < r(v_0), r(v_1), ..., r(v_{|V|-1}) >;$

---

5   Experimental Results

5.1   Experiment Setup

We have implemented two heuristics in C++ and conducted extensive experiments to evaluate the performance of the algorithms using widely different network configurations. The network used in our experiments has been generated using the Georgia Tech Internetwork Topology Models tool (GT-ITM)[2]. The network topology was randomly generated using the pure random model, where N nodes are randomly placed on a $s \times s$ square, and for each pair of nodes a link is generated with a probability σ, $(0 \leq \sigma \leq 1)$. Specifically, we choose the scale to be 1000, which means that the nodes of graph will be generated in a $1000 * 1000$ logical grid. We set N, the number of nodes, to be 100 and σ to be 0.8. The undirected graph such generated and used in our experiments contains 3997 links. The weight of a link is the Euclidean distance between the two nodes. The tool also provides another parameter β, which can be used to adjust the

---

[2] Interested readers are referred to [8,17] for details of the topology model and usage of the tool.

Algorithm 2: The pseudo code of one by one replica placement

/* **Procedure:** Main Routine   */
input : $G = (V, E)$, Q, M
output: a replica placement solution X

1 **begin**
2   *Find all-pairs shortest paths distance;*
3   *Build influencing sets and influenced sets;*
4   **for** $\forall v \in V$ **do**
5 | $IG\_R(v) = ID\_R(v) = \phi, r(v) = 0;$
6   **end**
7   *mark all nodes as unsatisfied;*
8   *calculate RDP and RPR for each node;*
9   **while** *there exists unsatisfied node* **do**
10   *Select u, an unsatisfied node which has the highest RDP value;*
11   *Select the node p, $r(p) == 0$, with the highest RPR value from $IG(u)$;*
12         $r(p) = 1;$
13   *and update all IG_R and ID_R sets related with p;*
14   **for** $\forall v \in V \&\& v$ *is unsatisfied* **do**
15   **if** $|IG\_R(v)| \geq m(v)$ **then**
16 | *mark node v as satisfied;*
17   **end**
18   **end**
19   **end**
20 **end**
21 **return** $X = < r(v_0), r(v_1), ..., r(v_{|V|-1}) >;$

---

ratio of long connections relative to short ones. In the current network generation configuration, we simply adopted the default value for β.

In the following reported experiments, different values of other parameters will be chosen for different evaluation purposes. Once the graph topology is generated, it poses constraints to the choices of values of M and Q. Therefore, we must first check for each node v if its m value is no greater than the cardinality of its influencing set $IG(v)$. To ensure the intrinsic constraints agree with each other, we can adjust them through trial and error. For example, we can reduce values of M. We can also test out larger $q(v)$ values. The larger $q(v)$ is, the larger the $IG(v)$ tends to be. Alternatively, we can even go back to the topology generation phase to assign a higher value to σ. This is because, for a node that demands m fault-tolerant requirement, the cardinality of its influenced set is at least m. If σ is too small, the chance for the node to get an influencing set with a cardinality larger than m is likely low. That is why σ was chosen as large as 0.8 for our network generation. If both M and Q for V have to be fixed but the graph does not meet the sufficient and necessary condition, then we must add link(s). Another way is to write validating routine to spot out cases where the value of m of a given node v is larger than the cardinality of the influencing set of v, then we can investigate the configuration of those m in practical environments for further action. For the simulation purpose of our concern, we can always reset m to a value smaller than the cardinality of the influencing set of v so that our algorithms will have a set of legitimate input to proceed.

5.2   Calculation the Super Lower Bound SLB

In order to quantitatively evaluate the effectiveness of the algorithms for different parameter configurations, ideally we should compare the replica cost results found by our heuristics with the cost provided by the optimal solutions. However, as our problem is NP-hard, we cannot have a polynomial time algorithm to achieve the optimal solution. Therefore, we relax the $\{0, 1\}$ constraints of X to the range of $[0, 1]$, consequently, we convert the original integer linear programming optimization problem into a linear programming optimization problem which can be efficiently solved. The optimal result computed from this relaxed linear programming optimization problem then will be used as the "super" lower bound for the solutions of the original optimization problem. This super lower bound serves as a comparison basis to evaluate the effectiveness of our solutions. Specifically, after we have calculated the cost due to the "super" lower bound (SLB) and the cost due to the feasible solution obtained from our heuristics (HSC), we use a cost approximation ratio ($\frac{HSC}{SLB}$) to measure the effectiveness of the solutions found by our heuristics. The smaller this ratio is, the closer the feasible solution found by our heuristics approximates the super lower bound of the relaxed optimal solution and the better the solution is supposed to be. So our approximation ratio is much strict and this evaluation approach has been commonly adopted by the research community [3, 16].

5.3   The Effectiveness of the Algorithms

Performance with Different Fault-Tolerant Requirements. In the first group of experiments, we fixed the storage cost and the QoS of every node respectively. Then we evaluate the effectiveness of the two algorithms by changing M in two different methods. In the first method the values of M of all nodes are identical; while in the second method M is taken from a discrete uniform distribution for different nodes.

The first method is relatively straightforward to experiment. For the second method, we generate random numbers for discrete uniform distribution of fault-tolerant level $m(v)$. When there are more than one combination for a value, we average them. For example, if we observe $m = 4$, the uniform distributions in three different ranges $[1, 7]$, $[2, 6]$, $[3; 5]$ will be selected. Both the replication results and the super lower bounds are averaged to get the final experimental results.

The experimental results are shown in sub-figures (a) and (b) on Fig. 2. From the figures, we observe that both algorithms can find very effective solutions. The relative cost (approximation ratio) of the feasible solution to the super optimal cost is below 1.7. The experimental results also show that when the value of m is assigned with the discrete uniform distribution, the corresponding performance

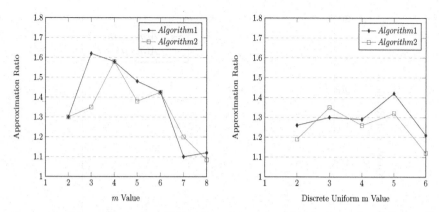

Approximation ratio under different m Approximation ratio under different discrete uniform m values.
values.

Fig. 2. Approximation ratio of two proposed heuristic algorithms with different fault-tolerant value (M) setups. (QoS $= 300$, storage cost $= 100$)

is better than that when m is assigned with a constant value. Since the super lower bound is lower than the integer lower bound of the original optimization problem, the performance of our heuristics are actually even more promising.

Performance with Different QoS Requirements. In the second group of experiments, we investigate the effects of different QoS requirements on the performance of the two heuristics. So we fix fault-tolerant requirement (m) and the storage cost (s) of each node for different experiments.

The results are charted in Fig. 3. From the figure, we observe that the total number of needed replica goes down when the QoS requirement is not so strict (value of QoS goes up). The reason is when QoS requirement goes down (the value of QoS increases under our setup), the chance for all the nodes to be satisfied by less replicas increases too and this will lead to overall sparser replica placement. The figure also indicates that Algorithm 2 is slightly more effective

Table 1. Number of replicas on Cartesian product of two sets: Q and M. Here, X represents that the graph cannot meet the requirements.

Value of m	Value of QoS		
	100	*300*	*500*
1	43	5	4
2	X	13	10
3	X	25	15
4	X	34	19
5	X	40	22

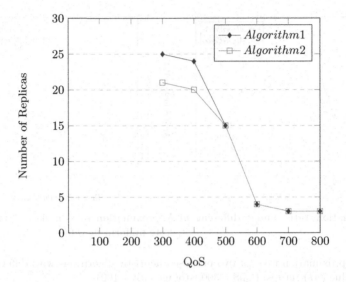

Fig. 3. Total number of replicas with different QoS requirements when fault-tolerant requirement and storage cost are fixed (m = 3, storage cost = 100).

than Algorithm 1 in terms of finding smaller replica numbers and thus less cost under the uniform storage cost setup.

Effect of Both QoS and Fault-Tolerant Requirements on Replica Placement. Furthermore, we conducted the third group of experiments to evaluate the conjunct influence of both QoS and m on the solution of replica placement using Algorithm 1. Table 1 indicates when both QoS and fault-tolerant requirements are very strict, there will be no replica placement solution for the given m and QoS due to that the requirements can't be met by the graph. From the experimental results we have the following observations. When QoS is fixed, our algorithms tend to return solutions that require more replicas, as M increases (higher fault tolerance requirement). The reason is that the larger M is, the more redundant replicas should be placed. On the other hand, if M is fixed, the larger QoS is (the QoS requirement is lower), the less replicas will be needed to meet the overall M constraints. This is because as QoS goes up, one replica could contribute to more nodes, which leads to an overall sparser solution. Table 1 shows how the number of replicas in our solutions changes with different QoS and fault-tolerant requirements.

6   Conclusion

Existing replica problem research only focuses on either QoS or fault tolerance. However, as more and more contemporary applications have become increasingly distributed and decentralized, the need for more sophisticated replica problems to consider both QoS and fault tolerance simultaneously is on the rise.

In this work, we propose a unique graph model to capture multiple attributes of replica cost, QoS and fault tolerance at each individual node and link level. Based on the proposed model, we formulate a novel optimal replica placement problem and prove that the problem is NP-hard. To find optimized solutions to the problem efficiently, we introduce two core concepts, influencing set and influenced set, to capture the relationship between each individual node and the nearby nodes that meet the individualized *QoS* requirements. We then present two criteria, Replica Demanding Priority and Replica Profit Ratio, which are metrics integrating the cost, QoS, fault-tolerant, and topology features, to measure how necessary some special replica configuration is needed by a node and given replica's contribution to other nodes. Based on the proposed criteria, we further design two specific heuristic algorithms under one unified algorithmic framework. Our preliminary experimental results show that the solutions found by our algorithms are very close to the optimal results.

References

1. Aral, A., Ovatman, T.: A decentralized replica placement algorithm for edge computing. IEEE Trans. Netw. Serv. Manage. **15**(2), 516–529 (2018)
2. Charikar, M., Guha, S., Tardos, É., Shmoys, D.B.: A constant-factor approximation algorithm for the k-median problem. J. Comput. Syst. Sci. **65**(1), 129–149 (2002)
3. Cheng, C., Wu, J., Liu, P.: QoS-aware, access-efficient, and storage-efficient replica placement in grid environments. J. Supercomput. **49**(1), 42–63 (2009)
4. Cornuéjols, G., Nemhauser, G., Wolsey, L.: The uncapicitated facility location problem. Cornell University Operations Research and Industrial Engineering, Technical report (1983)
5. Du, Z., Hu, J., Chen, Y., Cheng, Z., Wang, X.: Optimized QoS-aware replica placement heuristics and applications in astronomy data grid. J. Syst. Softw. **84**(7), 1224–1232 (2011)
6. Du, Z., Zhang, S., Bader, D.A., Hu, J.: A 2-approximation algorithm for QoS-aware and fault-tolerant replica placement. In: 2020 IEEE High Performance Extreme Computing Conference (HPEC). IEEE (2020)
7. Garey, M.R., Johnson, D.S.: Computers and Intractability: A Guide to the Theory of NP-Completeness. W. H. Freeman & Co., New York (1979)
8. Georgia Tech Internetwork Topology Models. http://www.cc.gatech.edu/projects/gtitm/. Accessed 4 Apr 2012
9. Ghanbari, H., Litoiu, M., Pawluk, P., Barna, C.: Replica placement in cloud through simple stochastic model predictive control. In: 2014 IEEE 7th International Conference on Cloud Computing, pp. 80–87. IEEE (2014)
10. Grace, R.K., Manimegalai, R.: Dynamic replica placement and selection strategies in data grids-a comprehensive survey. J. Parallel Distrib. Comput. **74**(2), 2099–2108 (2014)
11. Hashem, I.A.T., et al.: The role of big data in smart city. Int. J. Inf. Manage. **36**(5), 748–758 (2016)
12. Khan, S.U., Maciejewski, A.A., Siegel, H.J.: Robust CDN replica placement techniques. In: the 23rd IEEE International Parallel Distributed Processing Symposium (IPDPS), pp. 1–8. Rome, Italy (2009)

13. Lim, A., Rodrigues, B., Wang, F., Xu, Z.: k-center problems with minimum coverage. Theoret. Comput. Sci. **332**(1–3), 1–17 (2005)
14. Mills, K.A.: Algorithms for Optimal Replica Placement in Data Centers. Ph.D. thesis, The University of Texas at Dallas (2017)
15. Sahoo, J., Salahuddin, M.A., Glitho, R., Elbiaze, H., Ajib, W.: A survey on replica server placement algorithms for content delivery networks. IEEE Commun. Surv. Tutor. **19**(2), 1002–1026 (2016)
16. Tang, X., Xu, J.: QoS-aware replica placement for content distribution. IEEE Trans. Parallel Distrib. Syst. **16**(10), 921–932 (2005)
17. Waxman, B.M.: Routing of multipoint connections. IEEE J. Sel. Areas Commun. **6**(9), 1617–1622 (1988)
18. Yousafzai, A., Gani, A., Noor, R.M.: Availability aware continuous replica placement problem. arXiv preprint arXiv:1605.04069 (2016)
19. Yu, S.S., Yu, S.: Autonomous unmanned road vehicle for making deliveries (2015), US Patent App. 14/318,690

Neural Network Compression and Acceleration by Federated Pruning

Songwen Pei[1,2,3](✉), Yusheng Wu[1], and Meikang Qiu[4]

[1] University of Shanghai for Science and Technology, Shanghai 200093, China
swpei@usst.edu.cn, wysheng6088@163.com
[2] State Key Laboratory of Computer Architecture, Institute of Computing Technology, Chinese Academy of Sciences, Beijing 100190, China
[3] Shanghai Key Laboratory of Data Science, Fudan University, Shanghai 200433, China
[4] Department of Computer Science, Texas A&M University-Commerce, Texas 75428, USA
meikang.qiu@tamuc.edu

Abstract. In recent years, channel pruning is one of the important methods for deep model compression. But the resulting model still has tremendous redundant feature maps. In this paper, we propose a novel method, namely federated pruning algorithm, to achieve narrower model with negligible performance degradation. Different from many existing approaches, the federated pruning algorithm removes all filters in the pre-trained model together with their connecting feature map by combining the weights with the importance of the channels, rather than pruning the network in terms of a single criterion. Finally, we fine-tune the resulting model to restore network performance. Extensive experiments demonstrate the effectiveness of federated pruning algorithm. VGG-19 network pruned by federated pruning algorithm on CIFAR-10 achieves 92.5% reduction in total parameters and 13.58× compression ratio with only 0.23% decrease in accuracy. Meanwhile, tested on SVHN, VGG-19 achieves 94.5% reduction in total parameters and 18.01× compression ratio with only 0.43% decrease in accuracy.

Keywords: Model compression · Channel pruning · Federated pruning · Neural network · Pre-trained model

1 Introduction

Convolution neural networks (CNNs) have achieved prominent success in various recognition tasks, especially in computer vision, e.g., image classification [1], face recognition [2], traffic sign detection [3] and video analysis. In these fields, compared with traditional methods, recent CNNs [4–7] have achieved state-of-the-art performance.

However, CNNs have a huge number of computation and memory requirements, greatly imposing restrictions on their employment on resource-constrained platforms and applications such as robotics, portable devices, and

© Springer Nature Switzerland AG 2020
M. Qiu (Ed.): ICA3PP 2020, LNCS 12453, pp. 173–183, 2020.
https://doi.org/10.1007/978-3-030-60239-0_12

drones. To make CNNs available on resource-constrained devices, there are lots of studies on model compression which mainly contains following categories, namely, quantization [8], sparse or low-rank compressions [9], and pruning [10,11]. Network pruning is proved to be a valid way to reduce the network parameters and accelerate the training process. It mainly contains weight [9,12] and channel pruning [13–15]. Although these kinds of methods achieved considerable pruning ratio while maintaining the performance, the final network obtained by using single pruning standard still has a lot of redundant parameters. Therefore, we will investigate it because there are still rooms for further improvement on pruning neural network.

In this paper, we propose a method called federated pruning algorithm, a plain yet effective the scheme of pruning redundant parameters. It addresses the problem that the network still has redundant feature maps after single pruning. Our approach combines the weights with the importance of the channels to trim much less significant feature maps so as to obtain a more compact network. After pruning, the resulting narrower model requires to be fine-tuned to hold the similar level of accuracy on the baseline as much as possible.

This paper is organized as follows: related works in model acceleration and compression will be introduced in Sect. 2. Secondly the details of our approach will be described in Sect. 3. Effectiveness of the federated pruning algorithm will be illustrated and discussed by experimental comparisons in Sect. 4. Finally, the conclusion and future works will be given in Sect. 5.

2   Related Works

2.1   Weight Pruning

Based on the loss function of Hessian, weight pruning dates back to Optimal Brain Damage [16]. More recently, weights in neural network are pruned in [9,12,17], which results in small models. An iterative pruning way to abandon the small weights in which values are lower than the predefined threshold is proposed by [12]. A deep neural network compression pipeline is proposed by [9], it removes firstly the unimportant connections. Based on stochastic gate, sparse networks through L0-norm regularization is achieved in [18]. However, these methods have limited effect on actual acceleration, this approach results in sparse connectivity patterns. Therefore, it requires the support of sparse convolution libraries.

2.2   Channel Pruning

Pruning channels in the convolution layer is a very effective technique in order to get a more efficient and compressed model. Some heuristic methods contain pruning channels based on their corresponding filter weight norm [14] and average percentage of zeros in the output [13]. A channel selection strategy based on LASSO and a least square reconstruction algorithm to prune filters are proposed by [15]. The above channel pruning can obtain an effective network, which can achieved high ratio of compression and high performance of acceleration compared with the original network. However, there are still redundant feature maps in the resulting model for all the channel pruning methods, which affects the performance of compression and acceleration.

2.3   Neural Architecture Learning

Neural architecture automatically with reinforcement learning is proposed by some recent works. Studies on neural architecture search try to find the optimal network structures. For instance, the HAP problem which is useful for real-time systems is solved by [19]. Based on the heterogeneous scratchpad memory architecture, a genetic algorithm is introduced to perform data allocation to different memory units in [20]. In addition, the HSDefender (Hardware/Software Defender) technique is pointed out to perform protection and checking together by [21]. To achieve effective key distribution with a remarkable level of trust, an effective and practical data fusion trust system is proposed by [22,23]. Therefore, federated pruning algorithm learns network architecture through a single training process, which meets our goal of network compact.

3   Method

3.1   Federated Pruning

Channel pruned process mainly consists of the following steps: (1) Implementing the channel-level sparsity on the initial network; (2) Evaluating the significance of weights and feature maps by magnitude of weights and the scaling factors of individual channels; (3) Adopting a global pruning strategy to remove the insignificant weights, channels and corresponding feature maps; (4) Fine-tuning the resulting network with no loss of performance. The procedure of pruning is described in Fig. 1.

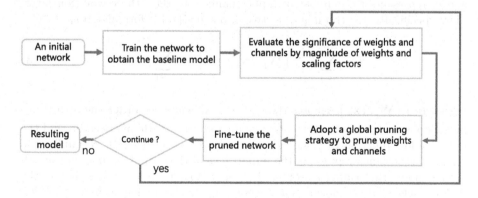

Fig. 1. The procedure of federated pruning

Evaluate the Importance of Weights. Pruning the insignificant connections with small weights in trained neural networks is proposed by [12], which inevitably results in a decrease in network performance. Therefore, the way we evaluate the importance of weights is remarkably essential. Inspired by previous work, the magnitude of these weights represents the importance of the feature map in current layer, so we get a global threshold in all layers by the pruning ratio p_w, which is defined as a certain percentile of all the weight values. All connections with weights below a threshold are removed from the network. The set of current pruned network parameters is represented by

$$\mathcal{W} = \{\mathbf{W}_1, \mathbf{W}_2, \cdots, \mathbf{W}_l, \mathbf{W}_{l+1}, \cdots, \mathbf{W}_L\} \tag{1}$$

where W_l and W_{l+1} denote the weight matrices on the lth and $l+1$ convolutional layer, respectively. We require further optimize the loss function $\mathcal{L}(\cdot)$ of the network. A set of training examples is viewed as

$$\mathcal{L}(\cdot) = \{\mathcal{X} = \{x_0, x_1, \cdots, x_N\}, \mathcal{Y} = \{y_0, y_1, \cdots, y_N\}\} \tag{2}$$

where $\mathcal{X}$ and $\mathcal{Y}$ denotes an input and a corresponding output of the training examples, respectively. N represents the number of images. Therefore, we retrain the current network to minimize the loss function $\mathcal{L}(\cdot)$ of the network so as to learn the final weights.

Channels Selection. Selecting the least representative channels are straightforward but the implementations are nontrivial. Rather than imposing sparsity on features or convolutional weight parameters, sparsity is induced on the N scaling factors γ_l by recent channel pruning methods [10]. Also, we use the channel-wise importance measure. Given a training set $\{(\mathbf{X}_n, \mathbf{Y}_n)\}_{n=1}^{N}$, where $\mathbf{x}_n$ denotes the n-th input features of training samples and $\mathbf{y}_n$ is the corresponding actual outputs. We assume that a network has L layers. The network can be parameterized by $\{\mathbf{W}_{l+1} \in \mathbb{R}^{c_{l+1} \times c_l \times k \times k}, 1 \leq l \leq L\}$ for the l-th convolution layer, where $\mathbf{W}_{l+1}$ denotes a matrix of connection weights, k is the kernel size and c_l represents the number of input channels in the l-th convolution layer. More specifically, the training objective of our method is formulated as:

$$L = \sum_{n=1}^{N} \mathcal{L}\left[f\left(\mathbf{W}_l^{(i)}; \mathbf{x}_n\right), \mathbf{y}_n\right] + \lambda \sum_{\gamma \in \Gamma}(|\gamma_l|) \tag{3}$$

where $f\left(\mathbf{W}_l^{(i)}; \mathbf{x}_n\right)$ denotes the network's outputs vector for the n-th training sample x_n, y_n denotes the corresponding ground-truth label vector, and the choice of loss function $\mathcal{L}(\cdot)$ depends on the specific learning task. γ_l implies the scaling factor which represents the importance of the output channel in the l-th layer denotes the scaling factor, and $|\gamma_l|$ is a sparsity-induced penalty on the scaling factors. It is known that $L1$-norm is widely used to achieve sparsity. Therefore, we also impose $L1$ regularization on the BN layer directly.

Our goal is to obtain a sparse γ_l. Namely, if $\gamma_l = 0$, then we can safely remove the corresponding feature maps because its outputs have no contribution to subsequent computation. Inspired by the recent work [10], we can effectively prune the insignificant channels with the relative value of scaling factors by removing all their incoming and outgoing connections of corresponding channel. As for the ratio of pruning, we discard the channels below the global threshold that we pre-define. The pruning process is summarized in Algorithm 1.

Algorithm 1. Federated pruning

Input: The baseline model $\mathbf{M}$ and its parameters $\mathbf{W}^*$; training set: $\{(\mathbf{x}_n, \mathbf{y}_n)\}_{n=1}^N$; pruning ratio: p_w,p_c; the training epochs: $epoch_{max}$; number of layers: L;
Output: The resulting model $\mathbf{M}$ and its parameters $\mathbf{W}^*$;
1: Get the index of all layers weights
2: Obtain the index of threshold by pruning ratio p_w;
3: **for** l=1, ..., L **do**
4:     Use global pruning strategy in each hierarchy
5:     Trim these connections with weights below the threshold
6: **end for**
7: Obtain the latest model $\mathbf{W}^*$ after fine-tuning network;
8: **for** $epoch = 1$; $epoch \leq epoch_{max}$; $epoch + +$ **do**
9:     **for** $l = 1$; $l \leq L$; $l + +$ **do**
10:         Implement the sparsity regularization for BN layers;
11:         Zeroize γ_l by ℓ_1-norm filter selection;
12:     **end for**
13: **end for**
14: Prune the network by pruning ratio p_c;
15: Fine-tune the resulting model with parameters $\mathbf{W}^*$ from $\mathbf{W}$;

---

4 Experiments

In this section, we aim to evaluate the performance of federated pruning algorithm with the traditional VGG-16 and VGG-19 models on the CIFAR and SVHN datasets.

4.1 Datasets and Models

CIFAR and SVHN. CIFAR-10 is compose of 50,000 training images and 10,000 testing images which draws from 10 categories of objects, while CIFAR-100 contains identical number of images drawing from 100 categories of objects. Each image in the CIFAR dataset is real color images in life with resolution 32×32. To verify the effectiveness of Federated pruning algorithm, we compare our method with [10,12] on the CIFAR dataset. On the other hand, SVHN integrates more than 600,000 digital images consisting of 32×32 colored digit images obtained from house numbers in Google Street View images. About data augmentation, the input images are normalized using standard deviations and channel means. We compare our method with [10,12] on SVHN dataset.

VGGNet. Federated pruning is evaluated on the popular convolution networks: VGG-16 and VGG-19. Traditional VGG-16 model is constructed by 13 convolutional layers and 3 FC layers, while 3 FC layers of the VGG-16 model are replaced with 1 FC layer. Specifically, VGG-19 model which contains 16 convolutional layers with batch normalization layer is adopted on all datasets. We compare the algorithm transformation on the VGG-19 network.

4.2   Implementation Details

Training. All the networks are normally trained from scratch as baselines. VGG-19 is trained on all the datasets. The network is trained on the CIFAR dataset with batch size of 64 for 180 epochs and 40 epochs on the SVHN dataset, respectively. Meanwhile, Criterion function takes the cross entropy loss and Optimization scheme adopts the Stochastic Gradient Descent (SGD). The initial learning rate for these two datasets is set to 0.1, and is divided by 10 at 30% and 70% of the total number of epochs, respectively. In addition, a weight decay of 10^{-4} is used to prevent networks from overfitting.

Pruning and Fine-Tuning. The network is pruned by assessing the significance of channels and weights for VGGNet. When we carry out the federated pruning algorithm to prune the network normally trained from scratch as baselines, a global pruning threshold for simplicity is adopted unlike in [14] where different layers are trimmed by a variety of ratios. The pruning threshold is determined by the percentile of all channels and weights, for instance, 30% or 80% channels and weights are pruned. A new narrower model is created after the pruning process. On the CIFAR dataset, the fine-tuning uses all settings kept the same as in normal training. On the SVHN dataset, the pruned model is fine-tuned for 50 epochs to restore original accuracy. In our experiments, all fine-tuning operations are adopted on the compact model to avoid loss of precision after pruning the networks.

4.3   Results and Analysis

Effect of Federated Pruning Algorithm on Network Performance. We compare federated pruning algorithm with other state-of-the-art models for VGG-19 for the CIFAR dataset, our method achieves the better performance. For example, network slimming [10] with fine-tuning achieves 8.81× compression ratio in Table 1, but our federated-wc with fine-tuning compresses 13.58× with only 0.23% accuracy decreased. Comparing to weight-pruned [12], when pruning 92.5% parameters of VGG-19, our federated-cw has only 0.06% accuracy drop, which is less than weight-pruned [12] (0.27%). However, we can observe that our federated-wc only achieves 3.70× compression ratio on the CIFAR-100, which is typically less than network slimming [10]. This is possibly the reason that more classes exist in the CIFAR-100. In addition, we test our federated pruning approach with the pruning ratio of 60% and 80% for the SVHN dataset. Same with

Table 1. Results on benchmark datasets. "Baseline" means the initial model without any other pruning operations. In column-1, Federated-CW and Federated-WC represent the different models due to removing insignificant information in the opposite order, respectively. "Acc" is the accuracy on test datasets for different models. The pruned ratio of parameters and compression ratio are also displayed in column-5&7.

Model	Datasets	Acc (%)	Parameters	Pruned (%)	Size	Compression ratio
VGG-19 (Baseline)	CIFAR-10	93.66	2.0×10^7	–	160.3 MB	1×
Weight-pruned [12]	CIFAR-10	93.39	6.0×10^6	70	80.2 MB	2×
Channel-pruned [10]	CIFAR-10	94.04	2.3×10^6	88.5	18.2 MB	8.81×
Federated-cw	CIFAR-10	93.60	6.8×10^5	96.6	18.2 MB	8.81×
Federated-wc	CIFAR-10	93.43	1.5×10^6	92.5	11.8 MB	13.58×
VGG-19 (Baseline)	CIFAR-100	73.12	2.0×10^7	–	160.3 MB	1×
Weight-pruned [12]	CIFAR-100	72.49	8.0×10^6	60	80.4 MB	2×
Channel-pruned [10]	CIFAR-100	73.62	4.9×10^6	84.5	39.2 MB	4.10×
Federated-cw	CIFAR-100	73.30	9.8×10^5	95.1	19.6 MB	8.20×
Federated-wc	CIFAR-100	72.28	5.4×10^6	73	43.4 MB	3.70×
VGG-19 (Baseline)	SVHN	95.76	2.0×10^7	–	160.3 MB	1×
Weight-pruned [12]	SVHN	95.74	4.0×10^6	80	80 MB	2×
Channel-pruned [10]	SVHN	95.64	2.2×10^6	89	17.8 MB	9.01×
Federated-cw	SVHN	95.62	6.7×10^5	96.7	8.9 MB	18.01×
Federated-wc	SVHN	95.33	1.1×10^6	94.5	8.9 MB	18.01×

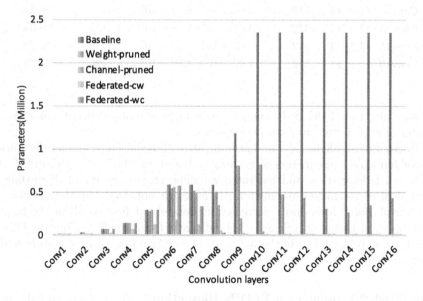

Fig. 2. Statistics of remaining parameters for each layer for VGG-19 on the CIFAR-10.

[10], we fine-tune the pruned model by the federated pruning algorithm. Table 1 also shows that federated pruning outperforms previous methods on the dataset, again. Federated-wc reduces more than 90% parameters of the model with only negligible (0.43%) accuracy loss, and obtains 18.01× compression ratio, which is much than that of weight-pruned [12] and network slimming [10]. These results

Table 2. Performance of several layers for the resulting model on CIFAR-10. In column-3&6, Channel denotes the decrement of feature maps compared with the baseline models. Ratio in the last column is the rate of channel reduction. FLOPs pruned is also shown in the column-5.

Layer	$h_l \times w_l$	Channels	FLOPs	FLOPs pruned	Channels	Ratio%
Conv1	224×224	64	8.67×10^7	4.74×10^7	29	54.7%
Conv2	224×224	64	1.85×10^9	–	64	0%
Conv3	112×112	128	9.25×10^8	5.78×10^7	120	6.3%
Conv4	112×112	128	1.85×10^9	–	128	0%
Conv5	56×56	256	9.25×10^8	1.08×10^7	253	1.2%
Conv6	56×56	256	1.85×10^9	4.34×10^7	250	2.3%
Conv7	56×56	256	1.85×10^9	7.44×10^8	153	40.2%
Conv8	56×56	256	1.85×10^9	1.67×10^9	25	90.2%
Conv9	28×28	512	9.25×10^8	8.85×10^8	22	95.7%
Conv10	28×28	512	1.85×10^9	1.82×10^9	7	98.6%
Conv11	28×28	512	1.85×10^9	1.84×10^9	1	99.8%
Conv12	28×28	512	1.85×10^9	1.84×10^9	4	99.2%
Conv13	14×14	512	4.62×10^8	4.59×10^8	4	99.2%
Conv14	14×14	512	4.62×10^8	4.58×10^8	5	99.0%
Conv15	14×14	512	4.62×10^8	4.54×10^8	9	98.2%
Conv16	14×14	512	4.62×10^8	4.39×10^8	26	94.9%

demonstrate that federated pruning can achieve a more compact model with similar or even better network performance.

In order to highlight federated pruning algorithm efficiency, we plot the statistics of remaining parameters of each layer based on VGG-19 in the Fig. 2. As expected, federated-cw and federated-wc show greater ability of discarding a great deal of redundant parameters. It is worth noticing that the number of the convolutional layers only decreases moderately at first on all methods, yet the number of parameters decreases significantly from Conv9 to Conv16. This is possibly due to the effect of the network extracting representative feature maps in the beginning.

Federated-Wc Impact on FLOPs Reductions. The purpose of federated pruning algorithm is to lower required the amount of computing resources. FLOPs for different layers in the model is determined to require a great deal of different computation due to the fact that the channels and sizes of input feature maps and convolution kernels are different. Amounts of computation of VGG-19 model are removed on the CIFAR-10 dataset, which is described in Table 2, e.g. convolutional layers with great FLOPs pruned: Conv10, Conv11, Conv12. FLOPs reductions are typically around 90%. To emphasize federated-wc's efficiency, from the Table 1, we can observe that federated-wc achieves 13.58× compression ratio, which is higher than that of other methods. This indicates that the resulting model obtained by making federated-wc to remove great many calculation is actually an effective compact model.

Comparison with Other Pruning Criteria. At present, evaluating the importance of a convolutional kernel or feature map on CNN has many common pruning criteria. Comparison results of VGG-16 models obtained by making these criteria are summarized in Fig. 3. We select the federated-wc method to compare the network performance with those criteria due to the better pruning effect of federated-wc pruning algorithm. Figure 3 illustrates that the accuracy of the resulting model with federated-wc pruning algorithm is higher and more efficient than that of other methods. Especially when the floating point operation is reduced to more than 70%, our method can implement the most optimal network performance.

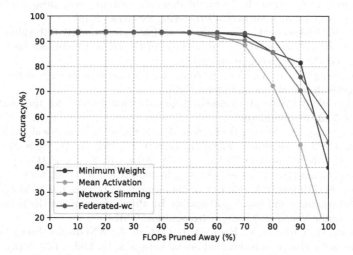

Fig. 3. Comparison of pruning criteria. The curve is influenced by network accuracy and model FLOPs for VGG-16 on the CIFAR-10.

5   Conclusion

In this paper, we propose a novel federated pruning algorithm by exploiting the weight and channel pruning to learn more compact CNNs required while maintaining high accuracy, which is rarely considered before. By measuring the importance of channels and weights, the basic model was pruned and compressed to get a more compact model. What's more, compared to the original model, fine-tuning can restore similar accuracy. On multiple datasets, we proved, obviously, the resulting model can have fewer parameters than networks that have been pruned by other criteria through our federated pruning algorithm, and also implement higher compression ratio with modest accuracy loss. In the future, we plan to work on how to apply our method to the deeper network, or combine our pruning strategy with other criteria, e.g., low-precision weights and matrix decomposition, to implement the higher acceleration and compression.

Acknowledgement. We would like to thank the anonymous reviewers for their invaluable comments. This work was partially funded by the National Natural Science Foundation of China under Grant (61975124, 61332009, and 61775139), the Shanghai Natural Science Foundation(20ZR1438500), the Open Project Program of Shanghai Key Laboratory of Data Science (2020090600003), and the Open Project Funding from the State Key Lab of Computer Architecture, ICT, CAS under Grant CARCH201807. Any opinions, findings and conclusions expressed in this paper are those of the authors and do not necessarily reflect the views of the sponsors.

References

1. Srivastava, R.K., Greff, K., Schmidhuber, J.: Training very deep networks. In: Advances in Neural Information Processing Systems (2015)
2. Schroff, F., Kalenichenko, D., Philbin, J.: Facenet: a unified embedding for face recognition and clustering. In Proceedings of the IEEE Conference on Computer Vision and Pattern Recognition (2015)
3. Pei, S., Tang, F., Ji, Y., Fan, J., Ning, Z.: Localized traffic sign detection with multi-scale deconvolution networks (2018)
4. Pei, S., et al.: 3DACN: 3D augmented convolutional network for time series data. Inf. Sci. **513**, 17–29 (2020)
5. Ma, N., Zhang, X., Zheng, H.-T., Sun, J.: ShuffleNet V2: practical guidelines for efficient CNN architecture design. In: Ferrari, V., Hebert, M., Sminchisescu, C., Weiss, Y. (eds.) Computer Vision – ECCV 2018. LNCS, vol. 11218, pp. 122–138. Springer, Cham (2018). https://doi.org/10.1007/978-3-030-01264-9_8
6. Sandler, M., Howard, A., Zhu, M., Zhmoginov, A., Chen, L.-C.: Mobilenetv 2: Inverted residuals and linear bottlenecks. In: Proceedings of the IEEE Conference on Computer Vision and Pattern Recognition (2018)
7. Rastegari, M., Ordonez, V., Redmon, J., Farhadi, A.: XNOR-Net: ImageNet classification using binary convolutional neural networks. In: Leibe, B., Matas, J., Sebe, N., Welling, M. (eds.) ECCV 2016. LNCS, vol. 9908, pp. 525–542. Springer, Cham (2016). https://doi.org/10.1007/978-3-319-46493-0_32
8. Zhou, A., Yao, A., Guo, Y., Xu, L., Chen, Y.: Incremental network quantization: Towards lossless cnns with low-precision weights. arXiv preprint arXiv:1702.03044 (2017)
9. Han, S., Mao, H., Dally, W.J.: Compressing deep neural networks with pruning trained quantization and huffman coding. arXiv preprint (2015)
10. Liu, Z., Li, J., Shen, Z., Huang, G., Yan, S., Zhang, C.: Learning efficient convolutional networks through network slimming. In: Proceedings of the IEEE International Conference on Computer Vision (2017)
11. Ye, J., Lu, X., Lin, Z., Wang, J.Z.: Rethinking the smaller-norm-less-informative assumption in channel pruning of convolution layers. arXiv preprint arXiv:1802.00124 (2018)
12. Han, S., Pool, J., Tran, J., Dally, W.: Learning both weights and connections for efficient neural network. In: Advances in Neural Information Processing Systems (2015)

13. Hu, H., Peng, R., Tai, Y.-W., Tang, C.-K.: Network trimming: a data-driven neuron pruning approach towards efficient deep architectures. arXiv preprint arXiv:1607.03250 (2016)
14. Li, H., Kadav, A., Durdanovic, I., Samet, H., Graf, H.P.: Pruning filters for efficient convnets. arXiv preprint arXiv:1608.08710 (2016)
15. He, Y., Zhang, X., Sun, J.: Channel pruning for accelerating very deep neural networks. In: Proceedings of the IEEE International Conference on Computer Vision (2017)
16. LeCun, Y., Denker, J.S., Solla, S.A.: Optimal brain damage. In: Advances in Neural Information Processing Systems (1990)
17. Guo, Y., Yao, A., Chen, Y.: Dynamic network surgery for efficient DNNs. In: Advances in Neural Information Processing Systems (2016)
18. Louizos, C.M.W., Kingma, D.P.: Learning sparse neural networks through l_0 regularization. arXiv preprint arXiv:1712.01312 (2017)
19. Qiu, M., Sha, E.H.-M.: Cost minimization while satisfying hard/soft timing constraints for heterogeneous embedded systems. In: ACM Transactions on Design Automation of Electronic Systems (TODAES) (2009)
20. Qiu, M., et al.: Data allocation for hybrid memory with genetic algorithm. IEEE Trans. Emerg. Top. Comput. **3**(4), 544–555 (2015)
21. Shao, Z., Xue, C., Zhuge, Q., Qiu, M., Xiao, B., Sha, E.H.-M.: Security protection and checking for embedded system integration against buffer overflow attacks via hardware/software. IEEE Trans. Comput. **55**, 443–453 (2006)
22. Qiu, H., Qiu, M., Zhihui, L., Memmi, G.: An efficient key distribution system for data fusion in v2x heterogeneous networks. Inf. Fusion **50**, 212–220 (2019)
23. Song, Y., Gao, S., Li, Y., Jia, L., Li, Q., Pang, F.: Distributed attention-based temporal convolutional network for remaining useful life prediction. IEEE Int. Things J. https://doi.org/10.1109/JIOT.2020.3004452

Scalable Aggregation Service for Satellite Remote Sensing Data

Jianwu Wang[1]([✉])(iD), Xin Huang[1], Jianyu Zheng[2], Chamara Rajapakshe[2],
Savio Kay[1], Lakshmi Kandoor[1], Thomas Maxwell[3], and Zhibo Zhang[2]

[1] Department of Information Systems, University of Maryland, Baltimore County,
Baltimore, MD 21250, USA
{jianwu,xinh1,savio1,k170}@umbc.edu
[2] Department of Physics, University of Maryland, Baltimore County, Baltimore,
MD 21250, USA
{jzheng3,charaj1,zhibo.zhang}@umbc.edu
[3] Goddard Space Flight Center, NASA, Greenbelt, MD 20771, USA
thomas.maxwell@nasa.gov

Abstract. With the advances of satellite remote sensing techniques, we
are receiving huge amount of satellite observation data for the Earth.
While the data greatly helps Earth scientists on their research, conducting data processing and analytics from the data is getting more and
more time consuming and complicated. One common data processing
task is to aggregate satellite observation data from original pixel level
to latitude-longitude grid level to easily obtain global information and
work with global climate models. This paper focuses on how to best
aggregate NASA MODIS satellite data products from pixel level to grid
level in a distributed environment and provision the aggregation capability as a service for Earth scientists to use easily. We propose three
different approaches of parallel data aggregation and employ three parallel platforms (Spark, Dask and MPI) to implement the approaches.
We run extensive experiments based on these parallel approaches and
platforms on a local cluster to benchmark their differences in execution
performance and discuss key factors to achieve good speedup. We also
study how to make the provisioned service adaptable to different service
libraries and protocols via a unified framework.

Keywords: Big data · Data aggregation · Remote sensing ·
Servicelization · Benchmark

1 Introduction

The advances in climate study in recent years have resulted in astronomical
growth of available climate data. There are two main sources for climate data:
climate simulation model and satellite remote sensing. The paper [15] from *Science* magazine estimates the total worldwide available climate data will increase
from less than 50 PB in 2015 to about 350 PB in 2030. Among projected data

© Springer Nature Switzerland AG 2020
M. Qiu (Ed.): ICA3PP 2020, LNCS 12453, pp. 184–199, 2020.
https://doi.org/10.1007/978-3-030-60239-0_13

size in 2030, climate model simulation results and satellite remote sensing data consist of about 188 PB (54% percent) and 155 PB (44% percent) respectively.

A basic data processing task in climate study is to aggregate satellite observation data from original pixel level to latitude-longitude grid level to easily obtain global information and work with global climate simulation models. It is because most global climate simulation models, such as the climate simulation models in Phase 6 of the Coupled Model Intercomparison Project, known as CMIP6 [12], conduct simulation by dividing the Earth into 3-dimensional (latitude, longitude, altitude) grids, and solving physics equations (including mass and energy transfer) within each grid and its interactions (including radiant exchange) with neighboring grids. By aggregating satellite observation data from pixel level to grid level, Earth scientists can conduct many studies with climate model simulation data and aggregated satellite observation data since they have the same granularity. For instance, a climate model can be evaluated via the comparison between its simulation results with satellite observation data.

This paper addresses two specific challenges in satellite data aggregation: 1) how to efficiently aggregate data from pixel level to grid level in a distributed environment, 2) how to provision data aggregation as services so Earth scientists can achieve data aggregation without downloading data to local machine. To address these two challenges, we discuss different approaches of parallel data aggregation and how different factors such as big data platform and sampling ratio affect the execution performance and aggregation results. We also discuss how to make provisioned services adaptable to different service libraries and protocols. The software implementations of our work is open-sourced at [2].

The contributions of this paper are fourfold. First, we propose three parallel data aggregation approaches and discuss their differences. All approaches can show good scalability in our experiments with 738 GB input data with the best speedup ratio as 97.03 when running on 10 distributed compute nodes. Second, we apply the above data aggregation approaches with three popular parallel platforms/techniques: Spark [3,9], Dask [4,18] and MPI [16]. We benchmark and analyze their performance differences through our experiments. Third, we apply sampling techniques in our data aggregation approaches to understand how sampling affects execution speedup and the correctness of the results. Our experiments show sampling only has less than 1% data loss and its affects on execution speedup is mixed. Fourth, to work with different service protocols/libraries such as REST [17] and ZeroMQ [7], we adopt the Stratus framework [6] for servicelization of the data aggregation capability. Users only need to change one parameter to switch from one service protocol/library to another.

The rest of the paper is organized as follows. The background is introduced in Sect. 2. The data aggregation logic is explained in Sect. 3. Section 4 contains our proposed three scalable data aggregation approaches. Section 5 describes the experiments on different scalable aggregation approaches, different parallel platforms, and different sampling ratios. The servicelization of our data aggregation capability is discussed in Sect. 6. Finally, we discuss related work in Sect. 7 and conclude our paper in Sect. 8.

2 Background

2.1 MODIS Satellite Remote Sensing Data

The MODIS (Moderate Resolution Imaging Spectroradiometer) is a key instrument on board NASA's Terra (launched in 1999) and Aqua (launched in 2002) satellite missions as part of the larger Earth Observation System (EOS). MODIS measures the reflection and emission by the Earth-Atmosphere system in 36 spectral bands from the visible to thermal infrared with near daily global coverage and high-spatial resolution (250 m to 1 km at nadir). These measurements provide a critical observational basis for understanding global dynamics and processes occurring on the land, in the oceans, and in the lower atmosphere. MODIS is playing a vital role in the development of validated, global, interactive Earth system models that are able to predict global change accurately enough to assist policy makers in making sound decisions concerning the protection of our environment.

MODIS atmosphere properties products are processed into three levels, i.e., Level 1 (L1), Level 2 (L2) and Level 3 (L3). The Level 1 products contain geolocation and the raw reflectance and radiance measurements for all 36 MODIS spectral bands, at 250 m, 500 m, or 1 km spatial resolutions. The Level 2 products contain the geophysical properties, such as cloud mask, cloud and aerosol optical thickness, retrieved from the Level 1 products. The retrieval process is usually based on sophisticated algorithms developed by the MODIS science teams. Because Level 2 products are derived from the Level 1 products, they usually have the same or similar spatial resolution. For example, Level 2 MODIS cloud properties products (product name "MOD06" for Terra and "MYD06" for Aqua) have a nominal spatial resolution of 1 km. The Level 3 processing produces Earth-gridded geophysical parameter statistics, which have been averaged (e.g., daily or monthly), gridded (e.g., 1° × 1° degree), or otherwise rectified or composited in time and space. The Level 3 MODIS Cloud Properties products contain hundreds of 1° × 1° global gridded Scientific Data Sets (SDSs) or statistics derived from the Leval 2 products. The Level 1 and Level 2 products are often called pixel products and the Level 3 products are often called gridded products. Many atmospheric/climate research studies are done using Level 3 data. All three levels of MODIS data are publicly available at [5].

2.2 Parallel Platforms

Spark [3] is one of the most popular big data platform. By following and extending the MapReduce paradigm [10], Spark embeds computation using high-level functions like Map, Reduce, CoGroup and Cross, and achieve parallelism by distributing input data among many parallel tasks of the same function. Spark works well with Hadoop distributed file system (HDFS) to achieve parallel computation on partitioned data. It supports multiple programming languages including Scala, Java and Python. For job scheduling, Spark employs a master process communicating with parallel worker processes, maintains a task queue and distributes the next task in the queue to a worker process after the worker process finishes its current assigned task.

Dask [4,18] is another scalable platform targeted for big data processing and analytics. Similar to Spark, a Dask application is composed as a task graph which can be distributed within one computer or a distributed computing environment. Dask employs a similar master-worker framework for task scheduling. Because Dask is implemented in Python, it is native to work with other Python libraries and packages such as Numpy, XArray, Pandas.

Message Passing Interface (MPI) [16] defines message-passing standard to achieve parallel computing within a distributed computing environment such as a cluster. As a dominant model used in high-performance computing, MPI defines how to distribute tasks to multiple compute nodes and CPU cores and communicate among the parallel processes on synchronization. Unlike Dask and Spark, MPI does not use master-worker architecture. Instead, each process is assigned with a rank number, and the communication and synchronization are done by sending/receiving messages to/from different ranked processes. MPI supports multiple programming languages including C, Fortran and Python.

3   MODIS Data Aggregation Logic

MODIS data aggregation from Level 2 to Level 3 requires collecting all relevant Level 2 data files and calculating corresponding values based on the target variable and the spatial and temporal ranges. In this paper, we focus on monthly global aggregation of one atmospheric variable, called Cloud Fraction. In this section, we will first explain how to do it for a single Level 2 cloud properties product file, namely MYD06/MOD06, then how to combine results from all relevant Level 2 files.

As shown in Fig. 1, the process of generating Level 3 data from Level 2 file involves four main steps. In the first step, it reads one file from MYD06/MOD06 and its corresponding file from MYD03/MOD03 and produces grid-level counts for both cloud pixels and total pixels.

In its first step, it reads 'Cloud_Mask_1km' variable from the MYD06/MOD06 file, and reads 'Latitude' and 'Longitude' variables from the MYD03/MOD03 file. Based on MYD/MOD manual and HDF file convention, each of these three variables is a 2030×1354 2D array. Also, their values of three arrays are 1-to-1 mapped, namely the corresponding longitude and latitude value of each cloud mask value in Cloud_Mask array can be found at Latitude array and Longitude array respectively using the same array index.

The second step of the process is sampling, which only takes partial data from the original 2D (2030×1354) arrays. Based on Earth's geographic coordinates, each latitude grid is about 111 km in distance and each longitude grid is about 0–111 km in distance depends on the longitude's value (0 km for $\pm 90°$ and 111 km for $0°$). It means each (lat, lon) grid covers up to 111 km $\times$ 111 km area (over 12k pixels). Also based on the nature of cloud coverage, if a pixel is cloudy, its surrounding pixels are also likely to be cloudy. To reduce the computing load, we

188   J. Wang et al.

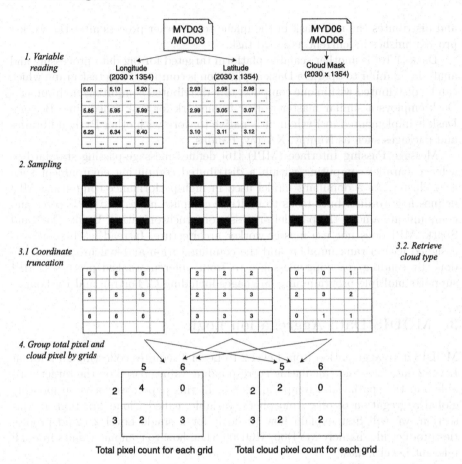

Fig. 1. Illustration of MODIS data aggregation from Level 2 to Level 3 for a single file.

could use different sampling ratios to only aggregate a portion of pixels at 1 km resolution. As shown in the figure, by using sampling ratio 2 for both longitude and latitude, the number of pixels to be aggregated is only 1/4 of original pixels.

The third step is to convert coordinate resolution from Level 2 to Level 3. The original latitude and longitude values in MOD03/MYD03 have precision as 5, while we only need to have integral latitude and longitudes in Level 3 data. So we need to convert coordinate resolution to integer by removing each value's floating part. For instance 25.12345° in Level 2 will be just 25° in Level 3.

The fourth step is to retrieve cloud condition information by doing bit operation. Based on the manual of MOD06/MYD06, there are four types of cloud condition which are encoded as two digits of the values in binary. So by doing bit operation, we can retrieve the values for cloud condition information. Out of the four possible values (0, 1, 2 and 3), only 0 value means the pixel is cloudy. The last step of the process is to calculate the total pixel count and cloud pixel count for each grid. It first group pixels into corresponding grids. Then it counts all pixels within each grid to get the total pixel table and only counts zero value pixels within each grid to get the cloud pixel table.

To get global aggregation results for a month, we need to do the same processing in Fig. 1 for all relevant files and eventually get results for every grid of global 180×360 longitude/latitude grids. Every 5 min, the satellite remote sensing retrieval algorithm obtains a snapshot data file, called *granule*, of the area the satellite covers at the time. Each variable measured by the satellite is saved as a 2D pixel array in the granule file. The 2D array size of the MODIS data we work with is 2030×1354. Because MODIS Level 2 product contains such as granule files every 5 min, the pixel number for a full day is about 800 million ($2030 \times 1354 \times 288$) and the number for a full month is 24.5 billion. The aggregation process will group these 24.5 billion values into 64,800 (180×360) grid bins based on their geographic locations and conduct corresponding calculations for the pixels within a grid bin. On average, each bin aggregates values from about 378,700 pixels.

4 Three Scalable MODIS Aggregation Approaches

In this section, we propose three scalable approaches for MODIS data aggregation, each in one subsection. Approach illustrations are in Figs. 2, 3 and 4. The main differences are task granularity and file/record count, where the numbers in parentheses, e.g., $(2 \times 288 \times 31)$, are the file/record number for the step.

4.1 File Level Scalable MODIS Aggregation

The overall logic of our file level scalable MODIS aggregation approach is shown in Fig. 2. The implementations of the same approach in Spark, Dask and MPI are slightly different because of their programming model difference.

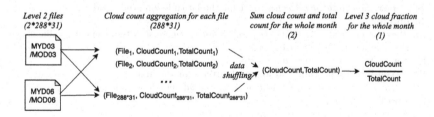

Fig. 2. Illustration of file level parallel aggregation.

We employed MapReduce model in our Spark based file level scalable MODIS aggregation. Each Map task calls the function shown in Algorithm 1. The function takes one MOD03/MYD03 file and one MOD06/MYD06 file as inputs and generates a 2D (180×360) array which which element contains the aggregated cloud pixel count and total pixel count for each grid. In the algorithm, output arrays are initialized in line 1, and data are read into three 2D (2030 × 1354) arrays in lines 2–4. Then line 5 applies the same sampling for all three variables. Line 6 operates on the two longitude/latitude arrays by truncating the floating part of the values. Then in the first for loop (lines 7–9), we can find the integral longitude-latitude grid indices for each pixel and increment the total pixel count by 1 for the grid. Then we check which pixel is cloudy in line 10 and use the second for loop (lines 11–14) to update cloudy pixel count based its grid location.

In this approach, the total (MOD03/MYD03, MOD06/MYD06) file pair number is 8928 (288 × 31). By first creating a list of 8928 file pairs and setting partition number to be 8928, Spark will generate 8928 Map tasks and one task for each file pair. These 8928 Map tasks run in parallel on distributed nodes. After receiving outputs from the Map phase, the two 2D (180 × 360) arrays of cloudy pixel count and total pixel count are simply added to two final 2D (180 × 360) arrays via a Reduce sum function. The cloud fraction ratio is calculated via dividing cloudy pixel counts by total pixel counts for each grid.

Algorithm 1: Data aggregation for each file: *aggregateOneFileData()*

Input: MYD06/MOD06 file path: *M06_file*; MYD03/MOD03 file path: *M03_file*
Output: 2D (180x360) array for cloud pixel count and total pixel count of each grid:
  cloud_pixel_count, total_pixel_count
1: Initialize *cloud_pixel_count* and *total_pixel_count* to all zero 2D (180x360) array
2: Read *Cloud_Mask_1km* variable from *M06_file* file and extract its cloud phase values to a 2D (2030x1354) array: *2D_pixel_array_cloud_mask*
3: Read *Latitude* variable values from *M03_file* file to a 2D (2030x1354) array: *2D_pixel_array_lat*
4: Read *Longitude* variable values from *M03_file* file to a 2D (2030x1354) array: *2D_pixel_array_lon*
5: Apply the same sampling ratio for all above three variables
6: Convert floating-point numbers in *2D_pixel_array_lat* and *2D_pixel_array_lon* to integral numbers
7: **for** each grid location in (*2D_pixel_array_lat, 2D_pixel_array_lon*) **do**
8:     Increment *total_pixel_count* by 1 for the latitude-longitude grid
9: **end for**
10: Retrieve element indices in *2D_pixel_array_cloud_mask* if the element's value shows the pixel is cloudy: *cloud_indices*.
11: **for** each index in *cloud_indices* **do**
12:     Get integral latitude and longitude values for the index
13:     Increment *cloud_pixel_count* by 1 for the latitude-longitude grid
14: **end for**
15: Output (*cloud_pixel_count, total_pixel_count*)

In our Dask and MPI implementations, we used similar Map function so that all 8928 tasks from all file pair combinations can be executed in parallel. After the tasks are done, results are integrated via a for loop.

There is a major difference among the above implementations using Spark, Dask and MPI. Both Spark and Dask support dynamic scheduling of tasks by distributing tasks in the queue to available worker processes. It is particularly useful for our aggregation application because the tasks' execution times can vary from 1 s to 15 s. Dynamic scheduling can achieve good load balance among worker processes. For MPI, we had to programmatically assign tasks to processes without the knowledge of loads for each process.

4.2   Day Level Scalable MODIS Aggregation

The overall logic of day level scalable MODIS aggregation approach is shown in Fig. 3. In day level scalable MODIS aggregation, each function is similar with the one in Algorithm 1 except each function processes one day data (288 files) via an additional for loop. So there will be 31 tasks to be processed in total. Our tests show the execution times for the tasks are also very different, varying from 380 s to 685 s. The implementations in Spark, Dask and MPI are similar with their file level implementations.

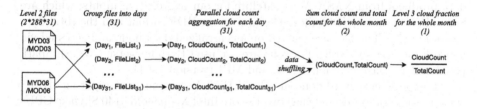

Fig. 3. Illustration of day level parallel aggregation.

4.3   Pixel Level Scalable MODIS Aggregation

The overall logic of pixel level scalable MODIS aggregation approach is shown in Fig. 4. In this approach, each function still processes one file pair like Algorithm 1. The difference lies in how to generate outputs. In stead of generating two 2D (180×360) arrays like the first two approaches, it outputs a list of all pixels in the input file. By looping through all pixels in the input, each pixel outputs a tuple of key-value pair. The key element is a tuple of grid information: (latitude, longitude) and the value element is a tuple of cloudy pixel count. If a pixel is cloudy, the element is $(1, 1)$, otherwise it is $(0, 1)$. Because each input file contains a 2D (2030×1354) array, the output list has 2,748,620 ($=2030 \times 1354$) elements.

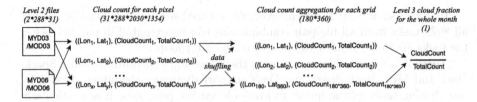

Fig. 4. Illustration of pixel level parallel aggregation.

In our Spark based implementation, we used *flatMap* to wrap the above function so that the elements of the output listed are flattened and shuffled based on their keys, and *reduceByKey* to group the results based on their grid information. We do not have implementations in Dask and MPI because they do not have similar higher-level functions like flatMap and reduceByKey.

5   Experiments

In this section, we explain the experiments we conducted to benchmark and evaluate the differences of using different parallel approaches, parallel platforms and sampling ratios. We choose January 2008 MODIS Level 2 cloud properties data products from NASA Aqua satellite for our aggregation inputs, which are 8928 MYD03 and MYD06 files. The files are in HDF format and the total data size is 738 GB. The files are located on a centralized data node and accessed via network file system (NFS). For software, we used Python (version 3.6.8), Spark (version 2.4), Dask (version 1.1.4), and MPI (version 1.4.1).

All experiments were done in a local High Performance Computing (HPC) cluster. Each computer node has two 18-core Intel Xeon Gold 6140 Skylake CPUs (36 cores in total) and 384 GB of memory. To make fair comparison, we allocate 2 CPU cores and 20 GB memory for each parallel process. By running on 2, 4, 6, 8, 10 and 12 nodes, we can have 36, 72, 108, 144, 180 and 216 parallel processes respectively.

5.1   Comparison Between Different Parallel Approaches

We listed the exact execution times in Table 1 and the speedup in Fig. 5. The bold numbers in the table are the shortest execution times for each distributed environment size. The speedups are calculated via dividing the execution times of serial versions of each approach by the execution times in Table 1. We note our serial version of file level aggregation approach is also used in calculating speedup of pixel level aggregation because it is difficult to implement serial version of key-based data shuffling. The table and figure show the performances of file level parallel aggregation are the best in most cases. We believe a big reason is the task granularity. By having 8928 tasks and each task processing one pair of files, file level parallel aggregation approach can keep all processes running in parallel with assigned tasks for distributed environments in our experiments. Day level parallel aggregation approach shows the worst performance. We believe one reason is many processes will be idling once the available process number is greater than day number (31). Another reason could be the difficulty to achieve load balance among parallel processes because its task granularity is much higher than file level approach. The performance of our spark pixel level parallel aggregation approach is close to spark file level approach. The biggest difference between these two approaches is the data record number to be shuffled after Map/flatMap phase: 8928 records in spark file level approach and 800 million records in pixel file level approach. We believe the reason that the huge data record number in the pixel level approach did not slow down the execution is Spark supports local partial aggregation at Map phase before data is shuffled across network if the following phase is reduceByKey function, which is similar to the Combiner in Hadoop.

Table 1. Execution time results (in Seconds) for scalability evaluation.

Node number	Spark			Dask		MPI	
	File level	Day level	Pixel level	File level	Day level	File level	Day level
2	**256.30**	803.86	279.81	769.82	890.84	788.10	306.51
4	**225.07**	603.66	269.81	423.76	635.48	479.56	428.45
6	**214.07**	729.08	245.80	422.94	595.03	326.90	434.04
8	197.07	464.00	**152.82**	335.16	593.53	205.74	329.25
10	181.68	690.66	208.77	462.80	632.09	**135.10**	418.66
12	164.30	507.79	216.07	364.17	571.87	**161.03**	303.57

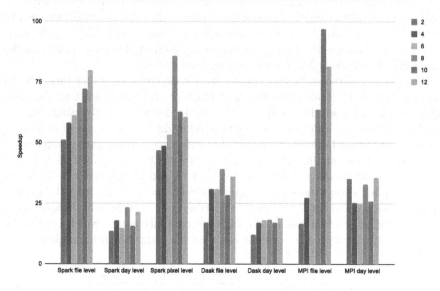

Fig. 5. Execution Speedup for scalability evaluation.

5.2   Comparison Between Different Parallel Platforms

From Table 1 and Fig. 5, we can see Spark achieves the best speedup on average among all three parallel platforms, especially for the file level approach. We believe the reason is Spark can manage task scheduling more efficiently and achieve better load balancing among parallel processes. For the comparison between MPI and Dask, the speedups done via MPI are better than those via Dask in most cases. We think it is due to less coordination overhead for MPI based parallelization. We also notice the best speed up is achieved by file level parallelization via MPI on its execution 10 nodes. It shows the advantage of MPI because of its low coordination overhead, especially if the static scheduling via the MPI approach happens to achieve relatively balanced work loads among different nodes.

5.3   Comparison Between Different Sampling Ratios

We further evaluated how different sampling ratios will affect the execution times and aggregation result quality. Because we conduct sampling for both longitude and latitude direction, sampling ratio of n means only reading 1 out of n^2 pixels. We tested our file level parallelization with Spark implementation on multiple nodes. Table 2 shows the execution times of different sampling ratios on different nodes where the baseline row is done without parallelization and Spark, and the no sampling column is for experiments without sampling. The bold numbers in the table are the shortest execution times for each environment size. From the table, we can see the execution times increase from no sampling to sampling, then decrease with higher sampling ratios. We believe the reason is the additional time for sampling operation could be longer than the time saved for downstream operations with less data, especially when sampling ratio is low. Further, the table shows the executions achieve good scalability for all sampling ratios. For sampling ratio as 5, all parallel executions take less time than corresponding no sampling executions and the largest speed up ratio is 2.586 when running on 4 nodes.

We also calculated data loss percentage for different sampling ratios based on Formula 1. It first calculates the percentage of absolute cloud fraction value difference for each grid, then computes the average for all grids. By using absolute cloud fraction value difference, not actual value difference, in the formula, we can avoid offsetting between positive value differences and negative differences. Our experiments show the data loss percentages are 0.1801%, 0.3021%, 0.4535% and 0.6167% when sampling ratios are 2, 3, 4 and 5, respectively. It shows higher sampling ratio causes a little more data loss percentage, but all data loss percentages are quite small (below 1%). Further, our experiments show the data losses are the same when running for different compute nodes, which verifies the correctness of our approach does not change with compute node numbers.

$$\frac{1}{180*360} \sum_{(-90,-180)}^{(90,180)} \left| \frac{CF(i,j)_{orig} - CF(i,j)_{samp}}{CF(i,j)_{orig}} \right| \tag{1}$$

Table 2. Execution time results (in Seconds) for different sampling ratios.

	No sampling	Sampling ratio			
		2	3	4	5
Baseline	**42,817.87**	133,830.33	267,660.87	86,225.44	55,484.36
2 nodes	1,254.83	5,395.07	2,509.40	1,510.31	**1,008.71**
4 nodes	1,460.57	3,312.09	1,272.82	1,045.13	**564.79**
6 nodes	598.71	2019.24	994.65	590.83	**424.14**
8 nodes	791.23	506.44	461.81	465.77	**415.51**
10 nodes	371.49	269.63	450.24	447.98	**320.76**
12 nodes	259.57	316.91	373.20	474.62	**254.23**

6 Big Data Service for MODIS Aggregation

To simplify on-demand MODIS data aggregation by users, we further provision the above scalable data aggregation capability as services. In this way, users do not need to have a distributed environment with proper back-end libraries installed and download large-scale MODIS data for aggregation. One challenge we face is how to make the services work with different service libraries/protocols, such as REST [17] and ZeroMQ [7]. To achieve this flexibility for our services, we employ an open-source framework called Stratus (Synchronization Technology Relating Analytic Transparently Unified Services) [6] developed by co-author Maxwell. We will explain the Stratus framework, and how our data aggregation services are implemented via Stratus.

The Stratus framework provides a workflow orchestration approach for incorporating Earth data analytic services as a unified solution. It defines a common set of API for workflow and request/response. It consists of a set of orchestration nodes and each implements a particular composition strategy on a particular technology and is designed to interface with other Stratus nodes. Then, an integration framework can be constructed by combining orchestration nodes. Currently available Stratus service handlers include endpoint, ZeroMQ [7], OpenAPI [1] and REST [17]. Stratus can support them by having a separate implementation of the same unified API for each specific library/protocol.

In order to expose a capability within the Stratus framework, that capability must be wrapped as a Stratus endpoint. A typical Stratus based analytics service architecture includes: 1) client application used by users to connect to remote server, 2) server application which accepts connections from the client, 3) endpoint which is a server-side operation for a certain task. In this architecture, a common request language can be established across all the supported endpoints. A common server is used by the client to submit service requests to any endpoint. Coordinating workflows composed of services developed by disparate teams requires the combination of multiple orchestration strategies, e.g., fan-out, publish-subscribe, distributed task queue, and request-reply.

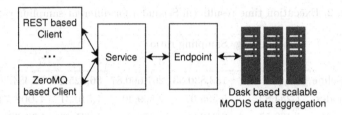

Fig. 6. Flexible MODIS data aggregation service.

Our implementation of MODIS aggregation servicelization using Stratus is illustrated in Fig. 6 and its endpoint and client side code are illustrated in Listings 1 and 2 respectively. For endpoint code, we only need to implement function *operation* to define an operation, and function *execute* to call the operation based on client-side inputs. For client-side code, request specifications are defined through a Python dictionary object. To switch from one service library/protocol to another, we only need to change the value for *type* parameter.

```
class XaOpsExecutable(Executable):
  #Definition of the operation
  def operate(self, M03_dir, M06_dir):
    if self.request['operation'][0]['name']=="
    cloudFraction":
      cf = modis.calculateCloudFraction(M03_dir, M06_dir)
    return cf
  #Executes the operation.
  def execute(self, **kwargs) -> TaskResult:
    inputSpec = self.request.get('input', [])
    cf = self.operate(inputSpec['path1'], inputSpec['path2
    '])
    result = xarray.DataArray(cf.tolist(), name='test')
    return TaskResult(kwargs, [result])
```

Listing 1. Endpoint for MODIS cloud fraction aggregation.

```
if __name__ == "__main__":
  settings = dict(stratus=dict(type="rest"))
  stratus = StratusCore(settings)
  client = stratus.getClient()
  requestSpec = dict(input=dict(path1="MYD03", path2="
    MYD06", operation=[dict(name="cloudFraction")])
  # Submit the request to the server and wait for the
    result
  task: TaskHandle = client.request(requestSpec)
  result: Optional[TaskResult] = task.getResult(block=True
    )
```

Listing 2. Client for MODIS cloud fraction aggregation.

7   Related Work

Benchmarking for Big Data Analytics. There have been many studies on benchmarking for big data analytics including [13,21]. Study [14] compared Spark and Dask in a deep learning satellite image application and their experiments show Spark achieves better performance. The experiments in [11] show no significant execution performance differences between Spark and Dask for their neuroimaging pipeline applications. The work at [8] compares OpenMP, OpenMP + MPI, and Spark for K-means clustering on distributed environment. Our study complements these studies by comparing not only different platforms but also different scalable data aggregation approaches, and analyzing the reasons for the differences.

Climate Analytics as Services. With the rapid increase of climate data, researchers have been studying how to apply service oriented architecture in climate analytics to achieve better efficiency and flexibility than the traditional way of downloading data to a local machine and then analyzing it. Book [20] collected recent studies on how to apply cloud computing for Ocean and Atmospheric Sciences. Among them, the Climate Analytics-as-a-Service (CAaaS) framework [19] is most similar to our work by addressing both service challenge and big data challenge. It differentiates two types of services for big climate data: 1) Analytic Services that run MapReduce style analytic jobs/workflows and 2) Persistent Services that manage the storage and access of data produced by Analytic Services. Our work in this paper falls in the Analytic Service category by supporting on-demand aggregation of satellite data. Our work further addresses the service library/protocol variety challenge by employing the Stratus framework so the same service side implementation can support different service libraries/protocols with simple configuration change at client side.

8   Conclusions

With astronomical growth of available climate data, we believe big data techniques and service techniques are promising to achieve scalable and on-demand analytics for climate data. In this paper, we study how to integrate these techniques for a fundamental climate data aggregation application. We proposed three different aggregation approaches and compared their performance in three different platforms. From the experiments, we conclude that, while we can achieve speedup using all approaches and platforms, defining proper task granularity and dynamic scheduling are key factors to enable good scalability and load balance. The best speed up ratio we can achieve is close to 100. Our experiments also show proper sampling ratio design could achieve execution speedup with little data loss. Last, we discussed how to leverage the Stratus framework to easily support different service libraries/protocols.

For future work, we will extend our satellite data aggregation capability for more climate variables (such as cloud top height), more statistics (such as standard deviation and histogram), more flexible spatial area selection. We also plan to expose the service on public cloud environments, such as Amazon Web Service, for users to use without requiring a distributed environment.

Acknowledgment. This work is supported by an NASA CMAC grant (80NSSC18K 0796) and an NSF CyberTraining grant (OAC-1730250). The execution environment is provided through the High Performance Computing Facility at UMBC.

References

1. OpenAPI Initiative (OAI). https://www.openapis.org. Accessed 28 May 2020
2. Scalable MODIS Data Aggregation Platform. https://github.com/big-data-lab-umbc/MODIS_Aggregation. Accessed 28 May 2020
3. Apache Spark Project (2020). http://spark.apache.org. Accessed 28 May 2020
4. Dask: Scalable analytics in Python (2020). https://dask.org/. Accessed 28 May 2020
5. NASA LAADS Distributed Active Archive Center (2020). https://ladsweb.modaps.eosdis.nasa.gov/search/. Accessed 28 May 2020
6. Stratus: Synchronization Technology Relating Analytic Transparently Unified Services (2020). https://github.com/nasa-nccs-cds/stratus/. Accessed 28 May 2020
7. ZeroMQ: An open-source universal messaging library (2020). https://zeromq.org/. Accessed 28 May 2020
8. Barajas, C., et al.: Benchmarking parallel k-means cloud type clustering from satellite data. In: Zheng, C., Zhan, J. (eds.) Bench 2018. LNCS, vol. 11459, pp. 248–260. Springer, Cham (2019). https://doi.org/10.1007/978-3-030-32813-9_20
9. Chambers, B., Zaharia, M.: Spark: The Definitive Guide: Big Data Processing Made Simple. O'Reilly Media Inc., Sebastopol (2018)
10. Dean, J., Ghemawat, S.: Mapreduce: simplified data processing on large clusters. Commun. ACM **51**(1), 107–113 (2008)
11. Dugré, M., Hayot-Sasson, V., Glatard, T.: A performance comparison of dask and apache spark for data-intensive neuroimaging pipelines. In: 2019 IEEE/ACM Workflows in Support of Large-Scale Science (WORKS), pp. 40–49. IEEE (2019)
12. Eyring, V., et al.: Overview of the coupled model intercomparison project phase 6 (cmip6) experimental design and organization. Geoscientific Model Development (Online), 9(LLNL-JRNL-736881) (2016)
13. Han, R., Lu, X., Xu, J.: On big data benchmarking. In: Zhan, J., Han, R., Weng, C. (eds.) BPOE 2014. LNCS, vol. 8807, pp. 3–18. Springer, Cham (2014). https://doi.org/10.1007/978-3-319-13021-7_1
14. Nguyen, M.H., Li, J., Crawl, D., Block, J., Altintas, I.: Scaling deep learning-based analysis of high-resolution satellite imagery with distributed processing. In: 2019 IEEE International Conference on Big Data (Big Data), pp. 5437–5443. IEEE (2019)
15. Overpeck, J.T., Meehl, G.A., Bony, S., Easterling, D.R.: Climate data challenges in the 21st century. Science **331**(6018), 700–702 (2011)
16. Pacheco, P.: Parallel Programming with MPI. Morgan Kaufmann, San Francisco (1997)
17. Pautasso, C., Wilde, E., Alarcon, R.: REST: Advanced Research Topics and Practical Applications. Springer, New York (2013). https://doi.org/10.1007/978-1-4614-9299-3

18. Rocklin, M.: Dask: parallel computation with blocked algorithms and task scheduling. In: Proceedings of the 14th Python in Science Conference, no. 130–136. Citeseer (2015)
19. Schnase, J.L.: Climate analytics as a service. In: Cloud Computing in Ocean and Atmospheric Sciences, pp. 187–219. Elsevier (2016)
20. Vance, T.C. , Merati, N., Yang, C., Yuan, M.: Cloud computing for ocean and atmospheric science. IEEE (2016)
21. Wang, L., et al. Bigdatabench: a big data benchmark suite from internet services. In: 2014 IEEE 20th International Symposium on High Performance Computer Architecture (HPCA), pp. 488–499. IEEE (2014)

Edge-Assisted Federated Learning: An Empirical Study from Software Decomposition Perspective

Yimin Shi[1,2], Haihan Duan[1,2], Yuanfang Chi[3], Keke Gai[4], and Wei Cai[1,2(✉)]

[1] The Chinese University of Hong Kong, Shenzhen, Shenzhen, China
{yiminshi,haihanduan}@link.cuhk.edu.cn
caiwei@cuhk.edu.cn
[2] Shenzhen Institute of Artificial Intelligence and Robotics for Society,
Shenzhen, China
[3] The University of British Columbia, Vancouver, Canada
yuanchi@ece.ubc.ca
[4] Beijing Insitute of Technology, Beijing, China
gaikeke@bit.edu.cn

Abstract. Federated learning is considered to be a privacy-preserving collaborative machine learning training method. However, due to the general limitation of the computing ability of the terminal device, the training efficiency becomes an issue when training some complex deep neural network models. On the other hand, edges, the nearby stationary devices with higher computational capacity, might serve as a help. This paper presents the design of a component-based federated learning framework, which facilitates the offloading of training layers to nearby edge devices while preserving the users' privacy. We conduct an empirical study on a classic convolutional neural network to validate our framework. Experiments show that this method can effectively shorten the time cost for mobile terminals to perform local training in the federated learning process.

Keywords: Federated learning · Deep learning · Mobile edge computing · Program decomposition · Distributed computing

1 Introduction

In a traditional cloud-centric approach, if the cloud wants to utilize the data from mobile terminal devices to train a Machine Learning (ML) model, local data of the terminal has to be directly uploaded to the cloud, which can hardly preserve the privacy of clients and will pose a burden on the backbone networks [7].

This work was supported by Project 61902333 supported by National Natural Science Foundation of China, by the Key Area R&D Program of Guangdong Province with grant No. 2018B030338001, by the Shenzhen Institute of Artificial Intelligence and Robotics for Society (AIRS).

Currently, Federated Learning (FL) has been introduced as a privacy-preserving distributed ML approach [10]. In a gradient-descent based FL, mobile terminals will locally train the given model with its privacy-sensitive dataset. Meanwhile, based on the current parameters, all terminals will return different sets of parameter gradients to the cloud [17]. This approach avoids directly sending the dataset to the untrusted cloud, which makes it possible to preserve user privacy and reduce network overhead.

Compare to the traditional cloud-centric approach, FL migrates part of the task from the cloud to the terminal. However, the computation resource of the mobile terminal device varies and is generally constrained. When some more complicated ML models like Deep Neural Network (DNN) [5] are trained with multimedia datasets, it may be too time-consuming for the terminal to complete the task. However, the computing power of the mobile edge computing (MEC) device is much stronger than that of the terminal. The distance and the communication cost between edge and terminal are relatively low, and it will not create a significant burden on the backbone network. Besides, the edge has higher security than the cloud [9]. Therefore, the edge may help solve this problem.

In terms of model training efficiency optimization, current works focus on the following two aspects: the optimization of the traditional FL procedure by improving the existing protocol and the optimization of the DNN training process by applying distributed methods. However, these works do not consider how to apply the distributed methods to accelerate the local update process of FL with the help of MEC.

In our work, we propose a component-based FL framework that can offload some of the training layers of the DNN model to the nearby edge while preserving the privacy of the terminal user. The local parameter updating process is further decomposed into inter-independent components by coarse-grained program decomposition, and each component can be separately deployed either on the edge or the terminal. Each layer component can independently complete the forward and backward propagation. We conduct an empirical study on a classic convolutional neural network (CNN). Experiments show that this framework can efficiently shorten the time cost of local training. The optimal strategy of component deployment and the effectiveness of user privacy preservation in different edge trusting situations are separately discussed.

The remaining sections of the paper are organized as follows. We review some related works in Sect. 2, and discuss the general method applied for layers decomposition of DNN in Sect. 3. Section 4 presents the architecture and the program design of the proposed component-based distributed training system. In Sect. 5, we discuss different trusting situations of the edge device and provide the optimal strategies which can efficiently preserve user privacy. Moreover, in Sect. 6, we present the results of a set of experiments that explore the efficiency of the system in different conditions. Finally, Sect. 7 concludes the paper and discusses the aspects that can be further studied on the basis of this work.

2   Related Work

In the field of federated learning, current researches on the optimization of FL focus on at least two aspects as follow: (1) User privacy preservation and (2) FL efficiency improvement. For users' privacy preservation, some existing protocols [1] have been proven to be effective. But some existing cloud storage protection method [13] is challenging to use directly in the FL process. For FL efficiency improvement, recent works discuss how to make user selection [11] for clients during the learning process, as well as how to determine the frequency of local up-dates and global aggregation [16]. Our work focuses on improving the efficiency of the local updates through the distributed method with the help of MEC, which is different from the above works and does not contradict each other.

In the field of software program decomposition, currently, there are two primary schemes: Fine-Grained Decomposition and Coarse-Grained Decomposition. Coarse-grained decomposition partitions the program into a set of functional-independent and stateless components. A web-based coarse-grained program decomposition platform has been developed in the current work [2] [3], which provides a good execution engine and API for software decomposition. In this work, we apply the coarse-grained decomposition to decompose the layers in a CNN into different functional components. The number of components is ensured to be sufficient to provide us with more possible component deployment strategies.

In the field of the distributed deep neural network, current works focus on the following aspects: (1) Improving the training efficiency of DNN through distributed parallel training method [4]. (2) Using distributed DNN to do the training process or forward propagation to protect user privacy [8,12,15]. However, the existing works do not carry out program design or deployment strategy under the framework of FL, do not guarantee user privacy, or propose a suitable decomposition schema. In this paper, we decompose the training process of DNN into more fine-grained components under the framework of FL and MEC.

3   Layers Decomposition

Different DNN models have a lot in common in their architectures: they are usually composed of multiple relatively independent layers with different functionalities. Furthermore, there exists a sequential relationship between the different layers, which provides the possibility to design a general DNN decomposing method: In the forward propagation process of DNN models, the Lth layer always receives the input from the $(L-1)^{th}$ layer and gives the output to the $(L+1)^{th}$ layer. In the process of backward propagation, the Lth layer receives the gradients from the $(L+1)^{th}$ layer, calculates gradients of the weight, and returns the gradients of its input to the $(L-1)^{th}$ layer.

The dependency relationship between adjacent layers must be considered when decomposing the training process of the DNN model. Each layer in the

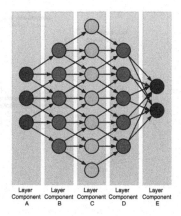

Fig. 1. Layer decomposition of DNN.

original DNN model can be decomposed as an independent component where the specific calculating procedure of the layer is implemented and encapsulated. Meantime, in the forward and backward propagation, the original layer-to-layer association can be achieved through mutual communication between devices. Besides, the intermediate results generated during the forward propagation process must be reasonably cached since these results will be used when performing the backward propagation.

In our work, in the local update process, we decompose the entire training procedure into different inter-independent coarse-grained layer components, as shown in Fig. 1. Each component is stateless and can be selected by instructions to execute a specific layer's forward or backward propagation procedure. Meanwhile, each component can be independently deployed on different devices and communicate with each other through the network protocol. The design of the layer component will be further discussed in detail in the next section.

4   System Design

4.1   Component-Based Program Design

The system is divided into two main components, which separately execute on the cloud and the client-side. The component executes on the client-side is responsible for the local updating and can be further decomposed into multiple more fine-grained components, which can be deployed either on the terminal or edge. As shown in Fig. 2, the components on the client generally interact with the terminalController component to cache intermediate data and exchange messages. Since the layer components have the possibility of migration at run time, there is no direct interaction between the layer components.

The entire procedure of FL can be decomposed into three sets of components: CloudController, TerminalController, and Layer Components.

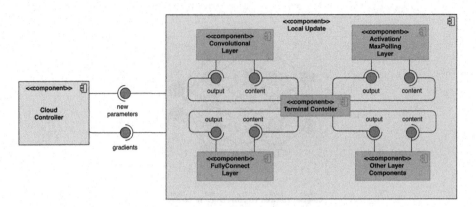

Fig. 2. Component diagram of the system.

CloudController Component. The cloudController component fixedly executes on the cloud and is responsible for the original cloud task, which includes initialization, parameter distribution, and gradient aggregation. When the initialization instruction is raised, the cloudController will perform cloud initialization and then send the initialization instruction to the terminalController component. When terminalController returns the gradient of weights, the cloudController will perform aggregation and distribute new parameters.

TerminalController Component. The terminalController fixedly executes on terminal devices. It is responsible for managing local privacy-sensitive datasets, caching intermediate results generated in the forward propagation of the DNN model, and determining which layer components to call next after the execution of the last layer component and gives the corresponding content. When the initialization instruction is raised, the terminalController will perform local initialization and inform the cloudController after its completion. In the forward propagation, when the intermediate output of the last layer component is sent back, the terminalController will cache the calculation results and redirect the control to the next layer component. In the backward propagation, it will store the sent back gradients from the last layer component and redirect the control.

Layer Components. Layer Components (Fig. 3) are a collective term for various layer components. Layer components in the general CNN model include and are not limited to convolutional layer components, fully connected layer components, activation function components, and pooling layer components. Since the communication cost and calculation cost of each layer is generally different, layer components should be established for each layer in a particular model. For example, if a CNN model is mainly composed of two convolutional layers and a fully connected layer, then it should consist of two independent convolutional

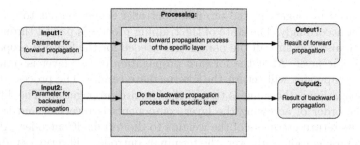

Fig. 3. The IPO diagram of Layer Components.

layer components with different hyper-parameters and only one fully connected layer type component. These components are independent of each other and are uniformly called by the terminalController.

4.2 Data Flow

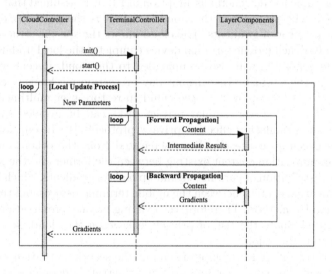

Fig. 4. The sequence diagram of the system.

Before the training process, the cloudController will issue initialization instructions to the terminalController. After the entire initialization is completed, in each round of training, the cloudController will send the current weight to the terminalController of each terminal. The local update process of each terminal can be divided into the forward propagation and the backward propagation.

In forward propagation, the terminalController sends content to each layer component accordingly. The state of the training process is passed through content, which also contains all the parameters needed for the forward propagation of this specific layer. After the forward propagation of this layer is completed, the layer component will return the intermediate results. The process of backward propagation is the same as that of the forward propagation. The difference is that the order of accessing the layer components is reversed, and the layer components return gradients of the weights to the terminalController. After the backward propagation is all over, the terminalController will return gradients of weights in this round of training to the cloudController. After cloudController aggregates the gradients returned by all terminals, it will update the existing weights and distribute them back to the terminalController to start the next iteration of training.

5   Privacy Preservation Analysis

Privacy preservation is one of the main features of FL. When using distributed methods for FL training, it is necessary to formulate specific component deployment strategies according to different situations to preserve user privacy.

First, the same as the general assumption of FL, it is assumed that the cloud is not trusted. The property of the cloud is that it is the publisher of training tasks and aggregator of gradients. The cloud knows the specific structure of the model, and after each mobile terminal device completes the local update process, the gradients generated will also be uploaded to the cloud. Recent works have proven that in the FL instances without any protection mechanism, algorithms like DLG [18] can be deployed on the cloud to restore user multimedia inputs at the pixel level. However, similar privacy risks can be avoided by gradient pruning [18] or formulating effective privacy protocols [1]. In our work, if the components execute on the cloud are separated from the other components, then the communication content existing between the former and the latter only includes the distribution of parameters and uploads of gradients, which is same as the communication content between the mobile terminal devices and the cloud in the traditional FL method. Therefore, the existing privacy preservation methods can still help eliminate the risk of privacy leakage on the cloud, so this is not the focus of our discussion.

Second, three different assumptions about the security situation of the edge device are made. The corresponding privacy preservation deployment strategies are discussed base on each assumption. Assumptions: (a)The edge is fully trusted (b)The edge is untrusted, but there is no direct communication or cooperation between the edge and the cloud (c)The edge is untrusted while there is direct communication or cooperative relationship between the edge and the cloud.

Case A. The edge device may be fully trusted in some cases; for example, the edge device is privately owned by an organization instead of being public. In this case, there is no additional privacy concern when doing the local update. The optimal strategy is which can maximize the overall efficiency.

Case B. The edge is untrusted, but it does not have direct communication or cooperation with the cloud. It means that there may exist potential attackers on the edge device, but the edge has different knowledge from the cloud: the edge does not have the specific knowledge about the entire structure of the DNN model. However, edge knows the input and the corresponding intermediate results and gradients of the layer components executing on it. In this way, it should be avoided to deploy the input layer component on edge, which will lead to direct leakage of user input. However, since the components are independent of each other and are stateless, layer components executing on edge will not be informed about the structure of the entire DNN model or the hyper-parameters, algorithms like DLG will not be able to obtain user privacy. Therefore, in this case, the optimal deployment strategy should be a strategy that maximizes the operating efficiency after fixing the input layer component on the mobile terminal.

Case C. The edge is not trusted, and there may exist direct communication or cooperation between the edge and the cloud, the potential attackers on edge and the cloud may share knowledge about the DNN model. In this case, directly giving the input of a specific layer component to the edge could be dangerous. Furthermore, we need to find new privacy preservation methods, such as designing new protocols, to ensure client privacy.

6   Experiments and Results

6.1   Experimental Settings

In our work, we decompose the FL process with classic CNN model LeNet [6] as the training model into two fixed components and seven free components (Fig. 5) as follow: Conv1 and Conv2 are layer components correspond to the convolutional layers in the order of the forward propagation, ReLu1 and ReLu2 are components for activation layers, Pool1 and Pool2 for pooling layers, FC for the remaining fully connected layers. The system is implemented by using the existing distributed software platform [2, 3] bases on program decomposition. Base on the different deployment strategies in Table 1, we conducted a series of experiments and discussions on the efficiency of local training.

We deploy our system and conduct all experiments by constructing virtual machines on the existing OpenStack-based cloud platform. If there is no further instruction, the default experiment configuration is as follows: the operating system of the edge is Ubuntu 18.04, whose operating environment is Node.js 12.18.0, with 8 CPU cores and 32 GB RAM. The operating system of the terminal is Ubuntu 18.04, whose operating environment is Firefox 77.0.1, with 4 CPU cores and 16 GB RAM. The screenshots of the log output on the edge and terminal during the local training process are shown in Fig. 6.

Table 1. Different components deployment strategies.

Strategy	Components on terminal	Conponents on edge	Strategy	Components on terminal	Conponents on edge
A	No Component	All Components	B	Conv1	Conv2, ReLu1, ReLu2, Pool1, Pool2, FC
C	ReLu1, ReLu2	Conv1, Conv2, Pool1, Pool2, FC	D	Pool1, Pool2	Conv1, Conv2, ReLu1, ReLu2, FC
E	FC	Conv1, Conv2, ReLu1, ReLu2, Pool1, Pool2	F	Conv1, ReLu1, ReLu2	Conv2, Pool1, Pool2, FC
G	Conv1, Pool1, Pool2	Conv2, ReLu1, ReLu2, FC	H	Conv1, FC	Conv2, ReLu1, ReLu2, Pool1, Pool2
I	ReLu1, ReLu2, Pool1, Pool2,	Conv1, Conv2, FC	J	Conv1, ReLu1, ReLu2, Pool1, Pool2	Conv2, FC
K	ReLu1, ReLu2, Pool1, Pool2, FC	Conv1, Conv2	L	Conv1, ReLu1, ReLu2, Pool1, Pool2, FC	Conv2
M	All Components	No Component			

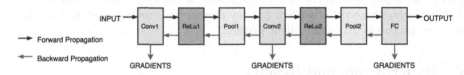

Fig. 5. LeNet architecture from a component perspective.

6.2   Run-Time Efficiency with Different Deployment Strategies

We observe the time cost of each iteration for the mobile terminal and the edge to complete the local update process through the distributed FL method under different deployment strategies in Table 1. It is defined that the time cost of each iteration starts when the new parameter distributed by the cloudController component, ends when the local update is completed and terminalController uploads the gradients to the cloudController. The batch size is set to 10 for each strategy. Figure 7 plots the execution curves of several representative deployment strategies in 150 iterations. The specific component deployment situation corresponding to each strategy is shown in Table 1. In Strategy M, all components are deployed on the terminal, which is similar to the traditional FL method. In Strategy A, all components are deployed on edge. The Strategy I and J are the optimal deployment strategies can be found under two different trusting assumptions (Case A and B) of edge devices, which can maximize the training efficiency.

As can be found from the figure, with same operating environment and hardware setting, The run-time efficiency of Strategy J and Strategy M is more unstable than Strategy A, and Strategy I. Especially in Strategy M, which is

Fig. 6. Screenshots of the log generate on both edge and terminal during the training process. The left figure is the console output of the Node.js server on edge. The right figure is the console output of the Chrome browser on the terminal.

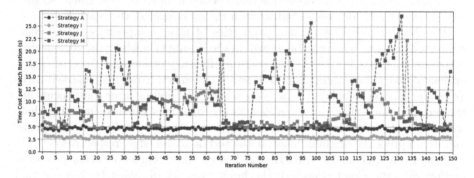

Fig. 7. Time cost of each iteration with different deployment strategies.

equivalent to the traditional FL deployment, the run-time efficiency of local training fluctuates greatly and periodically. This may be caused by the more stable execution in the Node.js environment at the edge compare to the browser engine at the terminal. Besides, it can also be seen that there exist significant differences in the average execution efficiency of different deployment strategies: compare with Strategy M, the average training efficiency of Strategy I and J increases by 74.8% and 36.5% respectively, which will be discussed in detail in the next section.

6.3   Average Efficiency with Different Deployment Strategies

Multiple sets of different component deployment strategies are tested with different batch sizes. In Fig. 8, the histogram shows the average time costs of each deployment strategy for an iteration with the batch size of 5, 10, and 20, which

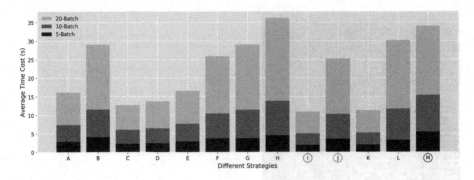

Fig. 8. Average time cost of different deployment strategies with different batch sizes. (Color figure online)

represented by dark blue, blue and light blue respectively. The specific deployment location of layer components in each strategy is shown in Table 1, we can see that there exists an apparent difference between different deployment strategies. Moreover, the difference widens with the increment of the batch size.

By observing the deployment properties of layer components in different strategies, the conclusion can be drawn. In the assumption where the edge device is completely trusted (Case A), the training efficiency is our only concern, so the optimal deployment strategy for all tested batch sizes is Strategy I (marked with green circle), which deploys the activation layer and polling layer components on the terminal and other layer components on edge. It is worth mentioning that some efficient compression techniques [14] can be further utilized to compress and decompress the original image at the terminal and edge, respectively.

When the edge is not trusted, and has no direct communication or cooperation with the cloud (Case B), the input layer locally must be kept on the terminal, so of all the strategies deploying Conv1 component on the terminal, Strategy J (marked with yellow circle) is the most efficient for all tested batch sizes. Strategy J deploys the input layer, activation layer, and pooling layer components on the terminal and other layer components on edge. The above two deployment strategies I and J are more efficient than the traditional federated learning deployment strategy M (marked with red circle). It can be inferred from the result that, since the communication costs of the layer components are relatively similar, when deploying some layer components with lightweight computation task (such as the activation layer component and the pooling layer component) on the terminal and deploy the computation-intensive layer components (such as convolutional layer component) on edge, the overall efficiency is higher.

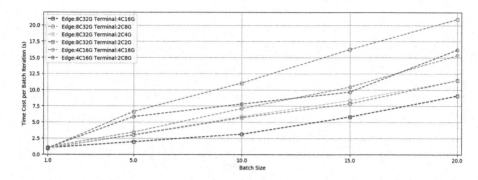

Fig. 9. Average time cost of a single iteration for different batch sizes when the hardware configuration of the edge and the terminal changes.

6.4 Effects of Different Hardware Resources on Average Training Efficiency

The effect of different hardware configurations on local training efficiency is also observed. In the following experiment, the layer components are deployed according to Strategy J, and the computing power resources of the edge and the terminal are adjusted. The average time cost of the batch iteration with different batch sizes is calculated, and results are shown in Fig. 9. When the batch size is equal to 1, with different computing power resources, the average training efficiency is basically the same. However, the difference in efficiency comes to be more significant with the gradual increment of the batch size. Among all the hardware configurations tested with the batch size of 20, the training efficiency comes to be the highest when the edge has 8 CPU cores and 32 GB RAM, and the terminal has 4 CPU cores and 16 GB RAM. On the contrary, the training efficiency comes to be the lowest when the edge has 4 cores with 16 GB RAM, and the terminal has 2 cores with 2 GB RAM. The training time cost of the latter is about 2.31 times the former.

It could be concluded that when the memory of the terminal device is below a certain threshold, which is about 2 GB in our experiment, the local training efficiency will be greatly reduced. Besides, changes in computing resources of edge devices will also significantly affect the local training efficiency. However, when the computing resources of the edge device remain unchanged, and the memory size of the terminal device is relatively sufficient, the computing resource of the terminal device has a limited effect on efficiency.

6.5 Effects of Different Terminal Operating Environments on Training Efficiency and Run-Time Stability

Different operating environments on the terminal device are tested to find their effects on the training efficiency further. We separately test the efficiency of the mobile terminal device using Chrome 83.0 browser and Firefox 77.0.1 browser

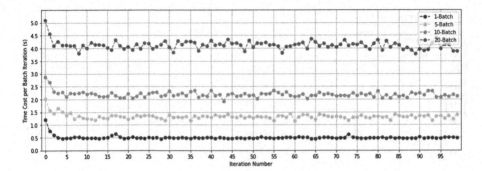

Fig. 10. Compare the time cost per iteration with different batch sizes when using the Chrome 83.0 browser in the Ubuntu 18.04 operating system as the terminal executing environment.

as the operating engine when the batch size is equal to 1, 5, 10 and 20 under the default hardware configuration with deployment strategy J. The results are shown in Fig. 10 and Fig. 11. It can be found that different terminal operating environments will make a noticeable difference in training efficiency and stability. In Fig. 10, when using Chrome as the browser, for all batch sizes, the time cost of each batch iteration is relatively stable. As the batch size increases, the fluctuation of the curve slightly increases but still not apparent. In Fig. 11, when using Firefox as the browser, it can be observed that for batch sizes like 1 and 5, the curve does not fluctuate significantly, but for larger batch sizes like 10 and 20, the curve fluctuates periodically and significantly, and there exist some unreasonably high peaks.

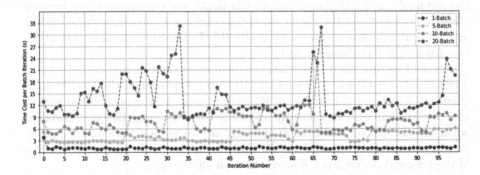

Fig. 11. Compare the time cost per iteration with different batch sizes when using the Firefox 77.0.1 browser in the Ubuntu 18.04 operating system as the terminal executing environment.

In fact, with the batch size increasing from 1 to 20, the standard deviation of the time cost of each iteration in the Chrome environment increases from 0.083

to 0.170, while for the Firefox, it increases from 0.318 to 4.884. Moreover, the latter's average time cost in different batch sizes is about 2.85 times the former. Since in some application scenarios of FL, the aggregation server may need to communicate with a large number of terminal devices at the same time in a single iteration of training, the global time cost of the specific iteration may depend more on terminals cost maximum time to complete the local training procedure instead of the average cost of all terminals. Therefore, the operating environment for the terminal should be reasonably chosen so that the local training of the terminal in each iteration is stable and has high average efficiency.

7   Conclusions

In this paper, we have proposed a component-based framework for FL, which can effectively improve the local training efficiency. Unlike the previous related works, we apply the coarse-grained program decomposition on the DNN model to allow the terminal to offload the training layers to the edge while preserving user privacy. An empirical study on a classic CNN is conducted to show the system's effectiveness and explore the optimal deployment strategies under different edge-trusting assumptions.

Several aspects that can be the future work: (1) apply program decomposition for more complex DNN models and find the corresponding optimal deployment strategies; (2) improve the system and provide API; (3) consider the impact of different network environments on training efficiency.

References

1. Bonawitz, K., et al.: Practical secure aggregation for privacy-preserving machine learning. In: Proceedings of the 2017 ACM SIGSAC Conference on Computer and Communications Security, pp. 1175–1191, October 2017
2. Cai, W., Zhu, C., Chi, Y.C., Leung, V.C.: Balancing cloud and mobile terminal: an empirical study on decomposed cloud gaming. In: 9th IEEE International Conference on Cloud Computing Technology and Science (CloudCom 2017), Hong Kong, China, 11 December–14 December 14 (2017)
3. Cai, W., Chi, Y., Zhou, C., Zhu, C., Leung, V.C.: UBCGaming: ubiquitous cloud gaming system. IEEE Syst. J. 12(3), 2483–2494 (2018)
4. Harlap, A., et al.: Pipedream: fast and efficient pipeline parallel DNN training. arXiv preprint arXiv:1806.03377 (2018)
5. LeCun, Y., Bengio, Y., Hinton, G.: Deep learning. Nature 521(7553), 436–444 (2015)
6. LeCun, Y., Bottou, L., Bengio, Y., Haffner, P.: Gradient-based learning applied to document recognition. Proc. IEEE 86(11), 2278–2324 (1998)
7. Lim, W.Y.B., et al.: Federated learning in mobile edge networks: a comprehensive survey. In: IEEE Communications Surveys & Tutorials (2020)
8. Mao, Y., Yi, S., Li, Q., Feng, J., Xu, F., Zhong, S.: A privacy-preserving deep learning approach for face recognition with edge computing. In: Proceedings USENIX Workshop Hot Topics Edge Computer (HotEdge), pp. 1–6 (2018)

9. Mao, Y., You, C., Zhang, J., Huang, K., Letaief, K.B.: A survey on mobile edge computing: the communication perspective. IEEE Commun. Surv. Tutorials **19**(4), 2322–2358 (2017)
10. McMahan, H.B., Moore, E., Ramage, D., Arcas, B.A.: Federated learning of deep networks using model averaging (2016)
11. Nishio, T., Yonetani, R.: Client selection for federated learning with heterogeneous resources in mobile edge. In: ICC 2019–2019 IEEE International Conference on Communications (ICC), pp. 1–7. IEEE, May 2019
12. Osia, S.A., et al.: A hybrid deep learning architecture for privacy-preserving mobile analytics. IEEE Internet of Things J. **7**(5), 4505–4518 (2020)
13. Qiu, H., Noura, H., Qiu, M., Ming, Z., Memmi, G.: A user-centric data protection method for cloud storage based on invertible DWT. IEEE Trans. Cloud Comput. **2020**, 1–12 (2019)
14. Qiu, H., Zheng, Q., Memmi, G., Lu, J., Qiu, M., Thuraisingham, B.: Deep residual learning based enhanced JPEG compression in the internet of things. IEEE Trans. Ind. Inform. **2020**, 1–10 (2020)
15. Wang, J., Zhang, J., Bao, W., Zhu, X., Cao, B., Yu, P.S.: Not just privacy: improving performance of private deep learning in mobile cloud. In: Proceedings of the 24th ACM SIGKDD International Conference on Knowledge Discovery & Data Mining, pp. 2407–2416, July 2018
16. Wang, S., et al.: Adaptive federated learning in resource constrained edge computing systems. IEEE J. Selected Areas Commun. **37**(6), 1205–1221 (2019)
17. Yang, Q., Liu, Y., Chen, T., Tong, Y.: Federated machine learning: concept and applications. ACM Trans. Intell. Syst. Technol. (TIST) **10**(2), 1–19 (2019)
18. Zhu, L., Liu, Z., Han, S.: Deep leakage from gradients. In: Advances in Neural Information Processing Systems, pp. 14747–14756 (2019)

A Dynamic Partitioning Framework for Edge-Assisted Cloud Computing

Zhengjia Cao[1,2], Bowen Xiao[1,2], Haihan Duan[1,2], Lei Yang[3], and Wei Cai[1,2(✉)]

[1] The Chinese University of Hong Kong, Shenzhen, Shenzhen, China
{zhengjiacao,bowenxiao,haihanduan}@link.cuhk.edu.cn
caiwei@cuhk.edu.cn
[2] Shenzhen Institute of Artificial Intelligence and Robotics for Society,
Shenzhen, China
[3] South China University of Technology, Guangzhou, China
sely@scut.edu.cn

Abstract. Edge computing is commonly adapted to reduce network delay and fluctuation in the traditional cloud-terminal architecture. Many research teams have been dedicated to algorithms and optimizations in the cloud-edge data cache and computation offloading schemes, while there is a blank in practical implementation to facilitate such a paradigm. In this work, we proposed a component-based framework that facilitates dynamic partitioning of a software program. We have designed and implemented the very first test-bed for further optimizing the components distribution strategy among cloud, edge, and terminal devices. Experiments have revealed the characteristics of components execution in the proposed architecture, showing that the system can improve computing performance under the real-world unstable network environments.

Keywords: Cloud computing · Edge computing · Software decomposition · Offloading · Distributed system

1 Introduction

With abundant computational resources as a guarantee, cloud computing is becoming the main direction of software evolution. The software in recent years, such as cloud-hosted augmented reality, online video editing, games, etc., often demands assistance from cloud computing technologies [3] in data storage and computing resources [14]. However, accessing the remote cloud server introduces inevitable network delay. As a supporting infrastructure of cloud computing, edge computing [6,16] utilizes hardware facilities at the edge of the network as

This work was supported by Project 61902333 supported by National Natural Science Foundation of China, by the Key Area R&D Program of Guangdong Province with grant No. 2018B030338001, by Project 61972161 supported by National Natural Science Foundation of China, by the Shenzhen Institute of Artificial Intelligence and Robotics for Society (AIRS).

M. Qiu (Ed.): ICA3PP 2020, LNCS 12453, pp. 215–229, 2020.
https://doi.org/10.1007/978-3-030-60239-0_15

relay nodes for reducing latency, pushing the cloud services closer to the terminal end to avoid backbone network communication as much as possible [2]. Due to the hardware limitation, the resources available at the edge node are far less powerful than cloud servers. Therefore, to achieve a balanced workload between the end devices as well as a reduced communication latency, it is necessary to distribute the calculation tasks between the mobile terminal, the edge node, and the remote cloud server according to the runtime status of the application.

In order to build software in the traditional cloud computing engineering implementation, the engineer should first analyze various functions in the program, identify the module's dependencies, code the software and finally deploy the application in the cloud server to achieve the functioning remote service. Such a classic engineering solution is generally called the server-terminal architecture [1]. However, heterogeneous computation distribution of cloud, edge, and terminal brings new challenges to the traditional computing framework. First, there is a tendency that the software in recent years is becoming more calculation-heavy. With the introduction of complex computing tasks like deep learning algorithms, decomposition may become the main revolution direction of reducing software complexity. Second, it is hard to predict the network environment of the terminal since more mobile devices are introduced into the system. The engineers can no longer assume a stable network condition between the end devices, making it hard to determine what kind of components are most suitable for running on the edge. Third, the types of terminal devices will continue to expand in the future as the Internet of Things (IoT) technologies grow, ranging from smart buttons to smart cars. Their computational performance may vary significantly. Therefore, the new types of software in the edge-computing system are now more computational and communication intensive. Thus, to fully utilize the computational capacity of the devices at each end of the distributed system, software should be able to be decomposed into components to achieve the optimal distribution of overall performance.

Decomposing software into a set of executable components and adjust this partition results dynamically according to the system environment is critical in maximizing the program's overall efficiency. In fact, various types of cloud applications that are widely used nowadays (e.g. the classic MapReduce [17] and deep neural networks [12]) have the improvement potential for software decomposition since it is possible to decompose the data modules and the neural network architectures in these programs. Although there have been many current studies on the algorithms and optimization for partitioning program components under simulated scenarios [15,19], few have done practical implementations and quantitative experiments on the partitioning strategies. Thus, it is essential to implement a system that facilitates the measurements and the performance testing of the current software under the edge-assisted cloud computing scenario.

In this paper, we have proposed and implemented a component-based framework for an edge computing system that allows dynamic partitioning of the software components. Our proposed system can help in further investigations of optimizing the components distribution strategies between the cloud, edge, and terminal devices by measuring the system performance of different types of applications in the deployed settings. Our experiments have also proved the effectiveness of our system by outperforming the traditional cloud-terminal system in the real-world unstable network environments. The outline of the paper is as follows. We reviewed the related works in Sect. 2. The design of our proposed system is shown in Sect. 3. Section 4 presents our implementation of our testing environment. In Sect. 5, we discussed different experiment settings and presented the corresponding results. Section 6 concludes the paper and discusses aspects for future improvements based on this work.

2 Related Work

2.1 Decomposition

Software decomposition is a way to decouple a large system by dividing software into executable components, which is closely related to the program slicing problem [18]. Software decomposition can be classified as functional decomposition [11,13] or micro-services decomposition [7,8]. Functional decomposition decomposes the program into a series of functions (methods) and constructs a working process tree by analyzing the execution process. Micro-services decomposition adopts the theory from Object-Oriented Programming model. In this work, we adopt the programming model from micro-services, deconstructing the applications in our system into different low-coupling instances in software engineering terms.

2.2 Performance Evaluation

Performance evaluation of decomposed software components is a crucial technique for evaluating the execution efficiency of heterogeneous components. Methods can be summarized as model-based analysis [9,11] and statistic prediction [13,20]. The model-based analysis builds a model on the static structure of the program code and estimates the system performance by combing the structure with the real-time execution state of each component. Statistic prediction infers performance evaluation results by maintaining a timeline table of all components at execution. However, the current performance evaluation methods have shown these drawbacks: For the model-based analysis, it is hard to construct a model that can precisely predict the performance. For the statistic prediction, it heavily relies on two assumptions: 1. Components' performance within one application is stable. 2. The consumption of components execution is based on static time complexity analysis. Our previous results [5] have proved that these two assumptions are not correct. Therefore, to address problems in performance evaluation, we have implemented an effective measuring system in our testing framework for component-based software.

2.3 Migration

To support the migration of software between different end devices, current popular solutions include virtual machine migration [9–11] and mobile agent migration [4,5]. Virtual machine migration should build a complete application including the system image. This kind of solution is easy to deploy in the cloud but hard for mobile devices and edge nodes, which is weak in computation ability to support the full capacity of running components. On the contrary, mobile agent solution can migrate corresponding decomposed heterogeneous resources in the components level without consuming much computational resources. In this paper, we adopt Mobile Agent as our migration implementation to studying the component distribution between the cloud, edge, and terminal.

3 System Design

Based on the cloud-terminal architecture proposed in the previous work [5], we continue to design an architecture that incorporates edge computing as follows.

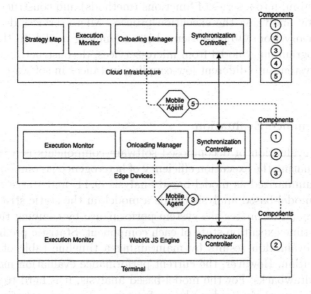

Fig. 1. System Architecture

Figure 1 describes the main components of the proposed system. Overall, the entire architecture can be divided into three layers. The first layer is the Cloud server layer with unlimited computing resources and storage of all the software components. The rectangle on the right represents the component pools of each layer, which is part of the storage that persists the executing code of the corresponding components. The dots with different numbers represent different components. The Edge server connects with the cloud and starts to request components. Then the cloud carries components to the edge's component pool and the terminal's component pool. If the edge detects the existence of a missing component on the terminal side, it can also pass this code to the terminal's component pool. In this process, the component performs inter-component communication through the synchronization controller to prevent unmatched computing status.

3.1 Constructing Components

Execution Monitor is a general system component that exists on each side of the system. It provides monitoring tools for measuring the resource usage of all the components executing on that end, including the runtime CPU cycle or memory usage of components. These data are collected to analyze the effect of the software components distributions. On the terminal-side, the performance recorder(e.g., an FPS recorder) is created to demonstrate the real-time user experience of the application. Since it best indicates the overall system's user satisfaction, we have been adapting this as the criteria for the system QoE in our experiments. *Synchronization Controller* in our system is designed to allow parameter updates in the application. Since in a remotely distributed system, the components are executed separately on different end devices, it is common to have the problem of data getting out of synchronization. Thus, in our system architecture, every component can only communicate with each other through the *Synchronization Controller*. Moreover, the parameter messages in our system need to be compiled and serialized into a JSON string before passing to its destination component as a message. *Onloading Manager* is only implemented in the cloud and the edge system, since all the application codes were originally stored on the cloud and the edge devices. When the system was first started, it would call the *Onloading Manager* to do the first preloading steps to have all the application components stored in the database on the cloud as well as on the edge ends. Since our system does not require any installation of application software on the terminal device, the *Onloading Manager* will load the software components to the terminal before the software was started or running as a backend service as the software is being used in the terminal end. *Onloading manager* achieves components code migration through *Mobile Agent*, a stringified message of the software components that can be passed between ends. The *Strategy Map* in the cloud is a configuration file to guide the distribution of each component. It needs to be loaded in the cloud server's initial stage since the cloud server will be loading the corresponding components to each end in the first preloading process. The dynamic partitioning of the components can be easily achieved by changing the configuration file of the *Strategy Map*. The *WebKit JS Engine* is a native Javascript Engine in browser kernel that is responsible for image rendering computation on the terminal-side.

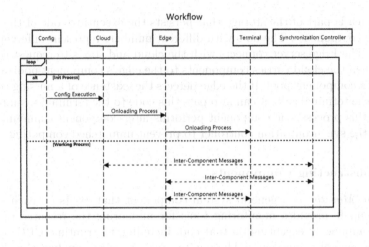

Fig. 2. Workflow

3.2 System Workflow

As shown in Fig. 2, after the application instance starts to run, the entire running
process is similar to an infinite state machine. In the initialization phase, the
cloud server will first launch the application instance in the cloud server to
start its execution. Meanwhile, the cloud server transmits the corresponding
component to the Edge server and the terminal server through the *Onloading
Manager* according to the pre-specified partition strategy in *Strategy Map*. After
the local calculation of each component finishes, each component in the cloud,
edge, or terminal will pass through the message passing in the *Synchronization
Controller*. In this way, we can ensure the proper execution of the software in
this distributed setting.

4 Test Bed Implementation

After evaluating the current system's needs and the experimental goals, we
decided to reconstruct the original cloud-terminal system to support offloading to the edge end. JavaScript is currently the only language supported by the
browser on all platforms. Using Node.js, a runtime package of JavaScript, we
wrote our server end program. This kind of combination allows us to implement the whole project within one programming language. In this setting, the
offloaded and migrated computing components can continue its execution in any
end without any modification to the code. On the terminal-side, we embedded
a Web Kit based browser to parse and execute our JavaScript code so that our
system can be well tested on any platform.

We have been using the JavaScript Socket.IO library to establish our connection between the end devices. The library enables us to build sockets allowing
real-time, bidirectional, and event-based communication between the terminal

and the servers. In our system implementation, each end device will establish two socket connections when the system is fired up: the cloud-terminal socket, the edge - terminal socket, and the edge-cloud socket. These sockets are responsible for component onloading, connection management, and components message passing. In our system, to enable the migration of the components, we have encapsulated the software components into separated JavaScript files wrapped in a function. The component will first be converted as an executable string in the system, and then use the built-in function *eval* in JavaScript to run corresponding tasks. The demonstration of our test bed is shown in Fig. 3.

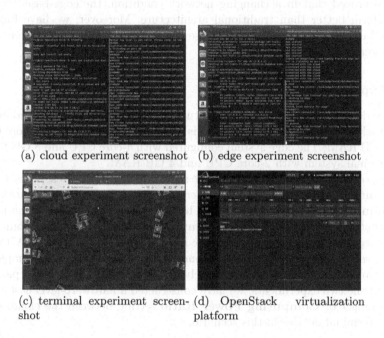

(a) cloud experiment screenshot   (b) edge experiment screenshot

(c) terminal experiment screen-   (d) OpenStack   virtualization
shot   platform

Fig. 3. System screenshots

5   Experiment

In this article, our experiment's main purpose is to prove the edge-based system's performance improvement compared with pure cloud-terminal architecture. We adopt the Tank Game Application illustrated in previous work [5] as the main testing benchmark.

The experiment part is constructed as follows. We first conducted single-end measurements, which means all components are all running on either the cloud, the edge, or the terminal device. The first experiment is to measure the effects of an increased computational burden. We have considered two aspects for the computational intensity: the iteration times of each component, and the component quantity in the system. After finishing the necessary measurements of computation ability, we have conducted *intra-cloud*, *intra-edge*, and *intra-terminal* experiments to determine the message passing ability of each ends. They have shown varying degrees of attenuation in FPS as the message length increase of message frequency increases. Combined with the conclusion above, we have tested the edge-based system with cloud-terminal architecture in different network configurations. Finally, we have compared different hardware computing power setups to discover the relationships between the system performance and the computational capacity of the edge device. Based on our measurements, we have proved that in a changing network condition, the edge-based system can perform better than traditional architecture. Moreover, we have found a marginal effect in the reward of increase the resources at edge to increase the system performance.

5.1   Experiment Settings

We have deployed our systems on Ubuntu 18.04 virtual machines created on the OpenStack platform. The cloud system is initially configured with 16-core CPU and 32 GB RAM; The edge system configured with 4-core CPU and 8 GB RAM; The terminal system with 2-core CPU and 4 GB RAM.

With our distributed system, we have mainly tested from two aspects: **Computational Cost** can be adjusted by *Iteration Times*, *component quantity*, and **Communication Cost** can be adjusted by *Communication Frequency* and *Message Length*. In our experiments for **Computational Cost**, no other communication except for those essential for basic inter-component invocation. With this setting, we can analyze the impact of components on computational complexity and communication complexity separately. After measuring the system parameters from the two experiments, we have constructed a hardware-level case study on the impact of **Computing Power Ratio** setting between the cloud, edge, and the terminal devices in this scenario.

Computational Cost: This experiment will consider factors that determine the computational intensity of the components. Considering our system design, we have chosen the *Iteration Number* and *Component Quantity* as our parameters for measurements of the system's performance on calculation tasks. *Iteration Number* is set to be the calculation task number of a single component, which is proportional to computation complexity. *Component quantity* increases the computational pressure by having more components running at the same time.

Iteration Number: To find out the execution characteristics of each end, we have designed our experiment with all the components running in the cloud, edge, and terminal end. We start with a small number of iteration (10 iterations) and then gradually increase the number of iterations to 16000. The result is shown in Fig. 4. From the above measurement cases, we can infer that the higher iteration number will decrease the FPS value in the cloud and edge ends. Cloud outperforms the other ends when the components have a smaller number of iterations. Moreover, cloud and edge execution of high iteration components are less stable comparing to the all-terminal execution, which shows almost no FPS change when the iterations number is getting higher. Generally, when the component has a relatively small size of iteration (e.g., 2000 iterations), it should be offloaded to cloud/edge for execution in order to give the best QoE.

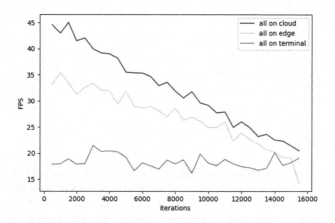

Fig. 4. FPS and iteration numbers at each end

Component Quantity: To further discover the execution characteristics of each end devices, we have experimented with different numbers of components in our system. For all experiments in this series, the iteration number was set to be 2000 iterations, since it was shown in the last experiment that a relatively small iteration is more suitable for cloud/edge execution. Component quantity indicates more components running in a single time period, which can be a measurement of concurrency. Similar to the iteration number experiment, we increase the number of components executing in the single-end system and record their average FPS. The results of Fig. 5 show that an increased number of components leads to a lower FPS value. Moreover, the cloud and edge server are more resistant to the increase in the component quantity than the terminal. Thus, we decided to offload components with more instances to the cloud or the edge side in our further experiments.

Based on the previous two experiments regarding the computing capacity of the end devices in our system, we have constructed a strategy map for further experiments. Components with a large number of instances and a smaller amount of iterations should be distributed to cloud/edge, and a small number of components with large iteration numbers should be left on the terminal.

To verify that our conclusion of different types of components execution on the single end system can show the characteristics of the distributed system, we have further conducted an experiment with components originally executing on the cloud migrating to different end systems and evaluate their performances. The result is shown in Fig. 6. The components with small size of iterations and

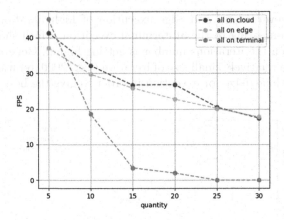

Fig. 5. FPS and quantity numbers at each end

large quantities are originally running at the cloud, yielding a high FPS. The FPS quickly drops when it was migrated to the terminal for execution after 500 time steps. It then further executed for 500 time steps, and then changed to edge for execution. The performance of edge is similar to the cloud in this current setting.

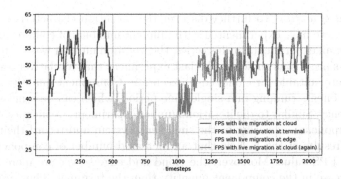

Fig. 6. Live migration of the components

Communication Cost: This experiment will modify the communication frequency and message length under fixed iteration number and component quantity. We aim at discovering the property of both intra-end and inter-end communication in the distributed system.

Intra-end Communications: The three graphs represent *intra-cloud*, *intra-edge* and *intra-terminal* in Fig. 7 respectively. For *intra-cloud* communications, it can be observed that for communication within the cloud, the FPS value will drop as the message length and communication frequency increases. Most of the FPS value is still acceptable for the gaming environment, which means that the cloud end is more suitable for executing components with long and frequent message passing within their instances. For *intra-edge* communications, we have observed a general decrease in overall FPS performance due to the limited computing resources compared to the cloud. However, the execution performance is still at a satisfying level. From the Fig. 7(2), it can be observed that *intra-edge* communication displays a similar property to the *intra-cloud* communication. Thus, the edge is also more suitable for components with larger communication frequencies and message length. For *intra-terminal* communications, it can be observed that transmitting data between components in the terminal greatly reduces the performance of the system. The FPS value dropped the most when the message length is long, and the communication is more frequent compared to the edge/cloud. This implies that the communication within the components at the terminal should be minimized as much as possible.

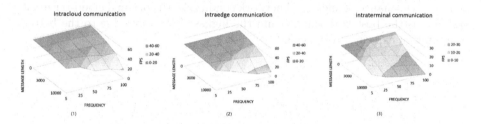

Fig. 7. FPS, message length and communication frequency

In this context, cloud/edge performs better when the message frequency and message length is growing. From the red area portion in the graph, which is the indicator of low-FPS, the *intra-terminal* communication is noticeable worse than the cloud/edge. It also proved that for the browser running on the terminal side, gathering all components is not a good choice, especially when the message communication is increasing. Additionally, we have also observed that comparing to the message length, and the communication frequency will have a more significant influence on the FPS value. For example, in the *intra-edge* setting, if we fix the communication frequency at 50 and only change the message length from 5000 to 10000, the FPS in this case drops from 42.13 to 38.99, whereas if we fix the message length at 5000 and change the communication frequency from 50 to 100, we can observe that the FPS drops to 30.67. This could be caused by the communication overhead of the system in our socket design.

Inter-end Communications: In this experiment, we have conducted experiments on cases where components communicate across heterogeneous environments. Good, fair and bad network were being discussed. Here are related settings: **Good network**, with network delay <30 ms and no package loss in the socket transmission. **Fair network**, with network delay at 60−80 ms and 1% package loss in the socket transmission. **bad network**, with network delay >120 ms and 5% package loss in the socket transmission.

Table 1. Strategy map for network experiments

	cloud_terminal configuration	MEC configuration
Cloud	'bot'*4, 'bot2'*4	'bot' *4
Edge	–	'bot1'*2 ,'bot2'*4
Terminal	'ui','exec','bot1'*2	'ui','exec'

The strategy map we adopted in the experiment is shown in Table 1. There are a total of 12 components running in our testing system, with iteration number 2000, message passing frequency 100, and message length 5000. From the results shown in Fig. 8, we can infer that the network condition has a direct impact on system performance. Under good network conditions, the original cloud-terminal system has the best performance and the highest FPS value due to the better computation capacity in the cloud.

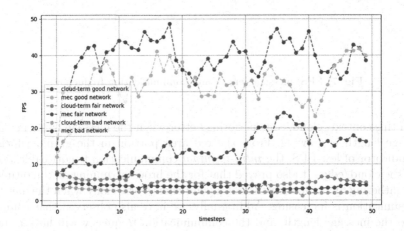

Fig. 8. FPS, message length and communication frequency between ends

When the network between the terminal and the cloud starts to become unstable, the edge-assisted systems perform much stable than the cloud-terminal system. The edge execution of communicative components can fully utilize the high-speed network between the edge and the other ends, resulting in a higher FPS. When the system is under a bad network condition with more package loss between the terminal and the cloud, the performance drops very quickly.

With this experiment, we have shown that with the assistance of edge, the system performance can be greatly improved when there is a fair but unstable network between the mobile device and the cloud. For general case components, our proposed edge-based system generally shows a better performance comparing to the current cloud-terminal system in an environment with network throttling. In component-wise communications, the message length and frequency can impact the overall performance. For our system, we found that the communication frequency will have a more significant influence on the overall FPS than the message length.

Computing Power Ratio: In this experiment, we altered the hardware settings (CPU core number, memory size) to achieve different computing power ratios between different end devices in the edge-assisted cloud computing system. In this experiment, our hypothesis is that increasing the number of resources that we deploy on the edge device will not be giving equal improvement to the system performance. We have adopted the strategy map discussed in the previous experiment settings. We adopted the fair network scenario (80 ms delay, 1% package loss in each transmission) to simulate common network situation in reality without extreme assumption on connection quality. We have tested our system using different sets of hardware. The computing power ratio settings of different experimental groups are shown in Table 2:

Table 2. Hardware configuration table

Cloud CPU cores	Edge CPU cores	Terminal CPU cores
16	8	2
16	4	2
8	4	2
8	2	1
4	2	1

We can infer from Fig. 9 that comparing with the original cloud-terminal configuration, adding edge to the system can have a great improvement on the system FPS. However, if we increase the CPU core number of the edge device from 4 cores to 8 cores, the improvement of FPS is not as significant as expected, which indicates a diminishing marginal utility in the hardware configurations for the edge end. This might be caused by the lack of edge offloading. The most computational intense components should be directed to the cloud for quicker computation. In this case, having a more powerful cloud will give a larger improvement to system performance.

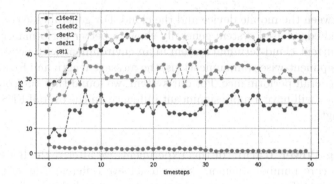

Fig. 9. FPS of different hardware settings w.r.t time steps

6   Conclusion

In this paper, we have proposed and implemented an edge-based dynamic partition testing platform. The proposed system can function as a monitoring tool for the execution of distributed applications in edge computing architecture. Our experiments on the system revealed the characteristics of program execution in different end devices. We have also shown that by properly offloading different types of program components, the edge-based system can improve the overall performance and is more resistant to the fluctuating network. For future works, we will focus on improving the current system from the following perspectives:

1. We shall realize automated component transferring between ends with the real-time data using cognitive algorithms.
2. We shall extend the current single edge system to include multiple edges in our architecture.

References

1. Balan, R.K., Satyanarayanan, M., Park, S.Y., Okoshi, T.: Tactics-based remote execution for mobile computing. In: Proceedings of the 1st International Conference on Mobile Systems, Applications and Services, pp. 273–286 (2003)
2. Bilal, K., Erbad, A.: Edge computing for interactive media and video streaming. In: 2017 Second International Conference on Fog and Mobile Edge Computing (FMEC), pp. 68–73. IEEE (2017)
3. Buyya, R., Yeo, C.S., Venugopal, S.: Market-oriented cloud computing: vision, hype, and reality for delivering it services as computing utilities. In: 2008 10th IEEE International Conference on High Performance Computing and Communications, pp. 5–13. IEEE (2008)
4. Cai, W., Chan, H.C., Wang, X., Leung, V.C.: Cognitive resource optimization for the decomposed cloud gaming platform. IEEE Trans. Circ. Syst. Video Technol. **25**(12), 2038–2051 (2015)
5. Cai, W., Chi, Y., Zhou, C., Zhu, C., Leung, V.C.: Ubcgaming: ubiquitous cloud gaming system. IEEE Syst. J. **12**(3), 2483–2494 (2018)

6. Chen, X., Jiao, L., Li, W., Fu, X.: Efficient multi-user computation offloading for mobile-edge cloud computing. IEEE/ACM Trans. Network. **24**(5), 2795–2808 (2015)
7. Chuang, S.N., Chan, A.T., Cao, J.: Dynamic service composition for wireless web access. In: Proceedings International Conference on Parallel Processing, pp. 429–436. IEEE (2002)
8. Chuang, S.N., Chan, A.T., Cao, J., Cheung, R.: Dynamic service reconfiguration for wireless web access. In: Proceedings of the 12th International Conference on World Wide Web, pp. 58–67 (2003)
9. Chun, B.G., Ihm, S., Maniatis, P., Naik, M., Patti, A.: Clonecloud: elastic execution between mobile device and cloud. In: Proceedings of the Sixth Conference on Computer Systems, pp. 301–314 (2011)
10. Chun, B.G., Maniatis, P.: Dynamically partitioning applications between weak devices and clouds. In: Proceedings of the 1st ACM Workshop on Mobile Cloud Computing & Services: Social Networks and Beyond, pp. 1–5 (2010)
11. Cuervo, E., et al.: Maui: making smartphones last longer with code offload. In: Proceedings of the 8th International Conference on Mobile Systems, Applications, and Services, pp. 49–62 (2010)
12. Fadlullah, Z.M., et al.: State-of-the-art deep learning: evolving machine intelligence toward tomorrow's intelligent network traffic control systems. IEEE Commun. Surv. Tutorials **19**(4), 2432–2455 (2017)
13. Kosta, S., Aucinas, A., Hui, P., Mortier, R., Zhang, X.: Thinkair: Dynamic resource allocation and parallel execution in the cloud for mobile code offloading. In: 2012 Proceedings IEEE Infocom, pp. 945–953. IEEE (2012)
14. Kumar, K., Lu, Y.H.: Cloud computing for mobile users: can offloading computation save energy? Computer **43**(4), 51–56 (2010)
15. Liu, J., Zhang, Q.: Adaptive task partitioning at local device or remote edge server for offloading in mec. arXiv preprint arXiv:2002.04858 (2020)
16. Rodrigues, T.G., Suto, K., Nishiyama, H., Kato, N.: A PSO model with VM migration and transmission power control for low service delay in the multiple cloudlets ECC scenario. In: 2017 IEEE International Conference on Communications (ICC), pp. 1–6. IEEE (2017)
17. Slagter, K., Hsu, C.H., Chung, Y.C.: An adaptive and memory efficient sampling mechanism for partitioning in mapreduce. Int. J. Parallel Program. **43**(3), 489–507 (2015)
18. Weiser, M.: Program slices: formal, psychological, and practical investigations of an automatic program abstraction method. Ph.D. thesis, University of Michigan (1979)
19. Yu, S., Wang, X., Langar, R.: Computation offloading for mobile edge computing: a deep learning approach. In: 2017 IEEE 28th Annual International Symposium on Personal, Indoor, and Mobile Radio Communications (PIMRC), pp. 1–6. IEEE (2017)
20. Zhang, X., Kunjithapatham, A., Jeong, S., Gibbs, S.: Towards an elastic application model for augmenting the computing capabilities of mobile devices with cloud computing. Mob. Netw. Appl. **16**(3), 270–284 (2011)

Deep Reinforcement Learning
for Intelligent Migration of Fog Services
in Smart Cities

Dapeng Lan[1(✉)], Amir Taherkordi[1], Frank Eliassen[1], Zhuang Chen[2(✉)],
and Lei Liu[3(✉)]

[1] Department of Informatics, University of Oslo, Oslo, Norway
{dapengl,amirhost,frank}@ifi.uio.no
[2] School of Computing Science, Simon Fraser University, Burnaby, Canada
zhuangchenuio@gmail.com
[3] State Key Laboratory of Integrated Services Networks, Xidian University,
Xi'an, China
tianjiaoliulei@163.com

Abstract. Fog computing plays a crucial role in future smart city applications, enabling services running along the cloud-to-thing continuum with low latency and high quality of service (QoS) requirements. However, the mobility of end users in smart city systems can result in considerable network performance and QoS degradation, hence interrupting fog services provisioning. *Service migration* is considered an effective solution to avoid service interruption and ensure service continuity, which can be carried out proactively or reactively. Existing work lacks intelligent and efficient migration solutions for fog services migrations. In this paper, we propose OCTOFOG, a fog services migration model and framework in the context of smart cities, featuring artificial intelligence for resource-efficient migration. We formulate proactive and reactive migration policies as an optimization problem, minimizing migration cost in terms of delay and energy consumption. We use a deep reinforcement learning (DRL) algorithm to solve the optimization problem to make fast migration decisions, using deep deterministic policy gradient (DDPG) based schemes. The evaluation results illustrate that OCTOFOG effectively reduces the total migration cost (*i.e.*, latency and energy).

Keywords: Fog computing · Smart city · Deep reinforcement learning · Service migration

1 Introduction

Internet of Things (IoT) applications have been evolving in various domains, of which smart city is a promising representative. Conventional smart city services are typically centralized, *i.e.*, data storage, data analytics and feedback

© Springer Nature Switzerland AG 2020
M. Qiu (Ed.): ICA3PP 2020, LNCS 12453, pp. 230–244, 2020.
https://doi.org/10.1007/978-3-030-60239-0_16

control solely rely on remote servers in the Cloud. However, traditional Cloud-centric approaches hardly fulfill the performance requirements of rapidly emerging smart city applications, such as virtual reality (VR)/augmented reality (AR), autonomous vehicles, and smart grid automation, all demanding high throughput, ultra low latency, and high availability and reliability [1–3]. In light of this, the fog computing and mobile edge computing (MEC) paradigms have been proposed to address the low latency and high throughput challenges by migrating cloud services to the proximity of local infrastructures in order to enable real-time applications and to elevate quality of service (QoS) [4–7].

However, the incorporation of fog platforms into smart city systems is constrained by several challenges. The mobility of end users and the limited coverage of fog servers can result in considerable network performance degradation, lower QoS support, and even interruption of fog services provision [8]. The latter concerns *ensuring service continuity*, which poses challenging issues such as task migration, content caching and context switching between fog nodes. One crucial issue is the spatiotemporal aspect of migration decisions, which can be *proactive* or *reactive*. The former refers decision making before the actual network-level handover based on, e.g., the stochastically predicted user mobility. In the latter, service migration decisions are made only after a network-level handover has happened.

The service migration problem has recently gained the attention of the research community [9]. Some models have been proposed to enable programming-level fog services migration, such as Foglet [10], multi-tier fog [11] and virtualization-based fog [12]. Moreover, Markov Decision Process (MDP) is a common technique to formulate the service migration problem and decision making [13]. Recently, the renaissance of research on artificial intelligence (AI), especially deep learning technologies can help solve the complex problem of migration optimization using MDP [14]. However, existing solutions still lack efficiency in resource usage and QoS support in service migration, such as low latency and resource optimization.

In this paper, we propose a comprehensive intelligent fog services migration model and framework, called OCTOFOG in the context of smart cities. The proposed approach is aimed to efficiently migrate services over fog platforms with respect to reactive and proactive decision models. OCTOFOG incorporates artificial intelligence to achieve resource-efficient and fast decision making in fog services migration. To this end, we formulate proactive and reactive migration policies as an optimization problem with the goal of minimizing migration cost in terms of delay and energy consumption. Our contributions are summarized as follows:

1. A fog service migration architecture, called OCTOFOG, in smart city systems considering both deep learning-based proactive and reactive service migrations and optimization of the total migration cost.
2. A mathematical service migration framework consisting of task and computation models, formulating the migration object function in terms of the cost of latency and energy consumption using DDPG schemes.

3. A thorough simulation evaluation of the OCTOFOG framework, proving the efficiency of the framework with the reduction of migration cost (*i.e.,* latency and energy).

The rest of the paper is organized as follows. In Sect. 2, we describe the related work. In Sect. 3, we explain the system model and also propose the intelligent migration and optimization model. We provide the solution to the optimization model in Sect. 4, while Sect. 5 explains the simulation setup and presents the numerical results. We conclude the paper and discuss the future work in Sect. 6.

2   Related Work

Generally, service migration refers to determining *when, where* and *how* to migrate fog services from a node to another fog node based on user mobility and the changes in demands or QoS [15]. These three questions raise several challenges such as frequency of migration, selection of candidate nodes for migration, stateful and stateless migrations, and so on. Considering when to migrate, there are two different aspects of migration that need to be carefully considered: the decision making mechanism for a migration, and the frequency of migration. The former can be performed reactively or proactively, as discussed in Sect. 1. The latter is related to the frequency of migrating fog services to the best possible fog node. It is not efficient to migrate fog services too frequently, as a migration may be resource demanding (*e.g.,* computation) and implicate a downtime period in which the migrated service is not available. Moreover, due to the delay-sensitive nature of mobile applications and instability of mobile connections, migration algorithms should be robust and efficient to make fast migration decisions. Not only the individual task requirements, such as latency and energy consumption, need to be satisfied, but also the global cost and QoS need to be considered [16]. The focus of this paper is on when to migrate a fog service.

There are a number of studies focusing on fog services migration, in terms of frameworks [10–12,17,18], techniques [13,19–21], and recent AI-based approaches [14,22,23]. In the rest of this section, we discuss each category in detail.

The work [10] presents Foglet, a programming model for geo-distributed computational fog nodes and end devices. The authors discuss the APIs and algorithms to handle the initial deployment and the possible migration of fog services. In [11], the authors propose a multi-tier fog architecture to facilitate the migration of multimedia services from the Cloud to fog nodes in order to prevent bandwidth outstripping and quality of experience (QoE) degradation from huge uploading/downloading multimedia traffic. A four-tier fog architecture is presented and exemplified with video service migration to highlight the benefits of service migration. In [12], the authors propose a virtualization-based fog architecture to support service migration among fog networks. The authors in [17] present a framework that supports a new seamless handover timing scheme among different computation fog nodes. It features a job pre-migration mechanism when the handover is expected to happen, which achieves a substantial latency reduction. In [18], the authors propose a multi-layer fog computing architecture with intelligence for smart cities to identify anomalous and hazardous events and to offer optimal responses and controls.

Some works presented migration techniques in fog computing environments. The work [24] analyzed the placement of the services using an ILP-based optimization approach and took into consideration QoS requirements, such as latency. Other works like [12] proposed VM migration in generic Fog architectures supporting virtual machine migration and discussed both the benefits and challenges associated with such a migration. Moreover, Markov Decision Process (MDP) is commonly used in fog computing to formulate the service migration problem and migration decision making [13,19], and [20].

The renaissance of research on artificial intelligence (AI), especially deep learning technologies, has paved the way for introducing intelligence to various disciplines, such as service migration in fog computing. Deep learning technologies have been considered an effective way to solve the complex problem of migration optimization. In [14], Tang et al. model the task migration strategy in a fog computing network as multiple dimensional MDP spaces, and utilize efficient deep reinforcement learning algorithms to reduce the space in order to achieve fast decision-making. A deep reinforcement learning algorithm is utilized to obtain the resource allocation policy in applications. Further, in [23], Dai et al. present a framework which leverages both deep reinforcement learning and blockchain to enable secure and intelligent resource management and orchestration in 5G wireless networks.

Discussion. As mentioned before, this paper is mainly focused on migration decision making with special consideration to the resource efficiency and latency. We achieve this by studying reactive and proactive migration scenarios together in order to optimize the migration process with respect to the above parameters. However, existing works in the categories of frameworks and techniques mostly lack an intelligent way to efficiently balance reactive and proactive service migration decision making. In addition, existing AI-based approaches are still in their infancy stage and this paper is aimed to practically explore AI techniques in intelligent fog services migration with a special focus on performance evaluation.

3   System Model

Most of existing fog computing frameworks include only centralized service management or orchestration [10,12]. In this case, the centralized manager needs to have a clear view of all the distributed components. However, with the explosion of IoT devices, it is very hard to manage resources in a centralized way. Some applications, such as VR/AR, need to transfer large amount of continuous data and some applications may need demanding QoS, such as low latency. In this case, the centralized system will face considerable pressure. Besides, the centralized way also faces the issue of single point of failure (SPOF). To tackle this, we propose a distributed fog intelligent framework, named OCTOFOG. The architecture consists of three layers: the cloud layer, the fog layer and the mobile user layer. Inspired by the octopus, an animal that 40% of its intelligence is in their main brain while 60% of its intelligence is distributed in their legs. They have very high intelligence combining the main brain and edge brain together. We apply the same philosophy to design the architecture of the OCTOFOG framework, as shown in Fig. 1. The cloud layer will have the main intelligence for resource management, service orchestration, data analytics and decision-making. The fog layer will assist the main intelligence and make local decisions. The main brain and the distributed edge brains will work collaboratively, which is referred to as distributed fog intelligence. The mobile user layer produces tasks, and parts of the tasks are offloaded to the corresponding fog nodes via the access networks.

We consider a smart healthcare system that monitors elder people's health status (*e.g.*, heart rate) and detects emergency situation (*e.g.*, falling down). As shown in Fig. 1, the users of the smart healthcare system may move around in the city. As a result, their fog domains change with their locations. Assume N users inside the same fog domain have access to the fog node. This set of users is denoted as $\mathcal{N} = \{1, 2, \ldots, N\}$. We assume each user has a computation-intensive task to perform, such as health condition prediction using a machine learning model. In this case, the tasks will be offloaded to the fog node in proximity.

The services provided by the system should not be interrupted if the users move. The tasks running on the current fog domain need to follow the users wherever they are. Figure 1 shows two ways to achieve this, i.e. proactive migration and reactive migration. For the reactive migration process, the task running on the current fog node will be temporarily interrupted and the user needs to initiate the task again when entering the next fog domain. For proactive migration, the user can inform the current fog node about his/her future location before leaving the current fog domain or the intelligent fog system predicts mobile user's future location, thus his/her task and the related context information can be moved to the fog node at the next location beforehand. This will shorten the latency between the mobile user devices and new fog nodes, and increase the quality of services.

3.1 Task Model

In the OCTOFOG framework, the task will be offloaded to the nearest fog node due to lack of computational resources. The computation-intensive task of a user n can be represented as $R_n \triangleq (B_n, D_n, \tau_n)$. Here B_n denotes the size of input data needed for computing R_n, including program codes, parameters and other context information. D_n denotes the number of CPU cycles required to compute task R_n. D_n is the same regardless of the hardware hosting a task. τ_n denotes the deadline for finishing task R_n. It is a crucial quality of service (QoS) requirement for smart healthcare systems. All the three parameters are task-specific and they can be estimated or obtained through task profiles.

We assume that a task cannot be divided into sub-tasks, which means that a task as a whole can either be migrated proactively or passively. We denote $\alpha_n \in$

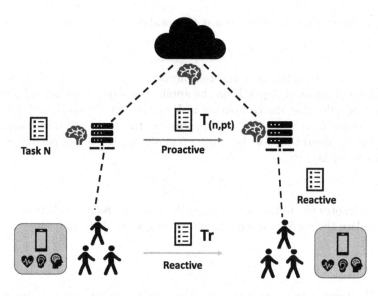

Fig. 1. The fog service migration scenario

$\{0,1\}$ the computation migration decision of task R_n and define the migration decision vector as $\mathcal{A} = [\alpha_1, \alpha_2, \ldots, \alpha_N]$. Here $\alpha_n = 0$ means reactive migration while $\alpha_n = 1$ means proactive migration.

3.2 Computation Model

There are two migration processes: the proactive migration process and the reactive migration process, as shown in Fig. 1.

Reactive Migration Model. Depending on the end-user application, some tasks may be migrated passively when there are not harsh deadlines. The reactive task migration process will be divided into two steps. First, the user will physically move from the current fog domain to the destination fog domain. Then, the user will offload the computation tasks to the new fog node. Here, we define T_r as the delay when the user moves from the old fog domain to the new fog domain. During this period, the computation-intensive tasks cannot run on the fog nodes. In this case, the user may offload the tasks to the Cloud or stop running the computation-intensive tasks. In this paper, we do not consider the former situation, but only consider the latter case. After the user arrives the new fog domain, the user can offload the computation tasks to the new fog node, which can be divided into three steps. Firstly, the task R_n needs to upload sufficient input data (*i.e.*, program codes, parameters, and context information) to the new fog node through the wireless access network. Then, the fog node allocates part of computational resources to execute the computing task. Finally, the new fog node returns the execution results to user n.

According to the procedures above, the total delay for reactive task migration is:

$$T_R = T_r + T_{n,r}^o \tag{1}$$

Here T_r denotes the delay when the user physically moves from an old fog domain to a new fog domain. It depends on the applications and scenarios. $T_{n,r}^o$ denotes the delay of offloading the computing task to the new fog node, which consists of three steps: uploading the task, processing the task and downloading the result. The required time for the first step of offloading the computing task (the transmission delay) is:

$$T_{n,rt}^o = \frac{B_n}{r_n} \tag{2}$$

where r_n denotes the uplink data rate in the wireless channel. Therefore, we can also calculate the corresponding energy consumption of the first step:

$$E_{n,rt}^o = P_n T_{n,rt}^o = \frac{P_n B_n}{r_n} \tag{3}$$

where P_n denotes the power consumption of the transmission state. For the second step of offloading the computing task, the latency is the processing delay in the fog node, which can be given as:

$$T_{n,rc}^o = \frac{D_n}{f_n} \tag{4}$$

where f_n stands for the allocated computational resource (such as CPU cycles per second) by the fog node to complete task R_n. In this step, we define the user end devices stay idle so that the power consumption of the idle state as P_n^i , the corresponding energy consumption is:

$$E_{n,rp}^o = P_n^i T_{n,rc}^o = \frac{P_n^i D_n}{f_n} \tag{5}$$

For the last step of offloading the computing task, the fog node will send back the result to the user n with task R_n. Therefore, the required time for downloading the processed result is:

$$T_{n,rb}^o = \frac{B_b}{r_b} \tag{6}$$

where B_b denotes the data size of the processed result and r_b defines the download data rate from the fog node to the user with task R_n. However, according to [25], the download data rate is very high in general and the data size of the results is quite small compared to the processed data, therefore we ignore the delay and energy consumption of this step in the rest of this paper. Therefore, according to (3), (4) and (5), the total delay and energy consumption for reactive task migration are:

$$T_{n,r}^o = \frac{B_n}{r_n} + \frac{D_n}{f_n} \tag{7}$$

and

$$E_{n,r}^o = \frac{P_n B_n}{r_n} + \frac{P_n^i D_n}{f_n} \tag{8}$$

Combining the latency (7) and energy (8) cost, the total cost of reactive task migration can be given as

$$C_{n,r}^o = I_n^t \left(T_r + T_{n,r}^o\right) + I_n^e E_{n,r}^o \tag{9}$$

where I_n^t and I_n^e denote the weight of time and energy cost of task R_n. The weights should satisfy $0 \le I_n^t \le 1, 0 \le I_n^e \le 1$ and $I_n^t + I_n^e = 1$. The weight may be different due to different kinds of tasks. In our paper, for the sake of simplicity, we assume the two weight parameters of task R_n remain the same during the migration process.

Proactive Migration Model. When the user moves out from the current fog domain, the system can forecast the destination fog domain or the system will receive notifications from the user about the future moving path. During this period, there are still some tasks running on the old fog node. With the proactive task migration, the old fog domain will automatically forward related running tasks to the new fog domain. The proactive migration will be divided into three steps: data transmission, processing the data, and sending back the results. First, the running tasks will offload the tasks from the old fog node to the new fog node through the wireless access network. We define the delay during first step transmission as $T_{n,pt}^o$. Then, the new fog node allocates part of the computational resource to execute the computing tasks. Finally, the new fog node returns the execution results to the user task R_n. According to the procedures above, the required time for the first step of offloading task (the transmission delay) is:

$$T_{n,pt}^o = \frac{B_n}{r_{np}} \tag{10}$$

where r_{np} denotes the transmission data rate between fog nodes in the wireless channel. Therefore, we can also calculate the corresponding energy consumption for the first step:

$$E_{n,pt}^o = P_n T_{n,pt}^o = \frac{P_n B_n}{r_{np}} \tag{11}$$

For the second step of offloading the task, the latency is the processing delay in the new fog node, which can be given as:

$$T_{n,pc}^o = \frac{D_n}{f_n} \tag{12}$$

During the first and the second steps mentioned above, we define the user end devices stay idle so that the corresponding energy consumption is:

$$E_{n,pc}^o = P_n^i (T_{n,pt}^o + T_{n,pc}^o) = P_n^i (\frac{B_n}{r_{np}} + \frac{D_n}{f_n}) \tag{13}$$

For the last step of offloading the task, the new fog node will send back the result to the user once the user has access to the new fog area. The result will be the same as Eq. 6. Due to the high speed of download rate and the small size of the result compared to the processed data, we ignore the delay and the energy consumption of this step.

According to (10), (11), (12) and (13), the delay and energy consumption cost of proactive migration are:

$$T_{n,p}^o = \frac{B_n}{r_{np}} + \frac{D_n}{f_n} \tag{14}$$

and

$$E_{n,p}^o = \frac{P_n B_n}{r_{np}} + P_n^i \left(\frac{B_n}{r_{np}} + \frac{D_n}{f_n} \right) \tag{15}$$

Combining the time (14) and energy (15) cost, the total cost of proactive migration can be given as

$$C_{n,p}^o = I_n^t T_{n,p}^o + I_n^e E_{n,p}^o \tag{16}$$

Therefore, we can get the sum cost of all users in the intelligent fog migration system that is expressed as:

$$C_{all} = \sum_{n=1}^{N} (1 - \alpha_n) C_{n,r}^o + \alpha_n C_{n,p}^o \tag{17}$$

where $\alpha_n \in \{0, 1\}$ denotes the migration decision of the intelligent fog system. $\alpha_n = 0$ means the tasks use reactive migration method, otherwise $\alpha_n = 1$ means proactive task migration.

3.3   Problem Formulation

According to the section above, we can formulate the migration process in OCTO-FOG as an optimization problem. The objective of this paper is to minimize the sum cost combining the reactive migration and proactive migration of the tasks, including the cost of transmission and execution delay, and energy consumption. With the constrain of maximum tolerable latency and consumption capacity, we can formulate the optimization problem as follows:

$$\min_{\mathcal{A}, f} \sum_{n=1}^{N} (1 - \alpha_n) C_{n,r}^o + \alpha_n C_{n,p}^o$$

$$\begin{aligned}
\text{s.t.} \quad & C1 : \alpha_n \in \{0, 1\}, \forall n \in \mathcal{N} \\
& C2 : (1 - \alpha_n) T_R + \alpha_n T_{n,p}^o \le \tau_n, \forall n \in \mathcal{N} \\
& C3 : \sum_{n=1}^{N} f_n \le F, \forall n \in \mathcal{N}
\end{aligned} \tag{18}$$

In (18), $\mathcal{A} = [\alpha_1, \alpha_2, \dots, \alpha_N]$ is the migration decision vector, while $f = [f_1, f_2, \dots, f_N]$ denotes computational resources allocation. The objective of this optimization function is to minimize the sum cost of the entire system. $C1$ means that the user tasks can choose either reactive migration or proactive migration process. $C2$ indicates that the time cost of the tasks should not exceed the maximum tolerated delay with either reactive migration or proactive migration. $C3$ guarantees that the sum of the required computation resources to be allocated for all user tasks can not exceed the entire computational resource F of the new fog node.

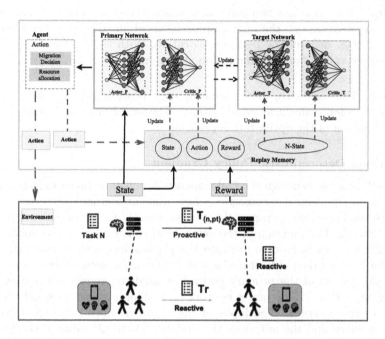

Fig. 2. Deep reinforcement learning empowered service migration

Therefore, this optimization problem can be solved by finding optimal values of migration decision vector $\mathcal{A}$ and computational resource allocation $\boldsymbol{f}$. However, as $\mathcal{A}$ is a binary variable, the feasible set and objective function of this optimization problem (18) is not convex. Besides, if the number of user tasks increases fast, the size of the problem (18) will increase dramatically. Therefore, it is NP hard to solve this non-convex problem extended from the Knapsack problem [26]. Therefore, we propose a new deep reinforcement learning-based method to find the optimal $\mathcal{A}$ and $\boldsymbol{f}$, instead of using conventional optimization methods.

4 Problem Solution

In this section, we introduce DRL-based techniques to solve the optimization problem (18). First, we define three elements in DRL, namely, specific state, action and reward. Then, we utilize the DDPQ method to estimate the action-value function.

4.1 Elements of DRL

State. The state in DRL is a space to reflect the environment. The state consists of two parts $S = (sc, ac)$ where sc denotes the sum cost of the entire system and ac denotes the computation capability of fog nodes: $ac = F - \sum_{n=0}^{N} f_n$.

Action. The action has two components in our system. The first component is $\mathcal{A} = [\alpha_1, \alpha_2, \dots, \alpha_N]$, the migration decisions of all the user tasks while $\boldsymbol{f} = [f_1, f_2, \dots, f_N]$, denotes the resource allocation. With the combination of these two possible values, we have the action vector: $[\alpha_1, \alpha_2, \dots, \alpha_N, f_1, f_2, \dots, f_N]$.

Reward. For each step, after executing each possible action a for a certain state s, the agent will get an award $R(s, a)$. The award function is related to the object function of the optimization problem.

4.2 Deep Deterministic Policy Gradient

The DRL process to design the tasks migration policy is shown in Fig. 2, which is based on the deep deterministic policy gradient method [27]. The primary network combines two deep neural networks, namely the actor network and the critic network. The function of the actor network is exploring the policy while the critic network is used to evaluate the performance and provides the critic value for learning the gradient of the policy. The target network that is regarded as an old version of the primary network is used to produce the target value for training Critic_P. The target network consists of a target actor network and a target critic network. The input of the target network is the N-state from replay memory and the output is the update of Critic_P value in the primary network. The replay memory includes current state, the selected action, reward, and the next state. These randomly sampled values can be used to train the primary network and the target network, which can reduce the effects of data correlation.

5 Simulation and Numerical Results

5.1 Simulation Setup

We evaluate the performance of the proposed DRL-empowered OCTOFOG service migration scheme through extensive simulations. We consider three fog nodes and 20 mobile users randomly distributed in a $1\,km \times 1\,km$ area in the environment. The computation capability of the fog nodes is randomly chosen from [4, 6, 8] GHz/sec. The transmission power and idle power of mobile user devices are set to $P_n = 500$ mW and $P_n^i = 100$ mW [28], respectively. We also assume that the data size of task offloading B_n (in kbits) complies with uniform distribution between (300,500). The number of CPU cycles D_n (in Megacycles) is with uniform distribution between (900, 1100). In addition, for convenience, the decision weight of each mobile user task is $I_n^t = I_n^e = 0.5$. The simulation is implemented using Python and TensorFlow.

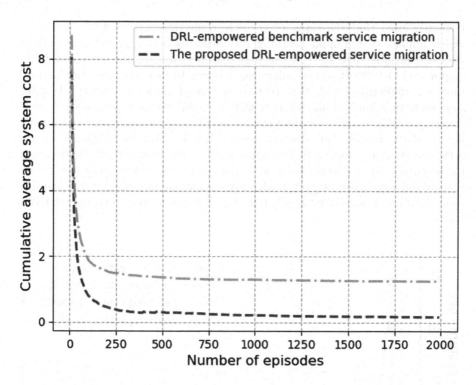

Fig. 3. Comparison of system cost under different schemes.

5.2 Numerical Results

The simulation results are shown in Figs. 3 and 4. OCTOFOG simultaneously performs comparison between proactive and reactive migrations in terms of latency and energy consumption.

We first compare different service migration schemes in terms of cumulative average system cost in Fig. 3. It shows clearly that the cumulative average system cost of OCTOFOG is lower than the benchmark service migration scheme. The reason is that OCTOFOG can select the most appropriate migration method to meet individual latency requirements and optimize the global cost of the fog nodes and mobile users. In addition, we observed that at the beginning of the learning process, the cost of each system for different scenarios was very high. As the number of episodes increases, the system cost reaches a relatively stable value after running 500 episodes iterations. In comparison, the benchmark scheme makes decisions without considering resource allocation, which results in higher transmission energy cost.

Figure 4 shows the convergence performance of OCTOFOG under different learning rates. First, the cumulative average system costs achieve convergence in all learning rates and larger learning rate tends to have faster convergence. Second, when the learning rate is 10^{-6}, the cumulative average system cost is obviously lower than the cases when the learning rate is 10^{-3}, 10^{-4}, and 10^{-5}, which implies that a small learning rate achieves better performance. Thus, we can conclude that 10^{-6} is the best learning rate for the proposed DRL-empowered OCTOFOG service migration scheme. In fact, choosing the learning rate is a challenging task that requires a careful trade-off, as large learning rate leads to a suboptimal solution while a small learning rate leads to slow convergence. Generally, we can set the learning rate according to the convergence degree of the model. Actually, the appropriate learning rate depends on the architecture of the model to be optimized and the state of the environment in the current optimization process. Furthermore, we observe that the system costs of different schemes are considerably high at the beginning of the learning process. With an increasing number of episodes, system costs become relatively stable.

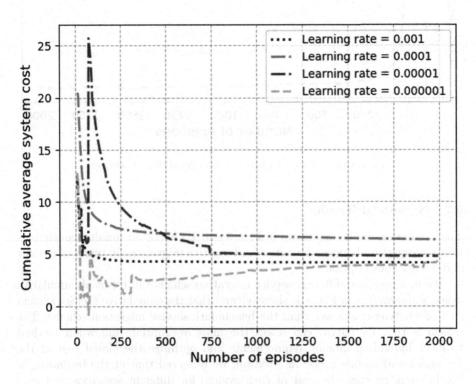

Fig. 4. Convergence performance under different learning rates.

6   Conclusion and Future Work

In this paper, we proposed OCTOFOG, an intelligent fog service migration framework for smart city applications. The proposed framework is empowered by artificial intelligence that enables resource-efficient realization of reactive and proactive service migration scenarios. To this end, we proposed a comprehensive framework to formulate reactive and proactive migration policies as an optimization problem, encompassing delay and energy cost for migration. To evaluate the proposed model, we simulated the optimization function using a DRL algorithm together with a deep deterministic policy gradient (DDPG) based scheme, which achieves fast convergence and better performance than the baseline function. We further investigated the learning parameters for our model. As future work, we will further investigate the modeling aspects of the proposed migration model to take into consideration other QoS parameters such as bandwidth and storage.

References

1. Yue, J., Xiao, M., Pang, Z.: Distributed fog computing based on batched sparse codes for industrial control. IEEE Trans. Ind. Inform. **14**(10), 4683–4691 (2018)
2. Lan, D., et al.: Latency analysis of wireless networks for proximity services in smart home and building automation: the case of thread. IEEE Access **7**, 4856–4867 (2019)
3. Chen, C., et al.: Delay-optimized v2v-based computation offloading in urban vehicular edge computing and networks. IEEE Access **8**, 18863–18873 (2020)
4. Dapeng, L., Taherkordi, A., Eliassen, F., Horn, G.: A survey on fog programming: concepts, state-of-the-art, and research challenges. In: DFSD 2019 (2019)
5. Feng, J., Yu, F.R., Pei, Q., Chu, X., Du, J., Zhu, L.: Cooperative computation offloading and resource allocation for blockchain-enabled mobile edge computing: a deep reinforcement learning approach. IEEE Internet Things J. **7**(7), 6214–6228 (2019). https://doi.org/10.1109/JIOT.2019.2961707
6. Qiu, H., Zheng, Q., Zhang, T., Qiu, M., Memmi, G., Lu, J.: Towards secure and efficient deep learning inference in dependable IoT systems. IEEE Internet Things J., 1 (2020)
7. Qiu, M., Sha, E.H.M.: Cost minimization while satisfying hard/soft timing constraints for heterogeneous embedded systems. ACM Trans. Des. Autom. Electron. Syst. **14**(2), 1–30 (2009)
8. Liu, L., Chen, C., Qiu, T., Zhang, M., Li, S., Zhou, B.: A data dissemination scheme based on clustering and probabilistic broadcasting in VANETs. Veh. Commun. **13**, 78–88 (2018)
9. Liu, L., Chen, C., Pei, Q., Maharjan, S., Zhang, Y.: Vehicular edge computing and networking: a survey. Mob. Netw. Appl. 1–24 (2020). https://doi.org/10.1007/s11036-020-01624-1
10. Saurez, E., et al.: Incremental deployment and migration of geo-distributed situation awareness applications in the Fog. In: DEBS 2016 (2016)
11. Rosario, D., et al.: Service migration from cloud to multi-tier fog nodes for multimedia dissemination with QoE support. Sensors **18**(2), 329 (2018)
12. Bittencourt, L.F., Lopes, M.M., Petri, I., Rana, O.F.: Towards virtual machine migration in fog computing. In: 2015 10th International Conference on P2P, Parallel, Grid, Cloud and Internet Computing (3PGCIC), pp. 1–8, November 2015
13. Zhang, W., et al.: Segue: quality of service aware edge cloud service migration. In: 2016 IEEE International Conference on Cloud Computing Technology and Science (CloudCom) (2016)

14. Tang, Z., Zhou, X., Zhang, F., Jia, W., Zhao, W.: Migration modeling and learning algorithms for containers in fog computing. IEEE Trans. Serv. Comput. **12**(5), 712–725 (2018)

15. Puliafito, C., et al.: Fog computing for the internet of mobile things: issues and challenges. In: 2017 IEEE International Conference on Smart Computing (SMART-COMP) (2017)

16. Bonomi, F., Milito, R., Zhu, J., Addepalli, S.: Fog computing and its role in the internet of things. In: Proceedings of the First Edition of the MCC Workshop on Mobile Cloud Computing, MCC 2012, pp. 13–16, ACM, New York (2012)

17. Ba, W., et al.: Follow me fog: toward seamless handover timing schemes in a fog computing environment. IEEE Commun. Mag. **55**(11), 72–78 (2017)

18. Tang, B., et al.: Incorporating intelligence in fog computing for big data analysis in smart cities. IEEE Trans. Ind. Inf. **13**(5), 2140–2150 (2017)

19. Ksentini, A., Taleb, T., Chen., M.: A markov decision process-based service migration procedure for follow me cloud. In: 2014 IEEE International Conference on Communications (ICC), pp. 1350–1354, June 2014

20. Yao, H., Bai, C., Zeng, D., Liang, Q., Fan, Y.: Migrate or not? exploring virtual machine migration in roadside cloudlet-based vehicular cloud. Concurrency Comput. Pract. Experience **27**(18), 5780–5792 (2015)

21. Sun, X., Ansari, N.: EdgeIoT: mobile edge computing for the internet of things. IEEE Commun. Mag. **54**(12), 22–29 (2016)

22. He, Y., Yu, F.R., Zhao, N., Leung, V.C.M., Yin, H.: Software-defined networks with mobile edge computing and caching for smart cities: a big data deep reinforcement learning approach. IEEE Commun. Mag. **55**(12), 31–37 (2017)

23. Dai, Y., Xu, D., Maharjan, S., Chen, Z., He, Q., Zhang, Y.: Blockchain and deep reinforcement learning empowered intelligent 5G beyond. IEEE Network **33**(3), 10–17 (2019)

24. Skarlat, O., Nardelli, M., Schulte, S., Dustdar, S.: Towards Qos-aware fog service placement. In: 2017 IEEE 1st International Conference on Fog and Edge Computing (ICFEC), pp. 89–96, May 2017

25. Chen, X., Jiao, L., Li, W., Fu, X.: Efficient multi-user computation offloading for mobile-edge cloud computing. IEEE/ACM Trans. Network. **24**(5), 2795–2808 (2016)

26. Jaszkiewicz, A.: On the performance of multiple-objective genetic local search on the 0/1 knapsack problem - a comparative experiment. IEEE Trans. Evol. Comput. **6**(4), 402–412 (2002)

27. Lillicrap, T.P.: Continuous control with deep reinforcement learning. arXiv preprint arXiv:1509.02971 (2015)

28. Cao, Y., Jiang, T., Wang, C.: Optimal radio resource allocation for mobile task offloading in cellular networks. IEEE Network **28**(5), 68–73 (2014)

A Novel Clustering-Based Filter Pruning Method for Efficient Deep Neural Networks

Xiaohui Wei, Xiaoxian Shen, Changbao Zhou, and Hengshan Yue[✉]

College of Computer Science and Technology, Jilin University,
Changchun 130012, China
weixh@jlu.edu.cn, {xxshen18,zhoucb18,yuehs18}@mails.jlu.edu.cn

Abstract. Deep neural networks have achieved great success in various applications, accompanied by a significant increase in the computational operations and storage costs. It is difficult to deploy this model on embedded systems. Therefore, model compress is a popular solution to reduce the above overheads. In this paper, a new filter pruning method based on the clustering algorithm is proposed to compress network models. First, we perform clustering with features of filters and select one for each category as a representative. Next, we rank all filters according to their impacts on the result to select configurable amounts of top features. Finally, we prune the redundant connections that are not selected. We empirically demonstrate the effectiveness of our approach with several network models, including VGG and ResNet. Experimental results show that on CIFAR-10, our method reduces inference costs for VGG-16 by up to 44% and ResNet-32 by up to 50%, while the accuracy can regain close to the original level.

Keywords: Clustering-based · Filter pruning · Deep neural networks

1 Introduction

Deep Convolution Neural Networks (CNNs) have excellent performance in fields such as computer vision, speech recognition, and natural language processing. Over the past few years, to achieve higher task accuracy, CNN models are usually designed to be deeper and wider with myriads of parameters to extract high-level features of the input data, which would bring significant computational and memory demands during inference. For example, a simple 16-layer VGGNet [1] model with more than 138 million parameters takes up more than

This work is supported by the National Natural Science Foundation of China (NSFC) (Grants No. 61772228, No. U19A2061), National key research and development program of China under Grants No. 2017YFC1502306 and Graduate Innovation Fund of Jilin University under Grants No. 101832018C026 No. 101832018C134.

© Springer Nature Switzerland AG 2020
M. Qiu (Ed.): ICA3PP 2020, LNCS 12453, pp. 245–258, 2020.
https://doi.org/10.1007/978-3-030-60239-0_17

500 MB of storage space and requires 30 billion floating-point operations. Therefore, for the deeper network, e.g., the champion network of ILSVRC 2015, 152-layer ResNet [2], the energy and storage overhead would be tremendous. For resource-constrained devices like cellphones and autonomous robotics, such a high demand for resources is prohibitive. Hence, it is impractical to deploy the large model into the resource-limited devices directly.

To deal with above problem, many model compression and acceleration methods have been proposed to reduce the storage and computation costs, such as efficient model design [3,4], knowledge distilling [5,6], matrix decomposition [7,8], network quantization [9,10], and network pruning [11–13]. By optimizing the convolution operations or network architectures, an efficient model design method accelerates the execution of the inference task. The knowledge distillation method transfers the useful information from the complex network to the simple network for memory and energy saving. The matrix decomposition method decomposes the complex matrix into several small matrixes to eliminate the informative parameters. Network quantization method is proposed to compress the model by decreasing the presentation precision of parameters, which requires special software or hardware for implementations. The network pruning approach removes redundant or non-informative elements in the network to decrease the energy and storage costs.

Compared with other compression or acceleration methods, the network pruning approach can easily and efficiently provide reasonable compression rates while minimally influencing the model accuracy. Thus, network pruning is one of the most popular methods to reduce network complexity [13,14]. Recent researches on pruning can be roughly divided into two categories, i.e., weight pruning [11] and filter pruning [12–15]. Weight pruning directly removes the unimportant connections in CNNs, generating a very non-structured sparse network model, which requires special software or hardware implementations for the resulting models. To avoid the limitations of non-structured pruning mentioned above, our work focuses on the filter pruning. This method directly discards the entire filters without introducing the sparsity.

In this paper, we propose a new filter pruning method combining high-correlated features sourcing from different reasonable operations. Compared with pioneer work, we make several contributions as following:

1. We propose a filter selection strategy based on the clustering algorithm to retain high-informative connections and eliminate redundancy. Owing to without introducing the sparsity of the CNNs model, it does not require additional specialized hardware and software support.
2. We reserve configurable amounts of high-impacted features through ranking all features of networks besides selecting representative features from the clustering results. In this way, the accuracy of the pruning network is maximally maintained.
3. Experimental results show that our work has better performance than existing methods. The statistics indicate that the parameters are reduced by about one half, and FLOP of VGG-16 and ResNet-32 are reduced by 44.2% and 50.7% respectively, but the accuracy only drops by less than 0.5%.

2   Related Work

Deep neural network weights have significant redundancy, and a wide variety of techniques have been proposed to exploit this attribute. [7] uses Singular Value Decomposition (SVD) in the nonlinear units to reduce the number of parameters by decomposing the complex matrix into several small matrixes. Additionally, quantization [9,10] can be used to reduce the model size and lower the computation overheads by reducing the number of bits of weight in many ways.

The early work [3,4] proposes new structure of CNN models named Squeeze-Net and MobileNet, respectively, to reduce the computations and parameters of the model. The other approach [5,6] is to reduce the convolutional overheads by transferring useful information from a model known as the teacher (usually of high capacity and performance) to another compact model known as the student.

Network pruning has been used both to reduce network complexity and to reduce over-fitting. Recent work [11] pays more attention to removing all connections whose weights are lower than the threshold. However, this method introduces the sparsity into the networks. To address this question, an acceleration method [12] is presented for CNNs to prune filters that are identified as having a small effect on the output accuracy.

3   Pruning Methods

In this section, we give a comprehensive description of our filter-level pruning approach. First, the overall framework of CNNs is presented. Then, we analyze the relevance and importance of filters by clustering and sorting. Finally, we select filters from the results of analysis to complete the model compress.

3.1   Framework

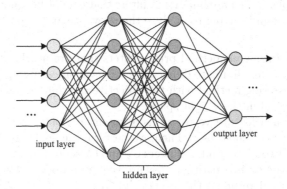

Fig. 1. The architecture of a neural network.

Typically, CNNs consist of input, output and the intermediate processing layers shown by Fig. 1. We define input channels, height and width of the input feature maps of ith convolutional layer as c_i, h_i and v_i, respectively. In CNNs, the input feature maps of a convolutional layer transform input $O_i \in R^{c_i \times h_i \times v_i}$ into output $O_{i+1} \in R^{c_{i+1} \times h_{i+1} \times v_{i+1}}$, where O_{i+1} serves as the input feature maps for the next convolutional layer. A convolutional layer convolves O_i with c_{i+1} 3D filters $F \in R^{c_i \times h_i \times v_i}$, in which one filter generates one feature map, resulting in c_{i+1} output feature maps O_{i+1}. The constitute of each 3D filter contains c_i 2D kernels $\gamma \in R^{k \times k}$. All the feature maps form the kernel matrix $W^i \in R^{m \times c_{i+1}}$, where $m = k^2 c_i$. Besides, $w_j^i, j = 1, \cdots c_{i+1}$, denotes the j_{th} feature in layer i, and each w_j^i corresponds to the j_{th} column of the kernel matrix $W^i = [w_1^i, \cdots w_{c_{i+1}}^i] \in R^{m \times c_{i+1}}$, as shown in Fig. 2.

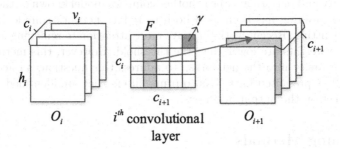

Fig. 2. Feature map corresponding to convolutional layer filters and related kernels in the next layer.

3.2   Filters Relevance and Importance Analysis

In over-sized deep neural network models, there are many redundant features with the similar weights resulting in filtering redundancy, which can be removed. And filters with large kernel weights sum tend to produce features maps with strong activations, which are important filters that cannot be ignored. Therefore, we use clustering and sorting to analyze the relevance and importance of filters.

Filter Clustering. Our method prunes filters that are nearly identical from a well-trained model to eliminate duplicative retrieval of feature maps. We use the hierarchical clustering method with the objective of grouping filter vectors that are nearly identical in the weight space.

Start with each feature vector v_i as a potential cluster and set the chosen cluster similarity threshold as τ. Then, we perform the agglomerative clustering by merging the two most similar clusters C_a and C_b so long as the mean of cosine similarity between their constituent feature vectors is above allowable threshold τ. The choice of τ determines the level of similarity, that is, the degree of feature sharing across a particular layer, ensuring features that are highly correlated are grouped. The clustering is convenient for selecting the representative filter next.

For a particular convolution layer of the model in Fig. 3, there are five filters 1, 2, 3, 4, 5. The clustering of them resulted in two clusters with members $\{1, 3, 5\}$, $\{2, 4\}$, resulting from the fact that cluster members have average similarity greater than τ.

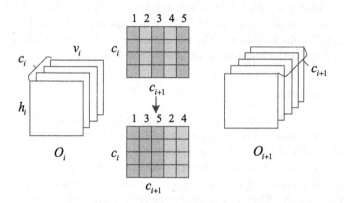

Fig. 3. The result of clustering filters based on a given threshold. 1, 3 and 5 constitute one cluster, and 2 and 4 constitute another. (Color figure online)

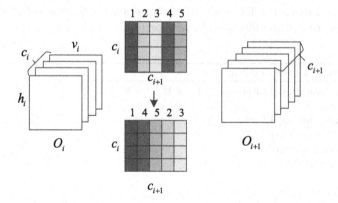

Fig. 4. The result of sorting the filter based on the value. (Color figure online)

Filter Sort. Our method also selects some important filters to minimize accuracy degradation. We measure the relative importance of a filter in each layer by calculating the sum of its absolute weights $\sum |F|$. Filters with large kernel weights greatly affect the inference accuracy as compared to the other filters in that layer.

As the sorting executes, we calculate the sum of the absolute weight values of each filter, using L1-norm as metric. Following, we sort filters by absolute weights sum for each layer. We sort the filters so as to select several top features to reserve.

For instance in Fig. 4, there are five filters 1, 2, 3, 4, 5 with different color depths. The filter has greater absolute weights sum with a higher color depth; otherwise, with a lower color depth. Their descending order is {1, 4, 5, 2, 3}, resulting from 1 (blackish green) the maximum and 3 (light green) the minimum value, according to the magnitude of absolute weights sum.

3.3   Filter Selection

The pruning method based on clustering algorithm is detailed in Algorithm 1, which takes a trained model as input and filter selection is performed at every layer of the model. As described in the pruning heuristic Algorithm 1, cluster-filters (line 3) is to cluster all the filters (columns of the kernel matrix W^i) into groups whose average similarity among cluster members is above a set threshold τ. One representative filter is randomly selected from each of the clusters. The statement aims at selecting characteristic filters to represent the repeated operations brought by similar filters to eliminate redundancy. Some filters with large kernel weights should be keeping, considering that they have a greater impact on the inference results and cannot be ignored. Select-filters (line 4) aims to sort filters by the weights sum and select filters with the larger sum values based on the sampling rate ε. The following step is to retain the filter selected above, and to prune the rest of the filters and their corresponding feature maps.

Algorithm 1. Filter pruning

1: Input: convolutional filters of ith layer W^i
2: Given: τ, ε
3: $I_1 = $ cluster_filters(W^i, τ)
   /**Filter classification**/
4: $I_2 = $ select_filters(W^i, ε)
   /**Filter sort**/
5: $I = I_1 \cup I_2$
   /**Merge filters**/
6: Initialize: W^i_{pruned} of the pruned model
7: $k \leftarrow 0$
8: **for** m **do** in I **do**
9:     copy: m^{th} column of W^i into k^{th} column of W^i_{pruned}
10:     $k \leftarrow k + 1$
11: **end for**

For instance in Fig. 5, there are five filters 1, 2, 3, 4, 5 with color codes: dark green, green, light green, light orange, and orange. The convolution of these filters with input O_i yields five output feature maps, which are connected to those in next layer by kernel matrix W^{i+1}. We use the color depth to represent weights sum and the color spectrum to represent the type of clusters, in which the darker color means larger values and different color spectrum means different clusters. The clustering of columns of W^i results in two clusters with members $\{1, 2, 3\}$ and $\{4, 5\}$. Then, we select one filter from each of the two clusters randomly and get $\{1, 4\}$. The following step is to select filters with large weights sum, as 1 and 5 are selected because of their darker color. Finally, we merge the results of clustering and the list of indices of I are then given as $\{1, 4, 5\}$.

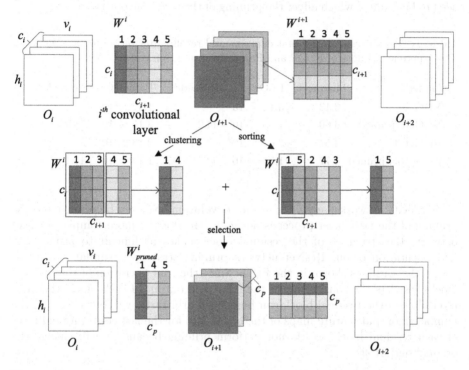

Fig. 5. The convolution layer is pruned according to our method, and the filters are divided into two categories: $\{1, 2, 3\}$ and $\{4, 5\}$ by the threshold value, in which $\{1, 4\}$ are selected as the representatives. At the same time, by numerical sorting, the weight value of 1 and 5 is larger. So, the filters 1, 4 and 5 are selected and the remaining filters are pruned.

There is no doubt that pruning filters will cause the degradation of network performance, but the performance can be retrained mostly by combining the clustered filters and top-ranked filters. We will validate our method in next section.

4 Experiments

We evaluate the pruning methods over two networks (VGG-16 and ResNet-34 on CIFAR-10) in this section. All experiments are implemented on Pytorch. CIFAR-10 is a small data set for identifying universal objects. It includes pictures in ten categories including airplanes, cars, birds, cats, deer, dogs, frogs, horses, boats and trucks. The size of each picture is 32×32, and there are 6000 images in each category, of which 50,000 are training pictures and 10,000 are test pictures. When pruning some filters of a convolutional layer, its corresponding feature map also need to be pruned, which affect the pruning of the next convolutional layer.

Table 1. Overall results. The best experimental results after retraining are reported. The experimental data of FLOP and Parameters are given respectively.

Model	Error(%)	FLOP	Pruned(%)	Parameters	Pruned(%)
VGG-16	6.13	3.14×10^8		1.47×10^7	
VGG-16-pruned	6.60	1.75×10^8	44.2	3.15×10^6	78.6
ResNet-34	4.57	3.67×10^9		2.13×10^7	
ResNet-34-pruned	4.77	1.81×10^9	50.7	1.19×10^7	44.1

For VGG networks, any filter in the convolution layer can be pruned and the pruning of the next layer proceeds normally. But for the more complex ResNet network, the structure of the residual block makes it difficult to prune. The implementation of our ResNet network pruning strategy is similar to that in [18] in the sense, as shown in the Fig. 6. Since the sum operation of the residual blocks restricts the consistency of the input and output of each block, the first-level filter in the residual block can be pruned arbitrarily without changing the number of output feature maps of the block. But for the last convolutional layer of each residual block, we do not perform pruning because it will change the original output size.

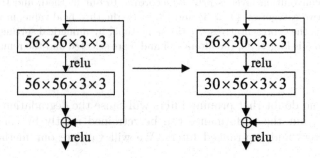

Fig. 6. Explanation of ResNet pruning strategy. For each residual block, we only prune the first convolutional layer, keeping the block output dimension unchanged. Left: before pruning; Right: after pruning.

4.1 VGG-16 on CIFAR-10

The slightly modified version of the VGG-16 model [1] that has 13 convolution layers and 2 fully connected layers, as shown in Table 2, can produce excellent results for CIFAR-10 datasets. It is also worth noting that the input to the linear layer is changed when the last convolutional layer is pruned, and the connections are also removed.

Table 2. VGG-16 on CIFAR-10 and the pruned model. Data of each layer before and after pruning.

	VGG-16			VGG-16-pruned			
	params	FLOP	Maps	params	FLOP	Maps	Prune(%)
Conv_1	1.79E+03	2.E+06	64	8.96E+02	9.18E+05	32	50
Conv_2	3.69E+04	3.78E+07	64	1.56E+04	1.60E+07	54	16
Conv_3	7.39E+04	1.89E+07	128	5.80E+04	1.48E+07	119	7
Conv_4	1.48E+05	3.78E+07	128	1.37E+05	3.51E+07	128	0
Conv_5	2.95E+05	1.89E+07	256	2.94E+05	1.88E+07	255	0
Conv_6	5.90E+05	3.78E+07	256	5.83E+05	3.73E+07	254	0
Conv_7	5.90E+05	3.78E+07	256	5.76E+05	3.69E+07	252	0
Conv_8	1.18E+06	1.89E+07	512	6.29E+05	1.01E+07	277	46
Conv_9	2.36E+06	3.78E+07	512	4.39E+05	7.02E+06	176	66
Conv_10	2.36E+06	3.78E+07	512	2.14E+05	3.42E+06	135	74
Conv_11	2.36E+06	9.44E+06	512	1.11E+05	4.43E+05	91	82
Conv_12	2.36E+06	9.44E+06	512	1.09E+05	4.36E+05	133	74
Conv_13	2.36E+06	9.44E+06	512	1.92E+05	7.67E+05	160	69

First, we show the number of non-redundant filters that exist for different thresholds τ and different sampling rates ε in Fig. 7. It can be seen that some convolutional layers in the VGG network easily extract features with high correlation, as examples of such layers are layer 1, 10, 11, and 12. It is observed that the convolutional layers in the later stage have more redundant filters than the convolutional layers in the earlier stage. One possible explanation is that the later convolutional layers have more filters. At the same time, it can be seen from the comparison that when the same threshold is set, the number of selected filters is appropriately increased if the filters with larger values are selected based on the sampling rate.

We prune each layer of the network independently to check its sensitivity. It can be seen from the Fig. 8 that the convolutional layer is more stable after adding some filters with larger value, resulting in the increase of accuracy. And it is found that the later-stage convolutional layers are more stable than the earlier-stage convolutional layers. Thus, we can determine the threshold and sampling rate required for network pruning according to the sensitivity of convolutional layer.

The statistical result comparing our method with others is shown in Table 3. When the threshold and sampling rate are 0.46 and 9% respectively, our method outperforms that Absolute filter sum approach [12], Network Sliming [13], Try-and-Learn [16], Redundant Features [14], and is able to prune more than 78.6% of the parameters resulting in 44.2% FLOP reduction and a competitive

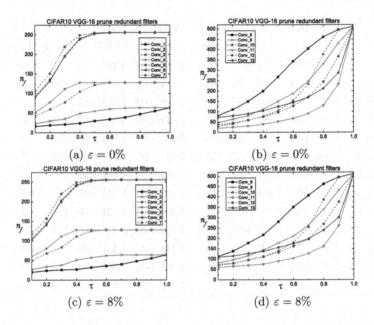

Fig. 7. Number of non-redundant filters (n_f) vs. cluster similarity threshold (τ) and sampling rate (ε) for VGG-16 trained on the CIFAR-10 dataset.

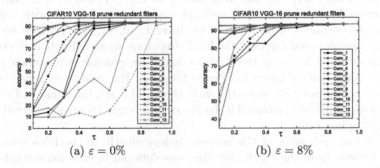

Fig. 8. The Accuracy vs. cluster similarity threshold (τ) and sampling rate (ε) for VGG-16 trained on the CIFAR-10 dataset.

classification accuracy. In addition, when the threshold and sampling rate are tune to 0.4 and 10% respectively, we are able to achieve less than 0.7% failure reduction. Consequently, we can infer that our pruning method performs better than other methods mentioned above.

Table 3. VGG-16 network performance evaluation after pruning. Data results for various technologies are reported.

VGG-16	Accuracy drop	FLOP pruned	Parameters pruned
Methods	(%)	(%)	(%)
Absolute filter sum [12]	0.40	34.2	64.0
Network Sliming [13]	-0.17	38.6	-
Try-and-Learn [16]	0.60	34.2	-
Redundant Features [14]	0.80	40.5	78.1
Ours($\tau= 0.46, \varepsilon= 9\%$)	0.47	44.2	78.6
Ours($\tau= 0.40, \varepsilon= 10\%$)	0.69	49.4	80.7

4.2   ResNet-34 on CIFAR-10

Simply increasing the depth will bring side effects (degeneration problem), so the ResNet network is proposed to solve overfitting. ResNet have four stages of residual blocks for feature maps with sizes of 56×56, 28×28, 14×14 and 7×7. A direct connection channel is added to the network as shown in Fig. 6. In this case, the neural network in this layer learns the residuals of the previous network output instead of learning the entire output.

There are also redundant filters in the ResNet network structure. As shown in Fig. 9, when we increase the sampling rate ε from 0% to 5%, which means adding the ranked important filters. It is observed that the the number of non-redundant filters do not change almost between (a)(b) and (c)(d), which proves the feasibility of the pruning method. Figure 10 shows the sensitivity of ResNet-34 layers to pruning and it can be observed that layers such as Conv 16 and 32 are more sensitive to filter pruning than other convolutional layers when the sample rate increase from 0% to 5%. In order to regain the accuracy by retraining the pruned model, we skip these sensitive layers while pruning.

Table 4. ResNet-34 network performance evaluation after pruning. Data results for various technologies are reported.

ResNet-34	Accuracy drop	FLOP pruned	Parameters pruned
Methods	(%)	(%)	(%)
Redundant Features [14]	0.34	41.7	28.7
Ours($\tau= 0.2, \varepsilon= 8\%$)	0.20	50.7	44.1
Ours($\tau= 0.3, \varepsilon= 3\%$)	0.15	40.9	28.0

The comparison result of retraining performance is shown in Table 4. With retraining, ResNet-34 is pruned with $\tau = 0.2$, $\varepsilon = 8\%$. The FLOP is reduced by 50.7% and the parameters is reduced by 44.1%, but the accuracy rate is only reduced by 0.2%. So, we can conclude that our pruning methods have competitive performance over [14] in terms of FLOP reduction.

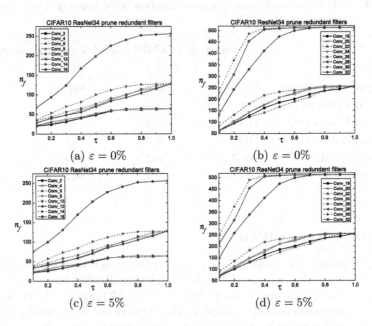

(a) $\varepsilon = 0\%$ (b) $\varepsilon = 0\%$

(c) $\varepsilon = 5\%$ (d) $\varepsilon = 5\%$

Fig. 9. Number of non-redundant filters (n_f) vs. cluster similarity threshold (τ) and sampling rate (ε) for ResNet-34 trained on the CIFAR-10 dataset.

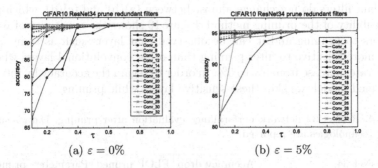

(a) $\varepsilon = 0\%$ (b) $\varepsilon = 5\%$

Fig. 10. The Accuracy vs. cluster similarity threshold (τ) and sampling rate (ε) for ResNet-34 trained on the CIFAR-10 dataset.

5 Conclusion

Modern CNNs usually have enormous amounts of features with similarity and have large inference costs. In this paper, a new filter pruning method based on clustering is proposed. The filter is selected from the clustering and sorting results to reduce the computational and memory overhead without introducing irregular sparsity. It achieves about 44.2% and 50.7% reduction in FLOP for VGG-16 and ResNet-32, respectively, with minor loss of accuracy. To recover the performance after pruning, models were retrained for a few iterations.

References

1. Simonyan, K., Zisserman, A.: Very deep convolutional networks for large-scale image recognition. arXiv preprint arXiv:1409.1556 (2014)
2. He, K., Zhang, X., Ren, S., Sun, J.: Deep residual learning for image recognition. In: Proceedings of the IEEE Conference on Computer Vision and Pattern Recognition, pp. 770–778 (2016)
3. Iandola, F.N., Han, S., Moskewicz, M.W., Ashraf, K., Dally, W.J., Keutzer, K.: Squeezenet: Alexnet-level accuracy with 50x fewer parameters and <0.5 MB model size. arXiv preprint arXiv:1602.07360 (2016)
4. Howard, A.G., et al.: Mobilenets: efficient convolutional neural networks for mobile vision applications. arXiv preprint arXiv:1704.04861 (2017)
5. Hinton, G., Vinyals, O., Dean, J.: Distilling the knowledge in a neural network. arXiv preprint arXiv:1503.02531 (2015)
6. Chen, T., Goodfellow, I., Shlens, J.: Net2net: accelerating learning via knowledge transfer. arXiv preprint arXiv:1511.05641 (2015)
7. Zhang, X., Zou, J., Ming, X., He, K., Sun, J.: Efficient and accurate approximations of nonlinear convolutional networks. In: Proceedings of the IEEE Conference on Computer Vision and Pattern Recognition, pp. 1984–1992 (2015)
8. Tai, C., Xiao, T., Zhang, Y., Wang, X., et al.: Convolutional neural networks with low-rank regularization. arXiv preprint arXiv:1511.06067 (2015)
9. Gupta, S., Agrawal, A., Gopalakrishnan, K., Narayanan, P.: Deep learning with limited numerical precision. In: International Conference on Machine Learning, pp. 1737–1746 (2015)
10. Vanhoucke, V., Senior, A., Mao, M.Z.: Improving the speed of neural networks on CPUs (2011)
11. Han, S., Pool, J., Tran, J., Dally, W.: Learning both weights and connections for efficient neural network. In: Advances in Neural Information Processing Systems, pp. 1135–1143 (2015)
12. Li, H., Kadav, A., Durdanovic, I., Samet, H., Graf, H.P.: Pruning filters for efficient convnets. arXiv preprint arXiv:1608.08710 (2016)
13. Liu, Z., Li, J., Shen, Z., Huang, G., Yan, S., Zhang, C.: Learning efficient convolutional networks through network slimming. In: Proceedings of the IEEE International Conference on Computer Vision, pp. 2736–2744 (2017)
14. Ayinde, B.O., Inanc, T., Zurada, J.M.: Redundant feature pruning for accelerated inference in deep neural networks. Neural Networks 118, 148–158 (2019)
15. He, Y., Liu, P., Wang, Z., Hu, Z., Yang, Y.: Filter pruning via geometric median for deep convolutional neural networks acceleration. In: Proceedings of the IEEE Conference on Computer Vision and Pattern Recognition, pp. 4340–4349 (2019)

16. Huang, Q., Zhou, K., You, S., Neumann, U.: Learning to prune filters in convolutional neural networks. In: 2018 IEEE Winter Conference on Applications of Computer Vision (WACV), pp. 709–718. IEEE (2018)

17. Krizhevsky, A., Sutskever, I., Hinton, G.E.: Imagenet classification with deep convolutional neural networks. In: Advances in Neural Information Processing Systems, pp. 1097–1105 (2012)

18. Luo, J.H., Wu, J., Lin, W.: Thinet: a filter level pruning method for deep neural network compression. In: Proceedings of the IEEE international Conference on Computer Vision, pp. 5058–5066 (2017)

Fast Segmentation-Based Object Tracking Model for Autonomous Vehicles

Xiaoyun Dong[1,2,3], Jianwei Niu[1,2,3,5], Jiahe Cui[1,2,3], Zongkai Fu[1,2,3], and Zhenchao Ouyang[1,2,4(✉)]

[1] Hangzhou Innovation Institution, Beihang University, Chuanghui Street #18, Binjiang, Hangzhou 310000, Zhejiang, China
ouyangkid@buaa.edu.cn
[2] State Key Laboratory of Virtual Reality Technology and Systems, Beijing, China
[3] Beijing Advanced Innovation Center for Big Data and Brain Computing (BDBC) Beihang University, Xueyuan Road #37, Haidian, Beijing 100191, China
[4] Nanhu Laboratory, Jiaxin 314000, Zhejiang, China
[5] Zhengzhou University Research Institute of Industrial Technology, Zhengzhou University, Zhengzhou 450001, China

Abstract. On-road object tracking is a critical module for both Advanced Driving Assistant System (ADAS) and autonomous vehicles. Commonly, this function can be achieved through single vehicle sensors, such as a camera or LiDAR. Consider the low cost and wide application of optical cameras, a simple image segmentation-based on-road object tracking model is proposed. Different from the detection-based tracking with bounding box, our model improves tracking performance from the following three aspects: 1) the Positional Normalization (PONO) feature is used to enhance the target outline with common convolutional layers. 2) The inter-frame correlation of each target used for tracking relies on mask, this helps the model reducing the influences caused by the background around the targets. 3) By using a bidirectional LSTM module capable of capturing timing correlation information, the forward and reverse matching of the targets in consecutive frames is performed. We also evaluate the presented model on the KITTI MOTS (Multi-Object and Segmentation) task which collected from out door environment for autonomous vehicle. Results show that our model is three times faster than Track RCNN with slightly drop on sMOTSA, and is more suitable for deployment on vehicular low-power edge computing equipment.

Keywords: Object tracking · Segmentation · Bi-LSTM · Autonomous vehicle · Deep learning

1 Introduction

Multi-object tracking (MOT) in out door environment is a basic and challenging computer vision task [1–3]. It has a wide range of application scenarios and actual requirement, such as security monitoring, industrial production, precision

Supported by Hangzhou Innovation Institution, Beihang University.

guidance, aerial warning and transportation [4]. In the era of big data, related face recognition systems, satellites, surveillance cameras, vehicle-mounted systems, etc. provide rich data sources for researches [5] and also greatly promote the development of related technologies.

Due to the unclear definition of tracking tasks when dealing with partially visible, occluded, or cropped targets, and the lack of well defined and organized dataset and evaluation metrics, previous track systems lack comprehensive evaluation. Until the emergence of large-scale labeled datasets [6–8] in recent years, multi-target tracking algorithms and models have been able to develop rapidly [9]. However, most of the MOT are designed for data collected from fixed scene or with static sensors.

With the rapid development of sensor technology, big data processing and artificial intelligence, autonomous driving technology that can improve transportation efficiency, traffic safety, and travel convenience, has received widespread attention in academia and industry in recent years [10]. At the same time, complex road environments, limited vehicle platform and highly random obstacles also pose challenges to autonomous vehicle perception systems. The MOT is one of the most basic and critical perception modules for autonomous vehicle [11,12].

The autonomous vehicle often equipped with rich sensors and can obtain information of the surrounding environment. Optical camera is the most commonly used sensors, and the vision-based MOT is widely studied [13]. Recent artificial intelligence solutions use deep learning modeling and big data-driven to build MOT model [14]. The convolutional neural network model relies on high-performance graphics cards, making the deployment of the model in a vehicle environment a challenge. Therefore, it is very necessary to study the tracking algorithm that balances performance and accuracy.

In this paper, a vision-based one stage MOT model is proposed for autonomous vehicle. Our tracking architecture is modified from Track R-CNN [3]. We first combine the Positional Normalization (PONO)[15] feature with a lightweight CNN backbone for refining object segmentation. The PONO combines first and second moments of features that can capture structural information. The bi-directional Long Short Term Memory (Bi-LSTM) module is then used to further enhance the association information of inter-frame targets. Finally, the segmentation masks of targets reduce the influence of the background, such as occluded targets and similar background. The final model can achieve better tracking performance while processing video stream data in a more efficient way. The contributions of this paper are summarized as follows[1]:

- We present a light-weight segmentation-based multi-object tracking model for autonomous vehicular platform.
- The model combines PONO features and Bi-LSTM modules to capture the spatial-temporal features of targets from continues frames for better tracking.

---

[1] We provide code online at https://github.com/XYunaaa/Fast-Segmentation-based-Object-Tracking-Model.

– We evaluate the presented model on KITTI Multi-Object and Segmentation (MOTS) dataset.

The rest of this paper organized as follows. The related work of deep learning based tracking models are summarized in Sect. 2. We present the segmentation-based tracking model architecture in Sect. 3, and evaluation results on KITTI MOTS dataset in Sect. 4. The final Section concludes the current model and possible improvements.

2   Related Works

Most of current object tracking algorithms follow the tracking-by-detection fashion which detect the targets frame by frame, and combine with matching algorithms to assign track ID for the same objects. The whole process mainly consists of three steps: detection, feature extraction/motion prediction and association. The detection module first find all possible targets in each frame. And then, the second module extracts a feature set for each target based on detection results. Finally, the association module matches targets with similar features or performs motion estimation based on inter-frame information.

The deep learning-based model mainly borrowed from object detection task directly [16–18], because the convolutional neural network models can automatically extract robust features for detection task instead of hand-made-craft features or simple statistical features. And then, the detection results from continuous frames are sent to different trackers, such as correlation filter [19–21], Kalman filter [22–24], clustering algorithm [25,26] or Hungarian algorithm [27], optical flow [28], etc., for ID matching.

As mentioned before, the deep learning models are mainly used for solve the detection process, both one-stage and two-stage models are used. The 2D bounding box (BBox) [22,23,29] is a common and efficient way to describe the results. In static scenes and top view applications, such as the security monitoring [30], face recognition, etc., the 2D BBox can capture the main outline of the targets. However, when dealing with occlusion targets and close-range targets, such as following a bus or waiting for traffic lights in a crowded traffic environment, the BBox may contain other objects in the background. This has a little effect on classification and detection tasks, but it will affect the characterization of the target, resulting in matching failure [3]. Segmentation-based tracking use masks for each target, and can significantly reduce the interference caused by the background. Moreover, splitting detection and feature extraction also affects the overall efficiency of the model.

The current solution avoids the usage of segmentation networks mainly for the following reasons. Segmentation models are usually very large, especially for the multi-channel results, and will affect processing efficiency. The segmentation-based annotation datasets are also very limited. Traditional features, such as HOG, SIFT, Daisy, etc., cannot generate from irregular masks.

To overcome the above problems, several studies perform end-to-end tracking, and use convolutional features for both detection and target encoding [3,8,31] to speed up the model efficiency. Moreover, combining with segmentation model, the convolutional encoding of the target mask is also more accurate.

The Siamese network is another CNN-based tracking solution, which consists of two parallel branches for different purposes. Namely, a searching branch response for high-level semantic representation at global area, and a detecting branch for low-level fine-grained representation at local area. The later branch is often offline trained for the predefined targets, and shares the weights with its twin branch. Fully convolutional Siamese trackers can achieve real-time performance, but due to not fully exploit semantic and objective information, it is often less accurate than other state-of-the-art works. Subsequent researches modified the Siamese network from different aspects, such as constructing cascaded region proposals based on convolution features from different layers of Siamese network [32], combining with Graph Convolutional Network (GCN) [33], increasing network depth [34] and refining match at different stage [35].

It is also important to consider temporal information in tracking tasks, and the most simple way is to combine with LSTM module [36]. TrackR-CNN [3] also deployed LSTM in their model, but found that there is no essential improvement. Other works [30,37] introduce the Bi-LSTM to explore target matching from both forward and reverse image sequences. Ning Wang et al. [38] even designed a novel unsupervised deep tracking framework to compute the consistency loss for network training from forward and backward image sequences.

Base on previous works, our model tries to optimize the data and detection results through PONO features and segmentation, and combine Bi-LSTM to capture temporal features.

3   Proposed Model

The architecture of the proposed model is presented in Fig. 1, the main workflow is borrowed from TrackR-CNN [3]. We also embed the whole detection, segmentation and target feature extraction and inter-frame matching into a whole CNN workflow. We first use a lightweight ResNet50 [39] backbone combining with the profile enhanced module of PONO to get the spatial association of the target from consecutive frames. Different lengths of subset sequences are also evaluated. A Bi-LSTM module that can extract temporal information is attached behind. This module learns the inter-frame relationship from frontward and backward, and each gate state exchanges information between the two parallel branches. The whole backbone combines spatio-temporal feature together for region proposal and generates mask, classification score and association vector for each target. All information is used for linking the inter-frame targets into tracks over time.

3.1   ResNet50 with PONO

To reduce total floating-point operations per second (Flops) of the model, we first use the ResNet50 as spatial-feature backbone. However, shallow networks

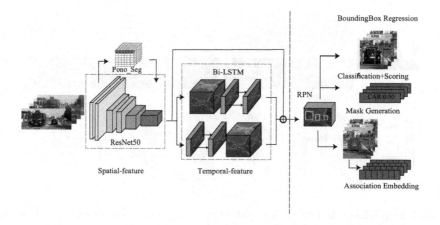

Fig. 1. The proposed segmentation-based on-road object tracking model.

will reduce the model fitting ability, especially on segmentation tasks. Therefore, we combine the Positional Normalization feature with ResNet50. Different from previous feature normalization (e.g., Batch, Group, Layer or Instance normalization), PONO not only benefit for network convergence, but also offer a clear structural information from the input images, as shown in Fig. 2(top). PONO achieves this by normalizing over the channels at any given pixel location, and the extracted statistics are position dependent and reveal structural information at this particular layer as Eq. 1.

$$
\begin{aligned}
u_{w,h,b} &= \tfrac{1}{C} \sum_{C}^{c=1} X_{w,h,c,b} \\
\sigma_{w,h,b} &= \sqrt{\tfrac{1}{C} \sum_{C}^{c=1} (X_{w,h,c,b} - u_{w,h,b})^2 - \epsilon} \\
PONO(X_{w,h,c}) &= \tfrac{X_{w,h,c} - u_{w,h,b}}{\sigma_{w,h,b}}
\end{aligned}
\tag{1}
$$

Where $u_{w,h,b}$ and $\sigma_{w,h,b}$ are the mean and standard deviation or the first and second order extension, respectively. $\epsilon = 10^{-5}$ is a small constant coefficient. The w, h, c, b are the width, height, channel and batch of input feature map of current layer, and X is activation. Commonly, after the normalization operation is deployed on each layer, the two extracted statistics are discarded. PONO treats them as single channel feature maps with same scale of input, and combine them to convolutional feature together as final output. It can be seen that the added amount of calculation is very small, however, the outline information is very significant and can be used to assist segmentation. The ResNet50 can be treat as texture encoding, while the PONO can be treat as outline encoding. Figure 2 shows an example of how to combine the common convolutional layer and a 3-layer residual module with PONO layer together.

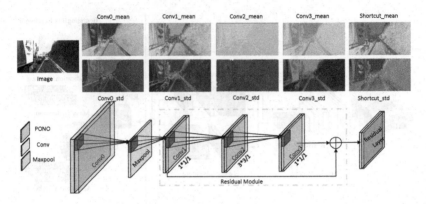

Fig. 2. Examples of the first (mean) and second (std) moments of the statistical PONO feature (top). The residual PONO module that combines residual convolutional and PONO layers (bottom).

3.2   Bi-LSTM

Building sequence model with RNN or LSTM module can capture temporal information, however, previous tracking tasks only considered forward order. However, if we treat the motion of vehicles and pedestrians as event, each frame is highly related with its context states (both forward and backward). Several human pose estimation and tracking tasks [37] use Bi-LSTM for encoding the temporal state of the target. Instead of using stacked LSTM, we use single Bi-LSTM unit to encode both forward and backward states of on-road objects as temporal-feature. We attached Bi-LSTM module behind the convolutional backbone, and to ensure the effectiveness of LSTM training, the nature order of frame sequence is used for training instead of random sampling.

3.3   Multi-task Region Proposal

As the previous two backbone modules generate spatio-temporal features, the final region proposal of our model directly borrow the multi-task network from TrackR-CNN [3,30,37]. It contains three individual branches, i.e., a mask-based segmentation, a classification and association vector generator. The fixed association vector is 128 generated from a fully connected layer from the mask, and is used for tracking with the Hungarian algorithm.

The whole model is trained on KITTI MOTS Challenge dataset[2]. This dataset is extended from KITTI tracking dataset which contains 21 labeled sequences. Among them, the sequence {0, 1, 3, 4, 5, 9, 11, 12, 15, 17, 19, 20} for training and sequence {2, 6, 7, 8, 10, 13, 14, 16, 18} for validation. Each time we feed more than one frame as model input as a sliding window for training and validation.

---

[2] https://www.vision.rwth-aachen.de/page/mots.

4   Experiments

We first quantify the background noise introduced by Bounding box to the target detection process in an open road environment. Figure 3 shows two examples of the bounding box based detection and mask based detection, it is easy to see that the bounding box introduces background noises and overlapped vehicles.

Fig. 3. The bounding box vs. mask: bounding boxes involve invalid background.

Take the MOTS dataset of KITTI for example. We evaluate the pixel noise introduced by bounding boxes by comparing with mask labels in Table 1. We evaluate the pixels of mask and bounding box for cars and pedestrians according to Eq. 2, respectively. Where the cumulative pixel (CP) is calculated by adding all the $mask/bbox$ of class (car/ped) according to the tracking ID (t) in each sequence. The mask and bounding box donate incremental pixels (megapixel) of each category of targets in the 21 sequences. Seq 17 does not contain cars, and Seq 3, 5, 6, 8, 18 and 20 do not contain pedestrians.

$$
\begin{aligned}
CP_{car} &= \sum_{t\in T}^{t=1} f_{\{mask,bbox\}}^{c=car} \\
CP_{ped} &= \sum_{t\in T}^{t=1} f_{\{mask,bbox\}}^{c=ped}
\end{aligned}
\tag{2}
$$

From Table 1, it is easy to see that the ratio of the mask/BBox for car ranges from 0.12 to 0.82, and the ratio of the mask/BBox for pedestrian ranges from 0.23 to 0.61. The average m/b for the 21 sequences is 0.28 for cars and 0.31 for pedestrians, respectively. The smaller the ratio, the greater the background noise introduced by BBox. And the pedestrians are affected more than cars, this may lead to poor tracking results for pedestrians when dealing with BBox. The overall ratio ranges from 0.13 to 0.65, and only 6 out of 21 sequences have a ratio larger than 50%.

We then train and evaluate our proposed model in the same way mentioned in Track R-CNN [3], the only difference is our hardware is a little different from theirs. All the testings are deployed on a GPU server with NVidia RTX2080ti@11.3 Gbps, with 32G memory and Intel Core i9-9900K CPU. During training, we can only set a maximum batch size of 7 due to the limitation of the GPU memory, the rest of the configuration is basically consistent with the original experiment [3].

The MOSTA (multi-object tracking and segmentation accuracy), MOTSP (mask-based multi-object tracking and segmentation precision) and sMOTSA

Table 1. Pixel compare: Bounding box vs. Mask (M=Million)

Seq	Cars			Pedestrian			Total		
	mask(M)	bbox(M)	m/b	mask(M)	bbox(M)	m/b	mask(M)	bbox(M)	m/b
0	2.15	2.89	0.75	0.06	0.10	0.56	10.08	25.01	0.40
1	21.00	28.71	0.73	0.21	0.39	0.54	30.43	63.15	0.48
2	2.14	2.91	0.74	0.17	0.33	0.52	7.41	24.09	0.31
3	2.69	3.32	0.81	-	-	-	3.56	6.07	0.59
4	2.57	3.62	0.71	0.12	0.22	0.55	6.31	16.99	0.37
5	3.07	3.90	0.79	-	-	-	5.69	12.88	0.44
6	3.15	4.15	0.76	-	-	-	5.86	11.55	0.51
7	28.56	36.55	0.78	0.09	0.19	0.49	38.51	59.06	**0.65**
8	2.48	3.20	0.77	-	-	-	4.54	8.86	0.51
9	20.54	27.65	0.74	0.03	0.06	0.57	33.01	67.40	0.49
10	1.71	2.28	0.75	0.05	0.07	**0.61**	7.00	12.88	0.54
11	16.45	22.06	0.75	0.10	0.20	0.51	19.99	3 5.88	0.56
12	0.13	0.16	0.78	0.01	0.02	0.53	0.61	1.51	0.41
13	0.20	0.24	**0.82**	2.14	4.12	0.52	14.01	36.30	0.39
14	2.60	10.46	0.25	0.21	0.56	0.38	3.74	29.52	**0.13**
15	12.28	46.73	0.26	1.97	8.43	**0.23**	17.76	130.93	0.14
16	2.53	21.11	**0.12**	6.33	23.93	0.26	13.48	94.41	0.14
17	-	-	-	4.40	13.80	0.32	10.92	67.46	0.16
18	12.04	42.21	0.29	-	-	-	13.75	65.25	0.21
19	11.81	38.42	0.31	16.73	54.19	0.31	77.66	352.53	0.22
20	26.35	96.36	0.27	-	-	-	51.41	229.78	0.22
All	50.19	176.99	**0.28**	16.73	54.19	**0.31**	142.81	647.55	**0.22**

(soft multi-object tracking and segmentation accuracy)[3] are introduced to evaluate the tracking model and is calculated as Eq. 3. Where TP is comprised of
hypothesized masks which are mapped to a ground truth mask, and $\widetilde{TP}$ is the
number of true positives. FP are hypothesized masks that are not mapped to
any ground truth mask. FN are the ground truth masks which are not covered
by any hypothesized mask. M denote the latest tracked predecessor of a ground
truth (mask), or $\emptyset$ if no tracked predecessor exists. The set IDS is defined as
the set of ground truth masks whose predecessor was tracked with a different
id. MOTSA is the mask-based multi-target tracking accuracy metric that only
consider successfully tracked targets. MOTSP is the mask-based multi-object
tracking and segmentation precision. sMOTSA measures both segmentation as
well as detection and tracking quality.

$$MOTSA = \frac{|TP| - |FP| - |IDS|}{|M|}$$
$$MOTSP = \frac{TP}{|TP|} \quad\quad (3)$$
$$sMOTSA = \frac{\widetilde{TP} - |FP| - |IDS|}{|M|}$$

The batch size of 5 is used during all training and testing, as we found the
enlarged batch size can slightly increase the model performance and has no
influence on FPS during testing. When the batch size is greater than 5, the
model performance will not continue to improve, as shown Table 2.

Table 2. Comparison of different batch sizes for training.

Batch		2	3	4	5	6	7
sMOTSA	Car	65	70.8	72.3	**72.6**	72.6	72.1
	Ped	29.7	31.9	34	**34.2**	34.1	34
FPS		10.29					

Table 3 illustrates the comparison of different model settings and model performances, two different backbones (Resnet50 and Resnet101) are used during testing, + means the backbone is integrated with PONO features in each module. The backbone is designed for charging the spatial feature, and is attached with a temporal module. Three different temporal modules are compared in our work, 3DCNN, LSTM and Bi-LSTM. In the end of the model, two tracking vector generation mechanisms are considered, i.e., generating from the RPN and from the mask.

We only evaluate the sMOTSA and FPS (Frame per Seconds) in the current stage. The $\sqrt{}$ donates the selected combination of modules for each setting, + donates the PONO feature is used, and $\sqrt{}/\sqrt{}$ means two repeatedly stacked modules. As we focus on tracking task, we randomly select a sequence during training and testing each time, and use the default frame order as model input.

Table 3. Comparison of different model settings.

	Backbone		Temporal Module			Vector		Performance		
								sMOTSA		FPS
	Resnet50	Resnet101	3DCNN	LSTM	Bi-LSTM	RPN	Mask	Car	Ped	
1		$\sqrt{}$	$\sqrt{}$			$\sqrt{}$		**76.2**	**46.8**	3.12
2	$\sqrt{}$		$\sqrt{}$			$\sqrt{}$		68.1	37.3	5.79
3	$\sqrt{}$+		$\sqrt{}$			$\sqrt{}$		70.7	30.2	10.29
4	$\sqrt{}$			$\sqrt{}$		$\sqrt{}$		61.4	39.3	8.82
5	$\sqrt{}$			$\sqrt{}\sqrt{}$		$\sqrt{}$		61.4	38.4	6.61
6	$\sqrt{}$				$\sqrt{}$0.05	$\sqrt{}$		68.5	42.4	**13.37**
7	$\sqrt{}$+				$\sqrt{}$0.05	$\sqrt{}$		72.3	34	10.36
8	$\sqrt{}$+				$\sqrt{}$0.8	$\sqrt{}$		73.7	37.3	10.36
9	$\sqrt{}$+				$\sqrt{}$0.8		$\sqrt{}$	74.9	43.6	10.15

The configuration1 is Track R-CNN [3] with a combination of Resnet101, 3DCNN, RPN-based tracking vector, it achieves the highest sMOTSA on both car and pedestrian, but can only reach 3.12 FPS on NVidia Titan XP@12G. This is far cry from the basic requirement of 10 Hz in autonomous system. By replacing the backbone with Resnet50, the FPS increases to 5.79 (nearly twice the original value), however, the sMOTSA drop to 68.1 an 37.3, respectively. Then, we add

PONO feature with Resnet50 backbone, the model FPS is surprisingly reached 10.29, and the car sMOTSA increased by about 2%, but pedestrian sMOTSA drops to 30.2. Which means that the PONO can heavily increase the backbone speed, and has benefit for large targets of cars.

Configuration 4, 5 and 6 show the performances of different temporal modules, except for the 3DCNN and LSTM, we also added a Bi-LSTM to capture the patterns from both toward and backward. Compare to configuration2, LSTM module shows higher processing efficiency for extracting temporal features than 3DCNN, but the sMOTSA is much lower than 3DCNN. The stack of two LSTM shows no better than single LSTM module, and increased processing time from 8.82 FPS to 6.61 FPS. Moreover, the Bi-LSTM shows similar performance on car, but much higher sMOTSA on pedestrian (42.4), and configuration6 can achieve the highest FPS at 13.37 among all configurations. This also donates that convolution-based Bi-LSTM can confirm the timing characteristics of the target through reverse information and greatly improve the efficiency of feature extraction and later tracking.

We further adjust the weight of the temporal module, as this is not a main module, we only consider two different conditions, i.e., 0.05 and 0.8, for Bi-LSTM. Considering the configuration6 and 7, when using Resnet50 with PONO, the sMOTSA of car further improves to 72.3 while the sMOTSA of pedestrian drops to 34. We then enlarged the weight of temporal module to 0.8 as in configuration8, both the sMOTSA of the car and pedestrian have increased significantly, and this process has not affect on FPS. In configuration9, the tracking vector is generated from mask instead of RPN, the tracking performance of sMOTSA is raises to 74.9 for car and 43.6 for pedestrian, and the FPS only drops 0.21. This is very close to Track R-CNN with a deeper backbone of Resnet101, but the FPS has increased more than three times, and can meet the needs of autonomous driving systems. Table 4 lists the detailed tracking indicators of the above-mentioned models with different configurations on cars and pedestrians.

Table 4. Details of tracking performances.

	Cars			Pedestrians		
	sMOTSA	MOTSA	MOTSP	sMOTSA	MOTSA	MOTSP
1	76.2	87.8	87.2	46.8	65.1	75.7
2	68.1	79.8	86.3	37.3	56.3	73
3	70.7	82.1	87	30.2	52.9	66.5
4	64.1	76	86.4	39.3	57.8	73.2
5	61.4	73.6	86.2	38.4	57.2	73.2
6	68.5	80.2	87	42.4	62.2	73.2
7	72.3	83.2	87.5	34	56.4	67
8	73.7	84.9	87.4	37.3	59.5	68.2
9	74.9	85.6	87.7	43.6	63.4	73.4

Figure 4 illustrates an example of ID matching based on the mask-based tracking vectors, the upper figure donates segmentation results at time T, while the bottom figure donates segmentation results at time $T+1$. After segmentation, we use the tracking vectors in each frame for matching track IDs with the famous Hungary algorithm. Figure 4(a) is the segmentation results in frame T and $T+1$. While Fig. 4(b), 4(c), 4(d), 4(e) and 4(f) donate the matching process for each targets. Due to the van and bus are ignored classes in KITTI tracking dataset, the model does not tracking on them during both training and testing. We also illustrated the calculated distances between the current target vectors (in $T+1$ frame) and last target vectors (in T frame), each target uses the same color of number and BBox. Our model successfully tracked four cars and a pedestrian, i.e., Car34, Ped83, Car44, Car52 and Car53, between the two frames.

(a) Segemention results (b) Matching for Car34

(c) Matching for Ped83 (d) Matching for unlabeled Car44

(e) Matching for Car53 (f) Matching for unlabeled Car52

Fig. 4. Evaluation of inter-frame ID matching.

This process can also be presented as matching matrix with distances information as shown in Fig. 5. When the element on the main diagonal is the smallest, and the distance of the same target (track ID) in the adjacent frames has the shortest distance, the model can achieve the best matching result based on the tracking vector distances. The vector distance between all correctly matched target frames is less than 1 (around 0.5).

Figure 6 illustrates four examples of 2D tracking results of our final model, the gray trajectory lines are ground truth calculated according to BBox center, while the trajectory line consistent with the same color of the mask is the predicted result of our model. It is easy to see that the predicted trajectories of the final model are very close to the ground truth in ideal occasions.

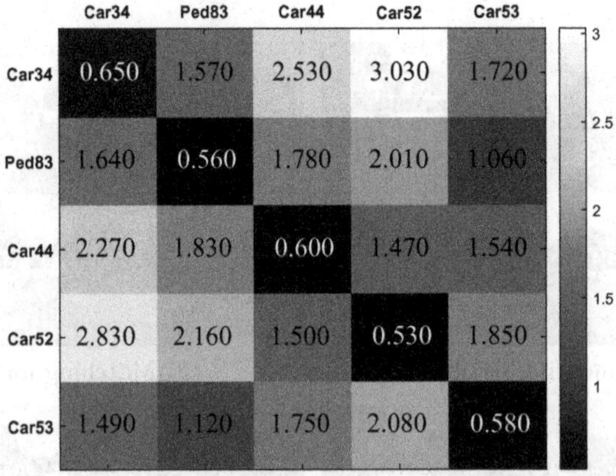

	Car34	Ped83	Car44	Car52	Car53
Car34	0.650	1.570	2.530	3.030	1.720
Ped83	1.640	0.560	1.780	2.010	1.060
Car44	2.270	1.830	0.600	1.470	1.540
Car52	2.830	2.160	1.500	0.530	1.850
Car53	1.490	1.120	1.750	2.080	0.580

Fig. 5. The matching distance matrix of inter-frame ID matching.

(a) Scenario 1

(b) Scenario 2

Fig. 6. Examples of 2D tracking results of the final model.

5   Conclusion

The cars and pedestrians are two main moving targets in open street environment. Accurate tracking of dynamic targets is an important module for perception system of autonomous vehicles, and can provide powerful information support for subsequent driving decision making and motion planning. To address accurate and efficient dynamic target tracking in dynamic open street area, this paper presented a novel segmentation-based object tracking model. We use lightweight backbone of Resnet50 with PONO module for generating the spatial features, and then combined with a Bi-LSTM for generating temporal feature from continuous inter-frame information in forward and reverse orders. The final tracking vector generated from segmentation mask also helps increase the tracking performance. Testing results on the public KITTI MOTS dataset show that the tracking performance of our model is slightly lower than Resnet101 based Track R-CNN on MOTSA, MOTSP and sMOTSA, but is three times faster and can be deployed for autonomous vehicles.

The current model still has some drawbacks, such as the tracking is relied on segmentation results, when the segmentation fails or is wrong, the entire system will fail. Moreover, the current tracking results on pedestrian is also not remarkable and safe enough for usage. We plan to solve those problems with re-identification strategies in the future.

Acknowledgment. This work has been supported by National Natural Science Foundation of China (61772060, 61976012), Qianjiang Postdoctoral Foundation (2020-Y4-A-001), and CERNET Innovation Project (NGII20170315).

References

1. Porzi, L., Hofinger, M., Ruiz, I., Serrat, J., Bulò, S.R., Kontschieder, P.: Learning multi-object tracking and segmentation from automatic annotations. arXiv preprint arXiv:1912.02096 (2019)
2. Osep, A., Voigtlaender, P., Weber, M., Luiten, J., Leibe, B.: 4D generic video object proposals. arXiv preprint arXiv:1901.09260 (2019)
3. Voigtlaender, P., et al.: Mots: multi-object tracking and segmentation. In: Proceedings of the IEEE Conference on Computer Vision and Pattern Recognition, pp. 7942–7951 (2019)
4. Li, X., Weiming, H., Shen, C., Zhang, Z., Dick, A., Van Den Hengel, A.: A survey of appearance models in visual object tracking. ACM Trans. Intell. Syst. Technol. (TIST) **4**(4), 1–48 (2013)
5. Ward, J.S., Barker, A.: Undefined by data: a survey of big data definitions. arXiv preprint arXiv:1309.5821 (2013)
6. Leal-Taixé, L., Milan, A., Reid, I., Roth, S., Schindler, K.: Motchallenge 2015: towards a benchmark for multi-target tracking. arXiv preprint arXiv:1504.01942 (2015)
7. Milan, A., Leal-Taixé, L., Reid, I., Roth, S., Schindler, K.: Mot16: a benchmark for multi-object tracking. arXiv preprint arXiv:1603.00831, 2016

8. Chen, Y., Jing, L., Vahdani, E., Zhang, L., He, M., Tian, Y.: Multi-camera vehicle tracking and re-identification on AI city challenge 2019. In: Proceedings of CVPR Workshops (2019)
9. Milan, A., Schindler, K., Roth, S.: Challenges of ground truth evaluation of multi-target tracking. In: Proceedings of the IEEE Conference on Computer Vision and Pattern Recognition Workshops, pp. 735–742 (2013)
10. Shi, W., Alawieh, M.B., Li, X., Yu, H.: Algorithm and hardware implementation for visual perception system in autonomous vehicle: a survey. Integr. **59**, 148–156 (2017)
11. Leal-Taixé, L., Milan, A., Schindler, K., Cremers, D., Reid, I., Roth, S.: Tracking the trackers: an analysis of the state of the art in multiple object tracking. arXiv preprint arXiv:1704.02781 (2017)
12. Ouyang, Z., Niu, J., Liu, Y., Guizani, M.: Deep CNN-based real-time traffic light detector for self-driving vehicles. IEEE Trans. Mob. Comput. **19**(2), 300–313 (2019)
13. Luo, W., et al.: Multiple object tracking: a literature review. arXiv preprint arXiv:1409.7618 (2014)
14. Ciaparrone, G., Sánchez, F.L., Tabik, S., Troiano, L., Tagliaferri, R., Herrera, F.: Deep learning in video multi-object tracking: a survey. Neurocomputing **381**, 61–88 (2020)
15. Li, B., Wu, F., Weinberger, K.Q., Belongie, S.: Positional normalization. In: Advances in Neural Information Processing Systems, pp. 1620–1632 (2019)
16. Ren, S., He, K., Girshick, R., Sun, J.: Faster R-CNN: towards real-time object detection with region proposal networks. In: Advances in Neural Information Processing Systems, pp. 91–99 (2015)
17. Redmon, J., Farhadi, A.: Yolov3: an incremental improvement. arXiv preprint arXiv:1804.02767 (2018)
18. Szegedy, C., Ioffe, S., Vanhoucke, V., Alemi, A.A.: Inception-v4, inception-ResNet and the impact of residual connections on learning. In: 31st AAAI Conference on Artificial Intelligence (2017)
19. Wang, L., Xu, L., Kim, M.Y., Rigazico, L., Yang, M.H.: Online multiple object tracking via flow and convolutional features. In: 2017 IEEE International Conference on Image Processing (ICIP), pp. 3630–3634. IEEE (2017)
20. Wang, Q., Zhang, L., Bertinetto, L., Hu, W., Torr, P.H.: Fast online object tracking and segmentation: a unifying approach. In: Proceedings of the IEEE Conference on Computer Vision and Pattern Recognition, pp. 1328–1338 (2019)
21. Zhao, D., Hao, F., Xiao, L., Tao, W., Dai, B.: Multi-object tracking with correlation filter for autonomous vehicle. Sensors **18**(7), 2004 (2018)
22. Bewley, A., Ge, Z., Ott, L., Ramos, F., Upcroft, B.: Simple online and realtime tracking. In: 2016 IEEE International Conference on Image Processing (ICIP), pp. 3464–3468. IEEE (2016)
23. Yu, F., Li, W., Li, Q., Liu, Y., Shi, X., Yan, J.: POI: multiple object tracking with high performance detection and appearance feature. In: Hua, G., Jégou, H. (eds.) ECCV 2016. LNCS, vol. 9914, pp. 36–42. Springer, Cham (2016). https://doi.org/10.1007/978-3-319-48881-3_3
24. Chen, J., Sheng, H., Zhang, Y., Xiong, Z.: Enhancing detection model for multiple hypothesis tracking. In: Proceedings of the IEEE Conference on Computer Vision and Pattern Recognition Workshops, pp. 18–27 (2017)
25. Tan, X., et al.: Multi-camera vehicle tracking and re-identification based on visual and spatial-temporal features. In: Proceedings of the IEEE Conference on Computer Vision and Pattern Recognition Workshops, pp. 275–284 (2019)

26. Zhang, S., et al.: Tracking persons-of-interest via adaptive discriminative features. In: Leibe, B., Matas, J., Sebe, N., Welling, M. (eds.) ECCV 2016. LNCS, vol. 9909, pp. 415–433. Springer, Cham (2016). https://doi.org/10.1007/978-3-319-46454-1_26

27. Kieritz, H., Hubner, W., Arens, M.: Joint detection and online multi-object tracking. In: Proceedings of the IEEE Conference on Computer Vision and Pattern Recognition Workshops, pp. 1459–1467 (2018)

28. Weinzaepfel, P., Revaud, J., Harchaoui, Z., Schmid, C.: Deepflow: large displacement optical flow with deep matching. In: Proceedings of the IEEE International Conference on Computer Vision, pp. 1385–1392 (2013)

29. Danelljan, M., Bhat, G., Khan, F.S., Felsberg, M.: Atom: accurate tracking by overlap maximization. In: Proceedings of the IEEE Conference on Computer Vision and Pattern Recognition, pp. 4660–4669 (2019)

30. Maksai, A., Fua, P.: Eliminating exposure bias and metric mismatch in multiple object tracking. In: Proceedings of the IEEE Conference on Computer Vision and Pattern Recognition, pp. 4639–4648 (2019)

31. Li, X., Ma, C., Wu, B., He, Z., Yang, M.H.: Target-aware deep tracking. In: Proceedings of the IEEE Conference on Computer Vision and Pattern Recognition, pp. 1369–1378 (2019)

32. Fan, H., Ling, H.: Siamese cascaded region proposal networks for real-time visual tracking. In: Proceedings of the IEEE Conference on Computer Vision and Pattern Recognition, pp. 7952–7961 (2019)

33. Gao, J., Zhang, T., Xu, C.: Graph convolutional tracking. In: Proceedings of the IEEE Conference on Computer Vision and Pattern Recognition, pp. 4649–4659 (2019)

34. Li, B., Wu, W., Wang, Q., Zhang, F., Xing, J., Yan, J.: Siamrpn++: evolution of siamese visual tracking with very deep networks. In: Proceedings of the IEEE Conference on Computer Vision and Pattern Recognition, pp. 4282–4291 (2019)

35. Wang, G., Luo, C., Xiong, Z., Zeng, W.: SPM-tracker: series-parallel matching for real-time visual object tracking. In: Proceedings of the IEEE Conference on Computer Vision and Pattern Recognition, pp. 3643–3652 (2019)

36. Milan, A., Rezatofighi, S.H., Dick, A., Reid, I., Schindler, K.: Online multi-target tracking using recurrent neural networks. In: 31st AAAI Conference on Artificial Intelligence (2017)

37. Kim, C., Li, F., Rehg, J.M.: Multi-object tracking with neural gating using bilinear LSTM. In: Proceedings of the European Conference on Computer Vision (ECCV), pp. 200–215 (2018)

38. Wang, N., Song, Y., Ma, C., Zhou, W., Liu, W., Li, H.: Unsupervised deep tracking. In: Proceedings of the IEEE Conference on Computer Vision and Pattern Recognition, pp. 1308–1317 (2019)

39. He, K., Zhang, X., Ren, S., Sun, J.: Deep residual learning for image recognition. In: Proceedings of the IEEE Conference on Computer Vision and Pattern Recognition, pp. 770–778 (2016)

A Data Augmentation-Based Defense Method Against Adversarial Attacks in Neural Networks

Yi Zeng[1], Han Qiu[2(✉)], Gerard Memmi[2], and Meikang Qiu[3]

[1] University of California San Diego, San Diego, CA 92122, USA
y4zeng@eng.ucsd.edu
[2] Telecom Paris, Institut Polytechnique de Paris, Palaiseau 91120, France
{han.qiu,gerard.memmi}@telecom-paris.fr
[3] Texas A&M University-Commerce, Texas 75428, USA
meikang.qiu@tamuc.edu

Abstract. Deep Neural Networks (DNNs) in Computer Vision (CV) are well-known to be vulnerable to Adversarial Examples (AEs), namely imperceptible perturbations added maliciously to cause wrong classification results. Such variability has been a potential risk for systems in real-life equipped DNNs as core components. Numerous efforts have been put into research on how to protect DNN models from being tackled by AEs. However, no previous work can efficiently reduce the effects caused by novel adversarial attacks and be compatible with real-life constraints at the same time. In this paper, we focus on developing a lightweight defense method that can efficiently invalidate full whitebox adversarial attacks with the compatibility of real-life constraints. From basic affine transformations, we integrate three transformations with randomized coefficients that fine-tuned respecting the amount of change to the defended sample. Comparing to 4 state-of-art defense methods published in top-tier AI conferences in the past two years, our method demonstrates outstanding robustness and efficiency. It is worth highlighting that, our model can withstand advanced adaptive attack, namely BPDA with 50 rounds, and still helps the target model maintain an accuracy around 80%, meanwhile constraining the attack success rate to almost zero.

Keywords: Adversarial Examples · Deep learning · Security · Affine Transformation · Data augmentation

1 Introduction

With the rapid development of the Deep Neural Networks (DNNs) in Computer Vision (CV), there are more and more real-world applications that rely on the DNN models to classify images or to make decisions [11]. However, in recent years, the DNN models are well known to be vulnerable to Adversarial Examples (AE) which threats the robustness of the DNN usage [25]. Basically,

© Springer Nature Switzerland AG 2020
M. Qiu (Ed.): ICA3PP 2020, LNCS 12453, pp. 274–289, 2020.
https://doi.org/10.1007/978-3-030-60239-0_19

the AEs can be generated by adding carefully designed perturbations that are imperceptible to human eyes but can mislead DNN classifiers with very high accuracy [8].

Today, several rounds of AE attack and corresponding defense techniques have been developed as shown in [18]. The initial research on adversarial attacks on DNN models such as Fast Gradient Sign Method (FGSM) [8] aims at generating AEs by directly calculating the model gradients with respect to the input images. Such methods are then defeated by the defense methods based on various kinds of methods such as model distillation [16]. Then, the improved AE attacks are proposed to combine the gradient-based approach with the optimization algorithm such as the CW [4] aims to find the input features that made the most significant changes to the final output to mislead the DNN models. Such an optimized gradient-based approach can defeat many previous defense methods including the model distillation. Later, advanced defense methods are proposed to mitigate such attacks by obfuscating the gradients of the inference process. Specifically, some data augmentation techniques are deployed in such defense methods such as image compression, image denoising, image transformation [21], etc. Then, such state-of-the-art methods are then defeated by more advanced attack methods such as Backward Pass Differentiable Approximation (BPDA) [2] that can effectively approximate the obfuscated gradients to defeat these defenses.

In this paper, we propose a novel defense method that combines several data augmentation techniques together to mitigate the adversarial attacks against on the DNN models. We propose our method, Stochastic Affine Transformation (SAT), by deploying the image translation, image rotation, and image scaling method together. Our method can be used as a preprocessing step on the input images which makes our solution agnostic on many DNN models. Firstly, our method has little influence on the DNN inference which can effectively maintain the classification accuracy of benign images. Then, intensive experimentation and comparison have been performed to show the improvement of our method compared with several previous state-of-the-art defense solutions. Moreover, our method is a lightweight preprocess-only step that can be used on resource-constrained use cases such as the Internet of Things (IoT) [20].

This paper includes two main contributions. (1) We design a data augmentation-based defense solution to mitigate the initial and optimized gradient-based adversarial attacks on DNN models. Our method combining several steps of data augmentation techniques can be used as a preprocessing step on input images that can effectively maintain the agnostic DNN model's accuracy. (2) Our method can also defeat the advanced adversarial attack method such as BPDA which outperforms many previous state-of-the-art defense solutions.

This paper is organized as follows. Section 2 discusses the background information of this research including the brief definition of the adversarial examples and the previous data augmentation-based defense solutions. Section 3 presents our threat model and defense requirements. Section 4 proposes our methodology including the algorithm and the design details. Section 5 illustrates the experimentation details and evaluation results comparing with the previous state-of-the-art solutions. We then conclude in Sect. 6.

2   Research Backgrounds

In this section, we briefly introduce the background of the AEs in DNNs, the related work on AEs, and the state-of-the-art preprocessing-based defense methods based on data augmentation techniques.

2.1   Adversarial Examples in Deep Neural Networks

AEs can be explained as imperceptible modified samples that force one or multiple DNN models outputs with wrong results. This was first highlighted by [25]. By denoting I an input image, an adversarial example generated from it can be denoted as $\widetilde{I} = I + \delta$, where δ is the adversarial perturbation. The target model, which conducts inference for classification tasks can be denoted as f, thus the problem of performing adversarial attacks on the target DNN model can be formulated as Eq. 1.

$$min\|\delta\|, \; s.t. \; f(\widetilde{I}) \neq f(I) \tag{1}$$

This equation can be interpreted as an optimization task that searches for a $\widetilde{I}$ based on I that can be misclassified by the target model, while keeping $\widetilde{I}$ visually as similar to I as possible. The aforementioned AE generation case is untargeted which aims to mislead the DNN classifier without a pre-set wrong label. As a targeted AE generation procedure aims to attack the DNN classifier to misclassify an input I with original label l as the pre-set wrong label l'. In the concern of real-life adoption of DNN models, both cases can result in serious outcomes if the models are not protected.

Since the time this vulnerability of DNNs has been discovered, various kinds of attacks have been proposed in the past few years to help the society to better understand the nature of AEs. To sum up, past work on adversarial attacks can be classified into two main approaches including initial gradient-based and optimized gradient-based. Fast Gradient Sign Method (FGSM) [8] is one of the most famous initial gradient-based adversarial attacks which calculates the model gradients based on the sign of the gradient of the classification loss concerning the input image. FGSM performs a one-step gradient update along the direction of the sign of gradient at each pixel under L_{inf} constraints to generate AEs. Later on, variations of FGSM were introduced to better searching for the optimum AE based on a single input. Such kind of methods includes I-FGSM [10] and MI-FGSM [6], aim at iteratively calculating the perturbations based on FGSM with a small step or with momentum.

Then, optimized gradient-based AE attacks are proposed to calculate the gradients based on adopting optimization algorithms to find optimal adversarial perturbations directly between the input images and output predicted labels [4]. Such kind of attack is especially powerful in a whitebox or graybox scenario by adopting optimization algorithms to enhance the gradient calculation. Various optimized gradient-based AE attacks were proposed in recent years including Jacobian-based Saliency Map Attack (JSMA [15]), DeepFool [14], LBFGS [25], Carlini & Wagner (CW [4]), and Backward Pass Differentiable Approximation (BPDA [2]). We should highlight the BPDA attack here, as it invalidates dozens of existing state-of-art defense approaches in recent evaluations [2]. The BPDA attack in a manner assumes that a defense function $g(\cdot)$ maintains the property $g(I) \approx I$ in order to preserve the functionality of the target model $f(\cdot)$. Then the adversary can use $g(I)$ on the forward pass and replace it with I on the backward pass when calculating the gradients.

2.2   Data Augmentation-Based Preprocessing Defense Solutions

Various defensive strategies have been proposed to defeat adversarial attacks. One direction is to train a more robust model from either scratch or an existing model. Those approaches aim to rectify AEs' malicious features by including AEs into the training set [28], processing all the training data [32], or revising the DNN topology [16]. However, training a DNN model is very time and resource-consuming, especially for real-life cases, where models are more complicated. Besides, in real-life, DNN models are packed as closed-source applications and cannot be modified, thus those methods are not applicable. Most of all, the adversary can still adaptively generate AEs for the new models [2].

A more promising direction is to preprocess the input data to eliminate adversarial influence without touching the DNN model. These solutions are more suitable in the concern of real-life cases, as it is feasible, efficient, and lightweight. Thus, The preprocessing based defense is within the scope of this paper, as they do not require any laborious work with the DNN models, which made them competitive with most of the real-life defense scenarios. Below we describe some previous works and their limitations:

Feature Distillation (FD) [12] designed a compression method based on the JPEG compression but modified the quantization step. The basic idea is to measure the importance of input features for DNNs by leveraging the statistical frequency component analysis within the DCT of JPEG. It demonstrated a huge improvement in defending adversarial attacks compared with the standard JPEG compression method [19].

SHIELD [5] aims to randomize the quantization step by tuning the window size and quantization factors in the JPEG compression method. In SHIELD, the Stochastic Local Quantization (SLQ) method is used to divide an image into 8×8 blocks and applies a randomly selected JPEG compression quality (tuning quantization factors) to every block. The advantage is that the authors randomized the selective quantization steps which make the defense process different for different input images and make the adversarial attacks more difficult.

Bit-depth Reduction (BdR) [31] performs a simple type of quantization that can remove small (adversarial) variations in pixel values from an image. In the evaluation of that work, it demonstrates a more effective result comparing to adversarial training. However, recently developed attacks are not within the scope of that work, namely the BPDA attack.

Pixel Deflection (PD) [17] aims to add similar natural noises that are not sensitive to the DNN model. The idea of deflection is to randomly sample a pixel from an image and replace it with another randomly selected pixel from within a small square neighborhood. This could generate an artificial noise that affects little on the DNN model but can disturb the adversarial perturbations. Then, a BayesShrink denoising process is followed to recover the image content before this image is feed into the DNN model. The results of such a method are convincing since it introduces randomness into the preprocessing step and does not require any modification on the DNN model. However, the robustness of this method is significantly reduced if the attackers have knowledge of the preprocess step [2].

3   Threat Model and Defense Requirements

3.1   Threat Model

Untargeted attacks and targeted attacks are two major types of adversarial attacks. Untargeted attacks try to mislead the DNN models to an arbitrary label different from the correct one. On the other hand, targeted attacks only considering succeed when the DNN model predicts the input as one specific label desired by the adversary [4]. In this paper, we only evaluate the targeted attacks. The untargeted attacks can be mitigated in the same way.

We consider a full whitebox scenario, where the adversary has full knowledge of the DNN model and the defense method, including the network architecture, exact values of parameters, hyper-parameters, and the details of the defense method. However, we assume the random numbers generated in real-time are perfect with a large entropy such that the adversary cannot obtain or guess the correct values. Such a targeted full whitebox scenario represents the strongest adversaries, as a big number of existing state-of-the-art defenses are invalidated as shown in [27].

As for the adversary's capability, we assume the adversary is outside of the DNN classification system, and he is not able to compromise the inference computation or the DNN model parameters (e.g., via fault injection to cause bit-flips [22] or backdoor attacks [7]). What the adversary can do is to manipulate the input data with imperceptible perturbations. In the context of computer vision tasks, he can directly modify the input image pixel values within a certain range. We use l_∞ and l_2 distortion metrics to measure the scale of added perturbations: we only allow the generated AEs to have either a maximum l_∞ distance of $8/255$ or a maximum l_2 distance of 0.05 as proposed in [2].

3.2   Defense Requirements

Various life-concerned vital tasks are already implemented with DNN in real-life, e.g., video surveillance [26], face authentication [13], autonomous driving [29], network traffic identification [33], etc. Most of those cases' inference is conducted either locally in distributed computing units or remotely with the help of cloud servers. In the case where inference procedures are conducted locally, resources are highly constrained, thus only lightweight designs of defense can protect the system without draw extra burden over those units. Moreover, for both cases, previous resolutions, e.g. adversarial training or extra modifications over the target model would be considerably costly for real-life cases (where each sample is of large size, thus any kind of retraining can be laborious). As mentioned in the previous section, most of the existing defense methods either too complicated to be compatible with real-life constraints [3,9,23] or not capable of effectively reduce the impact brought by adversarial attacks [17].

To cope with those constraints in real-life adoption of DNNs, we believe the following properties should be taken considered when designing novel adversarial defense methods:

- Accuracy-preserving: they should not affect much on the prediction accuracy of the DNN model on clean data samples that processed by those methods.
- Security: they should be capable to effectively reduce effects brought by adversarial perturbations.
- Lightweight: the defense method should not be too heavy to impact the devices' or units' performance or operations, considering the limited onboard computing capabilities and resources.
- Generalization Ability: the defense method should neither require modifications over the target DNN's structure nor require any kind of retraining.

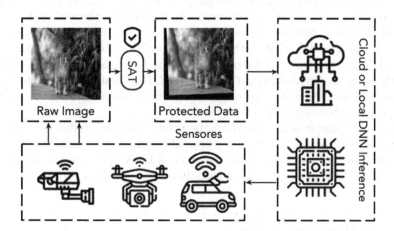

Fig. 1. A system overview of adopting SAT in the real-life use case.

Thus, for a better adaptation over nowadays real-life scenarios, we aim to design a preprocessing-only method to meet all those requirements. It has proved in our previous work [18], preprocessing-only adversarial defenses are competent enough to defense adversarial attacks, even for whitebox attacks.

4   Proposed Methodology

To better adapt to nowadays real-life scenarios' adoption of DNN, we present our efficient defense method against adversarial attacks that can conduct protections on the fly, termed Stochastic Affine Transformation (SAT). Thanks to the lightweight design of SAT, this defense should be more compatible with both cloud DNN inference as well as local or edge DNN inference respecting real-life scenarios. Figure 1 illustrates an overview of adopting the SAT method in real life. The details of SAT will be present in Sect. 4.1. The analysis of the three hyperparameters respecting the defense efficiency is illustrated in Sect. 4.2.

4.1   SAT Algorithm

Following the logic of adding randomness to the affine transformation without harming the classification accuracy [9, 18, 30], we propose a simple but effective way of image distortion as an adversarial example defense. The algorithm is designed based on combining several affine transformation methods. The details are illustrated in Algorithm 1.

Three basic affine transformations with randomized coefficients are bounded tother in this single procedure, namely, translation, rotation, and scaling. We add randomness to those three simple affine transformations so that the attacker cannot utilize a useful gradient to generate adversarial examples even acknowledges the details of this defense. Such is done by acquiring different coefficients that follow three uniform distribution for different samples. To be specific, there are three coefficients along with the raw image as the input of the SAT method. T is the translation limit, R is the rotation limit, and S is the scaling limit. The original input will first be randomly shifted away from its original coordinates according to δ_x and δ_y that both follow the uniform distribution in the range $(-T, T)$. Then, the data will be randomly rotated at a certain angle δ_r that follows the uniform distribution in the range $(-R, R)$. Finally, the distorted image will be acquired by scaling up or down δ_s times, where δ_s follows a uniform distribution in the range $(1 - S, 1 + S)$.

Since only simple affine transformations and random number generator are adopted in SAT, we believe SAT is compatible with both cloud DNN inference procedures as well as localized or edge devices DNN inference procedure. This lightweight design is also hardware friendly and can conduct protections on the fly.

4.2   SAT Hyper-parameters

As aforementioned, there are three essential coefficients in the SAT method. In this part, we did a thorough evaluation and analysis of those three coefficients respecting the efficiency of defending adversarial attacks.

We believe a higher variance between the original data and the protected data while maintaining a high classification accuracy can help more to defend adversarial attacks, which is proved in our previous work [18]. In this part, three different metrics are adopted to evaluate this variance, namely the l_2 norm, Structural Similarity (SSIM) index, and the Peak Signal-to-Noise Ratio (PSNR). The classification accuracy (ACC) is the priority of most classification tasks thus is as well taken considerate in this part.

ALGORITHM 1: Stochastic Affine Transformation

Input: original image $I \in \mathbb{R}^{h \times w}$
Output: transformed image $I' \in \mathbb{R}^{h \times w}$
Parameters: translation limit T; scaling limit S, rotation limit R.

1   $I' = O^{h \times w}$;
   /* 1.Translation */
2   $\delta_x \sim \mathcal{U}(-T, T)$;
3   $\delta_y \sim \mathcal{U}(-T, T)$;
4   $\Delta_x = \delta_x \times w$;
5   $\Delta_y = \delta_y \times h$;
6   **if** $(x + \Delta_x \in (0, w)) \wedge (y + \Delta_y \in (0, h))$ **then**
7 | $I'(x, y) = I(x + \Delta_x, y + \Delta_y)$;
8   **end**
   /* 2.Rotation */
9   $\delta_r \sim \mathcal{U}(-R, R)$;
10   $\Delta_r = \delta_r \times \pi/180$;
11   **for** (x_i, y_j) *in* $\{(x, y)|x \in (0, w), y \in (0, h)\}$ **do**
12 | $x'_i = -(x_i - \lfloor w/2 \rfloor) \times sin(\Delta_r) + (y_j - \lfloor h/2 \rfloor) \times cos(\Delta_r)$;
13 | $y'_j = (x_i - \lfloor w/2 \rfloor) \times cos(\Delta_r) + (y_j - \lfloor h/2 \rfloor) \times sin(\Delta_r)$;
14 | $x'_i = \lfloor x'_i + \lfloor w/2 \rfloor \rfloor$;
15 | $y'_j = \lfloor y'_j + \lfloor h/2 \rfloor \rfloor$;
16 | **if** $(x'_i \in (0, w)) \wedge (y'_j \in (0, h))$ **then**
17 | | $I'(x_i, y_j) = I(x'_i, y'_j)$;
18 | **end**
19   **end**
   /* 3.Scaling */
20   $\delta_s \sim \mathcal{U}(1 - S, 1 + S)$;
21   $h_{new} = \delta_s \times h$;
22   $w_{new} = \delta_s \times w$;
23   $I' = \text{reshape}(I', (h_{new}, w_{new}))$;
24   **if** $\delta_s > 1$ **then**
25 | $I'(x, y) = cropping(I', (h, w))$;
26   **end**
27   **if** $\delta_s < 1$ **then**
28 | $I'(x, y) = padding(I', (h, w))$;
29   **end**
30   **return** I';

---

l_2 is a widely adopted metric in deep learning domain to measure the amount of difference of two samples in the term of Euclidean distance, a higher l_2 indicates a greater difference. Equation 2 shows how l_2 can be computed.

$$l_2(I', I) = \sqrt{(I'_R - I_R)^2 + (I'_G - I_G)^2 + (I'_B - I_B)^2}/(h \times w \times 3) \quad (2)$$

Where I' and I are the two 3-channel (RGB) samples to be compared. h and w are the height and width of those samples respectively.

SSIM is a metric in the computer vision domain normally being adopted to measure the similarity between two images, where smaller SSIM reflects greater difference. To be specific, SSIM is based on three comparison measurements between two samples, namely luminance (l), contrast (c), and structure (s). Each comparison function is elaborated in Eqs. 3–5 respectively.

$$l(I', I) = \frac{2\mu_{I'}\mu_I + c_1}{\mu_{I'}^2 + \mu_I^2 + c_1} \tag{3}$$

$$c(I', I) = \frac{2\sigma_{I'}\sigma_I + c_2}{\sigma_{I'}^2 + \sigma_I^2 + c_2} \tag{4}$$

$$s(I', I) = \frac{\sigma_{I'I} + c_2/2}{\sigma_{I'}\sigma_I + c_2/2} \tag{5}$$

Where $\mu(\cdot)$ computes the mean of a sample, $\sigma^2(\cdot)$ computes the variance of a sample. $c1$ is equal to $(0.01 \times L)^2$, $c2$ is equal to $(0.03 \times L)^2$. Here L is the dynamic range of the pixel values. Finally, the SSIM can be acquired by computing the product of those three functions.

PSNR is most commonly used to measure the quality of reconstruction of lossy compression codecs, say compression, augmentation, or distortion, etc. A smaller PSNR indicates a greater difference between the two evaluating samples. PSNR can be defined via the mean squared error (MSE) between two comparing samples, which is explained in Eq. 6.

$$PSNR(I', I) = 20 \cdot log_{10}(255) - 10 \cdot log_{10}(MSE(I', I)) \tag{6}$$

Where $MSE(\cdot)$ computes the MSE between two inputs.

Table 1. Comparisons of different methods over ACC and amount of changes.

Defense	l_2 norm	SSIM	PSNR	ACC
SAT	**0.322**	**0.194**	**10.219**	**0.98**
FD [12]	0.1343	0.4310	18.050	0.97
SHIELD [5]	0.0405	0.8475	28.345	0.94
BdR [31]	0.0709	0.7730	23.010	0.92
PD [17]	0.0147	0.9877	37.100	0.97

We tried different values of T and S in the range $[0.01, 0.5]$. Different values of R is acquired in the range $[0, 40]$. Thus, each coefficient will test 11 values in the respecting range. As for different metrics, we will acquire 1331 ($11 \times 11 \times 11$) results from different combinations of those different values of coefficients. Figure 2 demonstrates the change of those four metrics' value when different T, S, and R are adopted.

In Fig. 2(a), the changes of ACC with those three coefficients varies is presented. The right side of Fig. 2(a) is the color-bar that reflecting the ACC attained with respecting combinations of those three coefficients. We can learn that lower T, S, and R can help the model maintain a high ACC. To ensure a high ACC, we set 95% ACC as a standard, thus those combinations with ACC below this standard would not be taken further considerations.

From Fig. 2(b) we can learn that the l_2 is not that sensitive with S and R comparing to T in their respecting range. Combining the information provided from Figs. 2(b), 2(c), and 2(d) we can as well acquire this similar analytical result for SSIM and PSNR. This phenomenon also reflects that the l_2, SSIM, and PSNR can all reflecting the scale of changes in a similar manner respecting affine transformations.

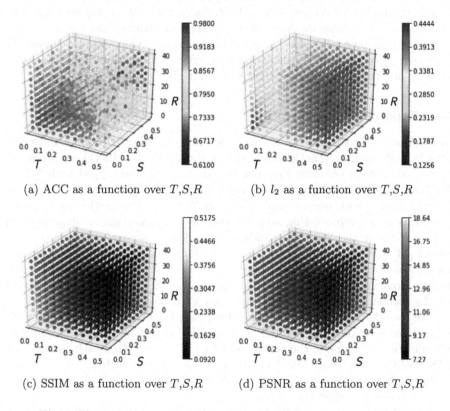

(a) ACC as a function over T,S,R

(b) l_2 as a function over T,S,R

(c) SSIM as a function over T,S,R

(d) PSNR as a function over T,S,R

Fig. 2. Metrics of reconstructed images under different values of T,S,R

By overlapping Fig. 2(a) with the other three figures, we can acquire a set of optimum coefficients that ensures high ACC and great variance at the same time. The set of coefficients for the following experiment is set as follows: $T = 0.16$, $S = 0.16$, and finally $R = 4$.

We compare the SAT method using those fine-tuned hyperparameters with other state-of-art adversarial defense methods, which presented in Table 1. As demonstrated, SAT can create a greater difference between raw samples and protected samples while maintaining a high ACC than other methods. We will evaluate whether this greater variance will help and how much will it help DNN models to defend adversarial attacks in the following section.

5   Experimentation and Evaluation

In this section, we conduct a comprehensive evaluation of the proposed technique. Various adversarial attacks are taken considered in this part: 4 kinds of standard adversarial attacks (FGSM, I-FGSM, LBFGS, and C&W) are conducted, advanced interactive gradient approximation attack, namely BPDA, is also conducted to evaluate the robustness of SAT. We compare SAT with four state-of-art defense methods publish in top-tier artificial intelligence conferences from the past two years. This section would be divided into three parts to elaborate on the settings of the experiment, efficiency over standard adversarial attacks, and the efficiency over BPDA respectively.

5.1   Experimental Settings

Tensorflow [1] is adopted as the deep learning framework to implement the attacks and defenses. The learning rate of the C&W and BPDA attack is set to 0.1. All the experiments were conducted on a server equipped with 8 Intel I7-7700k CPUs and 4 NVIDIA GeForce GTX 1080 Ti GPU.

SAT is of general-purpose and can be applied to various models over various platforms as a preprocessing step for computer vision tasks as illustrated in Fig. 1. Without the loss of generality, we choose a pre-trained Inception V3 model [24] over the ImageNet dataset as the target model. This state-of-the-art model can reach 78.0% top-1 and 93.9% top-5 accuracy. We randomly select 100 images from the ImageNet Validation dataset for AE generation. These images can be predicted correctly by this Inception V3 model.

We consider the targeted attacks where each target label different from the correct one is randomly generated [2]. For each different attack, we measure the classification accuracy of the generated AEs (ACC) and the attack success rate (ASR) of the targeted attack. To be noticed that the untargeted attacks are not within our scope in this work. A higher ACC or lower ASR indicates the defense is more resilient against the attacks.

For comparison, we re-implemented 4 existing solutions including FD [12], SHIELD [5], Bit-depth Reduction [31], and PD [17].

5.2   Evaluation on Defending Adversarial Attacks

We first evaluate the efficiency of our proposed method over standard adversarial attacks, namely FGSM, I-FGSM, C&W, and LBFGS. For FGSM and I-FGSM, AEs are generated under l_∞ constraint of 0.03. For LBFGS and C&W, the attack process is iterated under l_2 constraint and stops when all targeted AEs are found.

Table 2. Comparisons of different defense against attacks respecting ACC.

Defense	Clean	FGSM($\epsilon = .03$)	IFGSM($\epsilon = .03$)	C&W	LBFGS
Baseline	1.00	0.42	0.02	0.06	0.00
SAT	**0.98**	**0.61**	**0.85**	**0.78**	**0.96**
FD [12]	0.97	0.47	**0.87**	**0.84**	**0.97**
SHIELD [5]	0.94	0.49	0.84	**0.78**	0.92
BdR [31]	0.92	0.47	0.82	0.61	0.90
PD [17]	0.97	0.42	0.30	0.11	0.86

We measure the model accuracy (ACC) and attack success rate (ASR) with the protection of SAT and other defense methods.

The results are shown in Tables 2 and 3. For benign samples only, our proposed techniques have the smallest influence on the model accuracy comparing to past works. For defeating AEs generated by these standard attacks, the attack success rate can be kept around 0% and the model accuracy can be drastically recovered, which an efficiency against different kinds of adversarial attacks is demonstrated. To be noticed, previous work can only attain an accuracy of around 50% on samples attacked by the FGSM ($\epsilon = .03$). SAT can recover the accuracy to 0.61%.

Comparing to the Table 1 which compares different methods' capability of creating a variance between input and defended sample, the defense efficiency evaluated in this part shows a strong correlation with the amount of variance generated by the defense method. This has confirmed the previous conclusion in our previous work [18].

In a nutshell, the effectiveness of SAT against standard adversarial attack is demonstrated, as we can attain a state-of-art defense efficiency on all the evaluated attacks comparing to other methods.

Table 3. Comparisons of different defense against attacks respecting ASR.

Defense	IFGSM($\epsilon = .03$)	C&W	LBFGS
Baseline	0.95	1.00	1.00
SAT	**0.01**	**0.01**	**0.00**
FD [12]	**0.00**	0.06	**0.00**
SHIELD [5]	**0.01**	0.02	**0.00**
BdR [31]	0.02	0.2	**0.00**
PD [17]	0.6	0.11	0.02

The ASR evaluation of the FGSM is not shown here since the baseline ASR is 0.

5.3   Evaluation on Defending Advanced Adversarial Attacks

We then evaluate the effectiveness of SAT against the BPDA attack. Since the BPDA attack is an interactive attack, we record the ACC and ASR for each round for different defense methods.

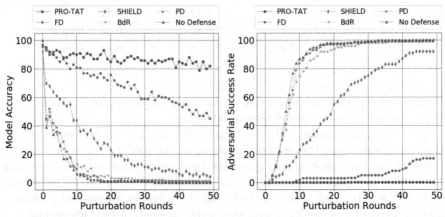

(a) ACC per round under BPDA attack.   (b) ASR per round under BPDA attack.

Fig. 3. ACC and ASR of various techniques under the BPDA attack.

The model prediction accuracy and attack success rate in each round are shown in Fig. 3(a) and 3(b), respectively. We can observe that after 50 attack rounds, all other three prior solutions except FD can only keep the model accuracy lower than 5%, and attack success rates reach higher than 90%. Those defenses fail to mitigate the BPDA attack. FD can keep the attack success rate lower than 20% and the model accuracy is around 40%. This is better but still not very effective in maintaining the DNN model's robustness.

In contrast, SAT is particularly effective against the BPDA attack. As our method can maintain an acceptable model accuracy (around 80% for 50 perturbation rounds), and restrict the attack success rate to 0 for all the record rounds. This result is as well consistent with the l_2, SSIM, and PSNR metrics compared in Table 1: the randomization effects in SAT cause greater variances between I' and I, thus invalidating the BPDA attack basic assumption, which is $I' \approx I$.

To sum up, the effectiveness of SAT against the BPDA attack is demonstrated, as a considerable improvement over the defense efficiency against the BPDA attack is shown comparing to previous work.

6   Conclusion

In this paper, we proposed a lightweight defense method that can effectively invalidate adversarial attacks, termed SAT. By adding randomness to the coefficients, we integrated three basic affine transformations into SAT. Compared with four state-of-art defense methods published in the past two years, our method clearly demonstrated a more robust and effective defense result on standard adversarial attacks. Moreover, respecting the advanced BPDA attack, SAT showed an outstanding capability of maintaining the target model's ACC and detain the ASR to 0. This result is almost 50% better than the best result achieved by previous work against full whitebox targeted attacks.

References

1. Abadi, M., et al.: Tensorflow: a system for large-scale machine learning. In: 12th {USENIX} Symposium on Operating Systems Design and Implementation ({OSDI} 16), pp. 265–283 (2016)
2. Athalye, A., Carlini, N., Wagner, D.: Obfuscated gradients give a false sense of security: circumventing defenses to adversarial examples. In: International Conference on Machine Learning, pp. 274–283 (2018)
3. Buckman, J., Roy, A., Raffel, C., Goodfellow, I.: Thermometer encoding: one hot way to resist adversarial examples (2018)
4. Carlini, N., Wagner, D.: Towards evaluating the robustness of neural networks. In: 2017 IEEE Symposium on Security and Privacy (SP), pp. 39–57. IEEE (2017)
5. Das, N., et al.: Shield: fast, practical defense and vaccination for deep learning using JPEG compression. In: Proceedings of the 24th ACM SIGKDD International Conference on Knowledge Discovery & Data Mining, pp. 196–204 (2018)
6. Dong, Y., Liao, F., Pang, T., Hu, X., Zhu, J.: Discovering adversarial examples with momentum. arXiv preprint arXiv:1710.06081 (2017)
7. Gao, Y., Xu, C., Wang, D., Chen, S., Ranasinghe, D.C., Nepal, S.: Strip: a defence against trojan attacks on deep neural networks. In: Proceedings of the 35th Annual Computer Security Applications Conference, pp. 113–125 (2019)
8. Goodfellow, I.J., Shlens, J., Szegedy, C.: Explaining and harnessing adversarial examples. arXiv preprint arXiv:1412.6572 (2014)
9. Guo, C., Rana, M., Cisse, M., van der Maaten, L.: Countering adversarial images using input transformations. In: International Conference on Learning Representations (2018)
10. Kurakin, A., Goodfellow, I., Bengio, S.: Adversarial examples in the physical world. arXiv preprint arXiv:1607.02533 (2016)
11. LeCun, Y., Bengio, Y., Hinton, G.: Deep learning. Nature **521**(7553), 436–444 (2015)
12. Liu, Z., et al.: Feature distillation: DNN-oriented JPEG compression against adversarial examples. In: 2019 IEEE/CVF Conference on Computer Vision and Pattern Recognition (CVPR), pp. 860–868. IEEE (2019)
13. Mao, Y., Yi, S., Li, Q., Feng, J., Xu, F., Zhong, S.: A privacy-preserving deep learning approach for face recognition with edge computing. In: Proceedings USENIX Workshop Hot Topics Edge Computing (HotEdge), pp. 1–6 (2018)

14. Moosavi-Dezfooli, S.M., Fawzi, A., Frossard, P.: Deepfool: a simple and accurate method to fool deep neural networks. In: Proceedings of the IEEE Conference on Computer Vision and Pattern Recognition, pp. 2574–2582 (2016)

15. Papernot, N., McDaniel, P., Jha, S., Fredrikson, M., Celik, Z.B., Swami, A.: The limitations of deep learning in adversarial settings. In: 2016 IEEE European Symposium on Security and Privacy (EuroS&P), pp. 372–387. IEEE (2016)

16. Papernot, N., McDaniel, P., Wu, X., Jha, S., Swami, A.: Distillation as a defense to adversarial perturbations against deep neural networks. In: 2016 IEEE Symposium on Security and Privacy (SP), pp. 582–597. IEEE (2016)

17. Prakash, A., Moran, N., Garber, S., DiLillo, A., Storer, J.: Deflecting adversarial attacks with pixel deflection. In: Proceedings of the IEEE Conference on Computer Vision and Pattern Recognition, pp. 8571–8580 (2018)

18. Qiu, H., Zeng, Y., Zheng, Q., Zhang, T., Qiu, M., Memmi, G.: Mitigating advanced adversarial attacks with more advanced gradient obfuscation techniques. arXiv preprint arXiv:2005.13712 (2020)

19. Qiu, H., Zheng, Q., Memmi, G., Lu, J., Qiu, M., Thuraisingham, B.:Deepresidual learning based enhanced JPEG compression in the internet of things. IEEE Trans. Ind. Inf. (2020)

20. Qiu, H., Zheng, Q., Zhang, T., Qiu, M., Memmi, G., Lu, J.: Towards secure and efficient deep learning inference in dependable IoT systems. IEEE Internet of Things J. (2020)

21. Qiu, M., Qiu, H.: Review on image processing based adversarial example defenses in computer vision. In: 2020 IEEE 6th International Conference on Big Data Security on Cloud (BigDataSecurity), pp. 94–99. IEEE (2020)

22. Rakin, A.S., He, Z., Fan, D.: Bit-flip attack: crushing neural network with progressive bit search. In: Proceedings of the IEEE International Conference on Computer Vision, pp. 1211–1220 (2019)

23. Shaham, U., Yamada, Y., Negahban, S.: Understanding adversarial training: increasing local stability of supervised models through robust optimization. Neurocomputing **307**, 195–204 (2018)

24. Szegedy, C., Vanhoucke, V., Ioffe, S., Shlens, J., Wojna, Z.: Rethinking the inception architecture for computer vision. In: Proceedings of the IEEE conference on Computer Vision and Pattern Recognition, pp. 2818–2826 (2016)

25. Szegedy, C., etal.: Intriguing properties of neural networks. arXiv preprint arXiv:1312.6199 (2013)

26. Tang, Y., Zhang, C., Gu, R., Li, P., Yang, B.: Vehicle detection and recognition for intelligent traffic surveillance system. Multimedia Tools Appl. **76**(4), 5817–5832 (2015). https://doi.org/10.1007/s11042-015-2520-x

27. Tramer, F., Carlini, N., Brendel, W., Madry, A.: On adaptive attacks to adversarial example defenses. arXiv preprint arXiv:2002.08347 (2020)

28. Tramèr, F., Kurakin, A., Papernot, N., Goodfellow, I., Boneh, D., McDaniel, P.: Ensemble adversarial training: attacks and defenses. arXiv preprint arXiv:1705.07204 (2017)

29. Wu, B., Iandola, F., Jin, P.H., Keutzer, K.: Squeezedet: unified, small, low power fully convolutional neural networks for real-time object detection for autonomous driving. In: Proceedings of the IEEE Conference on Computer Vision and Pattern Recognition Workshops, pp. 129–137 (2017)

30. Xie, C., Wang, J., Zhang, Z., Ren, Z., Yuille, A.: Mitigating adversarial effects through randomization. In: International Conference on Learning Representations (2018)

31. Xu, W., Evans, D., Qi, Y.: Feature squeezing: detecting adversarial examples in deep neural networks. In: 25th Annual Network and Distributed System Security Symposium, NDSS 2018, San Diego, California, USA, February 18–21, 2018. The Internet Society (2018)
32. Yang, Y., Zhang, G., Katabi, D., Xu, Z.: Me-Net: towards effective adversarial robustness with matrix estimation. In: International Conference on Machine Learning, pp. 7025–7034 (2019)
33. Zeng, Y., Gu, H., Wei, W., Guo, Y.: $deep-full-range$: A deep learning based network encrypted traffic classification and intrusion detection framework. IEEE Access **7**, 45182–45190 (2019)

User Recruitment with Budget Redistribution in Edge-Aided Mobile Crowdsensing

Yanlin Zhang[1,2], Peng Li[1,2(✉)], and Tao Zhang[3]

[1] College of Computer Science and Technology, Wuhan University of Science and Technology, Hubei, China
lipeng@wust.edu.cn
[2] Hubei Province Key Laboratory of Intelligent Information Processing and Real-time Industrial System, Hubei, China
[3] Department of Computer Science, New York Institute of Technology, New York, USA

Abstract. Mobile crowdsensing is an efficient data collection method using various mobile smart devices to collect data. Most of the existing mobile crowdsensing frameworks adopt a two-layer architecture model with only a cloud platform to recruit users. However new challenges arise: one is how to reduce the pressure and transmission cost of the cloud platform. Another is how to improve the coverage rate of some areas with low completion rates to ensure data quality. In this paper, we propose an original three-layer mobile crowdsensing framework composed of the cloud platform, edge nodes, and users. It transfers user recruitment and data processing to edge nodes, which offload the data of the cloud platform. Moreover, we propose the offline and online mechanisms based on users' reputations to solve user recruitment in the edge node. Furthermore, a budget redistribution (BRD) algorithm is proposed. It dynamically redistributes the budget according to the task completion rate of different edge nodes. Finally, we show the proposed mechanism was truthful, individual rationality, calculation efficiency, and budget feasibility. Extensive simulations on real data sets show the reliability and effectiveness of our proposed framework and algorithms.

Keywords: Mobile Crowdsensing · Edge computing · User recruitment · Budget redistribution

1 Introduction

With mobile smart devices have abundant embedded sensors and powerfully computing capabilities, using mobile smart devices to collect data has become an emerging method in data collection. This paradigm is called Mobile Crowd

This work is partially supported by the NSF of China (No. 61502359), the Hubei Provincial Natural Science Foundation of China (No.2018CFB424).

M. Qiu (Ed.): ICA3PP 2020, LNCS 12453, pp. 290–305, 2020.
https://doi.org/10.1007/978-3-030-60239-0_20

Sensing(MCS) [1,2], which has been applied in many fields, such as noise monitoring, traffic monitoring, health monitoring, and indoor positioning [3–5].

Most of the existing MCS frameworks are two-layer architecture, with a central cloud platform to select users, process the collected data, and reward users (see in Fig. 1a). However, the sense distance of the cloud platform to the area is generally relatively far away. When a user uploads data to the cloud platform, with the long transmission time and high latency, it is possible to increase the sensing cost of the user and combat the enthusiasm of users' participation in tasks. Additionally, to enhance the coverage rate of the unfinished Points of Interested (PoIs), the platform has to recruit numerous users to sense for improving the overall coverage of the task, which generates data redundancy and costs rising.

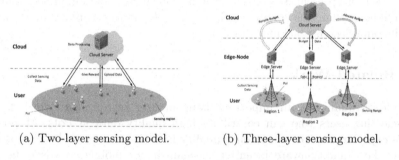

(a) Two-layer sensing model.   (b) Three-layer sensing model.

Fig. 1. Two different sensing models.

In this paper, considering the challenges of the two-layer architecture, we refer to the edge node as the middle layer and propose a three-layer architecture model including a cloud platform, edge nodes, and users (see in Fig. 1b). First, we divide a sensing area into several sub-areas, each of which has a corresponding edge node to sense. User recruitment and data processing functions are transferred from the cloud platform to the edge node [6], to reduce the pressure of the cloud platform. Then, we design an offline user recruitment mechanism based on users' reputation for each edge node, using diminishing marginal utility [7] to select users who meet the pre-set threshold to sense task. In reality, some users may leave in the middle of the task, resulting in different task completion rates in different areas. Hence, in an online algorithm, the edge node dynamically adjusts the recruitment threshold to add new users to sense the task. At the same time, the cloud platform will recycle all remaining budget from edge nodes and redistribute it based on the task completion rate of edge nodes. As for data quality, we utilize the reputation to quantify participation users in the recruitment of the edge node and believe that users with higher reputation have a higher probability of reliable data quality. We select users based on their prices and reputations, as well we utilize the k-Means clustering algorithm as a measure of user data quality.

Specifically, our contributions are summarized as follows:

- We propose a three-layer framework that includes cloud platform, edge nodes, and users to address low completion rates in the sparse areas of the traditional framework.
- We propose an offline mechanism and a dynamic budget online mechanism assisted by edge nodes and utilize reputation to quantify the users' long-term performance to improve data quality.
- We show the proposed mechanism was truthful, individual rationality, calculation efficiency and budget feasibility.
- Through extensive simulations, the results show that our proposed framework and mechanism are effective and efficient.

The remainder of this paper is organized as follows. Section 2 gives the related work. Section 3 presents system model and problem formulation. In Sect. 4, offline and online mechanisms are proposed, respectively. Section 5 gives the simulations results, and Sect. 6 concludes this paper.

2   Related Work

MCS data quality is an important issue in user recruitment, which focuses on selecting users who can contribute high quality data in PoIs. At present, there exist large works about data quality in MCS. Gao et al. [8] proposed an online data quality-aware incentive mechanism for mobile crowdsensing with extra rewards. Ding et al. [9] proposed a multi-objective optimization model of service node selection based on the trade-off between service quality and resource consumption in MCS. Wang et al. [10] proposed a multitask assignment scheme in MCS. They introduced a task-specific minimum sensing quality threshold to redefine the problem of multi-task assignment. Yang et al. [11] designed an unsupervised learning method to quantify the data quality and long-term reputation of participants and used outlier monitoring technology to filter out abnormal data items. Jin et al. [12] proposed a quality-aware amount reward mechanism that rewards participants who have made tremendous efforts to sense data.

User recruitment is a basic problem in MCS. Task publishers want to recruit the most effective users who can provide high-quality sensing services. Zhang et al. [13] proposed conflict-aware participant recruitment in MCS, which introduced participant ability and reputation evaluation indicators. Liu et al. [14] proposed a user recruitment method to improve data inference accuracy in sparse MCS. Wu et al. [15] proposed user characteristic aware recruitment for MCS. They divided different hot areas according to the user's historical information and choose participating users according to the user's reputation, willingness, etc. Karaliopoulos et al. [16] proposed optimal user choice engineering in mobile crowdsensing with bounded rational users, which decision tree was used to selected users. Wang et al. [17] proposed a novel personalized task-oriented worker recruitment mechanism for MCS based on users' preference.

Mobile edge computing (MEC) [18] is to sink computing power to distributed base stations and add computing, storage, processing, and other functions on the wireless network side. Xia et al. [19] proposed a framework for quality-aware sparse data collection with edge nodes. Ma et al. [20] usesd edge computing to address users' privacy protection and reputation updates in their work.

Unlike most of the existing research work, in this paper, we propose a three-layer sensing framework, which uses edge nodes to offload the data of the cloud platform, to reduce the pressure of the cloud platform. Then, we design a user recruitment mechanism for each edge node. The cloud platform can dynamically recycle and allocate the budget of edge nodes based on the task completion rate of edge nodes, so that each sub-area gets enough budget to recruit users.

3   System Model and Problem Formulation

This section describes the proposed three-layer architecture framework for MCS. It consists of three parts, included cloud platform, edge nodes, and users.

Cloud Platform: The cloud platform is in charge of the overall control of the sensing task. When the cloud platform publishes a task and gives a budget B of this task, the task contains a set of PoIs in the sensing area is $P = \{1, 2, ..., K\}$, where K is the number of the PoIs. The users should reach all the PoIs to complete the task. Each $k \in P$ needs to be covered d_k time(s), and the location of k is p_k^l. The deadline of the task is T and the set of edge nodes is $N = \{1, 2, ..., I\}$, where I is the number of edge nodes.

Edge Node: Each node $i \in N$ has a coverage range of R. If several PoIs are within this range, they constitute a subset of PoIs for the node i, denoted by P_i ($P_i \subseteq P$), and we assume that a PoI only belong to one edge node. According to the size of the edge node, it gets the budget B_i from the cloud platform

User: The candidate set of edge node i consists of users who are prepared to participate in the sensing task and within the range of edge node i, denoted by U_i. Each $j \in U_i$ is given a reputation value $u_{i,j}^r$ based on user's historical sensing record and has a request reward $u_{i,j}^c$. We assume that user j can only be recruited by one edge node i in the task. If selected, edge node i evaluate the data quality contributed by user j and gives the final reward $u_{i,j}^{c'}$.

3.1   Edge Node PoIs Allocation

For edge node i, the radius of the covered area is N_i^R and its location is N_i^l, so the set of PoI that edge node i needs to sense is P_i. The PoI $k \in P$ is divided into the nearest edge nodes i, and then is added to P_i and removed from P, $P_i = P_i \cup \{k\}$, $P = P \setminus \{k\}$. After the allocation, all the PoIs are sensed by the corresponding edge node, and the budget is allocated proportionally according to the number of PoIs that each edge node needs to sense, as $B_i = B \times \frac{|P_i|}{|P|}$.

3.2   Users' Sensing Ability

Edge node i recruits users based on the number of PoIs and given budget. When user j decides to participate in the sensing task and is within the coverage of edge-node i, constituting the candidate set U_i. The platform hopes to select the most suitable user set from the candidate to perform tasks, denoted by U_i^s. Each user participating in the edge node i is required to cover all the PoIs (reach the PoI and upload data). Based on user j's reputation, we could calculate the number of PoIs that user j can complete. We denote the user j's ability as follows

$$f(u_{i,j}^r) = \begin{cases} \alpha \times \sin(\dfrac{\pi}{2} \times u_{i,j}^r), & \text{if } u_{i,j}^r \geq \overline{u}^r \\ \beta \times \sin(\dfrac{\pi}{2} \times u_{i,j}^r), & \text{if } u_{i,j}^r < \overline{u}^r \end{cases}, \tag{1}$$

where α, $\beta \in [0,1]$, and they are weight coefficient. $f(u_{i,j}^r)$ represents the completion ability of user j with reputation value $u_{i,j}^r$, which is in the range of $[0,1]$, and $\overline{u}^r$ denotes the average reputation of the edge node i.

For edge node i, the total expected collected amount of data by U_i^s is defined as the utility function of the edge node i, as

$$V(U_i^s) = \sum_{k=1}^{|P_i|} \min(d_{i,k}, \sum_{j=1}^{|U_i^s|} E_{p_{i,k}}(u_{i,j}^r)), \tag{2}$$

where $d_{i,k}$ reprents request sensing data at PoI k, and $E_{p_{i,k}}(u_{i,j}^r)$ denotes user j expected sensing data at PoI k. $E_{p_{i,k}}(u_{i,j}^r) = b_j * f(u_{i,j}^r)$, and b_j means user j's sensing data when completes the task. As $f(u_{i,j}^r)$ denotes the completion ability of user j, the expected sensing data is proportional to the user j's completion ability.

We adopt diminishing marginal utility to describe the utility of selecting a user on the whole. When the set is empty, adding a user will get a huge utility, while there are enough users, the utility of adding a new user will be less.

Theorem 1. *The proposed Eq. (2) is a monotonic submodule function.*

Proof. Since $V(U_i^s) = \sum\limits_{k=1}^{|P_i|} \min(d_{i,k}, \sum\limits_{j=1}^{|U_i^s|} E(u_{i,j}^r))$, $\forall\ U_{i,1} \subseteq U_{i,2} \subseteq U_i$, there is $V(U_{i,1}) \leq V(U_{i,2})$. Meanwhile, $\forall u \in U \backslash U_{i,2}$, there is

$$V(U_{i,1} \cup \{u\}) - V(U_{i,1})$$
$$= \sum_{k=1}^{|P_i|} \min(\max(0, d_{i,k} - \sum_{j=1}^{|U_i^s|} E(u_{i,j}^r)), E(u))$$
$$\geq \sum_{k=1}^{|P_i|} \min(\max(0, d_{i,k} - \sum_{j=1}^{|U_i^s|} E(u_{i,j}^r)), E(u))$$
$$= V(U_{i,2} \cup \{u\}) - V(U_{i,2}).$$

Consequently, we obtain the utility of adding a user j into U_i^s, as

$$V_{u_{i,j}}(U_i^s) = V(U_i^s \cup \{u_{i,j}\}) - V(U_i^s).\qquad(3)$$

Therefore, the marginal utility of user j is that adding user j to participating set U_i^s to minus the utility of not adding the user j.

3.3   Data Estimation and Reputation Update

Many factors affect data quality, such as the accuracy of the sensor. The higher the accuracy, the higher the data quality. However, due to some user's wrong operations, low-quality data is inevitable. We add all user-contributed data at the same PoI k in the edge node i to a set $X_{i,k} = \{d_{i,k,1}, ..., d_{i,k,j}\}$, where $d_{i,j,k}$ means the contributed data by user j at PoI k. Using the K-Means algorithm to find the center point ω of the cluster, and calculating the offset value of each data point from the center point. We believe that the closer to the center point, the more reliable the data quality of the user j at the PoI. Thus, the data quality of the user j under the PoI k can be expressed as:

$$q_{i,j,k} = \left(\frac{1}{\frac{d_{i,j,k}}{\lambda_k} + \tau}\right) \Big/ \left(\sum_{j=1}^{|U_i^d|} \frac{1}{\frac{d_{i,j,k}}{\lambda_k} + \tau}\right),\qquad(4)$$

where $\lambda_k = \sum\limits_{j=1}^{|U_i|} d_{i,j,k}$ denotes the total offset distance at PoI k, τ is a constant, avoiding the problem of divisor being zero, and $q_{i,j,k} \in [0,1]$. According to the quality of data completed at all PoIs by user, the reputation update function is defined as

$$R(q_{i,j}) = \gamma \times \left(\frac{1}{1 + e^{1 - q_{i,j}}}\right),\qquad(5)$$

where $q_{i,j} = \sum\limits_{k=1}^{P_i}(q_{i,j,k} - \bar{q}_{i,k})$ represents the sum of offset between user j and the average data quality $\bar{q}_{i,k}$ of each PoI i. γ is a quality influencing factors. After users contribute low-quality data, their reputation will reduce faster, but reputation increasing while contributing high-quality data, with a slower rate.

3.4   Budget Redistribution

In the actual execution of the task, we divide T into n slots. For each slot t, the platform will investigate the task completion rate of each edge node, and then dynamically redistribute the budget of each edge node. The task completion rate of edge node i is denoted as

$$\varphi_i = 1 - (\|P_i^r - P_i^a\|_F)/(\|P_i^r\|_F),\qquad(6)$$

where P_i^r represents a set of required coverage frequency of edge node i, and P_i^a represents current actual coverage frequency of edge node i. $\|\cdot\|_F$ is the F norm.

During each time solt t, some users may be leave, denoted as U_i^d, and we calculate the amount of data that they actually contributed before leaving, as

$$V^a(U_i^d) = \sum_{k=1}^{|P_i|} \min(d_{i,k}, X_{i,k}).$$

(7)

Thence, there will be a gap between $V(U_i^d)$ and $V^a(U_i^d)$, denoted as

$$\delta_i = (V(U_i^d) - V^a(U_i^d))/\sum_{j=1}^{|U_i^d|} u_{i,j}^c.$$

(8)

where $V(U_i^d)$ denote as the edge node i's total expected collected amount of data. We could use Eq. (2) to calculate. Edge nodes will adjust the number of recruited users according to the gap. The cloud platform will redistribution the budget of each edge node i based on its task completion rate of φ_i.

Finally, when a task is completed, the remaining budget of the cloud platform will be used as a extra reward to motivate users. The cloud platform distributes rewards to all participating users in this task based on their reputation. The user finally obtains the actual reward is

$$u_{i,j}^{c'} = u_{i,j}^c + B \times \frac{P_i}{P} \times \frac{u_{i,j}^r}{\sum_{j=1}^{|U^s|} u_{i,j}^r}.$$

(9)

The user j's extra reward is determined by the proportion of the number of PoIs of the edge node i and the j's own reputation. The higher the reputation of the user j, the more rewards he/she gets.

3.5   Problem Formulation

Cloud platform expects each edge node to maximize the value $V^a(U_i^s)$ from selected users U_i^s under budget constraints. Therefore, our final optimization goal is defined as

$$\text{maximize: } \sum_{i=1}^{|N|} V^a(U_i^s)$$

$$\text{subject to: } \sum_{i=1}^{|N|} \sum_{j=1}^{|U_i^s|} u_{i,j}^{c'} \leq B, \forall i \in N, j \in U_i^s.$$

(10)

The above problem (10) is a set cover problem, which is NP-Hard. Then, we adopt a suboptimal solution to solve the problem. Only the user who meets the pre-set threshold request of the edge node i can be selected, called the marginal utility density threshold, denoted as $\Delta\rho_i$. The total amount of data that the edge node i need to be collected is $V(P_i)$, $\Delta\rho_i = V(P_i)/B_i$. In the same way, it can obtain the individual's marginal utility density threshold. In this way, we can help edge nodes quickly select the proper users to perform tasks.

4   Mechanism Design

In mechanism design, we propose the edge node user recruitment offline algorithm and online algorithm with a dynamical redistribution budget.

4.1   Offline Algorithm

In the offline algorithm, all the PoIs of the cloud platform is divided according to the distance of the PoI from the edge node, which gets a budget from the cloud platform. Users who want to participate in the task and are within the coverage of edge node i are U_i, from which edge node i select user with higher marginal utility value. If user j meets the pre-set threshold and the budget is enough to recruit him or her, then the user j will be recruited (see Line 6–11). Those selected users U_i^s begin sensing, eventually departure and upload sensing data (see Line 12).

Algorithm 1 Offline Algorithm

Input: B,T,P
Output: $\sum_{i=1}^{|N|} V(U_i^d)$
1: PoIs and budget allocation
2: **for** $i \in N$ **do**
3:      $\Delta\rho_i = V(P_i)/B_i$
4:      $(U_i, U_i^s, U_i^d) \leftarrow (\emptyset, \emptyset, \emptyset)$
5:      Add all arrived users to U_i
6:      **while** $U_i \neq 0$ **do**
7:          $u_{i,j} = argmax_{u \in U_i}(V_u(U_i^s))$
8:          **if** $\Delta\rho_i \leq V_u(U_i^s)/u^c$ and $B_i \geq u^c$ **then**
9:              $U_i^s = U_i^s \cup \{u\}$
10:             $B_i = B_i - u^c$
11:         $U_i = U_i \setminus \{u\}$
12: Users sensing and all leaved users add into U_i^d
13: **return** $\sum_{i=1}^{|N|} V(U_i^d)$

4.2   Online Algorithm

In the actual situation, due to users may leave the task halfway, resulting in different completion rate of the edge node. Some edge nodes complete the task rapidly with budgets left, while some edge nodes completion rate is low and the budget is insufficient, which needs to supplement the budget to recruit new users to promote the task completion rate. At each slot t, the cloud platform will observe the task completion of each edge node and reallocate the budget.

First of all, similarly the cloud platform divides PoIs and allocates budget for each edge node. Each edge node calculates the recruitment threshold and initializes the participant set(see Line 1–4). Then, we use the offline algorithm to select users (see Line 8). However, in each round, there may be users who leave the task halfway or complete the task and leave, denoted as U_i^d. The edge node i need to dynamically adjust the recruitment threshold $\Delta\rho_i$ according to the completion of the departure users. At the same time, the edge node i may not have enough budget to recruit new users.

Algorithm 2 Online Algorithm

Input: B,T,P
Output: $\sum_{i=1}^{|N|} V(U_i^d)$
1: PoIs and budget allocation
2: **for** $i \in N$ **do**
3:      $\Delta\rho_i = V(P_i)/B_i$
4:      $(U_i, U_i^s, U_i^d) \leftarrow (\emptyset, \emptyset, \emptyset)$
5: **while** $t <= T$ **do**
6:      **for** $i \in N$ **do**
7:          Add all arrived users to U_i
8:          Select users who meet the requirements
9:          Users who depart at time t are ΔU_i^d
10:          Remove ΔU_i^d from U_i^s and add ΔU_i^d to U_i^d
11:          $(\Delta\rho_i^{t+1}, B_i^{t+1}) \leftarrow \text{BRD}(\Delta U_i^d, U_i^d, B_i^t)$
12:          **if** $\Delta\rho_i == 0$ **then**
13:              Remove i from set N
14:              $B = B + B_i$
15:      $t = t + 1$
16: **return** $\sum_{i=1}^{|N|} V(U_i^d)$

Therefore, We use Algorithm 3 to adjust the recruitment threshold $\Delta\rho_i$ and carry out the budget recycle and redistribution (see Line 11). From line 12 to 14, if an edge node finally completes the task, it will be removed and the remaining budget will be recycled.

Algorithm 3 Budget Redistribution(BRD)

Input: $\Delta U_i^d, U_i^d, B_i^t$
Output: $\Delta\rho_i^{t+1}, B_i^{t+1}$
1: $\varphi_i = 1 - (\|P_i^r - P_i^a\|_F)/\|P_i^r\|_F$
2: $\varphi = 1 - (\|P^r - P^a\|_F)/\|P^r\|_F$
3: $\delta_i = (V(U_i^d) - V^a(U_i^d))/\sum_{j=1}^{|U_i^d|} u_{i,j}^c$
4: **if** $\varphi_i < \varphi$ **then**
5:      $B_i^{t+1} = \sum_{i=1}^{|N|} B_i^t \times (\frac{1-\varphi_i}{\sum_{i=1}^{|N|} 1-\varphi_i})$
6:      $= (g^{\delta_i}) \times (V(P_i) - V(U_i^d))/B_i^{t+1}$
7: **return** $(\Delta\rho_i^{t+1}, B_i^{t+1})$

In the budget reallocation algorithm, we will calculate the current overall completion rate φ and the completion rate φ_i of the edge node i. Similarly, we will use Eq. (2) to calculate δ_i (see Line 1–3). If the completion rate of edge node i is lower than the overall, the cloud platform will recycle the budgets of all edge nodes and then redistribute the budget according to the completion rate of each edge node (see Line 4–5). Then, the recruitment threshold $\Delta\rho_i^{t+1}$ of the next round of edge nodes is changed according to the current amount of remaining data $V(P_i) - V(U_i^d)$, budget B_i and the gap g^{δ_i} (see Line 6).

4.3   Theoretical Analysis

Next, we make a theoretical analysis of the proposed mechanism, including truthful, individual rationality, calculation efficiency and budget feasibility.

Theorem 2. *The proposed offline mechanism and the online mechanism are truthful.*

Proof. We assume that each user who participates in the task is rational but selfish, and the data he contributes will not exceed the edge node requires. Consequently, user u will only adopt two strategies in sensing. S1: The contributed data meets the requirements of the edge node. S2: The contributed data less than the requirements of the edge node.

We assume that reward of a user u adopting strategy S1 is r (S1), and reward adopting strategy S2 is r (S2), then the data contributed by user u adopting different strategies are V(r(S1)) and V(r(S2)), which represents the different marginal utility of different strategies. It's obviously that V(r(S1)) > V(r(S2)). Therefore, in an online algorithm, the edge node prefers to select users with higher marginal utility. Users who adopt strategy S1 will be given priority to select. Besides, adopting strategies S1 and S2 will cause different reputation values to change. The reputation value of adopting strategy S1 will increase while adopting strategy S2 will decrease. According to Eq. (9), then

$$(u_{i,j}^c + B \times \frac{P_i}{P} \times \frac{u_{i.j}^r(S1)}{\sum_{j=1}^{U^s} u_{i.j}^r}) > (u_{i,j}^c + B \times \frac{P_i}{P} \times \frac{u_{i.j}^r(S2)}{\sum_{j=1}^{U^s} u_{i.j}^r}). \tag{11}$$

The above formula indicates that the reward satisfies r(S1) > r(S2). The higher the reputation, the more extra rewards will be obtained, adopting strategy S1 will get more rewards than strategy S2. Hence, S1 is the best choice for users to get more rewards. If a user continues to choose strategies S2, his or her reputation will be close to 0, which eventually leads to extra reward tending to 0.

Theorem 3. *The proposed offline mechanism and the online mechanism are individual rationality.*

Proof. For any user u who is selected to participate in sensing task, the actual reward can be obtained by Eq. (9). Among that, budget $B \geq 0$, so $u_{i,j}^{c'} \geq u_{i,j}^c$ is always established, and the selected user's reward will be no less than his expected reward.

Theorem 4. *The proposed offline mechanism and the online mechanism are both in polynomial time.*

Proof. For the offline algorithm, each edge node i PoIs set is P_i, and need to choose an appropriate user set U_i^s from U_i to participate in the sensing task. The worst scenario is that all arriving users are recruited, which takes $O(|P_i| * |U_i|)$. Departure users U_i^d, a subset of U_i, $O(|U_i^d|) < O(|U_i|)$. Therefore, the computational complexity is $O(|N| * |P| * |U|)$ in N edge nodes.

For the online algorithm, it is divided into n time slots. In each round, the computational complexity is the same as the offline algorithm, and the total computational complexity is $O(n * |N| * |P| * |U|)$.

Theorem 5. *The proposed mechanism is budget feasibility.*

Proof. In each slot, the edge node recruits new users under budget constraints to ensure that the budget will not be exceeded. (see Line 8, Algorithm 1). In the BRD algorithm, the cloud platform will reallocate the budget while the edge node completion rate lower than average(see Line 4–5, Algorithm 3). The above can reflect the budget flexibility in our proposed algorithm.

5   Performance Evaluation

Fig. 2. Sensing area

In this section, we introduce the used data set, simulation settings and then give the analysis of simulation results.

5.1   Setup

We employ real data set to evaluate our offline and online recruitment algorithms, which is the Chicago taxi driving data in 2013. The relevant settings of the simulation are as follows. All the trajectory data are from the Chicago taxi driving data in 2013 with the sensing area in Fig. 2. In the simulation, 5 edge node servers are set up, using the five-point sampling method. The total number of PoIs are 100, and each PoI must be covered at least 5 times (users arrive and upload data). After the data preprocessing, 187 taxis were selected, using the taxi cost as the user request reward of the task, and the taxi travel time as the user perceives time. When it exceeds this time, it is considered that the user has left. The sensing budget is changed from 200 to 800 units. The total time is divided into 12 slots, and each slot is 60 units time.

To compare the proposed algorithm, we slightly changed the algorithm Online-QIM proposed in [8] to the QIM algorithm, which is equivalent to the three-layer architecture except for the edge nodes. We compare the task completion rate, the number of recruitment users, and the budget surplus in three algorithms.

- *Our-offline*: Our proposed offline algorithm, which divides the sensing area into multiple sub-areas according to the edge nodes. Each edge node allocates budget and recruits users separately according to the number of PoIs.
- *Our-online*: Our proposed online algorithm. Based on offline, the cloud platform dynamically redistribution the budget according to the task completion rate of different edge nodes.
- *QIM*: Online dynamic recruitment of users. Compared with our proposed algorithm, there is only one edge node (cloud platform), and the recruitment threshold is dynamically adjusted.

5.2   Results and Analysis

Our simulation results are shown in Figs. 3, 4 and 5 respectively, showing the changes in task completion rate, the actual final number of recruitment users, and the final budget surplus. We conducte a total of 10 rounds of sensing simulations and take the average.

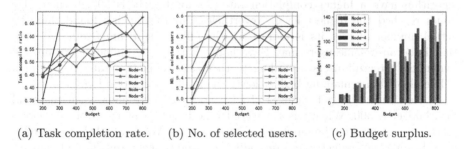

(a) Task completion rate.   (b) No. of selected users.   (c) Budget surplus.

Fig. 3. The offline algorithm's task completion rate changes at each edge node, the total number of people recruited, and the final budget.

Our Offline Algorithms: First, we perform a simulation analysis of the proposed offline algorithm. Figure 3 shows our proposed offline algorithm result about the task completion rate, the total number of selected users, and the budget surplus of 5 edge nodes when the total budget is varied from 200 to 800. Figure 3a represents the task completion rate of each edge node. When budget is 200, all edge nodes task completion rate of all edge nodes is relatively low, and each gap is relatively large. As the budget increases, the completion rate increases slightly, whereas the fluctuation is relatively large, which is in the range of [0.45, 0.65]. Figure 3b reveals the range of recruitment user quantity varies from 5 to 7 when the budget changes. In Fig. 3c, as the budget increases, the final budget surplus of each edge node growing, which means that the incentive rewards of participating users will rise.

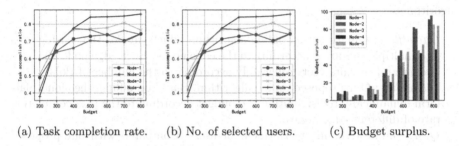

(a) Task completion rate.   (b) No. of selected users.   (c) Budget surplus.

Fig. 4. The onffine algorithm's task completion rate changes at each edge node, the total number of people recruited, and the final budget.

Our Online Algorithms: Figure 4 shows the change in the task completion rate, the total number of recruits, and the budget surplus of the same 5 edge nodes while the total budget changes from 200 to 800. Compared with the offline algorithm we proposed, as shown in Fig. 4a, the completion rate of the online algorithm remains stable when the budget is increased, which is in the range of [0.4, 0.85]. When the total budget is the same, the edge nodes of the online algorithm have a higher completion rate. As for budget is 500, both 5 edge nodes reached the highest level. Nevertheless, in Fig. 4b, it's found that the actual number of recruits is more than the offline algorithm (from 6 to 14), that is because some users may leave halfway, then the online algorithm will dynamically recruit users according to the task completion rate, resulting in online algorithm remaining budget lower than offline ones (see Fig. 3c and 4c).

Table 1 shows budget changes of edge nodes in 12 rounds when the total task budget is 300. When t is 1, each edge node gets a budget from the cloud platform according to the number of interest points. We can find that each edge node has a large budget gap, especially edge node 2 gets a 60.95 unit budget. However, in the next round, the budget of edge node 2 is changed to 28.77. This is because some edge nodes have low completion rates, which lead to budget recycle and allocation. In the same way, the budget of edge node 5 is changed from 7.59 to 17.32 in round 7 to 8. Finally, the budget for each edge node has a different degree of surplus. This shows that our dynamic budget redistribution is successful.

Table 1. Budget changes of edge nodes in 12 rounds when the total task budget is 300

Node	T								
	1	2	3		7	8	9		12
1	24.35	30.15	29.72	...	19.68	12.55	15.41	...	9.06
2	60.95	28.77	30.44	...	20.55	13.15	6.75	...	10.04
3	23.35	29.28	29.39	...	19.38	16.72	9.47	...	9.60
4	17.3	30.05	28.25	...	7.59	17.32	6.62	...	10.32
5	21.35	29.05	29.5	...	20.15	17.71	12.1	...	11.33

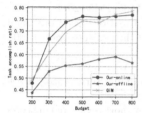 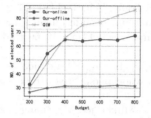 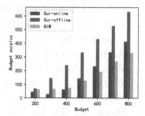

(a) Task completion rate.   (b) No. of selected users.   (c) Budget surplus.

Fig. 5. The offline, online and QIM in the task completion rate, the total number of recruits and budget surplus.

Figure 5 reveals the comparison between the proposed offline algorithm, online algorithm, and a slightly modified QIM algorithm in the task completion rate and the number of recruited users. From Fig. 5a, we can discover that the proposed online algorithm and the QIM algorithm task completion rate is basically equivalent, and finally hovers around 0.75. However, after the budget over 500, the gap of recruiters is gradually widening (see in Fig. 5b and 5c). It may be due to the low task completion rate, then QIM recruited more users to speed up the completion of the task. The proposed online algorithm divides a sensing area into multiple sub-areas, then the total number of people will be slightly less than just one sensing area. The offline algorithm does not dynamically adjust participating users, resulting in the task completion rate and the number of recruits is relatively low.

6   Conclusion

In this paper, we propose a three-layer sensing framework with a cloud platform, edge node, and user to address traditional cloud platform problem with low completion rate in local areas. Based on this, we propose offline and online edge node user recruitment mechanisms. In the online user recruitment mechanism, the cloud platform can dynamically reallocate the budget. Then, We theorize that the proposed mechanism has truthful, individual rationality, calculation efficiency, and budget feasibility. The simulation experiment shows that the proposed online algorithm is effective and reliable. However, it is still a single task user recruitment problem. We plan to expand to the multi-task user recruitment problem in the future.

References

1. Guo, B., et al.: Mobile crowd sensing and computing: the review of an emerging human-powered sensing paradigm. ACM Comput. Surv. **48**(1), 7 (2015)
2. Xiao, L., Chen, T., Xie, C., Dai, H., Poor, H.V.: Mobile crowdsensing games in vehicular networks. IEEE Trans. Veh. Technol. **67**(2), 1535–1545 (2017)
3. Wang, J., et al.: Real-time and generic queue time estimation based on mobile crowdsensing. Front. Comput. Sci. **11**(1), 49–60 (2017). https://doi.org/10.1007/s11704-016-5553-z
4. Mao, Y., You, C., Zhang, J., Huang, K., Letaief, K.B.: A survey on mobile edge computing: the communication perspective. IEEE Commun. Surv. Tutor. **19**(4), 2322–2358 (2017)
5. Zhang, X., Shu, L., Huo, Z., Mukherjee, M., Zhang, Y.: A short review of constructing noise map using crowdsensing technology. In: Romdhani, I., Shu, L., Takahiro, H., Zhou, Z., Gordon, T., Zeng, D. (eds.) CollaborateCom 2017. LNICST, vol. 252, pp. 37–43. Springer, Cham (2018). https://doi.org/10.1007/978-3-030-00916-8_4
6. Laoudias, C., Moreira, A., Kim, S., Lee, S., Wirola, L., Fischione, C.: A survey of enabling technologies for network localization, tracking, and navigation. IEEE Commun. Surv. Tutor. **20**(4), 3607–3644 (2018)
7. Layard, R., Mayraz, G., Nickell, S.: The marginal utility of income. J. Pub. Econ. **92**(8–9), 1846–1857 (2008)
8. Gao, H., Liu, C.H., Tang, J., Yang, D., Hui, P., Wang, W.: Online quality-aware incentive mechanism for mobile crowd sensing with extra bonus. IEEE Trans. Mob. Comput. **18**(11), 2589–2603 (2019)
9. Ding, S., He, X., Wang, J.: Multiobjective optimization model for service node selection based on a tradeoff between quality of service and resource consumption in mobile crowd sensing. IEEE Internet Things J. **4**(1), 258–268 (2017)
10. Wang, J., et al.: Multi-task allocation in mobile crowd sensing with individual task quality assurance. IEEE Trans. Mob. Comput. **17**(9), 2101–2113 (2018)
11. Yang, S., Wu, F., Tang, S., Gao, X., Yang, B., Chen, G.: On designing data quality-aware truth estimation and surplus sharing method for mobile crowdsensing. IEEE J. Sel. Areas Commun. **35**(4), 832–847 (2017)
12. Jin, H., Su, L., Nahrstedt, K.: Theseus: incentivizing truth discovery in mobile crowd sensing systems. In: Proceedings of the 18th ACM International Symposium on Mobile Ad Hoc Networking and Computing, pp. 1–10 (2017)
13. Zhang, L., Ding, Y., Wang, X., Guo, L.: Conflict-aware participant recruitment for mobile crowdsensing. IEEE Trans. Comput. Soc. Syst. **7**(1), 192–204 (2020)
14. Liu, W., Yang, Y., Wang, E., Wu, J.: User recruitment for enhancing data inference accuracy in sparse mobile crowdsensing. IEEE Internet Things J. **7**, 1802–1814 (2019)
15. Wu, D., Li, H., Wang, R.: User characteristic aware participant selection for mobile crowdsensing. Sensors **18**(11), 3959 (2018)

16. Karaliopoulos, M., Koutsopoulos, I., Spiliopoulos, L.: Optimal user choice engineering in mobile crowdsensing with bounded rational users. In: IEEE INFOCOM 2019-IEEE Conference on Computer Communications, pp. 1054–1062. IEEE (2019)
17. Wang, Z., et al.: Towards personalized task-oriented worker recruitment in mobile crowdsensing. IEEE Trans. Mob. Comput. (2020)
18. Marjanović, M., Antonić, A., Žarko, I.P.: Edge computing architecture for mobile crowdsensing. IEEE Access **6**, 10662–10674 (2018)
19. Xia, X., Zhou, Y., Li, J., Yu, R.: Quality-aware sparse data collection in mec-enhanced mobile crowdsensing systems. IEEE Trans. Comput. Soc. Syst. **6**(5), 1051–1062 (2019)
20. Ma, L., Liu, X., Pei, Q., Xiang, Y.: Privacy-preserving reputation management for edge computing enhanced mobile crowdsensing. IEEE Trans. Serv. Comput. **12**(5), 786–799 (2019)

Multi-user Service Migration for Mobile Edge Computing Empowered Connected and Autonomous Vehicles

Shuxin Ge[1], Weixu Wang[1], Chaokun Zhang[1], Xiaobo Zhou[1(✉)], and Qinglin Zhao[2]

[1] Tianjin Key Laboratory of Advanced Networking, College of Intelligence and Computing, Tianjin University, Tianjin 300350, China
{cecilge,weixuwang,xiaobo.zhou,zhangchaokun}@tju.edu.cn
[2] Faculty of Information Technology, Macau University of Science and Technology, Avenida Wei Long, Macau, Taipa, China
zqlict@hotmail.com

Abstract. Connected and autonomous vehicles (CAVs) are promising in improving driving safety and efficiency, which are usually empowered by mobile edge computing (MEC) to push computing and storage resources to the edge networks. By deploying vehicular services at the edge servers in close proximity to vehicles, the service latency can be greatly reduced. Due to the high mobility of vehicles, the services have to be migrated to follow the vehicles to achieve a balance between the service latency and the migration cost. Making service migration decisions for each vehicle independently will suffer from the interference among the vehicles. Moreover, trajectory prediction, which is crucial for service migration decisions, becomes intractable when the number of vehicles is large. In this paper, we investigate the multi-user service migration problem in MEC empowered CAVs, and formulate the service migration of all the vehicles as an optimization problem with the aim of minimizing the average latency, where the interference among different vehicles is taken into account. We then develop an efficient multi-user service migration scheme based on Lyapunov optimization, called ING, to solve the optimization problem in an online fashion without predicting the trajectories of the vehicles. Finally, a series of simulations based on real-world mobility traces of Rome taxis are conducted to verify the superior performance of the proposed ING algorithm as compared with the state-of-the-art solutions.

Keywords: Connected and autonomous vehicles · Mobile edge computing · Service migration · Mobility

This work is supported in part by the National Natural Science Foundation of China under Grant No. 61702365 and 61872451, in part by the Natural Science Foundation of Tianjin under Grant No. 18ZXZNGX00040 and 18ZXJMTG00290, and the Macao FDCT under Grant 0076/2019/A2.

M. Qiu (Ed.): ICA3PP 2020, LNCS 12453, pp. 306–320, 2020.
https://doi.org/10.1007/978-3-030-60239-0_21

1 Introduction

The development of automotive industry and the popularity of private cars are driving various issues, such as traffic congestion, accidents, and air pollution [1]. Connected and autonomous vehicles (CAVs) are promising to address these issues by improving vehicle safety and efficiency, as well as optimizing the road usage [2]. To ensure driving safety, which is the most important task of CAVs, a variety of novel vehicular services, e.g., surrounding vehicle perception, high definition (HD) mapping, are designed to control vehicle behavior with ultra-low latency [3]. In conventional vehicular networks, these services are usually deployed in the remote cloud. However, the high delay incurred by the long propagation and congestion of the backhaul networks does not satisfy the stringent latency requirement of CAVs [4].

Mobile edge computing (MEC) is regarded as a key enabler for CAVs by pushing computing and storage resources to the edge networks [5,6]. With MEC, the vehicular services can be deployed at the edge server, such that the requests from the vehicle can be served locally with low latency [7]. However, as the vehicle moves across different geographical locations due to its high mobility, the response delay will be increased and even become unacceptable if the vehicle is far away from the original edge server where the corresponding service is deployed [8]. A possible solution to maintain service continuity and to enjoy the benefits of MEC is to migrate the service to the nearest edge server, i.e., follow the vehicle, and then the service resumes exactly where it stopped before migration [9]. Obviously, frequent migration induces a huge amount of transmission cost, therefore a proper service migration strategy is needed to make a trade-off between the transmission cost and the service response delay [10].

Service migration in MEC has attracted a lot of research interest in recent years. There is many related works as shown in the following parts. Zhang et al. [6] use acyclic data flow graphs to model the service requests in edge server, which aims to minimize the latency in processing request. Ceselli et al. [11] propose a method to measure the long-term profit of migration, and make the service migration decision to maximize the long-term profit of migration with perfect knowledge of mobile user trajectory. This method does not work in practice where the user mobility is usually unavailable. By assuming the user moves in a straight line, i.e., following a one-dimensional (1-D) mobility pattern, the authors in [12] propose a method to predict user trajectory by utilizing Markov Decision Process (MDP) and then make the service migration decision based on it. Considering the more realistic case where the mobile user moves following a two-dimensional (2-D) mobility pattern, Wang et al. [13] propose a service migration strategy by predicting the user trajectory with a 2-D MDP that has a larger state space. It should be mentioned that most of the existing works design migration strategies from the perspective of a single user [14]. To the best of our knowledge, an efficient multi-user service migration strategy for MEC empowered CAVs is still missing.

Applying the single-user service migration strategy for each vehicle independently seems to be a potential solution, however, it encounters two key challenges. First, the interference among different vehicles may affect the optimality of the service migration strategy. Since the resources of the edge servers are limited, the service migration decisions made by different vehicles independently may affect each other and decrease the quality of experience (QoE), leading to resource scarcity and abuse [15]. Second, the performance of the single-user service migration strategy greatly relies on the accuracy of the trajectory prediction. However, as the number of vehicles increases, the accuracy of trajectory prediction of each vehicle may decrease rapidly and even become intractable due to the state space explosion.

To overcome these two challenges, in this paper, we propose an efficient multi-user service migration scheme for MEC empowered CAVs. Instead of designing service migration strategy from each vehicle independently, we formulate the service migration process of all the vehicles as a mixed integer nonlinear programming (MINP), where the interference among the vehicles is taken into account. Moreover, inspired by [16], we solve the formulated MINP problem in an online fashion by utilizing Lyapunov optimization, thus the trajectory prediction issue can be circumvented. The contributions of this paper are summarized as follows.

- We investigate the multi-user service migration problem in MEC empowered CAVs, and formulate the service migration of all the vehicles as an optimization problem with the aim of minimizing average latency. In this scheme, the interference among users is taken into account to facilitate the accurate characterization of service latency and energy consumption.
- The multI-user QoE-orieNted miGration (ING) algorithm utilized the Lyapunov optimization technique to decouple the migration process in different time slots. We first transform the original optimization problem into an online fashion without predicting user trajectory, and then divide it to a series of service-by-service migration sub-problems to reduce complexity. Furthermore, We also prove that the solution of ING is quite close to the optimal solution of the original problem under certain conditions.
- The effectiveness of the proposed multi-user service migration scheme was established on the basis of real-world taxi traces in Rome. A series of simulations are conducted to verify the effectiveness of the proposed multi-user service migration scheme in terms of service latency, energy consumption, and deadline guarantee rate.

The rest of this paper is organized as follows. Section 2 introduces the system model and formulates the multi-user service migration problem considered in this paper. The proposed ING algorithm is presented in Sect. 3 along with the theoretical analysis. Section 4 shows the performance evaluation and discussion. Finally, the conclusion is drawn in Sect. 5.

2 System Model and Formulation

2.1 System Overview

In this paper, we consider a multi-vehicles networks consists of N base stations (BSs) and V vehicles, as shown in Fig. 1. Each BS equipped with a edge

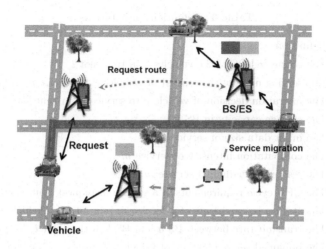

Fig. 1. The overview of the system considered in this paper. Vehicles request services from its connected BS constantly as it move. BS can directly provide services which the BS has been deployed or has been migrated. Otherwise, BS require route the request from other BS to provide services.

server which can provide vehicular services to the vehicles within its radio range. We use S_n to denote the maximum storage size of the edge server in BS n, $n = 1, 2, \cdots, N$. Moreover, the BSs are connected to each other via wireless communication. The V vehicles are moving among the coverage regions. As the BSs are densely deployed, the vehicle is usually in the coverage regions of multiple BSs. Typically, the vehicle will connect to the BS with the largest received signal strength indicator (RSSI).

There are M different vehicular services deployed in the edge servers. However, due to the limited storage capacity, one edge server can not accommodate all the services. Note that, since all the edge servers are connected with each other, the vehicle is able to access all the services once it is in the coverage region of arbitrary BS. Let λ_m, γ_m and D_m denote the input data size, the computation intensity (i.e. CPU cycles/bit) and the response deadline of service m, $m \in [1, M]$, respectively. Let $f_{m,v}, \theta_{m,v}$ denote the CPU cycle and storage requirements of a service request from vehicle v, $v \in [1, V]$ to service m, respectively.

We divide the total time into T time slots, where each time slot lasts for τ seconds. In time slot t, $t = 0, 1, 2, \cdots, T$, all the vehicles generates service requests according to some stochastic statistics. When the vehicle is in the coverage area of the BS deployed with corresponding service, the request can be served locally and thus the latency is minimized. As the vehicle moves, we have to decide whether migrate the service by trading off the latency and the energy consumption. Table 1 summarizes the key parameter notations in our paper.

Table 1. Definitions of notations

Notation	Definition
$L_t^B(v,n)$	The connected BS of the vehicles v in time slot t
$L_{m,t}^S(v,n)$	The serving node of the request of vehicles v to service m in time slot t
$L_{m,t}^M(v,n)$	The migration decision of vehicle v to service m in time slot t
S_n	The storage capacity in BS n
λ_m	The input data size of service m
γ_m	The computation intensity of service m
D_m	The response deadline of service m
$f_{m,v}$	The CPU cycle requirements of a service m request from vehicle v
$\theta_{m,v}$	The storage requirements of a service m request from vehicle v
$R(a,b)$	The transmit rate between BS a and BS b in time slot t
$d_{m,t}(v)$	The remained service latency of vehicle v for service m in time slot t
$D_{m,t}(v)$	The execution time of vehicle v for service m in time slot t
$E_{m,t}$	The total energy consumption for service m in time slot t

2.2 Service Placement Model

At the beginning of each time slot t, the networks operator mange service for each vehicle v to decide which BS n will provide service m to vehicle v in time slot $t+1$. We use a V-by-N matrix $\boldsymbol{L}_t^B = \{L_t^B(v,n)|n = 1,2,\cdots,N\}$ to denote the connectivity between vehicle v and BS n, where $L_t^B(v,n) = 1$ means vehicle v connects to BS n, and $L_t^B(u,n) = 0$ otherwise. Obviously, we have

$$\sum_{n=1}^N L_t^B(v,n) = 1, \forall t,v. \tag{1}$$

We define a V-by-N binary matrix $\boldsymbol{L}_{m,t}^S = \{L_{m,t}^S(v,n)|v = 1,2,\cdots,V;\ n = 1,2,\cdots,N\}$, where $L_{m,t}^S(v,n) = 1$ indicates the request of vehicle v to service m is served by server node n in time slot t, and $L_{m,t}^M(v,n) = 0$ otherwise. Since the request of vehicle V to service m can be served by only one server, or no request for service m, we have

$$\sum_{n=1}^N L_{m,t}^S(v,n) \leq 1, \forall t,v,m. \tag{2}$$

Another V-by-N binary matrix $\boldsymbol{L}_{m,t}^M = \{L_{m,t}^S(v,n)|v = 1,2,\cdots,V;\ n = 1,2,\cdots,N\}$ is used to present the migration decision in each time slot t, where $L_{m,t}^M(v,n) = 1$ denotes the service m of vehicle v migrate to BS n in time slot t, and $L_{m,t}^S(v,n) = 0$ otherwise. Thus, the server node in next time update by

$$L_{m,t+1}^S(v,n) = L_{m,t}^M(v,n) \tag{3}$$

Similar to (2), we have

$$\sum_{n=1}^{N} L_{m,t}^{M}(v,n) \le 1, \forall t, v, m. \tag{4}$$

As there is a one-to-one mapping between the vehicle and its service data, $L_{m,t}^{M}(v,n) = 1$ also indicates service data of vehicle v to service m is placed at BS n in time slot t. And the storage constraints for each BS n can be described as

$$\sum_{m=1}^{M} \sum_{v=1}^{V} \theta_{m,v} L_{m,t}^{M}(v,n) \le S_n, \forall t, n. \tag{5}$$

2.3   Service Latency

We use $R(a,b)$ to denote the transmission rate between BS a and BS b, which can be calculated by Routing Information Protocol (RIP). Since the transmit time between vehicle and BS is hard to calculate precisely (vehicles are always on the move), we use a constant value C to approximate it. Assuming vehicles v generate a request to service m at the beginning of time slot t_0. The request transmission latency is

$$d_{m,t_0}^{T}(v) = \frac{\lambda_m}{R(\pi_{t,v}^{B}, \pi_{m,t,v}^{S})} + C, \tag{6}$$

where

$$\pi_{t,v}^{B} = \arg\max_{n} L_t^{B}(v,n), \quad \pi_{m,t,v}^{S} = \arg\max_{n} L_{m,t}^{S}(v,n).$$

and the computation latency of vehicle v for service m is

$$d_{m,t}^{C}(v) = \frac{\lambda_m \gamma_m}{f_{m,v}}. \tag{7}$$

Note service m may be migrated from BS $\pi_{m,t,v}^{S}$ to BS $\pi_{m,t,v}^{M}$ in time slot t. The service migration latency in time slot t can be expressed as

$$d_{m,t}^{M}(v) = \frac{\theta_{m,v}}{R(\pi_{m,t,v}^{S}, \pi_{m,t,v}^{M})}. \tag{8}$$

We use $\pi_{m,t,v}^{M} = \arg\max_{n} L_{m,t}^{M}(v,n)$ to denote the migration target BS. In the case that $\pi_{m,t,v}^{S} = \pi_{m,t,v}^{M}$, service m is not migrated in time slot t, and the service migration latency is $d_{m,t}^{M}(u) = 0$. It should be mentioned that the request may not be completed within one time slot t. We use $d_{m,t}(v)$ and $D_{m,t}(v)$ to represent the remained service latency (i.e., how long will it take to complete the request) and the execution time (i.e., how long is the service been processed) in time slot t, which are

$$d_{m,t}(v) = \begin{cases} d_{m,t}^{T}(v) + d_{m,t}^{C}(v) + d_{m,t}^{M}(v), & t = t_0, \\ \max\{d_{m,t-1}(u) + d_{m,t}^{M}(u) - \tau, 0\}, & t > t_0, \end{cases}$$

and
$$D_{m,t}(v) = \min\{d_{m,t}(v), \tau\}. \tag{9}$$
respectively.

2.4 Energy Consumption

Correspondingly, the transmission energy consumption is
$$E_{m,i}^T(v) = pd_{m,t_0}^T(v), \tag{10}$$
where p is the transmission power. The computing energy consumption is
$$E_{m,t}^C(v) = \kappa f_{m,v}^3 \left[D_{m,t}(v) - d_{m,t}^T(v) - d_{m,t}^M(v)\right]. \tag{11}$$
where κ is the unit energy consumption when BS n computing with unit CPU cycle. And the migration energy consumption is
$$E_{m,t}^M(v) = pd_{m,t}^M(v). \tag{12}$$

Thus, the total energy consumption in time slot t can be calculated as
$$E_{m,t} = \sum_{v=1}^{V} \left[E_{m,t}^T(v) + E_{m,t}^C(v) + E_{m,t}^M(v)\right]. \tag{13}$$

2.5 Problem Formulation

Our goal is to minimize the average service latency while maintaining system energy budget by selecting the optimal migration strategy $\boldsymbol{L}_{m,t}^M$, which can be expressed as

$$\mathbf{P1}: \quad \min_{\boldsymbol{L}_{m,t}^M} \lim_{T \to \infty} \frac{1}{T} \sum_{t=0}^{T-1} \sum_{m=1}^{M} \sum_{v=1}^{V} D_{m,t}(v), \tag{14}$$

$$s.t. \quad \lim_{T \to \infty} \frac{1}{T} \sum_{t=0}^{T-1} \sum_{m=1}^{M} E_{m,t} \le \hat{E}, \tag{15}$$

$$d_{m,t}(v) \le D_m - \tau(t - t_0), \forall v, m, t,$$
$$(1), (2), (4), (5). \tag{16}$$

where constraint (15) indicates that the the long-term energy of the whole system does not exceed the energy constraint $\hat{E}$. The constraint (16) ensures the request has to be completed before its response deadline. The major challenge that impedes the derivation of the optimal solution to the above problem is the lack of future information, i.e., optimally solving **P1** requires complete offline information (e.g., user trajectory, request distribution) which is difficult to predict. Moreover, even if the future information is known in advance, **P1** is still a MINP problem that is NP-hard. Denote O^* as the optimal average delay achieved by **P1**.

3   Online Service Migration Algorithm

3.1   Algorithm Design

In this subsection, we present the details of proposed ING algorithm. We first transform **P1** into a series of solvable sub-problems by utilizing Lyapunov optimization, and then we can obtain the sub-optimal service migration decisions without trajectory prediction.

For notational simplicity, we first set

$$x_m(t) = E_{m,t} - \frac{\omega_m \lambda_m \gamma_m \mathbf{E}\{f_{m,v}^2 | v\} \hat{E}}{\sum\limits_{m=1}^{M} \omega_m \lambda_m \gamma_m \mathbf{E}\{f_{m,v}^2 | v\}}, \tag{17}$$

which is used to denote the drift between energy consumption for service m in time slot t and expected energy consumption for service m. ω_m is the priority of service m. We then construct a energy queue to follow the long-term energy constraint as

$$e_m(t+1) = \max\{e_m(t) + x_m(t), 0\}.$$

The Lyapunov function is defined as $L(x) \triangleq \frac{1}{2}x^2$, representing the "congestion level" in energy deficit queue. A small value of $L(e_m(t))$ implies that the queue backlog is small, i.e., the virtual queue has strong stability. Let $\triangle_1^e$ denote the one-slot Lyapunov drift, which is $\triangle_1^e = L(e_m(t+1)) - L(e_m(t))$, and we have

$$\triangle_1^e = \frac{1}{2} \sum_{m=1}^{M} [e_m^2(t+1) - e_m^2(t)]$$

$$\leq \frac{1}{2}x_m^2(t) + \sum_{m=1}^{M} e_m(t)x_m(t) \leq MU + \delta \sum_{m=1}^{M} e_m(t), \tag{18}$$

where U and δ are the upper bound of $\frac{1}{2}x_m^2(t)$ and the expectation of $x_m(t)$, respectively. Thus, the optimal migration decision in each time slot can be obtained by solving **P2** as follows

$$\mathbf{P2}: \min_{L_{m,t}^M} \ Q \sum_{m=1}^{M} \sum_{v=1}^{V} D_{m,t}(v) + \sum_{m=1}^{M} e_m(t)E_{m,t},$$

$$s.t. \qquad (1), (2), (4), (5), (16), \forall t, v,$$

where Q is used to adjust the trade-off between latency and energy consumption. Similarly, we also set

$$y_{t,n}(m) = \sum_{v=1}^{V} \theta_{m,v} L_{m,t}^M(v,n) - \frac{\omega_m \mathbf{E}\{\theta_{m,v} | v\} S_n}{\sum\limits_{m=1}^{M} \omega_m \mathbf{E}\{\theta_{m,v} | v\}}. \tag{19}$$

which is used to denote the drift between resource requirement for service m in time slot t and resource expected to be allocated for service m. Without resources constraints, the migration decision for each service m can be regarded as independently. We construct another virtual queen $S_{t,n}(m)$ to indicate how far the current resource usage deviates from the total resource budget, and decompose **P2** into M sub-problem.

$$S_{t,n}(m+1) = \max\{S_{t,n}(m) + y_{t,n}(m), 0\}, \qquad (20)$$

We define a Lyapunov drift-plus-penalty function $\triangle_M^s = L(S_{t,n}(M)) - L(S_{t,n}(0))$ to solve the real-time problem

$$\triangle_M^s = \frac{1}{2}\sum_{m=1}^{M}[S_{t,n}^2(m+1) - S_{t,n}^2(m)]$$

$$\leq \frac{1}{2}\sum_{m=1}^{M}y_{t,n}^2(m) + \sum_{m=1}^{M}[S_{t,n}(m) - S_{t,n}(0)]y_{t,n}^2(m)$$

$$\leq MB + \frac{M(M-1)}{2}y_{t,n}^2(m) \leq M^2B, \qquad (21)$$

where B is the upper bound of $\frac{1}{2}y_{t,n}^2(m)$. With Lyapunov optimization, the underlying objective is to minimize a supremum bound on the following drift-plus-cost expression in each time slot for all services

$$\sum_{n=1}^{N}\triangle_M^s + Q'(\triangle_1^e + Q\sum_{m=1}^{M}\sum_{v=1}^{V}D_{m,t}(v))$$

$$\leq NM^2B + Q'MU + Q'QO^*. \qquad (22)$$

Thus, **P2** can be solved by solving M sub-problem **P3**.

$$\mathbf{P3} : \min_{L_{m,t}^M} \ Q'[Q\sum_{v=1}^{V}D_{m,t}(v) + e_m(t)E_{m,t}(v)]$$

$$+ \sum_{n=1}^{N}S_{t,n}(m)\sum_{v=1}^{V}\theta_{m,v}L_{m,t}^S(v,n),$$

$$s.t. \qquad (1),(2),(4),(16), \forall t,m,v, \qquad (23)$$

where $Q' \geq 0$. By solving **P3** using particle swarm optimization (PSO) algorithm, we obtain the near-optimal solution of **P1**, and the whole process is summarized in Algorithm 1. For each service and each time slot, we can solve **P3** with PSO algorithm, which the details are omitted here due to space limitation. After this step, the energy queue of each BS is updated according to (18) (Lines 4), as well as resource queue which is updated by (20) (Lines 6). Finally, the solution in the time slot t of **P3** becomes the initial state in the time slot $t + 1$ (Lines 7).

Algorithm 1. ING Algorithm

Input: $:e_m(0) \leftarrow 0, S_{t,n}(0) \leftarrow 0, \boldsymbol{L}^B_{m,0}, \boldsymbol{L}^S_{m,0}, \hat{E}, S_n$
Output: : service migration decision $\boldsymbol{L}^M_{m,t}$
1: **for** $t = 0$ to $T - 1$ **do**
2: **for** $m = 1$ to M **do**
3:    obtain $\boldsymbol{L}^M_{m,t}$ by solving **P3**;
4:    update $e_m(t+1)$ according to (3.1);
5: **end for**
6: update $S_{t,n}(m+1)$ according to (20);
7: update $\boldsymbol{L}^S_{m,t+1}$ according to (3);
8: **end for**

3.2   Performance Bound

Theorem 1. *By letting δ go to zero, the average system service latency satisfies*

$$\lim_{T \to \infty} \frac{1}{T} \sum_{t=0}^{T-1} \sum_{m=1}^{M} \sum_{v=1}^{V} D_{m,t}(v) < O^* + \frac{MU}{Q} + \frac{NM^2 B}{QQ'}. \tag{24}$$

the energy consumption and resource utilization in BS n in time slot t satisfy

$$\lim_{T \to \infty} \frac{1}{T} \sum_{t=0}^{T-1} \sum_{m=1}^{M} E_{m,t} < \hat{E} + \sqrt{2QO^* + 2MU}, \tag{25}$$

$$\sum_{M=1}^{M} \sum_{v=1}^{V} \theta_{m,v} L^S_{m,t}(v,n) \le S_n + \sqrt{2M^2 B + \frac{2Q'O^*}{N}}, \tag{26}$$

respectively.

Proof. Summing the inequality (22) over $t = 0, 1, \cdots, T-1$ and then dividing the result by T, it is easy to have (24). Meanwhile, (25) can be obtained according to

$$\frac{1}{T} \sum_{t=0}^{T-1} [\sum_{m=1}^{M} E_{m,t} - \hat{E}] \le \frac{1}{T} \sum_{t=0}^{T-1} \sum_{m=1}^{M} [e_m(t+1) - e_m(t)]$$

$$= \frac{\sqrt{2\triangle^e_{T-1}}}{T} \le \frac{\sqrt{2MU + QO^*}}{T}.$$

Similarly, (26) can be obtained from

$$\sum_{m=1}^{M} \sum_{v=1}^{V} \theta_{m,v} L^M_{m,t}(v,n) - S_n \le \sqrt{2\triangle^s_M} \le \sqrt{2M^2 B + \frac{2Q'O^*}{N}}.$$

Theorem 1 shows that the difference between the average service latency with ING and the optimal solution of **P1** is no more than $O(1/Q, 1/Q')$. Meanwhile, the energy consumption and resources utilization are within a bounded deviation $O(Q)$ and $O(Q')$ compared to the given constraints, respectively.

Table 2. Service Parameters

Type	D_m	λ_m	$\gamma_m(K)$	ν_m	θ_m(MB)	ω_m
Emergency stop	0.1	3200	36	10	[1,3]	0.9
Collision risk	0.1	4800	40	10	[1,3]	0.9
Accident report	0.5	4800	28	2	[2,3]	0.9
Parking	0.1	1200	80	1	[3,5]	0.5
Traffic control	1	1200	45	1	[4,5]	0.8
Platoon	0.5	4800	88	2	[4,6]	0.6
Face detection	0.5	3200	50	1	[5,10]	0.3

4   Performance Evaluation

4.1   Simulation Setup

In this section, we evaluate the performance of ING algorithm with simulations. We simulate a 500m×500m area served by 9 BSs that are regularly deployed on a grid networks. The vehicle trajectories are collected from the real-world Rome taxi traces obtained on 2014 [17]. Each active taxi requests a set of services that is independent of the services of the other taxis. Seven typical vehicular services, i.e., emergency stop, collision risk, accident report, parking, traffic control, platoon, and face detection, are deployed on the edge servers [18]. The parameters of the services are listed in Table 2. The constant value C, unit energy consumption κ and transmission power p are set to 0.02sec, 10^{-26} and 0.5 W respectively. The transmission rate between BS is generated by the random networks model within [5, 20] Mbps. The available computation capability for each service and storage in each BS n follow uniform distributed within [500,1000] Kcycles/bit and [1,5] GB, respectively.

We compare ING with four benchmarks.

- **Always migration (AM)**: The service is always migrated to the BS that is nearest to the vehicle.
- **No migration (NM)**: The service is deployed in the original BS and never migrate to other BS.
- **Partial Dynamic Optimization Algorithm (PDOA)** [14]: This is a single-user service migration algorithm which picks up a fraction of the services according to the priority queue, and migrate these services to reduce the negative effect of the trajectory prediction error.
- **Dynamic Markov Decision Process (DMDP)** [13]: This is also a single-user service migration algorithm, where the trajectory is predicted using MDP, and optimal service migration decision is made to minimize the specific cost, which regarded as energy consumption in this experiment. Note that both PDOA and DMDP are applied to each vehicle independently in our simulations.

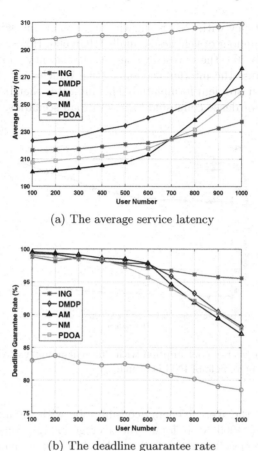

(a) The average service latency

(b) The deadline guarantee rate

Fig. 2. The average service latency and deadline guarantee rate with different number of vehicles, where the energy constraint per vehicle of ING is 50 mJ.

4.2 Results and Discussions

The average service latency with different number of vehicles are shown in Fig. 2(a). As can be seen from Fig. 2(a), as the number of vehicles increases from 100 to 1000, the average service latency with NM is the highest. This is reasonable as the services never migrate to follow the vehicles. On the other hand, when the number of vehicles is smaller than 700, the lowest average service latency can be achieved with AM since the services are always successfully migrated to the nearest BSs to the corresponding vehicles. However, as the number of vehicle exceeds 700, the average service latency with AM increases exponentially. This is because the edge servers run out of resources with too many vehicles, and the interference among the vehicles affects the service migration decisions. The average service latency with PDOA is quite close to that with AM, and the average service latency with DMDP is the second highest. It is found that the average service latency with ING is slightly larger than that with both AM and PDOA when the number of vehicles is smaller than 600. However, as the number of vehicles exceeds 700, the lowest average service latency can be achieve with ING. This is because the interference among the vehicles is taken into account in the optimization problem formulated. This indicates that ING is suitable for CAVs, especially for dense congested areas in urban scenarios.

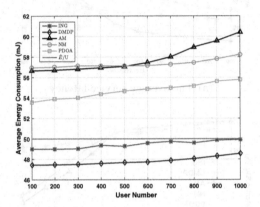

Fig. 3. The average energy consumption with different number of vehicles, where the energy constraint per vehicle of ING is 50 mJ.

Figure 3 shows the average energy consumption with the number of vehicles varies from 100 to 1000. It is found the average energy consumption of AM and NM are quite close to each other, which is also larger than the other three methods. This is because both the migration and data transmission consumes a lot of energy. The average energy consumption with DMDP is the lowest since its objective is to minimize the average energy consumption. The average energy consumption with ING lies between that with DMDP and PDOA, which is quit close to the energy constraint per vehicle $\hat{E}/V$ specified in **P1**.

Figure 4 presents the average service latency of ING with different energy constraints per vehicle $\hat{E}/V$, where the number of vehicles is 500. It is surprising to observe that the average service latency of ING decreases as $\hat{E}/V$ increases, and finally become the smallest among the 5 methods after $\hat{E}/V$ exceeds 54 mJ. This indicates a trade-off between the service latency and the average energy consumption can be achieved by ING. For some latency-sensitive services, we can further decrease the latency by sacrificing some energy consumption. However, the performance gain by increasing the energy constraints becomes stable when $\hat{E}/V$ is large enough.

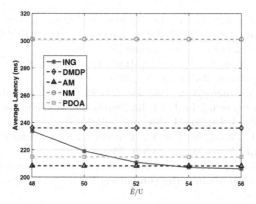

Fig. 4. The average service latency with different $\hat{E}/V$, where the number of vehicle is 500.

5   Conclusion

In this paper, we studied the multi-user service migration problem for MEC empowered CAVs. We formulate the service migration process as a MINP optimization problem to minimize the average service latency, where the interference among the vehicles is taken into account. An online algorithm, ING, is proposed to solve the MINP problem based on Lyapunov optimization, and therefore the trajectory prediction of the vehicles is no longer needed. We also showed the solution of the ING algorithm is quite close the optimal solution of the MINP problem formulated by theoretical analysis. The simulations results based on the real world taxi trajectory in Rome demonstrated that the proposed solution is superior than the state-of-the-art solutions. It is shown that ING is suitable for CAVs, especially for dense congested areas in urban scenarios. As a centralized solution, ING may suffer from the single point failure problem. An efficient decentralized solution is left as a future study.

References

1. Shah, S.A.A., Ahmed, E., Imran, M., Zeadally, S.: 5g for vehicular communications. IEEE Commun. Mag. **56**(1), 111–117 (2018)
2. Shi, W., Jie, C., Quan, Z., Li, Y., Xu, L.: Edge computing: vision and challenges. IEEE Internet Things J. **3**(5), 637–646 (2016)
3. Zhang, Y., Wang, C., Wei, H.: Parking reservation auction for parked vehicle assistance in vehicular fog computing. IEEE Trans. Veh. Technol. **68**, 3126–3139 (2019)
4. Xu, J., Chen, L., Zhou, P.: Joint service caching and task offloading for mobile edge computing in dense networks. In: IEEE Conference on Computer Communications, April 2018, pp. 207–215 (2018)
5. Mao, Y., You, C., Zhang, J., Huang, K., Letaief, K.B.: A survey on mobile edge computing: the communication perspective. IEEE Commun. Surv. Tutor. **19**(4), 2322–2358 (2017)

6. Zhang, C., Zheng, R., Cui, Y., Li, C., Wu, J.: Delay-sensitive computation partitioning for mobile augmented reality applications. In: IEEE/ACM International Symposium on Quality of Service, June 2020
7. Ge, X., Tu, S., Mao, G., Wang, C., Han, T.: 5G ultra-dense cellular networks. IEEE Wirel. Commun. 23(1), 72–79 (2016)
8. Elsayed, S.A., Abdelhamid, S., Hassanein, H.S.: Proactive caching at parked vehicles for social networking. In: IEEE International Conference on Communications, Kansas City, MO, USA, 20–24 May, pp. 1–6 (2018)
9. Wang, S., Xu, J., Zhang, N., Liu, Y.: A survey on service migration in mobile edge computing. IEEE Access 6, pp. 23 511–23 528 (2018)
10. Ouyang, T., Zhou, Z., Chen, X.: Follow me at the edge: mobility-aware dynamic service placement for mobile edge computing. IEEE J. Sel. Areas Commun. 36(10), 2333–2345 (2018)
11. Ceselli, A., Premoli, M., Secci, S.: Mobile edge cloud network design optimization. IEEE/ACM Trans. Netw. 25(3), 1818–1831 (2017). https://doi.org/10.1109/TNET.2017.2652850
12. Ksentini, A., Taleb, T., Min, C.: A Markov decision process-based service migration procedure for follow me cloud. In: IEEE International Conference on Communications (2014)
13. Wang, S., Urgaonkar, R., Zafer, M., He, T., Chan, K., Leung, K.K.: Dynamic service migration in mobile edge computing based on Markov decision process. IEEE/ACM Trans. Netw. 27(3), 1272–1288 (2019)
14. Yu, X., Guan, M., Liao, M., Fan, X.: Pre-migration of vehicle to network services based on priority in mobile edge computing. IEEE Access 7, 3722–3730 (2019)
15. Machen, A., Wang, S., Leung, K.K., Ko, B.J., Salonidis, T.: Live service migration in mobile edge clouds. IEEE Wirel. Commun. 25(99), 2–9 (2018)
16. Sun, Y., Zhou, S., Xu, J.: EMM: energy-aware mobility management for mobile edge computing in ultra dense networks. IEEE J. Sel. Areas Commun. 35(11), 2637–2646 (2017)
17. Bracciale, L., Bonola, M., Loreti, P., Bianchi, G., Amici, R., Rabuffi, A.: CRAWDAD dataset roma/taxi (v. 2014–07-17). https://crawdad.org/roma/taxi/20140717
18. Nasrin, W., Xie, J.: Sharedmec: sharing clouds to support user mobility in mobile edge computing. In: IEEE International Conference on Communications, May 2018, pp. 1–6 (2018)

A Precise Telecom Customer Tariff Promotion Method Based on Multi-route Radial Basis Kernel Fuzzy C-means Clustering

Chenghao Wei[✉], Timur Valiullin, and Long Hao

Big Data Institute, College of Computer Science and Software Engineering, Shenzhen University, Shenzhen 518000, China
chenghao.wei@szu.edu.cn

Abstract. Precise tariff promotion for existing customers is important for telecom operators to increase their revenue. The basis of successful marketing campaign mainly relies on the analysis of users' product usage behavior. Unfortunately, traditional segmentation methods cannot discover complex usage behavior of users, which may result in a low conversion rate during customer tariff promotion activity. To address this problem, we introduce a novel soft clustering method called Multi-Route Radial Basis Kernel Fuzzy C-Means (MRRBKFCM) for telecom customer segmentation based on their mobile internet usage data. The kernel function in the proposed algorithm was taken as the radial basis kernel function, the iterative formula for the recalculation of cluster centers was obtained by gradient method and particle swarm optimization (PSO) algorithm. Compared with the performance of the traditional Kernel Fuzzy C-means (KFCM) and Fuzzy C-means (FCM), the effectiveness of the proposed algorithm has been validated by the public UCI data sets. The customers of GuanAn branch, SiChuan telecom company are used as the input for the proposed customer segmentation algorithm. We launched the promotion campaign according to the outputs of our method for city district of Guan An, Lin Shui and Yue Chi. 31285 users has been used for customer promotion campaign. Compared with the average conversion rate 5.81% using traditional marketing strategy, our method reaches to 19.11%, 17.42% and 7.99% for each district respectively, it greatly reduces marketing cost and increases the conversion rate.

Keywords: Telecom tariff precise promotion · Customer segmentation · Radial basis kernel function · Soft clustering · Multi-route optimisation

---

This work was supported by National Natural Science Foundation of China (61972261). We would like to thank anonymous reviewers for their valuable work.

M. Qiu (Ed.): ICA3PP 2020, LNCS 12453, pp. 321–336, 2020.
https://doi.org/10.1007/978-3-030-60239-0_22

1 Introduction

With the advent of 5G, competition in telecommunication market is becoming increasingly fierce. More than 70% of the revenue of telecom companies comes from their existing customers. Therefore, customer maintenance, customer preservation and appreciation of existing customers become the top priority. Without well-performed customer segmentation based on their product usage behavior, it would consume a lot of extra resources, and obviously reduce operators' income and profits. But there are two main difficulties during tariff promotion campaigns. The first difficulty is that the limited resources of telecom operators cannot be used for the tariff promotion of each customer. It is impossible to achieve the goal by using the traditional strategy, which mainly relies on experienced marketing department staff to filter customers list by limited product usage indicators. The tariff promotion for customers was held by selling calls from telecom company staff. Judged by company manual and personal experience, such an approach would inevitably lead to inefficiency and harass customers. The second difficulty is that results of general data mining algorithms cannot be easily applied. Although telecom enterprises carried out data mining, especially using hard clustering algorithm for customer segmentation, such as partition-based clustering, hierarchical clustering, density-based clustering and model-based clustering. However, the obtained results themselves did not provide high efficiency at the business level. Actually, key attribute of telecom customers is their mobile internet usage. Based on which different customer usage behavior patterns can be discovered. It is important to identify the tariff package that will satisfy their requirements. By carefully analysing the tariff package distribution of each group with similar pattern, we could discover suitable tariff package promotion for customers.

Clustering method has been applied to generate the customer promotion list [1]. However, each customer cannot simply belong to one cluster due to complex usage behavior [2], it is necessary to establish a soft membership degree, that is, each existing customer needs a coefficient to measure its belonging degree to a different segments. Fuzzy C-means (FCM) [3–5] can be used as a tool for soft clustering, which was used for fraud detection by telecom company [6]. However, with the development of telecom business, customers' features are expanding rapidly. Traditional FCM does not have good performance with non-linear data. Its derivatives were introduced with kernel function [7–9].

Since the membership matrix and cluster centers contain gradient information in the iteration process, it may fall into local minimal points. Based on the mobile internet traffic usage data, we introduce the Mutli-Route Radial Basis Kernel Fuzzy C-Means (MRRBKFCM) clustering algorithm to complete the soft clustering task and to avoid local minima. With a better performance, it can be used for customer segmentation that can finally help operators to increase the conversion rate during tariff package promotion campaign. This method uses inner product operation of kernel function in high-dimensional characteristic space to complete the distance calculation between samples and cluster centers. At the same time, particle swarm optimization (PSO) algorithm was introduced to find the alternative path of optimization to avoid local minima problem. PSO provides multiple search path, it could reduce the total number of iterations to improve the performance of the algorithm.

In Sect. 2, two parts of related works are introduced, the first one is preliminary of telecom user segmentation and precise marketing, while the second one is the basics of FCMs and its successors. In Sect. 3, we discussed kernel fuzzy clustering problems and the novel MRRBKFCM algorithm. In Sect. 4, we report experimental results that demonstrate the feasibility and effectiveness of the proposed method with UCI data sets. Second part of the section is dedicated to the analysis of the clustering results for the real world data provided by China Telecom SiChuan corporation. Finally, the conclusion is provided in Sect. 5.

2 Related Work

2.1 Telecom Customer Segmentation

Nowadays, telecom operators pay great attention to their high value customers. The statistical rule of telecom company points out that 80% of their incomes come from the top 20% of profitable customers and 80% of the costs are incurred on the 20% of unprofitable customers [10,11]. However, how to find those incipient high value customers is a crucial issue for operators. In marketing application, customers can be divided according to their customer life time value (LTV) [12,13], namely, incipient customers, value-enhancing customers, stable customers and detained customers [14].

Incipient customers are new to operators. Due to their short time of service usage, such kind of customers are still at product exploratory stage. Apparently, their consumption demands are not clear. Average Revenue Per Customer (ARPC) [12] of those customers keeps increasing, some may be accompanied by network traffic depression, network traffic overload or even network traffic saturation. As their contract time goes by, some of those customers will turn to value-enhancing customers. Both their consumption demands and their consumption characteristics are clear. Later, some value-enhancing customers will become stable customers. Accurate telecom tariff package promotion not only enhances operators' income, but also increases users' brand loyalty. Efficient tariff package promotion depends on well-performed customer segmentation. It divides a large customer group into several smaller ones. In each customer group after the division, there is a high degree of similarity within the same group, but there is a large difference between groups.

Data mining clustering methods is a way to achieve the aim. In order to make churn management successful for telecom operators, density-based clustering algorithms have been used as a tool for evaluating their loyalty and churn rate [15]. For analyzing mobile phone users' behaviors, clustering based on their call detail records (CDR) has been done with K-means method in [16] for marketing campaign. In addition, segmentation of telecom customers based on customer value by decision tree model was proposed in [17]. A hybrid data mining model for customer classification with genetic algorithm (GA) as a decision maker was proposed in [18] for searching a target customer, who is most likely to purchase a new service from the telecom operator. A cluster algorithm based on k-means and multi-variable quantum shuffled frog leaping algorithm was proposed in [19], which was used as a customer segmentation method in the telecom customer marketing.

2.2   Kernel Fuzzy C-Means Overview

K-means is one of the most popular clustering algorithms. It partitions data objects into k clusters, where the number of clusters k is chosen in advance according to the application purpose [20]. Extension of basic K-means called the K-modes algorithm, which used a simple matching dissimilarity measure to deal with categorical data [21]. For a given data set $D = \{x_1, \ldots, x_i, \ldots, x_n\}, x_i \in \mathrm{R}^d$, K-means algorithm tried to group data set D into k clusters, where $C = \{c_1, \ldots, c_j, \ldots, c_k\}, c_j \in \mathrm{R}^d$. The basic objective function of K-means [20] is given as follows:

$$J_{Kmeans} = \sum_{j=1}^{k} \sum_{x \in c_j} \|x - \mu_j\|^2, \tag{1}$$

where $\mu_j = \frac{1}{|c_j|} \sum_{x \in c_j} x_i$. But, this model and its successors are inappropriate for real application without clear boundaries between clusters. To tackle this issue, the researchers introduced the fuzzy theory into clustering algorithms, which can assign data object partially to multiple clusters. The degree of membership in the fuzzy clusters depends on the closeness of the data object to the cluster centers. Fuzzy C-means presented a soft clustering strategy to handle this issue. Unlike the cost function J_{Kmeans}, the FCMs was modified with a degree of membership function [5].

$$J_{FCM} = \sum_{i=1}^{n} \sum_{j=1}^{k} u_{ij} \|x_i - c_j\|^2 \tag{2}$$

where u_{ij} is the ith sample's degree of membership to the jth cluster. Fuzzy clustering algorithms are more flexible than non-fuzzy algorithms, which is operable in real application [22]. FCMs algorithm uses the value of $0-1$ to measure the membership degree u_{ij} of each sample to different cluster centers. According to the fuzzy partition concept introduced, the sum of membership degrees of each sample data to all clusters is equal to 1.

$$u_{ij} = \frac{1}{\sum_{j=1}^{k} \left(\frac{\|x_i - c_j\|}{\|x_i - c_k\|} \right)^{\frac{2}{m-1}}} \tag{3}$$

$$c_j = \frac{\sum_{i=1}^{n} u_{ij}^m x_i}{\sum_{i=1}^{n} u_{ij}^m}, \tag{4}$$

The optimization process of J_{FCM} is achieved by iteratively updating Eq. (3) and Eq. (4), where m (m > 1) is a scalar termed the weighting exponent and controls the fuzziness of the resulting clusters. It will converge to a stable value until relative change in the centroid values becomes small. Details of the algorithm are described in Algorithm 1.

Algorithm 1 Basic Fuzzy C-means Algorithm

Input: data set:$D = \{x_1, \ldots, x_i, \ldots x_n\}, x_i \in \mathrm{R}^d$, cluster number: k, a real number: m, minimum threshold: ε.

Output: The cluster $C = \{c_1, \ldots, c_j, \ldots, c_k\}$

1: Initialise the membership matrix $U = [u_{ij}]$ as $U^{(0)}$

2: Repeat:

3: At the h-th iteration, calculate the center of $c_k^{(h)}$ by $\dfrac{\sum\limits_{i=1}^{n} u_{ij}^m x_i}{\sum\limits_{i=1}^{n} u_{ij}^m}$

4: Update the $U^{(h+1)}$ by $u_{ij} = \dfrac{1}{\sum\limits_{q=1}^{k} \left(\dfrac{\|x_i - c_j\|}{\|x_i - c_q\|}\right)^{\frac{2}{m-1}}}$

5: Until: $\left\| U^{(h+1)} - U^h \right\| \leq \varepsilon$ then stop; otherwise return to step 3

Kernel method [7] uses kernel function to perform inner product operation in high-dimensional feature space, which can project data from the nonlinear low-dimensional feature space into linear high-dimensional feature space without knowing the specific non-linear mapping form, which greatly improves the separability between different patterns [9]. The cost function of kernel fuzzy C-Means and the corresponding membership matrix can be defined as follows:

$$J_{\mathrm{KFCM}} = 2 \sum_{i=1}^{n} \sum_{j=1}^{k} u_{ij}^m (1 - K(x_i, c_j)) \qquad (5)$$

$$u_{ij} = \frac{\left(\dfrac{1}{K(x_i,x_i)+K(c_j,c_j)-2K(x_i,c_j)}\right)^{\frac{1}{(m-1)}}}{\sum\limits_{q=1}^{k} \left(\dfrac{1}{K(x_i,x_i)+K(c_q,c_q)-2K(x_i,c_q)}\right)^{\frac{1}{(m-1)}}}, \qquad (6)$$

where initial centroid matrix $C = [c_q]_{k \times d}$ and $U = [u_{ij}]_{n \times d}$ is fuzzy membership matrix. Here, $0 \leq u_{ij} \leq 1$ is a confusion value. In this paper, we use radial basis kernel function, for any given two vectors v and w, we can calculated the kernel values as the following equation.

$$K_r(v,w) = \exp(-\gamma \|v - w\|^2), \gamma > 0 \qquad (7)$$

Based on the Eq. (7), we derived $K_r(x_i, x_i)$ equals to 1. With this result, Eq. (6) will be calculated by the following membership matrix.

$$u_{ij} = \frac{(\frac{1}{1-K_r(x_i,c_j)})^{\frac{1}{(m-1)}}}{\sum\limits_{q=1}^{k}(\frac{1}{1-K_r(x_i,c_q)})^{\frac{1}{(m-1)}}}. \tag{8}$$

Minimizing Eq. (5) with constraint of U, we will obtain the new center.

$$c_j = \frac{\sum\limits_{i=1}^{n} u_{ij}^m K_r(x_i, c_j) x_i}{\sum\limits_{i=1}^{n} u_{ij}^m K_r(x_i, c_j)} \tag{9}$$

3   Multiple Cost Function Optimization for Kernel Fuzzy Clustering Model

In Radial Basis Kernel Fuzzy C-means (RBKFCM) algorithm, Eq. (2) is obtained based on the cost function J_{KFCM}, which is used for updating the value of center c_j. The explicit iterative expression was calculated by lagrange multiplier and it will be the maximum point. But, there are two main drawbacks during optimization.

- From the point of view of model optimization, unsupervised learning algorithms often fall into multiple local minima, which may also lead to the reduction of generalization performance.
- It is not quite a big issue for them to determine the number of clusters due to the limited number of tariff package. The core of this issue is the acquisition of membership matrix and applied them in tariff package promotion campaign.

In order to avoid local minima during the optimization process, multi-route optimization for RBKFCM is introduced. In the process of optimization, both lagrange multiplier iteration value and PSO optimization value are used to determine the cluster center of the next iteration by comparing the converged degree.

3.1   Multi-route Radial Basis Kernel Function

The MRRBKFCM algorithm's object function includes a kernel function for high dimensional distance evaluation and fuzzy membership function. Then, the lagrange multiplier method is used to obtain the iteration formula of clustering center. Based on the iterated formula, a particle swarm optimization algorithm is introduced for second stage optimization, the route with small value of cost function is selected as the one for the next iteration. With such optimization strategy, we implement multi-route clustering optimization and the objective function Eq. (5) with RBF kernel function.

$$J_{\text{RBKFCM}}(U,C) = 2 \sum_{i=1}^{n} \sum_{j=1}^{k} u_{ij}^{m}(1 - \text{K}_r(x_i, c_j)) \tag{10}$$

The membership confusion matrix is

$$u_{ij} = \frac{\left(\frac{1}{1-\text{K}_r(x_j,c_i)}\right)^{\frac{1}{m-1}}}{\sum\limits_{q=1}^{k}\left(\frac{1}{1-\text{K}_r(x_j,c_q)}\right)^{\frac{1}{m-1}}} \tag{11}$$

Unlike gradient descent optimization method, each PSO particle is a feasible solution to the cost function. The direction and position of particle are determined by current position and flight speed of all particles. The particles modify their flight speed according to their own *pbest* and global *gbest*. The *pbest* is the best position obtained by particle search in past generations, and the global *gbest* is the best position obtained by all particles. The updating formulas of particle velocity and position are as the following equation:

$$v_{id}(t+1) = wv_{id}(t) + s_1 r_1[p_{id}(t) - c_{iq}(t)] + s_2 r_2[g_{id}(t) - c_{id}(t)], \tag{12}$$

and

$$c_{id}(t+1) = c_{id}(t) + v_{id}(t+1). \tag{13}$$

where, s_1 and s_2 are acceleration factors, and there are equal to 2, r_1 and r_2 are random number in the range value of [0,1], w is the inertial factor. Suppose $c_i(t)$ is the cluster center generated for the first iteration of MRRBKFCM algorithm. From the Eq. (11), we can find $u_{ij}(t+1)$. Then, we can obtain the value of $c_i^1(t+1)$ with $c_i(t)$ and $u_{ij}(t+1)$ by using the Eq. (10) and the corresponding cost function J_{RBKFCM}^1. By using the PSO algorithm, a series of J_{MRRBKFCM} are obtained for selection of the possible centers at the next stage, and the best J_{RBKFCM}^2 will be chosen. Compared with J_{RBKFCM}^1 and J_{RBKFCM}^2, the smallest value's corresponded C is the used as the next iteration C. The process of calculation is provided in Algorithm 2.

4   Experiment Results

In this study, we compared the performances of the FCM, GKFCM and MRRBKFCM in terms of five measures using eight UCI data sets in Table 1. The parameters of MRRBKFCM s_1 and s_2 are set to 2, w is set to 0.5, initial position of 20 particles are generated by 20 cluster center matrixes and γ is set to 2^{-4}. The five measures, Adjust Rand Index (ARI), Normalized Mutual Information (NMI), F1, Precision (PE) and Recall (RE), are used for performance comparison. The algorithm with larger values of ARI, NMI, F1, PE, and RE

Algorithm 2 Multi-Route RBF Kernel Fuzzy C-Means

Input: dataset:$D = \{x_1, \ldots, x_i, \ldots x_n\}, x_i \in \mathbb{R}^n$, cluster number: k, a real number: m, the minimum threshold: ε, s_1 and s_2 are 2, the inertia coefficient is 0.5.

Output: The cluster: $C = \{c_1, \ldots, c_j, \ldots c_k\}$

1: Initialization of input clustering centers matrix $C(t)$ at iteration time t=0
2: Calculate the fuzzy membership degree matrix $U(t+1)$ with data set D by using Eq. (11)
3: Calculate the descent gradient clustering center $C^1(t+1)$ with $U(t+1)$ and $C(t)$ by using Eq. (9)
4: Calculate the first cost function value $J^1_{RBKFCM}(t+1)$ with $U(t+1)$ and $C(t+1)$ by using Eq. (10)
5: Calculate the value of $C^2(t+1)$ using PSO algorithm
6: Calculate the second cost function value $J^2_{RBKFCM}(t+1)$ with $U(t+1)$ and $C^2(t+1)$ by using Eq. (10)
7: If($J^1_{RBKFCM}(t+1) < J^2_{RBKFCM}(t+1)$)
8: $C(t+1) = C^1(t+1)$ and $J_{RBKFCM} = J^1_{RBKFCM}(t+1)$
9: Else
10: $C(t+1) = C^2(t+1)$ and $J_{RBKFCM} = J^2_{RBKFCM}(t+1)$
11: If($J_{RBKFCM}(t+1) - J_{RBKFCM}(t) \le \varepsilon$)
12: Stop the process and output confusion matrix u_{ij} and the cluster C
13: Else
14: Return to Step 2

---

indicates better performance. As can be seen from the table, our method shows the best performance in terms of experiments with data set D1, D3, D4, D5, D6 and D7. Although the MRBRKFCMs has lower ARI, F1 and NMI values compared with FCMs and KFCMs using data set D2, D8, it still reaches to 0.85, 0.94 and 0.91. The average performance of our algorithm is better than the other two methods.

4.1 Analysis of Segmentation Result Obtained Using Our Method

Table 2 shows a customer data description of telecom company. It contains two types of data. The first one is basic customer information, e.g., user age, while the second one is users' behavior information, e.g., user's mobile internet usage. We applied our algorithm to the real world data presented by a telecom company to perform clustering of 31285 GuanAn citizens, SiChuan province. Telecom company offers four tariff packages to their users which are 99RMB, 129RMB, 169RMB and 199RMB, therefore, we choose 4 as the number of clusters. The given data contains daily observations of users' mobile internet usage for three months with the sample rate of half hour. The input data for the algorithm is calculated by averaging customers' mobile network usage of three month to a single day. By analysing each cluster center that represents the behavior of one of four distinguished groups, we can learn the usage pattern of the whole group, thus we can perform more efficient promotion campaign to enhance the conversion rate.

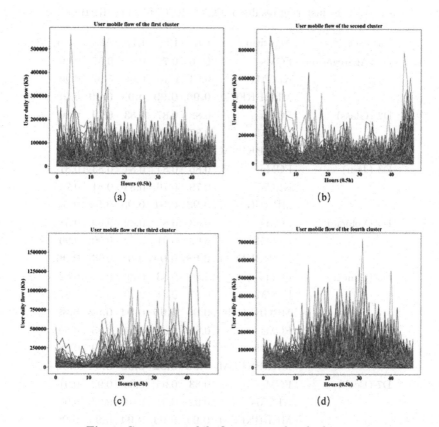

Fig. 1. Customer mobile flow usage of each cluster

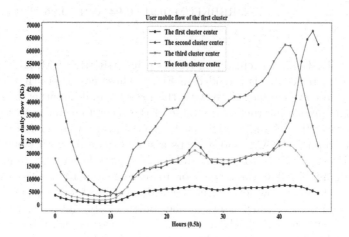

Fig. 2. Centers of different clusters

Table 1. The clustering results of FCMs, KFCMs and MRRBKFCM.

Data set	Algorithm	F1	PE	RE	ARI	NMI
D1 (Aggregation)	FCMs	0.86	0.76	0.65	0.57	0.49
	KFCMs	0.94	0.75	0.68	0.97	0.90
	MRRBKFCM	**0.99**	**0.99**	**0.99**	**0.99**	**0.98**
D2 (Glass)	FCMs	0.88	0.87	0.83	0.79	0.80
	KFCMs	0.87	0.79	0.69	**0.89**	0.93
	MRRBKFCM	**0.89**	**0.80**	**0.83**	0.85	**0.92**
D3 (Jain)	FCMs	0.86	0.84	0.84	0.83	0.81
	KFCMs	0.79	0.76	0.83	0.81	0.74
	MRRBKFCM	**0.93**	**0.91**	**0.93**	**0.94**	**0.92**
D4 (Vehicle)	FCMs	0.83	0.89	0.82	0.89	0.68
	KFCMs	0.92	0.91	0.92	0.90	0.90
	MRRBKFCM	**0.98**	**0.98**	**0.98**	**0.95**	**0.90**
D5 (Cancer)	FCMs	0.89	0.83	0.89	0.84	0.82
	KFCMs	0.83	0.81	0.79	0.73	0.87
	MRGKFCMs	**0.95**	**0.95**	**0.94**	**0.92**	**0.89**
D6 (Bupa)	FCMs	0.82	0.83	0.84	0.83	0.80
	KFCMs	0.82	0.81	0.84	0.89	0.98
	MRRBKFCM	**0.93**	**0.92**	**0.93**	**0.91**	**0.92**
D7 (Ecoli)	FCMs	0.83	0.95	0.93	0.92	0.91
	KFCMs	0.92	0.91	0.94	0.93	0.96
	MRRBKFCM	**0.94**	**0.94**	**0.94**	**0.94**	**0.96**
D8 (Haberman)	FCMs	0.93	0.95	**0.98**	0.93	**0.94**
	KFCMs	**0.95**	0.92	0.93	0.93	0.93
	MRRBKFCM	0.94	**0.94**	0.94	**0.95**	0.91

Figure 1 illustrates the clustering results using MRBRKFCM, hourly behavior of each cluster center is provided in Fig. 2. Those customers that belong to the first cluster are characterized with the lowest mobile internet usage along the day. The mobile internet usage value of the second cluster center starts to increases after 8:00 PM and quickly reaches to the peak by the end of the day, it reduces at around 5:00 AM, and it keeps a stable value between 10:00 AM and 4:00 PM. Users of the third cluster show a high mobile internet usage during daytime, the mobile flow value keeps on increasing from 10:00 AM to 8:00 PM. It quickly reduces to low level after 8:00PM, possibly those customers generally sleep at an early time. Users of the fourth cluster illustrate a low daily mobile internet usage, it slowly increases in the early morning and reaches to peak at roughly 11:00 AM, which becomes a stable value between 2:00 PM and 8:30 PM, it slowly decreases after 10:00PM. To sum up, customers belonging to

Table 2. Telecom customer dataset description

Feature	Description	Range
User age	Customer online age (month)	0–357
Contract length	Customer current package age (month)	0–236
Length of incoming call	Total incoming calls length per month (mins)	0–1396
Number of incoming call	Total incoming calls number per month (number)	0–1140
User mobile flow	User mobile flow per 0.5 h (KB)	0–1.58

the first cluster show very low mobile traffic usage all day, customers of the second cluster have a high mobile usage tendency at night time. The third cluster customers generally illustrate a very high mobile internet usage during daytime and it keeps a low level at night time. The fourth cluster users show a steady low mobile internet usage along the day.

It is not enough to generate a precise marketing strategy by only considering the cluster centers pattern change. Figure 3 illustrates the package tariff distribution in each cluster. The customer number of the first cluster, the second cluster, the third cluster and the fourth cluster are 19476, 1851, 1695 and 8263 respectively. In the first cluster, Fig. 3(a) shows that 40.5% and 33.9% of its users use the tariff package 99 RMB and 129 RMB, respectively. The cluster center of

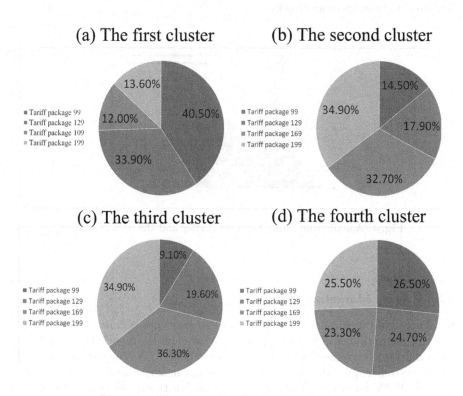

Fig. 3. Tariff package distribution in each cluster

the first cluster in Fig. 2 illustrates a very low mobile flow, whose average value below 5 MB per half hour. According to the service of telecom company, a user could have two or three SIM cards, some SIM cards may not be used. The center of the second cluster in Fig. 2 illustrates the highest mobile flow at the middle of night. Those users with the tariff package 169 and the tariff package 199 accounts 67% of the total users, also those customers' age is mainly distributed between 23 and 40 as shown in Fig. 4. The age distribution of the third cluster is between 20 and 40 while the age distribution of the fourth cluster is between 25 and 50, which is shown in the Fig. 5. Those users illustrate a good incipient additional value for telecom company. There are 268 users with 99 tariff package and 331 users with 129 tariff package in the second cluster. Great attention should be paid to those customers for tariff package promotion, especially with mobile internet package at night time usage. Customers of the third cluster illustrates high mobile flow at day time. There are 9.1% and 19.6% of tariff package 99 and tariff package 129 in the third cluster with 155 and 332 users, respectively. Those customers have high mobile flow at day time, company can offer them additional flow package promotion with discount at day time. The mobile flow usage of users in the fourth cluster decrease after 8:00 PM. The distribution of package tariff in the fourth cluster is roughly equal. The fourth cluster has 8263 users with common mobile internet usage pattern, the company should pay an attention to customer care works.

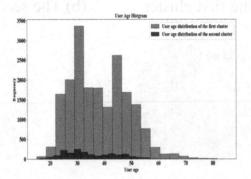

Fig. 4. Age distribution of the first cluster and the second cluster

4.2 Precise Marketing Strategy

In China Telecom company, there are three main types of marketing strategies that are promotion of ITV package, flow package and tariff package. Table 3 provides different strategies for users of different clusters. We suggest taking action before customer quits the service, for example, check the user's broadband flow usage and ITV service usage of low value users. For those users with 169 and 199 tariff package, we choose Personal selling and Customer care with satisfaction

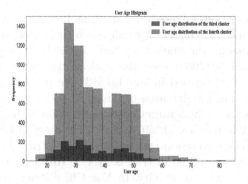

Fig. 5. Age distribution of the third cluster and the fourth cluster

Table 3. Marketing strategy

Cluster 1	Cluster 2	Cluster 3	Cluster 4
T1 : Check user's broadband flow usage	T1 : Nighttime flow package marketing	T1 : Daytime flow package marketing	T1 : 129 Package promotion
T2 : Check user's ITV usage and customer care	T2 : Nighttime flow package marketing	T2 : Daytime flow package marketing	T2: 169 Package promotion
T3 : Personal selling of 199 package	T3 : Check user's broadband flow usage	T3 : Personal selling of 199 package	T3 : Personal selling of 199 package
T4 : Customer care and satisfaction survey	T4 : Customer care and satisfaction survey	T4 : Customer care and satisfaction survey	T4 : Customer care and satisfaction survey

T1 99 tariff fee; T2 129 tariff fee; T3 169 tariff fee; T4 199 tariff fee

survey respectively. Considering a nighttime usage pattern of customers from the second cluster, we can offer nighttime flow package promotion for those 99 and 129 tariff users, whose age is between 25 and 35 in Fig. 4. Compared with the second cluster, users of the third cluster mainly use the internet during the daytime. Our promotion activities are mainly focused on daytime products or mobile internet packages. In the fourth cluster, users with different tariff packages are equally distributed, we will consider tariff package upgrading strategy, users with 99 package will be suggested to be promoted to 129 package tariff and so on. According to clustering results, each user will have a membership probability value to all clusters. We choose two clusters with the highest membership probability, and depending on his current tariff package we apply the corresponding marketing strategies from Table 4. For example, a 99 tariff user has membership value of 0.45 in the second cluster and membership value of 0.21 in the third cluster. According to Table 3, a nighttime flow package will be first suggested to this user then followed by a daytime flow package. Table 4 provides three users' MRRBKFCM clustering results with their corresponding marketing strategy. We recommend the marketing staff to use the strategy in the Table 3. For example, the 260100179691 user, the staff can try to do customer care and satisfaction survey as described in the Table 3. We have used this strategy to do customer promotion for city area of Guan An(GA), Lin Shui(LS) and Yue Chi(YC). The number of 3925 target customer has been chosen for promotion. Marketing staff got in touch with 607 out of 1359 users in Guan An area with period of 45 days. The conversion rate in Guan An district is 19.11% (116/607). The conversion rate in Lin Shui district is 17.42% (61/350), which is quite high for a short 15 days campaign activity. In Yue Chi district, the conversion rate is 7.99% (98/1226) for 13 days campaign activity. Compared with the average conversion rate 5.81% of traditional marketing strategy, our method has a better performance as shown in Table 5.

Table 4. Clustering results of MRRBKFCM with marketing strategy

UserID	Package	Cluster 1	Cluster 2	Cluster 3	Cluster 4	Cluster	Strategy
260100179691	199	0.07	0.46	0.31	0.16	2	C2/T4|C3/T4
260100253818	129	0.30	0.27	0.22	0.22	1	C1/T2|C2/T2
260100267208	99	0.12	0.41	0.29	0.19	2	C2/T1|C3/T1

Table 5. Precise promotion comparisons

City area	Total users	Contacted users	Period time	99–129	129–169	169–199	Flow	ITV	Upgrades	Rate
Our promotion strategy										
GA	1359	607	45	37	14	0	35	30	116	19.11%
LS	805	350	15	8	3	1	20	14	61	17.42%
YC	1761	1226	13	10	9	0	31	35	98	7.99%
Traditional promotion strategy										
GA	1869	1526	90	70	0	0	0	0	70	4.50%
LS	2330	1507	90	94	0	0	0	0	90	6.23%
YC	2175	1694	90	114	0	0	0	0	114	6.72%

5   Conclusion

In this paper, we introduce a novel multi-route radial basis kernel fuzzy C-means (MRRBKFCM) to avoid local minima optimization problems for soft clustering. In terms of clustering efficiency, experiment results with UCI datasets illustrate the advantage of the MRRBKFCM compared to the traditional fuzzy C-means and Gaussian kernel fuzzy C-means. The MRRBKFCM algorithm has been used to perform customer segmentation to increase the efficiency of the tariff promotion campaign and achieves high conversion rate. A deep analysis of customers behavior is completed based on their daily mobile internet usage patterns. Meanwhile, corresponding precise marketing strategy has been applied to users in GuanAn. Implementation of the proposed algorithm allowed to increase the conversion rate of marketing campaign.

References

1. Wei, C.P., Chiu, I.T.: Turning telecommunications call details to churn prediction: a data mining approach. Expert. Syst. Appl. **23**(2), 103–112 (2002)
2. Chiu, C.Y., Chen, Y.F., Kuo, I.T., Ku, H.C.: An intelligent market segmentation system using k-means and particle swarm optimization. Expert. Syst. Appl. **36**(3), 4558–4565 (2009)
3. Bezdek, J.C., Ehrlich, R., Full, W.: FCM: the fuzzy c-means clustering algorithm. Comput. Geosci. **10**(2), 191–203 (1984)
4. Pal, N.R., Bezdek, J.C.: On cluster validity for the fuzzy c-means model. IEEE Trans. Fuzzy Syst. **3**(3), 370–379 (1995)
5. Pal, N.R., Pal, K., Keller, J.M., Bezdek, J.C.: A possibilistic fuzzy c-means clustering algorithm. IEEE Trans. Fuzzy Syst. **13**(4), 517–530 (2005)
6. Subudhi, S., Panigrahi, S.: Use of possibilistic fuzzy c-means clustering for telecom fraud detection. In: Proceedings of the International Conference on Computational Intelligence in Data Mining, pp 633–641 (2017)
7. Hofmann, T., SchLkopf, B., Smola, A.J.: Kernel methods in machine learning. Ann. Stat. **36**(3), 1171–1220 (2008)
8. Ding, Y., Fu, X.: Kernel based fuzzy c-means clustering algorithm based on genetic algorithm. Neurocomputing **188**(5), 233–238 (2015)
9. Girolami, M.: Mercer kernel-based clustering in feature space. IEEE Trans. Neural Netw. **13**(3), 780–784 (2002)
10. Duboff, R.S.: Marketing to maximize profitability. J. Bus. Strateg. **13**(6), 10–13 (1992)
11. Dwyer, F.R. Customer lifetime valuation to support marketing decision making. J. Dir. Mark. **3**(4), 6–13 (1989)
12. Berger, P.D., Nasr, N.I.: Customer lifetime value: marketing models and applications. J. Interact. Mark. **12**(1), 17–30 (1998)
13. Donkers, B., Verhoef, P.C., Jong, M.G.D.: Modeling CLV: a test of competing models in the insurance industry. Quant. Mark. Econ. **5**(2), 163–190 (2007)
14. Gloy, B.A., Akridge, J.T., Preckel, P.V.: Customer lifetime value: an application in the rural petroleum market. Agribus **13**(3), 335–347 (1997)
15. Karahoca, A., Kara, A.: Comparing clustering techniques for telecom churn management. In: Proceedings of the 5th WSEAS International Conference on Telecommunications and Informatics, pp. 27–29 (2006)
16. Lin, Q., Wan, Y.: Mobile customer clustering based on call detail records for marketing campaigns. In: International Conference on Management and Service Science, pp. 1–4 (2009)
17. Han, S.H., Lu, S.X., Leung, S.C.H.: Segmentation of telecom customers based on customer value by decision tree model. Expert Syst. Appl. **39**(4), 3964–3973 (2012)
18. Ahn, H., Ahn, J.J., Oh, K.J., Kim, D.H.: Facilitating cross-selling in a mobile telecom market to develop customer classification model based on hybrid data mining techniques. Expert Syst. Appl. **38**(5), 5005–5012 (2011)
19. Cheng, C.: A novel cluster algorithm for telecom customer segmentation. In: 16th International Symposium on Communications and Information Technologies, pp. 324–329 (2016)

20. Jain, A.K.: Data clustering: 50 years beyond K-means. Pattern Recognit. Lett. **31**(8), 651–666 (2010)
21. Huang, Z.X.: Extensions to the K-means algorithm for clustering large data sets with categorical values. Data Min. Knowl. Discov. **2**(3), 283–304 (1998)
22. Asokan, G., Mohanavalli, S.: Fuzzy clustering for effective customer relationship management in telecom industry. In: International Conference on Computational Science, Engineering and Information Technology, pp. 571–580 (2011)

Clustering by Unified Principal Component Analysis and Fuzzy C-Means with Sparsity Constraint

Jikui Wang[1] (ID), Quanfu Shi[2], Zhengguo Yang[1] (ID), and Feiping Nie[3] (✉) (ID)

[1] College of Information Engineering, Lanzhou University of Finance and Economics, Lanzhou 730020, Gansu, China
[2] Degree Management and Graduate Office, Lanzhou University of Finance and Economics, Lanzhou 730020, Gansu, China
[3] School of Computer Science and Center for OPTical IMagery Analysis and Learning (OPTIMAL), Northwestern Polytechnical University, Xi'an 710072, Shaanxi, China
feipingnie@gmail.com

Abstract. For clustering high-dimensional data, most of the state-of-the-art algorithms often extract principal component beforehand, and then conduct a concrete clustering method. However, the two-stage strategy may deviate from assignments by directly optimizing the unified objective function. Different from the traditional methods, we propose a novel method referred to as clustering by unified principal component analysis and fuzzy c-means (UPF) for clustering high-dimensional data. Our model can explore underlying clustering structure in low-dimensional space and finish clustering simultaneously. In particular, we impose a L_0-norm constraint on the membership matrix to make the matrix more sparse. To solve the model, we propose an effective iterative optimization algorithm. Extensive experiments on several benchmark data sets in comparison with two-stage algorithms are conducted to validate effectiveness of the proposed method. The experiments results demonstrate that the performance of our proposed method is superiority.

Keywords: Machine learning · Artificial Intelligence · Principal component analysis · Fuzzy c-means clustering · Sparsity constraint

1 Introduction

Clustering is a hot research topic in machine learning, which is widely used in many fields to solve different application problems, including pattern recognition [1] and image segmentation [2, 3]. Researchers have proposed many clustering algorithms in the past decades. Among all of these algorithms, KMeans [4] is the most famous method due to its low complexity when dealing with different kinds of data sets. However, KMeans cannot find representative samples of the clusters. To deal with the problem, k-medoids is proposed to minimize the average dissimilarity of all objects of the data set to the

Submit to ICA3PP 2020 Special Session on Artificial Intelligence and Security (AIS).

nearest medoid. There are other several KMeans variant algorithms [5–8], which attempt to eliminate the limitations issues. Huang [9–12] extended the KMeans algorithm for clustering categorical data.

Later, Fuzzy C-Means (FCM) is proposed, which introduces the fuzzy sets to make the clustering method more powerful. However, FCM still exists several drawbacks, including the limitation to convex clusters, the sensitivity to outliers and initialization. Therefore, many extensions and variants of FCM clustering are advanced [13–16]. Kernel based FCM [13] optimizes FCM, based on the genetic algorithm optimization which is combined of the improved genetic algorithm and the kernel technique to optimize the initial clustering center and to guide the categorization. Modified probabilistic FCM [14] jointly considers the typicality as well as the fuzzy membership measures to model the bias field and noise. Conditional spatial FCM [15] improves the robustness of FCM through the incorporation of conditioning effects imposed by an auxiliary variable corresponding to each pixel. Generalized entropy-based possibilistic FCM [16] utilizes the functions of distance in the fuzzy, probabilistic and entropy terms of the clustering objective function to decrease the noise effects on the cluster centers.

Nowadays, the dimensionality of the data sets is always very high. Due to the curse of dimensionality, it is usually computationally expensive. To alleviate the problem, a lot of dimensionality reduction techniques are proposed, which aim to extract the most important features and remove the noisy data, such as Principal Component Analysis (PCA) [17], Locality Preserving Projections (LPP) [18], orthogonal locality preserving projection (OLPP) [19, 20] and linear discriminant analysis (LDA) [21]. However, using PCA, LPP, and OLPP to construct the projection matrix only relies on feature information and ignores label information which is important to reveal the relations between classes and features [22, 23].

For clustering high-dimensional data, the state-of-the-art methods always project the original data into an optimal subspace by a projection matrix beforehand, and then conduct the final clustering assignments by a concrete clustering method. However, the results obtained by a two-stage strategy may deviate from the assignments by directly optimize the unified objective function. Motivated by the idea, we design an objective model which combines principal component analysis and fuzzy c-means. Moreover, we introduce a flexible sparse constraint imposed on the membership matrix. Our model takes the advantages of two classical methods. Through adjusting the parameter, the model can find the optimal projection matrix for clustering task.

The rest of this paper is organized as follows. In Sect. 2, we briefly discuss some related work. Our method is introduced in Sect. 3. In Sect. 4, we present the experiment results on several real world data sets. At last, the conclusion is given in Sect. 5.

2 Related Works

In this section, we briefly review Principal Component Analysis and Fuzzy c-means.

2.1 Principal Component Analysis

Given a data set $X = [x_1, x_2, \cdots, x_n] \in \mathbb{R}^{d \times n}$, where d denotes the number of dimensionality and n is the number of data sample. Principal Component Analysis (PCA) is the most famous dimensionality reduction method, which is aimed to find a low-rank matrix to best approximate the give data matrix in the sense of Euclidean distance. Suppose the samples are centered beforehand, PCA is to solve the following problem:

$$\min_{Rank(Z)=k} \|X - Z\|_F^2. \tag{1}$$

According to the full rank decomposition, any matrix $Z \in \mathbb{R}^{d \times n}$ with rank k can be decomposed as $Z = WV^T$, where $W = \mathbb{R}^{d \times k}$, $V = \mathbb{R}^{n \times k}$. After denoting an identity matrix by I, The problem (1) can be rewritten as:

$$\min_{W^T W = I} \left\|X - WV^T\right\|_F^2. \tag{2}$$

Taking the derivative with respect to V and setting it to zero, we have $V = X^T W$. Thus the problem becomes:

$$\max_{W^T W = I} Tr\left(W^T XX^T W\right). \tag{3}$$

The columns of the optimal solution W to problem (3) are the k eigenvectors of $X^T X$ corresponding to the k largest eigenvalues.

2.2 Fuzzy C-Means Clustering

As one of the most popular fuzzy clustering techniques, fuzzy c-means clustering (FCM) is to minimize the following objective function:

$$\sum_{i=1}^{n} \sum_{k=1}^{c} \|x_i - b_k\|_2^2 \theta_{ik}^m \tag{4}$$
$$s.t. \theta_i \geq 0, \theta 1_c = 1,$$

where $\theta \in \mathbb{R}^{n \times c}$ is the membership matrix and whose elements are nonnegative and their sum equals to one on each row. b_k denotes k-th center. The only difference between FCM objective function and KMeans objective function is to introduce a parameter m which expresses the fuzziness of a given sample to each of the clusters. When m tends towards infinity, the objective function of FCM identical to that of KMeans. The fuzzy c-means algorithm is more robust for ambiguity than KMeans.

3 Clustering by Unified Principal Component Analysis and Fuzzy C-Means with Sparsity Constraints

Can we perform subspace learning and clustering simultaneously? Motivated by the idea, we propose a novel method, which can conduct subspace learning and clustering simultaneously. The followings describe the details of the proposed method.

3.1 Formulation

For clustering high-dimensional data, the traditional two-stage method includes two steps, for instance:

Step1: Dimension reduction by PCA

Denote a centering matrix $H = 1 - \frac{1}{n}\mathbf{11}^T$, the objective function of PCA can be represented as:

$$\max_{W^T W=I} Tr(W^T XHX^T W). \tag{5}$$

The columns of the optimal solution W to problem (5) are formed by k eigenvectors of $X^T HX$ corresponding to the k largest eigenvalues.

Step2: Clustering by FCM

Then to finish clustering by solving the following question:

$$\sum_{i=1}^{n} \sum_{k=1}^{c} \|Y_i - b_k\|_2^2 \theta_{ik}^m \tag{6}$$
$$s.t. \theta_i \geq 0, \theta \mathbf{1}_c = 1,$$

where $Y = W^T X$ denotes the data in low-dimensional subspace.

This two-stage method does not consider the interaction of dimension reduction and clustering.

By using a combination of model (5) and model (6), we propose a novel method referred to clustering by unified principal component analysis and fuzzy c-means (UPF) for clustering high-dimensional data. Our model can explore underlying clustering structure in low-dimensional space and finish clustering simultaneously. In particular, we impose a L_0-norm constraint on the membership matrix to make the matrix more sparse. The model of UPF is defined as follows:

$$\max_{W,B,\theta} Tr(W^T XHX^T W)$$
$$-\lambda \sum_{i=1}^{n} \sum_{k=1}^{c} \|W^T x_i - b_k\|_2^2 \theta_{ik}^m \tag{7}$$
$$s.t. W^T W = I, \|\theta_i\|_0 = K, \theta_i \geq 0, \theta_i \mathbf{1}_c = 1,$$

where $W \in \mathbb{R}^{d \times d'}$ is a projection matrix and which can project the data from the originally d-dimensional space to the d'-dimensional space. $\theta \in \mathbb{R}^{n \times c}$ is the membership matrix constrained by the L_0-norm, whose elements are nonnegative and their sum equals to one on each row. θ_i denotes the i-th row of θ, θ_{ik} denotes an element of θ. The hyper parameter $m > 1$ controls how fuzzy the cluster will be and $K \in \mathbb{N}_+$ denotes the number of nonzero elements in θ_i, which controls the sparseness of membership matrix. As is shown that t PCA and FCM are unified into UPF model, we use the coefficient $\lambda \geq 0$ to control the importance of the two models.

Note that the term $Tr(W^T XHX^T W)$ attempts to extract useful information which is important to PCA. Meanwhile, the second term $\lambda \sum_{i=1}^{n} \sum_{k=1}^{c} \|W^T x_i - b_k\|_2^2 \theta_{ik}^m$ attempts to assign all the points to their centers in low-dimensional subspace by FCM. In this way, the model (7) not only takes the impact of PCA on FCM, but also involves the effect of FCM on PCA.

3.2 Optimization

In this subsection, we provide an efficient iterative method to solve the problem (7), whose convergence has already been proved in [24]. More concretely, we alternatively update a group of optimization variables while keeping other optimization variables fixed. The details of solving the problem (7) is represented as follows.

Step 1: Solving W, B while fixing θ

Now, the model (7) can be formulated into:

$$
\max_{W,B} Tr\left(W^T XHX^T W\right) \\
-\lambda \sum_{i=1}^{n} \sum_{k=1}^{c} s_{ik} \left\| W^T x_i - b_k \right\|_2^2 \\
s.t.\, W^T W = I,
\tag{8}
$$

where, $s_{ik} = \theta_{ik}^m$. We took the derivative of the objective function of model (8) with respect to b_k:

$$
\frac{\partial \left(Tr\left(W^T XHX^T W\right) - \lambda \sum_{i=1}^{n} \sum_{k=1}^{c} s_{ik} \left\| W^T x_i - b_k \right\|_2^2 \right)}{\partial b_k}
$$

$$
= 0 - 2\lambda \left(\sum_{i=1}^{n} s_{ik} W^T x_i - \sum_{i=1}^{n} s_{ik} b_k \right)
$$

$$
= -2\lambda \left(\sum_{i=1}^{n} s_{ik} W^T x_i - \sum_{i=1}^{n} s_{ik} b_k \right),
$$

and then make the above derivative equal to zero,

$$
-2\lambda \left(\sum_{i=1}^{n} s_{ik} W^T x_i - \sum_{i=1}^{n} s_{ik} b_k \right) = 0
$$

$$
\Rightarrow \sum_{i=1}^{n} s_{ik} W^T x_i - \sum_{i=1}^{n} s_{ik} b_k = 0
$$

$$
\Rightarrow b_k = W^T \sum_{i=1}^{n} x_i s_{ik} \Big/ \sum_{i=1}^{n} s_{ik}.
$$

Let $y_k = \sum_{i=1}^{n} x_i s_{ik} \Big/ \sum_{i=1}^{n} s_{ik}$, then,

$$
b_k = W^T y_k.
\tag{9}
$$

Now, the problem (8) can be formulated into:

$$
\max_{W} Tr\left(W^T XHX^T W\right) \\
-\lambda \sum_{i=1}^{n} \sum_{k=1}^{c} s_{ik} \left\| W^T x_i - W^T y_k \right\|_2^2. \\
s.t.\, W^T W = I
\tag{10}
$$

In order to solve the problem (10), we introduce the following lemma:

Lemma 1. Let $W \in \mathbb{R}^{d \times d'}$ be a transformation matrix and d_{ii}^k, d_{kk}^i, M and $\tilde{M}$ are defined as follows:

$$d_{ii}^k = \sum_{k=1}^c s_{ik},$$

$$d_{kk}^i = \sum_{i=1}^n s_{ki},$$

$$M = XD^k X^T - 2XSY^T + YD^i Y^T,$$

$$\tilde{M} = \left(M + M^T\right) \big/ 2, \tag{11}$$

where the degree matrix D^k is a diagonal matrix whose i-th diagonal element is D_{ii}^k, and the degree matrix D^i is a diagonal matrix whose k-th diagonal element is D_{kk}^i. The following equality holds:

$$\sum_{i=1}^n \sum_{k=1}^c s_{ik} \left\| W^T x_i - W^T y_k \right\|_2^2 = Tr\left[W^T \tilde{M} W \right] \tag{12}$$

Let $\tilde{X} = W^T X$ and $\tilde{Y} = W^T Y$, by definition:

$$\begin{aligned}
&\sum_{i=1}^n \sum_{k=1}^c s_{ik} \|\tilde{x}_i - \tilde{y}_k\|_2^2 \\
&= \sum_{i=1}^n \sum_{k=1}^c s_{ik} (\tilde{x}_i - \tilde{y}_k)^T (\tilde{x}_i - \tilde{y}_k) \\
&= \sum_{i=1}^n \sum_{k=1}^c s_{ik} (\tilde{x}_i^T \tilde{x}_i - 2\tilde{y}_k^T \tilde{x}_i + \tilde{y}_k^T \tilde{y}_k) \\
&= \sum_{i=1}^n d_{ii}^k \tilde{x}_i^T \tilde{x}_i + \sum_{k=1}^c d_{kk}^i \tilde{y}_k^T \tilde{y}_k - 2 \sum_{i=1}^n \sum_{k=1}^c s_{ik} \tilde{y}_k^T \tilde{x}_i
\end{aligned} \tag{13}$$

An observation will simplify the first and the second term of the above expression:

$$\sum_{i=1}^n d_{ii}^k \tilde{x}_i^T \tilde{x}_i = Tr(D^k \tilde{X}^T \tilde{X}) = Tr[\tilde{X} D^k \tilde{X}^T]$$

$$\sum_{k=1}^c d_{kk}^i \tilde{y}_k^T \tilde{y}_k = Tr(D^i \tilde{Y}^T \tilde{Y}) = Tr[\tilde{Y} D^i \tilde{Y}^T] \tag{14}$$

We further have for the third term:

$$\sum_{i=1}^n \sum_{k=1}^c s_{ik} \tilde{y}_k^T \tilde{x}_i$$

$$= \sum_{i=1}^{n} \left(\sum_{k=1}^{c} s_{ik} \tilde{y}_k^T \right) (\tilde{X} e_i)^T$$

$$= \sum_{i=1}^{n} e_i S Y^T \tilde{X} e_i$$

$$= Tr\left[S \tilde{Y}^T \tilde{X} \right] = Tr\left[\tilde{X} S \tilde{Y}^T \right], \tag{15}$$

where Denoting by e_i is the i-th canonical vector. Putting these expressions together, we have:

$$\sum_{i=1}^{n} \sum_{k=1}^{c} s_{ik} \left\| W^T x_i - W^T y_k \right\|_2^2$$

$$= \sum_{i=1}^{n} \sum_{k=1}^{c} s_{ik} \left\| \tilde{x}_i - \tilde{y}_k \right\|_2^2$$

$$= \sum_{i=1}^{n} d_{ii}^x \tilde{x}_i^T \tilde{x}_i + \sum_{k=1}^{c} d_{ii}^y \tilde{y}_i^T \tilde{y}_i - 2 \sum_{i=1}^{n} \sum_{k=1}^{c} s_{ik} \tilde{y}_k^T \tilde{x}_i$$

$$= Tr[\tilde{X} D^x \tilde{X}^T] + Tr[\tilde{Y} D^y \tilde{Y}^T] - 2Tr[\tilde{X} S \tilde{Y}^T]$$

$$= Tr[W^T (X D^x X^T - 2 X S Y^T + Y D^y Y^T) W]$$

$$= Tr[W^T M W] = Tr[W^T ((M + M^T)/2) W]$$

$$= Tr[W^T \tilde{M} W].$$

Thus, the Lemma 1 is proved. □

Now, the model (10) can be written as follows:

$$\max_{W} Tr(W^T X H X^T W) - \lambda Tr(W^T \tilde{M} W)$$

$$= \max_{W} Tr[W^T X (H - \lambda \tilde{M}) X^T W]$$

$$s.t. \ W^T W = I.$$

Therefore the Optimal W is formed by the d' eigenvectors corresponding to the d' largest eigenvalues of $X(H - \lambda \tilde{M})X^T$. Then we update B with Eq. (9) immediately.

Step 2: Solving θ while fixing W, B

Because the constraints exist on every row, therefore, we further divide the problem (7) into n independent sub-problems and impose a L_0-norm constraints on the membership vector for each sample. Therefore, Now, the model (7) can be formulated into:

$$\min_{\theta^i} \sum_{k=1}^{c} \left\| W^T x_i - b_k \right\|_2^2 \theta_{ik}^m$$

$$s.t. \ \theta_i \geq 0, \theta_i 1_c = 1, \|\theta_i\|_0 = K \tag{16}$$

It is obvious that $\|\theta_i\|_0 = K$ makes the membership vector with the sparsity K, which avoids the incorrect or invalid clustering partitions from outliers. To make the problem (16) more simple, we need to do some transformations as follows:

$$\min_{\theta^i} \sum_{k=1}^{c} h_{ik}\theta_{ik}^m$$

$$s.t. \ \boldsymbol{\theta}_i \geq 0, \ \boldsymbol{\theta}_i 1_c = 1, \ \|\boldsymbol{\theta}\| =\|_0 K \qquad (17)$$

where $h_{ik} = \left\| W^T x_i - b_k \right\|_2^2 \in \mathbb{R}$. We define the following Lagrangian function:

$$J(\theta^i, \alpha_i) = \sum_{k=1}^{c} h_{ik}\theta_{ik}^m + \alpha_i \left(\sum_{j=1}^{c} \theta_{ij} - 1 \right) \qquad (18)$$

Let θ_{ik}^* denotes the optimal solution of θ_{ik}, α_i^* denotes the optimal solution of α_i. According to KKT condition, we have:

$$\frac{\partial J(\theta^i, \alpha_i)}{\partial \theta_{ik}^*} = md_{ik}\theta_{ik}^{m-1} + \alpha_i^* = 0 \qquad (19)$$

$$\frac{\partial J(\theta^i, \alpha_i)}{\partial \alpha_i^*} = \sum_{j=1}^{c} \theta_{ij}^* - 1 = 0 \qquad (20)$$

Thus, we obtain the optimal solution of θ_{ik} as follows:

$$\theta_{ik}^* = h_{ik}^{\frac{1}{1-m}} \bigg/ \sum_{s=1}^{c} h_{is}^{\frac{1}{1-m}}. \qquad (21)$$

Substituting the Eq. (21) into problem (17), its optimal value arrives at:

$$\sum_{k=1}^{c} h_{ik}\theta_{ik}^m = \left(\sum_{k=1}^{c} h_{ik}^{\frac{1}{1-m}} \right)^{1-m}$$

$$= 1 \bigg/ \left(\sum_{k=1}^{c} (\tfrac{1}{h_{ik}})^{\frac{1}{m-1}} \right)^{m-1} \qquad (22)$$

It is obvious that the minimum depends on h_{ik}, the smaller the better. Consequently, we select first K smallest elements as well as their corresponding membership values in θ^i, meanwhile, setting the membership values of the rest elements as zeros. Intuitively, the above optimization is summarized in Algorithm 1.

Algorithm 1: Algorithm to solve Problem (7).

Input : Data matrix $X \in \mathbb{R}^{d \times n}$,The number of clusters c

Parameter: reduced dimension number d', K and λ

Output: projection matrix W , the membership matrix θ

Initialization: Random initialization θ_i , which satisfied $\theta_i \geq 0$, $\theta_i 1_c = 1$, $H = I - \frac{1}{n} 11^T$

1. Calculate each $y_k = \sum_{i=1}^{n} x_i \theta_{ik} / \sum_{i=1}^{n} \theta_{ik}$

2. **WHILE** not converge **DO**

3. Calculate $D_{ii}^x = \sum_{j=1}^{c} \theta_{ij}$, $D_{kk}^y = \sum_{i=1}^{n} \theta_{ik}$

4. Calculate $M = XD^x X^T - 2X\theta Y^T + YD^y Y^T$

5. Calculate $M = (M + M^T) / 2$

6. Update W , which is formed by the d' eigenvectors corresponding to the d' largest eigen-values of $X(H - \lambda \tilde{M}) X^T$

7. Update $b_k = W^T y_k$

8. Calculate $h_{ik} = \| W^T x_i - b_k \|_2^2$

9. Update $\theta_{ik} = h_{ik}^{\frac{1}{1-m}} / \sum_{s=1}^{K} h_{is}^{\frac{1}{1-m}}$, we select first K smallest h_{ik} as well as their corresponding

membership values in θ_i , meanwhile, setting the membership values of the rest elements as

zeros.

10. **END WHILE**

11. **RETURN** projection matrix W and the membership matrix θ

3.3 Complexity Analysis

Given a data matrix $X \in \mathbb{R}^{d \times n}$, where d denotes the number of dimensionality, n is the number of data sample, c denotes the number of centers, the computational complexity of UDF is mainly divided into five parts.

1) We need $O(nd^2)$ to calculate $\tilde{M}$.
2) We need $O(nd^2)$ to update the matrix W.
3) We need $O(ndd')$ to project the original data to low-dimensional space by W.
4) We need $O(ndc)$ to update the centers matrix B.
5) We need $O(ndc)$ to update the matrix θ.

Considering that $c \ll n$, $d \ll n$ and $d' \ll d$, the overall computational complexity is $O(nd^2 t)$, where t is the number of iterations. It's notable that the UPF is linear scaled algorithm with respect to n.

3.4 Convergence Analysis

In this section, we prove the convergence of the proposed UDF iterative procedure given in Algorithm 1.

Theorem 1. The Algorithm 1 monotonically increases the objective of the problem (7) in each iteration and converges to the local optimum.

Proof. We utilize an alternately iterative optimization method to solve the problem (7), which includes two steps. According to the Algorithm 1, after i-th iteration, we can obtain $W = W^{(t)}, B = B^{(t)}$ and $\theta = \theta^{(t)}$. In the next iteration, we first fix $\theta = \theta^{(t)}$ and update W and B. Let $W^{(t+1)}$ denotes the optimal solution of W, $B^{(t+1)}$ denotes the optimal solution of B in step 1. Thus, we have the following inequality:

$$
\begin{aligned}
&Tr\left(W^{(t+1)T}XHX^TW^{(t+1)}\right) \\
&-\lambda \sum_{i=1}^{n}\sum_{k=1}^{c}\left\|W^{(t+1)T}x_i - b_k^{(t+1)}\right\|_2^2 \theta_{ik}^{(t)m} \\
&\geq Tr\left(W^{(t)T}XHX^TW^{(t)}\right) \\
&-\lambda \sum_{i=1}^{n}\sum_{k=1}^{c}\left\|W^{(t)T}x_i - b_k^{(t)}\right\|_2^2 \theta_{ik}^{(t)m}.
\end{aligned}
\tag{22}
$$

In step 2, we fix $W = W^{(t+1)}, B = B^{(t+1)}$ and update θ with the sparsity constrains K. It is obvious that the minimum depends on h_{ik}, the smaller the better. Algorithm 1 selects first K smallest elements as well as their corresponding membership values in θ^i, meanwhile, setting the membership values of the rest elements as zeros. Let $\theta^{(t+1)}$ denotes the optimal solution of θ. According to the above analyses, we have the following inequality:

$$
\begin{aligned}
&Tr\left(W^{(t+1)T}XHX^TW^{(t+1)}\right) \\
&-\lambda \sum_{i=1}^{n}\sum_{k=1}^{c}\left\|W^{(t+1)T}x_i - b_k^{(t+1)}\right\|_2^2 \theta_{ik}^{(t+1)m} \\
&\geq Tr\left(W^{(t+1)T}XHX^TW^{(t+1)}\right) \\
&-\lambda \sum_{i=1}^{n}\sum_{k=1}^{c}\left\|W^{(t+1)T}x_i - b_k^{(t+1)}\right\|_2^2 \theta_{ik}^{(t)m}.
\end{aligned}
\tag{23}
$$

Combining formula (22) and (23), the following inequality holds:

$$
\begin{aligned}
&Tr\left(W^{(t+1)T}XHX^TW^{(t+1)}\right) \\
&-\lambda \sum_{i=1}^{n}\sum_{k=1}^{c}\left\|W^{(t+1)T}x_i - b_k^{(t+1)}\right\|_2^2 \theta_{ik}^{(t+1)m} \\
&\geq Tr\left(W^{(t)T}XHX^TW^{(t)}\right) \\
&-\lambda \sum_{i=1}^{n}\sum_{k=1}^{c}\left\|W^{(t)T}x_i - b_k^{(t)}\right\|_2^2 \theta_{ik}^{(t)m}.
\end{aligned}
\tag{24}
$$

Because $\lambda \geq 0$ and $W^TW = I$, the following inequality holds:

$$
Tr(W^TXHX^TW) - \lambda \sum_{i=1}^{n}\sum_{k=1}^{c}s_{ik}\left\|W^Tx_i - W^Ty_k\right\|_2^2
$$

$$\leq Tr(\boldsymbol{W}^T \boldsymbol{XHX}^T \boldsymbol{W}) \leq Tr(\boldsymbol{XHX}), \tag{25}$$

therefore, the objective function of model (7) has an upper bound. Thus, the Theorem 1 is proved.   □

3.5 Discussion of UPF

Consider the limiting case when $\lambda \to 0$, UPF is equivalent to the two-stage strategy of performing the two popular methods PCA and FCM in sequence.

Proof. When $\lambda \to 0$, as is shown that the step 1 of UDF will amount to computing the d' maximal eigenvalues of the matrix $\boldsymbol{XHX}^T$ and the associated eigenvectors, which is a standard PCA. Then, the projection matrix $\boldsymbol{W}$ projects the original data to the low-dimensional d'-space, and update B simultaneously. In step 2, UDF perform a standard FCM to solve θ in low-dimensional space. Next iteration, the $\boldsymbol{W}$ would never be changed, and B, θ would be never changed too. Therefore, the iterations are completed.   □

Thus, we have proved that when $\lambda \to 0$, UPF is equivalent to PCA+FCM.

4 Experiments

In this section, extensive experiments on several real world benchmark data sets are conducted to evaluate the effectiveness of our proposed UPF method. Concretely, we tested the clustering abilities between UPF、PCA+KMeans and PCA+FCM on several benchmark data sets. Our experiments are conducted on a Windows7 computer with a 2.7 GHz AMD A12-9800B R7 CPU and 8-GB RAM, MATLAB 2012a.

4.1 Data Sets

We have tested the clustering ability of the proposed method on six benchmark, which are IRIS, Breast, Dermatology, COIL20, ORL and Yale, and they are described as follows.

IRIS. This is perhaps the best known database to be found in the pattern recognition literature. The data set contains 3 classes of 50 instances each, where each class refers to a type of iris plant. One class is linearly separable from the other 2; the latter are not linearly separable from each other.

BREAST. The dataset consists of 699 observations, of which 683 are complete. Discounting repetitions, there are 213 distinct observations about malignant tumors and 236 distinct observations about benign ones.

Dermatology. This data set contains consists of 366 different objects, 34 attributes, 33 of which are linear valued and one of them is nominal.

COIL20. The dataset is constructed by 1440 gray-scale images of 20 objects (72 images per object). The objects are placed on a motorized turntable against a black background and their Images are taken at pose intervals of $5°$.

ORL. This dataset [25] consists of 40 different subjects, 10 images per subject and each image is resized to 32×32 pixels. The images are taken against a dark homogeneous background with the subjects in an upright, frontal position.

Yale. The dataset [26] contains 165 gray-scale images of 15 individuals. There are 11 images per subject, one per different facial expression or configuration, and each image is resized to 32×32 pixels.

4.2 Evaluation Metrics

In our experiments, we adopt clustering accuracy (*Acc*) as evaluation metrics. For the metrics, the higher value indicates the better clustering quality. Let r_i be the clustering result and s_i be the ground truth label of x_i. *Acc* is defined as:

$$Acc = \frac{\sum_{i=1}^{n} \delta(s_i, map(r_i))}{n}.$$

Here n is the total number of samples. $\delta(x, y)$ is the delta function that equals one if $x = y$ and equals zero otherwise. $map(r_i)$ is the best mapping function that utilizes the Kuhn-unkres algorithm to permute clustering labels to match the ground truth labels.

4.3 Experiments Results

In the clustering experiment, we set the number of clusters to be the ground truth in each data set and $m = 1.1$. The parameter λ can be tuned by a grid-search strategy from 0.01 to 10 with step 0.1. Specifically, we tuned parameter K for different data sets also by a grid-search strategy from 1 to c. For Dermatology data sets, we compared the performance by setting reduced dimensions to the range of 1 to 34, while for all other three data sets, we set the scale to be 1 to 100. For all other methods, we ran 20 times with the same initialization and wrote down the best performance in each dimension. As for our proposed UPF, we ran 20 times and recorded the best results. The comparison results are shown in Table 1.

As is shown in Table 1. that UPF outperforms the two-stages methods on all data sets.

Table 1. Comparison results on IRIS, BREAST, Dermatology, Yale, COIL20, and ORL in terms of accuracy.

Date sets	PCA+KMeans	PCA+FCM	UPF
IRIS	88.93 ± 0.018	89.33 ± 0.013	**96.67 ± 0.033**
Breast	60.09 ± 0.000	60.09 ± 0.000	**94.99 ± 0.162**
Dermatology	79.45 ± 0.045	83.03 ± 0.030	**94.81 ± 0.084**
Yale	40.42 ± 0.012	46.28 ± 0.010	**50.91 ± 0.037**
COIL20	62.94 ± 0.017	63.76 ± 0.013	**77.71 ± 0.050**
ORL	54.25 ± 0.010	59.20 ± 0.008	**83.50 ± 0.086**

4.4 Convergence Analysis

The value scales of the objective functions on different data sets vary greatly, in order to put all convergence curves in one picture, we normalize the objective function values beforehand. The convergence curves of UPF on IRIS, BREAST, Dermatology, Yale, COIL20, and ORL data sets are illustrated in the following Fig. 1.

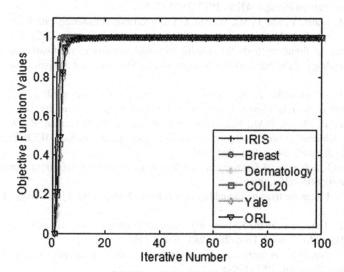

Fig. 1. The convergence curves of UPF on IRIS, BREAST, Dermatology, Yale, COIL20, and ORL

Form Fig. 1, UDF converges when it runs less than 20 times on all the six benchmark data set, so the convergence speed of the proposed algorithm is very fast.

5 Conclusion

In this paper, we have proposed a novel clustering algorithm, named UPF, which introduces a model to combine FCM clustering and PCA dimensionality reduction. UPF method not only completes unsupervised subspace learning via the optical subspace projection matrix but also completes clustering simultaneously. Moreover, we impose a L_0-norm constraint on the membership matrix to make the model more robust and sparse. UPF model can be solved by the proposed tractable and skillful iterative optimization method and guarantees the optimality and convergence. Theoretical analyses and extensive experiments on several public data sets demonstrate the superior performance of our models.

Acknowledgements. The work was partial supported by National Natural Science Foundations of China (61962012), Xing-Long scholar project of Lanzhou University of Finance and Economics, and Gansu Provincial Institutions of Higher Learning Innovation Ability Promotion Project (2019B − 97).

References

1. Woźniak, M., Graña, M., Corchado, E.: A survey of multiple classifier systems as hybrid systems. Inf. Fusion **16**, 3–17 (2014)
2. Wang, L., Pan, C.: Robust level set image segmentation via a local correntropy-based KMeans clustering. Pattern Recogn. **47**(5), 1917–1925 (2014)
3. Gong, M., Liang, Y., Shi, J., Ma, W., Ma, J.: Fuzzy c-means clustering with local information and kernel metric for image segmentation. IEEE Trans. Image Process. **22**(2), 573–584 (2013)
4. Macqueen, J.: Some methods for classification and analysis of multivariate observations. In: Proceedings of the 5th Berkeley Symposium on Mathematical Statistics and Probability (1967)
5. Arthur, D., Vassilvitskii, S.: KMeans++: the advantages of careful seeding. In: Proceedings of the 18th Annual ACM-SIAM Symposium on Discrete Algorithms (2007)
6. Kanungo, T., Mount, D.M., Netanyahu, N.S., Piatko, C.D., Silverman, R., Wu, A.Y.: An efficient KMeans clustering algorithm: analysis and implementation. IEEE Trans. Pattern Anal. Mach. Intell. **24**(7), 881–892 (2002)
7. Frahling, G., Sohler, C.: A fast K-means implementation using coresets. Int. J. Comput. Geom. Appl. **18**(6), 605–625 (2008)
8. Elkan, C.: Using the triangle inequality to accelerate KMeans. In: Proceedings of the ICML (2003)
9. Huang, Z.: Extensions to the KMeans algorithm for clustering large data sets with categorical values. Data Min. Knowl. Disc. **2**(3), 283–304 (1998)
10. Huang, Z., Ng, M.K.: A fuzzy k-modes algorithm for clustering categorical data. IEEE Trans. Fuzzy Syst. **7**(4), 446–452 (1999)
11. Huang, J.Z., Ng, M.K., Rong, H., Li, Z.: Automated variable weighting in K-means type clustering. IEEE Trans. Pattern Anal. Mach. Intell. **27**(5), 657–668 (2005)
12. Li, M.J., Ng, M.K., Cheung, Y., Huang, J.Z.: Agglomerative fuzzy K-means clustering algorithm with selection of number of clusters. IEEE Trans. Knowl. Data Eng. **20**(11), 1519–1534 (2008)
13. Yi, D., Xian, F.: Kernel-based fuzzy c-means clustering algorithm based on genetic algorithm. Neurocomputing **188**, 233–238 (2015)
14. Aparajeeta, J., Nanda, P.K., Das, N.: Modified possibilistic fuzzy c-means algorithms for segmentation of magnetic resonance image. Appl. Soft Comput. **41**, 104–119 (2016)

15. Adhikari, S.K., Sing, J.K., Basu, D.K., Nasipuri, M.: Conditional spatial fuzzy c-means clustering algorithm with application in MRI image segmentation. Adv. Intell. Syst. Comput. **340**, 539–547 (2015)
16. Askari, S., Montazerin, N., Zarandi, M.H.F.: Generalized possibilistic fuzzy c-means with novel cluster validity indices for clustering noisy data. Appl. Soft Comput. **53**, 262–283 (2017)
17. Svante, W., Kim, E., Paul, G.: Principal component analysis. Chemometr. Intell. Lab. Syst. **2**, 37–52 (1987)
18. He, X., Niyogi, P.: Locality preserving projections. In: Proceedings of the NIPS (2003)
19. Cai, D., He, X., Han, J., Zhang, H.-J.: Orthogonal laplacianfaces for face recognition. IEEE Trans. Image Process. **15**(11), 3608–3614 (2006). A Publication of the IEEE Signal Processing Society
20. Nie, F., Wei, Z., Li, X.: Unsupervised feature selection with structured graph optimization. In: Proceedings of the 13th AAAI Conference on Artificial Intelligence (2016)
21. Welling, M.: Fisher linear discriminant analysis. Department of Computer Science, University of Toronto. (2005)
22. Nie, F., Wang, X., Huang, H.: Clustering and projected clustering with adaptive neighbors. In: Proceedings of the ACM SIGKDD International Conference on Knowledge Discovery and Data Mining (2014)
23. Chang, X., Yang, Y.: Semisupervised feature analysis by mining correlations among multiple tasks. IEEE Trans. Neural Netw. Learn. Syst. **28**, 2294–2305 (2016)
24. Bezdek, J.C., Hathaway, R.J.: Convergence of alternating optimization. Neural Parallel Sci. Comput. **11**(4), 351–368 (2003)
25. Samaria, F.S., Harter, A.C.: Parameterisation of a stochastic model for human face identification. In: Proceedings of the 2nd IEEE Workshop on Applications of Computer Vision (1994)
26. Belhumeur, P.N., Hespanha, J.P., Kriegman, D.J.: Eigenfaces vs. fisherfaces: Recognition using class specific linear projection. Yale University New Haven United States, Technical report (1997)

A Hierarchical-Tree-Based Method for Generative Zero-Shot Learning

Xizhao Wang, Zhongwu Xie, Weipeng Cao$^{(\boxtimes)}$, and Zhong Ming

College of Computer Science and Software Engineering, Shenzhen University,
Shenzhen 518060, China
caoweipeng@szu.edu.cn

Abstract. It is currently a popular practice to use the class semantic information and the conditional generative adversarial network (CGAN) technique to generate visual features for the unseen classes in zero-shot learning (ZSL). However, there is currently no good ways to ensure that the generated visual features can always be beneficial to the prediction of the unseen classes. To alleviate this problem, we propose a hierarchical-tree-based method for constraining the generation process of CGAN, which can tune the generated visual features based on the multi-level class information. Moreover, to enhance the mapping ability of the model from the visual space to the semantic space, we add a multi-expert module to the traditional single mapping channel, which helps the model to mine the mapping relationship between the visual space and the semantic space. Extensive experimental results on five benchmark data sets show that our method can achieve better generalization ability than other existing generative ZSL algorithms.

Keywords: Zero-shot learning · Hierarchical tree · Generative adversarial networks

1 Introduction

In real-world applications, especially in medical image recognition and wild animal recognition scenarios, sometimes there are no training samples for some classes due to the difficulty of obtaining data or the high cost of labeling them [13]. We call these classes unseen classes, and the corresponding is seen classes, which refer to the classes that have corresponding labeled samples and can be directly used for model training. To enable the model to accurately predict unseen classes, zero-shot learning (ZSL) [22,31] was proposed, which aims to recognize these classes without any training samples with the help of their side information (i.e., the description information of the classes). Generally, the

This work was supported by National Natural Science Foundation of China (61836005, 61976141, 61732011), and the Opening Project of Shanghai Trusted Industrial Control Platform (TICPSH202003008-ZC).
X. Wang and Z. Xie—Joint first authors.

© Springer Nature Switzerland AG 2020
M. Qiu (Ed.): ICA3PP 2020, LNCS 12453, pp. 352–364, 2020.
https://doi.org/10.1007/978-3-030-60239-0_24

side information corresponding to these classes is encoded into semantic vectors of the same dimension, and related techniques include: attribute [7], gaze [14], word2vec [28], etc.

According to the different testing settings, the existing ZSL algorithms can be divided into two categories: conventional zero-shot learning (CZSL) and generalized zero-shot learning (GZSL). For CZSL, testing samples only contain the unseen classes samples; for GZSL, the testing samples include both the seen and unseen classes samples. It is clear that the testing setting of GZSL is closer to the real world than that of CZSL [30]. At present, there are two main strategies to realize ZSL: embedding strategy and generative strategy. The embedding strategy follows the assumption that the distributions of the seen classes and the unseen classes in both the visual space and the semantic space are similar [13], and the main implementation is to transfer the mapping relationship learned from the seen classes to the recognition of the unseen classes. The advantage of this strategy is that it is easy to implement, but its disadvantage is that it may cause the model to suffer from the hubness problem [24] and bias problem [33].

Inspired by generative adversarial networks (GAN) [10], many GAN-based generative ZSL algorithms have been proposed in recent years [8,17,31]. In general, these algorithms aim to use random noises and the semantic vectors to generate pseudo samples and then transform the ZSL task into the general supervised learning task. This generative strategy is straightforward but suffers from the following weakness: the generators of these algorithms are easy to generate unrepresentative visual features, especially for the unseen classes. To alleviate this problem, in [17], the authors proposed to use a regressor (the same to the module R in Fig. 1), which can enhance the correlation between the generated features and the known semantic vectors to a certain extent. But it can not guarantee the quality of these features [9], especially for the unseen classes, because the information used to constrain the generator is still too scarce.

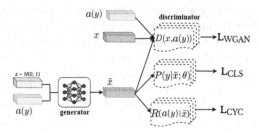

Fig. 1. The framework of our method

To solve the above problems, we propose a novel constraint method based on the class hierarchy tree to guide the generator to generate more useful visual features for the current ZSL task. Specifically, we first use the K-means technique to build a class hierarchy tree based on the semantic vectors corresponding to the classes and then use the multi-level class information to evaluate the quality of the generated features. In other words, the generated features can be evaluated from multiple dimensions and the corresponding evaluation loss can be used to guide the update of the generator. Moreover, to enhance the mapping ability of the model from the visual space to the semantic space, we add a multi-expert module, which can embed a variety of different structures or algorithms.

The main contributions of this paper can be summarized as follows: (1) We propose to use a class hierarchy tree to constrain the generator of the generative ZSL algorithms for the first time, which can make it generate more representative visual features for model training; (2) We propose to add a multi-expert module, which is beneficial to mine the relationship between the visual space and the semantic space; (3) Extensive experiments on five benchmarks verify that our method outperforms the previous state-of-the-art approaches.

2 Preliminaries

2.1 Generative Adversarial Networks (GAN)

GAN [10] can be used to generate images (or text) from random noise, and its architecture usually contains two modules: generator and discriminator. The input of the generator is random noise z that follows a specific distribution (e.g., $z \sim N(0,1)$). During training, the generator and the discriminator are trained in an adversarial manner, that is, the generator tries its best to generate fake samples to fool the discriminator, while the discriminator does its best to distinguish samples as whether they come from the generator or the original training data set. When the discriminator cannot distinguish the real samples from the generated samples, the model training is completed. At this point, we think that the generator has learned the distribution of the original samples.

In recent years, GAN has received extensive attention, and some notable work includes: in [19], the authors proposed that the label information can be added to the input of the generator and the discriminator, so that the model can better capture the data distribution under the specific class constraints. Different from [19,26] and [25] use sentence descriptions as the auxiliary constraints. In [3], the authors proposed to use the Wasserstein distance to alleviate the problem of unstable model training and model collapse in GAN. In [11], the authors proposed an improved WGAN, which can accelerate its training speed. At present, it has become a popular practice to integrate Wasserstein distance into conditional GAN (CGAN), and various CWGAN algorithms and applications have been proposed [6,32].

2.2 Generative Methods for Zero-Shot Learning

In recent years, the generative strategy has been widely concerned in the field of ZSL, because once one can generate training samples for unseen classes in some way, then the ZSL tasks become the general supervised learning tasks. Related representative work includes: in [20], the authors used the conditional variable auto-encoder to generate the unseen classes samples. Using CGAN to generate training samples for the unseen classes is currently a hot research topic. In this learning paradigm, the input of the generator is usually the semantic vectors corresponding to the unseen classes and the random noise following a specific distribution. Here the semantic vectors play the role of constraining the output generated by CGAN. In [31], the authors pointed out that it is easier to generate low-dimension visual features through the generator than directly generating pixel-level images, and their experimental results showed that using the generated features can also improve the generalization ability of the model. Similar work also includes [8,17], etc. Note that the generator in our method also generates visual features rather than the pixel-level images.

2.3   Constraints are Necessary

In [31], the authors demonstrated that it is difficult for CWGAN to ensure that the generated features are beneficial to the final model decision. To optimize this problem, they proposed f-CLSWGAN, which uses a classifier trained based on the seen classes samples to evaluate the quality of the generated features and uses the corresponding evaluation loss to constrain the update of the generator. Their experimental results show that even if only additional constraints are imposed on the seen classes, this method can still effectively improve the generalization ability of the final model. In [8], the authors pointed out that if we do not constrain the generated features for the unseen classes during the update of the generator, it will cause the distribution of the generated unseen classes features to be far from their true distribution. To solve this problem, they proposed Cycle-CLSWGAN, which can alleviate the above problem by reconstructing the generated features back to their corresponding semantic vectors. Similar work also includes AFC-GAN [17]. The difference between AFC-GAN and Cycle-CLSWGAN is that the regressor used in the former is pre-trained, and its internal parameters will not be updated throughout the training process, while the latter will update these parameters.

3   The Details of Our Method

3.1   Notations

In this study, $\mathcal{X}$ represents the visual space, $\mathcal{A}$ represents the visual space, and $\mathcal{Y}$ represents the label space. In ZSL, generally the original data set ($\mathcal{D}$) will be divided into the seen classes set $\mathcal{D}_S$ and the unseen classes set $\mathcal{D}_U$, where $\mathcal{D}_S = \{x, y, a(y) | x \in \mathcal{X}_S, y \in \mathcal{Y}_S, a(y) \in \mathcal{A}_S\}$, $\mathcal{D}_U = \{x, y, a(y) | x \in \mathcal{X}_U, y \in \mathcal{Y}_U, a(y) \in \mathcal{A}_U\}$, and $\mathcal{Y} = \mathcal{Y}_S \cup \mathcal{Y}_U$, $\mathcal{Y}_S \cap \mathcal{Y}_U = \emptyset$.

In the experiment, we used the testing setup of GZSL, that is, the testing samples can come from both the unseen and seen classes. Therefore, $\mathcal{D}_S$ will also be divided into two subsets: the subset for training $\mathcal{D}_S^{Tr}$ and the subset for testing $\mathcal{D}_S^{Te}$. So the original data set was finally divided into three parts: $\mathcal{D}_S^{Tr}$, $\mathcal{D}_S^{Te}$, and $\mathcal{D}_U^{Te}$.

3.2   The Basic Idea of Our Method

The basic idea of our method is to optimize the constraints of CWGAN to make the generated features more conducive to the final model decision. The difficulty here is how to evaluate the quality of the generated features during the training process, and then tune the generator based on the evaluation feedback. In [17], the authors pointed out that the generated features should be able to be reconstructed into their corresponding semantic vectors. However, the method they proposed mainly helped the prediction of the seen classes [9], but the constraints on the generated unseen classes features are insufficient.

Inspired by their work, we propose a hierarchical-tree-based constraint method to provide more information about the unseen classes for the generator in CWGAN. Specifically, first, we use the K-means technique to cluster the semantic vectors corresponding to the seen and unseen classes to form a class hierarchy tree (as shown in Fig. 2). Note that the higher-level ancestor nodes are the high-level abstraction of the concepts of the child nodes.

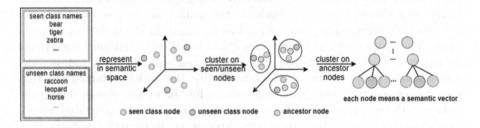

Fig. 2. Tree-building process.

During training, the model first maps the generated features to the semantic space and then uses the above class hierarchy tree to evaluate their quality and obtain the corresponding loss. This loss will provide the generator with additional constraint information. The class hierarchy tree can evaluate whether the generated features conform to the semantic features of the classes from multiple dimensions, thereby guiding the generator to generate better visual features. Moreover, the multi-level tree structure also gives us the opportunity to use multi-expert modules (i.e., the mapping channels) to enhance the feature extraction ability of the model (as shown in Fig. 3), which helps the model to better mine the relationship between the visual space and the semantic space.

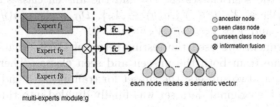

Fig. 3. The architecture of the regressor used in our method.

3.3   Details of Our Method

The basic framework used in our method is CWGAN, where the input of the generator is the random noise (i.e., z, $z \sim N(0,1)$) and the semantic vector (i.e., a) corresponding to each specific class, and its output is the generated visual features. The loss of the generator can be obtained as follows:

$$\mathcal{L}_G = -E[D(G(z,a),a)] - \eta E[log P(y|G(z,a)) - \beta E[t|R(G(z,a))] \quad (1)$$

where the first term is the Wasserstein loss [3], the second item is the classification loss for evaluating the quality of the generated features, and the last item is the reconstruction loss. η and β are the trade-off parameters. t is the identifier corresponding to the hierarchical tree.

The loss of the discriminator can be calculated as follows:

$$\mathcal{L}_D = E[D(G(z,a),a)] - E[D(x,a)] + \lambda E[(\|\nabla_{\hat{x}} D(\hat{x},a)\|_2 - 1)]^2 \quad (2)$$

where the last term refers to the penalty loss, which is also known as the Lipschitz constraint [11], in which $\hat{x} = \mu x + (1 - \mu)\tilde{x}, \mu \sim U(0,1)$. Here λ is a penalty parameter, which was set to 10 in our study.

The global optimization objective of our model is:

$$\min_{\mathcal{G}} \max_{D} \mathcal{L}_{GAN} = E[D(x,a)] - E[D(G(z,a),a)]$$
$$-\lambda E[(\|\nabla_{\hat{x}} D(\hat{x},a)\|_2 - 1)]^2 - \eta E[log P(y|G(z,a)) - \beta E[t|R(G(z,a))]] \quad (3)$$

After training, one can use the generator to augment the original training data set of the ZSL task, thereby improving the generalization ability of the model. It is worth mentioning that to improve the prediction accuracy of the ZSL model, we adopted the ensemble learning framework designed in [17] to make predictions. Specifically, we trained classifiers f_v and f_s based on the training data provided by visual features and labels, and the training data provided by semantic vectors and labels, respectively. The final prediction results of the model are determined by these two classifiers:

$$y = argmax\,[f_v(x) + \hat{\lambda} f_s((g(x))] \quad (4)$$

where $\hat{\lambda}$ is the trade-off parameter.

The learning framework of our method is shown in Fig. 1, which contains a generator, a discriminator (D), a classifier (P), and a regressor (R). The training process of the regressor can be summarized as Algorithm 1. The training methods of other modules are the same as traditional methods.

4   Experiments

4.1   Datasets

In the experiment, we selected five benchmark data sets to evaluate the performance of our method, including Animals with Attributes1 (**AWA1**) [16], Animals with Attributes2 (**AWA2**) [30], Caltech-UCSD Bird-200-2011 (**CUB**) [29],

Oxford Flowers (**FLO**) [21], and SUN Scene Recognition (**SUN**) [23]. The details of the data sets are shown in Table 1.

Algorithm 1. The training process of the regressor

1: **Input:** Training data $\mathcal{D}=(X_S^{Tr}, A_S, A_U)$; the number of nodes at each level in the class hierarchy tree, which is represented by array arr, and $arr[0]$ represents the number of leaf nodes; the learning rate α; and the trade-off parameters $\beta_0, \beta_1, ...,$ β_n.

2: **Output:** The parameters (i.e., θ_r) of the regressor.

3: **Step1**: K-means clustering
   if $len(arr) == 1$:
       return A;
   $temp\_att = A$;
   while $i < len(arr) - 1$ do:
       $node = arr[i]$;
       kmean = Kmeans(n_clusters=$node$).fit($temp\_att$);
       $cluster\_center$ = kmean.cluster_center_;
       save $cluster\_center$;
       $temp\_att = cluster\_center$;

4: **Step2**: Build the framework of the regressor R according to the architecture of each expert module.

5: **Step3**: Let T_{ij} represent the a_j of the i-th layer.
   while not done do:

6:     $pred_0, pred_1, ..., pred_n = R(x|x \in jth\ class)$;

7:     $\mathcal{L} = \beta_i \sum_{i=0}^{n} \|pred_i - T_{ij}\|^2$;

8:     $\theta_r = \theta_r - \alpha \nabla_{\theta_r}(L)$;

9: **return** θ_r;

Table 1. The details of the experimental data sets

Dataset	AWA1	AWA2	CUB	FLO	SUN
Images	30475	37322	11788	8189	14340
Attributes	85	85	312	1024	102
Seen classes	40	40	150	82	645
Unseen classes	10	10	50	20	72

To make a fair comparison with other ZSL algorithms, we used the same division method as [30] for the AWA1, AWA2, CUB, and SUN data sets; for FLO, we used the division method mentioned in [26]. A pre-trained ResNet-101 [12] model was chosen as the feature extractor for all the ZSL algorithms.

4.2   Implementation Details

As shown in Fig. 1, our learning framework mainly includes three functional modules: generator, discriminator, and regressor. In our experiments, for the generator and the discriminator, we used the multilayer perceptron (MLP) with a single hidden layer, in which the number of nodes in the hidden layer is 4096 and the activation function is LeakyReLU [18]. The output layer of the generator contains 2048 nodes and uses ReLU as its activation function; while the output layer of the discriminator has one output node but does not use any activation function. For the regressor, we used the neural network with a single hidden layer as the basic expert module (three expert modules in total), the activation function of the hidden layer is LeakyReLU, and the activation function of the output layer is ReLU. For AWA1, AWA2, and CUB, the number of hidden nodes corresponding to these three expert modules are 2048, 3072, and 4096, respectively; for SUN, they are 7168, 8189, and 9216, respectively; for FLO, they are 3072, 4096, and 5120, respectively. In this study, Adam [15] was chosen as the optimization algorithm and the learning rate was set to $1e-4$. In addition, the level of the class hierarchy tree was set to 3.

In Eq. (1), the last term can be expanded as follows:

$$\beta_i E[t|R(G(z,a))] = \beta_{seen} \|pred_0 - a^s\|^2 + \beta_{unseen} \sum_{i=0}^{2} \|pred_i - T_{ij}\|^2, \quad (5)$$

where β_{seen} is a constant, β_{unseen} is a vector, and T_{ij} refers to the ancestor of a_j when $i > 0$ and refers to the a_j itself when $i = 0$. The first term is used for constraining the generated seen classes features to reconstruct to their corresponding semantic vectors, while the last term is used for constraining the generated unseen classes features to reconstruct to their ancestors and their corresponding semantic vectors. The values of these parameters in our experiments are shown in Table 2.

Moreover, the value of the $\hat{\lambda}$ in Eq. (4) was set to 2 in SUN, and was set to 1.5 in other data sets.

Table 2. The parameter settings of the regressor.

Dataset	AWA1	AWA2	CUB	FLO	SUN
Node number in HT	[50, 20, 5]	[50, 20, 5]	[200, 20, 5]	[102, 50, 10]	[717, 300, 50]
β_{seen}	1,	1,	1,	1,	10,
β_{unseen}	[0.1, 1, 1]	[0.1, 1, 1]	[0.1, 1, 1]	[1,1,1]	[0.1, 5, 5]

Note: HT means the class hierarchy tree. $\beta_{unseen} = [\beta_0, \beta_1, \beta_2]$ correspond to the weight of loss evaluation from bottom to top of the class hierarchy tree.

4.3 Experimental Results

In this study, we used the harmonic mean $\mathcal{H}$ to evaluate the performance of the proposed algorithm, which can be obtained as follows:

$$\mathcal{H} = \frac{2 \times A_s \times A_u}{A_s + A_u} \tag{6}$$

where A_s and A_u are the top-1 accuracy of the model on the seen classes and the unseen classes, respectively. Note that the larger the value of $\mathcal{H}$, the better the generalization ability of the model.

We compared our method with the other seven popular ZSL algorithms (i.e., SJE [2], ESZSL [27], ALE [1], DEVISE [9], GAZSL [33], f-CLSWGAN [31], and AFC-GAN [17]), and the corresponding results are shown in Table 3.

Table 3. Experimental results of our method and other ZSL algorithms on benchmarks

Method	AWA1			AWA2			CUB			FLO			SUN		
	As	Au	$\mathcal{H}$	As	Au	$\mathcal{H}$	As	Au	$\mathcal{H}$	As	Au	$\mathcal{H}$	As	Au	$\mathcal{H}$
SJE [2]	74.6	11.3	19.6	73.9	8.0	14.4	59.2	23.5	33.6	47.6	13.9	21.5	47.6	13.9	21.5
ESZSL [27]	75.6	6.6	12.1	77.8	5.9	11.0	68.3	12.6	21.0	56.8	11.4	19.0	11.0	27.9	15.8
ALE [1]	16.8	76.1	27.5	81.8	14.0	23.9	62.8	23.7	34.4	61.6	13.3	21.6	33.1	21.8	26.3
DEVISE [9]	68.7	13.4	22.4	74.7	17.1	27.8	53.0	23.8	32.8	44.2	9.9	16.2	27.4	16.9	20.9
GAZSL [33]	84.2	29.6	43.8	86.9	35.4	50.3	61.3	31.7	41.8	77.4	28.1	41.2	39.3	22.1	28.3
f-CLSWGAN [31]	61.4	57.9	59.6	68.9	52.1	59.4	57.7	43.7	49.7	73.8	59.0	65.6	42.6	36.6	39.4
AFC-GAN [17]	66.8	58.2	62.2	-	-	-	59.7	53.5	56.4	80.0	60.2	68.7	36.1	49.1	41.6
ours	72.4	61.5	66.5	77.3	60.9	68.1	59.0	54.7	56.8	78.6	65.0	71.2	38.0	47.9	42.4

From Table 3, one can observe that our model can achieve the highest $\mathcal{H}$ on all the benchmark data sets, which implies that our method has better generalization ability than others. Moreover, our model has higher prediction accuracy for the unseen classes (i.e., A_u) than other algorithms on most of the data sets (4/5, i.e., AWA1, AWA2, CUB, and FLO), which indirectly reflects that our strategy does improve the quality of the generated unseen classes features.

4.4 Parameter Sensitivity Analysis

To explore the impact of the three hyper-parameters (η, $\hat{\lambda}$, and β_{unseen}) in our method on model performance, here we use the one-variable-at-a-time method to analyze the sensitivity of the model to them. As mentioned in Section III, η is used to weigh the loss generated by the classifier, and its impact on the performance of the model on AWA2 and SUN is shown in Fig. 4(a). It can be observed from Fig. 4(a) that the value of η should not be set too large. Generally, as long as its value is fixed at 0.01, the model can achieve good performance. $\hat{\lambda}$ is used to weigh the prediction results of the two classifiers in Eq. (4), and

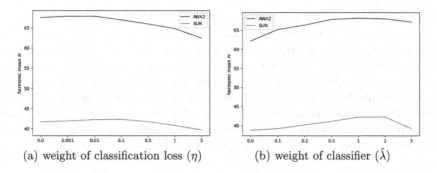

(a) weight of classification loss (η)   (b) weight of classifier ($\hat{\lambda}$)

Fig. 4. Parameter sensitivity

its impact on model performance is shown in Fig. 4(b). It can be observed from Fig. 4(b) that setting the value of $\hat{\lambda}$ to 1.5 is a good choice.

β_{unseen} is an important parameter in building the class hierarchy tree, and its influence on the generalization ability of the final model is shown in Fig. 5(a)–Fig. 5(b). The elements in β_{unseen} represent the weights of the reconstruction loss in building the tree from the leaf nodes to the root node, and the length of β_{unseen} is the height of the tree. From Fig. 5(a)–Fig. 5(b), one can observe that when the height of the tree is 2 or 3, the performance of the model can be significantly improved, which again verifies the effectiveness of our method.

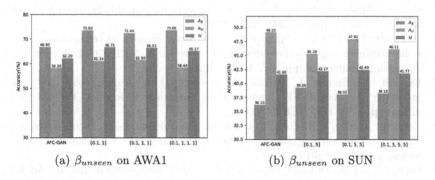

(a) β_{unseen} on AWA1   (b) β_{unseen} on SUN

Fig. 5. The recognition results of the model with different β_{unseen} on AWA1 and SUN

5   Conclusions

To improve the generalization ability of the generative ZSL algorithms, we designed a hierarchical-tree-based method to enhance the quality of the generated features in this paper. Specifically, our method can evaluate the quality of the generated visual features from multiple levels and tune the generator through the corresponding loss feedback. Moreover, to enhance the mapping ability of the model from the visual space to the semantic space, we added the multi-expert module to the traditional single mapping channel to realize the multi-level extraction and transformation of visual features, which is conducive to the model for mining the relationship between the visual features and the semantic features. We evaluated the effectiveness of the proposed algorithm on five benchmark data sets and the experimental results show that our model can achieve higher prediction accuracy than the other seven popular ZSL algorithms. In the future, we will consider incorporating non-iterative algorithms [4,5] to our method to improve the training efficiency of its classifier.

References

1. Akata, Z., Perronnin, F., Harchaoui, Z., Schmid, C.: Label-embedding for attribute-based classification. In: Proceedings of the IEEE Conference on Computer Vision and Pattern Recognition, pp. 819–826 (2013)
2. Akata, Z., Reed, S., Walter, D., Lee, H., Schiele, B.: Evaluation of output embeddings for fine-grained image classification. In: Proceedings of the IEEE Conference on Computer Vision and Pattern Recognition, pp. 2927–2936 (2015)
3. Arjovsky, M., Chintala, S., Bottou, L.: Wasserstein generative adversarial networks. In: Proceedings of the 34th International Conference on Machine Learning, Proceedings of Machine Learning Research, vol. 70, pp. 214–223 (2017)
4. Cao, W., Gao, J., Ming, Z., Cai, S., Shan, Z.: Fuzziness-based online sequential extreme learning machine for classification problems. Soft Comput. **22**(11), 3487–3494 (2018). https://doi.org/10.1007/s00500-018-3021-4
5. Cao, W., Wang, X., Ming, Z., Gao, J.: A review on neural networks with random weights. Neurocomputing **275**, 278–287 (2018)
6. Ebenezer, J.P., Das, B., Mukhopadhyay, S.: Single image haze removal using conditional wasserstein generative adversarial networks. In: 2019 27th European Signal Processing Conference, pp. 1–5. IEEE (2019)
7. Farhadi, A., Endres, I., Hoiem, D., Forsyth, D.: Describing objects by their attributes. In: 2009 IEEE Conference on Computer Vision and Pattern Recognition, pp. 1778–1785. IEEE (2009)
8. Felix, R., Kumar, V.B., Reid, I., Carneiro, G.: Multi-modal cycle-consistent generalized zero-shot learning. In: Proceedings of the European Conference on Computer Vision, pp. 21–37 (2018)
9. Frome, A., et al.: DeViSe: a deep visual-semantic embedding model. In: Advances in Neural Information Processing Systems, pp. 2121–2129 (2013)
10. Goodfellow, I., et al.: Generative adversarial nets. In: Advances in Neural Information Processing Systems, pp. 2672–2680 (2014)
11. Gulrajani, I., Ahmed, F., Arjovsky, M., Dumoulin, V., Courville, A.C.: Improved training of wasserstein GANs. In: Advances in Neural Information Processing Systems, pp. 5767–5777 (2017)
12. He, K., Zhang, X., Ren, S., Sun, J.: Deep residual learning for image recognition. In: Proceedings of the IEEE Conference on Computer Vision and Pattern Recognition, pp. 770–778 (2016)

13. Jiang, H., Wang, R., Shan, S., Chen, X.: Transferable contrastive network for generalized zero-shot learning. In: Proceedings of the IEEE International Conference on Computer Vision, pp. 9765–9774 (2019)
14. Karessli, N., Akata, Z., Schiele, B., Bulling, A.: Gaze embeddings for zero-shot image classification. In: Proceedings of the IEEE Conference on Computer Vision and Pattern Recognition, pp. 4525–4534 (2017)
15. Kingma, D.P., Ba, J.: Adam: a method for stochastic optimization. arXiv preprint arXiv:1412.6980 (2014)
16. Lampert, C.H., Nickisch, H., Harmeling, S.: Learning to detect unseen object classes by between-class attribute transfer. In: 2009 IEEE Conference on Computer Vision and Pattern Recognition, pp. 951–958. IEEE (2009)
17. Li, J., Jing, M., Lu, K., Zhu, L., Yang, Y., Huang, Z.: Alleviating feature confusion for generative zero-shot learning. In: Proceedings of the 27th ACM International Conference on Multimedia, pp. 1587–1595 (2019)
18. Maas, A.L., Hannun, A.Y., Ng, A.Y.: Rectifier nonlinearities improve neural network acoustic models. In: Proceedings of ICML, vol. 30, p. 3 (2013)
19. Mirza, M., Osindero, S.: Conditional generative adversarial nets. arXiv preprint arXiv:1411.1784 (2014)
20. Mishra, A., Krishna Reddy, S., Mittal, A., Murthy, H.A.: A generative model for zero shot learning using conditional variational autoencoders. In: Proceedings of the IEEE Conference on Computer Vision and Pattern Recognition Workshops, pp. 2188–2196 (2018)
21. Nilsback, M.E., Zisserman, A.: Automated flower classification over a large number of classes. In: 2008 Sixth Indian Conference on Computer Vision, Graphics and Image Processing, pp. 722–729. IEEE (2008)
22. Palatucci, M., Pomerleau, D., Hinton, G.E., Mitchell, T.M.: Zero-shot learning with semantic output codes. In: Advances in Neural Information Processing Systems, pp. 1410–1418 (2009)
23. Patterson, G., Xu, C., Su, H., Hays, J.: The sun attribute database: beyond categories for deeper scene understanding. Int. J. Comput. Vis. **108**(1–2), 59–81 (2014)
24. Radovanović, M., Nanopoulos, A., Ivanović, M.: Hubs in space: popular nearest neighbors in high-dimensional data. J. Mach. Learn. Res. **11**(Sep), 2487–2531 (2010)
25. Reed, S., Akata, Z., Yan, X., Logeswaran, L., Schiele, B., Lee, H.: Generative adversarial text to image synthesis. arXiv preprint arXiv:1605.05396 (2016)
26. Reed, S.E., Akata, Z., Mohan, S., Tenka, S., Schiele, B., Lee, H.: Learning what and where to draw. In: Advances in Neural Information Processing Systems, pp. 217–225 (2016)
27. Romera-Paredes, B., Torr, P.: An embarrassingly simple approach to zero-shot learning. In: International Conference on Machine Learning, pp. 2152–2161 (2015)
28. Socher, R., Ganjoo, M., Manning, C.D., Ng, A.: Zero-shot learning through cross-modal transfer. In: Advances in Neural Information Processing Systems, pp. 935–943 (2013)
29. Wah, C., Branson, S., Welinder, P., Perona, P., Belongie, S.: The caltech-ucsd birds-200-2011 dataset (2011)
30. Xian, Y., Lampert, C.H., Schiele, B., Akata, Z.: Zero-shot learning—a comprehensive evaluation of the good, the bad and the ugly. IEEE Trans. Pattern Anal. Mach. Intell. **41**(9), 2251–2265 (2018)
31. Xian, Y., Lorenz, T., Schiele, B., Akata, Z.: Feature generating networks for zero-shot learning. In: Proceedings of the IEEE Conference on Computer Vision and Pattern Recognition, pp. 5542–5551 (2018)

32. Zheng, M., et al.: Conditional wasserstein generative adversarial network-gradient penalty-based approach to alleviating imbalanced data classification. Inf. Sci. **512**, 1009–1023 (2020)

33. Zhu, Y., Elhoseiny, M., Liu, B., Peng, X., Elgammal, A.: A generative adversarial approach for zero-shot learning from noisy texts. In: Proceedings of the IEEE Conference on Computer Vision and Pattern Recognition, pp. 1004–1013 (2018)

Fast Computation of the Exact Number of Magic Series with an Improved Montgomery Multiplication Algorithm

Yukimasa Sugizaki[1(⊠)] and Daisuke Takahashi[2]

[1] Graduate School of Science and Technology, University of Tsukuba, 1-1-1
Tennodai, Tsukuba, Ibaraki 305-8573, Japan
sugizaki@hpcs.cs.tsukuba.ac.jp
[2] Center for Computational Sciences, University of Tsukuba, 1-1-1 Tennodai,
Tsukuba, Ibaraki 305-8577, Japan
daisuke@cs.tsukuba.ac.jp

Abstract. The numbers of magic series of large orders are computed on
Intel Xeon Phi processors with an improved and optimized Montgomery
multiplication algorithm. The number of magic series can be efficiently
computed by Kinnaes' formula, of which the most time-consuming element is modular multiplication. We use Montgomery multiplication for
faster modular multiplication, and the number of operations is reduced
through procedural simplifications. Modular addition, subtraction, and
multiplication operations are vectorized by using the following instructions: Intel Advanced Vector Extensions (AVX), Intel Advanced Vector
Extensions 2 (AVX2), and Intel Advanced Vector Extensions 512 (AVX-
512). The number of magic series of order 8000 is computed on multiple
nodes of an Intel Xeon Phi processor with a total execution time of
1806 days. Results are compared with salient studies in the literature to
confirm the efficacy of the approach.

Keywords: Magic series · Magic square · Montgomery
multiplication · Intel Xeon Phi processor

1 Introduction

A magic square of order m is an $m \times m$ array composed of distinct integers
$1, 2, \ldots, m^2$ where the sums of the columns, rows, main diagonal, and antidiagonal take the same number $\sigma = \frac{1}{m} \sum_{i=1}^{m^2} i = m(m^2 + 1)/2$. Magic squares have
intrigued both professional and recreational mathematicians for a long time.
Indeed, the history of magic squares can be traced back to ancient China and
they have many applications in combinatorics, computer science, cryptography,
and physics [1–4]. Enumerations and number approximations of magic squares
and their derivations have been accomplished [4–11].

A magic series of order m is a set of m distinct integers in $\{1, 2, \ldots, m^2\}$ whose
sum is σ. Since magic squares are composed of magic series, we can estimate the

© Springer Nature Switzerland AG 2020
M. Qiu (Ed.): ICA3PP 2020, LNCS 12453, pp. 365–382, 2020.
https://doi.org/10.1007/978-3-030-60239-0_25

number of magic squares from the number of magic series [7], and thus magic series are also of interest. In this paper, we denote the exact number of magic series of order m as $M(m)$.

As for related works of computing $M(m)$, Kraitchik [12] counted $M(m)$ for up to $m = 7$ by classifying them by their properties. Bottomley [7,13], Trump [7], and Gerbicz [7,14] independently computed $M(m)$ for up to $m = 150$ by using similar recursive algorithms. Kinnaes [7,15] computed $M(m)$ for up to $m = 1000$ based on the relationship that $M(m)$ appears as a coefficient of the q-binomial coefficient. Endo [16] computed $M(4000)$ with Kinnaes' formula and Montgomery multiplication using 16 GPUs. We also computed $M(4000)$ with similar methods in this paper. The results by Endo and us are submitted privately to Trump, which exactly match each other. However, Endo carried out a naive Montgomery multiplication without the improvements we propose in this paper.

Methods for approximating $M(m)$ are also proposed. Bottomley [13,17] put forward an approximation formula for $M(m)$, of which the relative error is as low as 10^{-7} for $m = 100$ [7]. Quist [17] improved Bottomley's empirical formula by adding an analytical correction term, which reduced the relative error to as low as 10^{-11} for $m = 1000$.

It is conjectured, but not yet proved, that $M(m)$ is an even number except for $M(1) = 1$. However, it is not known how many further correction terms are needed for the approximation formulas to estimate the number of large-order magic series. One of the motivations to compute the numbers of magic series of large orders is to investigate these problems.

To the best of our knowledge, Kinnaes' formula is the fastest algorithm for computing the exact number of magic series but an implementation of this formula with optimized Montgomery multiplication using Intel Xeon Phi processors has not yet been reported. Importantly, the most time-consuming operation in Kinnaes' formula is modular multiplication, which plays an important role in various areas such as cryptography. Therefore, developing a faster way of computing Kinnaes' formula has the potential to benefit many applications. In this paper, we present a fast way of computing the number of magic series by using Kinnaes' formula on Intel Xeon Phi processors, with an improved Montgomery multiplication algorithm.

The remainder of the paper is organized as follows. We introduce the magic series and Kinnaes' formula in Sect. 2. Modular addition, subtraction, and multiplication algorithms and their improved implementations are discussed in Sect. 3. We evaluate and compare the execution time of our program with that of Kinnaes in Sect. 4. Next, the numbers of magic series of large orders are computed using Intel Xeon Phi 7250 processors in parallel and the results are compared with those in prior studies in Sect. 5. The achievements of this paper are summarized in Sect. 6. Algorithms presented herein are proved in Appendix.

2 Magic Series

2.1 Definition

A magic square of order m is an $m \times m$ array composed of distinct integers $1, 2, \ldots, m^2$ where the sums of the columns, rows, main diagonal, and antidiagonal take the same number $\sigma = \frac{1}{m} \sum_{i=1}^{m^2} i = m(m^2 + 1)/2$. The magic square of order 3, which is unique, is shown in Fig. 1.

2	9	4
7	5	3
6	1	8

Fig. 1. Magic square of order 3

A magic series of order m is a set of m distinct integers in $\{1, 2, \ldots, m^2\}$ whose sum is σ. For example, if $m = 3$, then $\sigma = 15$ and therefore all possible magic series of order three are $\{1, 5, 9\}$, $\{1, 6, 8\}$, $\{2, 4, 9\}$, $\{2, 5, 8\}$, $\{2, 6, 7\}$, $\{3, 4, 8\}$, $\{3, 5, 7\}$, and $\{4, 5, 6\}$, which implies $M(3) = 8$. From the definition, we can discern that magic squares consist of magic series.

2.2 Kinnaes' Formula

Kinnaes [15] proposed a formula for $M(m)$ which is quick to execute:

$$rM(m) = \binom{m^2}{m} + 2 \sum_{i=1}^{\lfloor r/2 \rfloor} \frac{1}{\omega^{im^2(m-1)/2}} \prod_{j=1}^{m} \frac{\omega^{i(m^2-m+j)} - 1}{\omega^{ij} - 1}, \qquad (1)$$

where r is an integer which is larger than $m^2(m-1)/2$ and of which the smallest prime factor is larger than m, and ω is a primitive r-th root of unity. That is, $\omega, \omega^2, \omega^3, \ldots, \omega^{r-1}$ are all not equal to one, and $\omega^0 = \omega^r = 1$.

Though a root of unity is easily found in the imaginary domain, computation of imaginary numbers may result in numerical errors. To circumvent this problem, Kinnaes [15] performed the computation over a finite field $\mathbb{Z}/N\mathbb{Z}$, where N is a prime number. In this case, r must be a factor of $N - 1$ so that there exists a root of unity. Finally, Kinnaes used the Chinese remainder theorem to reconstruct $M(m)$ from $M(m) \bmod N$ for several values of N. In this paper, we follow the same procedure as Kinnaes.

Algorithm 1. Kinnaes' method for finding ω over $\mathbb{Z}/N\mathbb{Z}$ [15]

Input: N positive integer, r factor of $N-1$
Output: a primitive r-th root of unity ω
1: **for** z in $\{2, 3, \ldots, N-1\}$ **do**
2:      **for** d in $\{d \; : \; d \mid (N-1), \; r \mid d\}$ **do**
3:          **if** $z^d \bmod N = 1$ **then**
4:              **return** $z^{d/r} \bmod N$
5:          **end if**
6:      **end for**
7: **end for**

Algorithm 2. Proposed method for finding ω over $\mathbb{Z}/N\mathbb{Z}$ for prime r

---
Input: N positive integer, r prime factor of $N-1$
Output: a primitive r-th root of unity ω
1: **for** z in $\{2, 3, \ldots, N-1\}$ **do**
2:      $\omega \leftarrow z^{(N-1)/r} \bmod N$
3:      **if** $\omega \neq 1$ **then**
4:          **return** ω
5:      **end if**
6: **end for**

2.3   Finding the Primitive r-th Root of Unity over $\mathbb{Z}/N\mathbb{Z}$

Before computing Eq. (1), we first have to find the value of ω. For this purpose, Kinnaes [15] used Algorithm 1, of which the time complexity is $\mathcal{O}(N\sqrt{N})$. When r is prime, Algorithm 2 finds ω in time complexity $\mathcal{O}(N \log N)$. We note that, for many values of m, the smallest value of r is prime, as described later. We use Algorithm 2 if r is prime, or else Algorithm 1. Refer to Theorem 1 in Appendix for the proof of the algorithm.

Algorithm 3. Montgomery multiplication [19]

Input: $0 \leq A, B < N < \beta$, $\gcd(\beta, N) = 1$, $\mu = -N^{-1} \bmod \beta$
Output: $C = AB\beta^{-1} \bmod N$, $0 \leq C < N$
1: $t \leftarrow AB$
2: $q \leftarrow \mu t \bmod \beta$
3: $C \leftarrow (t + qN)/\beta$
4: **if** $C \geq N$ **then**
5:      $C \leftarrow C - N$
6: **end if**
7: **return** C

Algorithm 4. Radix-β interleaved Montgomery multiplication [19,21,22]

Input: $A = \sum_{i=0}^{e-1} a_i \beta^i$, $0 \leq a_i < \beta$, $0 \leq A, B < N < \beta^e$, $\gcd(\beta, N) = 1$,
    $\mu = -N^{-1} \bmod \beta$
Output: $C = AB\beta^{-e} \bmod N$, $0 \leq C < N$
  1: $C_0 \leftarrow 0$
  2: **for** i **from** 0 **to** $e - 1$ **do**
  3:    $t_i \leftarrow C_i + a_i B$
  4:    $q_i \leftarrow \mu t_i \bmod \beta$
  5:    $C_{i+1} \leftarrow (t_i + q_i N)/\beta$
  6: **end for**
  7: **if** $C_e \geq N$ **then**
  8:    $C \leftarrow C_e - N$
  9: **else**
10:    $C \leftarrow C_e$
11: **end if**
12: **return** C

---

3   Fast Implementation of Kinnaes' Formula on an Intel Xeon Phi Processor

The most time-consuming operations in Eq. (1) are $\omega^{i(m^2-m+j)} \bmod N$ and $\omega^{ij} \bmod N$ for $j = 1, 2, \ldots, m$. As per Kinnaes [18], these powers can be computed with modular multiplication, which is usually slow on modern processors because it normally involves division. However, Montgomery multiplication [19], shown in Algorithm 3, is known to avoid this.

In our implementations, we utilize vector units of processors so that we can compute multiple modular multiplications in parallel. The Intel Advanced Vector Extensions 512 (AVX-512) instruction set, which the Intel Xeon Phi processor supports, and the Intel Advanced Vector Extensions 2 (AVX2) instruction set involve 32-bit $\times$ 32-bit $\rightarrow$ 64-bit unsigned integer multiplication instruction pmuludq [20]. In order to fully utilize the bit width, β is normally set to 2^{32}.

However, our preliminary experiments show that only $M(m)$ of $m \leq 1290$ can be computed with $N < \beta = 2^{32}$. Therefore, we use radix-β interleaved Montgomery multiplication [19,21,22], which is shown in Algorithm 4. The basic idea of this algorithm is to split numbers into base-β digits and conduct modular multiplication of each digit from bottom to top, which is similar to long multiplication. Nevertheless, if we choose $\beta = 2^{32}$ in this algorithm, there will be some overflow when computing unsigned 64-bit integer addition. Unfortunately, the AVX-512 and AVX2 instruction sets do not have carry bits. Though there is a technique to detect the overflow, it requires additional comparison and arithmetic instructions, which reduces the performance. Therefore, we use $\beta = 2^{31}$ in Algorithm 4. Because $N < 2^{62}$ is sufficient for at least $m \leq 10000$, we choose $e = 2$ along with $\beta = 2^{31}$ here.

We note that the first $e - 1$ moduli β in Algorithm 4 can be replaced by 2β. Because $2\beta = 2^{32}$ and 32-bit $\times$ 32-bit $\rightarrow$ 32-bit integer multiplication instruc-

Algorithm 5. Radix-2^{31} interleaved Montgomery multiplication of 62-bit numbers [23]

Input: $\beta = 2^{31}$, $A = a_1\beta + a_0$, $B = b_1\beta + b_0$, $N = N_1\beta + N_0$, $0 \le a_i, b_i, N_i < \beta$,
    $0 \le A, B < N < 2^{62}$, $\gcd(\beta, N) = 1$, $\mu = -N^{-1} \bmod \beta$
Output: $C = AB\beta^{-2} \bmod N$, $0 \le C < N$
  1: $s_0 \leftarrow a_0 b_0$
  2: $s_1 \leftarrow a_0 b_1$
  3: $q \leftarrow s_0 \mu \bmod \beta$
  4: $t \leftarrow (qN_0 + s_0)/\beta + qN_1 + s_1$
  5: $s_2 \leftarrow a_1 b_0 + t \bmod \beta$
  6: $s_3 \leftarrow a_1 b_1 + \lfloor t/\beta \rfloor$
  7: $r \leftarrow s_2 \mu \bmod \beta$
  8: $u \leftarrow (rN_0 + s_2)/\beta + rN_1 + s_3$
  9: **if** $u \ge N$ **then**
 10:     $C \leftarrow u - N$
 11: **else**
 12:     $C \leftarrow u$
 13: **end if**
 14: **return** C

---

tion pmulld [20] truncates overflowed bits, we can eliminate $e - 1$ bitwise AND instructions. Refer to Theorem 2 in Appendix for the proof.

Takahashi put forward a modular squaring algorithm in [23] which uses Montgomery multiplication specialized for the case where $e = 2$. The modular squaring algorithm can be easily extended to modular multiplication, as shown in Algorithm 5. However, we note that the operations $s_2 \leftarrow a_1 b_0 + t \bmod \beta$ and $s_3 \leftarrow a_1 b_1 + \lfloor t/\beta \rfloor$ in Algorithm 5 can be replaced by $s_2 \leftarrow a_1 b_0 + t$ and $s_3 \leftarrow a_1 b_1$, which reduces a bitwise AND operation, a shift operation, and an addition. Refer to Theorem 3 in Appendix for the proof. By employing this technique, the computation becomes approximately eight percent faster on an Intel Xeon Phi processor.

We also note that the first modulus β when calculating q in Algorithm 5 can be replaced by 2β in a similar way to the above. We need $N < 2^{61}$ instead of $N < 2^{62}$ to avoid overflow here, which is sufficient for at least $m \le 10000$. Though the number of primes becomes larger by approximately two percent by this change of condition, the computation becomes faster by approximately twelve percent.

Prior studies [24,25] independently discovered that the final conditional subtraction in Algorithm 5 is not necessary when $N < \beta/4$, that is, $N < 2^{60}$ instead of $N < 2^{61}$ in our case. Though this technique accelerates the computation by approximately one percent on an Intel Xeon Phi processor, it increases the number of primes by approximately two percent. Therefore, we choose not to use this technique in our computation.

The resultant Montgomery multiplication algorithm with the foregoing improvements, which we use to compute Eq. (1), is shown in Algorithm 6.

Algorithm 6. Radix-2^{31} interleaved Montgomery multiplication of 61-bit numbers

---

Input: $\beta = 2^{31}$, $A = a_1\beta + a_0$, $B = b_1\beta + b_0$, $N = N_1\beta + N_0$, $0 \le a_i, b_i, N_i < \beta$,
$\quad 0 \le A, B < N < 2^{61}$, $\gcd(\beta, N) = 1$, $\mu = -N^{-1} \bmod \beta$
Output: $C = AB\beta^{-2} \bmod N$, $0 \le C < N$

1: $s_0 \leftarrow a_0 b_0$
2: $s_1 \leftarrow a_0 b_1$
3: $q \leftarrow s_0 \mu \bmod 2\beta$
4: $t \leftarrow (qN_0 + s_0)/\beta + qN_1 + s_1$
5: $s_2 \leftarrow a_1 b_0 + t$
6: $s_3 \leftarrow a_1 b_1$
7: $r \leftarrow s_2 \mu \bmod \beta$
8: $u \leftarrow (rN_0 + s_2)/\beta + rN_1 + s_3$
9: **if** $u \ge N$ **then**
10: $\quad C \leftarrow u - N$
11: **else**
12: $\quad C \leftarrow u$
13: **end if**
14: **return** C

As described earlier, Algorithms 5 and 6 involve conditional subtraction $u - N$ when $u \ge N$, which is normally slow due to branching. However, the operation can be replaced by $\min(u, u - N)$ (where u and N are unsigned 64-bit integers) with the complementary arithmetic of wrap-around two [23]. We also vectorize modular addition and subtraction operations for Eq. (1) in a similar way.

Nevertheless, though the AVX-512 instruction set includes unsigned 64-bit integer min instruction pminuq, the Intel Advanced Vector Extensions (AVX) and the AVX2 instruction sets do not [20]. Therefore, we use the blendvpd d, a, b, c instruction from the AVX instruction set, which selects b_i if the most significant bit of c_i is set, or else a_i, and stores the result in d_i, where subscript i denotes the i-th 64-bit lane of a register [20]. Because $-N \le u - N < N$ since $0 \le u < 2N$, and the most significant bit of a signed value indicates that the value is negative, the blendvpd d, a, b, c instruction where $b_i = u$ and $a_i = c_i = u - N$ yields the same result as $d_i = \min(u, u - N)$. As the Intel Streaming SIMD Extensions 4.1 (SSE4.1) instruction set also involves the blendvpd instruction [20], this technique can also be implemented on older processors.

We create our program to support running with multiple processor threads in multiple computation nodes in parallel. Because the terms in the summation of Eq. (1) do not depend on each other, we can realize an "embarrassingly" parallel computation by splitting the terms into independent chunks. For in-node multithreading, we utilize OpenMP [26] to create threads, and guided scheduling is used to distribute the chunks to the threads. That is, chunk sizes are initially large, and they decrease gradually between iterations to manage load imbalances. Each thread then computes the specified range of the summation for four or eight prime values at a time, depending on the vector length of the underlying processor. Almost the same procedure is applied for inter-node parallelization.

The computation results are finally gathered into the master thread and written to a file. Then the reconstruction of the number of magic series with the Chinese remainder theorem is done by a separate script written in the Python language.

3.1   Choosing Parameters for Kinnaes' Formula

Equation (1) can be computed with multiple values of r, which enables us to use larger primes. However, it only reduces the number of operations by less than one percent. In addition, using a larger value of r increases the number of operations. Therefore, we use the smallest value of r through the computation of $M(m)$ for each m. We note that the smallest values of r are prime for approximately 59 percent of $2 \leq m \leq 10000$.

Fermat's little theorem states that $x^{N-1} \equiv 1 \pmod{N}$ for prime N and positive integer $0 < x < N$. For a positive integer r which divides $N - 1$, it holds that $\left(x^{(N-1)/r}\right)^r = x^{N-1} \equiv 1 \pmod{N}$, so $x^{(N-1)/r}$ is an r-th root of unity which is not necessarily primitive. We denote such N as $N - 1 = qr$ and thus $N = qr + 1$, where q is a natural number. The Dirichlet prime number theorem [27] states that there are infinitely many primes of this form. As we use the Chinese remainder theorem, using large primes reduces the number of computations. Therefore, we choose as large values of N and q as possible such that $N < 2^{61}$ and thus $q < (2^{61} - 1)/r$.

The magic series of order m is a set of m distinct integers $\{a_1, a_2, \ldots, a_m\}$ chosen from a set $\{1, 2, \ldots, m^2\}$ to satisfy the sum condition $\sum_{i=1}^{m} a_i = \frac{1}{m} \sum_{i=1}^{m^2} i = m(m^2+1)/2 = \sigma$. Suppose that we choose $m-1$ distinct integers $a_1, a_2, \ldots, a_{m-1}$ from $\{1, 2, \ldots, m^2\}$, and let $a_m = \sigma - \sum_{i=1}^{m-1} a_i$. Then, $\{a_1, a_2, \ldots, a_m\}$ is a magic series of order m if a_m satisfies $1 \leq a_m \leq m^2$, and is distinct from $a_1, a_2, \ldots, a_{m-1}$. Therefore, the upper bound of $M(m)$ is $\binom{m^2}{m-1}$.

From the above, if we choose t primes $N_1, N_2, \ldots, N_t$ such that $P = \prod_{i=1}^{t} N_i \geq \binom{m^2}{m-1}$ and compute $M(m) \bmod N_i$ for $i = 1, 2, \ldots, t$, then we can reconstruct $M(m)$ from $M(m) \bmod N_i$ for $i = 1, 2, \ldots, t$ using the Chinese remainder theorem. That is,

$$M(m) = \left(\sum_{i=1}^{t} (M(m) \bmod N_i)\, u_i P_i \right) \bmod P,$$

where $P_i = P/N_i$ and $u_i P_i \equiv 1 \pmod{N_i}$.

4   Performance Results

4.1   Calculation of Parameters

We first calculate the sufficient prime values of N which are smaller than 2^{61} by using the smallest value of r for each m. Next, we find r-th roots of unity ω for each N by using Algorithm 2 or 1, depending on whether r is prime or not, respectively. The programs used here are implemented in the Python language.

Table 1 compares the execution times of the programs on a single thread of an Intel Core m3-7Y32 processor. Our Algorithm 2 is approximately ten times faster than Kinnaes' Algorithm 1 at least for the selected values of m.

Table 1. Execution times to compute ω

m	r	Time (sec)	
		Algorithm 1	Algorithm 2
1000	499500011	10.46	0.982
2000	3998000021	15.75	1.166
3000	13495500019	14.11	1.289
4000	31992000011	13.12	1.459
5000	62487500003	16.48	1.539
8000	255968000021[a]	13.37	2.121

[a]The smallest value of r for $m = 8000$ is actually 255 968 000 003, which is not prime and thus cannot be used for our Algorithm 2. Therefore, we use the smallest prime value of r here to compare the execution time.

4.2 Computation of Kinnaes' Formula

Before computing $M(m)$ for large m, we evaluated the execution times of our program in comparison to Kinnaes' on the processors listed in Table 2. In this preliminary experiment, we computed $M(200)$ using the same parameters (r and N) for both programs.

Table 2. Processor specifications

Processor	Number of cores	Number of threads	Base frequency (GHz)	Vector width (bit)
Intel Xeon Phi 7250	68	272	1.40	512
Intel Core i7-7820X	8	16	3.60	512
Intel Xeon E5-2670 v3	12	24	2.30	256
Intel Core i7-3770K[a]	4	8	3.50	256

[a] The computations of large orders with Kinnaes' program are done by using this machine, which is of Trump. During the computation, Trump overclocked the machine to 4.264 GHz, and the vector units are not used explicitly [7, 18, 28].

Kinnaes' program [18], which does not use Montgomery multiplication, is written in C++, and expects to be run with a single thread. Therefore, we invoked the same number of processes as the number of threads of the underlying processor to compute $M(m)$ mod N for multiple values of N by using Kinnaes' program. Our program, which uses Montgomery multiplication, is written in

C with OpenMP, and thus can be run with an arbitrary number of threads automatically.

For a fair comparison, we implemented modular multiplication operations both with and without Montgomery multiplication in our program, along with implementing both vector and scalar versions for each. In the vector versions, the AVX-512 intrinsic functions are used on Intel Xeon Phi 7250 and Intel Core i7-7820X, and the AVX and the AVX2 functions are used on Intel Xeon E5-2670 v3. The versions without Montgomery multiplication as well as Kinnaes' program use a modulus operator for modular multiplication, which results in limiting $N < 2^{31}$, while the versions with Montgomery multiplication allow $N < 2^{61}$. Therefore, it can be stated that the versions with Montgomery multiplication are impeded in this preliminary experiment.

We used commands `icc` and `icpc` of Intel C++ Compiler version 19.0.3.199 to compile C and C++ source codes, respectively. To utilize vector instructions and processor architecture-specific optimization, `-xknl`, `-xskylake-avx512`, and `-xhaswell` compiler options are added for Intel Xeon Phi 7250, Intel Core i7-7820X, and Intel Xeon E5-2670 v3, respectively. Furthermore, the `-O2` option, which increases the optimization level of the compiler, and the `-qopenmp` option, which enables OpenMP functionality, are added for all processors.

According to the results shown in Table 3, Kinnaes' program on our machines is approximately 3.7 times faster than originally recorded by Kinnaes, which indicates the progressive performance gain of modern processors. Also, our scalar implementations without Montgomery multiplication are approximately 6.7 times faster than Kinnaes' program, which shows that our implementation is better optimized for the processors used. In addition, by employing Montgomery multiplication, vector implementations become approximately 3.6 times faster. Furthermore, vectorizing the program with Montgomery multiplication makes it approximately 6.8 times faster. It turns out that, with all the improvements above, the vectorized implementation with Montgomery multiplication on Intel Xeon Phi 7250, which we employ in the next section, is the fastest.

Table 3. Execution times to compute $M(200)$

Processor	Time with and without Montgomery multiplication (sec)				Time of Kinnaes' program (sec)
	Vector		Scalar		
Intel Xeon Phi 7250	2.243	8.041	27.30	26.89	74.47
Intel Core i7-7820X	6.432	18.52	34.43	43.57	60.92
Intel Xeon E5-2670 v3	12.96	57.12	40.58	48.19	62.91
Intel Core i7-3770K	N/A	N/A	N/A	N/A	247.5 [28]

Table 4. Execution times to compute $M(m)$

m	Total execution time (day)		
	Kinnaes' [7,15,28]	Endo's [16]	Ours
500	2.5	N/A	0.001
1000	9.5	N/A	0.05
2000	N/A	N/A	1.67
3000	N/A	N/A	13.7
4000	N/A	480	57
5000	N/A	N/A	164
8000	N/A	N/A	1806

5   Computation of the Number of Magic Series of Large Orders

We computed $M(m)$ for up to $m = 1000$ and $M(2000)$, $M(3000)$, and $M(4000)$ on a three-node cluster of Intel Xeon Phi 7250 processors. In addition, we computed $M(5000)$ and $M(8000)$ on the Oakforest-PACS system at the Joint Center for Advanced High Performance Computing (JCAHPC), which also consists of Intel Xeon Phi 7250 processors. Our results up to $m = 1000$ exactly match all preceding results put forward by Kraitchik [12], Bottomley [13], Trump [7], and Kinnaes [7,15]. Furthermore, our result of $M(4000)$ matches exactly with that put forward by Endo [16].

Table 4 shows and compares total execution times according to Kinnaes, Endo, and our program (in the latter two cases, sums of execution times with each processor are listed). By virtue of the improved performance of modern processors, and the improved algorithms and program, our computation for $m = 1000$ is approximately 190 times faster than Kinnaes'. Endo used NVIDIA GeForce RTX 2080 Ti GPU [16], of which the theoretical peak performance of a 32-bit integer operation is 11.8 Top/s, while that of Intel Xeon Phi 7250 is 3.05 Top/s. Therefore, our computation is approximately 33 times faster than Endo's, with respect to theoretical peak performance.

5.1   Comparison with Approximation Values

Bottomley [13,17] proposed an approximation formula for $M(m)$:

$$M(m) \simeq M_B(m) = \left(\frac{1}{\pi}\sqrt{\frac{3}{e}}\right)\frac{(em)^m}{m^3 - \frac{3}{5}m^2 + \frac{2}{7}m}.$$

Though this formula is based on statistical theory, the correction coefficients are empirical [17]. Quist [17] proposed another approximation formula with an analytical correction:

Table 5. Comparison of $M_B(m)$ and $M_Q(m)$ approximations with the exact $M(m)$

m	$M_B(m)$, $M_Q(m)$, and $M(m)$
100	$9.0430\underline{063647} \times 10^{236}$
	$9.0430070\underline{857} \times 10^{236}$
	$9.0430073681 \times 10^{236}$
500	$1.148464538\underline{0726} \times 10^{1558}$
	$1.148464537\underline{0536} \times 10^{1558}$
	$1.1484645373362 \times 10^{1558}$
700	$3.665277783\underline{935} \times 10^{2286}$
	$3.665277781\underline{732} \times 10^{2286}$
	$3.665277782060 \times 10^{2286}$
1000	$6.591829227\underline{4998} \times 10^{3424}$
	$6.591829225\underline{1991} \times 10^{3424}$
	$6.5918292254015 \times 10^{3424}$
2000	$1.863197694\underline{12995} \times 10^{7460}$
	$1.863197693\underline{93775} \times 10^{7460}$
	$1.86319769394489 \times 10^{7460}$
3000	$2.188751577\underline{452287} \times 10^{11723}$
	$2.188751577\underline{346787} \times 10^{11723}$
	$2.188751577349272 \times 10^{11723}$
4000	$1.367859861\underline{798863} \times 10^{16134}$
	$1.367859861\underline{760870} \times 10^{16134}$
	$1.367859861761525 \times 10^{16134}$
5000	$5.621300608\underline{742761} \times 10^{20654}$
	$5.621300608\underline{641406} \times 10^{20654}$
	$5.621300608642784 \times 10^{20654}$
8000	$7.776376616382\underline{1194} \times 10^{34686}$
	$7.776376616326\underline{1906} \times 10^{34686}$
	$7.7763766163266560 \times 10^{34686}$

The underlined numbers are the first incorrect digits in each approximation.

$$M(m) \simeq M_Q(m) = \left(\frac{1}{\pi} \sqrt{\frac{3}{e}} \right) e^m m^{m-3} \left(1 + \frac{3}{5m} + \frac{31}{420m^2} \right).$$

We computed the lower and upper bounds of each approximation formula by using the MPFI [29] library, which does arbitrary-precision interval arithmetic. The common digits of each bound as well as exact numbers are listed in Table 5. We note that Quist [17] did not correctly round lower digits of the approximations in their table, and that our exact results match well with the approximations.

From the above, we conclude that our results of $M(2000)$, $M(3000)$, $M(4000)$, $M(5000)$, and $M(8000)$ are correct. As a reference, we list the trailing 20 digits of the results in Table 6. Interested readers are invited to contact the authors for the full results.

Table 6. Trailing digits of $M(m)$

m	Trailing 20 digits of $M(m)$
1000	03477582982638400562
2000	62405569331413763196
3000	52086897891336347386
4000	24973137583787600694
5000	91587228772436547066
8000	57498574781607879608

6   Conclusions

This paper computed the numbers of magic series by using Kinnaes' formula. For a faster computation of $M(m)$ on the Intel Xeon Phi 7250 processor, we improved the Montgomery multiplication algorithm and vectorized modular operations. In addition, we computed $M(2000)$, $M(3000)$, $M(4000)$, $M(5000)$, and $M(8000)$ on multiple nodes of an Intel Xeon Phi 7250 processor in parallel, the results of which match well with the prior studies. Note especially that the computation of $M(m)$ for m larger than 4000 had not been achieved before.

Appendix   Proofs of Algorithms

Theorem 1. *Let N be a prime number, r be a prime factor of $N-1$, and z be a positive integer such that $0 < z < N$. Then, $\omega = z^{(N-1)/r} \bmod N$ is a primitive r-th root of unity in $\mathbb{Z}/N\mathbb{Z}$ if $\omega \not\equiv 1 \pmod{N}$.*

Proof. Let $\mathrm{ord}(z)$ be an order of an integer z in $\mathbb{Z}/N\mathbb{Z}$ for prime N. In other words, $\mathrm{ord}(z)$ is the smallest positive integer which is greater than 0 such that $z^{\mathrm{ord}(z)}$ is congruent to 1 modulo N [30,31]. We show that the order of ω is r if $\omega \not\equiv 1 \pmod{N}$.

It holds that $\mathrm{ord}(\omega) = \mathrm{ord}\left(z^{(N-1)/r}\right) = \mathrm{ord}(z)/\gcd\left(\frac{N-1}{r}, \mathrm{ord}(z)\right)$ [30]. Since Lagrange's theorem states that $\mathrm{ord}(z)$ divides $N - 1$, there exists an integer s which divides $N - 1$ and satisfies $\mathrm{ord}(z) = \frac{N-1}{s}$. Then, it holds that

$$
\operatorname{ord}(\omega) = \frac{N-1}{s} \Big/ \gcd\left(\frac{N-1}{r}, \frac{N-1}{s}\right)
$$

$$
= \frac{N-1}{s} \Big/ \gcd\left(\frac{N-1}{rs} \cdot r, \frac{N-1}{rs} \cdot s\right)
$$

$$
= \frac{N-1}{s} \cdot \frac{rs}{N-1} \cdot \frac{1}{\gcd(r,s)}
$$

$$
= \frac{r}{\gcd(r,s)}.
$$

Because $\gcd(r,s)$ is equal to 1 or r since r is prime, the order of ω is 1 or r.

Now, if the order of ω is 1, then $\omega = \omega^{\operatorname{ord}(\omega)} \equiv 1 \pmod{N}$. The contraposition of this statement shows that the order of ω is r if $\omega \not\equiv 1 \pmod{N}$.

Theorem 2. *Replacing the first $e-1$ moduli β with 2β does not change the result of Algorithm 4.*

Proof. Denote variables after substitution with superscript $\prime$. We show that $C'_e = C_e$.

If $i = 0$, then $t'_0 = a_0 B = t_0$ and hence $q'_0 = \mu t'_0 \bmod 2\beta = \mu t_0 \bmod 2\beta$. Therefore, there are two cases where $q'_0 < \beta$ and $q'_0 \geq \beta$. $q'_0 < \beta$, and thus $q'_0 = q_0$, is the same case as Algorithm 4, so $C'_1 = C_1$. As for $q'_0 \geq \beta$, and thus $q'_0 = q_0 + \beta$,

$$
C'_1 = (t'_0 + q'_0 N)/\beta
$$
$$
= (t_0 + q_0 N + \beta N)/\beta
$$
$$
= C_1 + N.
$$

Assume that C'_i is equal to C_i or $C_i + N$ where $1 \leq i \leq e-2$.

If $C'_i = C_i$, then

$$
t'_i = C'_i + a_i B
$$
$$
= C_i + a_i B
$$
$$
= t_i
$$

and hence $q'_i = \mu t'_i \bmod 2\beta = \mu t_i \bmod 2\beta$. Therefore,

$$
C'_{i+1} = (t'_i + q'_i N)/\beta
$$
$$
= \begin{cases} (t_i + q_i N)/\beta = C_{i+1} & \text{if } q'_i < \beta \text{ and thus } q'_i = q_i \\ (t_i + q_i N + \beta N)/\beta = C_{i+1} + N & \text{if } q'_i \geq \beta \text{ and thus } q'_i = q_i + \beta \end{cases}
$$

On the other hand, if $C'_i = C_i + N$, then

$$
t'_i = C'_i + a_i B
$$
$$
= C_i + N + a_i B
$$
$$
= t_i + N
$$

and hence

$$q_i' = \mu t_i' \bmod 2\beta$$
$$= \mu(t_i + N) \bmod 2\beta$$
$$= \{(\mu t_i \bmod 2\beta) + (\mu N \bmod 2\beta)\} \bmod 2\beta.$$

Here, μt_i is equal to q_i or $q_i + \beta$ modulo 2β, and μN is equal to -1 or $(-1 + \beta)$ modulo 2β since $\mu = -N^{-1} \bmod \beta$. Therefore, q_i' becomes

$$q_i' = \begin{cases} q_i - 1 & \text{if } \mu t_i = q_i \bmod 2\beta, \ \mu N = -1 \bmod 2\beta \\ q_i - 1 + \beta & \text{if } \mu t_i = q_i \bmod 2\beta, \ \mu N = (-1 + \beta) \bmod 2\beta \\ q_i + \beta - 1 = q_i - 1 + \beta & \text{if } \mu t_i = (q_i + \beta) \bmod 2\beta, \ \mu N = -1 \bmod 2\beta \\ q_i + \beta - 1 + \beta = q_i - 1 & \text{if } \mu t_i = (q_i + \beta) \bmod 2\beta, \ \mu N = (-1 + \beta) \bmod 2\beta \end{cases}$$

modulo 2β. Therefore,

$$C_{i+1}' = (t_i' + q_i' N)/\beta$$
$$= \begin{cases} (t_i + N + q_i N - N)/\beta = C_{i+1} & \text{if } q_i' = q_i - 1 \\ (t_i + N + q_i N - N + \beta N)/\beta = C_{i+1} + N & \text{if } q_i' = q_i - 1 + \beta \end{cases}$$

By mathematical induction, it holds that C_{i+1}' is equal to C_{i+1} or $C_{i+1} + N$ for $i = 1, 2, \ldots, e - 2$.

If $i = e - 1$, then C_{e-1}' is equal to C_{e-1} or $C_{e-1} + N$. As for $C_{e-1}' = C_{e-1}$, this is the same case as Algorithm 4, so $C_e' = C_e$. As for $C_{e-1}' = C_{e-1} + N$,

$$t_{e-1}' = C_{e-1}' + a_{e-1} B$$
$$= C_{e-1} + N + a_{e-1} B$$
$$= t_{e-1} + N$$

and hence

$$q_{e-1}' = \mu t_{e-1}' \bmod \beta$$
$$= (\mu t_{e-1} + \mu N) \bmod \beta$$
$$= (q_{e-1} - 1) \bmod \beta.$$

Therefore,

$$C_e' = (t_{e-1}' + q_{e-1}' N)/\beta$$
$$= (t_{e-1} + N + q_{e-1} N - N)/\beta$$
$$= (t_{e-1} + q_{e-1} N)/\beta$$
$$= C_e,$$

assuming that $q_{e-1}' N = q_{e-1} N - N$ does not overflow.

From $0 \le C_i < 2N$, it follows that $0 \le C_i' < 3N$. Therefore, to avoid overflow in 32-bit registers of processors, it is required that $3N < 2^{64}$. From the prerequisite of Algorithm 4, $N < \beta^e$ and hence $N < 2^{62} = 2^{64}/4 < 2^{64}/3$ when $e = 2$ and $\beta = 2^{31}$. Thus, the condition is inherently satisfied.

Theorem 3. *Replacing* $s_2 \leftarrow a_1 b_0 + t \bmod \beta$ *by* $s_2 \leftarrow a_1 b_0 + t$, *and* $s_3 \leftarrow a_1 b_1 + \lfloor t/\beta \rfloor$ *by* $s_3 \leftarrow a_1 b_1$ *does not change the result of Algorithm 5 when* $N < 2^{61}$.

Proof. Denote variables after substitution with superscript $\prime$. We show that $u' = u$.

It holds that

$$
\begin{aligned}
s_2' &= a_1 b_0 + t \\
&= (a_1 b_0 + t \bmod \beta) - t \bmod \beta + t \\
&= s_2 + t - t \bmod \beta.
\end{aligned}
$$

Therefore,

$$
\begin{aligned}
r' &= s_2' \mu \bmod \beta \\
&= (s_2 + t - t \bmod \beta)\mu \bmod \beta \\
&= \{s_2 \mu \bmod \beta + (t - t \bmod \beta)\mu \bmod \beta\} \bmod \beta.
\end{aligned}
$$

Since $t - t \bmod \beta$ eliminates the remainder of t divided by β, it holds that $(t - t \bmod \beta)\mu \bmod \beta = 0$, and hence

$$
\begin{aligned}
r' &= (s_2 \mu \bmod \beta + 0) \bmod \beta \\
&= s_2 \mu \bmod \beta \\
&= r.
\end{aligned}
$$

Furthermore, since

$$
\begin{aligned}
s_3' &= a_1 b_1 \\
&= (a_1 b_1 + \lfloor t/\beta \rfloor) - \lfloor t/\beta \rfloor \\
&= s_3 - \lfloor t/\beta \rfloor,
\end{aligned}
$$

it holds that

$$
\begin{aligned}
u' &= (r'N_0 + s_2')/\beta + r'N_1 + s_3' \\
&= (rN_0 + s_2 + t - t \bmod \beta)/\beta + rN_1 + s_3 - \lfloor t/\beta \rfloor \\
&= (rN_0 + s_2 + t - t \bmod \beta)/\beta + rN_1 + s_3 - (t - t \bmod \beta)/\beta \\
&= (rN_0 + s_2)/\beta + rN_1 + s_3 \\
&= u.
\end{aligned}
$$

References

1. Biggs, N.L.: The roots of combinatorics. Historia Math. **6**(2), 109–136 (1979). https://doi.org/10.1016/0315-0860(79)90074-0
2. Nordgren, R.P.: On properties of special magic square matrices. Linear Algebra Appl. **437**(8), 2009–2025 (2012). https://doi.org/10.1016/j.laa.2012.05.031

3. Cammann, S.: The evolution of magic squares in China. J. Am. Orient. Soc. **80**(2), 116–124 (1960). https://doi.org/10.2307/595587
4. Xin, G.: Constructing all magic squares of order three. Discrete Math. **308**(15), 3393–3398 (2008). https://doi.org/10.1016/j.disc.2007.06.022
5. Beeler, M.: Appendix 5: The Order 5 Magic Squares (1973). (Privately Published)
6. Pinn, K., Wieczerkowski, C.: Number of magic squares from parallel tempering Monte Carlo. Int. J. Mod. Phys. C **9**(4), 541–546 (1998). https://doi.org/10.1142/S0129183198000443
7. Trump, W.: Magic Series. http://www.trump.de/magic-squares/magic-series
8. Beck, M., van Herick, A.: Enumeration of 4×4 magic squares. Math. Comput. **80**, 617–621 (2011). https://doi.org/10.1090/S0025-5718-10-02347-1
9. Ripatti, A.: On the number of semi-magic squares of order 6 (2018). arXiv: 1807.02983
10. Kato, G., Minato, S.: Enumeration of associative magic squares of order 7 (2019). arXiv: 1906.07461
11. Libis, C., Phillips, J.D., Spall, M.: How many magic squares are there? Math. Mag. **73**(1), 57–58 (2000). https://doi.org/10.1080/0025570X.2000.11996804
12. Kraitchik, M.: Mathematical Recreations, 2nd revised edn. Dover Publications (2006)
13. Bottomley, H.: Partition and composition calculator. http://www.se16.info/js/partitions.htm
14. Gerbicz, R.: Robert Gerbicz's Home Page. https://sites.google.com/site/robertgerbicz
15. Kinnaes, D.: Calculating exact values of $N(x, m)$ without using recurrence relations (2013). http://www.trump.de/magic-squares/magic-series/kinnaes-algorithm.pdf
16. Endo, K.: Private Communication (2019)
17. Quist, M.: Asymptotic enumeration of magic series (2013). arXiv: 1306.0616
18. Kinnaes, D.: Private Communication (2019)
19. Montgomery, P.L.: Modular multiplication without trial division. Math. Comput. **44**, 519–521 (1985). https://doi.org/10.1090/S0025-5718-1985-0777282-X
20. Intel Corporation: Intel 64 and IA-32 Architectures Software Developer's Manual. https://software.intel.com/en-us/articles/intel-sdm
21. Koç, Ç.K., Acar, T., Kaliski Jr., B.S.: Analyzing and comparing Montgomery multiplication algorithms. IEEE Micro **16**(3), 26–33 (1996). https://doi.org/10.1109/40.502403
22. Bos, J.W., Montgomery, P.L., Shumow, D., Zaverucha, G.M.: Montgomery multiplication using vector instructions. In: Lange, T., Lauter, K., Lisoněk, P. (eds.) SAC 2013. LNCS, vol. 8282, pp. 471–489. Springer, Heidelberg (2014). https://doi.org/10.1007/978-3-662-43414-7_24
23. Takahashi, D.: Computation of the 100 quadrillionth hexadecimal digit of π on a cluster of Intel Xeon Phi processors. Parallel Comput. **75**, 1–10 (2018). https://doi.org/10.1016/j.parco.2018.02.002
24. Dussé, S.R., Kaliski Jr., B.S.: A cryptographic library for the Motorola DSP56000. In: Damgård, I.B. (ed.) EUROCRYPT 1990. LNCS, vol. 473, pp. 230–244. Springer, Heidelberg (1991). https://doi.org/10.1007/3-540-46877-3_21
25. Walter, C.D.: Montgomery's multiplication technique: how to make it smaller and faster. In: Koç, Ç.K., Paar, C. (eds.) CHES 1999. LNCS, vol. 1717, pp. 80–93. Springer, Heidelberg (1999). https://doi.org/10.1007/3-540-48059-5_9
26. OpenMP Architecture Review Boards: OpenMP. https://www.openmp.org

27. Selberg, A.: An elementary proof of Dirichlet's theorem about primes in an arith-
    metic progression. Ann. Math. **50**(2), 297–304 (1949). https://doi.org/10.2307/
    1969454
28. Trump, W.: Private Communication (2019)
29. Revol, N., Rouillier, F.: Motivations for an arbitrary precision interval arithmetic
    and the MPFI library. Reliable Comput. **11**(4), 275–290 (2005). https://doi.org/
    10.1007/s11155-005-6891-y
30. Adams, W.W., Goldstein, L.J.: Introduction to Number Theory. Prentice-Hall
    (1976)
31. Childs, L.N.: A Concrete Introduction to Higher Algebra, 3rd edn. Springer, New
    York (2009). https://doi.org/10.1007/978-0-387-74725-5

I am Smartglasses, and I Can Assist Your Reading

Baojie Yuan, Yetong Han, Jialu Dai, Yongpan Zou[✉], Ye Liu,
and Kaishun Wu

College of Computer Science and Software Engineering, Shenzhen University,
Shenzhen, China
{yuanbaojie2018,hanyetong2017}@email.szu.edu.cn,
{yongpan,ly,wu}@szu.edu.cn

Abstract. Human reading states reflect people's mental activities and are closely related with learning behaviors. Consequently, it is of great value to identify reading states in many aspects. For example, it can provide timely assistance in learning processes and boost the learning efficiency of learners in a personalized way. Conventionally, researchers usually make use of EEG and fMRI to recognize human reading states. However, these methods have shortcomings of high device cost, low portability and bad user experience. In this paper, we design a real-time reading states detection system named ETist with commodity wearable glasses that can identify four reading states including attention, browsing, mind wandering and drowsiness via tracking eye movement. Through our experiments, we demonstrate that this system can recognize four fine-grained states with an average accuracy of 84.0%, which is applicable in a wide area of applications.

Keywords: Eye tracking · Reading states · Wearable devices

1 Introduction

Reading is a mainstream vision-based information acquisition channel and shows increasing significance due to the rapid update of knowledge. However, reading states show vital impact on the reading efficiency [1]. It is obvious that people needs to spend less time in understanding materials when they are concentrated [2]. On the contrary, the reading efficiency decreases vastly when a reader is distracted by other disturbers [3]. Motivated by such common cases, we ask such a question *What if we can identify a reader's reading states?* Theoretically, we envision that identifying a reader's reading states brings about several potential benefits as follows. On one hand, this information is helpful for readers to adjust themselves into the best state timely when they are reminded about being distracted. On the other hand, readers can get timely assistance when they are stuck in difficulties. For instance, for educators, knowing students' learning states provides guidance of improving teaching methods and optimizing contents organization.

© Springer Nature Switzerland AG 2020
M. Qiu (Ed.): ICA3PP 2020, LNCS 12453, pp. 383–397, 2020.
https://doi.org/10.1007/978-3-030-60239-0_26

Researchers in psychology and cognitive science have conducted some research work related to recognizing reading states. In general, previous work around this topic can be categorized into three different kinds by supporting technologies, namely, electroencephalogram (EEG)-based [4,5], functional magnetic resonance imaging (fMRI)-based [6–8] and eye tracking based [9–15]. EEG and fMRI-based approaches detect reading states by monitoring electrophysiological signals varying with brain activities. However, both of them have certain limitations. They require special devices which are too bulky and expensive to carry. The another kind of technique we utilize to recognize cognitive states of users to provide learning strategies and the like is eye-tracking. The rational behind this method is that different reading states induce unique eye movement patterns and allows for discrimination between reading states by analyzing data [16]. Nevertheless, researches before only focus on studying a certain specific state and its impact on reading, instead of identifying different states and providing corresponding assistance. In this paper, we propose a real-time reading cognitive detection system called ETist with commodity wearable glasses that can recognize four fine-grained reading states via tracking eye movements, namely *attention*, *browsing*, *mind wandering* and *drowsiness*. We design ETist system with a sequence of functional blocks that takes eye movements data as input and outputs human reading states. Each part of ETist system is carefully designed for achieving high performance and better user experience. We have also carefully designed experiments to evaluate the performance of ETist under different settings and the final results show that our system can achieve an average accuracy of 84.0% in recognizing four reading states in common reading scenarios. Finally, we also introduce a case study with ETist system to demonstrate that this system can provide appropriate support to readers based on the recognition results of reading states.

Contributions: In summary, our main contributions can be concluded as follows:

- To the best of our knowledge, this is the first attempt to detect fine-grained cognitive reading states and build a reading assistant system which is convenient for users in real time only using eye-tracking glasses.
- In order to identify the user's reading mental state according to the reading material in real time, we propose line and page segmenting algorithms to segment eye-tracking data according to the reading page.
- By combining with reading states recognition and POIs identification, we build a prototype system and conducts case-study experiments to evaluate its performance. The results show that, compared with traditional methods, our system can effectively assist a user's reading process.

2   Related Work

2.1   EEG-Based/FMIR-Based Cognitive State Recognition

As EEG signal could reflect the electrical activity of scalp, it can detect the human brain activities. EEG is commonly used to detect the cognitive state.

Berka *et al.* [4] found that the EEG signal contains some cognitive state information such as task engagement level and mental workload. Lan *et al.* [5] propose a classification system by using machine learning approaches to estimate the reading states. However, in order to use EEG to detect the cognitive state, it necessitates to stick the electrode slices on user's head or wear a wireless sensor headset. Compared with the eye tracking method, EEG is not portable and comfortable for user to use. fMRI can also detect the cognitive state by measuring the brain activities. Mather *et al.* [6] explain how fMRI reveals reading states of humans. Wang *et al.* [7] detect cognitive state by using fMRI classifiers. However, this technology can't be widely deployed because of the highly price of the fMIR device. Compared with fMRI, eye tracking prevails by being portable, affordable and comfortable to use.

2.2 Eye Tracking-Based Reading States Recognition

Eye tracking has been intensively used by psychologists to study basic cognitive processes during reading and other types of information processing such as pattern of information processing [10,13], effects of instructional strategies [14], effects of learning strategies [15]. Particularly, some researches use fusing-data for attention detection when reading or other information processing activities. For example, Li *et al.* [11] use eye tracking in conjunction with webcam and mouse dynamics to identify cognitive states. Bixler *et al.* [12] fuse eye gaze data and physiology data to achieve detecting of mind wandering when reading. There are some researches using eye-tracking techniques to detect driver cognitive distraction [17] and drowsiness [18] in real time. However, those works with fusing data need additional hardware, which brings burden to a user and causes other problems. and most researches pay more attention on detecting single state like mind wandering rather than recognizing multiple states. To the best of our knowledge, this is the first work to detect fine-grained cognitive reading states in real time only with low-cost eye-tracking glasses.

3 System Overview and Methodology

3.1 Concepts of Reading States

As aforementioned, we focus on studying four reading cognitive states, namely, attention, browsing, mind wandering and drowsiness. There are several reasons behind this. First, studies have shown that attention and mind wandering are two basic mental states in reading tasks [19]. Mind wandering usually happens when attention involuntarily shifts from reading to other unrelated thoughts, which has been proved to have negative effect on reading comprehension [2]. This suggests that it's possible to improve reading efficiency by attempting to hinder mind wandering in time and reduce its negative effects. Therefore, it is necessary to detect these two states for the purpose of helping people better focus on the task at hand. Second, for a fine-grained identification of reading states, we further divide the two basic states into four fine-grained states. In the following, we shall clarify them based on psychological literatures.

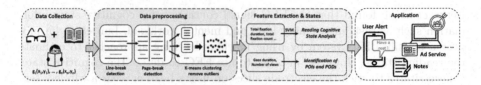

Fig. 1. An overview of the system architecture of ETist

- **Attention**: it is a kind of behavioral and cognitive process during which people concentrate on a discrete aspect of information while ignoring other perceivable information [20]. In our work, it corresponds to a state of focused reading.
- **Browsing**: it is a quick examination of the relevance of a number of objects which may or may not lead to a closer examination or acquisition/selection of (some of) these objects [21]. In this work, it corresponds to the state of fast and fluent reading.
- **Mind wandering**: It's a kind of involuntarily shift from thoughts related about reading to other unrelated thoughts. Mind wandering offen causes poor comprehension during reading [2].
- **Drowsiness**: it is a state of strong desire for sleep [22], which can make people stop reading. It can be defined as high-level mind wandering.

3.2   System Overview

Figure 1 shows the architecture of our ETist system which consists of three main parts, namely, data collection, data preprocessing and reading states recognition. When users are doing reading, the smartglasses with two eye cameras and one world camera collect eye movements data containing real-time coordinates of the eye and corresponding timestamps.

Due to the large number of fixation points and the existence of many invalid points, we propose line-break and page-break algorithms to locate the page where each fixation point is, and then use k-means algorithm to remove invalid outliers in preprocessing block. Following that, we perform feature extraction and selection for reading process within different pages, and further recognize reading states with machine learning methods. Based on the above results, our system can be applied in different areas such as advertising and education.

3.3   Methodology

Data Preprocessing. The data processing is composed of three steps. The smartglasses output three-dimensional space coordinates of eye movements at a certain rate. As mentioned above, each eye movement data point can be denoted by $g = (x, y, z)$, of which x, y and z represent three-dimensional space coordinates, respectively. However, it is to be noted that, the raw data output by the

(a) The reading direction is from the end of the bottom line of left page to the beginning of the top line of the right page

(b) The reading direction is from the end of the bottom line of right page to the beginning of the top line of the left page

Fig. 2. An example of reading track of page breaks

hardware system do not contain information about which line and page they belong to. So the first step is to segment raw eye-tracking data by the line-break algorithm and store each segment in an array. For a sequence of eye gaze points in the ith line, they can be denoted by $L_i = (g_{i1}, g_{i2}, ..., g_{il})$ where l is the total number of points in a line. Similarly, we utilize $P_j = (L_{j1}; L_{j2}; ...; L_{jn})$ to represent the whole data points within jth page. Then, we use a page-break detection algorithm to segment L in order to get lines which are in the same page. This will help reduce the amount of data processed, and can more accurately locate what was being read at the time. After handling pages and lines in reading materials, the third step is to remove outliers from the obtained data. In addition, because the user can not get the information while scanning, we also need to remove the scan points in the data. Through analysis, we adopt the optimized k-means metholdolgy to remove outliers.

Detection of Line Breaks. We segment raw eye-tracking data into parts corresponding to reading lines. We use $(g_1, g_2, \cdots, g_l)$ to represent a sequence of eye gaze points. Since the reading direction for a line break is from the end of text line to the beginning of a new line, which is against the regular reading direction, an eye gaze g_l must satisfy the first condition. Moreover, because the minimum spacing of two lines is line-spacing, g_l also meets the second condition. In a word, a eye gaze g_l is recognized as a line break if the following two conditions are satisfied.

Condition 1: Considering g_k is the beginning point of new line, the eye gaze direction from the g_i to g_l is opposite to the direction from g_l to g_k $(i < l < k)$.

Condition 2: Let y_o, y_n represent the y-coordinate of the beginning point of current line and new line respectively. The absolute value of y_n minus y_o must larger than the line-spacing.

Detection of Page Breaks. In this paper, we mainly focus on the page breaks when a user reads in sequence (shown in Fig. 2). There are mainly two scenarios of page break detection. When a user reads in sequence, the reading direction for a page break is from the bottom line of current page forwards to the top line of a new page. Since the book is divided into left page and right page, there are two kinds of page breaks. One kind is a user reads from left page to the right page (shown in Fig. 2a) and the direction of the eye movement is from the end of the bottom line of left page to the beginning of the top line of the right page. Another kind is the opposite, as shown in Fig. 2b. We use $x(i)$, $y(i)$ to represent the x-coordinate and y-coordinate of the eye-tracking data i respectively. In order to remove the noise point, $x(i)$, $y(i)$ in line j must satisfy:

$$|x(i) - x(i-1)| < P \quad (P > 0) \tag{1}$$

$$|y(i) - y(i-1)| < T \quad (T > 0) \tag{2}$$

The threshold P must smaller than the margin between the two pages and larger than the spacing between two words. Since the reading direction is from the bottom to top and the value of y will decrease until it reaches the first line, the $y(i)$ must satisfy:

$$y(i) - y(i-2) < -C \quad (C > 0) \tag{3}$$

$$y(i) < \min(y) + \frac{\max(y) - \min(y)}{4} \tag{4}$$

where $\min(x)$, $\min(y)$, $\max(x)$, $\max(y)$ represent the minimum, maximum value of x, y lying in reading lines. As for the first kind of page breaks, the value of x will increase because the reading direction is from the left page to the right page. Moreover, $x(i)$ belongs to a new line. Therefore, $x(i)$ must satisfy:

$$x(i) > \max(x) + d \tag{5}$$

$$x(i+1) - x(i) < P \tag{6}$$

where d is half of the margin between two pages. As for the second kind of page breaks, the value of x will decrease because the reading direction is from the right page to left page. In addition, $x(i)$ will lie in a new line. Therefore, $x(i)$ must satisfy:

$$x(i) < \min(x) - (\max(x) - \min(x)) \tag{7}$$

$$x(i+1) - x(i) < P \tag{8}$$

When the system detects the page breaks, it will clear the maximum and minimum value of the x and y and recalculate on the eye-tracking data which lies on the new page.

Feature Extraction. After performing data preprocessing, we extract a set of features for reading cognitive states classification and POIs detection. At the first stage, we extract several base eye movement features, which is fixations, saccades, long fixations and recessions by analyzing the gaze point data. Then we calculate the duration of all reading states using time stamp. Specifically, we embed each of the classification models utilized in the learning stage which include support vector machine (SVM), random forest (RF) and K nearest neighbor (KNN) during feature selection. Then we pick out the feature subset that achieves optimal average accuracy as the final selected features.

Gaze Feature Extraction. Human reading states are closely related with different eye movement states which can be classified into fixation, long fixation, saccade, and revisit according to Lai *et al.* [23]. We identify these movements based on the following metrics.

- *Fixation*: the eyes stay within a certain spatial range for a duration between 100 ms and 300 ms.
- *Long fixation*: the eyes stay within a certain spatial range for a duration more than 300 ms, which indicates a deep thinking state.
- *Saccade*: the eyes move rapidly from one fixation point to another and the duration of eye gaze of less than 100 ms.
- *Revisit*: the order in which the eyes read is the opposite of the order in which the text is arranged.

After identifying different kinds of base eye movements, we further extract eye-movement features covering temporal, spatial and counting aspects. During our experiments, we find that when a user is in a mind wandering state, the depths of his fixation points are greater than the distances from the eye to the reading material plane. when the user is in attention state and drowsiness reading cognitive state, the depth of his fixation points is roughly the same as the distance between the eye to the reading material plane. Figure 3 shows the depth of user's fixation point in attention state and mind wandering state, respectively. Therefore, the z-axis coordinate value of the gaze point is an important feature. And finally the features that we picked were total fixation duration, gaze duration, total revisited fixation, duration, total saccade duration, total reading time, total fixation count, revisited fixation count, number of blinks and saccade count based on our experiments.

POIs feature extraction. In this section, we will introduce the features that we choose to recognize the POIs of users when they are reading and the reason why we choose these features. As we can see, when a user is interested or has difficulty in reading a piece of material, he/she will spend more time on this part to obtain more detailed information. As a result, his/her gaze duration in the corresponding reading area is longer than that in other areas. Second, repetition is the first principle of all learning [24]. When a user has difficulty in reading a piece of content, the user may tend to repeatedly read the content to deepen the understanding of the content. Therefore, the user's gaze duration in this area

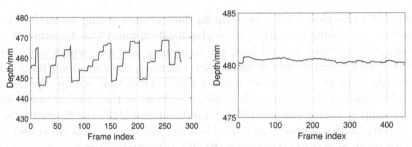

(a) The depth of user's fixation point (b) The depth of user's fixation point
in the state of attention   in the state of distraction

Fig. 3. The depth of user's fixation point in different states

will be longer than in other areas and the number of revisiting times will be
longer as well and we decide these two aspects as features to detect POIs. In
the process of features extraction of POIs, we perform a linear fit on each line of
every fixation point's centroid to obtain a fitting function $f(x)$ first. Since there
are generally less than 20 fixation points in each line, calculation will be greatly
reduced. When the user revisit certain part of material, the user may read the
previous content in the same line, or the content before this line. So the gaze
point $g = (x_i, y_i, z_i)$ must satisfy (9) or (10):

$$x_i - x_{i-1} < \alpha, \|y_i - y_{i-1}\| < \beta \tag{9}$$

$$y_{i-1} - y_i > \beta \tag{10}$$

The value of α is three-quarters of the space between two words, and the value
of β is the line spacing of reading content. When user end up revisiting the
contents, the returned gaze point $g = (x_i, y_i, z_i)$ must satisfy:

$$x_{i+1} - x_i < \alpha, \|y_i - y_{i+1}\| < \beta \tag{11}$$

After obtaining the coordinates of the gaze point that the user normally reads
for the first time, we need to know the serial number of lines the user revisit and
then increase the revisit number of it. If the gaze point $g = (x_i, y_i, z_i)$ satisfies
the (12), it is decided to be on this line.

$$\|f(x_i) - y_{i+1}\| < \beta \tag{12}$$

Then we calculate the number of fixation points n in each line of the reading
material. Since the sampling frequency of smart glasses 30 Hz, the duration of
gaze t can be calculated by (13).

$$t = n * \frac{1}{30} \tag{13}$$

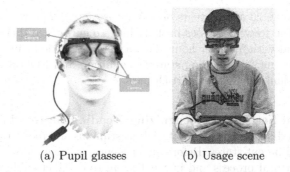

(a) Pupil glasses   (b) Usage scene

Fig. 4. Hardware of ETist

Classification. We use four kinds of classifiers including SVM, Random Forest, kNN and Ensemble Learning for data training. As a result, we find out that Random Forest classifier performs the best since it provides a high average classification accuracy of 84.0% and is more stable than other machine learning methods in English reading scenario. As for real-time system, we choose Random Forest as our classifier.

4   Implementation and Experiments

4.1   System Implementation

Hardware: The hardware of ETist system mainly include Pupil eye-tracking smartglasses [25], as shown in Fig. 4 and a data processing unit (*e.g.*, a PC). When a user performs reading with smartglasses, the eye camera captures eye movements with a resolution of 640×480 at a sampling rate of 30 fps. Each frame is sent to the connected PC for image processing to extract coordinates of the user's pupils. In our experiments, we utilize a Lenovo desktop with a Intel Core i5-4590 CPU and 8 GB RAM which runs software to capture and process images.

Software: The software of Pupil smartglasses is responsible for two-fold functions. On one hand, it controls the smartglasses to capture images and also extract coordinates of one pupil. On the other hand, as an open-source software, it provides powerful libraries for developers to design customized systems and realize personalized functions. Based on the open-source software, we implement our algorithms for page/line segmenting, outliers removal, and eyes' movements recognition.

4.2   Experiment Design

To assess the performance of our ETist system, we recruit a total number of 50 participants covering different grades of undergraduates and graduates in our campus. Among them, 30 participants are males and the others are females, with ages between 20 and 30 years old. What is more, 25 participants are short-sighted and have to wear glasses in daily life. To encourage their participation, each one is paid with ¥30 per hour. The system evaluation can be divided into two main parts, namely, reading states recognition and POIs detection. In the following, we shall give details of each part.

Experimental Setting of Recognizing Reading States. In the first part, we make use of different reading materials as stimulus and request participants to read them at different states. Before experiments, we give them an introduction of the experimental process and instruct them to conduct calibration following the procedure provided by the software. Knowing about our experiments, they will do some warm-up practice to help them be more familiar with the whole process. For different states, each participant is assigned with 100 pages of reading materials selected from different sources with different difficulty. In the following, we shall give some details about settings of evaluating different reading states.

– **Attention** The reading materials are selected from reading comprehension articles of CET 6[1]. To make them focused, each reading article is attached with several questions and participants are requested to answer them after finishing reading. Additional bonus will be based on their performance of answering questions.
– **Browsing** The reading materials of this setting are primary English compositions which can be easily understood by university students. Different from the above, participants are not required to answer questions after finishing reading.
– **Mind wandering** We refer to the experimental setup of Ylias et al. [26]. The reading materials in this setting are same with those in testing attention. The differences lie in two aspects. On one hand, during experiments, popular TV shows are played according to participants' preferences with the purpose of distracting them. On the other hand, the experiments are arranged between 16:00 pm and 17:00 pm during which people are more likely distracted.
– **Drowsiness** We refer to the experimental setup of Hu et al. [27]. To obtain the real state of drowsiness, we arrange experiments in midnights during which participants will first sleep at 23:00 pm for about 2.5 h, and then be alarmed to get up for experiments according to our survey of their daily schedules. They need to finish two pages of reading materials which are the same with above setting.

After finishing each page, participants will report their true states during experiments and then have a rest for about 3 mins before beginning a new repetition. If the self-report results differ from experimental setting, we discard corresponding data and restart experiments. The total experiments last for about 4 months.

---

[1] College English Test Band 6.

Experimental Setting of Detecting POIs. To evaluate the accuracy of detecting POIs, we request each participant to read 100 page of materials which are selected from comprehension articles of CET 6 and have different contents. During reading each article, participants are requested to mark their POIs on the materials. It is noted that the POIs marked by participants may be some key words, phrases, or short sentences. The ETist syetem detects POIs every 30 s which usually cover about two lines. If the marked POIs belong to the detected lines by smartglasses, it is considered as correct hit.

5   Performance Evaluation

5.1   Performance of Recognizing Reading States

We first evaluate the performance of recognizing reading states with training and testing data collected from the same person, which is referred as the identical-person case. For each participant, we use 5-fold cross-validation method to obtain the corresponding average confusion matrix, in which each row represents the ratios that a certain reading state is classified as others. Obtaining the confusion matrix of each participant, we further average them to get the results of different states. Figure 5a shows the result of identical-person tests. As we can see, the minimum and average accuracy of detecting reading cognitive state is 76.0% and 84.0%, respectively. As for the lower accuracy of recognizing drowsiness, we hypothesize that it is because three participants have experience of working over night. To verify our assumption, we exclude their data from our data set and classify data again. The final accuracy of reading drowsiness state is about 79.8%, with an increase of 3.8% compared with the former one.

Then we evaluate ETist's performance in cross-person scenario where classification models are trained and tested with samples collected from different persons. We use 200 data samples of one person as training data set to train a

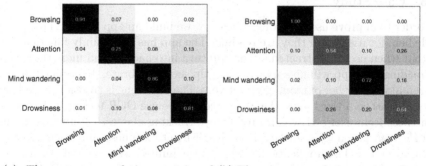

(a) The average confusion matrix of recognizing reading states in identical-person tests

(b) The average confusion matrix of recognizing reading states in cross-person tests

Fig. 5. The performance of recognizing different states in tests

model and test it with another person's data. Since different people have different reading abilities and habits, it is impossible to use data of one person to train the data of others. However, we can find people who have similar English reading abilities to achieve this goal. In all of participants, we find two persons have the most similar English reading abilities according to the score of the CET. The result is shown in Fig. 5b in the form of confusion matrix. As you can see, the accuracy of reading cognitive state detection is 70.0%, with a decrease of 14.0%. We hypothesize that although they have similar English reading abilities, there are some different in their reading habits, which leads to different reading tracks. However, we envision that with more training data, the recognition performance of ETist in cross-person scenario can be increased to above 70.0%.

5.2 Performance of Detecting POIs

To evaluate the performance of detecting POIs, we make use of *precision* and *recall* as evaluation metrics. During the experiment, the number of POIs user marked is 56, and the number of POIs system detected is 80, among which the number of marked POIs is 32. As we can see, the total number of missed points is 24, of which 20 points are difficult points. This is because experimenters think that some difficult points of these such as persons' names can be skipped without affecting the reading quality. As for the remaining points, they are users' POIs, but experimenters did not read them repeatedly. Consequently, the system did not detect those points. In addition, the number of false detections of the system is 48, the main reason is that readers understand the difficulties through repeated reading, which leads to false detections of the system. Since points of interest and difficulties in the reading content detected by the system will not interfere with the user's reading, the most important indicator for the detection of reading POIs is the recall rate, which is 57.1%. After excluding the difficulties that do not affect the understanding of reading, the recall rate is 88.9%.

5.3 Case Study

In order to improve user's reading efficiency, we build an app called ETist to provide help for user when they are reading. The app has two mainly function. One is that when user is distracted or the app turn into background more than 2 min in reading, the app will give user a warm tip such as vibrate, determined by user to remind him or her of reading attentively. The other one is translation. When a user focuses on a word more than five seconds, we use OCR technique to extract sentences from user's view captured by eye-tracking glasses. Afterwards, we use google translation to translate from English into Chinese and then send the translation to ETist application. Owing to the limited resolution of eye-tracking glasses, a fraction of words can not be extracted accurately with OCR. However, with high-definition cameras, we believe that this problem can be solved easily. To further evaluate the performance of ETist in boosting up reading efficiency with better user experience, we conduct a real-world case study with all the participants. According to researches before, reading efficiency is reflected by two

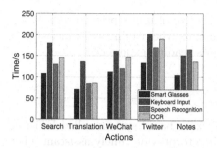

Fig. 6. The comparison of reading efficiency with different methods

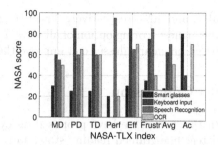

Fig. 7. The NASA scores of our method

main factors, namely, the understanding level and the reading time. Therefore, it can be roughly quantified by the following equation [28].

$$E = R \times C \tag{14}$$

where E, R and C represent reading efficiency, reading speed and understanding level, respectively. In each experiment, the experimenter indicated that he could fully understand the reading content, so we can set the value of C were 100%. Then the main factor that influencing reading efficiency of the experimenter is their reading time. To choose some traditional way for comparison, we conduct a survey by asking all the participants such a question *"what actions will you conduct when you encounter problems in reading?"*. By collecting their answers, we can get a ranking list of different choices according to the frequencies. The top five choices are searching online, translation, turning to friends via instant message applications(*e.g.*, wechat), posting on community websites, and making notes.

In the user-study experiments, we assign each participant with 100 pages of reading materials in each of which an uncommon geographical word is intentionally inserted into a certain position in the first three lines. In each session, participants conduct a certain kind of action mentioned above when they encounter the uncommon word. As comparison, we consider three other alternative ways to perform different actions when reading difficulties, namely, inputting via a keyboard, speech recognition and OCR text recognition. The reading speed is measured by the time consumed for finishing the whole page. Figure 6 shows the corresponding result. As we can see, the experimenter has the shortest reading time when using our ETist eye tracking assistant system in the same operation type. What is more, we also compare the user experience of using different methods. The approach that we used for this evaluation is NASA-TLX [29]. NASA-TLX is a multi-dimensional scale designed to obtain workload estimates from one or more operators while they are performing a task or immediately afterwards. The NASA-TLX score are shown in Fig. 7 which including scores that measure user's like level. From the average NASA-TLX score, Smart glasses score 35 points less than keyboard input, 42.5 points less than speech recognition, 23.3 points less than OCR text recognition, and users' reading load has been reduced by 56%,

61%, and 46%, respectively. From the perspective of user's preference, ETist is the highest scoring option of all ways. In summary, the reading assistance system based on smart glasses can improve the user's reading experience.

6   Conclusion

In this paper, we have proposed an eye-tracking based system called ETist to detect fine-grained reading states in real time to provide timely assistant when users are reading. Experiments have been conducted to evaluate the recognition performance of such a system. Results show that ETist can identify human reading states with an average accuracy of 84.0%. Moreover, we have also developed an application based on the output of ETist as a case study which aims to provide assistance for a reader stuck by vocabularies. We expect that more applications related with human mental states can be built upon this system and provide broader applications in the near future.

Acknowledgment. This research was supported in part by the China NSFC Grant (61802264, 61872248), Guangdong NSF 2017A030312008, Shenzhen Science and Technology Foundation (No. JCYJ20180305124807337, No. ZDSYS20190902092853047), Guangdong Science and Technology Foundation (2019B111103001, 2019B020209001), GDUPS (2015), and Natural Science Foundation of SZU (No. 860-000002110537).

References

1. Smallwood, J., et al.: Subjective experience and the attentional lapse: task engagement and disengagement during sustained attention. Consciousness Cognit. **13**(4), 657–690 (2004). https://doi.org/10.1016/j.concog.2004.06.003
2. Feng, S., D'Mello, S., Graesser, A.C.: Mind wandering while reading easy and difficult texts. Psychon. Bull. Rev. **20**(3), 586–592 (2012). https://doi.org/10.3758/s13423-012-0367-y
3. Smallwood, J., McSpadden, M., Schooler, J.W.: When attention matters: the curious incident of the wandering mind. Memory Cogn. **36**(6), 1144–1150 (2008)
4. Berka, C., et al.: EEG correlates of task engagement and mental workload in vigilance, learning, and memory tasks. Aviat. Space Environ. Med. **78**(5), B231–B244 (2007)
5. Lan, T., Adami, A., Erdogmus, D., Pavel, M.: Estimating cognitive state using EEG signals. In: Signal Processing Conference, 2005 13th European, pp. 1–4. IEEE (2005)
6. Mather, M., Cacioppo, J.T., Kanwisher, N.: How fMRI can inform cognitive theories. Perspect. Psychol. Sci. **8**(1), 108–113 (2013)
7. Wang, X., Hutchinson, R.A., Mitchell, T.M.: Training fMRI classifiers to discriminate cognitive states across multiple subjects. In: NIPS, pp. 709–716 (2003)
8. Chen, Y., Fu, S., Iversen, S.D., Smith, S.M., Matthews, P.M.: Testing for dual brain processing routes in reading: a direct contrast of chinese character and pinyin reading using fmri. J. Cogn. Neurosci. **14**(7), 1088–1098 (2002)
9. Rayner, K.: Eye movements and attention in reading, scene perception, and visual search. Quart. J. Exp. Psychol. **62**(8), 1457–1506 (2009)

10. Rayner, K., Chace, K.H., Slattery, T.J., Ashby, J.: Eye movements as reflections of comprehension processes in reading. Sci. Stud. Read. **10**(3), 241–255 (2006)
11. Li, J., Ngai, G., Leong, H.V., Chan, S.C.: Multimodal human attention detection for reading from facial expression, eye gaze, and mouse dynamics. ACM SIGAPP Appl. Comput. Rev. **16**(3), 37–49 (2016)
12. Bixler, R., Blanchard, N., Garrison, L., D'Mello, S.: Automatic detection of mind wandering during reading using gaze and physiology. In: Proceedings of the 2015 ACM on international Conference on Multimodal Interaction, pp. 299–306 (2015)
13. Brusnighan, S.M., Folk, J.R.: Combining contextual and morphemic cues is beneficial during incidental vocabulary acquisition: semantic transparency in novel compound word processing. Read. Res. Quart. **47**(2), 172–190 (2012)
14. Johnson, C.I., Mayer, R.E.: An eye movement analysis of the spatial contiguity effect in multimedia learning. J. Exp. Psychol. Appl. **18**(2), 178 (2012)
15. Tsai, M.J., Hou, H.T., Lai, M.L., Liu, W.Y., Yang, F.Y.: Visual attention for solving multiple-choice science problem: an eye-tracking analysis. Comput. Educ. **58**(1), 375–385 (2012)
16. Marshall, S.P.: Identifying cognitive state from eye metrics. Aviat. Space Environ. Med. **78**(5), B165–B175 (2007)
17. Liang, Y., Reyes, M.L., Lee, J.D.: Real-time detection of driver cognitive distraction using support vector machines. IEEE Trans. Intell. Transp. Syst. **8**(2), 340–350 (2007)
18. Devi, M.S., Bajaj, P.R.: Driver fatigue detection based on eye tracking. In: First International Conference on Emerging Trends in Engineering and Technology. ICETET 2008, pp. 649–652. IEEE (2008)
19. Schooler, J.W.: Zoning out while reading: evidence for dissociations between experience and metaconsciousness jonathan w. schooler, erik d. reichle, and david v. halpern. Thinking and seeing: Visual metacognition in adults and children, 203 (2004)
20. James, W.: The Principles of Psychology, vol. 1. Cosimo Inc. (2007)
21. Hjørland, B.: Theoretical clarity is not 'manicheanism': a reply to Marcia bates. J. Inf. Sci. **37**(5), 546–550 (2011)
22. Shen, J., Barbera, J., Shapiro, C.M.: Distinguishing sleepiness and fatigue: focus on definition and measurement. Sleep Med. Rev. **10**(1), 63–76 (2006)
23. Lai, M.-L., et al.: A review of using eye-tracking technology in exploring learning from2000 to 2012. Educ. Res. Rev. **10**, 90–115 (2013)
24. Bruner, R.F.: Repetition is the first principle of all learning. Available at SSRN 224340 (2001)
25. Pupil. Pupil eye-tracking glasses. 30 March 2017. https://pupil-labs.com/
26. Ylias, G., Heaven, P.C.L.: The influence of distraction on reading comprehension: a big five analysis. Personality Individ. Differ. **34**(6), 1069–1079 (2003)
27. Shuyan, H., Zheng, G.: Driver drowsiness detection with eyelid related parameters by support vector machine. Expert Syst. Appl. **36**(4), 7651–7658 (2009)
28. Jodai, H.: Reading rate and comprehension. Online Submission 2011
29. Hart, S.G.: Nasa-task load index (nasa-tlx); 20 years later. In: Proceedings of the Human Factors and Ergonomics Society Annual Meeting, vol. 50, pp. 904–908. Sage Publications Sage CA: Los Angeles (2006)

CHEAPS2AGA: Bounding Space Usage in Variance-Reduced Stochastic Gradient Descent over Streaming Data and Its Asynchronous Parallel Variants

Yaqiong Peng⑩, Haiqiang Fei⑩, Lun Li$^{(\boxtimes)}$⑩, Zhenquan Ding⑩,
and Zhiyu Hao⑩

Institute of Information Engineering, Chinese Academy of Sciences,
Beijing 100093, China
{pengyaqiong,feihaiqiang,lilun,dingzhenquan,haozhiyu}@iie.ac.cn

Abstract. Stochastic Gradient Descent (SGD) is widely used to train a machine learning model over large datasets, yet its slow convergence rate can be a bottleneck. As a remarkable family of variance reduction techniques, memory algorithms such as SAG and SAGA have been proposed to accelerate the convergence rate of SGD. However, these algorithms need to store per training data point corrections in memory. The unlimited space usage feature is impractical for modern large-scale applications, especially over data points that arrive over time (referred to as streaming data in this paper). To overcome this weakness, this paper investigates the methods that bound the space usage in the state-of-the-art family of variance-reduced stochastic gradient descent over streaming data, and presents CHEAPS2AGA. At each step of updating the model, the key idea of CHEAPS2AGA is always reserving N random data points as samples, while re-using information about past stochastic gradients across all the observed data points with limited space usage. In addition, training an accurate model over streaming data requires the algorithm to be time-efficient. To accelerate the model training phase, CHEAPS2AGA embraces a lock-free data structure to insert new data points and remove unused data points in parallel, and updates the model parameters without using any locking. We conduct comprehensive experiments to compare CHEAPS2AGA to prior related algorithms suited for streaming data. The experimental results demonstrate the practical competitiveness of CHEAPS2AGA in terms of scalability and accuracy.

Keywords: Machine learning · Stochastic gradient descent · Streaming data · Parallel algorithm · Multicore

1 Introduction

1.1 Motivation

The key problem of machine learning tasks is to fit a model over the training data points and use the model to make predictions. The model is an objective

M. Qiu (Ed.): ICA3PP 2020, LNCS 12453, pp. 398–414, 2020.
https://doi.org/10.1007/978-3-030-60239-0_27

function characterized by decision variables $\mathbf{w}$. Given a training dataset $\mathcal{S}$ consisting of n data points, the objective function is typically in the form of a finite-sum: $\mathcal{R}_{\mathcal{S}}(\mathbf{w}) = \frac{1}{n}\sum_{i=1}^{n} f_i(\mathbf{w})$, where $f_i(\mathbf{w})$ is the loss of $\mathbf{w}$ on the i-th data point in $\mathcal{S}$ and it is often convex. The goal is to find the vector $\mathbf{w}^*$ that minimizes $\mathcal{R}_{\mathcal{S}}(\mathbf{w})$. $\mathbf{w}^*$ is typically called Empirical Risk Minimizer (ERM). Gradient descent methods are widely used to pursue ERM for a given machine learning task.

Gradient descent methods are commonly classified into two forms: batch learning and incremental learning [1,2]. Batch learning algorithms use all the training data points to compute a gradient at each step of gradient descent, where the gradient computation cost increases with the size of data points. In contrast, incremental learning algorithms sample a single data point randomly from the training dataset to compute a gradient at each step of gradient descent. As the most popular incremental gradient descent methods, Stochastic Gradient Descent (SGD) [2,17] is more suitable for large datasets than batch algorithms due to low computation cost. However, the variance of standard SGD can only go to zero if decreasing step sizes are used in the updates for model decision variables [8], typically leading to a sub-linear convergence rate that can be non-trivially slower than the linear convergence rate of batch learning algorithms. To overcome the weakness, variance-reduced SGD methods come into play [4,8,18], which incorporate a correction term to approximate a full-data gradient at each update step and thus achieve linear convergence. The variance reduction techniques either periodically compute full gradients at pivot data points [8,10,19], or store per data point corrections in memory [3,4,18]. We call the latter as memorization algorithms like [6].

Most literatures on gradient descent methods assume that training data points are available beforehand [7,16]. However, in practical scenarios, training data points arrive as an endless sequence over time. We refer to these dynamically observed data points as streaming data in this paper. To the best of our knowledge, the best existing variance-reduced SGD over streaming data named STRSAGA [9] belongs to the memorization algorithms. Therefore, STRSAGA requires unlimited space usage. In addition, STRSAGA is a one-threaded algorithm. Because fitting an accurate model over streaming data requires the optimization algorithm to be time-efficient, it is essential to use parallelization technology to accelerate the training phase of the model.

1.2   This Work

Our goal in this paper is for a streaming variance-reduced SGD that is competitive with STRSAGA in terms of accuracy, while minimizing the running time and space usage. To achieve this goal, we focus on the parallelization on a single multicore machine, and present CHEAPS2AGA, a lock-free parallelization of STRSAGA with limited space usage. We show how to retain the accuracy of STRSAGA over the totally observed data points by only reserving N random data points as samples, while re-using information about past stochastic gradients across all the observed data points with limited space usage. In addition, CHEAPS2AGA embraces a lock-free data structure to insert new data points and remove unused data points in parallel. To guarantee the high time-efficiency, individual threads update the model without using any locking in CHEAPS2AGA.

1.3   Main Results

To show the practical competitiveness of CHEAPS2AGA, we conduct comprehensive experiments to evaluate CHEAPS2AGA and its competitive algorithms. The experimental results show that the accuracy of CHEAPS2AGA is comparable to that of STRSAGA and outperforms that of a prior asynchronous parallel SGD algorithm suited for processing streaming data. In addition, with an increasing number of threads, the model update time of CHEAPS2AGA is significantly reduced, indicating that it performs well in terms of scalability.

1.4   Contributions

In summary, this paper makes the following contributions:

- CHEAPS2AGA: a lock-free parallelization of variance-reduced SGD with limited space usage over streaming data.
- An experimental evaluation to show the competitiveness of CHEAPS2AGA with the related work in practice.

1.5   Organization

Section 2 overviews the most related work, and Sect. 3 provides a foundation for CHEAPS2AGA. Section 4 presents the main technical claims for CHEAPS2AGA. Section 5 shows the practical effectiveness of CHEAPS2AGA by comparing it to the state-of-the-art related work on representative machine learning problems. Section 6 concludes the paper.

2   Related Work

In this section, we first overview the recent progress on SGD algorithm and variance-reduced improvements, followed by their asynchronous parallel variants.

2.1   Stochastic Gradient Descent and Variance-Reduced Improvements

Gradient descent methods are widely used to solve the optimization problem of the finite sum form in machine learning domain [12,21]. Among the various form of the gradient descent methods, SGD [17] is the most popular one due to its low computation cost. However, SGD yields high variance because it computes the gradient at a single data point for each step of gradient descent, leading to slow convergence rate. Therefore, variance-reduced improvements are proposed to solve this problem, which incorporate a correction term that approximates a full-data gradient [3,4,8,18]. SAG is a representative variance-reduced SGD algorithm by storing the last gradient computed for each data point as the correction and using their average to approximate the full-data gradient [18]. Afterwards, Defazio et al. improve SAG by eliminating a bias of the update step, and present SAGA [4]. Daneshmand et al. find that SAGA can obtain a faster convergence rate to reach statistical accuracy on training dataset with gradually increased size [3]. Motivated by this finding, they present a novel algorithm called DYNASAGA [3]. SVRG is another type of variance-reduced SGD algorithm [8], which alternates between computing a full-data gradient and random-data gradient instead of storing historical gradients like SAG-style algorithm.

All the related works mentioned above are offline algorithms. To the best of our knowledge, STRSAGA is the best-performing variance-reduced stochastic gradient descent over streaming data [9]. It is designed based on DYNASAGA. STRSAGA assumes the memory resource is unlimited and needs to store all the observed data points and the last gradient computed for each of them. SSVRG is another streaming version of variance-reduced SGD algorithm [5], which is designed based on SVRG. SSVRG processes streaming data in a single pass, and it does not store historical gradients computed at any data point. Therefore, SSVRG consumes limited space usage. Unfortunately, SSVRG is significantly less accurate than STRSAGA. In this paper, we pursue an algorithm that matches the accuracy of STRSAGA and consumes limited space usage.

2.2 Asynchronous Parallel Variants of Stochastic Gradient Descent and Variance-Reduced Improvements

To meet the requirement of large-scale applications, many parallel implementations of SGD and its variance-reduced improvements are proposed by researchers [7,11,16,20]. HOGWILD! is a lock-free approach to parallelizing SGD [16], where each thread updates the model without using any locking. Zhang et al. further improve the performance of HOGWILD! on *Non-Uniform Memory Access* (NUMA) system [20]. The lock-free update scheme is extended to implement the asynchronous parallel variants of variance-reduced SGD [7,11]. For example, Reddi et al. present a unifying framework for many variance-reduced SGD, and propose an asynchronous parallel realization of SVRG [7], where each thread updates the essential parameters associated with the model without using any locking. Note that the effectiveness of the lock-free update scheme adopted by HOGWILD! depends on the assumption that most update steps only modify small parts of the decision variable. Unfortunately, this assumption cannot exist in SAGA-like algorithms because they use the average of historical gradients in an update step and this parameter relates to the whole parts of the decision variable. To solve this problem, Leblond et al. transform the dense SAGA update step to a sparse format, and present ASAGA, an asynchronous parallel implementation of SAGA that adopts the lock-free update scheme based on the transformation [11]. The transformation scheme used in ASAGA needs to first traverse the whole training data points and compute the probability for each dimension that the regarding value is non-zero in a training data point. This pretreatment is impractical over streaming data. In our previous work, we overcome the challenges of parallelizing the SAGA-like algorithm over streaming data in a lock-free manner, and present LFS-STRSAGA based on STRSAGA [15] for the settings where memory is not limited. To reduce data access conflicts in LFS-STRSAGA, each thread independently maintains the essential parameters associated with the data points that are ever picked by them for model update on the fly. In this paper, we further explore how to parallelize variance-reduced SGD over streaming data with limited space usage in a lock-free manner.

3   Preliminaries and Challenges

In this section, we describe some foundations for the design of CHEAPS2AGA. First, we specify the data stream setting and problem definition considered in this paper. Then, we present an overview of memory algorithms, a state-of-the-art family of variance reduction technologies for SGD. Last, we discuss the challenges for the lock-free parallelization of variance-reduced SGD with limited space usage on streaming data.

3.1   Settings and Problem Statement

We consider the data stream setting where training data points arrive over time. Let X_i be the set of training data points arriving between time step $i - 1$ and i $(i = 1, 2, 3, \ldots)$, and $S_i = \cup_{j=1}^{i} X_j$ $(i = 1, 2, 3, \ldots)$ denote the data points that have been seen till time step i. Our goal is to produce a model and the model's empirical risk is close to the ERM over the data points S_i at the end of each time step i.

3.2   Memory Algorithms: A Family of Variance Reduction Technologies for SGD

Optimization of an empirical or regularized convex risk function is a general problem that is pervasive in machine learning. Given a sample set S consisting of n training data points and a mathematical model characterized by a vector of weights $\mathbf{w} \in \mathbb{R}^d$ (referred to as decision variable), the empirical risk function measuring the average loss of $\mathbf{w}$ over S is: $\mathcal{R}_S(\mathbf{w}) = \frac{1}{n} \sum_{\mathbf{x} \in S} f_{\mathbf{x}}(\mathbf{w})$. The best we can do is to find a vector of weights $\mathbf{w}_S^*$ that minimizes $\mathcal{R}_S(\mathbf{w})$, namely $\mathbf{w}_S^* = \arg \min_{\mathbf{w} \in \mathcal{F}} \mathcal{R}_S(\mathbf{w})$. To solve the problem, SGD starts with a known initial vector of weights $\mathbf{w}^0$, and updates the vector repeatedly by taking the form: $\mathbf{w}^{t+1} \leftarrow \mathbf{w}^t - \eta \nabla f_i(\mathbf{w}^t)$ $(t \geq 0)$. Here, $\nabla f_i(\mathbf{w}^t)$ is the gradient computed at the data point (indexed by i) that is picked randomly in iteration $t + 1$, and $\mathbb{E} \nabla f_i(\mathbf{w}^t) = \nabla \mathcal{R}_S(\mathbf{w}^t)$. $\nabla \mathcal{R}_S(\mathbf{w}^t)$ is the full-data gradient. The variance of $\nabla f_i(\mathbf{w}^t)$ tends to be high. To better approximate the full-data gradient, variance-reduced SGD methods are proposed to incorporate a correction term into the update form of SGD. A state-of-the-art family of variance-reduced SGD algorithms, called memorization algorithms, correct the updates for the iterate sequence $\mathbf{w}^t$ ($t \geq 0$) by taking the form: $\mathbf{w}^{t+1} \leftarrow \mathbf{w}^t - \eta g_i(\mathbf{w}^t)$, $g_i(\mathbf{w}^t) = \nabla f_i(\mathbf{w}^t) - \overline{\alpha}_i$ with $\overline{\alpha}_i = \alpha_i - \overline{\alpha}$, where $\overline{\alpha} = \frac{1}{n} \sum_{j=1}^{n} \alpha_j$. $\overline{\alpha}_i$ are variance correction terms, which satisfy the condition $\mathbb{E}\overline{\alpha}_i = 0$ and thus guarantee unbiased updates $\mathbb{E}g_i(\mathbf{w}^t) = \nabla \mathcal{R}_\mathcal{S}(\mathbf{w}^t)$. Compared to $\nabla f_i(\mathbf{w}^t)$, the variance of $g_i(\mathbf{w}^t)$ is reduced. Different memorization algorithms take different update forms on the memory α_j. In this paper, we focus on the update rules of SAGA algorithm and its extensions.

Like SGD, the SAGA algorithm starts with a known initial vector of weights $\mathbf{w}^0$, and initializes α_i to $\nabla f_i(\mathbf{w}^0)$ for each i. Given the value of $\mathbf{w}$ and each α_i at the end of iteration t, the updates for iteration $t + 1$ in SAGA are as follows:

1. Select an index i uniformly at random.
2. Update $\mathbf{w}$ by taking the form: $\mathbf{w}^{t+1} \leftarrow \mathbf{w}^t - \eta(\nabla f_i(\mathbf{w}^t) - \alpha_i + \frac{1}{n} \sum_{j=1}^{n} \alpha_j)$.
4. Update α_i to $\nabla f_i(\mathbf{w}^t)$, and keep α_j unchanged for each $j \in [1, n], j \neq i$.

The key idea of SAGA is reusing historical gradients to reduce the variance of single-data gradient at each update step, thus yielding a better approximation of full-data gradient and higher convergence rate. To further improve the convergence rate, Daneshmand et al. present DYNASAGA, which gradually increases the size of effective sample set based on the phenomenon that the convergence rate of variance-reduced methods scale favourably for smaller sample sizes [3]. Therefore, different from that SAGA always samples a random data point from the whole training dataset, DYNASAGA starts with a small sample set, and adds a new data point into the sample set every other iteration. With the novel sample scheme, DYNASAGA achieves the statistical accuracy on an n-sample in $2n$ iterations, instead of $n log n$ iterations like SAGA. Both SAGA and DYNASAGA are offline algorithms, which know the whole training dataset in advance. Jothimurugesan et al. extend the idea of DYNASAGA to the streaming data scenario, and present STRSAGA [9]. Similar to DYNASAGA, STRSAGA extends the effective sample set in a controlled manner, namely that it adds a new data point every other iteration. However, STRSAGA chooses a new data point from a dynamically changing buffer which is used to store arriving data points, instead of a known training dataset in DYNASAGA. For any time the buffer is empty, STRSAGA performs further steps of SAGA on the current sample set.

3.3 Challenges for Parallelizing Memorization Variance-Reduced SGD with Limited Space Usage over Streaming Data

Here, we discuss the challenges for the lock-free parallelization of variance-reduced SGD with limited space usage on streaming data.

Challenge 1. *How to sample from the endless sequence of data points with limited space usage for gradient computation in parallel?*

Variance-reduced SGD needs to sample a data point from training dataset at each step of model update. To bound the space usage, we need to store part of the observed data points, while following the random sample principle. To achieve this goal, we use an array to reserve at most N random data points as the effective sample set, and each thread performs updates in alternation. Every other iteration a thread tries to place a new data point into the effective sample set and directly forces an update at this data point if successful. If the effective sample set is full, the insertion of a new data point will replace an existing data point from the effective sample set at random. In alternation, the thread samples a data point randomly from the effective sample set. In addition, we need to carefully design the lock-free synchronization scheme to coordinate threads with data sampling and replacement.

Challenge 2. *How to process the historical gradients with limited space usage?*

SAGA-like algorithms need to maintain the last gradient computed for each data point for approximating full-data gradients. To limit the space usage of historical gradients on streaming data, we only reserve the last gradients computed for the data points existing in the effective sample set. In addition, to reduce the data access conflicts among concurrent threads, each thread maintains a local version of per data point corrections. Section 4.2 describes the details.

Challenge 3. *How to reclaim the space usage of the replaced data points?*

When a data point is removed from the effective sample set, it will not be used in the subsequent updates. To save the memory space, the memory allocated to store the unused data points and their correction information will be reclaimed. The main problem is that a data point may be currently used by other threads at the point of time that the data point is replaced by a thread. Thus, when a data point is removed from the effective sample set, the corresponding memory cannot be reclaimed immediately. To solve the problem, we adopt a time-stamped memory reclamation scheme to safely reclaim the memory of any unused data point and its correction information. Section 4.3 describes the details.

4   CHEAPS2AGA: Lock-Free Parallelization of Variance-Reduced Gradient Descent with Limited Space Usage over Streaming Data

In this section, we describe the design of CHEAPS2AGA in detail, including the lock-free update rules and time-stamped memory reclamation scheme. As a warm up, let us first look into the main building blocks used to support the key ideas of CHEAPS2AGA.

4.1   Building Blocks

We focus on the parallelism on a single multicore system with shared memory. CHEAPS2AGA runs with $2n$ deterministic threads. Among the running threads, there are n master threads, and each master thread has a specific slave thread. A master thread is in charge of updating the model decision variables based on the effective sample set, which always includes at most N data points. A slave thread is in charge of reclaiming the memory of the data points removed by its master thread from the effective sample set and their corrections.

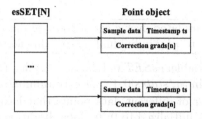

Fig. 1. The effective sample set data structure

The Effective Sample Set Data Structure. The main building block of
CHEAPS2AGA is a lock-free data structure for storing the data points in the
effective sample set and their corrections. Figure 1 shows a diagram for the data
structure. The data structure is an array with N cells. Each cell is a reference to
a *Point* object consisting of a data point, the corrections maintained by different
threads for this data point, and the time at which the data point is removed from
the effective sample set.

Thread-Local State. In Sect. 3.3, we mention the safety problem of the memory
reclamation for the data point removed from the effective sample set. Before
reclaiming the memory of a removed data point, we need to first ensure that
it will be never accessed by any thread. In CHEAPS2AGA, Given a removed
data point, if a master thread starts an update iteration after the data point is
removed, the thread cannot visit it. Based on this observation, we maintain an
array *states* to store the local state of each thread. A thread-local state records
the start time of the current update iteration performed by the owner thread (if
it does), and maintains a list of removed *Point* objects for memory reclamation.

With the support of two building blocks mentioned above, we implement the
lock-free update scheme and the time-stamped memory reclamation scheme in
CHEAPS2AGA. We will present the details in the following subsections.

4.2 Lock-Free Update Scheme

Table 1. Symbols used in Algorithm 1

$e(p)$	Indexes of the coordinates on which the values of the data point indexed by p are non-zero
η	Step size
x_v	Value of the vector x (e.g. $\mathbf{w}$, g, and α) on the coordinate indexed by v
e_v	Among the data points that have been picked by the local thread to update the decision variable, fraction of the data points that are non-zero on the v-th coordinate

---

Algorithm 1. Model updated over a set of training data points S_i that arrived in time steps 1 to i by individual master threads, $i > 0$

Input: w: decision variable; $esSET[0, 1, ..., N-1]$: effective sample set; $\widetilde{S}$: number of data points in S_i; $states[0, 1, ..., n-1]$: running states of all the master threads; $\widetilde{T}$: number of different data points that have been ever picked by any master thread, initialized to 0; t: global iteration count, initialized to 1 at each time step; $thread\_id$: id of current master thread; $next\_p$: recording the index of the next data point picked by the current thread, initialized to -1; sum: sum of local historical gradients, initialized to 0^d and maintained incrementally (d: number of features in a gradient).

Output: solution over S_i.

1: **for** ($t := $ FAA($\&t$, 1); $t \leq \rho$; $t := $ FAA($\&t$, 1)) **do**
2:     $size := 0$ // count the number of different data points that have been ever picked
3:     $states[thread\_id] \rightarrow ts := newTimestamp()$ // record the start time of iteration t
4:     **if** t is even **then** // try to place a new data point into the effective sample set
5:         **if** $next\_p \neq -1$ and $next\_p < \widetilde{S}$ **then** // pick the data point reserved in the previous even iteration
6:             $pi := next\_p, next\_p := -1$
7:             $size := pi + 1$
8:         **else**
9:             **if** $\widetilde{T} < \widetilde{S}$ **then** // try to pick a new data point
10:                 $pi := FAA(\&\widetilde{T}, 1)$
11:                 **if** $pi < \widetilde{S}$ **then**
12:                     $size := pi + 1$
13:                 **else**
14:                     $next\_p := pi$
15:                 **end if**
16:             **end if**
17:         **end if**
18:     **end if**
19:     **if** $size = 0$ **then** // if t is odd or fail to pick a new data point in the even iteration number, directly sample a data point from $esSET$ at random
20:         $size := min(\widetilde{T}, \widetilde{S})$
21:         sample $pi \sim$ Uniform$[0, min(size, N))$
22:         $p := esSET[pi]$ // read $esSET[pi]$ atomically
23:     **else**
24:         pick the data point indexed by pi from S_i; initialize a Point object p and update sum with the data point
25:         $old\_p := SWAP(\&esSET[pi \textbf{ mod } N], p)$ // place p into the effective sample set
26:         **if** $old\_p !=$ NULL **then** // deal with a replaced data point

27:          $old\_p \to ts := newTimestamp()$
28:          add old_p to $states[thread\_id].memory\_reclamation\_list$
29:       **end if**
30:    **end if**
31:    compute a gradient g at $p \to data$ based on the current state of $\mathbf{w}$, and $\alpha := p \to grads[thread\_id]$
32:    **for** $v \in e(p)$ **do**
33:       $\mathbf{w}_v := \mathbf{w}_v - \eta(g_v - \alpha_v + \frac{sum_v}{e_v \cdot size})$
34:    **end for**
35:    $sum := sum + g - \alpha$
36:    $p \to grads[thread\_id] := g$ // update the local historical gradient for p
37: **end for**
38: $states[thread\_id] \to ts := \top$ // $\top$: maximum time-stamp, indicating the current thread is not performing an update iteration

---

Algorithm 1 shows the lock-free update scheme adopted by CHEAPS2AGA, which uses the basic framework of our previous work [15]. Table 1 introduces the symbols used in Algorithm 1. At each time step, CHEAPS2AGA performs ρ update steps, which are evenly dispatched to each master thread. Suppose a master thread performs a model update in the global t-th iteration, where the iteration number is obtained by performing an atomic FAA[1] operation on a global counter t (line 1). If t is even, the thread picks a new data point (if exist) with index pi (lines 10–12), and places the data point into a random location of $esSET$ (lines 24–25), followed by using this fresh data point and its correction to update the model (lines 31–34). Note that the placement of a new data point accompanies the removal of an old one existing in $esSET$. In addition, different data points may be placed into the same location of $esSET$ at the same time. To ensure each removed data point to be perceived by the thread that removes it, each master thread uses the $SWAP$[2] atomic operation to place a data point into $esSET$ (line 25). When a master thread replaces an existing data point with a new data point, it adds the removed data point to a local list for its slave thread to periodically conduct memory reclamation (lines 26–29). When a slave thread finds a data point will be never used by any master thread, it reclaims the memory of the $Point$ object containing the data point and its corrections. If t is odd or no available data points ($\tilde{T} \geq \tilde{S}$) for the master thread to pick, the thread directly selects a data point from $esSET$ for the model update at random (lines 20–22). If the number of data points existing in $esSET$ is less than N, $esSET$ is equivalent to the effective sample set of STRSAGA. Otherwise, $esSET$ is equivalent to the set of N random data points from the effective sample set of STRSAGA, and sampling from $esSET$ is equivalent to sampling from the effective sample set of STRSAGA at random. Therefore, the sample

---

[1] FAA(**addr**, v): assuming the content stored in **addr** are a, this operation returns a and stores $a + v$ at **addr**.

[2] SWAP(**addr**, v): assuming the content stored in **addr** are a, this operation returns a and stores v at **addr**.

scheme used in CHEAPS2AGA follows the semantic of the alternation sample scheme used in STRSAGA.

A model update performed by a master thread in CHEAPS2AGA requires three parameters: 1) gradient computed at a chosen data point based on the current state of decision variable, 2) last gradient computed at the chosen data point, and 3) the average of last gradient computed by the thread at different data points. To reduce the computation expenses, we follow the rules in our previous work [15]: the model update form only considers the coordinates v on which the values of the chosen data point are non-zero (lines 32–34), and uses a coefficient $\frac{1}{e_v}$ to making the update unbiased on each coordinate v. The definition of e_v is given in Table 1. To ensure high performance, model updates do not use any locking.

4.3   Memory Reclamation

Algorithm 2. Memory reclamation process conducted by individual slave threads

Input: $states[0, 1, ..., n-1]$: running states of all the master threads; $thread\_id$: id of current slave thread, which is equal to id of its master thread.
Output: reclaim the memory of some removed data points.
1: **for** each $Point$ object $point$ in $states[thread\_id].memory\_reclamation\_list$ **do**
2:     **if** $point \rightarrow ts < \min_{0 \leq i \leq n-1}(states[i] \rightarrow ts)$ **then**
3:         remove $point$ from $states[thread\_id].memory\_reclamation\_list$
4:         free($point$)
5:     **end if**
6: **end for**

Here, let us discuss how to reclaim the memory of a removed data point safely. As shown in Algorithm 1, a master thread accesses a data point in two ways: (1) pick the data point as a new one; (2) sample the data point from the effective sample set. A master thread picks a new data point by using a FAA atomic operation to obtain its index, ensuring that each data point can be picked as a new data point only once. In addition, a master thread cannot sample a data point that has been removed from the effective sample set. Therefore, given a removed data point, if a master thread starts an update iteration after the data point is removed, the thread cannot access it. Based on the rule, we design a time-stamped memory reclamation scheme, which is conduced periodically by each slave thread in background. Algorithm 2 shows the details. When a slave thread starts to conduct memory reclamation process, it traverses the memory reclamation list of its master thread (lines 1–6), and checks whether a visited $Point$ object is removed before the start time of the current iterations performed by all the master threads (line 2). If a $Point$ object matches the precondition, the slave thread removes it from the list and reclaims its memory (lines 3–4).

Analysis for Space Usage. The space complexity of CHEAPS2AGA mainly lies on the decision variable, effective sample set data structure, thread-local states, and the number of *Point* objects in the memory reclamation lists of all the master threads. Obviously, the space usage of the first three instances is bounded by the number of components in the decision variable, $O(N \cdot (a \cdot d_1 \cdot n + b \cdot d_2 + c))$, and $O(n)$ respectively $(a, b, c$: constants; d_1: number of features in a gradient; d_2: number of features in a training data point; N: a predefined size of the effective sample set; n: number of master threads). In this paper, we assume that a master thread cannot be suspended infinitely during an operation. As time passes, master threads continuously start new update steps. Thus, slave threads limit the number of *Point* objects in the memory reclamation lists of all the master threads by taking the memory reclamation rule at lines 2–5 of Algorithm 2. Combining with the fact that the space usage of a *Point* object is bounded by $O(a \cdot d_1 \cdot n + b \cdot d_2 + c)$, the space usage of CHEAPS2AGA is limited. In our future work, we consider supporting thread failure scenario in CHEAPS2AGA: if a thread fails or it is suspended infinitely during an iteration, we need to avoid the case that the memory of any removed data point will be never reclaimed due to the thread failure.

5 Experimental Results

In this section, we evaluate the practical competitiveness of CHEAPS2AGA with the serial streaming data algorithm STRSAGA [9] and a state-of-the-art asynchronous parallel SGD algorithm HOGWILD! [16] through a set of experiments on real world datasets streamed in under various arrival distributions.

5.1 Study Cases

SGD algorithms are generally used to solve convex problems. For our experiments, we study a popular convex problem in machine learning, namely *logistic regression*. We use the RCV1 [13] and URL [14] datasets. The details of the datasets are provided in Table 2. Similar to [9,15], we convert these static training data into streaming data that follows the skewed and Poisson arrivals over the course of 100 time steps respectively.

Table 2. Datasets for logistic regression and step sizes used in experiments

Dataset	Number of training data points	Number of features	Density	Steps size η
RCV1	677399	47236	0.15%	0.5
URL	2396130	3231961	0.004%	0.05

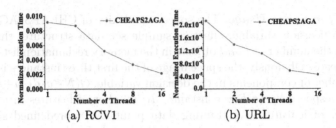

Fig. 2. Normalized update time of CHEAPS2AGA for logistic regression under Poisson arrivals on (a) RCV1 and (b) URL

5.2 Scalability

We first evaluate the scalability of CHEAPS2AGA by testing the model update time over the course of 100 time steps in all workloads with different number of master threads. The update time is normalized to the update time of STR-SAGA. Because CHEAPS2AGA and HOGWILD! have different computation complexity, we do not compare their execution time. Figure 2 shows the results under Poisson arrivals. With an increasing number of threads, the model update time of CHEAPS2AGA is significantly reduced, indicating that it performs well in terms of scalability. Because the datasets are sparse and CHEAPS2AGA conducts sparse processing on the average of historical gradients, even the model update time of single-threaded CHEAPS2AGA is non-trivially lower than that of STRSAGA. The experimental results under skewed arrivals have the similar figures.

5.3 Sub-optimality

Here, we evaluate the accuracy of different algorithms for logistic regression on RCV1 and URL datasets. The accuracy is evaluated in terms of the sub-optimality over the set of data points $\mathcal{S}_i$ that have been observed till each time step i like the related work [3,9,15]. The metric for the sub-optimality of an evaluated algorithm A is defined as follows:

$$SUBOPT_{\mathcal{S}_i}(A) = \mathcal{R}_{\mathcal{S}_i}(\mathbf{w}) - \mathcal{R}_{\mathcal{S}_i}(\mathbf{w}^*_{\mathcal{S}_i}) \tag{1}$$

where $\mathbf{w}$ is the solution produced by A over $\mathcal{S}_i$ and $\mathbf{w}^*_{\mathcal{S}_i}$ is the optimal solution over $\mathcal{S}_i$. Lower sub-optimality indicates better accuracy due to that it means the empirical risk of the solution produced by A is closer to the ERM.

Figures 3 and 4 show the sub-optimality of different algorithms for all the workloads under skewed arrivals and Poisson arrivals, respectively. To make the comparison fair, we implement a natural streaming data version of HOGWILD! (referred to as S-HOGWILD!), which performs ρ model update steps at each time step. Note that STRSAGA is a one-threaded algorithm. The number of master threads specified by the figures is configured in CHEAPS2AGA and S-HOGWILD!. Although CHEAPS2AGA only reserves N random data points as

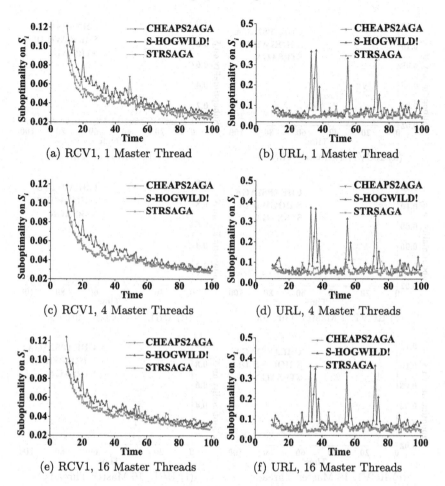

Fig. 3. Sub-optimality for logistic regression under skewed arrivals

the effective sample set dynamically, it reuses past stochastic gradients across all the observed data points by incrementally maintaining their average. Therefore, as shown in Figs. 3 and 4, CHEAPS2AGA with 1 master thread almost matches STRSAGA and significantly outperforms S-HOGWILD! for the most cases in terms of sub-optimality. With an increasing number of threads, the sub-optimality of CHEAPS2AGA is very close to that of STRSAGA for the URL dataset because this dataset is very sparse. For the RCV1 dataset, the sub-optimality of CHEAPS2AGA with 16 threads is bigger than that of STRSAGA, but still lower than that of S-HOGWILD!. This is because RCV1 is more dense than URL and more update conflicts among threads will occur.

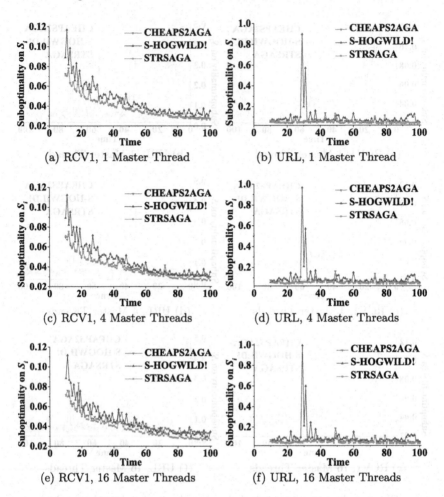

Fig. 4. Sub-optimality for logistic regression under Poisson arrivals

6 Conclusions

This paper investigates the methods that bound the space usage in the state-of-the-art family of variance-reduced stochastic gradient descent over streaming data, and presents CHEAPS2AGA. At each step of updating the model, the key idea of CHEAPS2AGA is always reserving N random data points and their corrections as samples, while re-using information about past stochastic gradients across all the observed data points by combining them into a structure with limited space usage. In addition, training an accurate model over streaming data requires the algorithm to be time-efficient. To accelerate the model training phase, CHEAPS2AGA embraces a lock-free data structure to deal with the arrival of new data points and the removal of unused data points in parallel, and updates the model parameters without using any lock. We conduct comprehensive experiments to evaluate the practical competitiveness of CHEAPS2AGA. The experimental results show that CHEAPS2AGA performs well in terms of scalability and accuracy.

Acknowledgements. This work was supported by National Natural Science Foundation of China under grant No.61702499. The authors would like to thank the anonymous reviewers for their useful suggestions which helped to improve the quality of the paper.

References

1. Bertsekas, D.P.: Nonlinear Programming, 3rd edn. Athena Scientific, Nashua (2016)
2. Bottou, L., Cun, Y.L.: Large scale online learning. In: Advances in Neural Information Processing Systems, vol. 16, pp. 217–224 (2004)
3. Daneshmand, H., Lucchi, A., Hofmann, T.: Starting small: learning with adaptive sample sizes. In: Proceedings of the 33rd International Conference on International Conference on Machine Learning, ICML 2016, pp. 1463–1471 (2016)
4. Defazio, A., Bach, F., Lacoste-Julien, S.: SAGA: a fast incremental gradient method with support for non-strongly convex composite objectives. In: Proceedings of the 27th International Conference on Neural Information Processing Systems, NIPS 2014, pp. 1646–1654 (2014)
5. Frostig, R., Rong, G., Kakade, S.M., Sidford, A.: Competing with the empirical risk minimizer in a single pass. In: Proceedings of the 28th Annual Conference on Learning Theory, COLT 2015, pp. 728–763 (2015)
6. Hofmann, T., Lucchi, A., Lacoste-Julien, S., McWilliams, B.: Variance reduced stochastic gradient descent with neighbors. In: Proceedings of the 28th International Conference on Neural Information Processing Systems, NIPS 2015, vol. 2, pp. 2305–2313. MIT Press, Cambridge (2015)
7. Reddi, S.J., Hefny, A., Sra, S., Poczos, B., Smola, A.J.: On variance reduction in stochastic gradient descent and its asynchronous variants. In: Advances in Neural Information Processing Systems, vol. 28, pp. 2647–2655 (2015)
8. Johnson, R., Zhang, T.: Accelerating stochastic gradient descent using predictive variance reduction. In: Proceedings of the 26th International Conference on Neural Information Processing Systems, NIPS 2013, pp. 315–323 (2013)
9. Jothimurugesan, E., Tahmasbi, A., Gibbons, P., Tirthapura, S.: Variance-reduced stochastic gradient descent on streaming data. In: Advances in Neural Information Processing Systems, vol. 31, pp. 9906–9915 (2018)
10. Konečný, J., Richtárik, P.: Semi-stochastic gradient descent methods. arXiv:1312.1666 (2013)
11. Leblond, R., Pedregosa, F., Lacoste-Julien, S.: ASAGA: asynchronous parallel SAGA. In: Proceedings of the 20th International Conference on Artificial Intelligence and Statistics, AISTATS 2017, pp. 46–54 (2017)
12. Lee, J., et al.: Wide neural networks of any depth evolve as linear models under gradient descent. In: Advances in Neural Information Processing Systems, vol. 32, pp. 8572–8583. Curran Associates, Inc. (2019)
13. Lewis, D.D., Yang, Y., Rose, T.G., Li, F.: RCV1: a new benchmark collection for text categorization research. J. Mach. Learn. Res. **5**, 361–397 (2004)
14. Ma, J., Saul, L.K., Savage, S., Voelker, G.M.: Identifying suspicious URLs: an application of large-scale online learning. In: Proceedings of the 26th Annual International Conference on Machine Learning, ICML 2009, pp. 681–688 (2009)

15. Peng, Y., Hao, Z., Yun, X.: Lock-free parallelization for variance-reduced stochastic gradient descent on streaming data. IEEE Trans. Parallel Distrib. Syst. **31**(9), 2220–2231 (2020)

16. Recht, B., Re, C., Wright, S., Niu, F.: HOGWILD!: a lock-free approach to parallelizing stochastic gradient descent. In: Advances in Neural Information Processing Systems, vol. 24, pp. 693–701 (2011)

17. Robbins, H., Monro, S.: A stochastic approximation method. Ann. Math. Stat. **22**(3), 400–407 (1951)

18. Roux, N.L., Schmidt, M., Bach, F.: A stochastic gradient method with an exponential convergence rate for finite training sets. In: Proceedings of the 25th International Conference on Neural Information Processing Systems, NIPS 2012, vol. 2, pp. 2663–2671 (2012)

19. Shah, V., Asteris, M., Kyrillidis, A., Sanghavi, S.: Trading-off variance and complexity in stochastic gradient descent. arXiv:1603.06861 (2016)

20. Zhang, H., Hsieh, C., Akella, V.: HOGWILD++: a new mechanism for decentralized asynchronous stochastic gradient descent. In: Proceedings of the 16th International Conference on Data Mining, ICDM 2016, pp. 629–638 (2016)

21. Zhao, P., Chen, P., Wang, S., Lin, X.: Towards query-efficient black-box adversary with zeroth-order natural gradient descent. In: Proceedings of the 34th AAAI Conference on Artificial Intelligence, AAAI 2020, pp. 6909–6916 (2020)

A Quantum Computer Operating System

Reid Honan[✉][iD], Trent W. Lewis[iD], Scott Anderson, and Jake Cooke[iD]

Flinders University, Adelaide, SA 5042, Australia
reid.honan@flinders.edu.au
https://www.flinders.edu.au/

Abstract. The growth of classical computers can be credited in no small part to the development of operating systems designed to provide program portability between different hardware configurations. Quantum computing should learn from classical computing history in order to expedite their development. The approach presented in this paper results in basic parallel execution of quantum programs. The approach utilises the information required for sequential execution including program dependancies, program mappings and qubit connectivities reshaping it to present novel data in the form of a serializability conflict graph. Utilising this graph the quantum system can easily determine parallelisable programs as well as how to parallelise them. This premiere system has been implemented and an attempted comparison has been made to other simulators. Because this system requires no new information, the comparison demonstrates that the cost of employing this strategy is largely similar to employing current sequential techniques, with the added novel benefit of parallel execution.

Keywords: Operating system · Quantum computing · Serializability graph

1 Introduction

Quantum computing stands as the next evolution in computing paradigms, able to augment current computing with a fundamentally new approach. Current work in quantum computing focuses on either creating the physical hardware (from a range of possibilities) [5,6,8,10,12,13,20] or looks at utilising this new paradigm through software applications [26,29]. Work has been done on optimising quantum programs [1,16,25] and mapping their resource allocation [11,14,15,23,28,31] so that it can be executed on the quantum hardware. Using the current state of the art implementations, quantum computers only execute a single program before moving onto the next program. This approach is reminiscent of batch processing in classical computing however it fails to actively utilise the complete power of the quantum hardware.

The current brute force cost to optimise and map a quantum program is so expensive that to map multiple quantum programs together is seemingly intractable. The approach presented in this paper tackles this problem and

© Springer Nature Switzerland AG 2020
M. Qiu (Ed.): ICA3PP 2020, LNCS 12453, pp. 415–431, 2020.
https://doi.org/10.1007/978-3-030-60239-0_28

presents an approach which results in simple parallel execution of quantum programs. The approach utilises the information required for sequential execution including program dependancies, program mappings and qubit connectivities, reshaping it to present novel data.

2 Definitions

2.1 Operating System Definition

Across the development lifetime of Operating Systems the exact implementation and feature support has differed, though they all support a base set of requirements. The requirements expected of all operating systems was outlined by Metzner [17] in Fig. 1.

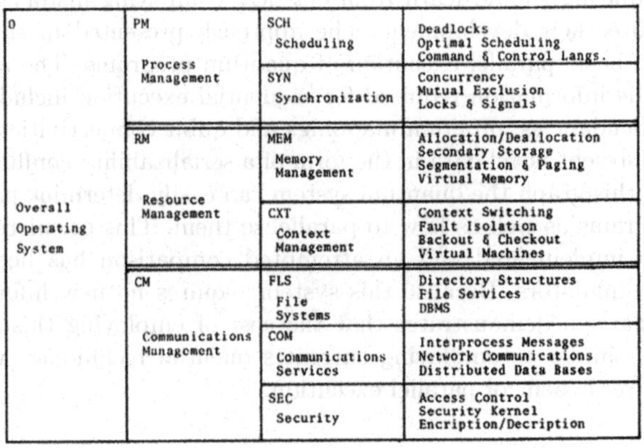

0	PM	SCH	Deadlocks
	Process Management	Scheduling	Optimal Scheduling Command & Control Langs.
		SYN Synchronization	Concurrency Mutual Exclusion Locks & Signals
Overall Operating System	RM	MEM Memory Management	Allocation/Deallocation Secondary Storage Segmentation & Paging Virtual Memory
	Resource Management	CXT Context Management	Context Switching Fault Isolation Backout & Checkout Virtual Machines
	CM	FLS File Systems	Directory Structures File Services DBMS
	Communications Management	COM Communications Services	Interprocess Messages Network Communications Distributed Data Bases
		SEC Security	Access Control Security Kernel Encription/Decription

Fig. 1. Operating system hierarchy [17]

2.2 Quantum Computer Definition

A universal definition which encompasses all implementations of quantum computers is difficult because of the varying hardware implementations. While the physical implementations differ, quantum computers must be able to manipulate quantum bits (qubits) and maintain these bits from corrupting due to decoherence or other sources of error and noise. To this end, the Divencenzo criteria [6] outline a hardware independent definition of a quantum computer as:

1. A scalable physical system with well characterized qubits
2. The ability to initialize the state of the qubits to a simple fiducial state
3. Long, relevant decoherence times
4. A "universal" set of quantum gates
5. A qubit-specific measurement capability.

When implemented, these criteria provide the basic building blocks of a quantum computer capable of supporting a quantum operating system.

3   Limitations of Quantum Computers

Some components of a classical operating system cannot be transferred to a quantum operating system due to current limitations with quantum computing. Qubits are currently limited to small execution times due to decoherence and other sources of error [21,34,35]. This limited execution time results in a lack of long term quantum memory ensuring that no quantum file systems can be implemented [21,34,35]. Due to the low execution time it is currently advised to treat quantum programs as singular blocks, this approach removes the context switching and limits the synchronisation to allocating qubits to programs without currently considering shared qubits.

While quantum networking [34] expands the capability and functionality of a quantum computer in a similar fashion to current computing systems, it is not required for a quantum computer as per the definition found in Sect. 2.2. Quantum networking has therefore been intentionally omitted from the approach presented below.

4   Validity of Concurrency

It is well established that quantum computers are capable of executing a quantum program on their hardware, depending on error and connectivity [21,35]. What is typically overlooked is the ability for a quantum computer to perform parallel computations. The basic structure of a quantum program is a circuit similar to Fig. 2.

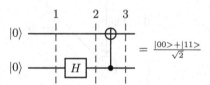

Fig. 2. An example Quantum Circuit, specifically an example of generating a basic entanglement state, commonly known as a Bell State

The circuit is read from left to right, with the input states specified on the left of each horizontal line. The circuit then manipulates those input states through either rotation gates like the *Hadamard (H)* gate above or controlled gates which act similar to boolean if statements. All quantum gates can be defined in terms of a unitary matrix [21] which acts upon the qubit state vector:

$$Hadamard = \frac{1}{\sqrt{2}} \begin{bmatrix} 1 & 1 \\ 1 & -1 \end{bmatrix} \tag{1}$$

$$ControlNot = \begin{bmatrix} 1 & 0 & 0 & 0 \\ 0 & 1 & 0 & 0 \\ 0 & 0 & 0 & 1 \\ 0 & 0 & 1 & 0 \end{bmatrix} \qquad (2)$$

Finally, the result of the circuit is found on the right hand side of the circuit. This circuit (Fig. 2) is a toy example of entanglement but serves the purpose of an example quantum circuit with which to demonstrate concurrency.

There are two main approaches to concurrency on a quantum computer, either the system attempts to execute multiple instances of the same circuit and thus minimise the computation time or the system attempts to execute multiple distinct programs at the same time. Regardless of the approach, the concept hinges on the ability to execute more than one program at once. Quantum hardware is unique in it's design as it only allows certain physical qubits to interact with each other [10], this is typically expressed in a connectivity graph like Fig. 3. The connectivity graph in Fig. 3 comes from an ibmq_5_yorktown - ibmqx2 v2.0.5 [10], and includes the error rates on the connections.

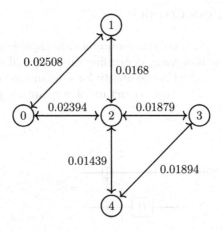

Fig. 3. Connectivity graph for ibmq_5_yorktown - ibmqx2 v2.0.5 [10].

This connectivity graph is only half of the problem, this example also requires the *programs* activity graph. The activity graph specifies the qubits required by the process, and the communication between them that the algorithm requires is denoted as an edge. The activity diagram is generated by parsing the instruction set of the process and generating a node for every qubit specified, and a directed edge for every controlled operation specified. An activity graph of Fig. 2 is included as Fig. 4.

Ignoring the problem of finding the optimal qubit mapping for the purposes of this example, it can be assumed that the system could choose qubits 0 and 1 to host the quantum program. Note that the change between Figs. 5 and 6 still leaves a configuration of 3 connected qubits. These 3 connected qubits can

Fig. 4. Activity graph generated from quantum circuit in Fig. 2

then be used to host an additional 2 qubit quantum program, thus allowing for concurrent hosting of quantum programs (Fig. 7).

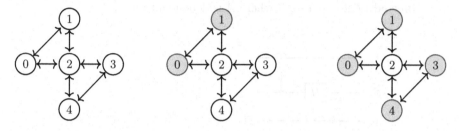

Fig. 5. Initial system   **Fig. 6.** Single map   **Fig. 7.** Double map

The quantum circuit of the two split quantum programs is found in Fig. 8, with the overall result of:

$$\frac{|00000> +|11000> +|00011> +|11011>}{2} \tag{3}$$

Upon measurement of the qubits, the resultant bitstring (for example: $|00011>$) should be segregated into the relevant results for each quantum program (for example: $|00>$ and $|11>$). Provided the qubit measurement is fine enough, in accordance with DiVincenzo criteria #5 [6], it is simple enough to measure the individual qubits for that program alone, instead of measuring the entire system and then subdividing.

5   Scheduling and Memory Management

Determining which programs can and should be parallelised through some form of a scheduling algorithm is key to achieving parallelisation. The proposed solution to this is to construct an intricate and complicated data structure which enables relative simplicity in the algorithms. The data structure is built using 5 seperate graphs which are combined into a single multi-layered graph with links between the layers where relevant. The layers are as follows:

1. Qubit Connectivity Map (Fig. 9)
    – Nodes = Qubits,
    – Edges = Connections between qubits.
2. Program Mappings (Fig. 10)
    – Nodes = Program Mapping,
    – Edges = N/A (This graph by default does not include edges because each mapping is individual, however edges will be added later in this paper).
3. Program Dependency Map (Fig. 11)
    – Nodes = Quantum Programs,
    – Directed Edges = $A \rightarrow B$ where A is dependant on B.

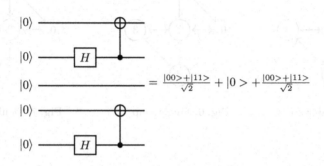

Fig. 8. Split quantum circuit

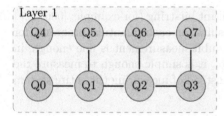

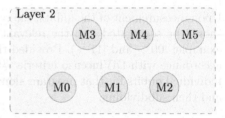

Fig. 9. An example of the qubit connectivity map

Fig. 10. An example of the Mapping graph

These layers form the base design of the data structure, with the links connecting the layers as:

1. Connection between the program mappings (Fig. 10) and the program dependency map (Fig. 11) where the mapping belongs to that program. This should result in a $1..n$ mapping. See Fig. 12
2. Connection between the qubit connectivity map (Fig. 9) and the program mappings (Fig. 10) where the qubits belong to the program mapping. This should result in a $n..n$ mapping. See Fig. 13.

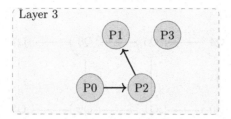

Fig. 11. An example of the process dependency graph

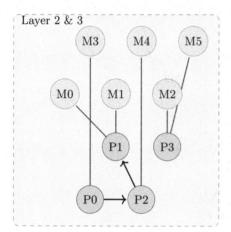

Fig. 12. An example of the programs and their mappings

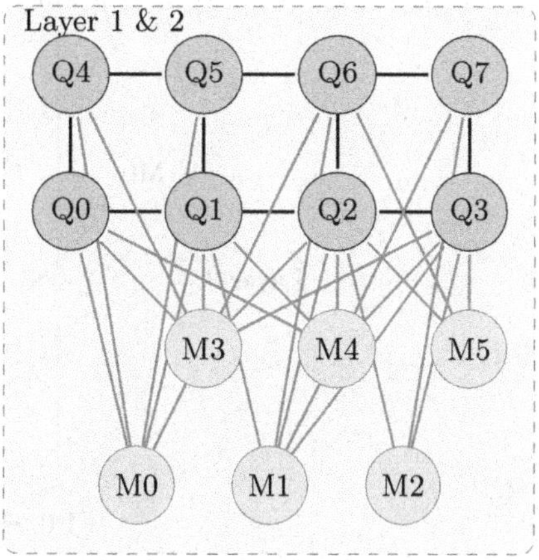

Fig. 13. An example of the mappings connected to their resources.

Overall this data structure combines all the required information for parallelism in a single concise structure. All of the information mentioned above was also required in some capacity for sequential execution. An example of this data structure can be found in Fig. 14, with 5 graphs connected together. While care has been taken to simplify the example, the complexity of the data structure cannot be ignored.

Storing the nodes from Layer 3 in a priority queue allows for simple application of a scheduler algorithm, for simplicity priority based on arrival time also known as First In First Out (FIFO) is recommended. When the scheduler determines the next program to execute, the edges between layers 3 and 2 can be followed to attempt each of the applicable mappings (preferably starting with 0 cost (perfect mapping) and increasing from there). Mapping cost is determined from the amount of swaps and tweaks required to force the program to execute using that mapping. Mappings which perfectly mirror the Activity graphs cost 0 to apply, with each swap increasing the cost by a fixed amount [31]. In the sequential mode, the first mapping will always fit and execute, before the system moves onto the next program.

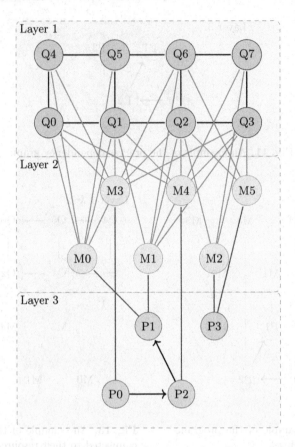

Fig. 14. An example of the multigraph data structure. Program dependency graph represented as the red nodes (P*) and black directed edges, program mappings represented as the green nodes (M*) with no edges, qubit connectivity map with the blue nodes (Q*), programs and the connection to their mappings represented with red edges and the mappings and which qubits they lock represented with orange edges. (Color figure online)

In the parallel mode, the next mapping may not fit (due to other programs currently consuming resources) and may require checking further, less efficient mappings or even halting until other programs free their resources (assigned qubits). Using the multi-layer data structure, unassigned layer 1 nodes (representing qubits) can be investigated as to whether mappings that will use them are able to be executed (all required nodes available). This approach allows the system to actively seek to execute more programs, however it ignores the current position of the scheduler. This can lead to issues of starvation for the next scheduled process because the nodes continue seeking processes which can run now instead of waiting for the required memory. A resolution to this is to have

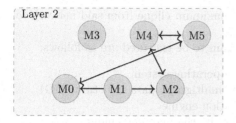

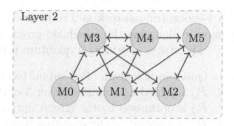

Fig. 15. An example of the conflict graph where edges mean they **can** be executed concurrently

Fig. 16. An example of the conflict graph where edges mean they **cannot** be executed concurrently

the scheduler 'reserve' nodes which will not be allocated to other programs and will wait for the scheduled task.

The parallel option mentioned above requires a large amount of execution and thread management to allow for the free qubits to be searched for applicable programs to execute. A better approach is to employ the use of a serialization conflict graph (Figs. 15 and 16) commonly found in database query processing [3]. This graph is used to determine when programs (or queries in databases) can be run in parallel or if they must be executed sequentially [3] by reviewing what resources are required by the program (query) and whether other programs (queries) require the same resources [3].

In the case of a quantum computer, currently the only resource is the qubits. By treating the programs as single indivisible blocks, the program can be considered as 'locking' the relevant qubits until it has completed its execution. By considering which programs lock which qubits (found within the qubit connectivity and Mapping graph), edges can be introduced into layer 2 which denote which programs do not lock the same qubits and can thus be executed concurrently.

Specifically in the layered data structure, the conflict graph [3] will inform which mappings can be executed together which may include multiple possible mappings of the same program. If the edges dictate two mappings which cannot be executed together (Fig. 16), then the number of edges expands greatly and determining the parallel mappings to execute becomes much more complicated. Alternatively using the edges to indicate two mappings which can be executed together (Fig. 15) results in smaller edge sets, while determining which mappings to execute refines to a maximum clique problem [7] of the 1-egocentric network (a network we pull out by selecting a node and all of its connections) [7] for the scheduled mapping.

6   Overall System Design

The overall execution of the operating system is to continuously execute the next program according to the scheduler. The executing program then generates a 1-egocentric network and retrieves the maximum clique from said network to execute alongside the scheduled program.

The stages of adding a quantum program to be executed are as follows:

1. Quantum Program (P_Q) submitted to operating system
2. P_Q is added as a node in layer 3 of the multigraph data structure (MG)
3. P_Q is optimised by the circuit optimisation engine
4. P_Q is mapped to the quantum hardware by the mapping engine
   - As each mapping is discovered it is added as a node in layer 2 of MG and connected to the program node in layer 3 of MG
   - As each mapping is discovered the layer 2 node in MG is connected to the relevant layer 1 qubits.
5. All mappings are now added to the conflict graph.

The stages of removing a finished quantum program are:

1. Return the results to the program for the user
2. Return to the layer 1 program node from the completed layer 2 mapping node
3. Delete all layer 2 children (mappings)
   - Remove all layer 3 qubit links from the layer 2 mapping nodes
   - Remove the layer 2 mapping from the layer 2 conflict graph.

If there is no programs to be executed it is recommended that the scheduler sleeps for a small period to allow for new programs to be added to the queue. The cost of the above implementation is largely reliant on the performance of the three largest system bottlenecks explored below.

6.1   System Bottlenecks

The largest bottlenecks with this system are all designed to be modular and can be replaced with a better algorithm/library when they are discovered. The system bottlenecks are:

1. Quantum program optimisation engine
2. Quantum program mapping engine - Subgraph isomorphism problem [31]
3. Parallel execution problem - Maximum clique problem [7].

Quantum Program Optimisation Engine. Quantum program optimisation is the process of eliminating redundant operations and attempting to simplify the overall program [16]. For example if a quantum program performs a NOT operation followed by a second NOT operation (see Fig. 17 and Eq. (4)) then the NOT operations cancel each other resulting in the same output as though they were never computed. These redundant operations result in time wasted and increases the amount of error in the system.

$$\begin{bmatrix} 0 & 1 \\ 1 & 0 \end{bmatrix} \times \begin{bmatrix} 0 & 1 \\ 1 & 0 \end{bmatrix} = \begin{bmatrix} 1 & 0 \\ 0 & 1 \end{bmatrix} \tag{4}$$

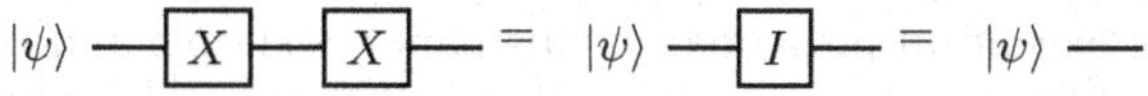

$$|\psi\rangle -\boxed{X}-\boxed{X}- \;=\; |\psi\rangle -\boxed{I}- \;=\; |\psi\rangle -$$

Fig. 17. Redundant quantum circuit

The optimisation process is not compulsory as sub-optimal programs can still be mapped and executed, however optimised programs should result in faster executions and/or a higher accuracy [16]. The balancing act with this process is to ensure that the benefit gained from optimising the program is at least as long as the time taken to optimise.

Quantum Program Mapping Engine. Mapping a quantum program to quantum resources is an ongoing research question [11,14,15,23,28,31]. At the core of the problem is the NP-Hard subgraph isomorphism problem [31]. Locating a mapping that works can usually be accomplished with relative speed, finding the optimal mapping is a different story. A better mapping results in a more accurate result and a faster execution time [31]. Multiple attempts have been made to resolve this problem but the current leading research stems from Siraichi et al. [31]. This process has to be completed for every quantum program that is entered into the system. Any delay taken in the mapping stage affects each quantum program, such that a 10 s delay to map a process results in a 100 s delay to map 10 processes.

Siraichi et al. [31] present a method which does not require any underlying knowledge of the qubit connectivity map and does not require any underlying structure within the qubit connectivity map. Siraichi [31] approaches the problem by mapping the program onto the hardware one qubit at a time, placing the qubit in response to already allocated qubits. This approach has the added benefit of providing all possible mappings as a result, which can then be used in the scheduling and memory management of the system. The algorithm [31] suffers from exponential growth during the search, where allocating qubits can result in growth which approximates $\prod_{k=0}^{q-1}(n-k)$ where n is the number of qubits in the computer and q is the number of qubits being mapped. In practice this limit is typically lower because of the relationship between qubits limiting the applicable matches.

Parallel Execution Problem. The conflict graph approach greatly simplifies the search for programs which can be executed in parallel and the maximal set of programs that can be executed in parallel. Following the conflict graph approach (assuming an edge represents 2 mappings that can be executed in parallel), the task is to determine a clique (a set of nodes where all the nodes are connected through edges) such that the clique is maximal. It is also recommended to find a maximum clique so as to parallelise as many programs as possible.

Because there exists a scheduled node which we must include, a 1 or 1.5 egocentric network [7] may be employed to reduce the search space to only those nodes which are neighbours of the scheduled node.

The parallel execution presented in this paper is simple and basic, akin to a structure built of blocks. A more advanced parallelism can be employed through more specific application of the locks. This approach yields better parallelism, however improper implementation will result in exploitable attack vectors. If the system considers a measurement as the trigger to free a qubit then programs can be properly interwoven, however this requires partial mapping and assumes that the qubit is not reused by the program. By incorrectly employing the above advanced approach, the system is then open to CWE-416: Use After Free [19] and CWE-200: Exposure of Sensitive Information to an Unauthorized Actor [18] attacks.

7   Results and Discussion

In an attempt to demonstrate the capability of this operating system approach, a software simulator (entitled Glade) is in active development which supports the parallel execution of quantum programs. This Glade simulator is being developed in accordance with traditional operating system design [22,30,32,33], treating the programs as objects which contain all their relevant data. Qubits are assigned to the programs and are used for the lifetime of the program, this approach enables the usage of virtual memory addressing by the programs.

The Glade simulator has been tested against three other simulators which only support sequential execution, Qiskit [9], Q# (Qsharp) [2] and Rigetti [27]. These simulators were provided with the same randomly generated quantum circuits of varying size (number of qubits) and depth (number of gates). The simulators were investigated using the Python Memory Profiler [24] to demonstrate the simulators memory usage contrasted with their execution time. The results have been summarised in Table 3, The full graphs are available upon request. The simulators were tested by running 50 quantum programs (composed of up to 10 qubits ($\mu = 4.3$ qubits, $\sigma = 2.147$)) each run 10 times (for a total of 500 programs).

Due to inconsistencies in the quantum circuit language between simulators, some instructions have had to be added or composed of multiple smaller instructions to enable equal testing of all the simulators. All the simulators support a completely connected qubit connectivity graph thus enabling all qubits to be entangled with all others, this does simplify the previously introduced 3 layer graph structure by removing the mappings and simply allocating qubits to programs.

As shown in the feature matrix (Table 2), the simulators differ considerably in their design. This leads to difficulties comparing two simulators together, each built to their own specification and for their own purpose. Qiskit and Q# were developed to research and develop quantum computer programming. Due to this design choice, while the program optimisation and execution are extremely rapid they seem to forget simulation of the hardware. Rigetti conversely was designed to fully simulate a quantum computer and everything that entails. To that end, Rigetti utilises server programs to handle the simulation load.

Table 1. Computer/simulation details   **Table 2.** Quantum simulator feature matrix

Feature	Value
CPU	Intel i7-6770HQ
RAM	16 GB
OS	Ubuntu 19.10
Python	Version 3.7.5
Glade	Version 2.0
Qiskit	Version 0.16.1
Q#	Version 0.10.2002.2610
Rigetti QVM	Version 0.16.0
Rigetti Quilc	Version 0.17.0

Feature	Glade	Qiskit	Q#	Rigetti
Program execution	✓	✓	✓	✓
Program optimisation	✗	✓	✓	✓
Hardware emulation	✓	✗	✗	✓
Program mapping	✓	✗	✗	✓
Multithreaded execution	✓	✓	✓	✓
Decoherence emulation	✓	✗	✗	✗

It should be noted that Qiskit and Q# load multiple circuits and even multiple shots of the same circuits across multiple execution threads. This leads to a large performance benefit over literal sequential execution. Rigetti constructs an entire digital 10 qubit quantum computer simulator for each circuit, leading to the large memory usage and time spent. The Glade simulator requires each of the circuits to be loaded as individual jobs, with a seperate duplicate circuit loaded for each different job of the circuit. In other words, if a circuit was to be executed 10 times 10 circuits need to be loaded. The Glade simulator was then capable of simulating up to 5 executions simultaneously.

The Glade system employs on a largely optimal simulation of qubits and requires a large amount of threads (1 per qubit and a handful of overall system threads, e.g. 256 qubits results in 267 threads) in order to attempt to simulate decoherence. Because of the parallel nature of the Glade system it requires more time to begin processing quantum programs as different structures are initialised and loaded for use, while the other simulators can purely begin computation. The system defaults to providing 256 qubits for the user, however that value can be specified by the user to accommodate smaller or larger values.

Comparing Glade with the other simulators shows that while our execution time and memory usage is similar to the Qiskit and Q# simulators, alternatives like Rigetti vastly exceed those measurements. This is due in part to the extremely lean C++ programming of Glade and the simplified nature of our qubit representation. Due to the numerous issues listed above regarding accurate comparisons, the Table 3 results are included not as completely accurate comparisons, but as a means to highlight the system in comparison to other more mature systems.

The simulations and by extension the memory graphs were produced on an Intel Skull Canyon NUC with details according to Table 1.

Table 3. Quantum simulator results summary

Simulator	Execution time	Maximum memory usage
Glade	7.9 s	323.52 MiB
Rigetti	43778.3 s	4941.87 MiB
Qiskit	2.2 s	310.15 MiB
Q#	1.7 s	319.40 MiB

8   System Scaling

Now that the system has been specified and explored, the final inquiry is how well the system scales. In pursuit of an answer to this inquiry, the scaling of a quantum computer has been seperated into vertical and horizontal scaling. Vertical scaling considers the system still running on a single quantum computer but of a larger size. Leaving horizontal scaling to consider keeping the quantum computer the same size, but expanding over multiple quantum computers to extend the computing power available.

8.1   Vertical Scaling

The designs and algorithms discussed within this paper will continue to produce accurate results for any sized quantum computer. However due to inbuilt and inherent inefficiencies in the discussed algorithms, the cost of deploying them on larger quantum computers will continue to increase in accordance with the details of the quantum computer.

8.2   Horizontal Scaling

An alternative method for increasing the size of the system is to employ multiple distinct quantum computers under the purview of a single instance of the system. Each quantum computer provides a connectivity graph of its qubits and can be instructed individually. One can consider all n quantum computers as a single graph of n connected components. Alternatively you can store it as n graphs each of 1 connected component. Searching a single disconnected graph is faster than searching each individual graph, however the cost of considering all n graphs is greatly increased from considering them each individually.

Without access to a quantum network connection a quantum program can only be split over multiple distinct quantum computers provided that the activity graph is made of multiple connected components with each component on a separate quantum computer.

9   Related Works

This is not the first time that quantum computer operating systems have been mentioned in literature, that honour belongs to Corrigan-Gibbs et al. [4].

Though their work is the first to analyze the potential of quantum computer operating systems, their work is "necessarily (and shamelessly) speculative (p. 1)" [4]. Corrigan-Gibbs et al., spend their paper discussing the end uses of a quantum operating system, skipping past the actual design and implementation of the system. Due to this skipping of detail, their aspirational paper reads better as an Application Programming Interface (API) reference than an outline of a new system. Whilst the work presented here demonstrates the first attempt to fully describe the parallelisation of quantum programs on quantum hardware. It also appears that no further work has been done on this project since the publication of that paper.

10   Conclusion

Quantum computing is an ever developing field of research and exploration. This article has introduced and developed a wide variety of concepts and implementations to be further developed and discussed. The concept of a quantum operating system requires development in numerous areas (specifically the system bottlenecks), it is expected that this field will continue to develop and mature alongside the development of the actual hardware. As explored above, careful utilisation of existing information clearly provides the ability for parallel processing. The implementations provided in this article produce the correct result, though may suffer from extensive overheads and processing times surrounding the system bottlenecks.

The next phase of this research is to further develop the basic parallelism introduced in this paper moving towards a more specialised 'jigsaw' approach compared to the basic 'block' approach presented in this paper. This next phase will better consider different length programs and dynamic qubit allocation.

References

1. Barenco, A., et al.: Elementary gates for quantum computation. Phys. Rev. A **52**, 3457–3467 (1995). https://doi.org/10.1103/PhysRevA.52.3457
2. cjgronlund: Microsoft Quantum Documentation and Q# API Reference - Microsoft Quantum. https://docs.microsoft.com/en-us/quantum/, Library Catalog: docs.microsoft.com
3. Connolly, T.M., Begg, C.E.: Database Systems: A Practical Approach to Design, Implementation, and Management. Always Learning, 6th edn. Pearson, Boston (2015). Global edn., oCLC: 894742602
4. Corrigan-Gibbs, H., Wu, D.J., Boneh, D.: Quantum operating systems. In: Proceedings of the 16th Workshop on Hot Topics in Operating Systems, HotOS 2017, pp. 76–81. ACM Press, Whistler (2017). https://doi.org/10.1145/3102980.3102993, http://dl.acm.org/citation.cfm?doid=3102980.3102993
5. D-Wave: The D-Wave 2000Q Quantum Computer Technology Overview

6. DiVincenzo, D.P.: IBM: the physical implementation of quantum computation. Fortschr. Phys. **48**(9–11), 771–783 (2000). https://doi.org/10.1002/1521-3978(200009)48:9/11⟨771::AID-PROP771⟩3.0.CO;2-E. arXiv:quant-ph/0002077
7. Golbeck, J.: Analyzing the Social Web, First edn. Morgan Kaufmann is an imprint of Elsevier, Waltham (2013)
8. IBM: IBM Q Experience. https://quantumexperience.ng.bluemix.net/qx/experience
9. IBM: Qiskit. https://qiskit.org/
10. IBM: IBM Quantum Experience (2020). https://quantum-computing.ibm.com
11. Javadi-Abhari, A., et al.: Optimized surface code communication in superconducting quantum computers. In: Proceedings of the 50th Annual IEEE/ACM International Symposium on Microarchitecture, MICRO-50 2017, pp. 692–705 (2017). https://doi.org/10.1145/3123939.3123949, arXiv:1708.09283
12. Kalai, G.: How quantum computers fail: quantum codes, correlations in physical systems, and noise accumulation, p. 17 (2011)
13. Lee, C.: Bigger is better: quantum volume expresses computer's limit (2017). https://arstechnica.com/science/2017/05/quantum-volume-one-number-to-benchmark-a-quantum-computer/
14. Lin, C.C., Sur-Kolay, S., Jha, N.K.: PAQCS: physical design-aware fault-tolerant quantum circuit synthesis. IEEE Trans. Very Large Scale Integr. (VLSI) Syst. **23**(7), 1221–1234 (2015). https://doi.org/10.1109/TVLSI.2014.2337302
15. Maslov, D., Falconer, S.M., Mosca, M.: Quantum circuit placement. IEEE Trans. Comput.-Aided Des. Integr. Circ. Syst. **27**(4), 752–763 (2008). https://doi.org/10.1109/TCAD.2008.917562. arXiv:quant-ph/0703256
16. Treinish, M.: "Building a Compiler for Quantum Computers" - Matthew Treinish (LCA 2020) (2020). https://www.youtube.com/watch?v=L2P501Iy6J8
17. Metzner, J.: Structuring operating systems literature for the graduate course. ACM SIGOPS Oper. Syst. Rev. **16**(4), 10–25 (1982). https://doi.org/10.1145/850726.850728. http://portal.acm.org/citation.cfm?doid=850726.850728
18. MITRE: CWE - CWE-200: Exposure of Sensitive Information to an Unauthorized Actor (4.0). https://cwe.mitre.org/data/definitions/200.html
19. MITRE: CWE - CWE-416: Use After Free (4.0). https://cwe.mitre.org/data/definitions/416.html
20. Moore, S.K.: IBM edges closer to quantum supremacy with 50-qubit processor. IEEE Spectrum. https://spectrum.ieee.org/tech-talk/computing/hardware/ibm-edges-closer-to-quantum-supremacy-with-50qubit-processor
21. Nielsen, M.A., Chuang, I.L.: Quantum Computation and Quantum Information. Cambridge University Press, Cambridge, New York (2010). 10th Anniversary edn
22. Nutt, G.J.: Operating Systems, 3rd edn. Pearson Addison-Wesley, Boston (2004). International edn., oCLC: 248877291
23. Pedram, M., Shafaei, A.: Layout optimization for quantum circuits with linear nearest neighbor architectures. IEEE Circ. Syst. Mag. **16**(2), 62–74 (2016). https://doi.org/10.1109/MCAS.2016.2549950. http://ieeexplore.ieee.org/document/7476978/
24. Pedregosa, F.: Memory-profiler: a module for monitoring memory usage of a Python program. https://github.com/pythonprofilers/memory_profiler
25. Pius, E.: Automatic parallelisation of quantum circuits using the measurement based quantum computing model, p. 73 (2010)
26. Proos, J., Zalka, C.: Shor's discrete logarithm quantum algorithm for elliptic curves. arXiv:quant-ph/0301141 (2003)

27. Rigetti: Home. Library Catalog: rigetti.com. https://rigetti.com/
28. Shafaei, A., Saeedi, M., Pedram, M.: Qubit placement to minimize communication overhead in 2D quantum architectures. In: 2014 19th Asia and South Pacific Design Automation Conference (ASP-DAC), pp. 495–500. IEEE, Singapore (2014). https://doi.org/10.1109/ASPDAC.2014.6742940, http://ieeexplore.ieee.org/document/6742940/
29. Shor, P.W.: Polynomial-time algorithms for prime factorization and discrete logarithms on a quantum computer. SIAM J. Comput. **26**(5), 1484–1509 (1997). https://doi.org/10.1137/S0097539795293172. arXiv:quant-ph/9508027
30. Silberschatz, A., Galvin, P.B., Gagne, G.: Operating System Concepts, 8th edn. Wiley, Hoboken (2010). International Student Version edn., oCLC: 465175615
31. Siraichi, M.Y., dos Santos, V.F., Collange, C., Pereira, F.M.Q.: Qubit allocation as a combination of subgraph isomorphism and token swapping. Proc. ACM Program. Lang. **3**(OOPSLA), 1–29 (2019). https://doi.org/10.1145/3360546
32. Stallings, W.: Operating Systems: Internals And Design Principles. Pearson International Edition, 6th edn. Pearson/Prentice Hall, Upper Saddle River (2009). International edn., oCLC: 255116549
33. Tanenbaum, A.S.: Modern Operating Systems, 3rd edn. Pearson/Prentice Hall, Upper Saddle River (2009). International edn., oCLC: 254320777
34. Van Meter, R.: Quantum Networking. ISTE Ltd./John Wiley and Sons Inc., Hoboken (2014)
35. Yanofsky, N.S., Mannucci, M.A.: Quantum Computing for Computer Scientists. Cambridge University Press, Cambridge, New York (2008). oCLC: ocn212859032

Dynamic Knowledge Graph Completion with Jointly Structural and Textual Dependency

Wenhao Xie[1,2(✉)], Shuxin Wang[2], Yanzhi Wei[1,2], Yonglin Zhao[1,2],
and Xianghua Fu[2(✉)]

[1] College of Computer Science and Software Engineering,
Shenzhen University, Shenzhen, China
LeSamourai@163.com

[2] College of Big Data and Internet, Shenzhen Technology University, Shenzhen, China
fuxianghua@sztu.edu.cn

Abstract. Knowledge Graph Completion (KGC) aims to fill the missing facts in Knowledge Graphs (KGs). Due to the most real-world KGs evolve quickly with new entities and relations being added by the minute, the dynamic KGC task is more practical than static KGC task because it can be easily to scale up the KGs by add new entities and relations. Most existing dynamic KGC models are ignore the dependency between multi-source information and topology-structure so that they lose very much semantic information in KGs. In this paper, we proposed a novel dynamic KGC model with jointly structural and textual dependency based on deep recurrent neural network (DKGC-JSTD). This model learns embedding of entity's name and parts of its text-description to connect unseen entities to KGs. In order to establish the relevance between text description information and topology information, DKGC-JSTD uses deep memory network and association matching mechanism to extract relevant semantic feature information between entity and relations from entity text-description. And then using deep recurrent neural network to model the dependency between topology-structure and text-description. Experiments on large data sets, both old and new, show that DKGC-JSTD performs well in the dynamic KGC task.

Keywords: Knowledge Graph Completion · Dependency · Deep recurrent neural network

1 Introduction

Knowledge graphs (KGs), such as FreeBase [1] and DBPedia [2], are usually represented in highly structured form, in the form of (h, r, t), to describe the hundreds of millions of real-world facts, where h represents some head entity and r represents some relationship that connects h to some tail entity t. At present, KGs have been widely used in many fields of artificial intelligence. Despite their usefulness and popularity, KGs are often incomplete. Based on this observation, researchers aim to improve the integrity of KGs by predicting missing parts of triple. This task is often called Knowledge Graph Completion (KGC). According to whether the entities and relations in the triples belong to KGs. KGC

© Springer Nature Switzerland AG 2020
M. Qiu (Ed.): ICA3PP 2020, LNCS 12453, pp. 432–448, 2020.
https://doi.org/10.1007/978-3-030-60239-0_29

task can be divided into two categories: static KGC task and dynamic KGC task. Static KGC task mainly depends on the connectivity of existing KGs and can be able to discover hidden relations between existing entities.

Furthermore, most dynamic KGC models simply fuse the semantic information of each word in the entity description information or extract the semantic information related to the triple using the cosine similarity as additional semantic information of the entity. In order to extract the semantic feature information in entity description under different relational semantic spaces, DKGC-JSTD designed a novel feature extraction model (SAIE-DMMA), SAIE-DMMA can be divided into three steps: Semantic Space Transformation (SST), Semantic Association Information Extraction (SAIE) and Target Semantic Fusion (TSF). The specific details of each step will be explained later.

In the dynamic KGC task, most models are described the relevance between topology information and external text information based on the translation principle, which interprets relations as the translations operating on entities. But this may lose a lot of potential semantic dependency information. Thereby, we use a deep bidirectional GRU model [3] with multi-type cells to build deeper semantic dependency information between topology-structural and external text-description information, and use attention mechanism to strengthen relationship among entity, relations words and potential related information.

To summarize, our contributions lie in the following three points:

- We propose a novel feature extraction model (SAIE-DMMA) that not only simply fuse the semantic information of each word but also uses deep memory network and association pairing mechanism to extract association information between target entity and relationship, then designs a two-channel full convolutional neural network and multi-attentional mechanism based on relations to fuse semantic features in the entity description.
- In order to get more dependencies information between topology-structural and external multi-resource information in the KGs, we firstly use deep bidirectional GRU model with multi-type cells to build deeper semantic dependency between topology-structural and text-description, then strengthen potential relevance between entities and relationships based on attention mechanism.
- Empirically our proposed model (DKGC-JSTD) is effective, which has the best or comparative results compared with the existing models.

2 Related Works

2.1 Static KGC Models

The most fundamental and widely used model is a translation-based Representation Learning (RL) model called TransE [4]. TransE assume that the correct triples (h,r,t) need to satisfy $h + r \approx t$, i.e., relationship can be seen as the translation process from the embedding of head entity to the embedding of tail entity.

Although TransE works well for the one-to-one relationship, it fails to model more complex (e.g., one-to-many) relations. In order to have better performance in complex relations, other models consider that entities should have different vectors under different relations. To preserve the mapping properties of complex relations, TransH [5] interprets a relation as a translating operation on a hyperplane. TransR [6] first projects entities from entity space to corresponding relation space and then builds translations between projected entities. TransD [7] constructs a dynamic mapping matrix for each entity-relation pair by considering the diversity of entities and relations simultaneously. TorusE [8] defines the principle of TransE on Lie group. A torus, which is one of the compact Lie groups, can be chosen for the embedding space to avoid regularization. With the extensive application of deep learning in the field of natural language processing, many models begin to use deep learning technology to KRL. ConvE [9] uses 2D convolution over embeddings and multiple layers of nonlinear features to model KGs. CapsE [10] employs capsule network to model the entries in the triple at the same dimension. RSNs [11] integrate recurrent neural networks with residual learning to efficiently capture the long-term relational dependencies of entities within and between KGs.

2.2 Dynamic KGC Models

Due to the limitations of the static KGC algorithms, many researchers have begun to study dynamic KGC techniques. Jiang [12] present a novel time-aware KGC model (TAE) that is able to predict links in a KG using both the existing facts and the temporal information of the facts. TEKE [13] incorporates the rich textual content information to expand the semantic structure of the knowledge graph. DKRL [14] takes advantages of entity descriptions to learn knowledge representations. SSP [15] models the strong correlations between triples and the textual correlations by performing the embedding process in a semantic improvements against the state-of-the-art baselines. [16] propose a novel model (RDRL) based on description-embodied knowledge representation learning framework, which is able to take advantages of both fact triples and entity description. [17] introduce an embedding-based approach which leverages a weakly aligned multilingual KG for semi-supervised cross-lingual learning using entity descriptions. [18] proposes an accurate text-enhanced knowledge graph representation framework, which can utilize accurate textual information to enhance the knowledge representations of a triple. [19] propose a novel deep architecture to utilize both structural-embeddings and text-embeddings first using a weighted sum and then calculates the L_n distance between the translated head entity and tail entity. ConMask [20] is a novel open-world KGC model that uses relationship-dependent content masking, fully convolutional neural networks, and semantic averaging to extract relationship-dependent embeddings from the textual features of entities and relations in KGs.

In the dynamic KGC task, most models are describe the relevance between topology information and external multi-source information based on TransE. But this may lose a lot of potential semantic dependency information. Therefore, DKGC-JSTD builds the dependency between topology-structural and external text-description information by using deep bidirectional GRU model with multi-type cells, thereby enhancing the correlation between internal structure information and external multi-source information. In addition, we designed a novel feature extraction model (SAIE-DMMA) for DKGC-JSTD.

3   Model

In this section, we introduce the overall architecture and the modelling decisions of the DKGC-JSTD. To illustrate how this model works, we first introduce a practical example as well as the top-ranked target entity inferred by the DKGC-JSTD:

Example Task: Complete triple (Edward Waring, birthplace, ?), where Edward Waring is absent from the KG.

Snippet of Entity Description: "...In 1776, Edward Waring married Mary Oswell. They lived for a while in Shrewsbury but the town was not to Mary's liking and the couple moved to Waring's estate at Plealey in Pontesbury....".

Predicted Target Entity: Shrewsbury.

In this example, if we want to find the birthplace of the famous mathematician Edward Waring in the entity description, then we certainly can't read the entire text description about him from beginning to end. Instead, we can infer his birthplace by looking for the associated information between the target entity and the relationship. Here, Edward Waring lived in Shrewsbury for a while after getting married. So we might infer that Edward Waring was born in Shrewsbury and Shrewsbury would be the (correct) target entity. Here and throughout the present work, we denote the missing entity as the target entity, which can be either the head or the tail of a triple.

We decompose the above reasoning process into three steps: 1) Find semantic information related to the current task, 2) Indirect reasoning based on context and related textual information, 3) Parse the relevant information obtained by reasoning into the appropriate target entity. The DKGC-JSTD model is designed to simulate the above three steps, thus DKGC-JSTD can be divided into three parts:

- Semantic Space Transformation (SST), which remove semantic information irrelevant to the current relations.
- Semantic Association Information Extraction (SAIE), which extracts association semantic information between entities and relations, and integrate contextual information and related word information into association information embedding.
- Target Semantic Fusion (TSF), which chooses a target entity by computing a colligation score between target entity candidates in the KG, the degree of association between the extracted textural embedding and the structural embedding of the local KG, and other textual features.

DKGC-JSTD uses deep bidirectional GRU model with multi-type cells to build deeper semantic dependency information between internal topology-structural information and external text-description information, and uses different types of cell units to control different properties of entities and relations. Finally, it compares the degree of association between the extracted textural embedding and existing entities in the KG to resolve a ranked list of target entities. The overall structure of DKGC-JSTD is illustrated in Fig. 1. Later subsections describe the model in detail.

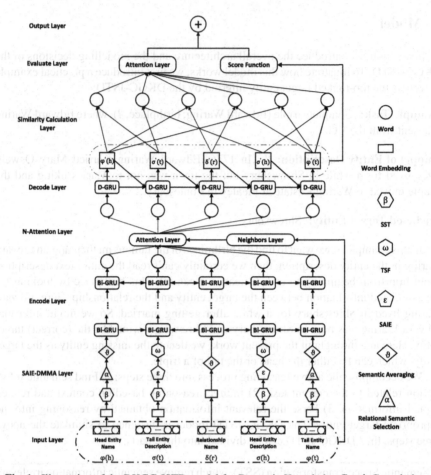

Fig. 1. Illustration of the DKGC-JSTD model for Dynamic Knowledge Graph Completion.

3.1 Semantic Space Transformation (SST)

According to TransD [7], an entity contain different aspects of semantic information and various relations may focus on different aspects of entities. In order to obtain the semantic information of entity under the specific relations, we use the projection operation to mask the irrelevant semantic information of entity. The basic idea of projection operation is illustrated in Fig. 2. For each triple (h, r, t), head/tail entities in the entity space are projected into r-relation space as h' and t' with operation M_{re}. The relation-specific projection can make various relations in the head/tail entities description text that actually hold the relation (denoted as colored circles) close with each other, and also get far away from those that do not hold the relation (denoted as colored triangles). In addition, the dashed line between entity and r-relation represents the compactness between entity and relation, and the more dashed lines there are, the closer related between entity and relation. In dynamic KGC task, we extend this idea to entity description.

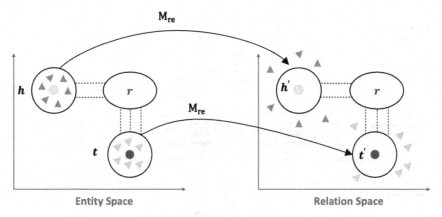

Fig. 2. Simple illustration of SST

Semantic Space Transformation process in DKGC-JSTD, an entity named symbol object is represented by three vectors. The first one captures the meaning of entity, the second one is to capture the semantics between target entity and relation, the third one is used to construct mapping matrices. An relation named symbol object is represented by two vectors, the first one captures the semantic information of the relation, the second one is used to construct mapping matrices. For example, given a triplet (h, r, t), its vectors are h, h_d, h_p, r, r_p, t, t_d, t_p, where subscript p marks the projection vectors, subscript d marks the entity description text vectors, h, h_p, t, $t_p \in \mathbb{R}^n$, h_d, $t_d \in \mathbb{R}^{n \times k}$, r, $r_p \in \mathbb{R}^m$, For each triplet (h, r, t), we set two mapping matrices M_{rhs}, $M_{rts} \in \mathbb{R}^{m \times n}$ to project entities and entities description from entity space to relation space. They are defined as follows:

$$M_{rhs} = r_p h_p^T + \sigma^2 \left(h r^T \right) \cdot I^{m \times n} \tag{1}$$

$$M_{rts} = r_p t_p^T + \sigma^2 \left(t r^T \right) \cdot I^{m \times n} \tag{2}$$

In addition, we use covariance matrices to describe the association information between entities and relations. As we initialize each mapping matrix with an covariance matrix, we add the $\sigma^2 \left(h r^T \right) \cdot I^{m \times n}$ to M_{rhs}, and add the $\sigma^2 \left(t r^T \right) \cdot I^{m \times n}$ to M_{rts}, identify matrix is used for dimension adjustment.

3.2 Semantic Association Information Extraction (SAIE)

In DKGC-JSTD, we use similar ideas to select the most relevant words for a given relation, and then extract association information between target entity and relation. through deep memory network and association pairing mechanism. The overall structure of SAIE is illustrated in Fig. 3. We formally define association information as:

$$\varepsilon(\varphi(e), \tau(r), \tau(t)) = M_{\varphi(e)} \circ \text{SAIE}\left(M_{\varphi(e)}, M_{\tau(r)}, M_{\tau(t)} \right) \tag{3}$$

Where e is an entity, r is a relationship, t is the target entity, φ and τ are the description and the name mapping functions respectively that return a word vector representing the

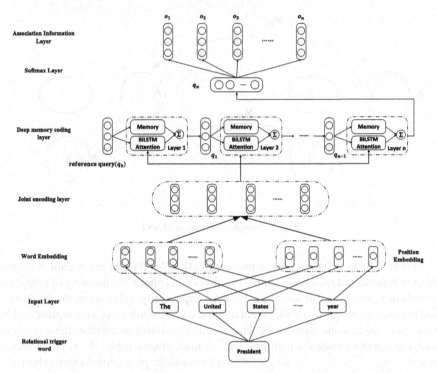

Fig. 3. Illustration of the Semantic Association Information Extraction (SAIE)

description or the name of an entity or relationship. $M_{\varphi(e)} \in \mathbb{R}^{|\varphi(e)| \times k}$ is the description matrix of e in which each row represents a k dimensional embedding for a word in $\varphi(e)$ in order. $M_{\tau(r)} \in \mathbb{R}^k$ represents the semantic synthesis matrix of all words in the title of relationship $\tau(r)$. $\circ$ is row-wise product, and SAIE calculate association semantic weight for each row, i.e., the embedding of each word, in $M_{\varphi(e)}$.

Before extracting the associated semantic information, we first use the simple cosine similarity function to find the word in the description text that is closest to the relationship [20] and denote as relation trigger word. Then we use the target entity and relation as a reference pair q_0, and use the relation trigger word and other words in the entity description as query pairs d_n. Finally, we leverage deep memory network to perform association pair extraction task. The formula of deep memory coding layer in SAIE is as follows:

$$g_i = ReLU\left(W_{att}\left[BiLSTM\left(D\right) \oplus q\right] + b_{att}\right) \tag{4}$$

$$\hat{a} = \sum_{i=1}^{n} softmax(g_i) \times d_i \tag{5}$$

$$\hat{q} = \tanh(W_m q + b_m) \tag{6}$$

$$q = \hat{q} + \hat{a} \tag{7}$$

Where *BiLSTM (D)* are two standard LSTM cells in opposite directions with input D, $D = \{d_1, \ldots d_t, d_{t+1}, d_n\}$ is the input sequence of the model and $d_n = [r_{max}; p_n; l_n]$, p_n is the word vector of each word in the entity description, l_n is the position marker of each word relative to the relation trigger word, q is the intermediate product of each layer encoding and its initial value is $q_0 = \{r, t, 0\}$. W_{att}, W_m are weight matrix, b_{att}, b_m are bias coefficient. Finally, SAIE uses the softmax layer to score:

$$o_i = softmax(p_i \times d_i + \hat{q}) \tag{8}$$

$$SAIE(M_{\varphi(e)}, M_{\tau(r)}, M_{\tau(t)}) = [o_1, o_2, o_3, \ldots \ldots o_n] \tag{9}$$

3.3 Target Semantic Fusion (TSF)

In this work, we decide to use two-channel fully convolutional neural network and multi-head attention mechanism based on relation as the target semantic fusion structure. In addition, due to the semantic information of relation is diverse and uncertain, we use multi-head attention mechanism based on relation to dynamically learn the uncertain representation of semantic information of relations. Figure 4 shows the overall architecture of the target semantic fusion process. The formula of multi-attentional mechanism based on relation is as follows:

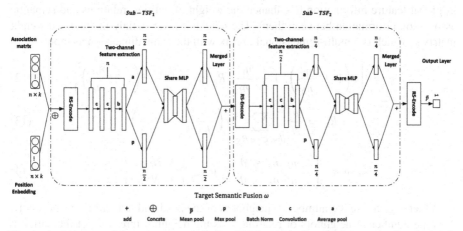

Fig. 4. Architecture of the target sematic fusion (TSF). n is the length of the entity description, k is the dimension of the word vector.

$$RHT = \frac{1}{2}\left[r_p(t - h)^T + r(t_p - h_p)^T\right] \tag{10}$$

$$Z = RHT * Multi\_head(A_1, A_2, \ldots \ldots A_h) \tag{11}$$

Where *RHT* is a selection factor for relational semantic, *Multi_head* is multi-head attention function [21], A_h is the calculation result in the h th semantic space. h, r, t are head entity, relation, and tail entity respectively, h_p, r_p, t_p correspond to the mapping matrix of head entity, relation and tail entity respectively. The TSF module we designed can be regarded as a target sematic fusion function ω, which generates a k dimensional embedding using the output of association matrix $\varepsilon(\varphi(e), \tau(r), \tau(t))$, where e is either a head or tail entity from triples, and t is the target entity corresponding to e.

Note that the TSF used here is different from the one that typically used in encoding and decoding tasks [22]. Rather than reconstructing the input, as is typical in machine translation. The goal of target semantic fusion is to extract the embedding w.r.t given relation, therefore we do not have the decoding operations. In addition, we extract the semantic association information of different levels of the entity description by half after each Sub-TSF layer. This is because the semantic features information of adjacent words in the entity description may have semantic relevance, so they will form new semantic representation information after feature fusion. Therefore, we use TSF to perform stepwise feature fusion on the semantic association information.

3.4 Neighbors Layer

Although we use bidirectional GRU [3] networks to learn the internal semantic relevance of each triplet, this only considers the local feature information of entities and relations, and it does not capture global feature information of entities and relations in KGs. Therefore, we consider using the neighbor subgraph structure of entities and relations as global feature information to enhance the weight of entities, relations and semantic association information between them. The neighbor layer mainly maintains a weight matrix φ, which is a multi-dimensional matrix with the same dimension as input:

$$\varphi = \left[\delta\left(1 + \frac{n_{hr}}{n_h}\right), 1 + \frac{n_{td}}{n_t}, R_g, 1 + \frac{n_{hd}}{n_h}, \delta\left(1 + \frac{n_{tr}}{n_t}\right)\right] \quad (12)$$

$$R_g = \frac{\mu}{N} \sum_{(h_n, t_n) \in R_N} \sigma\left(\frac{1}{|h_n + r - t_n|}\right) \quad (13)$$

$$\delta = \begin{cases} 1, & n_h/n_t \le \vartheta \\ \delta, & n_h/n_t > \vartheta \end{cases}, \quad \mu = \begin{cases} 1, & N \le \tau \\ \mu, & N > \tau \end{cases} \quad (14)$$

Where n_h, n_t are the number of neighbors of the head and tail entities respectively, N is the number of neighbors of relation, ϑ is the neighbor threshold of the entity, τ is neighbor threshold of the relation, δ is neighbor weight of the entity, μ is neighbor weight of the relation, $R_N = \{(h_{n1}, t_{n1}), (h_{n2}, t_{n2}), \ldots, (h_{ni}, t_{ni})\}$ is a binary set of relational neighbors, n_{hr}, n_{tr} are the number of neighbors that have the same relation path as the relation r in head/tail entity neighbor subgraph structure respectively. n_{hd}, n_{td} are the number of entity neighbors appearing in the head/tail description text information respectively, σ is sigmoid function.

3.5 Semantic Averaging/Relational Semantic Selection

Although we can use target semantic fusion to generate all the associated embedding between entities and relations used in DKGC-JSTD, such a process would result in a large number of parameters. Therefore, in order to avoid complex calculations caused by a large number of parameters, we use simple semantic averaging function $\vartheta(M) = \frac{1}{k_e}\sum_i^{k_e} M_{[i,:]}$ instead of target semantic fusion function to form entities or relations embedding, where $M \in \mathbb{R}^{k_e \times k}$ is the input embedding matrix from the entity $\varphi(\cdot)$ or relation name $\delta(\cdot)$. In addition, we use simple relational semantic selection function $R(E) = RHT * E$ to deal with the problem of semantic selection of entities in relational space, where $E \in \mathbb{R}^{k_e \times k}$ is the input entity $\varphi(\cdot)$.

4 Experiments

In this section, we evaluate the DKGC-JSTD model with dynamic and static entity prediction tasks, and compare the performance against several existing methods.

4.1 Settings

Training parameters were set empirically but without fine-tuning. We set the word embedding size $k = 200$, maximum entity description content $n = 512$, word position embedding size $k_{pe} = k$. The word embeddings are from the publicly available pre-trained 200-dimensional GloVe embeddings [23]. The number of Sub-TSF layers $k_{fcms} = 2$ where the Sub-TSF layer has a RS-Encode with $h = 8$ parallel attention modules. The parameters of the neighbor layer are $\vartheta = 5$, $\tau = 8$, $\delta = \mu = 1.5$. Max-pooling and mean-pooling in each Sub-TSF layer has a pool size and stride size of 2.

4.2 Data Sets

We will apply four data sets to the task of dynamic KGC and static KGC. The statistics of these KGs are summarized into Table 1.

Table 1. Data set statistics.

Data set	Entities	Rel.	Triples		
			Train	Valid	Test
FB15k	14951	1345	483142	50000	59071
FB20k	19923	1345	472860	48991	90149
DBPedia50k	49900	654	32388	399	10969
DBPedia500k	517475	654	3102677	10000	1155937

4.3 Dynamic Entity Link Prediction

For the dynamic KGC task, we generated a test set KG_{test} from the 10% of entities that were held out of KG_{train}. This test set has relations that connect the test entity to the entity in KG_{train}. For example, given an entity-relation partial triple (that was seen in the test set, but not seen during training), our goal is to predict the correct target entity in the KG_{train}.

In order to reduce the excessive cost of calculating scores for all entities in the KG. We applied the method of target entity filtering [20] to all KGC models. As a naive baseline we include the target filtering baseline method in Table 2, which assigns random scores to all the entities that pass the target filtering. DKGC-JSTD (no-extra) is a simple model that utilizes Semantic Averaging to replace SAIE-DMMA. DKGC-JSTD (no-dep) is a separation model that removes deep Bi-GRU coding modules. DKGC-JSTD (no-neib) is a separation model that removes the neighbor layer. In order to show our model (DKGC-JSTD) can perform entity link prediction. We compared our model with Jointly (dep) model, Jointly (A-LSTM) model, DKRL (2-layer CNN) model, SSP model, RDRL model, KDCoE-cross model and ConMask model proposed in Sect. 1 and Sect. 2. We use mean rank (MR), mean reciprocal rank (MRR) and HITS@k as evaluation metrics. Mean rank measures the average rank of correct entities/relations. HITS@k measures if correct entities/relations appear within the top-k elements. Mean reciprocal rank measures the sum of reciprocal of the rank of correct entities/relations.

The experimental results show that when we perform the entity link prediction for the larger data set (DBPedia500k data) and the smaller data set (DBPedia50k). The mean rank, mean reciprocal rank, and HITS@10 indicators of the DKGC-JSTD model are higher than Jointly (desp), Jointly (A-LSTM), DKRL (2-layer CNN), SSP, RDRL, KDCoE-cross and ConMask which considers entity description text. For example, on the experiment of DBPedia50k, where we assume that head entities are predicted through relations and tail entities, the HITS@10 of DKGC-JSTD model is higher about 5% to 36% than other eight models.

In addition, we find that SAIE-DMMA layer in DKGC-JSTD do boost MRR performance by at least 23% on both data sets compared to the extraction-free DKGC-JSTD (no-extra). Interestingly, the performance boost on the larger data set (DBPedia500k) is more significant than the smaller (DBPedia50k), which indicates that the SAIE-DMMA process can find more useful semantic information between entities and relations. On the other hand, we discovered that the HITS@10 of DKGC-JSTD model is higher about 7% to 13% than DKGC-JSTD (no-dep) model without considering dependence in both data sets, which indicates that uses deep bidirectional GRU model with multi-type cells can capture the dependency between external multi-source information and the local KG, thus enriching the association semantic information between them. In addition, we found that DKGC-JSTD (no-neib) dropped by about 6% on HIT@10, which can fully demonstrate that the subgraph structure information of entities and relations can further enhance the relevance between the structured information of existing knowledge graphs and the external text information.

Table 2. Dynamic Entity link prediction results on DBPedia50k and DBPedia500k. For Mean Rank (MR) lower is better. For HITS@10 and Mean Reciprocal Rank (MRR) higher is better.

Model	DBPedia50k Head			DBPedia50k Tail			DBPedia500k Head			DBPedia500k Tail		
	MR	HITS@10	MRR	MR	HITS@10	MRR	MR	HITS@10	MRR	MR	HITS@10	MRR
Target Filter Baseline	608	0.07	0.07	104	0.23	0.11	20654	0.01	0.01	3472	0.02	0.01
Jointly(desp)	545	0.08	0.09	106	0.22	0.08	8346	0.02	0.08	2036	0.12	0.07
Jointly(A-LSTM)	496	0.13	0.12	95	0.28	0.14	7956	0.04	0.09	1847	0.12	0.1
DKRL(2-layer CNN)	515	0.09	0.09	88	0.39	0.21	19473	0.01	0.01	2345	0.03	0.04
SSP	645	0.05	0.07	115	0.15	0.04	9739	0.02	0.06	2796	0.05	0.06
RDRL	379	0.13	0.14	94	0.32	0.14	7463	0.03	0.12	2453	0.03	0.03
KDCoE-cross	324	0.15	0.14	78	0.38	0.27	5214	0.08	0.19	875	0.23	0.18
AATE_E	134	0.22	0.19	56	0.45	0.34	4922	0.21	0.17	325	0.44	0.48
ConMask	103	0.37	0.39	38	0.75	0.62	2986	0.18	0.34	183	0.48	0.53
DKGC-JSTD(no-extra)	355	0.19	0.17	76	0.64	0.51	4266	0.13	0.16	682	0.36	0.42
DKGC-JSTD(no-dep)	215	0.31	0.34	65	0.65	0.58	3592	0.15	0.28	451	0.39	0.44
DKGC-JSTD(no-neib)	163	0.35	0.36	52	0.69	0.61	3255	0.18	0.31	322	0.45	0.49
DKGC-JSTD	**83**	**0.42**	**0.44**	**29**	**0.77**	**0.65**	**1864**	**0.22**	**0.37**	**155**	**0.52**	**0.56**

4.4 Static Entity Link Prediction

Because the dynamic entity link prediction task is less restrictive than the static entity link prediction task, it is possible for DKGC-JSTD to perform static entity link prediction tasks, even though it was not designed to do so. For static entity link prediction, we aim to predict a missing h(or$t$) for a given triple h, r, t by ranking all of the entities in the KG. Specifically, we randomly selected 90% of the triples in the KG and construct a training set using the selected triples. We made sure that these triples contained all entities and all relations. All other triples not included in training set as the test set. So in Table 3, we also compare the DKGC-JSTD model with other static KGC methods on the standard FB15k data set, FB20k data set and DBPedia500k data set. These methods are proposed in Sect. 1 and Sect. 2.

We find that DKGC-JSTD model sometimes outperforms static KGC methods on the static entity link prediction task. Especially on the FB20k data set. For example, the mean rank of DKGC-JSTD model is higher about 34% to 75% than other eight models. The HITS@10 of DKGC-JSTD model is higher about 2% to 3% than other eight models. This may be because the training subgraphs composed of randomly selected triple samples in the FB20k data set are sparse, resulting in many static KGC methods cannot fully learning the structural information in the KG.

4.5 Discussion

In this section, we elaborate on actual prediction results of some models and show some examples to emphasize the advantages and limitations of the DKGC-JSTD model.

Table 4 shows the performance of four KGC examples in four KGC models. In each case, four KGC models was provided the head and the relation and asked to predict the tail entity. Compared with the other three KGC models, DKGC-JSTD successfully ranks the correct entities within the top-3 results in most cases. But the third example does not perform well on DKGC-JSTD. Although Bakery_Music and Mediacorp_Suria have a business partnership, DBPedia indicates that it's parentCompany is actually Sony_Music_Entertainment, which ranked 5th. The reason for this error is because the relation trigger word for Bakery_Music was "recordLabel", which was not highly correlated to the relation name "parentCompany" from the model's perspective.

In addition, we can find an interesting phenomenon from the partial triple **<Vijay_(1989_film), writer, ?>**. The DKGC-JSTD model ranked the correct writer **B._Gopal** as the 1nd candidate, but the name "**B._Gopal**" does not actually appear in the film's description. This may be because the DKGC-JSTD model can accurately locate the relation words in the entity description, and infer the tail entity (**B._Gopal**) by capturing the semantic association information between the target entity and the relation in the training set.

Table 3. Static KGC on head and tail prediction. For HITS@10 higher is better. For Mean Rank (MR) lower is better.

| Model | FB15k | | | | FB20k | | | | DBPedia500k | | | |
| | Head | | Tail | | Head | | Tail | | Head | | Tail | |
	MR	HITS@10	MR	HITS@10	MR	HITS@10	MR	HITS@10	MR	HITS@10	MR	HITS@10
TransE	216	0.65	104	0.73	426	0.58	216	0.69	10032	0.14	2547	0.43
TransD	199	0.70	101	0.73	403	0.54	208	0.65	9874	0.16	2645	0.41
TransC	231	0.63	105	0.72	384	0.59	223	0.64	9912	0.15	2156	0.45
TorusE	188	0.72	115	0.70	412	0.61	192	0.73	12549	0.12	2914	0.39
ConvE	211	0.74	98	**0.77**	298	0.63	179	0.64	11779	0.18	2269	**0.45**
RotatE	171	0.76	108	0.69	265	0.63	188	0.72	9744	0.21	1952	0.42
CapsE	184	0.69	92	0.73	235	0.62	195	0.68	8452	**0.26**	2044	0.38
CrossE	**143**	**0.77**	80	0.75	203	0.65	143	0.71	1897	0.24	1789	0.43
DKGC-JSTD	153	0.75	**72**	0.72	**105**	**0.67**	**93**	**0.74**	**1579**	0.23	**1422**	0.31

Table 4. Entity prediction results on DBPedia50k data set. Top-3 predicted tails are shown with the correct answer in bold

DKGC-JSTD

Head	Relationship	Predicted Tails
Chakma_language	languageFamily	**Indo-Aryan language**, Hajong language, Language
Andrej_Pernecký	team	Nine, **ŠKF_Sereď**, A Clear and Present Danger
Bakery_Music	parentCompany	Mediacorp_Suria, Mediacorp, Toei_Company
Vijay_(1989_film)	writer	**B._Gopal**, David_Markham, Jeff Martin

KDCoE-cross

Chakma_language	languageFamily	English_language, **Indo-Aryan language**, Basque_language
Andrej_Pernecký	team	Ravi_Walia, Nikolai_Luzin, Andris_Bērziņš_(Latvian_President)
Bakery_Music	parentCompany	Machete_Music, Marvel_Entertainment, **Sony_Music_Entertainment**
Vijay_(1989_film)	writer	Prince_Rose, Malappuram, **B._Gopal**

AATE_E

Chakma_language	languageFamily	Tuareg_languages, Language, Filipino_language
Andrej_Pernecký	team	Alexandre_Kojève, Zetima, **ŠKF_Sereď**
Bakery_Music	parentCompany	Razor_&_Tie, Apple's_Way, **Sony_Music_Entertainment**
Vijay_(1989_film)	writer	Juventus_F.C, **B._Gopal**, Bananarama

ConMask

Chakma_language	languageFamily	Sami_languages, **Indo-Aryan language**, RegionalLanguage
Andrej_Pernecký	team	Luigi_Musso, Jacques_Lacan, **ŠKF_Sereď**
Bakery_Music	parentCompany	**Sony_Music_Entertainment**, Lorimar_Television, Ariola_Records
Vijay_(1989_film)	writer	Gardner_R._Withrow, Nottz, Daddy_Yankee

5   Conclusion and Future Work

In the present work we introduced a new dynamic KGC model DKGC-JSTD that uses SAIE-DMMA, relational semantic selection and semantic averaging to extract association embeddings from entity description, and then use deep bidirectional GRU model with multi-type cells and neighbor's layer to build deeper semantic dependency between internal topology-structural and external text information. We finally show that DKGC-JSTD can outperform existing methods on both dynamic entity link prediction tasks and static entity link prediction tasks.

DKGC-JSTD is only a KGC model for predicting entities. The goal for future work is to complete the existing knowledge graph through prediction relation.

Acknowledgements. This research is supported by the Scientific Research Platforms and Projects in Universities in Guangdong Province under Grants 2019KTSCX204.

References

1. Bollacker, K., et al.: Freebase: a collaboratively created graph database for structuring human knowledge. In: Proceedings of the 2008 ACM SIGMOD, International Conference on Management of Data, pp. 1247–1250 (2008). ACM
2. Auer, S., et al.: DBpedia: a nucleus for a web of open data. In: Aberer, K., et al. (eds.) The Semantic Web. Lecture Notes in Computer Science, vol. 4825, pp. 722–735. Springer, Berlin, Heidelberg (2007). https://doi.org/10.1007/978-3-540-76298-0_52
3. Schuster, M., Paliwal, K.K.: Bidirectional recurrent neural networks. IEEE Trans. Sig. Process. **45**(11), 2673–2681 (1997)
4. Bordes, A., et al.: Translating embeddings for modeling multi-relational data. In: Burges, C.J.C., et al.: Advances in Neural Information Processing Systems, pp. 2787–2795. Curran Associates, Inc. (2013)
5. Wang, Z., et al.: Knowledge graph embedding by translating on hyperplanes. In: Twenty-Eighth AAAI Conference on Artificial Intelligence. AAAI Press (2014)
6. Lin, Y., et al.: Learning entity and relation embeddings for knowledge graph completion. In: Twenty-Ninth AAAI Conference on Artificial Intelligence. AAAI Press (2015)
7. Ji, G., et al.: Knowledge graph embedding via dynamic mapping matrix. In: Meeting of the Association for Computational Linguistics & International Joint Conference on Natural Language Processing. AAAI Press (2015)
8. Ebisu, T., Ichise, R.: Toruse: knowledge graph embedding on a lie group. In: Thirty-Second AAAI Conference on Artificial Intelligence. AAAI Press (2018)
9. Dettmers, T., et al.: Convolutional 2D knowledge graph embeddings. In: Thirty-Second AAAI Conference on Artificial Intelligence. AAAI Press (2018)
10. Nguyen, D.Q., et al.: A capsule network-based embedding model for knowledge graph completion and search personalization. arXiv preprint arXiv:1808.04122 (2018)
11. Guo, L., Sun, Z., Hu, W.: Learning to exploit long-term relational dependencies in knowledge graphs. arXiv preprint arXiv:1905.04914 (2019)
12. Jiang, T., et al.: Towards time-aware knowledge graph completion. In: Proceedings of COLING 2016, the 26th International Conference on Computational Linguistics: Technical Papers, pp. 1715–1724 (2016)
13. Wang, Z., Li, J.Z.: Text-enhanced representation learning for knowledge graph. In: IJCAI, pp. 1293–1299 (2016)
14. Xie, R., et al.: Representation learning of knowledge graphs with entity descriptions. In: Thirtieth AAAI Conference on Artificial Intelligence, pp. 2659–2665 (2016)
15. Xiao, H., et al.: SSP: semantic space projection for knowledge graph embedding with text descriptions. In: Thirty-First AAAI Conference on Artificial Intelligence. AAAI Press (2017)
16. Dai, S., et al.: Learning entity and relation embeddings with entity description for knowledge graph completion. In: 2018 2nd International Conference on Artificial Intelligence: Technologies and Applications (ICAITA 2018). Atlantis Press (2018)
17. Chen, M., et al.: Co-training embeddings of knowledge graphs and entity descriptions for cross-lingual entity alignment (2018)

18. An, B., et al.: Accurate text-enhanced knowledge graph representation learning. In: Proceedings of the 2018 Conference of the North American Chapter of the Association for Computational (2018)
19. Xu, J., et al. Knowledge graph representation with jointly structural and textual encoding, 1318–1324. arXiv preprint arXiv:1611.08661 (2017)
20. Shi, B., Weninger, T.: Open-world knowledge graph completion. In: Thirty-Second AAAI Conference on Artificial Intelligence (2018)
21. Vaswani, A., et al.: Attention is all you need. In: Advances in Neural Information Processing Systems, pp. 5998–6008 (2017)
22. Zhou, B., et al.: Learning deep features for discriminative localization. In: Proceedings of the IEEE Conference on Computer Vision and Pattern Recognition, pp. 2921–2929 (2016)
23. Pennington, J., Socher, R., Manning, C.: Glove: global vectors for word representation. In: Proceedings of the 2014 Conference on Empirical Methods in Natural Language Processing (EMNLP), pp. 1532–1543 (2014)

MEFE: A Multi-fEature Knowledge Fusion and Evaluation Method Based on BERT

Yimu Ji[1,2,3,4,5], Lin Hu[1,3], Shangdong Liu[1,2,3,4,5]([✉]), Zhengyang Xu[1,3], Yanlan Liu[1,3], Kaihang Liu[1,3], Shuning Tang[1,3], Qiang Liu[1,3]([✉]), and Wan Xiao[3,4,6]([✉])

[1] School of Computer Science, Nanjing University of Posts and Telecommunications, Nanjing 210023, China
{lsd,liuqiang}@njupt.edu.cn
[2] Jiangsu High Technology Research Key Laboratory for Wireless Sensor Networks, Nanjing 210003, Jiangsu, China
[3] Institue of High Performance Computing and Bigdata, Nanjing University of Posts and Telecommunications, Nanjing 210003, Jiangsu, China
xiaow@njupt.edu.cn
[4] Nanjing Center of HPC China, Nanjing 210003, Jiangsu, China
[5] Jiangsu HPC and Intelligent Processing Engineer Research Center, Nanjing 210003, Jiangsu, China
[6] College of Educational Science and Technology, Nanjing University of Posts and Telecommunications, Nanjing 210023, China

Abstract. Knowledge fusion is an important part of constructing a knowledge graph. In recent years, with the development of major knowledge bases, the integration of multi-source knowledge bases is the focus and difficulty in the field of knowledge fusion. Due to the large differences in knowledge base structure, the efficiency and accuracy of fusion are not high. In response to this problem, this paper proposes MEFE (Multi-fEature Knowledge Fusion and Evaluation Method) based on BERT. MEFE comprehensively considers the attributes, descriptions and category characteristics of entities to perform knowledge fusion on multi-source knowledge bases. Firstly, MEFE uses entity category tags to build a category dictionary. Then, it vectorizes the category tags based on the dictionary and clusters the entities according to the category tags. Finally it uses BERT (Bidirectional Encoder Representation from Transformers) to calculate the entity similarity for the entity pairs in the same group. We calculate entity redundancy rate and information loss rate of knowledge base according to the fusion result, so as to evaluate the quality of the knowledge base. Experiments show that MEFE effectively improves the efficiency of knowledge fusion through clustering, and the use of BERT promotes the accuracy of fusion.

Keywords: Multi-source knowledge base · Knowledge fusion · Vectorization of category labels · BERT · Quality evaluation

1 Introduction

With the continuous development of the Internet and big data, a number of data are generated in people's daily lives, which contains rich and valuable resources. According

© Springer Nature Switzerland AG 2020
M. Qiu (Ed.): ICA3PP 2020, LNCS 12453, pp. 449–462, 2020.
https://doi.org/10.1007/978-3-030-60239-0_30

to the latest version of the white paper "Data Age 2025" released by IDC (Internet Data Center), the total amount of global data will reach 175ZB by 2025. It can be seen that the amount of data is explosively increasing in the information age we are in. We perform knowledge extraction on these large amounts of network data, and build a large knowledge base by taking entities as the center and entity relationships as the connections. These knowledge bases provide a very beneficial help for us to understand things.

In recent years, domestic and foreign academia and industry have established many knowledge bases and implemented many applications based on them. Typical foreign knowledge bases include DBpedia [1], WordNet, YAGO [2], Probase [3], NELL [4], etc. and domestic knowledge bases include Sogou Zhifang, zhishi.me [5] and Baidu Zhixin. The structure of these knowledge bases is different, and the knowledge in the knowledge base is complementary and repetitive, so making decisions based on a single knowledge base is not comprehensive enough. Knowledge fusion is to compare entities and match relationships in multi-source knowledge bases, verify the correctness of knowledge and organically combine knowledge, thereby constructing a more comprehensive large-scale knowledge base. However, due to the large number of entities and the large differences in the structure of each knowledge base, the efficiency and accuracy of knowledge fusion are not ideal. As to the efficiency of fusion calculation, this paper distributes entities into multiple groups of clusters through the vectorization of entity class labels. The entities in each cluster are fused independently, which will greatly improve the efficiency of fusion. Regarding to the accuracy of fusion calculation, we perform similarity calculation on the multiple features of the entity and use the BERT to calculate semantic similarity, which will greatly improve the accuracy of calculation. In summary, the contribution of this article lies in:

- This paper proposes a clustering method based on class label vectorization to improve the efficiency of multi-source knowledge base fusion.
- This paper proposes a method to calculate the multi-feature similarity of knowledge base entities based on BERT to improve the accuracy of multi-source knowledge base fusion.
- The algorithm proposed in this paper has been verified by experiments, and the results show that the algorithm has high efficiency and high accuracy.

2   Related Works

In recent years, scholars at home and abroad have carried out a lot of researches in the field of knowledge base alignment, and have achieved good results. The current research is mainly divided into paired entity alignment and collective entity alignment [6]. Paired entity alignment refers to calculating the similarity of multiple features such as the attributes of entities to determine whether the entities match. Collective entity alignment mainly uses the relationship between entities in the knowledge base to match. The research in this paper belongs to paired entity alignment. In paired entity alignment, entity alignment is mainly based on similarity and machine learning. The alignment algorithm based on similarity is mainly from the perspective of text similarity, relationship structure similarity, or a combination of the two. The earliest appeared is a similarity-based entity alignment probability model. This model establishes the basis for the entity alignment problem. A large number of scholars also apply this method, but when the number of unmatched entities in the knowledge base is much larger than the number of matched entities, the generalization ability of the model is poor. PARIS [7] proposed an iterative algorithm based on the model. Although the accuracy and recall rate are high, the iterative calculation time is too long. Lacoste-Julien developed a large knowledge base alignment algorithm SiGMa [8] through feature matching of similar functions. This algorithm uses the idea of greedy algorithm to optimize the objective function. The advantage is that the F-measure is high and has good scalability. But it requires a certain amount of manual intervention, and there is no alignment of categories and relationships. The alignment algorithm based on machine learning regards entity alignment as a binary classification problem. According to the data set, it is divided into supervised learning, semi-supervised learning and unsupervised learning. Under the supervised learning method, Cohen and Richman [9] obtained an adaptive distance function from the training set, and proposed a scalable and adaptive entity name matching and clustering technique. Through clustering, similar entities can be clustered together to a large extent reduce the amount of alignment calculation, but too many parameters need to be adjusted and rely on the training set, which increases the complexity and decreases the efficiency. McCallum and Wellner [10] also proposed a distance function based on conditional random fields to cluster similar entities, but the model of conditional random fields is more complicated. In an unsupervised manner, Fuzhen [11] proposed an unsupervised entity alignment approach using both attribute triples and relation triples of KGs. They use these aligned entity pairs to train a relation embedding model such that they could use relation triples to further align the remaining entities. Lastly, they utilize a bivariate regression model to learn the respective weights of similarities measuring from the two aspects for a result combination. In the alignment of collective entities, embedding-based models in recent years have been used for entity alignment. Such models are built on top of a knowledge graph embedding model that learns entity embedding to capture the semantic similarity between entities in the same knowledge graph. Trisedya [12] proposed a model that exploits large numbers of attribute triples existing in the knowledge graphs and generates attribute character embedding. Then they use a transitivity rule to further enrich the number of attributes of an entity to enhance the attribute character embedding. Weixin Zeng [13] proposed a collective EA framework. They employ three representative features, structural, semantic and string signals, which are adapted to capture different aspects of the similarity between entities in heterogeneous KGs. In order to make collective EA decisions, they formulate EA as the classical stable matching problem, which is further effectively solved by deferred acceptance algorithm. Yan Zhuang [14] proposed a novel hybrid human-machine framework for large-scale KB integration. They first partition the entities of different KBs into many smaller blocks based on their relations. Then they construct a partial order on these partitions and develop an inference model which crowdsources a set of tasks to the crowd and infers the answers of other tasks based on the crowdsourced tasks.

3   Multi-fEature Knowledge Fusion and Evaluation Algorithm Based on BERT

3.1   Design of MEFE

The task of this paper is to perform knowledge fusion on the multi-source knowledge base, and calculate the entity redundancy rate and information loss rate of the knowledge base according to the fusion result, so as to evaluate the quality of the knowledge base. Assuming that entities and relationships are marked as E and R respectively, the entity information in the knowledge base can be expressed in the form of triplets <E, R, E>, such as <entity, attribute, attribute value>, <entity, category, category value> and <entity, description, value>. Different knowledge bases have different levels of entity information perfection. The information perfection level mainly includes the number of entities and the comprehensiveness of the entity information. For example, in the knowledge base A, there are n triples about the entity E1, and in the knowledge base B, there are m triples about the entity E2. When E1 = E2, n < m, it means that the information of this entity in knowledge base B is more comprehensive than that of knowledge base A. Through the knowledge fusion of multi-source knowledge base, this kind of information can be found and the knowledge base information expansion will make the knowledge base more comprehensive. In addition, for entities with inconsistent namespaces but referring to the same object, such as "pineapple" and "pineapple" entities, you can identify and link each other by comparing the multiple features of the entities.

3.2   Algorithm Overview

The core of the algorithm in this paper is to first distribute the entities into multiple clusters through the category label vectorization of the entities, and then calculate the entity similarity for the set with more than one entity in the clustering result. The entity similarity is composed of the attribute similarity and the similarity of the description text of the entity. The flow of the algorithm is shown in "Fig. 1".

Before the fusion, we convert the entities expressed in the form of triples in the knowledge base into the form of records, such as <entity name, attribute 1, attribute 2, ..., attribute n, description, category>. To facilitate alignment calculation, the entity data of the two knowledge bases are stored in the csv file, and the file header is the entity name, attribute, description, category, and knowledge base name. Due to the different structure of the knowledge base, the classification of entity categories will be different, and some manual intervention is required to normalize the classification of different knowledge bases, and then establish a category dictionary for the vectorization of entity category labels. According to a custom vector distance function, we distribute entities with similar categories into the same group of clusters, calculate the edit distance and semantic similarity of the entity names in each cluster, and divide the entity names into similar blocks. The attributes and description similarity of the entities in the block with the number of entities greater than 1 are calculated. Since there are multiple attributes of an entity, the similarity of each component of the attribute needs to be calculated first, and then each attribute is assigned a weight according to the indicator function of the attribute. As a result, the total similarity of the attribute and the description similarity

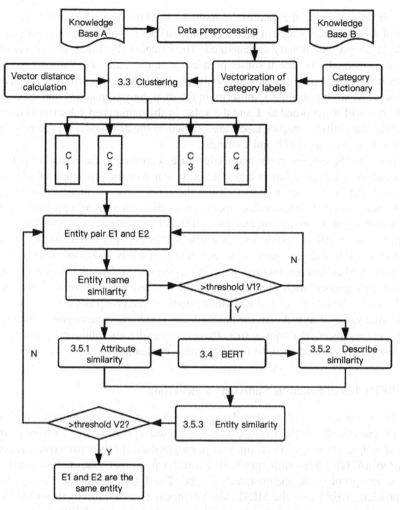

Fig. 1. Algorithm flow chart

are weighted and added to obtain the similarity of the entity. We set the threshold, which needs to be obtained through multiple experimental tests. If the similarity of the entity is higher than the threshold, it is determined that the entity is pointing the same entity in fact.

3.3 Vectorization and Clustering of Category Labels

There are often multiple categories of entities in the knowledge base, and the categories will have a good distinction between entities. Using category labels to quickly cluster entities in the knowledge base can greatly improve the efficiency of multi-source knowledge base fusion.

After preprocessing the data in the multi-source knowledge base, all the category labels of the entities are stored in the collection, and the set is cooperatively deduplicated. Finally a category dictionary is generated. The category dictionary is represented by a one-dimensional array, each subscript stores a unique category. We traverse all the entities in the csv file and map the category labels of the entities according to the category dictionary. Assuming that the x subscript in the dictionary is matched, the value of the x subscript will be recorded as 1 and the value of the unmatched subscript is recorded as 0. After the entity's category labels are mapped by the dictionary, the category labels vector will be composed of 0 and 1 strings.

Clustering the entities in the knowledge base according to the results of the above vectorization of category labels can effectively improve the efficiency of knowledge fusion calculation. In order to better cluster the entities, the calculation method of the custom category label vector in this paper is: when the same subscript position is both 1, the result is 1 and the remaining cases are all 0. For example, the category label vector of entity A is 10110100, and the category label vector of entity B is 11001001, then the calculation result of the category label vectors of A and B is 10000000. We traverse the collection of all entities and take the first entity as the standard to calculate the vectorized result of the category labels of all the remaining entities. When the result has m groups of subscripts with a value of 1, it is divided into the same group as the first entity in the cluster (the value of m is determined according to multiple experiments). Then all the entities in the cluster are removed from the original entity set, followed by analogy, and the clustering is stopped until the original entity set is empty.

3.4 BERT-Based Semantic Similarity Calculation

BERT is a language model based on Transformer network structure proposed by Google in 2018. The essence of BERT is to learn better feature representations for words through self-supervised learning of large amounts of corpus data. It solves two typical problems in the OpenAI GPT (Generative pre-trained transformer) model: the unidirectional information flow problem and the mismatch between Pre-Training and Fine-Tuning. For the first problem, BERT uses the MLM (Mask Language Model) mechanism to randomly cover 15% of the words in the sentence. These masked words let BERT predict, so that the self-attention learned by BERT can consider the model of bidirectional information. For the second problem, NSP (Next Sentence Prediction) is used to change the input of a single sentence in the model into the input of two sentences, so that the input of Pre-Training and Fine-Tuning are two sentences, which solves their mismatches. In the public data set, BERT has achieved the best performance in all tasks of NLP (Natural Language Processing).

BERT Network Structure. The BERT network structure adopts a multi-layer bidirectional Transformer structure [15]. Since the goal of BERT is to better train the feature representation of words, BERT only requires the Encoder layer of the Transformer structure. The network structure of BERT is shown in "Fig. 2". Where E1, ..., En are the encoded vectors of the sentence, Trm is the Transformer, T1, ..., Tn are the feature representation of the sentence output by BERT. The original input of BERT is to input two sentences. The two sentences are separated by a separator [SEP]. The beginning of the two sentences is represented by the characteristic character [CLS]. Token Embedding is first expressed as a vector for the input sentence, then Position Embedding is performed for each word, so as to remember the position of each word, and finally Segment Embedding for the sentence is used to mark which sentence each word belongs to. Combining the above three embedding features is the coding vector of the input sentence, so as to get E1, …, En.

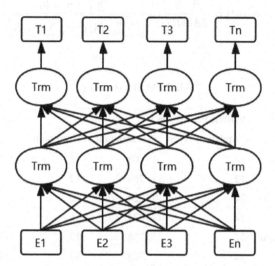

Fig. 2. BERT network structure diagram

Calculate Semantic Similarity. Text semantic similarity is one of the difficulties of NLP. Common similarity calculation methods include cosine similarity calculation, Manhattan distance, and Jaccard similarity coefficient. In this paper, the method of cosine similarity calculation is used to calculate the semantic similarity of text. Since the calculation of cosine similarity is based on vector space, the characterization of words is the key to determine the accuracy of semantic similarity calculation. In this paper, the pre-trained model in BERT is used to characterize the sentence. The test of the NLP task based on the public data set shows that the performance indicators of BERT are better than the best OpenAI GPT. Google provides multiple BERT pre-training models for different languages. Each model and some parameters of the model are shown in "Table 1". The model selected in this article is the Chinese BERT pre-training language model based on the whole word coverage (Whold Word Masking) released by Harbin Institute of Technology Xunfei joint laboratory. The BERT pre-training model based on full word coverage is an upgraded version of BERT released by Google in May 2019. The Chinese BERT pre-training model based on full word coverage applies the full word coverage technology to Chinese. The model uses Chinese Wikipedia data for pre-training and it is better than the Google provided in the NLP task. In this paper, the vector representation of the input sentence is obtained according to the model, and the cosine similarity of the two vectors is calculated according to the vector space. The cosine similarity is the semantic similarity of the calculated sentence. Assuming that the vectors of the two input words after using BERT are A and B, Ai and Bi are the components of the A and B vectors, and the semantic similarity calculation formula is as follows:

$$S = \cos(\theta) = \frac{A \cdot B}{\|A\| \|B\|} = \frac{\sum_{i=1}^{n} A_i * B_i}{\sqrt{\sum_{i=1}^{n} (A_i)^2} * \sqrt{\sum_{i=1}^{n} (B_i)^2}} \tag{1}$$

Table 1. Model statistics.

Model	Various Parameters			
	Number of layers	Hidden unit	Head number	Total parameters
BERT-base-uncased	12	768	12	110 MB
BERT-base-cased	12	768	12	110 MB
BERT-large-uncased	24	1024	16	340 MB
BERT-large-cased	24	1024	16	340 MB
BERT-large-ml-cased	12	768	12	110 MB
BERT-base-chinese	12	768	12	110 MB

3.5 Entity Similarity Calculation

Entities in the knowledge base are distributed according to the similarity of category labels to generate multiple sets of clusters. The entities in each cluster perform entity similarity calculation separately, including entity attribute similarity calculation and entity description similarity calculation. In order to calculate the similarity better, this paper uses the editing distance and the semantic similarity calculation to calculate the similarity together.

Attribute Similarity. Before calculating the attribute similarity, this paper first counts the frequency of each attribute in the knowledge base A and the total number of attributes of the library. Then we divides the frequency of the attribute by the total number of attributes to record the result as W (W is the weight of the attribute). An attribute with a higher frequency indicates that there are more entities with the attribute, which also indicates that the attribute is more universal. In this paper, the attribute that is more universal is considered as the more important attribute, and the corresponding weight should be greater. An entity will have multiple attributes. When calculating the attribute similarity, it is necessary to iteratively calculate the similarity of each pair of attribute components in the entity pair. The flow chart for calculating the similarity of the individual attribute components of the entity pair E1 and E2 is shown in "Fig. 3".

Thresholds k1 and k2 will be adjusted continually to the optimal value according to the experiment. According to the number of recorded unmatched attributes, the voting method is used to determine whether the attribute similarity of the entity pair needs to be calculated. That is, when the number of unmatched attributes is greater than the number of matched attributes, the attribute similarity of the entity pair does not need to be calculated. Otherwise it needs to be calculated. In this paper, when the attribute names of attribute N and attribute M do not match, we record attribute N and attribute M as the missing information values of entity E2 and entity E1, respectively. If in the end the entity pair is calculated as the same entity by similarity, it can track the missing value of the attribute information of each entity E1 and E2 basing on this recorded value. When the attribute names of attribute N and attribute M are successfully matched, the edit similarity or semantic similarity of the attribute pair is taken as the similarity S1. Similarly, the similarity S2, ..., Sn of other attributes can be calculated. We distribute the weight values to the successfully matched attributes, normalize these weight values to make their sum be 1, and set the normalized weight to W1, ..., Wn. Let the entity attribute similarity be AS.

$$AS = \sum_{k=1}^{n} S_K \times W_K \qquad (2)$$

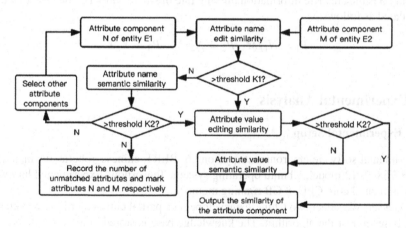

Fig. 3. Attribute similarity calculation

Description Similarity. The description of the entity is a short text that briefly explains the entity. Because the length of the short text is generally short, we use the BERT-based method to encode the short text. Then we calculate the cosine similarity of the generated vector, so as to obtain the description similarity of the entity pair. Let the description similarity of entities be ES.

Total Entity Similarity. Weights X1 and X2 are assigned to the attribute similarity and description similarity of the entities respectively, and the weights are taken as the optimal values based on multiple experiments. Let the total similarity of entities be ES.

$$ES = X_1 \times AS + X_2 \times ES \tag{3}$$

3.6 Quality Evaluation of Knowledge Base

In the fusion process of multi-source knowledge bases, knowledge sources from different sources can be used for knowledge complementation and knowledge discovery, so that the knowledge information in the knowledge base is more comprehensive and accurate. In this paper, the knowledge base is evaluated through two indicators: lack of entity attribute information and knowledge base redundant information. The evaluation can intuitively discover the deficiencies in knowledge base information.

After the fusion calculation of knowledge base A and knowledge base B, the algorithm will output the redundant entity records and entity attribute information of knowledge base A and knowledge base B respectively. Assuming that for the knowledge base A, the total number of entities is M, the number of redundant entities is N1, the number of entities with missing attribute information is N2, the total number of attributes of entity E is X, and there are Y attribute components missing, then P1 is the entity redundancy rate of knowledge base A, P2 is the entity information missing rate of knowledge base A, and P3 is the attribute information missing rate of single entity E. Then each index is calculated as follows:

$$P_1 = N_1/M \quad P_2 = N_2/M \quad P_3 = Y/X \tag{4}$$

4 Experimental Analysis

4.1 Experimental Setup

Experimental software environment: python3.6, BERT, chinese_roberta_wwm_ext_L-12_H-768_A-12 model, Ubuntu operating system, Tensorflow. Experimental hardware environment: 2 core CPU, 8 GB running memory.

Data set: We select two Chinese encyclopedia's partial entry records as two knowledge bases to test the algorithm. The knowledge base is stored in the form of csv file, the file header is the entity name, attribute, category, entity description information. First of all, 200 entities from each knowledge base are used for parameter optimization experiments to determine the optimal value of each parameter. Then all the entities in the knowledge base are used as the test set of the algorithm, and the aligned entity information is marked manually. The statistics of the parameter tuning data set are shown in Table 2, and the statistics of the algorithm test set are shown in Table 3.

Parameter: The threshold of this article includes the threshold of the category word vector, the threshold of the entity name clustering, the edit distance threshold, the semantic similarity threshold, the entity similarity threshold, the weight of the total attribute similarity and the description similarity the weight of.

Table 2. Statistics of parameter tuning experiment data.

Data set	Total quantity	
	Number of entities	Number of categories
knowledge base A	200	152
knowledge base B	200	84
total	400	236

Table 3. Algorithm test set data statistics.

Data set	Total quantity	
	Number of entities	Number of categories
Knowledge base A	1000	212
Knowledge base B	1000	124
Total	2000	336

4.2 Experimental Process and Results

Before conducting the algorithm performance test experiment, we adjust the various parameter values in the algorithm through parameter optimization experiments so that each parameter value reaches the global optimization. We take initial values of other parameters to observe the effect of different values of entity similarity threshold on alignment results. Among them, the threshold value of the category word vector is 1, the threshold value of entity name clustering is 0.95, and the edit similarity threshold is 0.96, the threshold of semantic similarity is 0.95, the weight of total attribute similarity is 0.6, and the weight of description similarity is 0.4. When the threshold of the similarity of entity names is different, the alignment results are shown in Table 4. It can be seen from the alignment results that when the entity similarity threshold is lower than 0.90, although the recall rate can be maintained at 100%, the accuracy rate will continue to decline. When the entity similarity threshold is 0.97, F-Measure can reach the maximum value. When the knowledge bases of different sources are merged, by comparing the aligned entity information, we will find the attribute information of an entity miss. This is a factor that mainly affects the similarity of entities. In the experiment, the calculation of entity similarity is reduced by voting Number of trips. The experiment found that when the average entity has 1–2 attributes missing, the entity similarity falls within the interval 0.97–0.98. When it is higher than 0.98, in fact, the entity pairs of the same entity will be classified as different entities due to the lack of some information of an entity, which results in the recall rate reduction. So the entity similarity threshold takes the value of this interval. After the entity similarity threshold is determined, several other experiments are fine-tuned for other parameters. Finally, after multiple experiments, when the category word vector threshold is 1, the entity name clustering threshold is 0.9, and the edit distance threshold is 0.93, the threshold of semantic similarity is 0.95, the threshold of entity similarity is 0.97, the weight of total attribute similarity is 0.5, and the weight of description similarity is 0.5, the experimental effect is the best.

Table 4. Results of parameter tuning experiment.

Entity similarity threshold	Evaluation index		
	Accuracy/ %	Recall rate/ %	F-measure/ %
0.90	62.50	100.0	76.92
0.97	98.50	94.20	96.30
0.98	98.80	90.30	94.35
0.99	100.0	70.00	82.35

According to the results of the parameter tuning experiment, the global optimal value of each parameter is selected to test the algorithm. The algorithm input in this paper is two knowledge bases, where the knowledge base is stored in csv files, and the two csv files are merged in the data preprocessing stage into a csv file. We add columns to mark the knowledge base to which the entity belongs. We calculate the category labels of all entities, establish a category dictionary and generate word vectors for each entity category through the mapping of the category dictionary. Then we gather the result of the category word vector operation as 1 into a cluster. According to the entity name for all entities in the same cluster Semantic similarity clustering, entities with semantic similarity greater than 0.9 are clustered together. After the clustering is completed, the entity pairs in each cluster are input into the algorithm. The attribute similarity and description similarity of the entities are calculated by using the BERT and the editing distance, and finally the weights are added to obtain the total similarity of the entities. The results of the algorithm output are the similarity results of the entity pairs, the redundant entities of each knowledge base and the information missing values of the entities in the knowledge base. We analyze the results in the following two aspects.

(1) The manually labeled test set contains 128 pairs of aligned entity pairs. The algorithm outputs the similarity information of the aligned entity pairs and all entity pairs. The similarity information results of the entity pairs are output in a csv file format, including the entity name of the entity pair, the knowledge base name, total attribute similarity, description similarity, entity similarity and whether the entity belongs to the same entity. We manually calibrate the results of the alignment entities output by the algorithm, and find that there are 120 pairs of aligned entity pairs, of which 118 pairs are correctly aligned. Based on the results, we calculate the accuracy rate, recall rate, and F-Measure of the algorithm. The algorithm in this paper is also compared with the alignment results of PARIS [7]. The comparison results are shown in Table 5. The accuracy of PARIS is higher than that of the method in this paper. The main reason is that the method in this paper will be affected by the lack of entity information. However, the F-Measure of MEFE is higher than that of PARIS. In terms of efficiency, if the entities of the knowledge base are not

clustered according to the aligned category labels, the entity pair that needs to be aligned is the product of the number of two knowledge base entities. According to the experimental results, the entity pairs to be aligned are calculated as 2000 pairs through the entity clustering, which is a reduction of 500 times compared to the calculation of entity pairs of the original scale.

Table 5. Comparing results.

Method	Evaluation index		
	Accuracy/ %	Recall rate/ %	F-measure/ %
MEFE	98.33	93.75	95.98
PARIS	99.70	90.60	94.90

(2) On the basis of the redundant entities of each knowledge base and the information missing value of the entities in the knowledge base, the redundancy rate and the entity information missing rate of the knowledge base are calculated separately. On the one hand, these two indicators can be used to evaluate the quality of the knowledge base. On the other hand, the redundant entities of the knowledge base can be eliminated, and the entity information can be expanded according to the lack of entity information, so that the entity information is more complete and the knowledge base has a higher quality.

5 Summary

This paper studies the knowledge fusion and evaluation algorithm of the multi-source knowledge base. On the one hand, it uses the category word vectors of the knowledge base entities to cluster the entities, which greatly reduces the number of entities in the knowledge base that need to be aligned and calculated, thereby improving the overall performance of the algorithm. On the other hand, BERT is used to calculate semantic similarity to improve the accuracy of the alignment algorithm. Finally, in the knowledge fusion algorithm, the knowledge base quality can be evaluated to a certain degree by calculating the redundancy rate of each knowledge base and the information missing rate of the knowledge base entities. The experimental results show that the algorithm proposed in this paper has greatly improved the accuracy rate, but the category difference of the knowledge sources from different sources is too large, which will affect the recall rate. In future research, we will comprehensively consider the relationship between the entities in the knowledge base and the structure of the knowledge base to further improve the performance of the knowledge fusion algorithm.

Acknowledgments. This work was supported by the National Key R&D Program of China (2017YFB1401300, 2017YFB1401302), Outstanding Youth of Jiangsu Natural Science Foundation (BK20170100), Key R&D Program of Jiangsu (BE2017166), Natural Science Foundation of the Jiangsu Higher Education Institutions of China (No. 19KJB520046), Natural Science Foundation of Jiangsu Province (No. BK20170900), Innovative and Entrepreneurial talents projects of Jiangsu Province, Jiangsu Planned Projects for Postdoctoral Research Funds (No. 2019K024), Six talent peak projects in Jiangsu Province, the Ministry of Education Foundation of Humanities and Social Sciences (No. 20YJC880104), NUPT DingShan Scholar Project and NUPTSF (NY219132) and CCF-Tencent Open Fund WeBank Special Funding (No. CCF-WebankRAGR20190104).

References

1. Lehmann, J., Isele, R., Jakob, M.: DBpedia: a largescale, multilingual knowledge base extracted from Wikipedia. Semant. Web **6**(2), 167–195 (2015)
2. Hoffart, J., Suchanek, F.M., Berberich, K.: YAGO2: a spatially and temporally enhanced knowledge base from Wikipedia. Artif. Intell. **194**, 28–61 (2013)
3. Wu, W., Li, H., Wang, H.: Probase: a probabilistic taxonomy for text understanding. In: Proceedings of the 2012 ACM SIGMOD International Conference on Management of Data, Scottsdale, USA, pp. 481–492 (2012)
4. Carlson, A., Betteridge, J., Kisiel, B.: Toward an architecture for never ending language learning. In: Proceedings of the 24th AAAI Conference on Artificial Intelligence, vol. 42, no. 4, pp. 1306–1313 (2010)
5. Niu, X., Sun, X., Wang, H., Rong, S., Qi, G., Yu, Y.: Zhishi.me - weaving chinese linking open data. In: Aroyo, L., et al. (eds.) ISWC 2011. LNCS, vol. 7032, pp. 205–220. Springer, Heidelberg (2011). https://doi.org/10.1007/978-3-642-25093-4_14
6. Solemn, L.G.: Feng Jianhua: overview of knowledge base entity alignment technology. Comput. Res. Develop. **53**(1), 165–192 (2016)
7. Suchanek, F.M., Abiteboul, S., Senellart, P.: PARIS: probabilistic alignment of relations, instances, and schema. Proc. VLDB Endow. **5**(3), 157–168 (2011)
8. Lacoste-Julien, S., Palla, K., Davies, A.: SIGMa: simple greedy matching for aligning large knowledge bases. In: Proceedings of the 2013 ACM SIGKDD Conference on Knowledge Discovery and Data Mining, pp. 572–580. ACM, New York (2013)
9. Cohen, W., Richman, J.: Learning to match and cluster large high-dimensional data sets for data integration. In: Proceedings of the 2002 ACM SIGKDD Conference on Knowledge Discovery and Data Mining, pp. 475–480. ACM, New York (2002)
10. McCallum, A., Wellner, B.: Conditional models of identity uncertainty with application to noun coreference. In: Proceedings of Advances in Neural Information Processing Systems, vol. 17, pp. 905–912. MIT Press, Cambridge, MA (2005)
11. He, F., et al.: Unsupervised entity alignment using attribute triples and relation triples. In: Li, G., Yang, J., Gama, J., Natwichai, J., Tong, Y. (eds.) DASFAA 2019. LNCS, vol. 11446, pp. 367–382. Springer, Cham (2019). https://doi.org/10.1007/978-3-030-18576-3_22
12. Trisedya, B.D., Qi, J., Zhang, R.: Entity alignment between knowledge graphs using attribute embeddings. In: AAAI-19, vol. 33, no. 01, pp. 297–304 (2019)
13. Zeng, W., Zhao, X., Tang, J.: Collective entity alignment via adaptive features. In: ICDE 2020, pp. 1870–1873 (2020)
14. Zhuang, Y., Li, G., Zhong, Z.: Hike: a hybrid human-machine method for entity alignment in large-scale knowledge bases. In: CIKM 2017, pp. 1917–1926 (2017)
15. Devlin, J., Chang, M.W., Lee, K.: Bert: pre-training of deep bidirectional transformers for language understanding. arXiv arXiv:1810.04805 (2018)

Comparative Analysis of Three Kinds of Laser SLAM Algorithms

Xin Liu[1](✉), Yang Lin[1], Hua Huang[1](✉), and Meikang Qiu[2]

[1] School of Information Engineering, Jingdezhen Ceramic Institute, Jingdezhen, China
117060100109@stu.jci.edu.cn, 378879@qq.com, jdz_hh@qq.com
[2] College of Computer Science and Software Engineering, Shenzhen University, Shenzhen, Guangdong, China
mqiu@szu.edu.cn

Abstract. With the development of artificial intelligence, the application of robots is also rapidly increasing. How to autonomously navigate and complete complex tasks for robots in an unknown environment is a hot spot in the research domain of simultaneous positioning and map construction (SLAM) algorithms. To better study and apply three common laser SLAM algorithms, by building a SLAM environment on the ROS robot platform, Hector SLAM, Gmapping, and Cartographer algorithms were used to conduct actual indoor mapping experiments. All three algorithms can achieve effective indoor two-bit mapping construction. By comparing and analyzing the three SLAM algorithms, the mapping accuracy of the Cartographer algorithm is significantly better than Hector SLAM and Gmapping algorithms. Meantime, the Cartographer algorithm has better robustness.

Keywords: Hector · Gmapping · Cartographer · SLAM

1 Introduction

SLAM [1, 2] is the focus of research in the field of mobile robots, and autonomous navigation is the key to mobile robots [3, 4] achieving autonomy and intelligence [5]. In life, mobile robots serve ground cleaning, shopping mall shopping guides, bank etiquette, etc. In industry, it is used in driverless [6], warehouse logistics, etc. In extreme environments, space exploration, rescue, and anti-terrorism missions are realized [7].

The earliest robots were mostly used in the industrial field, but in recent years, robots have gradually developed toward intelligence. Compared with traditional industrial robots, intelligent mobile robots have stronger autonomy and adaptability and can complete tasks better. The first autonomous mobile robot, Shakey, was born at Stanford University in the 1970s. It can complete tasks such as autonomous positioning, path planning, and navigation. This is also the world's first robot capable of autonomous movement. Since then, the application of robots has become more and more extensive, such as the British "Dragon Walker", which can replace people for dangerous jobs such as bomb removal; WillowGarage's PR2 robot, China's Cobos "ground protection" vacuum cleaner robots are all outstanding robots in housekeeping services.

© Springer Nature Switzerland AG 2020
M. Qiu (Ed.): ICA3PP 2020, LNCS 12453, pp. 463–476, 2020.
https://doi.org/10.1007/978-3-030-60239-0_31

For the surveying and mapping industry, robots used in the field of measurement have also achieved considerable results. In the past few years, Google has been devoted to indoor surveying and mapping technology and has launched Trekker and Cartographer [8] two backpack measuring robots. When a backpacker walks through a building, the onboard sensors automatically generate a floor plan of the building in real-time, and users can also mark points of interest on the map. In 2016, Zhonghaida integrated laser scanners, panoramic cameras, and mileage encoders broke through the laser-based SLAM technology and launched their HiScan-SLAM indoor mobile measurement system. Besides, robots perform measurement tasks such as indoor positioning, indoor space acquisition, and mapping.

Section 2 in the following reviews the related research of SLAM. Section 3 outlines the principles of the three kinds of laser SLAM algorithms [9]. The actual evaluation experiments of three algorithms are conducted in Sect. 4. Section 5 shows the summary of our paper and the future work.

2 Related Work

The SLAM problem is also known as the CML (Concurrent Mapping and Localization) problem, known as the "Holy Grail" in the field of autonomous mobile robot research. Since the development of the SLAM algorithm, a variety of SLAM systems have been formed, and all have achieved quite rich results. The earliest concept of SLAM was proposed by R. Smith and P. Cheeseman [10]. They established a statistical basis for describing the relationship between features and uncertain geometric objects. They believed that the position estimation of different road signs in the environment is highly correlated, and these correlations will continue to increase with continuous observations, using Kalman filtering to estimate robot poses and features in the environment; Dieter Fox [11] and Sebastian Thrun [12] propose to use probability to solve robot positioning problems, which has become the basis of the model of the probabilistic SLAM algorithm; Michael Montemerlo [13] of Carnegie Mellon University improved the data association and computational complexity of Kalman filtering, and proposed the Fast SLAM algorithm. This algorithm, based on particle filtering, not only greatly reduces the calculation but also helps to obtain accurate maps in large-scale, blurred landmark environments. Austin Eliazar of Duke University [14] a new method for creating maps, called distributed particle (DP) maps, that is, directly using sensor information to create occupancy maps, which avoids the problem of defining landmarks and corresponding data association. At the same time, a method of expanding to three-dimensional space is proposed; Daniel T. Savaria [15] fuses multiple sensors such as stereo photography vision system, odometer, and sonar for SLAM experiment, and uses SURF algorithm to describe map features. Yan Ma, Hehua Ju [16] improve the RBPF-SLAM algorithm by establishing the likelihood field LF to obtain the likelihood function and use laser, inertial navigation combined with odometer and other sensors to use it for lunar exploration vehicles.

3 Method Principle

The map is the main way we use to represent the environment. Mapping is a basic business in the surveying and mapping industry. At present, autonomous maps for robots mainly include raster maps [17], geometric feature maps [18], and topological maps. The representation of the map is usually based on the factors of actual situation of the robot's environment, the applicable conditions of the simultaneous positioning and mapping (SLAM) algorithm, and the needs of the robot. Choosing a suitable map representation method can not only meet the needs of building maps but also help us carry out other work based on maps, such as navigation and obstacle avoidance.

3.1 SLAM Algorithm Theory

SLAM usually contains several processes, of which the ultimate goal is to update the position estimation information of the robot. Since robot position information obtained through robot motion estimation usually has a large error, we cannot simply rely on robot motion to estimate robot position information. After obtaining the robot position estimate by using the robot motion equation, we can use the surrounding environment information obtained by the ranging unit to correct the robot position. The above correction process is generally achieved by extracting environmental features and then reobserving the position of the features after the robot moves. The core of SLAM is EKF which is used to estimate the exact position of the robot in combination with the above information. The features selected above are generally called landmarks. EKF will continue to estimate the position of the robot and the landmarks in the surrounding environment. The general process of SLAM is shown in Fig. 1.

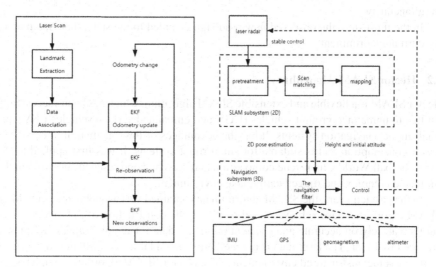

Fig. 1. Overview of the SLAM process block graph and Hector SLAM system framework

The robot's position will change when it moves. At this time, according to the observation of the robot position sensor, the feature points in the observation information are extracted, and then the robot estimate the current location and current environmental information combining the position of the currently observed feature point, the distance of the robot movement, and the position of the feature point observed before the robot movement through EKF.

In SLAM, we generally use the extended Kalman filter to estimate the state of the robot based on robot motion information and sensor measurement feature point information. Here, we will discuss in detail the specific steps of applying it to SLAM.

After obtaining the location and orientation of the signpost points and associating the signpost points, the SLAM process is divided into the following three parts:

1. Update the current state of the robot based on the robot motion information;
2. Update the estimated state based on the signpost information;
3. Add a new state to the current state;

The first step is relatively simple, it only needs to superimpose the output of the current controller to the state at the previous moment. For example, if the robot is currently located in (x, y, θ), and the current motion information is $(dx, dy, d\theta)$, then the current state of the first robot is:

$$(x = dx, x + dy, \theta + d\theta) \tag{1}$$

In the second step, we need to consider the signpost information. Based on the current state of the robot, we can estimate where the signpost should be, which is different from the measured position. The new information is the difference between the robot's estimated information and actual information based on the robot's state. At this time, according to the above new information, the method of each feature point is updated simultaneously.

In the third step, the observed new road sign is added to the state, that is, a map of the current environment.

3.2 Hector SLAM Algorithm

Hector SLAM is a flexible and extensible SLAM algorithm, which has been successfully applied to unmanned ground vehicles (UGV and unmanned surface vehicles USV) and small indoor navigation systems. It has the advantages of low calculation rate and low power consumption and is suitable for small robot systems. It was first applied to the robot football World Cup rescue competition and searched and rescued survivors in the simulated complex and rugged earthquake environment.

In the past ten years, the SLAM algorithm has obtained very valuable research. Most SLAM algorithms have achieved good results in the construction of 2D indoor maps, but they all rely on accurate and efficient odometer data for the high update frequency provided by Lidar. The laser data is not well utilized, and it is very difficult to construct maps for some unstructured environments. The Hector SLAM algorithm completes the SLAM task by using the laser data with high update frequency, rather than depending on the odometer.

Hector SLAM mainly includes two parts: 2D SLAM subsystem and navigation subsystem. Among them, the 2D SLAM system provides the posture information of the robot, and the task of drawing the environment map of the robot is completed by scanning matching; the navigation system uses a navigation filtering algorithm and fuses IMU or other sensor data to form a three-dimensional navigation scheme. The two subsystems are updated separately and only loosely coupled so that the two parts are synchronized. This article mainly discusses the 2D SLAM subsystem.

A navigation coordinate system is defined according to the principle of the right-handed coordinate system, with the initial position of the robot as the coordinate origin, the Z axis is vertically upward, and the X axis is directed to the direction of the initial yaw angle of the robot. The 3D coordinates of the robot can be expressed as $x = \left(\Omega^T, P^T, V^T \right)$ where $\Omega = (\phi, \theta, \psi)^T$ represents the robot's roll angle, pitch angle and yaw The Euler angle of the angle; $P = \left(p_x, p_y, p_z \right)^T$, $V = \left(v_x, v_y, v_z \right)^T$ represents the position information and velocity information of the robot. Inertial navigation measurement data is represented by $\mu = \left(\omega^T, a^T \right)^T$, where $\omega = \left(\omega_x, \omega_y, \omega_z \right)^T$ and $a = \left(a_x, a_y, a_z \right)^T$ respectively represent the robot's Angular velocity and acceleration information. Therefore, the robot's motion model can be written as the following nonlinear system:

$$\dot{\Omega} = E_\Omega \cdot \omega \tag{2}$$

$$\dot{P} = V \tag{3}$$

$$\dot{V} = R_\Omega \cdot a + g \tag{4}$$

R_Ω is the direction cosine matrix converted from the robot coordinate system to the navigation coordinate system; E_Ω is the mapping from the robot's angular velocity to the Euler angle. g is the acceleration of gravity. Here, we ignore the effect of the earth's rotation on the system.

Because the sensor has noise, it will cause the speed and position to deviate from the true value, so it may be necessary to fuse more sensor information. In Hector SLAM, this information is determined by the scan matcher, which is very suitable for the indoor environment.

3.3 Gmapping SLAM Algorithm

The Gmapping algorithm is currently the most widely used SLAM algorithm in robot neighborhood. The algorithm is a SLAM algorithm based on the principle of Rao-Blackwellized Particle Filter (RBPF). The map drawn is a grid map. The RBPF-SLAM algorithm was first proposed by Murph in 1999 to estimate the joint posterior probability $p(x_{1:k}, m|z_{1:k}, u_{1:k-1})$. That is, calculate the actual position of the robot (x) and create an environment map (m) when the odometer measurement value (u) and laser measurement value (z) are known. This posterior probability can be factored into:

$$p(x_{1:k}, m|z_{1:k}, u_{1:k-1}) = p(m|x_{1:k}, z_{1:k})p(x_{1:k}|z_{1:k}, u_{1:k-1}) \tag{5}$$

In this way, the SLAM problem is decomposed into estimating the robot's position based on the odometer data and laser data and then calculating the map based on the robot's position and laser data. Therefore, the construction process is very dependent on the pose of the robot. The biggest difference between the Gmapping algorithm and the Hector SLAM algorithm is the addition of odometer data.

According to the above factorization, the occupancy grid mapping algorithm and Inverse sensor model algorithm proposed by Thrun, Burgard, and Fox in 2006 is used to calculate $p(m|x_{1:k}, z_{1:k})$. Because the grid map m is composed of a limited number of grid cells, the posterior probability of the entire map can be approximated by the product of the edge probability: $p(m|x_{1:k}, z_{1:k}) = \prod_i p(m_i|x_{1:k}, z_{1:k})$. Applying a binary Bayesian filter to the construction of occupied raster maps to avoid the instability of the values around 0 and 1,

$$l_{k,i} = \log \frac{p(m_i|x_{1:k}, z_{1:k})}{1 - p(m_i|x_{1:k}, z_{1:k})} \tag{6}$$

From this the posterior probability of the map can be written as:

$$p(m_i|x_{1:k}, z_{1:k}) = 1 - \frac{1}{exp(l_{k,i})} \tag{7}$$

The posterior probability $p(x_{1:k}|z_{1:k}, u_{1:k-1})$ of the robot trajectory is solved according to particle filtering.

3.4 Cartographer Algorithm

Cartographer is a real-time simultaneous positioning and mapping (SLAM) library launched by Google in October 2006. Google designed a backpack, installed the sensor on the backpack, and used the operator to walk indoors to draw a two-dimensional grid map in real-time. Each frame of laser scan data obtained by lidar is inserted into a submap at the best-estimated position using scan matching which is only related to the current submap. Cartographer solves the error accumulation problem through periodic optimization instead of the particle filtering method. When a submap is generated and no new scan is inserted, a partial loop close is performed; after the submap is completed, the loop is automatically closed. If the laser scan is very close to the currently estimated pose, it will look for a consistent scan in the submap, and if it finds the best matching result with the current pose, it will be added to the optimization problem as a closed-loop constraint. The branch-and-bound and pre-computed grids are used to solve the global loopback problem: optimization is performed every few seconds. When the robot moves to a repeated location, scan matching will precede the generation of new scans to achieve closed-loop detection.

Cartographer is divided into local and global parts to draw a two-dimensional map. The robot's posture includes three parameters of translation (x, y) and rotation ξ_θ, expressed as $\xi = (\xi_x, \xi_y, \xi_\theta)$, and the laser observation value in the current posture is called scan. At the same time, the algorithm will add IMU data to estimate the direction of gravity to ensure that the laser position is parallel to the ground. Using a nonlinear optimization scheme to match a series of continuous laser observations (scans) with submaps (submap), this process is called scan matching.

Laser Data

The sub-graph structure is an iterative process in which laser data repeats matching in the sub-graph coordinate system. We write the laser point information as $H = \{h_k\}_{k=1,\ldots k}$, $h_k \in R^2$, the initial laser point is $0 \in R^2$. The measured pose of the robot is based on the laser position, so it is converted to the sub-picture coordinate system by the formula (8):

$$T_{\xi p} = \begin{pmatrix} cos\xi_\theta & -sin\xi_\theta \\ sin\xi_\theta & cos\xi_\theta \end{pmatrix} p + \begin{pmatrix} \xi_x \\ \xi_y \end{pmatrix} \tag{8}$$

Subgraph

A series of continuous laser dots are used to construct the subgraph. Subgraphs are expressed in the form of probabilistic grids. The constructed environment map can be regarded as a series of discrete grid points with certain values. The grid value is the probability that the grid will be occupied. We use the relevant pixels to approximate each grid point. When inserting the laser point into the grid, there are two cases of hits and miss in each grid. The grid closest to the endpoint of the laser point is hits, and the grid intersecting between the origin and endpoint of the laser point is a miss. If it is an unobserved grid, it will be assigned a probability value p_{hit} or p_{miss}, if it is an already observed grid, it will be updated with probability. In Fig. 2, the grid where the cross is located is hits, and the grid that intersects the laser beam is missing.

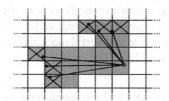

Fig. 2. Schematic diagram of hits and misses

$$odds(p) = \frac{p}{1-p} \tag{9}$$

$$M_{new}(x) = clamp\left(odds^{-1}(M_{old}(x) \cdot odds(p_{hit}))\right) \tag{10}$$

Ceres Scan Matching

Before inserting the laser point into the sub-graph, the scanning attitude should be optimized relative to the sub-graph. Here, the scan matching method based on Ceres is used to convert the scanning pose problem to a nonlinear least-squares problem.

$$\underset{\xi}{argmin} \sum_{k=1}^{K} \left(1 - M_{smooth}(T_\xi h_\xi)\right)^2 \tag{11}$$

In the above formula, T_ξ is the conversion coefficient to convert the laser information h_k from the laser coordinate system to the sub-picture coordinate system according to the scanning pose. $M_{smooth} : R^2 \to R$ is to use the bicubic interpolation method to smooth the probability value in the local subgraph, to eliminate the influence of the value outside the interval [0, 1].

The mathematical optimization of this smoothing function will usually have better accuracy than the resolution of the grid. Since the least squares problem is local optimization, a good initial estimate (initial value of pose) is needed, IMU is used to provide rotation variables for the initial value of pose. In the absence of IMU, the frequency or accuracy of scan matching can be improved.

4 Experiments and Evaluation

4.1 ROS Robot for Drawing

To realize the test of the SLAM algorithm, ROS (Robot Operating System) is needed. ROS is a highly flexible software architecture for writing robot software programs. It contains a large number of tool software, library code, and agreed protocols, and aims to simplify the difficulty and complexity of the process of creating complex and robust robot behaviors across robot platforms. The prototype of ROS comes from the STanford Artificial Intelligence Robot (STAIR) and Personal Robotics (PR) projects at Stanford University.

We use ROS Turtlebot2 is a low-cost robot development platform designed by Willow Garage, which is the second generation of the Turtlebot series. Its simple operation and strong scalability can not only meet the research and development needs of scientific research institutions but also provide a powerful robot application development platform for robot technology enthusiasts (Table 1).

Table 1. Turtlebot2 configuration parameters

Size	354 × 354 × 420 mm
Self-weight	6.3 kg
Maximum load	5 kg
Top speed	0.65 m/s
Rotation speed	180°/s
Chassis height	15 mm
Battery	2200 mAh Li-lon
Power interface	5 V/1 A, 12 V/1.5 A, 12 V/5 A, 19 V/1 A
Depth camera	Microsoft Kinect

Turltebot2 has a large size and a good load, which can meet our basic test requirements. And it is equipped with a Kinect depth camera, which provides a better test environment for the experiment.

4.2 Use of Kinect Depth Camera

Kinect has three lenses, and the center lens is an RGB color camera, used to collect color images. The left and right lenses are a 3D structured light depth sensor composed of an infrared emitter and an infrared CMOS camera, respectively, used to collect depth data (distance from the object in the scene to the camera). The color camera supports a maximum of 1280 * 960 resolution imaging, and the infrared camera supports a maximum of 640 * 480 imaging. Kinect is also equipped with focused tracking technology, and the base motor will rotate with the focused object. Kinect also has a built-in array microphone. The four microphones simultaneously pick up sounds, eliminate noise after comparison, and use the collected sounds for voice recognition and sound source localization.

4.3 Rplidar A2 Lidar

The system uses ubuntu16.04+ROS kinect, combined with RPLIDAR A2 officially supported by ROS to build a SLAM environment, which can very friendly support our three kinds of laser SLAM tests (Table 2).

Table 2. RPLIDAR A2 performance table

Project	Unit	Minimum value	Typical value	Maximum	Remarks
Ranging range	Meter	To be determined	0.15–12/18	To be determined	Measured based on white highly reflective objects
Scanning angle	Degree	Not applicable	0–360	Not applicable	
Range resolution	mm	Not applicable	$\frac{<0.5}{<1\% \text{ of actual distance}}$	Not applicable	$\frac{\text{Measuring objects within 1.5 m}}{\text{All ranges within the range}}$
Angle resolution	Degree	0.45	0.9	1.35	When scanning at 10 Hz
Single ranging time	Millisecond	Not applicable	0.25	Not applicable	
Measuring frequency	Hertz	2000	$\geq$4000	8000	
Scanning frequency	Hertz	5	10	15	

4.4 Comparative Analysis of Cartographer and Hector SLAM and Gmapping Algorithms

Comparative Analysis of The Real Environment

This article evaluates the Hector SLAM, Gmapping, and Cartographer experiments in real life. We used traditional measurement methods to measure the actual distance of objects including door frames, elevator openings, corridor width, flower beds, pillars and so on. Figure 3 is a schematic diagram of the experimental site drawn in CAD according to traditional surveying and mapping methods. Also based on MATLAB platform, the pixel distances of ground objects on the maps drawn by the three groups of SLAM algorithm were obtained respectively, and the errors between them and the traditional measurement methods were compared.

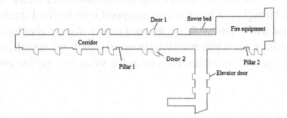

Fig. 3. Schematic diagram of the experimental scene

D-value is the difference between the length measured by the traditional method minus the length drawn by the algorithm. A is the traditional method to measure the length, and B is the algorithmic drawing length.

$$D = |A - B| \tag{12}$$

RMSD is a digital standard that measures the accuracy of observations. The RMSD of an estimator $\hat{\theta}$ concerning an estimated parameter θ is defined as the square root of the mean square error:

$$\mathrm{RMSD}\left(\hat{\theta}\right) = \sqrt{MSE\left(\hat{\theta}\right)} = \sqrt{E\left(\left(\hat{\theta} - \theta\right)^2\right)} \tag{13}$$

For an unbiased estimator, the RMSD is the square root of the variance, known as the standard deviation. The RMSD of predicted values $\hat{y}_t$ for times t of a regression's dependent variable y_t, with variables observed over T times, is computed for T different predictions as the square root of the mean of the squares of the deviations:

$$\mathrm{RMSD} = \sqrt{\frac{\sum_{t=1}^{T} \left(\hat{y}_t - y_t\right)^2}{T}} \tag{14}$$

$(\hat{y}_t - y_t)$ in RMSD is the D-value, which is obtained by dividing the sum of the squares of the D-value by the number of feature points and then squared to obtain the RMSD.

Mean is the mean of the n values $x_1, x_1, \ldots, x_n$. The arithmetic mean is the most commonly used and readily understood measure of central tendency in a data set. In statistics, the term average refers to any of the measures of central tendency. The arithmetic mean of a set of observed data is defined as being equal to the sum of the numerical values of every observation divided by the total number of observations. Symbolically, if we have a data set consisting of the values $a_1, a_2, \ldots a_n$, then the arithmetic mean A is defined by the formula:

$$A = \frac{1}{n} \sum_{i=1}^{n} a_i = \frac{a_1 + a_2 + \ldots + a_n}{n} \tag{15}$$

Table 3 and Fig. 4(a) show the comparison results of indoor features obtained by Hector SLAM mapping. Table 4 shows the comparison results of indoor features obtained by Gmapping algorithm mapping. Table 5 and Fig. 4(c) show the comparison results of indoor features obtained by Cartographer algorithm mapping. In Table 6 and Fig. 4, they are the comparison of the maximum difference, the minimum difference, and the medium error of the graphing results of the three algorithms, Hector SLAM, Gmapping, and Cartographer.

Table 3. Comparison of indoor features of Hector SLAM (unit: cm)

Serial number	Characteristic features	Length measured by traditional methods	Hector SLAM drawing length	D-value
1	Elevator door	101	100	1
2	Door 1	106	115	9
3	Door 2	90	90	0
4	Corridor	260	225	35
5	Flower bed	485	505	20
6	Pillar 1	113	110	3
7	Pillar 2	105	110	5
8	Fire equipment	33	40	7

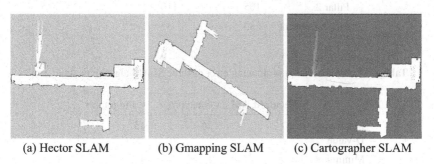

(a) Hector SLAM   (b) Gmapping SLAM   (c) Cartographer SLAM

Fig. 4. Three kinds of lidar algorithms to build map analysis

It can be seen from the comparison results of the real environment experiment: the maximum error of the Hector SLAM algorithm reaches 35 cm, which has a great relationship with the accumulation of errors in the scanning and matching process. Because Hector SLAM relies heavily on results which caused by laser scanning and matching

Table 4. Comparison of indoor features of Gmapping (unit: cm)

Serial number	Characteristic features	Length measured by traditional methods	Gmapping drawing length	D-value
1	Elevator door	101	98	3
2	Door 1	106	110	4
3	Door 2	90	89	1
4	Corridor	260	234	26
5	Flower bed	485	501	16
6	Pillar 1	113	112	1
7	Pillar 2	105	99	6
8	Fire equipment	33	49	16

Table 5. Comparison of indoor features of Cartographer (unit: cm)

Serial number	Characteristic features	Length measured by traditional methods	Cartographer drawing length	D-value
1	Elevator door	101	100	1
2	Door 1	106	115	9
3	Door 2	90	85	5
4	corridor	260	265	5
5	flower bed	485	500	15
6	Pillar 1	113	110	3
7	Pillar 2	105	110	5
8	Fire equipment	33	35	2

Table 6. Comparison unit of drawing error of three SLAM algorithms (unit: cm)

	Hector SLAM	Gmapping	Cartographer
Maximum D-value	35	26	15
Minimum D-value	0	1	1
Mean	10	9.125	5.625
RMSD	14.9583	12.5050	7.0267

and does not use odometer data, so it will affect the accuracy of Hector SLAM when the number of features in the environment is small. In this paper, the Gmapping mapping effect of the experimental site is better than Hector SLAM, and the accuracy can be controlled within 12 cm. At the same time, because Hector SLAM has very high requirements for the performance of the laser, it is required that the laser has high precision and high update frequency, but if the performance of the laser sensor used is slightly worse, it will affect the drawing effect. It can be seen from the experimental comparison data that Cartographer's mapping accuracy is the best, which can reach within 5 cm. This is because Cartographer does not adopt a particle filtering algorithm. The laser data scanning and matching uses SPA and BBS algorithms for optimization. Although the calculation amount is larger than Hector SLAM and Gmapping in principle, the speed is faster than expected. At the same time, Another major advantage of Cartographer is that it has loop detection, compared with Hector SLAM and Gmapping algorithms, during the experiment, it is found that Cartographer will correct the local map according to the global map. Therefore, when the local map has deviated, its closed-loop detection will pull the local map back to the global map for matching, which also makes Cartographer very robust.

5   Conclusion and Future Work

By comparing the real environment experiment of the three algorithms, the cumulative error caused by Cartographer algorithms significantly reduced, compared with the Hector SLAM and Gmapping algorithms, and its loopback detection function makes the algorithm very robust;

SLAM technology has great potention in areas such as mobile measurement, unmanned vehicles, model construction, and military reconnaissance, at the same time, accurate testing for SLAM is becoming increasingly important. Therefore, it is of great significance to study the algorithm application of the SLAM system, and it also has certain reference value for SLAM learning and deep research. SLAM is not only refers the aspect of lidar, and the related research of vision aspect will be carried out later.

Acknowledgment. This work is supported by the National Key Research and Development Plan of China under Grant No. 2016YFB0501801, National Natural Science Foundation of China under Grant No. 61170026, the National Standard Research Project under Grant No. 2016BZYJ-WG7-001, the Key Research and Development Plan of Jiang Xi province under Grant No. 20171ACE50022 and the Natural Science Foundation of Jiang Xi province under Grant No. 20171BAB202011, the science aánd technology research project of Jiang Xi Education Department under Grant Nos. GJJ180730, GJJ180727, GJJ181520, and the Science and Technology project of Jingdezhen under Grant Nos. 20182GYZD011-01, 20192GYZD008-01, 2019GYZD008-03, and the Open Research Project of the State Key Laboratory of Industrial Control Technology, Zhejiang University, China (No. ICT 20025).

References

1. Bonin-Font, F., Ortiz, A., Oliver, G.: Visual navigation for mobile robots: a survey. J. Intell. Robot. Syst. **53**, 263 (2008). https://doi.org/10.1007/s10846-008-9235-4

2. Kalogeiton, V.S., Ioannidis, K., Sirakoulis, GCh., Kosmatopoulos, E.B.: Real-time active SLAM and obstacle avoidance for an autonomous robot based on stereo vision. Cybern. Syst. **50**(3), 239–260 (2019)
3. Durrant-Whyte, H., Bailey, T.: Simultaneous localization and mapping: part I. IEEE Robot. Autom. Mag. **13**(2), 99–110 (2006). https://doi.org/10.1109/MRA.2006.1638022
4. Bailey, T., Durrant-Whyte, H.: Simultaneous localization and mapping (SLAM): part II. IEEE Robot. Autom. Mag. **13**(3), 108–117 (2006). https://doi.org/10.1109/MRA.2006.1678144
5. Jiang, B., Bishop, A.N., Anderson, B.D.O., Drake, S.P.: Optimal path planning and sensor placement for mobile target detection. Automatica **60**, 127–139 (2015)
6. Siagian, C., Chang, C.K., Itti, L.: Autonomous mobile robot localization and navigation using a hierarchical map representation primarily guided by vision. J. Field Robot. **31**(3), 408–440 (2014)
7. Aladrén, A., López-Nicolás, G., Puig, L., Guerrero, J.J.: Navigation assistance for the visually impaired using RGB-D sensor with range expansion. IEEE Syst. J. **10**(3), 922–932 (2016). https://doi.org/10.1109/JSYST.2014.2320639
8. Chainago, V.M., Jati, A.N., Setianingsih, C.: Development of non-platform mobile robot for simultaneous localization and mapping using ROS. In: 2019 IEEE International Conference on Aerospace Electronics and Remote Sensing Technology (ICARES), Yogyakarta, Indonesia, pp. 1–6 (2019). https://doi.org/10.1109/icares.2019.8914356
9. Yagfarov, R., Ivanou, M., Afanasyev, I.: Map comparison of lidar-based 2D SLAM algorithms using precise ground truth. In: 2018 15th International Conference on Control, Automation, Robotics and Vision (ICARCV), Singapore, pp. 1979–1983 (2018). https://doi.org/10.1109/icarcv.2018.8581131
10. Garulli, A., Giannitrapani, A., Rossi, A., et al.: Mobile robot SLAM for line-based environment representation. In: Proceedings of the 44th IEEE Conference on Decision and Control (2005)
11. Fox, D., Burgard, W., Thrun, S.: Markov localization for mobile robots in dynamic environments. J. Artif. Intell. Res. **11**, 391–427 (1999)
12. Thrun, S.: Probabilistic algorithms in robotics. AI Mag. **21**(4), 3–109 (2000)
13. Montemerlo, M., Thrun, S., Koller, D., Wegbreit, B.: FastSLAM 2.0: an improved particle filtering algorithm for simultaneous localization and mapping that provably converges (2003)
14. Eliazar, A.: DP-SLAM. Duke University, Durham (2005)
15. Savaria, D.T., Balasubramanian, R.: V-SLAM: Vision-based simultaneous localization and map building for an autonomous mobile robot. In: 2010 IEEE Conference on Multisensor Fusion and Integration for Intelligent Systems (MFI), Salt Lake City, UT. IEEE (2010)
16. Yan, M., Hehua, J., Pingyuan, C.: Research on localization and mapping for lunar rover based on RBPF-SLAM. In: 2009 International Conference on Intelligent Human-Machine Systems and Cybernetics, IHMSC 2009, Hangzhou, Zhejiang, pp. 306–311. IEEE (2009)
17. Kundu, A.S., Mazumder, O., Dhar, A., Bhaumik, S.: Occupancy grid map generation using 360° scanning xtion pro live for indoor mobile robot navigation. In: 2016 IEEE First International Conference on Control, Measurement and Instrumentation (CMI), Kolkata, pp. 464–468 (2016). https://doi.org/10.1109/cmi.2016.7413791
18. Ramaithitima, R., Whitzer, M., Bhattacharya, S., Kumar, V.: Automated creation of topological maps in unknown environments using a swarm of resource-constrained robots. IEEE Robot. Autom. Lett. **1**(2), 746–753 (2016). https://doi.org/10.1109/LRA.2016.2523600

Aspect-Level Sentiment Difference Feature Interaction Matching Model Based on Multi-round Decision Mechanism

Yanzhi Wei[1,2](✉), Xianghua Fu[2](✉), Shuxin Wang[2], Wenhao Xie[1,2], Jianwei He[1,2], and Yonglin Zhao[1,2]

[1] Shenzhen University, Shenzhen, China
workdempsey@foxmail.com
[2] Shenzhen Technology University, Shenzhen, China
fuxianghua@sztu.edu.cn

Abstract. Sentence matching is a key problem in natural language understanding, so the research on sentence matching can be applied to a large number of known natural language processing tasks, such as information retrieval, automatic question and answer, machine translation, dialogue system, paraphrase identification etc. In a series of natural language processing tasks, we need to rely on the participation and collaboration of the sentence matching model. The performance of the sentence matching model can greatly affect the final performance of these natural language processing tasks. We propose the Al-SFIM model, which improves the matching model from the perspective of word interaction. First, we propose sentiment attention mechanism based on the distribution of aspect-level sentiment difference to improve the interaction between cross-sentence words, and use the sentiment space position perception vector to improve the interaction between intra-sentence words, so that the model has the ability to perceive the subjective sentiment difference in the process of intra-sentence word interaction and cross-sentence word interaction. Then, we introduce a multi-round decision mechanism based on the accumulation of memory state, which iteratively updates the working memory state to make matching decisions in multiple rounds, so that the model can better understand the semantic of complex sentence. Experiment results show that the AL-SFIM model has made progress in sentence matching and has better matching performance for complex, long and incomprehensible sentences.

Keywords: Aspect-level · Sentiment · Matching · Word interaction

1 Introduction

In recent years, the Internet is playing an increasingly important role in the social development and people's life. On the Internet, there are various forms of data and information, and text information is a very important information carrier. There is already a huge amount of text data on the Internet. It has become a very challenging task to obtain useful information from the massive text data. Therefore, natural language processing

© Springer Nature Switzerland AG 2020
M. Qiu (Ed.): ICA3PP 2020, LNCS 12453, pp. 477–491, 2020.
https://doi.org/10.1007/978-3-030-60239-0_32

(NLP) is becoming more and more important in the field of computer science, and sentence matching is a very meaningful research direction in natural language processing, which is the basis of a series of natural language processing tasks.

Sentence matching is to analyze the relationship between two sentences. However, the existing sentence matching model only considers the objective information of the sentence and fails to capture the subjective sentiment differences between words in the process of word interaction, ignoring the impact of subjective sentiment information contained in the sentence on semantics. Moreover, models tend to make arbitrary "one-step" inferences about the relationships between sentences, so they do a poor job of matching complex, long and incomprehensible sentences. To solve these problems, we propose the aspect-level sentiment difference feature interaction matching model (AL-SFIM). From the perspective of word interaction matching, the process of word interaction is improved through the attention mechanism based on the distribution of aspect-level sentiment difference, and the method of word dependency calculation is improved by using the feature of aspect-level sentiment difference. Furthermore, the multi-round decision mechanism is added to the decision-making process of sentences relationship inference to improve the ability of the model to match complex, long and incomprehensible sentences.

Our main contributions can be summarized as follows: (1) We improved the word interaction process. The method of word dependency calculation is improved by using the feature of aspect-level sentiment difference; (2) We proposed the sentiment attention mechanism based on the distribution of aspect-level sentiment difference and sentiment space position perception vector, so the model can perceive the subjective sentiment of the sentence; (3) We used multi-round decision mechanism to improve the ability of the model to match complex, long and incomprehensible sentences.

We introduce the related work in Sect. 2 and illustrate the detailed framework of Al-SFIM model in Sect. 3. We describe experiments and analysis in Sect. 4. Finally, we conclude and discuss future work in Sect. 5.

2   Related Work

Many scholars have done a series of researches on sentence matching. The previous work can be roughly divided into two parts: one is sentence matching based on traditional feature engineering; the other is matching based on deep learning.

Traditional methods mainly rely on feature engineering, linguistic tools (such as syntax parsers, grammar parsers, dictionaries, etc.) and external knowledge. The disadvantage is that it requires a lot of time and energy, and it also encounters the problem of sparse training data. Xue [1] combined the translation language model with the maximum likelihood estimation by means of retrieval. Surdeanu [2] analyzed the influence of various types of features on sentence matching. Yih [3] used semantic features obtained from WordNet to enhance dictionary features. Tymoshenko [4] analyzed the influence of different syntactic and semantic structures extracted from shallow and deep syntactic parsers on sentence matching.

The traditional sentence matching process is complicated, requiring a large number of manually defined and extracted features to participate, so the definition and selection of features have high requirements. The generalization ability of the model is also poor, so different feature extraction schemes need to be designed according to different tasks. It takes a lot of time and energy to extract the features artificially, and not all the extracted features are effective. In addition, in the face of massive data, the traditional feature selection methods are difficult to find the implied feature, which may have excellent effect for some special samples.

The sentence matching model based on deep learning can overcome the problem of time-consuming feature engineering and has made a series of achievements in sentence matching. The application of deep learning method in sentence matching can be summarized into the following three stages: single semantic model, multiple semantic model, and direct modeling matching model.

The single semantic model simply encodes two sentences with the neural network, then calculates the matching score between the sentences. The single semantic models include DSSM [5], CDSSM [6], ARC-I [7], CNTN [8], and LSTM-RNN [9].

The multiple semantic model analyses sentences from the perspective of multiple grained and considers the local structure of phrases and sentences. Typical multiple semantic models include MV-LSTM [10], MultiGranCNN [11], uRAE [12], etc.

In the direct modeling matching model, the interaction of different words between sentences is considered more, the matching score between two words is calculated. Then the feature is extracted by neural network, and the connection between words is processed more finely. Typical models include Match Pyramid [13], DeepMatch [14], Match-SRNN [15], BiMPM [16], DIIN [17] and DRCN [18].

In general, the sentence matching method based on deep learning has more advantages than the traditional matching method based on traditional feature engineering, so the matching model based on deep learning is the trend of future research and development. Different matching models have their own advantages and disadvantages.

3   Method

3.1   Word Interaction Based on Aspect-Level Sentiment

Usually, there are two kinds of information contained in a sentence: objective information and subjective information; Objective information is a factual description of objective things. Subjective information is people's opinions on things, which can reflect their attitudes and preferences. On the Internet, the text about user's comments and communication often contains a wealth of subjective sentiment information. Mining these sentiment features in the text can help to better understand the semantic. The feature of sentiment polarity in sentence plays an important role in sentence understanding and is also beneficial to sentence matching. However, the subjective information expressed by the sentence-level sentiment polarity feature is still too crude. Therefore, it is necessary to carry out more fine-grained aspect-level sentiment analysis on sentences. Aspect-level sentiment refers to the fine-grained sentiment expressed by a word in sentence.

Thus, the semantic information of a sentence can be reflected more comprehensively, and the sentence matching task can be performed better. Combined with the characteristics of sentence matching task, we consider making use of aspect-level sentiment difference feature from two perspectives:

3.1.1 Cross-Sentence Word Interaction (CSWI)

The original dot-product attention mechanism was proposed by Luong [19].

$$\text{score}(s_t, h_i) = s_t^T h_i \tag{3.1}$$

For the original dot-product attention mechanism, given a sentence x, the attention weight between two words x_i and x_j in the sentence is calculated as follows:

$$Comp_{i,j} = Softmax(x_i \cdot x_j) \tag{3.2}$$

Then, the attention vector is obtained by weighting:

$$\tilde{x}_i = \sum_j Comp_{i,j} x_j \tag{3.3}$$

We refer to the word embedding space projection of word vector, and project different words into a sentiment space according to the aspect-level sentiment feature. In the sentiment space, the positions of different words express the polarity of sentiment. The closer the sentiment polarity of two words is, the closer the distance between two words in the sentiment space will be, and higher possibility that two words express similar semantics will be. As shown in the figure, word_2 is the center one. Because word_1 is closer to the center word, it is more likely to be semantically similar to the center word. Word_3 is further away, so less likely to be semantically similar to the central word (Fig. 1).

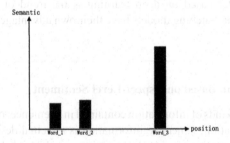

Fig. 1. The relationship between word distance and semantic similarity.

In order to introduce the feature of aspect-level sentiment difference in the process of attention calculating, it is necessary to add the dependent information of adjacent words in the sentiment space. In order to make use of both the original attention mechanism and the aspect-level sentiment difference features of sentences, we modify the dot-product attention mechanism by referring to the improved method of Guo [20] for the attention computing mechanism of words with different distances in the sentence. We modify the attention mechanism of dot-product by using the prior information based on gaussian distribution. Firstly, it is necessary to select a central word, and then assume that the semantic influence of words with different positions and distances in the sentiment space on the central word is following to the gaussian distribution, and then use the feature of aspect-level sentiment difference as the prior knowledge to correct the weight of other different words on the semantic influence of the central word.

The gaussian distribution is selected as the hypothesis basis in this paper. A series of experiments have been carried out by scholars [21] to explore the attention mechanism based on distance, which also shows that the gaussian distribution is a relatively reasonable hypothesis.

The variance of the standard gaussian distribution is $\sigma^2 = 1/(2\pi)$, and the probability density function is $\phi(d) = e^{-\pi d^2}$, where d is a random variable used to represent the sentiment difference between words. First, the aspect-level sentiment features of sentence are analyzed, and the sentiment difference information is calculated. Then, this function was added to the original dot-product attention mechanism calculation formula to correct the semantic influence of words at different positions in sentiment space, thus introducing aspect-level sentiment difference information in the process of attention weight calculation:

$$\tilde{x}_i = \sum_j \frac{\phi(d_{i,j}) \, com \, p_{i,j}}{Z_1} x_j = \sum_j \frac{e^{-d_{i,j}^2} \cdot e^{(x_i \cdot x_j)}}{Z_2} x_j = \sum_j \frac{e^{-d_{i,j}^2 + (x_i \cdot x_j)}}{Z_2} x_j$$

$$(3.4)$$

$$\tilde{x}_i = \sum_j Softmax\left(-d_{i,j}^2 + (x_i \cdot x_j)\right) x_j \qquad (3.5)$$

Where $Z_1 = \sum_k \phi(d_{i,k}) \, com \, p_{i,k}$ and $Z_2 = \sum_k e^{-d_{i,k}^2 + (x_i \cdot x_k)}$ are regularization factors. The above formula transforms the normal distribution into the gaussian distribution with bias terms, thus simplifying the additional product operation. Since the variance of the gaussian distribution is not necessarily the same as the standard normal distribution, it is necessary to introduce an additional variable w to restrict the above formula:

$$\tilde{x}_i = \sum_j Softmax\left(-wd_{i,j}^2 + (x_i \cdot x_j)\right) x_j \qquad (3.6)$$

3.1.2 Intra-sentence Word Interaction (ISWI)

In order to model the aspect-level sentiment difference information and the word position information, we consider introducing sentiment space position perception vector into the model. Based on the introduction of gaussian distribution in previous section, we construct the sentiment space position perception vector based on the assumption of gaussian distribution, so as to enable the model to perceive the interactive dependence caused by aspect-level sentiment differences between words. The calculation method of sentiment space position perception dependence based on gaussian distribution hypothesis is as follows:

$$Gaussian(n) = \exp\left(\frac{-n^2}{2\gamma^2}\right) \qquad (3.7)$$

Where, n represents the aspect-level sentiment feature of word and γ represents the position information of word in the sentence. In model training process, the sentence needs to be preprocessed, and each word in the sentence is represented by a vector. Therefore, matrix can be used to represent the dependent influence of sentiment space position perception on different dimensions of the word vector:

$$p(i, \gamma) \sim N\left(\text{Gaussian }(n), \sigma'\right) \tag{3.8}$$

Where, N represents the gaussian distribution, σ' represents the standard deviation, and $p(i, \gamma)$ represents the dependent effect on the i-th dimension of the word located in γ. The final result is the sentiment space position perception influence matrix of the whole sentence. Each column of matrix P represents the influence vector for a specific word, and each word in the sentence corresponds to a representation vector $p_j \in R^{d_p}$, which represents the sentiment space position perception relationship between words.

3.2 Multi-round Decision (MRD)

Considering the complexity of sentence matching task, especially for some complex sentences, it is difficult to perform matching in one step. For these complex sentences, it often requires multiple rounds of analysis and inference to fully understand the semantics of the sentences. Therefore, we further introduce the multi-round decision mechanism based on the accumulation of memory states, and combine it with the sentiment attention mechanism based on the distribution of aspect-level sentiment difference, so as to better match sentences.

This kind of multi-round inference has an important contribution to the correctness of the final decision. For example, in the field of MRC (Machine Reading Comprehension), many scholars have proposed many similar multi-round decision mechanisms, such as Hill [22], Shen [23], Xu [24], etc. In order to explore the application of multi-round inference mechanism in natural language inference, Liu [25] proposed a Stochastic Answer Networks, which maintains an additional working state and iteratively updates this state before making final inferential decision. Most of the previous sentence matching models directly infer the relationship between sentences. The Stochastic Answer Networks accumulates a state and updates it iteratively, and then decides the relationship between sentences in multiple rounds according to the accumulation of memory state, which enables the model to perform complex inference.

3.3 AL-SFIM Model

We propose the aspect-level sentiment difference feature interaction matching model (AL-SFIM). The overall structure of the model is shown in the following figure (Fig. 2):

(1) Word Representation Layer
In order to facilitate description, we describe that the two input sentences are P and H respectively, and sentence $P = \{p_0, p_1, \ldots, p_{m-1}\}$ has m words, and sentence $H = \{h_0, h_1, \ldots, h_{n-1}\}$ has n words. For each word in the sentence, the pre-trained word vector and sentiment space position perception vector should be obtained first. In

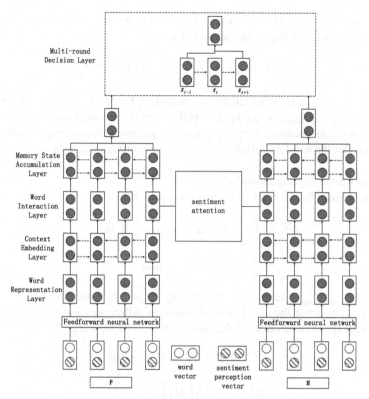

Fig. 2. AL-SFIM model.

this paper, the pre-trained word vector GloVe [26] is used as the initial representation of words, and the sentiment space position perception vector can be constructed by the method described in Subsect. 3.1. Then, the two vectors are concatenated to get the new vector. Finally, two independent double-layer forward neural networks are used to obtain the final representation $E^p \in \mathbb{R}^{d \times m}$ and $E^h \in \mathbb{R}^{d \times n}$ by combining the two vectors, where d represents the dimension of the vector.

(2) Context Embedding Layer

In this paper, the bidirectional LSTM network is used to encode contextual information for each word in the sentence, and according to the method of Goodfellow, maxout layer is used to compress the output of bidirectional LSTM network, and then the representation of $C^p \in \mathbb{R}^{2d \times m}$ and $C^h \in \mathbb{R}^{2d \times n}$ which contain contextual information of each word in the two sentences can be obtained.

(3) Word Interaction Layer

In this layer, the sentiment attention mechanism based on aspect-level sentiment difference distribution is used to perform word interaction, and the vectorization representation of word is used to complete attention weights calculation. This layer not only uses the sentiment attention mechanism to calculate the dependent weights of different words between two sentences, but also adds an additional mapping layer to regularize the attention weights according to the method of Liu [25]:

$$A_{\text{al}-sentiment} = dropout\left(f_{\text{al}-sentiment}\left(\hat{C}^p, \hat{C}^h\right)\right) \in \mathbb{R}^{m \times n} \qquad (3.9)$$

Where, $f_{\text{al}-sentiment}$ is the attention calculation based on the sentiment attention mechanism, $A_{\text{al}-sentiment}$ is the finally obtained attention weight matrix based on sentiment attention mechanism, $dropout$ is used for parameter smoothing, $\hat{C}^p$ and $\hat{C}^h$ are the results after the word vectors in the sentence pass through the mapping layer Re LU.

(4) Memory State Accumulation Layer
First, concatenate the information of the two sentences:

$$U^p = \left[C^p; \ C^h A_{\text{al}-sentiment}\right] \in \mathbb{R}^{4d \times m} \qquad (3.10)$$

$$U^h = \left[C^h; \ C^p A'_{\text{al}-sentiment}\right] \in \mathbb{R}^{4d \times n} \qquad (3.11)$$

Where, $A'_{\text{al}-sentiment}$ represents the transpose of the attention weight matrix $A_{\text{al}-sentiment}$ based on the distribution of sentiment differences. Finally, the memory of the two sentences can be obtained through the $BiLSTM$ network:

$$M^p = BiLSTM\left(\left[U^p; \ C^p\right]\right) \qquad (3.12)$$

$$M^h = BiLSTM\left(\left[U^h; \ C^h\right]\right) \qquad (3.13)$$

(5) Multi-round Decision Matching Layer
First, we need to calculate the initialized state s_0, which is obtained by summing the sentence memory:

$$s_0 = \sum_j \alpha_j M^h_j \qquad (3.14)$$

$$\alpha_j = \frac{\exp\left(\theta_2 \cdot M^h_j\right)}{\sum_{j'} \exp\left(\theta_2 \cdot M^h_{j'}\right)} \qquad (3.15)$$

In round t, the cumulative state can be calculated as follows:

$$s_t = GRU(s_{t-1}, x_t) \qquad (3.16)$$

$$x_t = \sum_j \beta_j M^p_j \qquad (3.17)$$

$$\beta_j = softmax(s_{t-1}\theta_3 M^p) \qquad (3.18)$$

In each round, a temporary decision of the round can be obtained:

$$P_t^r = softmax(\theta_4[s_t; x_t; |s_t - x_t|; s_t \cdot x_t]) \qquad (3.19)$$

Where, P_t^r represents the probability distribution of different categories. Finally, the temporary decisions in all rounds are used to get the final decision:

$$P^r = avg([P_0^r, P_1^r, \ldots, P_{T-1}^r]) \qquad (3.20)$$

For different matching tasks, the value range of r in above formula is different. For paraphrase identification, there are two kinds of matching results and two possible values. For natural language inference, there are three kinds of matching results and three possible values. The value range depends on specific matching task type.

4 Experiment

In this section, we will verify AL-SFIM model through experiments. We chose natural language inference and paraphrase identification tasks. First, we explored the performance of AL-SFIM model in these two classic sentence matching tasks, and then compared AL-SFIM model with some popular matching models. Finally, the performance and characteristics of AL-SFIM model are analyzed.

4.1 Datasets

The datasets selected in this paper are described as follows:

(1) Stanford natural language inference dataset (SNLI)
The Stanford natural language inference dataset (SNLI) is a natural language inference task dataset published by Stanford University. The dataset contains 570,000 pairs of inference sentences in English, each pair of which includes a premise sentence and a hypothesis sentence corresponding to the premise sentence.

(2) Paraphrase identification dataset of Quora website (Quora)
The Quora question pairs dataset is a paraphrase identification task dataset published by Quora website. The dataset contains 400,000 pairs of questions, which are collected from real questions asked by users on Quora.

(3) Multi-Genre NLI dataset (MultiNLI)
Multi-Genre inference dataset is a commonly used natural language inference dataset, the dataset consists of 433000 sentences which contain multiple types, and only half of these types appear in training set.

4.2 Experiment Design

In experiment, we use PyTorch framework to build AL-SFIM model, and the model will give corresponding sentence relations according to the input sentence pairs. The entire task can be abstracted to a classification problem, the number of classes is determined according to the given dataset, and it also determines the dimension of the prediction label of the final matching layer. The word vector is initialized by the 300 - dimension GloVe[1]. Where, the unknown word is set to 0, and the hidden layer size of LSTM network in the context embedding layer and memory state accumulation layer is set to 128, and the number of multi-round decision is set to 5. In addition, we refer to the method of Salimans [27] to do the parameters regularization, and the dropout rate is set to 0.2. Adamax optimization algorithm [28] is also adopted in this model. The initial learning rate of the optimizer is set to 0.002, which will be dynamically adjusted during training. For all datasets, we chose the best model parameters on the validation set and then evaluated them on the test set.

4.3 Results and Analysis

(1) Sentence Matching Accuracy In order to verify the sentence matching accuracy, a series of experiments are carried out on different datasets for different tasks. We conducted experiments on above datasets and use classification accuracy to illustrate model performance. At the same time, we also compared with some popular sentence matching models. The following table describes the sentence matching accuracy on different datasets (Table 1).

Table 1. Sentence matching accuracy for different datasets.

Dataset	Model	Accuracy
SNLI	PWIM	82.2
	InferSent	84.6
	SSE	85.5
	DecAtt	85.6
	BiMPM	86.9
	DIIN	88.0
	AL-SFIM	88.3
Quora	PWIM	83.4
	InferSent	86.6
	SSE	87.8

(*continued*)

---

[1] https://nlp.stanford.edu/projects/glove/.

Table 1. (*continued*)

Dataset	Model	Accuracy
	DecAtt	84.5
	BiMPM	88.1
	DIIN	89.0
	AL-SFIM	89.1
MultiNLI_m	InferSent	70.5
	DecAtt	71.9
	PWIM	72.2
	SSE	74.0
	BiMPM	77.0
	DIIN	78.8
	AL-SFIM	79.2
MultiNLI_u	InferSent	70.3
	DecAtt	71.3
	PWIM	71.6
	SSE	73.4
	BiMPM	76.7
	DIIN	77.8
	AL-SFIM	78.5

It can be seen from the experimental results that AL-SFIM model has achieved good matching accuracy for natural language inference and paraphrase identification on different datasets. Compared with other models, the accuracy of AL-SFIM model has been improved. Among them, SSE model is based on sentence interaction. It only uses the representation of sentences to carry out sentence matching and ignores word interaction information between two sentences. Therefore, it does not capture the relationship between words well enough. The performance of AL-SFIM model is also better than other complex models.

(2) Complex, Long and Incomprehensible Sentence Matching The AL-SFIM model perform semantic modeling of long and complex sentences more accurately by virtue of multi-round decision mechanism based on the accumulation of memory states, which also improved the matching accuracy. Similar multi-round inference mechanisms have been widely used in machine reading comprehension. In sentence matching, it is known that Liu [25] tried to explore the matching effect of the same model using single step inference and multiple steps inference. In order to further illustrate the MRD mechanism and verify the validity of multi-round decision component (MRD), we select some complex sentences from MultiNLI dataset, to explore the performance differences of sentence matching models. It can be seen from the experiment results that the matching performance of single round decision models are not ideal for complex and long sentences (Table 2).

Table 2. The performance of different models on complex sentences.

Dataset	Model	Accuracy
MultiNLI	SSE	67.6
	DecAtt	68.5
	InferSent	69.2
	PWIM	69.4
	AL-SFIM	73.8

(3) Sentiment Sentence Matching In order to qualitatively demonstrate the effectiveness of the sentiment attention mechanism, we also designed a new sentence matching task: Sentiment sentence matching (SSM). SSM is a variant task of sentence matching. The process and evaluation indexes of the task are basically the same as sentence matching task introduced above. The difference is that each sample sentence in the SSM task dataset must contain sentiment information. However, it is currently not possible to find a sentence matching dataset that all samples contain sentiment information. Therefore, we need to create a new dataset for the SSM task. In order to create the dataset that meet the requirements, we select the samples that meet the requirements from the original datasets introduced above to build the new sub-dataset. The experiment results are shown in the following Table 3.

Table 3. SSM accuracy of AL-SFIM model.

Dataset	Original test set	Sentiment test set
SNLI	88.3	90.3
Quora	89.1	91.2

By analyzing the results of the SSM expansion experiment, it can be seen that the model can better understand the subjective sentiment information contained in sentence by capturing the sentiment features, thus improving the model ability for semantic analysis and understanding. Sentiment attention mechanism can better capture the subjective sentiment dependence between words than the original attention mechanism, so that the model has the ability to perceive aspect-level sentiment difference in the process of intra-sentence word interaction and cross-sentence word interaction. Therefore, AL-SFIM model has a better matching performance for the sentences containing sentiment features.

5   Conclusions and Future Works

From the perspective of word interaction, we proposed a multi-round matching model AL-SFIM based on the distribution of aspect-level sentiment differences. The existing word interaction models generally use the attention mechanism to complete word interaction, but attention mechanism cannot capture the information of sentiment difference between words and ignores the effect of subjective sentiment information contained in sentence on semantic understanding. To solve this problem, we propose a sentiment attention mechanism based on the distribution of aspect-level sentiment differences to improve word interaction. The model has the ability to perceive subjective sentiment differences in the process of intra-sentence word interaction and cross-sentence word interaction. In addition, aiming at the poor matching performance of complex, long and incomprehensible sentences in the existing matching models, we also use the multi-round decision mechanism based on the accumulation of memory state for complex sentences. Through the multi-round decision mechanism based on the accumulation of memory state, the model can better understand the semantics of complex sentences. Experiments show that the AL-SFIM model proposed in this paper has made progress in sentence matching and has better matching performance for complex, long and incomprehensible sentences.

Nowadays, the sentence matching model based on deep learning has achieved good performance. Some existing researches focus on deepening the layer number of neural networks in order to obtain stronger feature encoding ability, but the results are not perfect. In the future, we can try to improve the network structure of matching model and explore new matching methods. In addition, the external domain knowledge plays an important role in sentence matching. Considering the development of knowledge graph technology in recent years, we can try to build external knowledge base by using knowledge graph in the future. Domain knowledge based on external knowledge will also be helpful to the decision process of sentence matching and may bring greater improvement to the matching model.

Acknowledgements. This research is supported by the Scientific Research Platforms and Projects in Universities in Guangdong Province under Grants 2019KTSCX204.

References

1. Xue, X., Jeon. J., Croft, W.B.: Retrieval models for question and answer archives. In: Proceedings of the 31st Annual International ACM SIGIR Conference on Research and Development in Information Retrieval, pp. 475–482 (2008)
2. Surdeanu, M., Ciaramita, M., Zaragoza, H.: Learning to rank answers to non-factoid questions from web collections. Comput. Linguist. **37**(2), 351–383 (2011)
3. Yih, W.-T., et al.: Question answering using enhanced lexical semantic models. In: Proceedings of the 51st Annual Meeting of the Association for Computational Linguistics, pp. 1744–1753 (2013)
4. Tymoshenko, K, Moschitti, A.: Assessing the impact of syntactic and semantic structures for answer passages reranking. In: Proceedings of the 24th ACM International on Conference on Information and Knowledge Management, pp. 1451–1460 (2015)

5. Huang, P.-S., et al.: Learning deep structured semantic models for web search using click-through data. In: Proceedings of the 22nd ACM International Conference on Information & Knowledge Management, pp. 2333–2338 (2013)
6. Shen, Y., et al.: A latent semantic model with convolutional-pooling structure for information retrieval. In: Proceedings of the 23rd ACM International Conference on Conference on Information and Knowledge Management, pp. 101–110 (2014)
7. Hu. B., et al.: Convolutional neural network architectures for matching natural language sentences. In: Advances in Neural Information Processing Systems, pp. 2042–2050 (2014)
8. Qiu, X., Huang, X.: Convolutional neural tensor network architecture for community-based question answering. In: Proceedings of the Twenty-Fourth International Joint Conference on Artificial Intelligence, pp. 1305–1311 (2015)
9. Palangi, H., et al.: Deep sentence embedding using long short-term memory networks: analysis and application to information retrieval. IEEE/ACM Trans. Audio Speech Lang. Process. 24(4), 694–707 (2016)
10. Wan, S., et al.; A deep architecture for semantic matching with multiple positional sentence representations. In: Thirtieth AAAI Conference on Artificial Intelligence, pp. 2835–2841 (2016)
11. Yin, W., Schütze, H.: MultiGranCNN: an architecture for general matching of text chunks on multiple levels of Granularity. In: Proceedings of the 53rd Annual Meeting of the Association for Computational Linguistics and the 7th International Joint Conference on Natural Language Processing, pp. 63–73 (2015)
12. Socher, R., et al.: Dynamic pooling and unfolding recursive autoencoders for paraphrase detection. In: Advances in Neural Information Processing Systems, pp. 801–809 (2011)
13. Pang, L., et al.: Text matching as image recognition. In: National Conference on Artificial Intelligence, pp. 2793–2799 (2016)
14. Lu, Z., Li, H.: A deep architecture for matching short texts. In: Advances in Neural Information Processing Systems, pp. 1367–1375 (2013)
15. Wan, S., et al.: Match-SRNN: modeling the recursive matching structure with spatial RNN. In: Proceedings of the Twenty-Fifth International Joint Conference on Artificial Intelligence, pp. 2922–2928 (2016)
16. Wang, Z., Hamza, W., Florian, R.: Bilateral multi-perspective matching for natural language sentences. In: Proceedings of the Twenty-Sixth International Joint Conference on Artificial Intelligence, pp. 4144–4150 (2017)
17. Gong, Y., Luo, H., Zhang, J.: Natural language inference over interaction space. In: 6th International Conference on Learning Representations (2018)
18. Kim, S., Kang, I., Kwak, N.: Semantic sentence matching with densely-connected recurrent and co-attentive information. In: The Thirty-Third AAAI Conference on Artificial Intelligence, pp. 6586–6593 (2019)
19. Luong, T., Pham, H., Manning, C.D.: Effective approaches to attention-based neural machine translation. In: Proceedings of the 2015 Conference on Empirical Methods in Natural Language Processing, pp. 1412–1421 (2015)
20. Guo, M., Zhang, Y., Liu, T.: Gaussian transformer: a lightweight approach for natural language inference. In: The Thirty-Third AAAI Conference on Artificial Intelligence (2019)
21. Im, J., Cho, S.: Distance-based self-attention network for natural language inference. arXiv Comput. Lang. (2017)
22. Hill, F., et al.: The Goldilocks principle: reading children's books with explicit memory representations. In: International Conference on Learning Representations (2016)
23. Shen, Y., et al.: An empirical analysis of multiple-turn reasoning strategies in reading comprehension tasks. In: International Joint Conference on Natural Language Processing, pp. 957–966 (2017)

24. Xu, Y., et al.: Multi-task Learning with sample re-weighting for machine reading comprehension. In: North American Chapter of the Association For Computational Linguistics, pp. 2644–2655 (2019)
25. Liu, X., Duh, K., Gao, J.: Stochastic answer networks for natural language inference. arXiv Comput. Lang. (2018)
26. Pennington, J., Socher, R., Manning, C.: Glove: global vectors for word representation. In: Proceedings of the 2014 Conference on Empirical Methods in Natural Language Processing (EMNLP), pp. 1532–1543 (2014)
27. Salimans, T., Kingma, D.P.: Weight normalization: a simple reparameterization to accelerate training of deep neural networks. In: Neural Information Processing Systems, pp. 901–909 (2016)
28. Kingma, D.P., Ba, J.: Adam: a method for stochastic optimization. In: International Conference on Learning Representations (2015)

Horus: An Interference-Aware Resource Manager for Deep Learning Systems

Gingfung Yeung[1], Damian Borowiec[1], Renyu Yang[2(✉)], Adrian Friday[1], Richard Harper[1], and Peter Garraghan[1]

[1] School of Computing and Communications, Lancaster University, Lancaster, UK
{g.yeung1,d.borowiec,a.friday,r.harper,p.garraghan}@lancaster.ac.uk
[2] School of Computing, University of Leeds, Leeds, UK
r.yang1@leeds.ac.uk

Abstract. Deep Learning (DL) models are deployed as jobs within machines containing GPUs. These DL systems - ranging from a singular GPU device to machine clusters - require state-of-the-art resource management to increase resource utilization and job throughput. While it has been identified that co-location - multiple jobs co-located within the same GPU - is an effective means to achieve this, such co-location incurs performance interference that directly debilitates DL training and inference performance. Existing approaches to mitigate interference require resource intensive and time consuming kernel profiling ill-suited for runtime scheduling decisions. Current DL system resource management are not designed to deal with these problems. This paper proposes Horus, an interference-aware resource manager for DL systems. Instead of leveraging expensive kernel-profiling, our approach estimates job resource utilization and co-location patterns to determine effective DL job placement to minimize likelihood of interference, as well as improve system resource utilization and makespan. Our analysis shows that interference cause up to 3.2x DL job slowdown. We integrated our approach within the Kubernetes resource manager, and conduct experiments in a DL cluster by training 2,500 DL jobs using 13 different models types. Results demonstrate that Horus is able to outperform other DL resource managers by up to 61.5% for resource utilization and 33.6% for makespan.

Keywords: Machine learning systems · Performance interference · Deep Learning · GPU scheduling · Cluster resource management

1 Introduction

Deep Learning (DL) is an increasingly important type of machine learning algorithm with significant potential for touching many aspects of society [12]. The rapid growth in the number of DL practitioners and the data they require in their computations has created a necessity for both individually powerful, as well as large-scale clusters of machines equipped with GPUs - specialized hardware accelerators - to facilitate the vast amounts of computation DL entails at

M. Qiu (Ed.): ICA3PP 2020, LNCS 12453, pp. 492–508, 2020.
https://doi.org/10.1007/978-3-030-60239-0_33

reduced training time. These systems, which we refer to as *DL systems*, use resource management frameworks to perform DL scheduling and job placement (i.e allocation of DL jobs onto GPUs). An important goal for DL systems is to maximize the effective utilization of these expensive resources through minimizing *makespan, job waiting,* and *job completion time (JCT)*.

DL systems, particularly clusters, experience issues associated with GPU underutilization and long queuing times [19]. One cause of such underutilization is that DL resource managers disallow *co-location* of multiple DL jobs within the same GPU [13,25]; a characteristic shared within other resource managers such as Kubernetes and Yarn that were originally designed for CPU-based workloads [16,35]. Instead, the majority of DL resource managers focus on reducing network latency and locality [3,13,25]. This inability to co-locate DL jobs within the same GPU results in reduced resource utilization, longer queuing times and reduced cost efficiency within DL systems.

Recent DL resource managers have been proposed that make placement decisions by consolidating DL jobs onto fewer machines to minimize workload and JCT [3,13,37]. However, while DL resource managers now exist that allow for co-location [29,37], there has been little attention drawn to the *performance interference* which arises between multiple DL jobs training within the same GPU. Performance interference (which we refer to as *interference*), results in slower training step time and overall epoch time. Previous work has demonstrated the existence of interference within DL systems, resulting in an 18% JCT slowdown [37]. Furthermore, as no existing DL resource manager considers the impact of interference in DL job co-location decisions, this can lead to poor placement of unsuitable jobs resulting in a higher makespan, increase in JCT, job eviction and job failures from GPU out-of-memory (OOM) errors [19].

We propose Horus, a DL system resource manager that maximises resource utilization and minimizes makespan whilst attempting to reduce JCT performance degradation by anticipating interference due to co-location. By leveraging DL model application features ascertained from [40], Horus is able to estimate the GPU utilization of DL jobs *prior to execution*, and make better placement decisions to determine suitable co-location combinations with the lowest interference. Our approach avoids the need to profile kernel patterns [6,26], modification of the GPU thread-block scheduler (commonly proprietary to GPU hardware manufacturers), and extensive online profiling of job execution in an isolated GPU at scheduler runtime; all of which are expensive and time consuming processes in terms of system development and job placement. Our core contributions are: **(1) Analysis of DL job co-location patterns**. We empirically measured 276 unique combinations of interference patterns from co-locating 13 prominent types of DL jobs comprising both vision and language models (Sect. 5.2). Our results demonstrate that co-locating DL jobs with high GPU utilization requirements leads to a 1.5X–3.2X JCT increase stemming from interference. Moreover, we observe that interference patterns significantly vary between different DL job combinations, and that equivalent GPU utilization can exhibit dissimilar JCT degradation patterns.

(2) Co-location DL resource manager. Via leveraging GPU utilization estimation, we construct a cost-based best-fit-decreasing model that determines suitable placement for DL job co-location (Sect. 4). Horus was integrated into Kuberentes [16] - an open source cluster manager, and was evaluated by submitting 2,500 DL jobs from different application domains into a GPU cluster under different workload patterns (Sect. 5). Results demonstrate that Horus achieves high resource utilization and scheduler performance, and outperforms various cluster resource managers - including other co-location approaches [35,37]) - with improvements up to 61.5% and 33.6% for resource utilization and makespan.

2 Background

2.1 Deep Learning

Deep Learning is a type of machine learning algorithm based on neural networks. DL models are formed by *deep neural networks* (DNNs), consisting of input, hidden, and output layers [12] and have two phases: *training* and *inference*. DL model training requires significant volumes of data [15] to iteratively minimize an error objective. Larger numbers of layers and units per layer results in higher number of floating point operations (FLOPs) to execute [32]. GPUs are frequently used to accelerate DL model training due to their ability to rapidly perform FLOPs execution using thousands of processing cores. For example, Nvidia GPUs define abstraction over a group of cores as streaming multiprocessors (SMs), which are used to execute GPU kernels. A complex DL model (i.e expressed by model depth and width) will result in greater number of FLOPs, and thus a higher GPU utilization in comparison to simpler DL models that leverage the same batch size and GPU architecture [1].

2.2 Deep Learning Resource Managers

Due to the growing number and scale of DL jobs that require training on TB-scale data, researchers and businesses leverage *DL systems*: powerful machines or clusters of machines to accelerate training. Users submit jobs to a DL system via a web portal or command line interface with specified job configurations (e.g. batch size, model, dataset) comprising one or more tasks that execute within a container. DL jobs are assigned resources and allocated onto machines through use of a *resource manager* to increase system resource efficiency to satisfy a specified Service Level Agreement (SLA). Existing DL system resource managers[1] have focused on a specific sub-set of objectives including minimizing makespan, JCT, as well as maximizing system resource utilization and energy-efficiency [4].

Recent studies of production DL systems have identified several challenges: low utilization of system resources reflected by an average GPU utilization of

---

[1] Which we refer to as DL resource managers.

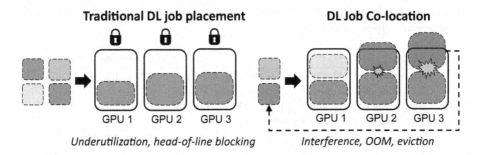

Fig. 1. Difference between traditional vs. DL resource managers: DL job co-location manifesting interference.

52% [19], and long queuing time for DL jobs between 4000 s–8000 s due to head-of-line blocking [13]. These challenges are exacerbated by DL systems leveraging non-preemptive traditional schedulers [16,35] that require DL jobs to hold exclusive access to a given GPU. This is particularly problematic in the context of schedulers due to its negative impact upon job throughput and makespan, system availability, and cost efficiency.

Improving upon established approaches, recent DL resource managers have been designed to address challenges of under-utilization and long queue times via improvements to network locality and bandwidth [3,13,25]. Another approach demonstrated to be effective for DL resource managers is enabling the *co-location* of DL jobs within the same GPU to execute simultaneously, improving overall system resource utilization [29,37]. However, few DL resource managers consider or capture the drawbacks of co-location when performing DL job placement decisions, including the manifestation of interference that might result (Fig. 1).

2.3   Deep Learning Interference

Interference occurs when multiple processes compete for limited resources within the same machine [8,23]. Interference of DL jobs co-located within the same GPU has been shown to result in 18% JCT degradation [37]. This is problematic when considering that DL jobs may train in the region of hours to days. Hence, in order for DL systems to fully exploit co-location, DL resource managers should consider the effects of interference when performing DL job placement.

Profiling DL job resource usage (notably GPU utilization) allows DL resource managers to minimize interference resultant from co-location. GPU interference differs from that of CPU interference because of their processor architectures. GPUs use a thread-block scheduler to schedule compute or memory intensive kernels to streaming multiprocessors, and leverage *single instruction multiple data* parallelism on many cores, whereas CPU uses *multiple instruction multiple data* parallelism on fewer cores. Precisely calculating interference for different co-located DL job combinations is challenging due to the diversity in model types and kernels that are implemented in different DL libraries [2].

Table 1. Studied DL models in Vision (Cifar10 [21]) and NLP (WikiText-2 [24], News-Commentary v14-en-zh [36]

Model	Type	Batch size
MobileNetV2 [27], GoogLeNet [31], ResNet [15], VGGNet [30], DenseNet [17], SqueezeNetV1 [18], ShuffleNetV2 [22], ResNeXt [38], MNASNet [33], PyramidNet [14], DualPathNetwork [7]	CNN	64, 128, 256
LSTM [11], Transformer [34]	RNN	16, 32, 64

While there exist several GPU resource managers that minimize interference [6,26], such approaches require extensive online profiling of job kernel access patterns at scheduling runtime. This is challenging as profiling DL job resource usage (e.g. instruction per cycle, DRAM throughput, compute efficiency) to infer interference by creating suitable performance profiles may extend DL job training within the regions of minutes to hours. Additionally it must be performed for every new DL model type submitted into the system. This results in considerable resource overhead when using profiling tools such as nvprof and nv-nsight-cu-cli, as well as increased DL system makespan. We believe an alternative approach is to understand how DL model types and model configurations result in varying DL job resource usage patterns, in order to infer interference for different co-location combinations. This would allow for DL resource managers to co-locate DL jobs more effectively through considering the impact of interference in placement decisions.

3   Deep Learning Interference Study

This section presents our analysis of different interference profiles for co-located DL jobs. While there have been prior studies into GPU interference [6,39], the majority of works use few or relatively simple DL models types and configurations (i.e. LeNet, Multilayer perceptron, MNIST). We conduct a micro-benchmark of different co-location combinations with heterogeneous configurations of prominent DL model types, and study their influence upon resource utilization and interference profiles.

3.1   Setup

A wide variety of DL jobs were deployed within a DL system (4 x Nvidia Geforce 1080, Intel i7-6850k, Nvidia Docker 2, CUDA Toolkit 10.0), and using the DL library frameworks AllenNLP [10] and Pytorch 1.1. Leveraging methods established within prior studies of GPU interference [6,39], micro-benchmarking was conducted by co-locating paired combinations of DL jobs within the same GPU device, and then measuring the corresponding JCT performance degradation

T_{deg} from interference during DL job training. Performance degradation is calculated as

$$T_{deg} = \frac{|T_{colo} - T_{solo}|}{T_{solo}} \quad (1)$$

where T_{colo} is the time taken for a co-located DL job to reach a fixed time epoch, and T_{solo} is the time taken for the same DL job training in isolation.

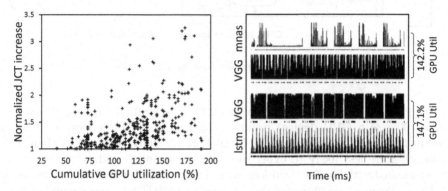

(a) Interference profiles from different DL job co-location combinations

(b) Kernel co-located jobs patterns with equiv. GPU utilization

Fig. 2. GPU utilization interference patterns

The DL job micro-benchmark comprises 13 unique DL model architectures including both convolution neural networks (CNNs) and recurrent neural networks (RNNs) from computer vision domains and natural language processing as shown in Table 1. Each model is then further modified with different model configurations such as residual blocks, projection dimensions, and scale parameters. DL models were selected due to their usage within previous DL resource managers [13, 25, 37] and prominence in the machine learning community. This provided a total of 24 unique DL model configurations, and 276 unique co-location combinations (300 when including DL jobs in isolation) for profiling. Each DL job is trained for a fixed set of five epochs to capture a stable performance profile. Analysis metrics were collected from monitoring the DL system, using `nvidia-smi` to collect statistics for GPU utilization, PCIe bandwidth and GPU memory usage.

3.2 Analysis

We found that co-located DL jobs that require high GPU utilization results in greater JCT slowdown from higher interference. For example, co-locating two VGG19 models (each requiring 90%+ utilization in isolation) results in over a 3.2X JCT increase. Figure 2a shows that for all co-located job combinations,

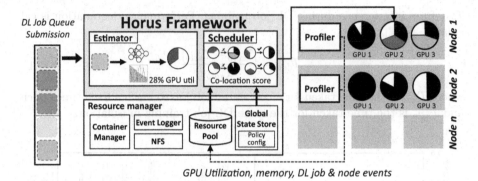

Fig. 3. Horus scheduler framework overview

GPU overcommitment (i.e. the cumulative GPU utilization requirement greater than 100% of a GPU device) results in an average JCT increase of 42%. In contrast, pairs of co-located DL jobs that individually require less than 50% utilization are less likely to experience overcommitment, and, as a result, can be co-located with similar sized jobs with minimal interference JCT increases of 1–10%. Moreover, we observed that the degree of JCT increase stemming from interference varies quite substantially, even for jobs with similar DL job utilization levels in isolation. This phenomena can be seen in kernel access patterns shown in Fig. 2b, whereby co-locating VGG with either LSTM (1 layer, hidden dimensions 128) or MNASNet (Depth 1.3) results in an equivalent GPU utilization of approximately 147%, however results in a 1.3X and 3.03X JCT increase. The reason for this behaviour is due to convolution kernels being intrinsically more compute intensive in contrast to GEMM kernels, as well as different memory transfer and compute patterns when contending for resources [6].

4   The Horus Framework

4.1   Overview

Horus is designed to operate within DL resource managers that leverage GPUs, and comprises two main components as shown in Fig. 3: the *Resource Estimator*, and *Resource Scheduler*. At submission time, the estimator calculates the GPU utilization of an executing, or incoming DL job, ascertained via online metric collection or prediction [40] (detailed in Sect. 4.2). The scheduler assigns DL jobs to GPUs by ranking their suitability to support co-location. Our approach attempts to greedily maximize GPU utilization to minimize makespan whilst attempting to avoid placement decisions that lead to severe interference causing JCT slowdown (detailed in Sect. 4.3).

Our resource manager can be deployed within a single DL system, or within a cluster integrated within existing resource managers such as Kubernetes [16]. In the context of clusters, Horus uses a shared-state, centralized architecture due to its suitability to handle long jobs (an common characteristic of DL job training), as well as providing a view of overall global cluster for high quality scheduling decisions [9,28]. A centralized cluster view is maintained through a shared centralised repository, with monitoring agents deployed in each machine reporting application and system utilization metrics for placement decisions.

4.2 Resource Estimator

Estimation of Expected GPU Utilization: While our framework does not modify any underlying DL libraries to function, it does require DL model utilization to be provided. Such utilization patterns can be provided via execution, or prediction. For Horus we have leveraged the prediction technique from Yeung et al. [40] to avoid executing each and individual incoming job j, which operates by traversing the computation graph, extracting model features and eventually estimating its GPU utilization, i.e. $\mathbb{E}(GUtil_j)$. Such features are well known and frequently modified by Machine Learning researchers and developers [27,32,33], hence it is relatively straight forward for practitioners to manually extract, or automatically collect such metrics via graph analysis tools such as TensorFlow profiler[2] and TorchScript[3].

Estimation of Expected GPU Memory: The only exception for straight forward metric collection the total job memory size (MiB) due to initialization and optimization of individual DL libraries. However, it is possible to estimate the minimum expected memory usage in bytes by considering the following four factors involved in both forward M^f and backward passes M^b [12]: (i) the batch size of data B, (ii) the number of activations A, (iii) number of gradients G and (iv) the number of parameters P. In addition to an initialization overhead δ, the overall estimated memory requirement for a given DL job j can be expressed as:

$$\mathbb{E}(GMem_j) = \mathbb{E}(M_j^f) + \mathbb{E}(M_j^b) + \delta = (B * A + P) + B * G + \delta \qquad (2)$$

The expectation of both GPU utilization and GPU memory will be used for node capacity check in the scheduler in case of tackling an incoming job.

4.3 Resource Scheduler

This section describes our scheduling approach to effectively co-locate DL jobs and handle potential placement issues from interference. We observe that in order to maximize GPU utilization, it is necessary to allow co-location of DL jobs onto the same GPU. Gandiva [37] employs a random trial-and-error strategy to co-locate DL jobs. In their approach, by monitoring the job in isolation and application throughput, a job is killed or migrated to another node randomly using an undefined threshold value and time period. In such an approach, it is

---

[2] https://github.com/tensorflow/tensorflow/tree/master/tensorflow/core/profiler.
[3] https://pytorch.org/docs/stable/jit.html.

possible for random job migration to be allocated with another incompatible job leading to equal or greater performance slowdown.

To alleviate the performance degradation stemming from co-location interference, the core of resource scheduling is to understand the compute resource requirement *prior* to job execution and incur as less overhead (cost) as possible. We design the cost to reflect the selection preference of a node mainly considering GPU memory usage and GPU utilization, respectively. We can therefore determine the most suitable placement based on per-node cost inference.

Cost Inference: We break down the cost of scheduling job j onto an individual GPU k (denoted by $j \to k$):

$$Cost_{j \to k} = C_{j \to k}^{GMem} + C_{j \to k}^{GUtil} + \epsilon_j^{jobType} \tag{3}$$

We add up the incurred cost of running job j regarding GPU memory usage and GPU utilization increase, followed by a calibration ϵ due to DL jobs type.

In particular, the cost of GPU memory $C_{j \to k}^{GMem}$ is referred to as a weighted proportion of GPU memory usage (Eq. 4), in light of the key implication – higher current GPU memory usage causes a higher chance of OOM and JCT slowdown.

$$C_{j \to k}^{GMem} = \omega * \frac{GMem_k^{used}}{GMem_k^{total}} \tag{4}$$

where $GMem_{total}$ is the total GPU memory of the device and ω is used to customize and indicate the performance impact.

As there exists a relationship between increased GPU utilization of co-located DL jobs and JCT slowdown (Sect. 3.2), we penalize the combinations of co-located DL jobs when over-commitment manifests – the total forthcoming GPU utilization (i.e., current GPU utilization $GUtil_k^{curr}$ and the estimated increment $\mathbb{E}(GUtil_j)$) exceeds 100%. Likewise, we add a penalty hyperparameter $\phi \in [1, 2]$ to tweak the slowdown impact due to GPU over-commitment:

$$GUtil_{j \to k} = GUtil_k^{curr} + \mathbb{E}(GUtil_j)$$

$$C_{j \to k}^{GUtil} = \begin{cases} \phi * |GUtil_{j \to k} - 100|, & \text{if } GUtil_{j \to k} > 100 \\ 100 - GUtil_{j \to k}, & \text{if } GUtil_{j \to k} \le 100 \end{cases} \tag{5}$$

The implication behind this *piecewise* cost setup is to pack a job so that the host node can approach to 100% GPU utilization without resource over-commitment. For instance, higher remaining GPU capacity results in a higher cost. On the other hand, once over-commitment occurs, the cost will be increasingly augmented, which indicates a reduced scheduling probability of the node, considering job's performance.

Similarly, as observed in Fig. 2b, different DL model architectures exhibit various degrees of JCT slowdown from interference, thus a *niceness* hyperparameter ν is added to accommodate this behaviour (Eq. 6. Our scheduling approach will favour a particular DL job types based on ν selection.

$$\epsilon_j^{jobType} = \begin{cases} \nu, & \text{if } C_{jobtype} \text{ is CNN} \\ 0, & \text{otherwise} \end{cases} \tag{6}$$

Algorithm 1. Best Fit Decreasing Job Scheduling Algorithm

Input: (J, S) // k jobs in the queue and current cluster state
Output: (scheduleStatus, NodeBindingResult res)

1: $\mathcal{J} \leftarrow$ DescendSort(J) // a job collection via descend sort by jobs' GPU util
2: $res \leftarrow$ dict()
3: **for** j in $\mathcal{J}$ **do**
4: **if** hasAllocatableResources(j, S) **then**
5:    $\mathcal{N} \leftarrow$ getFeasibleNodes(j, S) //capacity check(CPUs, Mems, GPU Mems)
6:    $\mathcal{G} \leftarrow getAllGPUs(\mathcal{N})$
7:    $\lambda \leftarrow j.requestedGPU$; $\sigma \leftarrow$ GetMinRequiredNodeNum(j),
8:    **if** LEN($\mathcal{N}$) $< \sigma$ **then**
9:      **continue**
10:    $\mathcal{C}_{\mathcal{G} \rightarrow j} \leftarrow$ GetCosts($\mathcal{G}$, j) // use Eq. 3 to calculate cost for each GPU
11:    $\mathcal{G}^{+} \leftarrow$ AscendSort($\mathcal{C}_{\mathcal{G} \rightarrow j}$).topK($\lambda$)
12:    $res \leftarrow$ Put(j, $\mathcal{G}^{+}$)
13: **if** Len(res) > 0 **then**
14: BindNodes(res, j)
15: **return** *ScheduleResults.SCHEDULED*, *res*
16: **else**
17: **return** *ScheduleResults.EMPTY*, Nil

Cost-Based Best-Fit-Decreasing Job Scheduling: Algorithm 1 outlines our scheduling solution. At each scheduling time, the scheduler will collect the current DL system state, and fetch job collection J from the queue. We adopt a best-fit-decrease like algorithm to prioritize jobs with larger GPU requests, avoiding their long-time starvation. To do so, we firstly sort the job collection $\mathcal{J}$ according to the pertaining GPU utilization (Line 1). We then iteratively attempt to find and bind resources for each job. Specifically, we firstly check the resource capacity and filter out available candidate nodes $\mathcal{N}$ that can satisfy all requirements of job j in terms of CPU, memory and GPU memory (Lines 4–9). GPU collection $\mathcal{G}$ is obtained from the pertaining candidate nodes. For instance, the GPU memory requirement is inferred by using Eq. 2 discussed in Sect. 4.2. We can further assess the incurred cost stemming from running the job j on each GPU of $\mathcal{G}$, separately, via the cost estimation in Eq. 3 (Line 10). To reduce possible interference, the best fit is to select the nodes with minimized impact in case of placing j. Hence the scheduler prefers nodes that host the selected GPUs ($\mathcal{G}^{+}$) with λ least costs (Lines 11–12), and finalizes placement decisions by sending binding requests to the scheduler (Lines 13–17).

Job Failover and Rescheduling: In some scenarios, it is possible for our approach (as well as other DL resource managers) to encounter issues associated with OOM errors due to co-located DL jobs exceeding the total GPU memory capacity resulting from incorrect estimation. We address this issue by using a separate thread to monitor job progress, and in the event of failure, jobs are resubmitted onto the scheduling queue. The scheduler will then update the DL job request with necessary GPU memory requirements, where GPU memory must be equal or greater than the job to be included in consideration.

In a worst case scenario whereby each DL job already fully utilizes GPU resources (i.e. no memory available for packing), our algorithm will pend waiting jobs until existing jobs terminate and release resources and so act similarly to traditional DL resource managers. However, we can leverage other priority or time-sharing primitives to reduce the waiting time.

Complexity Analysis: It is worth noting that at each scheduling cycle, the scheduler re-considers all available GPUs and nodes, given resource requirements are satisfied, including the number of GPUs, and the available GPU memory against the least expected memory calculated in Eq. 4. The time complexity of the capicity check procedure (Lines 4–5) is $\mathcal{O}(kNG)$ where k denotes the size of pending jobs, while N and G represent the number of nodes and GPUs in the DL system, respectively. In addition to $\mathcal{O}(NlogN)$ used for sorting and asymptotically $\mathcal{O}(N)$ in top-K selection, the overall complexity remains $\mathcal{O}(kNG)$. The scheduling can therefore scale well with the increment of either node or GPU scaling-out. Since our scheduler uses a placement algorithm for co-location is greedy, we do not guarantee algorithm optimality. Nevertheless, we find it is less invasive and less time consuming in improving scheduling effectiveness.

5   Evaluation

5.1   Experiment Setup

System: Horus was deployed onto a 12-GPU cluster with each node containing 4 x Nvidia RTX 2080 Ti GPUs, an AMD Ryzen 1920X 12 Core Processor (2 threads per core) with 10Gb Ethernet network, and 128 GB DDR4 memory. Each node was installed with Ubuntu Disco 19.04 and Nvidia driver 430.50. In our experiments, the DL libraries (AllenNLP [10], Pytorch 1.1) and CUDA toolkits responsible for DL job instantiation and execution were stored in a Docker container. Our cluster uses the Kubernetes 1.15.2 resource management framework due to its prominence within the resource management community. cAdvisor and DCGM were configured to extract data at 1 s and 250 ms intervals, respectively, as initial trial runs indicated that these parameters resulted in effective job throughput given our cluster configuration.

Comparative Algorithms: To evaluate the Horus scheduling co-location algorithm described in Sect. 4.3, we have designed and implemented two additional scheduling algorithms for comparison:

▷ *First in First Out (FIFO)*: Emulating slot-based approaches established in big data cluster schedulers such as Kubernetes [16] and YARN [35], FIFO assigns the next incoming DL job onto an idle GPU without job co-location. As threshold values are not defined in prior work, the timing period was set at 0–60 s so that all DL jobs achieve stable performance patterns, and we configure the performance threshold to 50% informed by interference patterns (Fig. 2a).

▷ *Opportunistic Bin Packing (OBP)*: Assigns DL jobs based on GPU memory availability. The algorithm assigns and co-locates DL jobs based on estimated memory requirements prior to submission via exploiting our memory estimation model described in Eq. 4. During job submission time, if a GPU with higher memory is available than estimated memory, then the scheduler will opportunistically allocate a schedule for that job to that GPU.

With the exception of FIFO, preemption was enabled for all algorithms. Our preemption strategy is triggered when a GPU is overcommitted (threshold defined by operator) or an idle GPU. This preemption allows previously co-located DL jobs experiencing interference to be rescheduled to another GPU. Horus have configured to operate job collection size $J = 15$ based on initial experiment runs to provide a sufficient number of candidate DL jobs for placement.

Workload: Experiments were conducted using a mixture of DL job types generated from Table 1, as well as new DL model configurations and model types (LSTM, Transformer), resulting in Horus being exposed to approximately 50% new DL jobs not used in predictor training. Selected models and datasets leveraged in our experiments are well established in micro-benchmarking DL resource managers [13,25,37]. Submitted DL jobs use a distribution between 3 min to 2 h following DL job sizes derived from JCT of production systems [13]. Jobs are characterized as short/long (<800 s or ≥800 s) and light/heavy (<60% or ≥60% GPU utilization). JCT was controlled by terminating jobs at specified epoch numbers to emulate JCT patterns of production systems, as well as train sufficient DL jobs in a reasonable time frame. For our experiments we focused on DL jobs requiring a single GPU for training, following established practice from other co-location DL schedulers [29]. This is because between 50%–86% of total production DL jobs have been shown to require a single GPU [5,13,19] and hence we have attempted to capture a broad spectrum of different job and model types. Our objective is to study changes in workload makespan and JCT due to interference from DL job co-location. Furthermore, locality—a key focus within prior DL cluster schedulers [13,25,37]—introduces a non-intuitive dimension of JCT heterogeneity even for jobs running in isolation, making it difficult to fairly measure potential trade-off gains between resource utilization against JCT increase when co-locating DL jobs.

Metrics: Algorithm effectiveness was measured using the following metrics: *Cluster GPU Resource Utilization*: GPU utilization of all devices, *Job Completion Time (JCT)*: End-to-end completion time for a DL job, commencing from the start of job execution and finishing at job completion. *Workload Makespan*: The total span-time to complete all DL jobs from en-queuing through to completion. Moreover, we have also measured parameters for general cluster resource utilization (CPU, memory, disk).

Experiment Runs: Each algorithm scheduled 100 DL jobs for each workload pattern five times each, successfully training a total of 1,500 DL jobs; equivalent to approximately 66 days of DL system GPU computation.

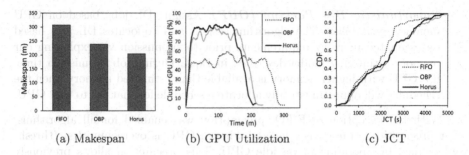

| (a) Makespan | (b) GPU Utilization | (c) JCT |

Fig. 4. Experiment results for comparing DL resource manager operation.

5.2   Results

Makespan: In all experiment runs, Horus was able to successfully schedule all DL jobs with the lowest makespan as shown in Fig. 4a. This is demonstrated by a makespan of 204 min, and is equivalent to a 33.6% improvement against FIFO, and a 22.2% improvement over OBP as shown in Table 2. The reason for a lower makespan is due to Horus being able to perform better placement decisions for co-locating DL jobs, by leveraging our GPU utilization estimator to avoid underutilization and OOM errors from overcommitment. We observe that OBP has the second lowest makespan for all experiments. FIFO has the highest makespan at 306.9 min due to longer queueing times for DL jobs waiting to acquire exclusive access to an idle GPU for training.

Table 2. Workload makespan, GPU utilization and JCT statistics.

Objective	Algorithm	Avg.	St. dev	Change
Makespan (mins)	FIFO	306.9	1.15	-
	OBP	238.6	4.9	22.2%
	Horus	**204.0**	**8.5**	**33.6%**
Utilization (%)	FIFO	43.1	16.7	-
	OBP	59.7	27.2	38.5%
	Horus	**69.6**	**26.9**	**61.5%**
JCT (s)	FIFO	1869.7	1054.3	-
	OBP	2277.5	1293.1	21.8%
	Horus	**2193.8**	**1307.3**	**17.3%**

Utilization: Horus is able to achieve high overall cluster resource utilization in all experiment runs as shown in Table 2 and Fig. 4b, reflected by an average 69% utilization in comparison to FIFO (43%) and OBP (60%). This is resultant of the Horus algorithm determining better co-location combinations for DL job placement to maximize GPU utilization from predicted memory and utilization requirements of DL jobs described in Sect. 4.3. While OBP achieve relatively high utilization compared to FIFO due their ability to perform co-location, OBP is able to achieve higher utilization as a result of its rapid scheduling cycle.

JCT: Figure 4c and Table 2 (JCT) shows the average JCT for DL jobs. We observe that FIFO achieves the fastest JCT at 1869.7 s, due to DL jobs acquiring exclusive GPU access disallowing co-location and thusly no interference. In contrast, we observe that all co-location algorithms experience JCT slowdown between 17.3%–21.8%. A note of particularly interest is Horus's ability to effectively co-locate and reduce makespan to achieve higher DL job throughput will paradoxically expose the DL cluster to greater interference and consequent JCT slowdown. Horus does however still achieve a lower JCT in comparison to OBP, and when considering our gains to resource utilization and makespan, we view this as an acceptable trade-off.

Scalability: Horus has been evaluated through empirical means in a DL cluster of comparable scale found in recent works [25, 29]. As discussed in Sect. 4.2, our algorithm complexity is a linear combination between the nodes and GPUs. Moreover, Horus was evaluated using established communication mechanisms in the Kubernetes framework for orchestrating job deployments and file system mounting, which has been demonstrated to scale to thousands of machines [16].

6 Related Work

DL Resource Managers: Gandiva [37] focuses primarily on improving the time-sharing, by enabling DL job *context-switching*, and extracting job throughput (e.g. minibatch per second) thus allowing 'random-and-trial' job co-location placement and eviction upon performance slowdown. Tiresias [13], focuses on improving average JCT and job starvation time. It does so by profiling network latency, consolidating distributed DL jobs and implementing a multi-level feedback queue, which adjusts job priorities. Optimus [25] implements a performance predictor model, which at runtime, adjusts the number of required parameter servers or workers. It assumes job convergence is predictable, which in many cases is difficult to ascertain [13]. All of the above DL cluster schedulers are complimentary to our work as they focus on addressing various challenges and scheduling objectives, related to locality, time-sharing and average JCT. Horus focuses on DL workload makespan and GPU utilization, as well as making placement decisions based on interference between co-located DL jobs in GPUs.

Interference-Aware Resource Managers: The study of GPU interference is an established area [6, 26], whereby approaches leverage heavy static profiling of GPU kernel access patterns within isolated machines to classify the workload types at the job submission time, identifying suitable placements. There also exist various cluster schedulers which reduce performance interference of heterogeneous CPU workloads [8, 9, 20]. As discussed in Sect. 2, these cluster schedulers are not designed to effectively handle GPU scheduling, particularly DL clusters due to differences in hardware, workload, and long queuing times

[19]. Horus builds upon these ideas, proposing a prediction discussed in Sect. 4.2 which complements other GPU interference-aware resource managers [29,39], by using lightweight profiling to characterize DL job utilization patterns.

GPU Interference Analysis: Interference analysis and fine-grained GPU kernel scheduling is an established field of research within the hardware architecture community. Researchers have proposed various solutions to mitigate kernel interference in GPU kernel scheduling [6]. All established DL cluster schedulers do not incorporate GPU kernel scheduling characteristics during job scheduling due to heterogeneous DL system hardware and DL frameworks. As an alternative to threadblock scheduling and kernel prediction, Horus uses application features to ascertain job utilization to alleviate GPU interference in DL systems, and provides a study focused on DL job interference due to co-location.

7 Conclusions

In this paper we have presented Horus, a resource management framework for Deep Learning systems that achieves high job throughput and resource efficiency via effective DL job co-location. Horus performs intelligent placement and co-location of DL jobs in GPUs by estimating job utilization patterns using model features without requiring DL library modification or heavy kernel profiling at scheduler runtime. From our analysis, we have empirically shown the diversity of interference profiles manifesting between co-located DL jobs, and can result in up to 3.2X increase within their completion time. We have created a resource estimator and scheduler integrated into existing DL cluster resource managers. Through experiments we have shown that Horus is capable of reducing makespan by up to 33.6% and improving cluster GPU utilization by 61.5%.

References

1. Nvidia Deep Learning Performance Guide, https://docs.nvidia.com/deeplearning/sdk/dl-performance-guide/index.html
2. Pytorch, https://pytorch.org/
3. Amaral, M., Polo, J., Carrera, D., Seelam, S., Steinder, M.: Topology-aware GPU scheduling for learning workloads in cloud environments. In: ACM SC (2017)
4. Bhuiyan, A., Guo, Z., Saifullah, A., Guan, N., Xiong, H.: Energy-efficient real-time scheduling of DAG tasks. ACM TECS **17**, 1–25 (2018)
5. Chaudhary, S., et al.: Balancing efficiency and fairness in heterogeneous GPU clusters for deep learning. In: ACM EuroSys 2020 (2020)
6. Chen, Q., Yang, H., et al.: Prophet: precise QoS prediction on non-preemptive accelerators to improve utilization in warehouse-scale computers. In: ACM SIGOPS Operating Systems Review (2017)
7. Chen, Y., Li, J., Xiao, H., Jin, X., Yan, S., Feng, J.: Dual path networks. In: Advances in Neural Information Processing Systems, pp. 4467–4475 (2017)
8. Delimitrou, C., Kozyrakis, C.: Paragon: QoS-aware scheduling for heterogeneous datacenters. In: ACM SIGPLAN Notices. ACM (2013)

9. Delimitrou, C., Kozyrakis, C.: Quasar: resource-efficient and QoS-aware cluster management. In: ACM ASPLOS (2014)
10. Gardner, M., Grus, J., Neumann, M., Tafjord, O., et al.: AllenNLP: a deep semantic natural language processing platform (2017)
11. Gers, F.A., Schmidhuber, J., Cummins, F.: Learning to forget: continual prediction with LSTM (1999)
12. Goodfellow, I., Bengio, Y., Courville, A.: Deep Learning. MIT Press, Cambridge (2016)
13. Gu, J., Chowdhury, M., Shin, K.G., Zhu, Y., et al.: Tiresias: a {GPU} cluster manager for distributed deep learning. In: USENIX NSDI (2019)
14. Han, D., Kim, J., Kim, J.: Deep pyramidal residual networks. In: IEEE CVPR, pp. 5927–5935 (2017)
15. He, K., Zhang, X., Ren, S., Sun, J.: Deep residual learning for image recognition. In: IEEE CVPR (2016)
16. Hightower, K., Burns, B., Beda, J.: Kubernetes: Up and Running: Dive into the Future of Infrastructure. O'Reilly Media Inc., Sebastopol (2017)
17. Huang, G., Liu, Z., Van Der Maaten, L., Weinberger, K.Q.: Densely connected convolutional networks. In: IEEE CVPR, pp. 4700–4708 (2017)
18. Iandola, F.N., Han, S., et al.: Squeezenet: alexnet-level accuracy with 50x fewer parameters and <0.5 mb model size. arXiv preprint arXiv:1602.07360 (2016)
19. Jeon, M., et al.: Analysis of large-scale multi-tenant GPU clusters for DNN training workloads. arXiv preprint arXiv:1901.05758 (2019)
20. Kambatla, K., Yarlagadda, V., Goiri, Í., Grama, A.: UBIS: utilization-aware cluster scheduling. In: IEEE IPDPS (2018)
21. Krizhevsky, A., Hinton, G., et al.: Learning multiple layers of features from tiny images. Tech. rep, Citeseer (2009)
22. Ma, N., Zhang, X., Zheng, H.T., Sun, J.: Shufflenet v2: practical guidelines for efficient CNN architecture design. In: ECCV (2018)
23. Mars, J., Tang, L., et al.: Bubble-up: increasing utilization in modern warehouse scale computers via sensible co-locations. In: IEEE/ACM MICRO (2011)
24. Merity, S., Xiong, C., Bradbury, J., Socher, R.: Pointer sentinel mixture models. arXiv preprint arXiv:1609.07843 (2016)
25. Peng, Y., Bao, Y., Chen, Y., Wu, C., Guo, C.: Optimus: an efficient dynamic resource scheduler for deep learning clusters. In: ACM EuroSys (2018)
26. Phull, R., et al.: Interference-driven resource management for GPU-based heterogeneous clusters. In: Proceedings of HDPC. ACM (2012)
27. Sandler, M., Howard, A., Zhu, M., Zhmoginov, A., Chen, L.C.: Mobilenetv 2: inverted residuals and linear bottlenecks. In: IEEE CVPR, pp. 4510–4520 (2018)
28. Schwarzkopf, M., Konwinski, A., Abd-El-Malek, M., Wilkes, J.: Omega: flexible, scalable schedulers for large compute clusters. In: ACM EuroSys (2013)
29. Shen, H., et al.: Nexus: a GPU cluster engine for accelerating DNN-based video analysis. In: ACM SOSP (2019)
30. Simonyan, K., Zisserman, A.: Very deep convolutional networks for large-scale image recognition. CoRR arXiv:1409.1556 (2014)
31. Szegedy, C., et al.: Going deeper with convolutions. In: Computer Vision and Pattern Recognition (CVPR) (2015)
32. Tan, M., Le, Q.V.: Efficientnet: rethinking model scaling for convolutional neural networks. arXiv preprint arXiv:1905.11946 (2019)
33. Tan, M., et al.: MNASNet: platform-aware neural architecture search for mobile. In: IEEE CVPR, pp. 2820–2828 (2019)

34. Vaswani, A., Shazeer, N., Parmar, N., et al.: Attention is all you need. In: NIPS (2017)
35. Vavilapalli, V.K., et al.: Apache hadoop yarn: yet another resource negotiator. In: ACM SoCC (2013)
36. (WMT19), A.M.T.: Shared task: machine translation of news. http://www.statmt.org/wmt19/translation-task.html
37. Xiao, W., et al.: Gandiva: introspective cluster scheduling for deep learning. In: USENIX OSDI (2018)
38. Xie, S., Girshick, R., Dollár, P., Tu, Z., He, K.: Aggregated residual transformations for deep neural networks. In: IEEE CVPR (2017)
39. Xu, X., et al.: Characterization and prediction of performance interference on mediated passthrough GPUs for interference-aware scheduler. In: HotCloud (2019)
40. Yeung, G.F., Borowiec, D., Friday, A., Harper, R., Garraghan, P.: Towards GPU utilization prediction for cloud deep learning. In: USENIX HotCloud (2020)

Attribute Bagging-Based Extreme Learning Machine

Xuan Ye[1], Yulin He[1,2,3(✉)], and Joshua Zhexue Huang[1,2,3]

[1] College of Computer Science and Software Engineering, Shenzhen University,
Shenzhen 518060, China
`yexuan@email.szu.edu.cn`, {`yulinhe,zx.huang`}`@szu.edu.cn`
[2] National Engineering Laboratory for Big Data System Computing Technology,
Shenzhen University, Shenzhen 518060, China
[3] Guangdong Laboratory of Artificial Intelligence and Digital Economy (SZ),
Shenzhen University, Shenzhen 518060, China

Abstract. Extreme learning machine (ELM) is a fast training scheme of
single-hidden-layer feedforward neural network. How to further improve
the prediction stability and accuracy of ELM in an ensemble learning way
becomes one of the hot research topics in the filed of supervised learning.
This paper proposes an attribute bagging-based ELM (AB-ELM) which
is an ensemble learning system for classification and regression tasks
by training the base ELMs on random samples of attributes instead of
the entire attribute set. AB-ELM uses the sampling with replacement
method to get the multiple randomized attribute subsets so as that the
different data subsets can be constructed for the training of base ELMs.
After obtaining a set of base ELMs, the weighted averaging method and
the weighted voting method are used to generate a combination output,
where the weight considers the information amount of training data subset. The relationship between the size of attribute subsets and the size
of base ELMs is also discussed in AB-ELM. On 4 classification and 4
regression data sets, we verify the training and testing performances of
AB-ELM in comparison with the classical ELM and the voting based
ELM (V-ELM). The experimental results show that AB-ELM obtains
the better prediction stability and accuracy than the classical ELM and
V-ELM and thus demonstrate the effectiveness of AB-ELM.

Keywords: Extreme learning machine · Ensemble learning · Attribute
bagging · Sampling with replacement · Randomized attribute subset

1 Introduction

With the development of science and technology, more and more data is produced all walks of life in recent years, such as medical big data [1]. The problems
of privacy and management in big data are also well worth studying [2,3]. The
data is large in quantity and dimension, which leads to a difficult task for model
training. For "Big Instance" data, a lot of related algorithms were proposed

© Springer Nature Switzerland AG 2020
M. Qiu (Ed.): ICA3PP 2020, LNCS 12453, pp. 509–522, 2020.
https://doi.org/10.1007/978-3-030-60239-0_34

and achieved a good performance. The concept of "Big Dimensionality" was proposed by Zhai et al., suggesting that higher dimensions also should be valued in industry and academia currently [4,5]. Therefore, how to deal with "Big Dimensionality" data effectively has become a very hot topic. Extreme Learning Machine (ELM), a single- hidden layer feedforward neural networks (SLFNs), was proposed by Huang et al. [6]. ELM is an algorithm which can be trained much faster and also produces a good generalization performance. It is widely used in various fields [7–9]. How to further improve the performance of ELM is also a hot topic. However, there are few algorithms related to effective processing of high dimensional data as well as improving the prediction stability and accuracy of traditional ELM at the same time. The research in this area is particularly important. In this paper, we propose an improved algorithm of ELM, which can not only handle the classification and regression problems of high dimensional data sets effectively, but also improve the prediction stability and accuracy of the classical ELM. It provides a method for high dimensional data processing.

Currently, there are many improved algorithms of ELM and the research of ensemble learning is the main direction [10]. Ensemble learning always have a better performance by combining multiple base learners rather than a single model. ELM can be trained faster because it avoids the need to iterate and updates the network weights. Compared with other ensemble learning algorithms, the ensemble learning of ELM requires less cost to be trained, which improves the train efficiency greatly. Liu et al. [11] proposed the earliest ensemble of ELM method called EN-ELM in 2010. EN-ELM introduces the strategy of cross-validation and ensemble learning in the training process. It generates Q training data sets and validation data sets by cross-validation and gets K base ELMs randomly. Therefore, EN-ELM has $Q \times K$ trained base ELMs for voting to get the final output. EN-ELM avoids the problem of overfitting and improves the generalization ability of the model. Online Sequential Extreme Learning Machine (OS-ELM) [12] is an incremental learning algorithm based on ELM. OS-ELM learns the new data set online without retraining the whole model and it also shows its good performance in the experiments. Lan et al. proposed EOS-ELM using the ensemble learning method based on OS-ELM [13], which further improved the stability and generalization ability of OS-ELM. Cao et al. [14] proposed another ensemble learning method based on ELM, called V-ELM. In V-ELM, each base ELM is trained by the entire data set and the majority voting method is used to get the final output. V-ELM avoids the instability caused by randomly generated weights in a single ELM and shows its effectiveness and stability. However, V-ELM needs more training time than ELM because it must train multiple ELMs with the entire data set. In dimension reduction area of ELM, a new multi-layer network structure was proposed for image recognition and data reduction by Yang te al. [15]. Kasun et al. proposed a dimension reduction framework referred to as ELM-AE and SELM-AE and both of them achieve a good performance in the experiment [16]. Deng et al. proposed a improved ELM called FSVD-H-ELM [5]. In order to process high dimension data, it embeds hidden nodes that are designed using singular value decomposition (SVD) into ELM. Although the above algorithms of ELM have shown their effectiveness, it still lacks the ensemble of ELM algorithm to reduce the dimension of data set as well as improve the prediction stability and accuracy of ELM.

In this paper, we propose an algorithm about ensemble learning of ELM called Attribute Bagging-Based Extreme Learning Machine (AB-ELM) to handle the supervised learning problems in classification and regression tasks. AB-ELM sets the size of attribute subset firstly. It uses the sampling with replacement method on the entire data set to get multiple randomized attribute subsets. Corresponding number of base ELMs are trained by these subsets. Then AB-ELM calculates the information amount of each attribute subset and the entire data set. The weights are normalized by the ratio of information amount between each attribute subset and the entire data set. How to choose an appropriate size of randomized attribute subsets and get the weights by the information amount are the key of AB-ELM. Compared with ELM and V-ELM, AB-ELM can be trained in parallel, which reduces the training cost greatly and achieves a better performance as well.

The rest of the paper is organized as follows: Sect. 2 briefs the overview of the classical ELM. In Sect. 3, we present the AB-ELM method and show the details of this algorithm. Section 4 shows the experimental results and analysis. At last, we give our conclusions and future works in Sect. 5.

2   Preliminaries

2.1   Extreme Learning Machine

Different from traditional neural network, ELM doesn't need to update parameters iteratively through back propagation algorithm and it generates weights between the input layer and the hidden layer as well as the biases of the hidden layer randomly [6,7]. The output of ELM is obtained by least square method, which leads to less time cost in model training. Compared with SVM and LS-SVM, ELM shows a better generalization performance [17]. The algorithm of ELM is as follows.

Giving N training samples $(\mathbf{x}_i, \mathbf{t}_i)$, where $\mathbf{x}_i = [x_{i1}, x_{i2}, \ldots, x_{in}]^T \in \mathbf{R}^n$, $\mathbf{t}_i = [t_{i1}, t_{i2}, \ldots, t_{im}]^T \in \mathbf{R}^m$. The weight $\mathbf{w}_i = [w_{i1}, w_{i2}, \ldots, w_{in}]^T$ between the input layer and the hidden layer of ELM and the bias $\mathbf{b}_i$ in the hidden layer are generated randomly. The output of SLFN with L hidden nodes can be formulated as

$$f_L(\mathbf{x}_j) = \sum_{i=1}^{L} \boldsymbol{\beta}_i g(\mathbf{w}_i \cdot \mathbf{x}_j + \mathbf{b}_i) = t_j, j = 1, 2, \cdots, N \tag{1}$$

where $\boldsymbol{\beta}_i \in \mathbf{R}^m$ is the link connecting the hidden layer to the output layer and $g(\mathbf{w}_i \cdot \mathbf{x}_j + \mathbf{b}_i)$ is the activation function, representing the output of the jth sample in the ith hidden node. Sigmoid function are choosen as the activation function in general. $\mathbf{w}_i \cdot \mathbf{x}_j$ is the inner product of $\mathbf{w}_i$ and $\mathbf{x}_j$. Equation (1) can be written as

$$\mathbf{H}_{(N \times L)} \cdot \boldsymbol{\beta}_{(L \times m)} = T_{(N \times m)} \tag{2}$$

where

$$\mathbf{H}(w_1, \cdots, w_L, b_1, \cdots, b_L, x_1, \cdots, x_N) = \begin{bmatrix} g(w_1, b_1, x_1) & \cdots & g(w_L, b_L, x_1) \\ g(w_1, b_1, x_2) & \cdots & g(w_L, b_L, x_2) \\ \vdots & \ddots & \vdots \\ g(w_1, b_1, x_N) & \cdots & g(w_L, b_L, x_N) \end{bmatrix}_{N \times L}$$

$\mathbf{H}$ is the output matrix of the hidden layer. $\boldsymbol{\beta} = [\beta_1, \cdots, \beta_L]^T$ is the output weight of ELM. $T = [t_1, \cdots, t_N]^T$ is the target matrix.

According to Eq. (1), the key of ELM algorithm is to get the $\boldsymbol{\beta}$. Huang et al. [7] shows that for a SLFNs, the input weights and the biases of hidden layer can be generated randomly. When the number of hidden layer nodes is less than the number of samples, Eq. (2) can be solved easily by

$$\tilde{\beta} = \mathbf{H}^\dagger T \tag{3}$$

where $\mathbf{H}^\dagger$ is the Moore-Penrose inverse of $\mathbf{H}$.

2.2   Voting Based ELM

Voting Based ELM (V-ELM) is an improved algorithm of ELM [14]. The input weights and the biases of ELM are randomly generated, resulting in its instability. V-ELM avoids that because it adopts an strategy of ensemble learning.

- In the training part of V-ELM, firstly $\mathbf{K}$, the number of base ELMs should be set and then each base ELM generates the input weights and biases randomly. At last, $\mathbf{K}$ base ELMs are trained by the training data set $(x_i^{train}, t_i^{train})_{train} \in \mathbf{R}^N$.
- In the testing part, V-ELM uses the testing data set $(x_i^{test}, t_i^{test})_{test} \in \mathbf{R}^{N^{test}}$ in $\mathbf{K}$ base ELMs to predict and get multiple outputs.

Assume that the number of classes is $\mathbf{D}$. For each testing sample x_{test} in k^{th} individual ELM, its output label can be indicated as c_k, where $c_k \in [1, 2, \ldots, D]$. The corresponding vector $\mathbf{S} = [S_1, S_2, \ldots, S_D]$ is created to store the outputs of $\mathbf{K}$ base ELMs, while S_i is the counts of label i in $\mathbf{K}$ base ELMs outputs. When k increases from 1 to $\mathbf{K}$, $\mathbf{S}$ is updated by Eq. (4).

$$S_{c_k} = S_{c_k} + 1 \tag{4}$$

The final output $\mathbf{C}_{test}$ of testing sample x_{test} in V-ELM is obtained by majority voting.

$$\mathbf{C}_{test} = \arg\max S_i, i = 1, 2, \ldots, D \tag{5}$$

3   Attribute Bagging Based ELM

3.1   Algorithmic Description

Attribute Bagging (AB) is an algorithm proposed by Bryll Robert [18]. AB selects the attributes randomly from the entire data set to obtain multiple attribute subsets. AB-ELM algorithm proposed in this paper combines AB method and the classical ELM. It also takes a new way of weight calculation in ensemble learning based on the information amount. We assume that the data set is $\mathbb{D}$, whose structure is shown in Table 1.

Table 1. Data set $\mathbb{D}$

	A_1	A_2	...	A_D
$\mathbf{x}_1$	x_{11}	x_{12}	...	x_{1D}
$\mathbf{x}_2$	x_{21}	x_{22}	...	x_{2D}
$\vdots$	$\vdots$	$\vdots$	$\ddots$	$\vdots$
$\mathbf{x}_N$	x_{N1}	x_{N2}	...	x_{ND}

First of all, we set $\mathbf{K}$ as the number of attribute subsets and $\mathbf{S}$ as the size of each attribute subset. We use the sampling with replacement method to get the multiple randomized attribute subsets $D_1, D_2, \ldots, D_K$. The dimension of D_k is $N \times S$ and $\mathbf{K}$ base ELMs are trained by these attribute subsets.

The weight is very important in some algorithms and the improved ELM algorithms about the weight have been investigated a lot [19,20]. In this paper, the weights of $\mathbf{K}$ base ELMs are calculated by the information amount. The information amount is an important element in the data set, which represents how much information of each attribute contains in the entire data set. AB-ELM defines the information amount of the entire data set is

$$H(\mathbb{D}) = \sum_{d=1}^{D} H(A_d) \tag{6}$$

where

$$H(A_d) = -\frac{1}{N} \sum_{n=1}^{N} \widetilde{f}_{-n}^{(d)}(x_{nd}), d = 1, 2, \ldots, D \tag{7}$$

$$\widetilde{f}_{-n}^{(d)}(x_{nd}) = \frac{1}{N} \sum_{m=1, m \neq n}^{N} \frac{1}{\sqrt{2\pi} h} \exp[-\frac{1}{2}(\frac{x_{nd} - x_{md}}{h})^2], n = 1, 2, \ldots, N \tag{8}$$

$H(A_d)$ is the information amount of attribute A_d and $\widetilde{f}_{-n}^{(d)}(x_{nd})$ is the probability density function of sample x_{nd} corresponding to A_d. h is the window width of $\widetilde{f}_{-n}^{(d)}(x_{nd})$. In this paper, $h = \frac{1}{\sqrt{N}}$.

The information amount of D_k is $H(D_k)$. The information amount's ratio of D_k to the entire data set $\mathbb{D}$ is

$$V_k = \frac{H(D_k)}{H(\mathbb{D})}, k = 1, 2, \ldots, \mathbf{K} \tag{9}$$

So the weight of ELM corresponding to k^{th} attribute subset is

$$w_k = \frac{V_k}{\sum_{k=1}^{\mathbf{K}} V_k}, k = 1, 2, \ldots, \mathbf{K} \tag{10}$$

At last, AB-ELM uses the same attribute extraction method as the training attribute subsets selection to get multiple test subsets. It also uses these with $\mathbf{K}$ pretrained base ELMs to obtain prediction outputs $E_1, E_2, \ldots, E_{\mathbf{K}}$. The final output of AB-ELM is calculated by weighted averaging in the regression task or by weighted voting in the classification task.

$$\mathbf{E} = \sum_{i=1}^{\mathbf{K}} w_i E_i \tag{11}$$

Compared with V-ELM, AB-ELM has certain novelty. First of all, AB-ELM takes the sampling with replacement method to obtain multiple randomized attribute subsets so as to reduce the dimension of the data set. In addition, AB-ELM takes a new way to calculate the weights of each base ELM, which helps the ensemble of ELM to achieve a good performance.

3.2   Parameter Selection

In AB-ELM, there are several parameters requiring to be set manually: the nodes in hidden layer, the number of base ELMs, the size of randomized attribute subsets. In ELM, the optimal number of hidden nodes is usually located in a small region [14]. With different numbers of hidden nodes selected from this area, the changes of network performance are very small. This paper chooses the hidden nodes corresponding to the best performance as the final nodes.

This paper also gives a suggestion for choosing the number of base ELMs $\mathbf{K}$. If the data set has Q attributes and the size of each randomized attribute subsets is $S(S < Q)$, the probability of each attribute which is not selected is

$$P = (1 - \frac{S}{Q})^{\mathbf{K}} \tag{12}$$

where $\frac{S}{Q}$ is the ratio of each attribute subset size to the entire data set. We want to minimize P, so we can set P as a minimum threshold, such as $P = 10^{-9}$ and get a suggestion about the number of base ELMs by the Eq. (12).

4   Experimental Results and Analysis

4.1   Experimental Setting

The experiments are carried out in four regression datasets and four classification datasets. The information of these datasets are shown in Table 2.

Table 2. Four classification datasets and four regression datasets

Data sets	Task	Attributes	Training	Testing
Multiple Features	Classification	649	1800	200
Smartphone	Classification	561	4596	1148
Semeion	Classification	256	1275	318
Musk	Classification	168	381	95
Communities and Crime	Regression	128	1596	398
Gas Sensor Array Drift	Regression	129	11128	2782
Residential Building	Regression	105	298	74
UJIIndoorLoc	Regression	529	16839	4209

Multiple Features, Smartphone, Semeion and Musk are the datasets used in the classification task. Communities and Crime, Gas Sensor Array Drift, Residential Building and UJIIndoorLoc are the regression datasets. All of those are from UCI Machine Learning Repository [21]. In order to avoid problems caused by different attribute dimension, this paper adopts max-min normalization method for each dataset to scale the value to $[0, 1]$. Root mean square error (RMSE) and Accuracy are used to measure the performance of algorithms in this paper. All of these experiments are run in MATLAB R2018a with Intel Core(TM) i5-7400 3.00 GHz CPU and 16 GB RAM. We use 10-fold crossing-validation to train and validate the algorithm. We can know the method of ELM parallel computation is also proposed to deal with large-scale data [22,23]. In AB-ELM, each individual ELM is able to be trained in parallel in principle. Therefore, we also use Parallel Computing Toolbox in MATLAB to train AB-ELM in parallel so as to reduce the training time.

4.2   Convergence of AB-ELM

In this section, we verify the convergence of AB-ELM through experiments. First of all, four ratios of the attribute subsets size to the entire data set size are set (30%, 50%, 70%, 90%). We want to know the performance in AB-ELM about small attribute subsets size. The experimental results are shown in Fig. 1 and Fig. 2. According to Fig. 1 and Fig. 2, we can see when K is small, AB-ELM is unstable. When $K > 10$, AB-ELM has reached convergence mainly and achieved a good performance on 8 datasets. It verifies that AB-ELM is feasible and effective.

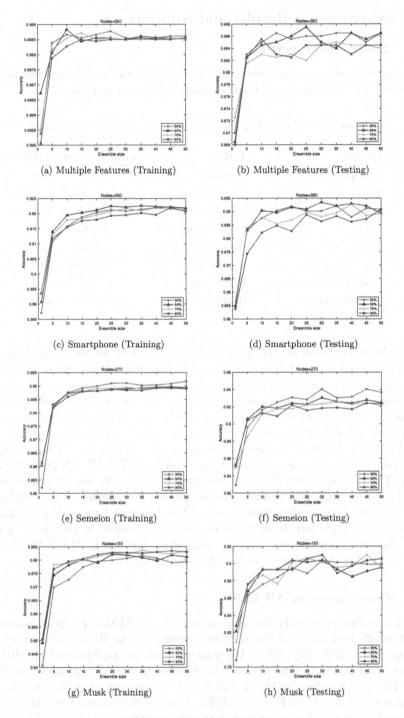

(a) Multiple Features (Training)   (b) Multiple Features (Testing)

(c) Smartphone (Training)   (d) Smartphone (Testing)

(e) Semeion (Training)   (f) Semeion (Testing)

(g) Musk (Training)   (h) Musk (Testing)

Fig. 1. Classification task

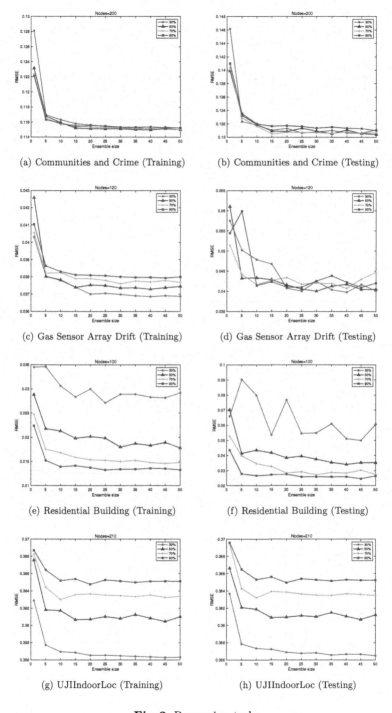

(a) Communities and Crime (Training)   (b) Communities and Crime (Testing)

(c) Gas Sensor Array Drift (Training)   (d) Gas Sensor Array Drift (Testing)

(e) Residential Building (Training)   (f) Residential Building (Testing)

(g) UJIIndoorLoc (Training)   (h) UJIIndoorLoc (Testing)

Fig. 2. Regression task

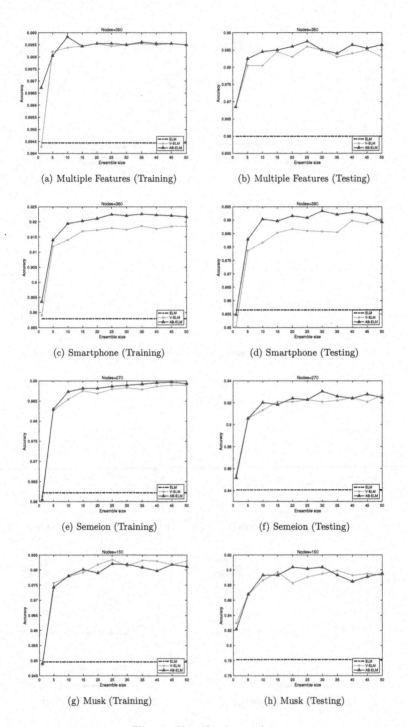

(a) Multiple Features (Training)

(b) Multiple Features (Testing)

(c) Smartphone (Training)

(d) Smartphone (Testing)

(e) Semeion (Training)

(f) Semeion (Testing)

(g) Musk (Training)

(h) Musk (Testing)

Fig. 3. Classification task

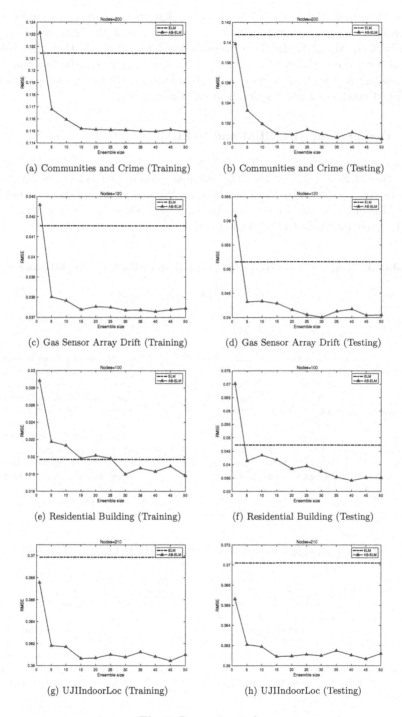

(a) Communities and Crime (Training)

(b) Communities and Crime (Testing)

(c) Gas Sensor Array Drift (Training)

(d) Gas Sensor Array Drift (Testing)

(e) Residential Building (Training)

(f) Residential Building (Testing)

(g) UJIIndoorLoc (Training)

(h) UJIIndoorLoc (Testing)

Fig. 4. Regression task

In the classification task, Fig. 1 shows that the accuracy of four ratios in experiments is very close. It means that small size of attribute subset is still work well in AB-ELM. In the regression task, Fig. 2 shows that the training results with the ratio of 50% differ from the optimal results by less than 0.01 on Residential Building dataset and UJIIndoorLoc dataset. It all proves that AB-ELM produces a good generalization performance.

4.3   Comparison with ELM and V-ELM

In this section, we evaluate the performance of AB-ELM, ELM and V-ELM. From Sect. 4.2, we can see that the suitable ratio of the randomized attribute subsets size to the entire data set is 50%. We choose 50% ratio in AB-ELM. V-ELM is only suitable for the classification task. Therefore, only AB-ELM and ELM are compared in the regression task.

Table 3. Comparison between ELM, V-ELM and AB-ELM in classification task

Datasets	Algorithms	Nodes	Size of each subset	Number of subsets	Training time(s)		Accuracy(%)	
					Serial	Parallel	Training	Testing
Multiple Features	ELM	350	649	/	0.2074	/	99.44%	96.00%
	V-ELM	350	649	50	10.0425	/	**99.85%**	98.30%
	AB-ELM	350	325 (50%)	50	4.2832	**3.6966**	**99.85%**	**98.65%**
Smartphone	ELM	380	561	/	0.3579	/	88.79%	85.65%
	V-ELM	380	561	45	11.5416	/	91.85%	88.88%
	AB-ELM	380	281 (50%)	45	11.2773	**10.0483**	**92.20%**	**89.21%**
Semeion	ELM	270	256	/	0.1173	/	96.22%	84.06%
	V-ELM	270	256	45	5.8261	/	98.89%	92.09%
	AB-ELM	270	128 (50%)	45	3.3438	**1.5757**	**98.97%**	**92.78%**
Musk	ELM	150	168	/	0.0137	/	94.96%	78.13%
	V-ELM	150	168	50	0.5731	/	**98.30%**	89.29%
	AB-ELM	150	84 (50%)	50	0.4291	**0.2337**	98.11%	**89.53%**

Table 4. Comparison between ELM and AB-ELM in regression task

Datasets	Algorithms	Nodes	Size of each subset	Number of subsets	Training time(s)		RMSE	
					Serial	Parallel	Training	Testing
Communities and Crime	ELM	200	128	/	0.0931	/	0.1214	0.1408
	AB-ELM	200	64 (50%)	50	1.6559	**1.1600**	**0.1150**	**0.1304**
Gas Sensor Array Drift	ELM	120	129	/	0.1598	/	0.0415	0.0515
	AB-ELM	120	65 (50%)	50	4.9718	**4.3956**	**0.0374**	**0.0405**
Residential Building	ELM	100	105	/	0.0089	/	0.0197	0.0473
	AB-ELM	100	53 (50%)	50	0.2207	**0.1245**	**0.0178**	**0.0352**
UJIIndoorLoc	ELM	210	529	/	0.8301	/	0.3699	0.3702
	AB-ELM	210	265 (50%)	50	17.0761	**13.6745**	**0.3610**	**0.3612**

The experimental results between AB-ELM, ELM and V-ELM are shown in Fig. 3 and Fig. 4. We also train AB-ELM in parallel, and its cost would be much less than V-ELM. For better analysis, the experimental results are also shown in the Table 3 and Table 4.

- In Fig. 3, AB-ELM shows a good performance on four classification datasets when the attribute subsets size is 50% of the entire dataset size. The results show that the performance of AB-ELM is better than ELM on 8 datasets. The accuracy of AB-ELM is close to V-ELM and even higher than V-ELM.
- Figure 4 shows that in the regression task the RMSE of AB-ELM is lower than ELM. It proves that AB-ELM performs better than ELM with same parameters. Although AB-ELM is less stable when K is small, it finally reaches convergence on four regression datasets and keeps stable on a small range.
- From Table 3 and Table 4, it shows when an appropriate K is chosen, the accuracy of AB-ELM is higher than that of ELM or V-ELM and the RMSE of AB-ELM is lower than that of ELM. It demonstrates the effectiveness of AB-ELM.

5   Conclusions and Future Works

This paper proposes a new improved ELM method based on attribute bagging to deal with the high dimension data and further improves the prediction stability and accuracy of ELM. AB-ELM reduces the dimension of the data by randomly attribute subsets selected and then it will be trained by those subsets in an ensemble learning way. AB-ELM also proposed a new way of weight calculation which considers the information amount of each attribute data set. The output of AB-ELM is generated by the weighted averaging or weighted voting method. Finally the experimental results demonstrate the feasibility and effectiveness of AB-ELM. AB-ELM performs better than ELM and V-ELM.

Acknowledgement. This paper was supported by Scientific Research Foundation of Shenzhen University for Newly-introduced Teachers (2018060), National Natural Science Foundation of China (61972261), and Open Foundation of Key Laboratory of Impression Evidence Examination and Identification Technology, Ministry of Public Security, People's Republic of China (HJKF201901).

References

1. Siuly, S., Zhang, Y.: Medical big data: neurological diseases diagnosis through medical data analysis. Data Sci. Eng. **1**(2), 54–64 (2016)
2. Wang, J., Chen, X.: Efficient and secure storage for outsourced data: a survey. Data Sci. Eng. **1**(3), 178–188 (2016)
3. Vargas-Solar, G., Zechinelli-Martini, J.L., Espinosa-Oviedo, J.A.: Big data management: what to keep from the past to face future challenges? Data Sci. Eng. **2**(4), 328–345 (2017)
4. Zhai, Y., Ong, Y.S., Tsang, I.W.: The emerging "big dimensionality". IEEE Comput. Intell. Mag. **9**(3), 14–26 (2014)

5. Deng, W.Y., Bai, Z., Huang, G.B., et al.: A fast SVD-hidden-nodes based extreme learning machine for large-scale data analytics. Neural Netw. **77**, 14–28 (2016)
6. Huang, G.B., Zhu, Q.Y., Siew, C.K.: Extreme learning machine: a new learning scheme of feedforward neural networks. In: IEEE International Joint Conference on Neural Networks, vol. 2, pp. 985–990 (2004)
7. Huang, G.B., Zhu, Q.Y., Siew, C.K.: Extreme learning machine: theory and applications. Neurocomputing **70**(1–3), 489–501 (2006)
8. Deng, J., Frühholz, S., Zhang, Z., et al.: Recognizing emotions from whispered speech based on acoustic feature transfer learning. IEEE Access **5**, 5235–5246 (2017)
9. Huang, Z., Yu, Y., Gu, J., et al.: An efficient method for traffic sign recognition based on extreme learning machine. IEEE Trans. Cybern. **47**(4), 920–933 (2016)
10. Huang, G.B., Wang, D.H., Lan, Y.: Extreme learning machines: a survey. Int. J. Mach. Learn. Cybern. **2**(2), 107–122 (2011)
11. Liu, N., Wang, H.: Ensemble based extreme learning machine. IEEE Signal Process. Lett. **17**(8), 754–757 (2010)
12. Liang, N.Y., Huang, G.B., Saratchandran, P., et al.: A fast and accurate online sequential learning algorithm for feedforward networks. IEEE Trans. Neural Netw. **17**(6), 1411–1423 (2006)
13. Lan, Y., Soh, Y.C., Huang, G.B.: Ensemble of online sequential extreme learning machine. Neurocomputing **72**(13–15), 3391–3395 (2009)
14. Cao, J., Lin, Z., Huang, G.B., et al.: Voting based extreme learning machine. Inf. Sci. **185**(1), 66–77 (2012)
15. Yang, Y., Wu, Q.M.J., Wang, Y.: Autoencoder with invertible functions for dimension reduction and image reconstruction. IEEE Trans. Syst. Man Cybern.: Syst. **48**(7), 1065–1079 (2016)
16. Kasun, L.L.C., Yang, Y., Huang, G.B., et al.: Dimension reduction with extreme learning machine. IEEE Trans. Image Process. **25**(8), 3906–3918 (2016)
17. Huang, G.B., Zhou, H., Ding, X., et al.: Extreme learning machine for regression and multiclass classification. IEEE Trans. Syst. Man Cybern. Part B (Cybern.) **42**(2), 513–529 (2011)
18. Bryll, R., Gutierrez-Osuna, R., Quek, F.: Attribute bagging: improving accuracy of classifier ensembles by using random feature subsets. Pattern Recogn. **36**(6), 1291–1302 (2003)
19. Zong, W., Huang, G.B., Chen, Y.: Weighted extreme learning machine for imbalance learning. Neurocomputing **101**, 229–242 (2013)
20. Li, K., Kong, X., Lu, Z., et al.: Boosting weighted ELM for imbalanced learning. Neurocomputing **128**, 15–21 (2014)
21. Dua, D., Gra, C.: UCI machine learning repository. University of California, School of Information and Computer Science, Irvine (2019). http://archive.ics.uci.edu/ml
22. Van Heeswijk, M., Miche, Y., Oja, E., et al.: GPU-accelerated and parallelized ELM ensembles for large-scale regression. Neurocomputing **74**(16), 2430–2437 (2011)
23. He, Q., Shang, T., Zhuang, F., et al.: Parallel extreme learning machine for regression based on MapReduce. Neurocomputing **102**, 52–58 (2013)

A Semi-supervised Joint Entity and Relation Extraction Model Based on Tagging Scheme and Information Gain

Yonglin Zhao[1,2(✉)], Xudong Sun[1], Shuxin Wang[2], Jianwei He[1,2], Yanzhi Wei[1,2], and Xianghua Fu[2(✉)]

[1] College of Computer Science and Software Engineering, Shenzhen University, Shenzhen 518061, China
1810272055@email.szu.edu.cn
[2] College of Big Data and Internet, Shenzhen University of Technology, Shenzhen 60590, China
fuxianghua@sztu.edu.cn

Abstract. Joint entity and relation extraction, completing the entity recognition and relation extraction simultaneously, can better integrate the information between two tasks and reduce the errors of each task. The methods based on tagging scheme treat the joint extraction as a sequence labeling task and have achieved outstanding results. However, those tag-based methods are insufficient in making full use of the information between entities and relations. It maybe the reason why they show a relatively poor precision. In this paper, we propose a novel semi-supervised approach that combines the tagging scheme with an information gain module. The information gain module is a combination of distant supervision and attention mechanism, which is used for obtaining prior information of candidate entities and relations. We believe that it is important for the joint extraction results to make much of the links among them. Our tagging scheme adds distant supervision to tag entity words and relational words in the given sentences, and combines with attention mechanism to improve their weight. It effectively helps our bias objective function improve model performance. Experiments on public dataset show that our methods are better than most of the existing pipelined and joint learning methods.

Keywords: Attention · Distant supervision · End-to-end · Information gain · Tagging scheme

1 Introduction

Extraction of entities and relations is the most fundamental task of information extraction (IE) in the field of neural language processing (NLP). Extracting useful information from massive unstructured text and structuring it into a format

---

Y. Zhao and X. Fu—This research is supported by the Scientific Research Platforms and Projects in Universities in Guangdong Province under Grants 2019KTSCX204.

M. Qiu (Ed.): ICA3PP 2020, LNCS 12453, pp. 523–537, 2020.
https://doi.org/10.1007/978-3-030-60239-0_35

available for downstream work is the significance of information extraction. It is an essential issue in automatic construction of knowledge base.

Conventional extraction approaches are based on the pipelined method which treat this task as two separated portion, named entity recognition (NER) [18,26] and relation classification (RC) [21]. There are several disadvantages of pipelined methods: (1) Error propagation, which means the errors of entity recognition will affect the following relation classification performance; (2) Ignoring the relationship between the two subtasks. (3) Generating unnecessary redundant information. Those methods classify the relation type after pairing the identified entities, so that those unrelated entity pairs bring redundant information and promoting errors.

Unlike pipelined methods, joint extracting approaches model the triplets directly. Joint learning frameworks usually include statistical methods [11,17,20] and neural network-based methods [3,10,23]. The statistical methods are seriously dependent on complex feature engineering and unable to capture global features, while the neural network-based methods learn the relations automatically which may not appear in the given sentence. However, most of the existing neural network-based methods extract entities and relation types separately and achieve joint extraction via parameter sharing rather than joint decoding. Recently, Zheng [30] proposed a sequence tagging-based neural network model which treats the joint extraction as a sequence labeling task. Their method achieved excellent results on the public dataset. But it shown poor precision even preformed an excellent F1 score in this task. We think that their methods cannot detect the entity and relation words in the given sentence perfectly and make full use of the information between those words in tagging scheme.

In this paper, we propose a semi-supervised joint extraction model, BiLSTM-DS-Att. It integrates the tagging scheme with an information gain module. The information gain module is used for obtaining prior information of candidate entities and relations which is a combination of distant supervision [16] and attention mechanism. We use distant supervision to generate the position information of candidate entity words and relation words in the given sentence and further use word level attention to increase the weight of these words. Our methods are proved better than most of the existing pipelined and joint learning methods in public dataset.

2   Related Work

Named entity recognition is a basic subtask of NLP. Traditional NER methods based on statistical machine learning mainly include: Hidden Markov Model (HMM), Maximum Entropy (ME), Support Vector Machine (SVM), and Conditional Random Fields (CRF) [13,19]. In recent years, neural network models such as RNN and LSTM [7,9] have been widely used for NER, which are characterized by strong sequence modeling capability to capture context information. Meanwhile, neural networks-based methods [7,22,31] have been being the mainstream methods of RC instead of handcrafted feature-based methods [21]. In our information gain model, we use Stanford CoreNLP[1] [14] to complete the task of NER.

---

[1] Stanford CoreNLP can be downloaded at: https://stanfordnlp.github.io/CoreNLP/.

The sequence tagging is one of the most common problem in NLP. It tags each of word in the given sequence with predefined labels. In recent years, the end-to-end model based on LSTM [5] has been successfully applied to various tagging tasks such as: Named Entity Recognition [9], CCG Supertagging [25], and Chunking [29] etc. In this paper, we adopted the new sequence tagging category called BIES [30], which converts the joint extraction involving sequence labeling task and classification task into a sequence labeling task absolutely. The tagging scheme is shown as Fig. 2. Their method better integrates the information between those two tasks and reduces the errors of each task. Most joint methods apply bidirectional LSTM to encode the input sentences, but decoding methods are always largely different. For examples, Lample [9] uses CRF layers to decode tag sequences, while Vaswani [25] and Katiyar [8] applies LSTM layer to produce the results.

Distant supervision [16, 28] is an efficient approach that automatically generates labeled data for RE. Various improved methods based on distant supervision have been widely used in relation extraction tasks. For example, aiming at the problem of error propagation caused by NLP tools in traditional relation extraction model, Zeng [28] proposed a convolutional neural network based on piece-wise (PCNN) and proposed a model based on multi-example learning. Moreover, Qin [33] designed a new deep reinforcement learning framework to improve the robustness of distant supervision. In our method, we adopt a combination of distant supervision and attention mechanism instead of using the results of distant supervision directly. Besides, we use a new mark strategy that only consider the relation type appearing in the given sentence.

In this paper, we propose an end-to-end model based on the tagging scheme BIES [30], while add the intermediate information of the input sentences obtained by distant supervision and attention mechanism to training in common. We apply bidirectional LSTM to encode the input sentences and LSTM layer to produce the tag sequence [8, 25], and connect encoder layer and decoder layer through an information gain module which is a combination of distant supervision and attention mechanism.

3   Method

Our end-to-end model is composed of a Bi-LSTM encoder layer, a LSTM decoder layer and a information gain module. Its structure is shown in Fig. 1(a).

3.1   Bi-LSTM Encoder Layer

In NLP sequence tagging tasks, Bi-LSTM encoding approach has been proved to be effective in capturing the semantic information of every word [1, 7]. In this paper, we use the basic Bi-LSTM to encoding the input sentences. Each given sentence will be converted to a embedding vector $W = w_1, ..., w_t, w_{t+1}, ..., w_n$ by word2vec[2] [15], where $w_t \in R^d$ is a d-dimensional word embedding, n is the length of input sentence.

---

[2] Word2vec can be downloaded at: https://code.g.oogle.com/archive/p/word2vec/.

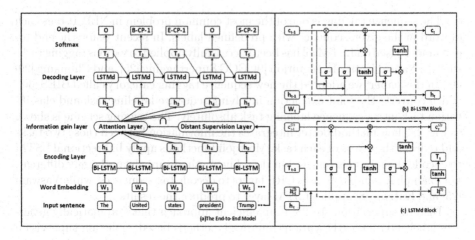

Fig. 1. The architecture of our end-to-end model

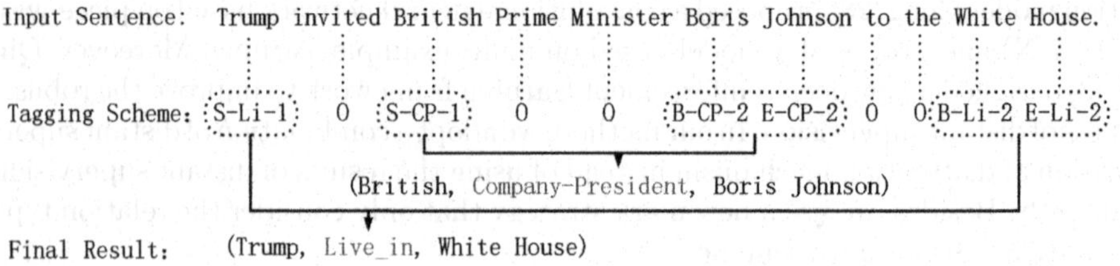

Input Sentence: Trump invited British Prime Minister Boris Johnson to the White House.

Tagging Scheme: S-Li-1   O S-CP-1   O   O   B-CP-2 E-CP-2 O O B-Li-2 E-Li-2

(British, Company-President, Boris Johnson)

Final Result: (Trump, Live_in, White House)

Fig. 2. An example of end-to-end model based on tagging scheme, where "Li" is a short for "Live_in" and "CP" is a short for "Country-President"

Encode layer finally outputs a hidden representation $H = [h_1, ..., h_t, h_{t+1}, ...h_n]$, where h_t represents the encoding information the t-th word. Its structure is shown in Fig. 1(b). And specific operations are defined as follows:

$$i_t = \sigma(W_{xi}x_t + W_{hi}h_{t-1} + W_{ci}c_{t-1} + b_i) \tag{1}$$

$$f_t = \sigma(W_{xf}x_t + W_{hf}h_{t-1} + W_{cf}c_{t-1} + b_f) \tag{2}$$

$$z_t = \tanh(W_{xc}x_t + W_{hc}h_{t-1} + b_c) \tag{3}$$

$$c_t = i_t \bigotimes u_t + f_t \bigotimes c_{t-1} \tag{4}$$

$$o_t = \sigma(W_{xo}x_t + W_{ho}h_{t-1} + W_{co}c_t + b_o) \tag{5}$$

$$h_t = o_t \bigotimes \tanh(c_t) \tag{6}$$

Where i is the input gate, f is the forget gate, o is the output respectively, b is the bias term, c is the cell memory, and $W_{(*)}$ is the parameter matrix.

3.2   Information Gain Module

The information gain module is a combination of distant supervision and global attention mechanism. We use distant supervision to automatically extract the potential relationship of the entities and generate position information of those words. The position information is recorded by a vector λ. Further more, we use word level attention to increase the weight of these words. Specifically, we firstly adopt the Stanford CoreNLP [14] tools to extract entities in the given sentence. Then using distance supervised classifier[3] trained by Google to extract the relation types between the entities. Because the extracted relation type may not exist in the given sentence partially result from the inaccuracy of distant supervision, so we set that the position of this relation word will be marked only if the corresponding relation type indeed appears in the given sentence. On the contrary, only entity words will be marked. We compare the embedding of relation types to the original sentence vector W.

In order to make the attention mechanism can fully utilize the position information obtained by distant supervision to achieve the effect of information gain, we uphold a hyperparameter matrix λ whose dimension is equal to the length of input sentence.

$$\lambda = [1, ..., \varepsilon, ..., \varepsilon, ..., 1, ..., \varepsilon, ...] \tag{7}$$

Where ε is an undetermined hyperparameter. The value of ε will be set through experimental tuning.

Firstly, we initialize a n-dimensional matrix with all values of 1. Then according to the position information of each entity words and relation words, we convert the value of corresponding position in the matrix λ to ε. In the given sentence "Trump invited British Prime Minister Boris Johnson to the White House", there are four entities "Trump", "British", "Boris Johnson" and "White House". Through distant supervision method we can get that the relation type between "Trump" and "White House" is "$Live_in$", and it is "$Company_president$" between "British" and "Johnson". However, both of them are not exit in the given sentence, so the corresponding matrix λ is defined as $\lambda = [\varepsilon, 1, \varepsilon, 1, 1, \varepsilon, \varepsilon, 1, 1, \varepsilon, \varepsilon]$. Finally, we adopt the output feature of encoder layer $H = [h_1, h_2, ...h_n]$ and attention mechanism to calculate the attention probability. The specific calculation process is defined as follows:

$$h_{ti} = U_a \tanh(U_b h_t + U_c h_i + b_a) \tag{8}$$

$$a_{ti} = \frac{\exp(h_{ti})}{\sum_{i=1}^{n} \exp(h_{ti})} \tag{9}$$

$$h_t' = \lambda \sum_{i=1}^{n} a_{ti} h_i \tag{10}$$

---

[3] Google distance supervised API can be downloaded at: https://developers.google.com/knowledge-graph.

Where h_t h_i are the feature representation of word t and word i, h_{Nt} is the hidden unit of word t about word i, U_a, U_b and U_c represent different weight matrices, b_a is the bias of attention mechanism, h_t' is a new feature representation of word t, and λ_t is the t-th element of matrix λ. In the Formula (10), We use the parameter matrix multiply the output feature matrix of the general attention module to get a new feature representation. This new feature representation is the input of decoder layer.

3.3   LSTM Decoder Layer

We use LSTM neural network to decode and generate the tag sequence [30]. When detecting the label of a word W_t, the input of the decoding layer are: the outputh_t' of the attention module, former predicted tag embedding T_{i-1}, former cell value $c_{t-1}^{(2)}$, and the former hidden vector in decoding layer $h_{t-1}^{(2)}$. Its structure is shown in Fig. 1(c). And the specific operations are defined as follows:

$$f_t^{(2)} = \sigma(W_{wf}^{(2)})h_t' + W_{hf}^{(2)}h_{t-1}^{(2)} + W_{tf}T_{t-1} + b_f^{(2)}) \tag{11}$$

$$z_t^{(2)} = \tanh(W_{wc}^{(2)}h_t' + W_{hc}^{(2)}h_{t-1}^{(2)} + W_{tc}T_{t-1} + b_c^{(2)}) \tag{12}$$

$$c_t^{(2)} = f_t^{(2)}c_{t-1}^{(2)} + i_t^{(2)}z_t^{(2)} \tag{13}$$

$$o_t^{(2)} = \sigma(W_{wo}^{(2)}h_t' + W_{ho}^{(2)}h_{t-1}^{(2)} + W_{co}^{(2)}c_t^{(2)} + b_o^{(2)}) \tag{14}$$

$$h_t^{(2)} = o_t^{(2)}\tanh(c_t^{(2)}) \tag{15}$$

$$T_t = W_{ts}h_t^{(2)} + b_{ts} \tag{16}$$

The final softmax layer computes normalized entity tag probabilities based on the tag predicted vector T_t:

$$y_t = W_yT_t + b_y \tag{17}$$

$$p_t^i = \frac{\exp(y_t^i)}{\sum_{j=1}^{N_t}\exp(y_t^i)} \tag{18}$$

where W_y is the softmax matrix, $N_t = 2*4*|R|+1$, where N_t is the total number of tags and $|R|$ is the size of predefined relation set.

3.4   Optimization

We train our model with RMSprop [32] method to maximize the log-likelihood of the training data. The objective function can be defined as:

$$L_{ner} = \max\sum_{j=1}^{|S|}\sum_{t=1}^{L_j}(\log(p_t^{(j)} = y_t^{(j)}|x_j,\theta)\cdot I(O)$$
$$+ \alpha\cdot\log(p_t^{(j)} = y_t^{(j)}|x_j,\theta)\cdot(1-I(O))) \tag{19}$$

where $|S|$ is the size of training set, L_j is the length of sentence x_j, $y_t^{(i)}$ is the label of word t in sentence x_j and $p_t^{(j)}$ is the normalized probabilities of tags which defined in Formula (18). In addition, $I(O)$ is a transforming function to distinguish the loss of tag "O" and relational tags which can indicate the results. $I(O)$ is defined as follows:

$$I(O) = \begin{cases} 1, \text{ if } tag = {'O'} \\ 0, \text{ if } tag \neq {'O'} \end{cases}$$

Where α is a bias parameter. A larger α means that the relation tag has a greater influence on the model.

4 Experiences

In this section, We will introduce our experimental settings, show and analyze our experimental results, and discuss some critical experimental parameters.

4.1 Experimental Settings

Dataset. To evaluate the performance of BiLSTM-DS-Att model and directly compare with LSTM-LSTM-Bias model, we use the New York Times (NYT[4]) [2] corpus as our experimental data. NER in NYT dataset is achieved by using NLP tools of Stanford University and further aligned with freebase database. In this paper, we employ two versions of NYT. One of them was used by Zeng [28] and the other was used by Lin [12]. Statistics of them are shown in Table 1.

Table 1. Statistics of two versions of NYT datasets

Dataset	Relations	Sentences	Triples
S-NTY-training data	24	Unknown	353,000
S-NTY-testing data	24	Unknown	3,880
B-NTY-training data	54	522,611	281,270
B-NTY-testing data	54	172,448	96,678

Evaluation. We adopt standard Precision, Recall and F1 score to evaluate the results. A triple is regarded as correct when its relation type and the head off-sets of two corresponding entities are both correct, and the ground-truth relation mentions are given and "None" label is excluded as Ren [20] and Zheng [30] did. We create a validation set by randomly sampling 15% data from test set and use the remaining data as evaluation based on Ren's suggestion. We run 15 times for each experiment then report the average results. On this basis, we analyze the bias parameters and hyperparameters ε, then observe the influence of these two parameters on the F1 value of the model.

---

[4] NYT dataset can be downloaded at: http://iesl.cs.umass.edu/riedel/ecml/.

Hyperparameters. Our model is composed of a Bi-LSTM encoding layer, an information gain module and a LSTM decoding layer with a bias objective function. The word embeddings used in the encoding part are initialed by running word2vec [15] on NYT training corpus. Primary parameter settings are shown in Table 2. The values of the bias parameter α and information gain parameter ε are determined by experiment. We will introduce it later in the analysis and discussion section.

Table 2. Primary parameter settings of our model

Parameters	Descriptions	Values
D	The dimension of word embedding	300
Num_En	Num of hidden layer nodes in encoding layer	300
Num_De	Num of hidden layer nodes in decoding layer	600
ρ	Embedding dropout ration	0.5
α	Bias parameter	10
ε	Information gain parameter	1.5

4.2   Experimental Results

Baselines. We compare our method with several classical triple extraction methods: the pipelined methods, the jointly extracting methods, as well as the end-to-end methods based on sequence tagging scheme.

The pipelined methods used in this paper are listed as follows: (1) DS-logistic [16] which integrates the advantages of supervised information extraction and unsupervised information extraction features; (2) LINE [24] which goes for any type of information networks; (3) FCM [4] which integrates word embeddings and lexicalized context for relation extraction tasks.

The jointly extracting methods are: (4) DS-Joint [11] which accomplishes joint extraction of entities and relations by structured perceptron based on manually annotated dataset; (5) MultiR [6] which is a classic distant supervised method which based on multi-instance learning algorithms in order to deal with the noisy training data; (6) CoType [20] which jointly embeds entity mentions, relation mentions, type labels and text features into intelligible representation.

We also compare our model with three typical end-to-end tagging models: (7) LSTM-CRF [9] is a entity recognition algorithm which uses a bidirectional LSTM to encode the input sentences and predicts the entity tag sequence by the conditional random fields. (8) LSTM-LSTM [25] decodes the tag sequence using a LSTM layer rather than CRF. (9) LSTM-LSTM-Bias [30] treat the joint extraction task as a sequence labeling task. In our method, we use the same bias function with it.

Results. The results of different methods are shown as Table3. We can see that our method achieves almost 2% improvement in F1 score over the best end-to-end tagging method LSTM-LSTM-Bias [30]. In addition, it shows that the jointly extracting methods are superior to pipelined methods in general, and the tagging methods are superior to most of the jointly extracting methods. It also declares the significant influence of error propagation and redundant information, as well as the validity of our method.

Table 3. The predicted results of different methods on extracting both entities and their relations.

Model	Precision	Recall	F1
DS+Logistic [16]	25.8	39.3	31.1
LINE [24]	33.5	32.9	33.2
FCM [4]	55.3	15.4	24.0
DS-Joint [11]	57.4	25.6	35.4
MultiR [6]	33.8	32.7	33.3
CoType [20]	42.3	**51.1**	46.3
LSTM-CRF [9]	**69.3**	31.0	42.8
LSTM-LSTM [25]	68.2	32.0	43.6
LSTM-LSTM-Bias [30]	61.5	41.4	49.5
BiLSTM-Att-DS (our model)	63.4	43.1	**51.3**

Compared with traditional pipelined and jointly extracting methods, the precision of end-to-end models have been significantly improved. But our methods achieve an outstanding balance between the precision and recall. It achieves 1.8% improvement in F1 score and 1.9% improvement in precision score over the best end-to-end tagging method LSTM-LSTM-Bias. It shows the effectiveness of our model on modeling and predicting entities and relations. The improvements may be result form that these end-to-end models all use the Bi-LSTM to encode input sentences and different neural networks to decode tagging sequences. In addition, those methods based on neural networks can better fit the data so that they can learn the common implicit features of training data well. And in our model, we focus on the impact of the connections among candidate entities and relation type belong to the same triple on the extraction results. It shows that the priori information obtained by distant supervision and attention can really help the bias objective function to improve the extraction accuracy.

Case Study. We observe the extraction results of our end-to-end model and select several typical examples to demonstrate the strengths and weaknesses of our methods. The results are shown in Table4. In this paper, we focus on extracting triples which consist of two entities and their relation type. It regards a triple is correct only when the head offsets of two entities their relation type are both correct. $E1$ and $E2$ respectively represent the performance of our approach on predicting single entity. The instance of $E1$ is correct only when the head offset of the first entity is correct, and the same to $E2$. Regardless of relation type, in this section, we treat the instance of $(E1, E2)$ is correct if the head offsets of two entities are both correct. Each example consists of four rows, the first row is the input sentence tagged artificially, the second and the third rows are the results extracted by model LSTM-LSTM and LSTM-LSTM-Bias, and the last one is obtained by our BiLSTM-Att-DS model.

Sentence1 is a example that the position of two interrelated entities is far away from each other in the given sentence, which make it more difficult to determine the relation between them. Compared with LSTM-LSTM and LSTM-LSTM-Bias, our model not only applies a bias objective function but integrates an information gain module which can enhance the relevance between entities. So, in this case, both of LSTM-LSTM-Bias and BiLSTM-Att-DS models can extract two related entities, while LSTM-LSTM can only predict one entity of "Canada" and cannot detect entity "The Rocky Mountains". Furthermore, a large number of experiments show that the performance of our model is better than that of LSTM-LSTM-Bias in the extraction of distant entities.

Sentence2 is a confused case which contains three interrelated entities. In our methods, we use distant supervision to detect the relations between candidate entities in advance. Therefore, our methods can better consider the connections among entities and their relation in given sentence and distribute different weights to those words according to their importance on extraction results. So, our model can extract two related entities correctly. LSTM-LSTM and LSTM-LSTM-bias can predict two entities, while they are not the correct entity pair owning this relation.

Sentence3 represents the situation that models can correctly detect the entities' head offset, however, the relational role of each entity is wrong. LSTM-LSTM considers both "Steve Jobs" and "Apple" as entity $E1$, while cannot find the corresponding entity $E2$. LSTM-LSMT–Bias and our model both can find the entities pair $(E1, E2)$, but it reverses the roles of "Steve Jobs" and "Apple". It shows that this tagging scheme can better detect the interrelated entity pair, but there is still room for improvement in distinguishing the relation type between them.

5   Analysis and Discussion

Our approach tends to use relational tags to strengthen connections between entities. To further study the influence of bias objective function, we analyze the proportion of predicted single entities whose corresponding entity cannot be found and compare it with LSTM-CRF, LSTM-LSTM methods. The results are shown in Fig. 5. Different from LSTM-LSTM and LSTM-CRF which not pay much attention to relational labels, LSTM-LSTM-Bias and our BiLSTM-Att-DS which based on bias objective function get a lower ratio on those single entities. It main that

Table 4. Results from different models

Standard S1	$[TheRockyMountains]_{E2contain}$, which are composed of many small mountains, are famous tourist attractions and the most famous mountains in $[Canada]_{E1contain}$.
LSTM-LSTM	The Rocky Mountains, which are composed of many small mountains, are famous tourist attractions and the most famous mountains in $[Canada]_{E1contain}$.
LSTM-LSTM-Bias	$[TheRockyMountains]_{E2contain}$, which are composed of many small mountains, are famous tourist attractions and the most famous mountains in $[Canada]_{E1contain}$.
BiLSTM-Att-DS	$[TheRockyMountains]_{E2contain}$, which are composed of many small mountains, are famous tourist attractions and the most famous mountains in $[Canada]_{E1contain}$.
Standard S2	$[Headlights]_{E2contain}$ are considered as the eyes of the $[vehicle]_{E1contain}$.
LSTM-LSTM	Headlights are considered as the eyes of the $[vehicle]_{E1contain}$.
LSTM-LSTM-Bias	Headlights are considered as the $[eyes]_{E2contain}$ of the $[vehicle]_{E1contain}$.
BiLSTM-Att-DS	$[Headlights]_{E2contain}$ are considered as the eyes of the $[vehicle]_{E1contain}$.
Standard S3	$[SteveJobs]_{E2CF}$, whe co-founder of $[Apple]_{E1CF}$, announced that he is taking another medical leave, raising new concerns about the company's future.
LSTM-LSTM	$[SteveJobs]_{E1CF}$, the co-founder of $[Apple]_{E1CF}$, announced that he is taking another medical leave, raising new concerns about the company's future.
LSTM-LSTM-Bias	$[SteveJobs]_{E1CF}$, the co-founder of $[Apple]_{E2CF}$, announced that he is taking another medical leave, raising new concerns about the company's future.
BiLSTM-Att-DS	$[SteveJobs]_{E1CF}$, the co-founder of $[Apple]_{E2CF}$, announced that he is taking another medical leave, raising new concerns about the company's future.

Standard S_i represents the gold standard of sentence i. The blue one is the correct result, and the red one is the wrong result. $EiCF$ in case "3" is short for $Ei_{Company-Founder}$.

those methods can better associate the entities. And our method combining distant supervision and attention mechanism can further strengthen the connection between entities when compared with LSTM-LSTM-Bias.

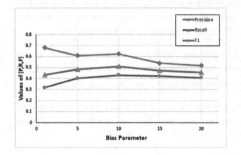

Fig. 3. The results on different bias parameter α when ε is set to 1

Fig. 4. The results on different ε when α is set to 10

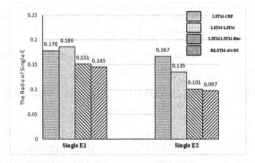

Fig. 5. The ratio of predicted single entities for each method. The higher of the ratio means that the more entities are left

In order to further analyze the effect of the bias objective function, we change the bias parameter α from 1 to 20, and the different extraction results are shown in Fig. 3. If α is too small, it will affect the recall scores. If α is too large, the precision will decrease. When $\alpha = 10$, our model can balance the precision and recall well, and can achieve the best F1 scores. Meanwhile, in order to avoiding the influence of the value of hyperparameter ε on the prediction result, we set it to 1.

And in order to analyze the effect of ε on final extraction results, we change the hyperparameter ε from 1 to 1.9. The different extraction results are shown in Fig. 4. If ε is too small, it will make an impact on the recall, and the precision will decrease if ε is too large which is similar to bias parameter α. When $\varepsilon = 1.5$, BiLSTM-Att-DS model can balance the precision and recall, and obtain the best F1 scores. And all the comparison tests in this section are performed with the value of α set to

10. With the change of ε, the fluctuation range of precision and F1 are relatively larger. It mains that our method can effectively improve extraction effect by taking an appropriate value for ε. The reason may be that this combination of distant supervision and attention can effectively enhance the accuracy of entity words prediction, at the same time pay much attention to strengthen links between them to promote entity pairs and relations prediction.

6 Conclusion

In this paper, we propose a semi-supervised method to improve the result of joint extraction. We put forward a combined method of distant supervision and attention mechanism and further combine it with end-to-end tagging model. Our methods are proved better than most of the existing pipelined and joint learning methods. However, it still have some deficiencies in the recognition of overlapping relations. In future work, we would like to investigate a learnable attention hyperparameter ε and attempt to improve the tagging scheme so that a word can have several tags to detect the overlapping relations.

References

1. Alzaidy, R., Caragea, C., Giles, C.L.: BI-LSTM-CRF sequence labeling for keyphrase extraction from scholarly documents. In: The World Wide Web Conference, ser. WWW 2019, pp. 2551–2557. Association for Computing Machinery, New York (2019)
2. Bollacker, K., Cook, R., Tufts, P.: Freebase: a shared database of structured general human knowledge, pp. 1962–1963 (January 2007)
3. Christopoulou, F., Miwa, M., Ananiadou, S.: A walk-based model on entity graphs for relation extraction (February 2019)
4. Gormley, M.R., Yu, M., Dredze, M.: Improved relation extraction with feature-rich compositional embedding models. arXiv preprint arXiv:1505.02419 (2015)
5. Hochreiter, S., Schmidhuber, J.: Long short-term memory. Neural Comput. **9**(8), 1735–1780 (1997)
6. Hoffmann, R., Zhang, C., Ling, X., Zettlemoyer, L., Weld, D.S.: Knowledge-based weak supervision for information extraction of overlapping relations. In: Proceedings of the 49th Annual Meeting of the Association for Computational Linguistics: Human Language Technologies, vol. 1. Association for Computational Linguistics, pp. 541–550 (2011)
7. Huang, Z., Xu, W., Yu, K.: Bidirectional LSTM-CRF models for sequence tagging. arXiv preprint arXiv:1508.01991 (2015)
8. Katiyar, A., Cardie, C.: Investigating LSTMs for joint extraction of opinion entities and relations. In: Proceedings of the 54th Annual Meeting of the Association for Computational Linguistics (Long Papers), vol. 1, pp. 919–929 (2016)
9. Lample, G., Ballesteros, M., Subramanian, S., Kawakami, K., Dyer, C.: Neural architectures for named entity recognition. arXiv preprint arXiv:1603.01360 (2016)
10. Li, P., Mao, K.: Knowledge-oriented convolutional neural network for causal relation extraction from natural language texts. Expert Syst. Appl. **115**, 512–523 (2019)

11. Li, Q., Ji, H.: Incremental joint extraction of entity mentions and relations. In: Proceedings of the 52nd Annual Meeting of the Association for Computational Linguistics (Long Papers), vol. 1, pp. 402–412 (2014)
12. Lin, Z., et al.: A structured self-attentive sentence embedding. arXiv preprint arXiv:1703.03130 (2017)
13. Luo, G., Huang, X., Lin, C.-Y., Nie, Z.: Joint entity recognition and disambiguation. In: Proceedings of the 2015 Conference on Empirical Methods in Natural Language Processing, pp. 879–888 (2015)
14. Manning, C.D., Surdeanu, M., Bauer, J., Finkel, J.R., Bethard, S., McClosky, D.: The Stanford CoreNLP natural language processing toolkit. In: Proceedings of 52nd Annual Meeting of the Association for Computational Linguistics: System Demonstrations, pp. 55–60 (2014)
15. Mikolov, T., Sutskever, I., Chen, K., Corrado, G.S., Dean, J.: Distributed representations of words and phrases and their compositionality. In: Advances in Neural Information Processing Systems, pp. 3111–3119 (2013)
16. Mintz, M., Bills, S., Snow, R., Jurafsky, D.: Distant supervision for relation extraction without labeled data. In: Proceedings of the Joint Conference of the 47th Annual Meeting of the ACL and the 4th International Joint Conference on Natural Language Processing of the AFNLP, vol. 2. Association for Computational Linguistics, pp. 1003–1011 (2009)
17. Miwa, M., Sasaki, Y.: Modeling joint entity and relation extraction with table representation. In: Proceedings of the 2014 Conference on Empirical Methods in Natural Language Processing (EMNLP), pp. 1858–1869 (2014)
18. Nadeau, D., Sekine, S.: A survey of named entity recognition and classification. Lingvisticae Investigationes 30(1), 3–26 (2007)
19. Passos, A., Kumar, V., McCallum, A.: Lexicon infused phrase embeddings for named entity resolution. arXiv preprint arXiv:1404.5367 (2014)
20. Ren, X., et al.: CoType: joint extraction of typed entities and relations with knowledge bases. In: Proceedings of the 26th International Conference on World Wide Web, pp. 1015–1024 (2017)
21. Rink, B., Harabagiu, S.: UTD: classifying semantic relations by combining lexical and semantic resources. In: Proceedings of the 5th International Workshop on Semantic Evaluation, pp. 256–259. Association for Computational Linguistics (2010)
22. Santos, C.N.D., Xiang, B., Zhou, B.: Classifying relations by ranking with convolutional neural networks. arXiv preprint arXiv:1504.06580 (2015)
23. Takanobu, R., Zhang, T., Liu, J., Huang, M.: A hierarchical framework for relation extraction with reinforcement learning. In: Proceedings of the AAAI Conference on Artificial Intelligence, vol. 33, pp. 7072–7079 (2019)
24. Tang, J., Qu, M., Wang, M., Zhang, M., Yan, J., Mei, Q.: Line: Large-scale information network embedding. In: Proceedings of the 24th International Conference on World Wide Web, pp. 1067–1077 (2015)
25. Vaswani, A., Bisk, Y., Sagae, K., Musa, R.: Supertagging with LSTMs. In: Proceedings of the 2016 Conference of the North American Chapter of the Association for Computational Linguistics: Human Language Technologies, pp. 232–237 (2016)
26. Yadav, V., Bethard, S.: A survey on recent advances in named entity recognition from deep learning models. arXiv preprint arXiv:1910.11470 (2019)
27. Yang, B., Cardie, C.: Joint inference for fine-grained opinion extraction. In: Proceedings of the 51st Annual Meeting of the Association for Computational Linguistics (Long Papers), vol. 1, pp. 1640–1649 (2013)

28. Zeng, D., Liu, K., Chen, Y., Zhao, J.: Distant supervision for relation extraction via piecewise convolutional neural networks. In: Proceedings of the 2015 Conference on Empirical Methods in Natural Language Processing, pp. 1753–1762 (2015)
29. Zhai, F., Potdar, S., Xiang, B., Zhou, B.: Neural models for sequence chunking. In: Thirty-First AAAI Conference on Artificial Intelligence (2017)
30. Zheng, S., Wang, F., Bao, H., Hao, Y., Zhou, P., Xu, B.: Joint extraction of entities and relations based on a novel tagging scheme. arXiv preprint arXiv:1706.05075 (2017)
31. Zheng, S., Xu, J., Zhou, P., Bao, H., Qi, Z., Xu, B.: A neural network framework for relation extraction: learning entity semantic and relation pattern. Knowl.-Based Syst. **114**, 12–23 (2016)
32. Zou, F., Shen, L., Jie, Z., Zhang, W., Liu, W.: A sufficient condition for convergences of Adam and RMSProp. In: Proceedings of the IEEE Conference on Computer Vision and Pattern Recognition, pp. 11127–11135 (2019)
33. Qin, P., Xu, W., Wang, W.Y.: Robust distant supervision relation extraction via deep reinforcement learning. arXiv preprint arXiv:1805.09927 (2018)

Research Progress of Zero-Shot Learning Beyond Computer Vision

Weipeng Cao[1], Cong Zhou[1], Yuhao Wu[1], Zhong Ming[1], Zhiwu Xu[1(✉)],
and Jiyong Zhang[2]

[1] College of Computer Science and Software Engineering, Shenzhen University,
Shenzhen, China
xuzhiwu@szu.edu.cn

[2] School of Automation, Hangzhou Dianzi University, Hangzhou, China

Abstract. Traditional machine learning techniques, including deep learning, most assume that the classes of testing samples belong to the subset of training samples. However, there are many scenarios that conflict with this assumption in the real world, that is, the classes of testing samples have never been seen in model training. To improve the generalization ability of the model in these cases, zero-shot learning (ZSL) was proposed, which can mine the mapping relationship between the features and the labels of the seen class samples and then transfer it to the prediction of unseen classes. Most of the existing ZSL algorithms or applications are concerned with computer vision problems. In fact, the above difficulties and the demand for ZSL also exist in other fields, but there is currently a lack of relevant research progress review. To make up for this gap, this paper reviews the latest research progress of ZSL beyond computer vision, introduces the general concepts of ZSL, classifies the mainstream models, and refines three issues worthy of study. This study is expected to provide ZSL-based solution guidance for researchers and engineers beyond the field of computer vision.

Keywords: Zero-shot learning · Deep learning · Transfer learning

1 Introduction

In recent years, with the rapid development of machine learning techniques, especially deep learning, related algorithms have made breakthroughs in many fields, such as computer vision and natural language processing. Most of these algorithms assume that the application environment is in a closed set state, that is, the classes of the testing samples must be the classes that have been seen during the model training. However, the real world is actually an open set state, that is, sometimes the classes of testing samples are never seen by the model.

This work was supported by National Natural Science Foundation of China (61836005) and the Opening Project of Shanghai Trusted Industrial Control Platform (TICPSH202003008-ZC).

M. Qiu (Ed.): ICA3PP 2020, LNCS 12453, pp. 538–551, 2020.
https://doi.org/10.1007/978-3-030-60239-0_36

We call the classes that one can see during the model training as the seen classes, which corresponds to the unseen classes. Most of the traditional machine learning algorithms are devoted to the prediction of the seen classes but there is no way to predict the unseen classes. To improve the prediction performance of the model for the unseen classes, zero-shot learning (ZSL) was proposed [33], which refers to the technology that can make the model accurately predict the unseen classes [34]. According to different testing settings, existing ZSL algorithms can be divided into two categories: traditional ZSL and generalized ZSL [44]. The difference between them is that the testing samples of the former can only come from unseen classes, while the testing samples of the latter can come from both the unseen classes and the seen classes.

One of the differences between ZSL and traditional machine learning is the construction of the training data set. In ZSL, in addition to labels (i.e., the classes), it is often necessary to provide the side information corresponding to the labels. The side information is usually the semantic coding of the classes and their attributes, which can be extracted manually or by automatic techniques such as word2vec [28]. After the training data set is constructed, the ZSL algorithm learns the general knowledge in the problem domain by mining the relationship between the features of the seen class samples, class information, and the side information, and then applies it to the prediction of unseen classes. For example, for image classification problems, in the training phase, the ZSL algorithm first learns the mapping function between the visual space corresponding to the seen classes images and the semantic space corresponding to the classes; in the testing phase, given a testing sample, the ZSL model can predict the semantic feature corresponding to its class according to its visual features, then compare the semantic features corresponding to the original side information to find the closest one, and finally map back to the corresponding class to complete the prediction.

The advantages of ZSL include: (1) It can improve the generalization ability and practicality of machine learning models. For a well-trained ZSL model, it can make an accurate prediction even if it encounters unseen classes, which is critical for real-world applications; (2) it greatly eases the dependence of machine learning algorithms on labeled data. In many scenarios, the cost of collecting a large number of training samples and labeling them is very expensive, such as medical image recognition and wild animal recognition scenarios. For these cases, sometimes we can obtain some prior knowledge related to the classes in advance, which can be used to construct the side information of the classes, and then one can use ZSL to train a model with good generalization ability. ZSL has attracted extensive attention in recent years, and many related algorithms and applications have been proposed [34,50].

However, most of the existing work is oriented to the field of computer vision. In fact, the problem to be solved by ZSL is a common problem in the real world. In other words, ZSL can be applied to solve problems other than computer vision. Actually, there have been many research achievements on this issue in recent years [13,29,35]. However, to the best of our knowledge, there is currently no relevant survey to introduce the research progress of ZSL beyond computer vision. To fill this gap, this paper introduces ZSL from the following four aspects (also shown in Fig. 1):

(1) General concepts: the history of ZSL, commonly used data sets, evaluation standards, and the strategies for constructing the side information;
(2) Mainstream algorithms: categories and characteristics of mainstream ZSL models;
(3) Research progress: mainly focus on the notable work of ZSL beyond computer vision;
(4) Future trends: discussion on the future research direction of ZSL.

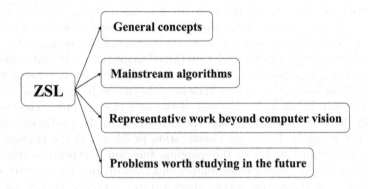

Fig. 1. The structure of this survey

Compare with other existing reviews on ZSL [34,50], our paper has the following highlight: This is the first review of research progress beyond computer vision that provides researchers and engineers in a wider field with a basic introduction to ZSL and representative application examples, which is expected to provide them with valuable guidelines for solving a wider range of real engineering problems.

The remainder of this paper will be organized as follows. In Sect. 2, we introduce several general concepts of ZSL, such as its evaluation standards; we group some notable ZSL models into five categories in Sect. 3; Sect. 2 and Sect. 3 can help readers understand the research status of ZSL from a macro perspective. In Sect. 4, we pay attention to the representative work of ZSL beyond computer vision. In Sect. 5, we summarize three worthy research directions for researchers. We conclude this paper in Sect. 6.

2   Introduction to General Concepts of ZSL

In this section, we briefly introduce the history of ZSL, the commonly used data sets, the evaluation criteria, and the constructive methods of the side information.

2.1   The Proposal and Development of ZSL

Before the concept of ZSL was formally proposed, some scholars had certain assumptions on the recognition of unseen classes. Representative work includes: zero-data learning of new tasks [22], ZSL with semantic output codes [33], unseen class learning by between-class attribute transfer [20], and ZSL through cross-modal transfer [39].

Specifically, zero-data learning of new tasks was proposed by Larochelle et al. in 2008 [22]. Its main goal is to apply the learned knowledge to the prediction of classes or tasks without training data, and to provide related semantic information for them, which is similar to the definition of ZSL. The authors also proposed the input space-based method and the model space-based method for dealing with this problem, which inspires the ZSL to use the side information of classes for classification.

In 2009, Palatucci et al. [33] proposed to use the semantic output coding classifier and the label base containing semantic knowledge to realize ZSL, which is the first time to propose the concept of "zero-shot learning". The model is mainly classified by the semantic codes information of class labels in the knowledge base. It compares the semantic codes information corresponding to the testing sample with the semantic codes information of the known classes in the knowledge base to distinguish the seen classes and the unseen classes.

Based on the semantic coding, Lampert et al. [20] proposed an unseen class learning method based on the between-class attribute transfer. The contributions of their work include: (1) Provided a unified training framework for most of the current ZSL methods; (2) Established a benchmark data set for ZSL (i.e., "Animals with attributes"); (3) Introduced the concept of "attribute" into ZSL. Based on this framework, two classic ZSL models are derived: the direct attribute prediction model (DAP) and the indirect attribute prediction model (IAP).

In 2013, Socher et al. [39] proposed a ZSL algorithm with a cross-modal transfer function, which transformed the ZSL into a subspace learning problem. The core idea of this method is to map the training images and their labels into the same subspace, and then use similarity measurement techniques to determine the labels of the testing samples.

Later, with the rapid development of deep learning, researchers began to use related technologies to realize the evolution from the low-level visual features to the deep-level visual features, and then better mining the mapping relationship between the visual space and the semantic space. For example, one can use the deep convolutional neural network proposed by Alex et al. in 2012 [19] and the word2vec technique proposed by Mikolov et al. in 2013 [28] to extract the deep visual features from the training data and obtain the semantic vectors corresponding to the labels, respectively, and then train the model based on the existing ZSL algorithm.

The classes of testing samples in conventional ZSL are completely different from those in the training phase, which deviates from the real-world rules because the classes of the testing samples in the real world should include both the seen classes and the unseen classes. To make the ZSL model more consistent with the real world, generalized ZSL (GZSL) was proposed [7], which has no restrictions on the types of testing samples. Therefore, GZSL is more difficult but also more practical, which has become one of the current research hotspots.

2.2 Data Sets and Evaluation Criteria

At present, ZSL is mainly used to solve computer vision problems and some representative data sets for this scenario include: ImageNet [9], Animals with Attributes (AwA) [7], Caltech-UCSD-Birds200-2111 (CUB) [42], Attribute Pascal and Yahoo (aPY) [11], SUN attribute [36], Oxford Flowers (FLO) [30], etc.

In the early stage, there was no unified standard to divide the seen classes and the unseen classes for a given data set, resulting in an unfair phenomenon in the performance evaluation of the ZSL algorithm. To alleviate this problem, Xian et al. proposed a standard data set segmentation method [46] in 2017, which unifies the benchmarks of model evaluation by unifying the evaluation protocol and the data segmentation. In addition, due to the uneven distribution of the classes in ZSL, the traditional mean average precision (MAP) can not reflect the performance of the ZSL algorithm well, so the class average accuracy was proposed [44] and has become one of the most commonly used evaluation indicators, which can be obtained using the following formula:

$$M = \frac{\sum_{i=1}^{K} A_u^i}{K} \tag{1}$$

where K is the number of the unseen classes and A_u^i refers to the prediction accuracy of the model on the i-th unseen class.

Moreover, to better evaluate the performance of the GZSL algorithm, the harmonic mean $\mathcal{H}$ was proposed [46] and has become one of the most commonly used evaluation indicators, which can be obtained as follows:

$$\mathcal{H} = \frac{2 \times A_s \times A_u}{A_s + A_u} \tag{2}$$

where A_s and A_u are the top-1 accuracy of the model on the seen classes and the unseen classes, respectively.

2.3 Side Information

Side information is used to describe the auxiliary information of the classes such as their attributes. It serves as a bridge between the seen classes and the unseen classes, making it possible to use the seen samples to train a ZSL model that can predict the unseen classes. Therefore, side information is an important part of ZSL, which can usually be obtained through two methods: human annotation and text-based learning [34,50].

Methods based on human annotation can be further divided into the attribute-based method and the non-attribute-based method. As a kind of prior knowledge, attributes can semantically represent specific classes and reflect their characteristics, so that they can be used to distinguish different classes in a data set. Commonly used attribute representation can be divided into binary attributes and continuous attributes. Binary attributes are used to describe whether a specific class or object has a certain attribute, if it is, the value of this dimension is 1, otherwise it is 0. Continuous attributes are generally used to describe the possibility of a specific class or object having a specific attribute, which is usually expressed in the form of real values. The non-attribute-based method directly uses the names of the classes as their semantic information description to construct the semantic vectors. Methods based on human annotation need the help of human's prior knowledge, so it inevitably has certain subjectivity. Moreover, when there are many classes, this method is time-consuming and expensive. The advantage of this method is that it is helpful to get high accuracy of the ZSL model, and to some extent, it can improve the interpretability of the results.

Text-based learning methods mainly use machine learning algorithms to obtain mapping models and then map classes or their descriptions to corresponding vectors. According to different auxiliary information, this kind of method can be divided into two categories: label embedding and text embedding. Label embedding mainly uses natural language processing models, such as word2vec, to represent the class labels in vectorization. The similarity between word vectors can be used as a reference for classification. The text embedding method needs to obtain the description text of the classes, and convert the text description into the corresponding semantic vector through text encoding models. The text-based methods can effectively reduce the labor cost, but the disadvantage is that the data sources often have noise and the results are less interpretable.

3   Categories and Characteristics of Mainstream Models

Inspired by the classification method of [44], here we divide the existing notable ZSL models into five categories: intermediate attribute classifiers models, compatibility models, hybrid models, transductive models, and generative models. The details of these five categories and the corresponding notable algorithms are shown in Table 1.

3.1   Intermediate Attribute Classifiers Models

In this learning paradigm, attributes are the key information that the ZSL model uses to make decisions. Specifically, given a testing sample, the ZSL model first predicts the attribute of its class and then selects the most probable class according to the similarity of the attribute to the attributes of the known classes. The consistency model and the hybrid model are also derived from the intermediate attribute classifiers models.

At present, the existing intermediate attribute classifiers models can be divided into two categories: the direct attribute prediction model (DAP) and the indirect attribute prediction model (IAP) [21]. In the training phase, DAP

Table 1. The details of the five categories of ZSL and the corresponding notable algorithms

Categories	Notable algorithms
Intermediate attribute classifiers models	DAP [26], IAP [21], etc.
Compatibility models	DEVISE [14], ALE [1], SJE [2], ESZSL [38], SAE [18], LATEM [43], CMT [39], etc.
Hybrid models	SSE [51], CONSE [31], SYNC [6], GFZSL [41], etc.
Transductive models	GFZSL-tran [41], DSRL [49], etc.
Generative models	f-xGAN [45], cycle-CLSWGAN [12], AFC-GAN [24], etc.

trains the attribute classifier. In the testing phase, one can directly obtain the attribute feature estimate by inputting the attributes of the testing sample into the model even if the testing class is an unseen class. The difference between IAP and DAP is that the attribute classifier of IAP cannot directly obtain the attribute feature estimate. IAP needs to input the class label of the sample and its attribute indication vector to indirectly obtain the attribute feature estimate.

3.2 Compatibility Models

The compatibility models are models that map the input and output to a subspace and then judge the compatibility of the input and output mapping vectors in the subspace and determine the class label. The compatibility models can be divided into linear compatibility models and non-linear compatibility models according to whether the compatibility function is linear. Some notable linear compatibility models include: deep visual-semantic embedding model (DEVISE) [14], attribute label embedding model (ALE) [1], structured joint embedding (SJE) [2], embarrassingly simple zero-shot learning (ESZSL) [38], semantic auto-encoder embedding (SAE) [18], etc. Some representative non-linear compatibility models include: latent embedding models (LATEM) [43], cross-mode migration model (CMT) [39], etc.

Compatibility models need to train the class label embedding function so that the class label can be accurately embedded in the feature space. However, the compatibility models rely heavily on the quality of auxiliary information. The learning ability of the linear compatibility models is often limited by the linear function, so its expression ability is not as good as the nonlinear compatibility models.

3.3 Hybrid Models

The hybrid models use a hybrid combination of feature subspace mappings corresponding to the training class labels to represent the mapping of the testing samples in the feature subspace, and then obtain the class label estimation of the testing sample based on the similarity between the input sample mapping and the testing class label mapping. Some notable hybrid models include: semantic similarity embedding model (SSE) [51], convex combination semantic embedding model (CONSE) [31], synthesized classifiers (SYNC) [6], generative framework for zero-shot learning (GFZSL) [41], etc.

The training mechanism of the hybrid models is similar to that of the IAP, so they also rely heavily on the similarity between the training classes and the testing classes, and the robustness of their models is relatively weak.

3.4 Transductive Models

Transductive models [40] use the class labels of the training classes and the side information of the testing classes to determine the class labels of the testing samples, and then add the testing samples with their predicted class labels to the original training data set. Then, continue to learn the new decision rules based on the augmented data set, and use the updated model to continue the above steps until all testing samples are labeled. Representative transductive models include GFZSL-tran [41], discriminative semantic representation learning (DSRL) [49], etc.

Transductive models belong to an online learning paradigm that can continue to update themselves during the testing phase. The disadvantage of this kind of method is that their training process needs a lot of calculation and the final model performance dependents on the initial accuracy of the model.

3.5 Generative Models

Recently, it is popular to combine generative adversarial networks (GAN) [17] with ZSL to produce generative ZSL models. Using the side information of the classes as the constraint of the GAN model can enable its generator to generate features related to the specific class, which in turn can make the model better distinguish different classes.

In particular, if the corresponding pseudo samples or features can be generated according to the side information of the unseen classes, the ZSL task can be converted into traditional supervised learning. Some notable work includes: f-xGAN [45], cycle-CLSWGAN [12], alleviating feature confusion GAN (AFC-GAN) [24], etc.

4   Application Progress of ZSL Beyond Computer Vision

ZSL has been widely used in the field of computer vision and some researchers have reviewed these works well [34,50]. Different from the existing surveys, here we focus on the research progress of ZSL in natural language processing and other fields.

4.1 Application of ZSL in Natural Language Processing

Nakashole et al. [29] combined ZSL with the bilingual dictionary induction to realize that one can translate an uncommon language (e.g.., Portuguese) into a common language (e.g.., English) through a third relatively relevant language (e.g.., Spanish) when only a small seed dictionary is used. The authors claimed that using ZSL to train the translation model can achieve high accuracy even with a small amount of labeled data.

In [13], Ferreira et al. proposed a complete semantic analyzer based on the word embedding and ZSL techniques. This semantic analyzer does not require annotated contextual data, ontology description of the target domain, and general word embedding features, which reduces the cost of manual annotation and can obtain performance comparable to the use of manual annotation data.

Pasupat et al. [35] used the ZSL technique to extract entities of specified categories in web pages. Most of the traditional methods are effective for ordinary categories, but they are relatively weak when facing categories with few samples. ZSL does not rely on multiple web pages to get entities, it only needs a single web page to get the target entities. Moreover, it can extract the target entities from the semi-structured data on the web page without complete category information.

In [27], Ma et al. used a ZSL framework to solve the problem that current named entity recognition methods cannot detect unseen entities. They proposed a label embedding method that combines prototype information and hierarchical information to learn the pre-trained label embedding. In this way, the above problems can be alleviated to some extent.

Funaki et al. [16] proposed an image-based cross-language document retrieval method, which can take images in two languages as the target, and deduce the common semantic subspace connecting two languages through generalized canonical correlation analysis, to realize document retrieval between different languages. This method can reduce the cost of manually creating a corpus when there is no or only a small number of parallel corpora.

4.2 Application of ZSL in Other Fields

In addition to applications in computer vision and natural language processing, ZSL has also been used in other fields. For example, for the human activity recognition problem, Zheng et al. [52] proposed that existing experience can be used to identify unseen human activities through knowledge transfer methods, and then Cheng et al. [8] realized the recognition of unseen activity categories based on the semantic description by using ZSL.

In the field of knowledge representation, Xie et al. [47] proposed the DKRL model to deal with the task of entity classification where at least one entity in the triple is not in the knowledge graph.

Robyns et al. [37] realized the identification of unseen physical layer devices by using ZSL. ZSL can also complete tasks such as generating Emoji expressions for unseen images [10], neural decoding of fMRI images [3], and identifying unseen molecular compounds [22].

5   Future Trends

Here we propose three open problems worthy of study.

5.1   Smart Side Information Construction Method

As mentioned in Sect. 2.3, the side information generated by manual labeling has the advantages of high accuracy and strong interpretability, but its disadvantages are high cost and low efficiency. The automatic text-based learning method can overcome these shortcomings, but the quality of the side information obtained by this method is difficult to guarantee. Combining these two methods to design a smart side information construction method will be a problem worthy of study. It is expected that the new method can obtain abundant side information quickly and accurately.

5.2   Generalized Zero-Shot Learning

As mentioned in 2.1, GZSL is more in line with the rules of the real world, that is, the testing samples can come from both the seen and unseen classes. Since the unseen classes samples have never been seen during model training, it is very difficult for the model to accurately predict them. Moreover, in this case, the three challenging issues in the ZSL field (i.e., domain shift problem [15], hubness problem [23], and the semantic gap problem [25]) will become more difficult [7]. How to improve the generalization ability of the ZSL model under the evaluation setting of GZSL is one of the current research hotspots. Although some researchers have put forward some enlightening algorithms [7,31,48], there is still a long way to go before GZSL can be truly applied. How to improve the generalization ability of the GZSL model is one of the most worthy research issues.

5.3   Combination with Other Technologies

Most fields are facing the problem of open set learning, how to combine the idea of ZSL with the existing algorithms in these fields to improve their generalization ability is a direction worth exploring. For example, some researchers have combined ZSL with reinforcement learning to optimize the strategies of agents to deal with unknown environments [32]. Moreover, one can also learn from the advantages of other technologies to improve the performance of the ZSL algorithm. For example, neural networks with random weights (NNRW [4,5]) have extremely fast training speeds, how to combine them with the existing ZSL algorithms to improve the training efficiency of the latter is an interesting research direction.

6   Conclusions

In this paper, we first introduce the research motivation, development history, and fundamental concepts of ZSL, and then classify the current mainstream ZSL models into five categories, which enable readers to have a macro understanding of the field of ZSL. Furthermore, we focus on the representative work of ZSL beyond computer vision, which is the biggest difference from other existing surveys. Moreover, we have refined three valuable research issues to provide direction for researchers. This paper is expected to provide guidance on open set learning for researchers and engineers in a wider range of fields.

In the future, we will give more details for the difficult issues mentioned in this paper, and add more representative work of ZSL beyond computer vision.

References

1. Akata, Z., Perronnin, F., Harchaoui, Z., Schmid, C.: Label-embedding for image classification. IEEE Trans. Pattern Anal. Mach. Intell. **38**(7), 1425–1438 (2016)
2. Akata, Z., Reed, S., Walter, D., Lee, H., Schiele, B.: Evaluation of output embeddings for fine-grained image classification. In: 2015 IEEE Conference on Computer Vision and Pattern Recognition (CVPR), pp. 2927–2936 (2015)
3. Caceres, C.A., et al.: Feature selection methods for zero-shot learning of neural activity. Front. Neuroinformatics **11**, 41 (2017)
4. Cao, W., Hu, L., Gao, J., Wang, X., Ming, Z.: A study on the relationship between the rank of input data and the performance of random weight neural network. Neural Comput. Appl. **32**, 1–12 (2020)
5. Cao, W., Wang, X., Ming, Z., Gao, J.: A review on neural networks with random weights. Neurocomputing **275**, 278–287 (2018)
6. Changpinyo, S., Chao, W.L., Gong, B., Sha, F.: Synthesized classifiers for zero-shot learning. In: Proceedings of the IEEE Conference on Computer Vision and Pattern Recognition, pp. 5327–5336 (2016)
7. Chao, W.-L., Changpinyo, S., Gong, B., Sha, F.: An empirical study and analysis of generalized zero-shot learning for object recognition in the wild. In: Leibe, B., Matas, J., Sebe, N., Welling, M. (eds.) ECCV 2016, Part II. LNCS, vol. 9906, pp. 52–68. Springer, Cham (2016). https://doi.org/10.1007/978-3-319-46475-6_4
8. Cheng, H.T., Sun, F.T., Griss, M., Davis, P., Li, J., You, D.: Nuactiv: recognizing unseen new activities using semantic attribute-based learning. In: Proceeding of the 11th Annual International Conference on Mobile Systems, Applications, and Services, pp. 361–374 (2013)
9. Deng, J., Dong, W., Socher, R., Li, L.J., Li, K., Fei-Fei, L.: Imagenet: a large-scale hierarchical image database. In: 2009 IEEE Conference on Computer Vision and Pattern Recognition, pp. 248–255. IEEE (2009)
10. Dinu, G., Lazaridou, A., Baroni, M.: Improving zero-shot learning by mitigating the hubness problem. In: ICLR (Workshop) (2014)
11. Farhadi, A., Endres, I., Hoiem, D., Forsyth, D.: Describing objects by their attributes. In: 2009 IEEE Conference on Computer Vision and Pattern Recognition, pp. 1778–1785 (2009)

12. Felix, R., Vijay Kumar, B.G., Reid, I., Carneiro, G.: Multi-modal cycle-consistent generalized zero-shot learning. In: Ferrari, V., Hebert, M., Sminchisescu, C., Weiss, Y. (eds.) ECCV 2018, Part VI. LNCS, vol. 11210, pp. 21–37. Springer, Cham (2018). https://doi.org/10.1007/978-3-030-01231-1_2

13. Ferreira, E., Jabaian, B., Lefèvre, F.: Zero-shot semantic parser for spoken language understanding. In: Sixteenth Annual Conference of the International Speech Communication Association (2015)

14. Frome, A., et al.: Devise: a deep visual-semantic embedding model. In: Advances in Neural Information Processing Systems 26, pp. 2121–2129 (2013)

15. Fu, Y., Hospedales, T.M., Xiang, T., Fu, Z., Gong, S.: Transductive multi-view embedding for zero-shot recognition and annotation. In: Fleet, D., Pajdla, T., Schiele, B., Tuytelaars, T. (eds.) ECCV 2014, Part II. LNCS, vol. 8690, pp. 584–599. Springer, Cham (2014). https://doi.org/10.1007/978-3-319-10605-2_38

16. Funaki, R., Nakayama, H.: Image-mediated learning for zero-shot cross-lingual document retrieval. In: Proceedings of the 2015 Conference on Empirical Methods in Natural Language Processing, pp. 585–590 (2015)

17. Goodfellow, I., et al.: Generative adversarial nets. In: Advances in Neural Information Processing Systems, pp. 2672–2680 (2014)

18. Kodirov, E., Xiang, T., Gong, S.: Semantic autoencoder for zero-shot learning. In: 2017 IEEE Conference on Computer Vision and Pattern Recognition (CVPR), pp. 4447–4456 (2017)

19. Krizhevsky, A., Sutskever, I., Hinton, G.E.: Imagenet classification with deep convolutional neural networks. In: Advances in Neural Information Processing Systems, pp. 1097–1105 (2012)

20. Lampert, C.H., Nickisch, H., Harmeling, S.: Learning to detect unseen object classes by between-class attribute transfer. In: 2009 IEEE Conference on Computer Vision and Pattern Recognition, pp. 951–958 (2009)

21. Lampert, C.H., Nickisch, H., Harmeling, S.: Attribute-based classification for zero-shot visual object categorization. IEEE Trans. Pattern Anal. Mach. Intell. 36(3), 453–465 (2014)

22. Larochelle, H., Erhan, D., Bengio, Y.: Zero-data learning of new tasks. In: AAAI'08 Proceedings of the 23rd National Conference on Artificial Intelligence - Volume 2, pp. 646–651 (2008)

23. Lazaridou, A., Dinu, G., Baroni, M.: Hubness and pollution: delving into cross-space mapping for zero-shot learning. In: Proceedings of the 53rd Annual Meeting of the Association for Computational Linguistics and the 7th International Joint Conference on Natural Language Processing (Volume 1: Long Papers), pp. 270–280 (2015)

24. Li, J., Jing, M., Lu, K., Zhu, L., Yang, Y., Huang, Z.: Alleviating feature confusion for generative zero-shot learning. In: Proceedings of the 27th ACM International Conference on Multimedia, pp. 1587–1595 (2019)

25. Li, Y., Wang, D., Hu, H., Lin, Y., Zhuang, Y.: Zero-shot recognition using dual visual-semantic mapping paths. In: 2017 IEEE Conference on Computer Vision and Pattern Recognition (CVPR), pp. 5207–5215 (2017)

26. Liang, K., Chang, H., Shan, S., Chen, X.: A unified multiplicative framework for attribute learning. In: Proceedings of the IEEE International Conference on Computer Vision, pp. 2506–2514 (2015)

27. Ma, Y., Cambria, E., Gao, S.: Label embedding for zero-shot fine-grained named entity typing. In: Proceedings of COLING 2016, the 26th International Conference on Computational Linguistics: Technical Papers, pp. 171–180 (2016)

28. Mikolov, T., Chen, K., Corrado, G., Dean, J.: Efficient estimation of word representations in vector space. arXiv preprint arXiv:1301.3781 (2013)
29. Nakashole, N., Flauger, R.: Knowledge distillation for bilingual dictionary induction. In: Proceedings of the 2017 Conference on Empirical Methods in Natural Language Processing, pp. 2497–2506 (2017)
30. Nilsback, M.E., Zisserman, A.: Automated flower classification over a large number of classes. In: 2008 Sixth Indian Conference on Computer Vision, Graphics & Image Processing, pp. 722–729. IEEE (2008)
31. Norouzi, M., et al.: Zero-shot learning by convex combination of semantic embeddings. arXiv preprint arXiv:1312.5650 (2013)
32. Oh, J., Singh, S., Lee, H., Kohli, P.: Zero-shot task generalization with multi-task deep reinforcement learning. In: Proceedings of the 34th International Conference on Machine Learning-Volume 70, pp. 2661–2670. JMLR. org (2017)
33. Palatucci, M., Pomerleau, D., Hinton, G.E., Mitchell, T.M.: Zero-shot learning with semantic output codes. In: Advances in Neural Information Processing Systems 22. vol. 22, pp. 1410–1418 (2009)
34. Pang, Y., Wang, H., Yu, Y., Ji, Z.: A decadal survey of zero-shot image classification. SCIENTIA SINICA Informationis 49(10), 1299–1320 (2019)
35. Pasupat, P., Liang, P.: Zero-shot entity extraction from web pages. In: Proceedings of the 52nd Annual Meeting of the Association for Computational Linguistics (Volume 1: Long Papers), pp. 391–401 (2014)
36. Patterson, G., Hays, J.: Sun attribute database: Discovering, annotating, and recognizing scene attributes. In: 2012 IEEE Conference on Computer Vision and Pattern Recognition, pp. 2751–2758 (2012)
37. Robyns, P., Marin, E., Lamotte, W., Quax, P., Singelée, D., Preneel, B.: Physical-layer fingerprinting of lora devices using supervised and zero-shot learning. In: Proceedings of the 10th ACM Conference on Security and Privacy in Wireless and Mobile Networks, pp. 58–63 (2017)
38. Romera-Paredes, B., Torr, P.: An embarrassingly simple approach to zero-shot learning. In: Proceedings of The 32nd International Conference on Machine Learning, pp. 2152–2161 (2015)
39. Socher, R., Ganjoo, M., Sridhar, H., Bastani, O., Manning, C.D., Ng, A.Y.: Zero-shot learning through cross-modal transfer. In: ICLR (Workshop) (2013)
40. Song, J., Shen, C., Yang, Y., Liu, Y., Song, M.: Transductive unbiased embedding for zero-shot learning. In: Proceedings of the IEEE Conference on Computer Vision and Pattern Recognition, pp. 1024–1033 (2018)
41. Verma, V.K., Rai, P.: A simple exponential family framework for zero-shot learning. In: Ceci, M., Hollmén, J., Todorovski, L., Vens, C., Džeroski, S. (eds.) ECML PKDD 2017, Part II. LNCS (LNAI), vol. 10535, pp. 792–808. Springer, Cham (2017). https://doi.org/10.1007/978-3-319-71246-8_48
42. Wah, C., Branson, S., Welinder, P., Perona, P., Belongie, S.: The caltech-UCSD birds-200-2011 dataset (2011)
43. Xian, Y., Akata, Z., Sharma, G., Nguyen, Q., Hein, M., Schiele, B.: Latent embeddings for zero-shot classification. In: 2016 IEEE Conference on Computer Vision and Pattern Recognition (CVPR), pp. 69–77 (2016)
44. Xian, Y., Lampert, C.H., Schiele, B., Akata, Z.: Zero-shot learning-a comprehensive evaluation of the good, the bad and the ugly. IEEE Trans. Pattern Anal. Mach. Intell. 41(9), 2251–2265 (2018)
45. Xian, Y., Lorenz, T., Schiele, B., Akata, Z.: Feature generating networks for zero-shot learning. In: Proceedings of the IEEE Conference on Computer Vision and Pattern Recognition, pp. 5542–5551 (2018)

46. Xian, Y., Schiele, B., Akata, Z.: Zero-shot learning - the good, the bad and the ugly. In: 2017 IEEE Conference on Computer Vision and Pattern Recognition (CVPR), pp. 3077–3086 (2017)
47. Xie, R., Liu, Z., Jia, J., Luan, H., Sun, M.: Representation learning of knowledge graphs with entity descriptions. In: Thirtieth AAAI Conference on Artificial Intelligence (2016)
48. Xie, Z., Cao, W., Wang, X., Ming, Z., Zhang, J., Zhang, J.: A biologically inspired feature enhancement framework for zero-shot learning. arXiv preprint arXiv:2005.08704 (2020)
49. Ye, M., Guo, Y.: Zero-shot classification with discriminative semantic representation learning. In: Proceedings of the IEEE Conference on Computer Vision and Pattern Recognition, pp. 7140–7148 (2017)
50. Zhang, L.N., Zuo, X., Liu, J.W.: Research and development on zero-shot learning. Acta Autom. Sin. **46(1)**(46), 1 (2020)
51. Zhang, Z., Saligrama, V.: Zero-shot learning via semantic similarity embedding. In: Proceedings of the IEEE International Conference on Computer Vision, pp. 4166–4174 (2015)
52. Zheng, V.W., Hu, D.H., Yang, Q.: Cross-domain activity recognition. In: Proceedings of the 11th International Conference on Ubiquitous Computing, pp. 61–70 (2009)

An Optimization of Deep Sensor Fusion Based on Generalized Intersection over Union

Lianxiao Meng[1,2], Lin Yang[2], Gaigai Tang[1,2], Shuangyin Ren[2(✉)], and Wu Yang[1]

[1] Information Security Research Center of Harbin Engineering University, Harbin, China
[2] National Key Laboratory of Science and Technology on Information System Security, Institute of System Engineering, Chinese Academy of Military Science, Beijing, China
renshuangyin@126.com

Abstract. 3D object detection is a major topic in unmanned driving and robotics, which is suffering from the low accuracy recently. We found that the loss function of 3D object detection network is the main cause to the low accuracy. For this, we proposed an optimized realization of deep sensor fusion network model (DSFN) based on Generalized Intersection over Union (GIoU). In DSFN, the designed backbone network is used to fuse point cloud features and image features, making full use of heterogeneous sensor information. Specifically, we introduced the GIoU as the loss function of the backbone network. We evaluated our model on KITTI dataset which is resulted from a LIDAR-camera setup. Compared with similar models, our model shows a higher accuracy.

Keywords: Deep Sensor Fusion Network (DSFN) · 3D bounding box estimation · Loss function · Deep learning · Camera and Lidar

1 Introduction

In recent years, artificial intelligence (AI) has developed rapidly, where the unmanned driving (UD) has been paid more attention by major enterprises, scholars and even general public. There are two quite different ways to achieve the goal of UD: one is a progressive method adopted by traditional enterprises, starting from the existing assisted driving system, and gradually increasing automatic steering to actively prevent collisions and other functions, to achieve conditional UD, and finally to complete UD when the costs and related technologies reach certain requirements. The other is represented by high-tech IT enterprises, they choose "one step" way to directly reach the ultimate goal of driverless driving. But the technical route chosen by the latter is more challenging and risky.

This research is supported by the National Natural Science Foundation of China (Grant No.61931017 and No.61831007).

M. Qiu (Ed.): ICA3PP 2020, LNCS 12453, pp. 552–562, 2020.
https://doi.org/10.1007/978-3-030-60239-0_37

Therefore, innovative algorithms and efficient and robust systems are needed to support it. Under this demand, object detection and positioning is particularly important, because it is equivalent to the intelligent unmanned system can "see" all directions, and provides a lot of useful information for the decision-making planning of UD system.

Nowadays, the latest development of convolutional neural networks has enabled 2D detection in complex environments. However, there is an urgent demand for precise detection in 3D environment, leading to an open challenge of 3D object detection. Thus, in this context, We propose an optimized realization of DSFN by using $GIoU$ (see Fig. 1) based on PointFusion [1], that can give a $6 - DoF$ pose and the 3D bounding box dimensions by combining point cloud and RGB information, along with identifying objects of interest in the scene.

Bounding box regression is one of the most basic components in many 2D/3D computer vision tasks. The tasks of target location, multi-object detection, target tracking and instance level segmentation all depend on the associated bounding box regression. The main trend of using deep neural network to improve application performance is to propose a better framework backbone [2] or a better strategy to extract reliable local characteristics [3].

Main contribution of our model as follows:

- The model we based on is validated on the KITTI 3D object detection dataset [4] with two indicators, namely, object classification accuracy and 3D bounding box accuracy. From many research works, we found that the object classification task had achieved a good enough performance, so we paid more attention on 3D bounding box accuracy while ensuring the object classification accuracy. For achieving a better 3D bounding box regression, we optimized the loss function by using IoU and $GIoU$ [5] in backbone network.

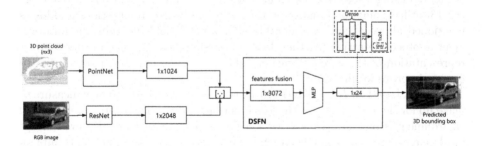

Fig. 1. An overview of the dense DSFN architecture. DSFN has two feature extractors: a PointNet variant that processes raw point cloud data, and a CNN (ResNet) that extracts visual features from an input image. Our fusion backbone network that directly regresses the box corner locations.

The rest of this article is as follows. The Sect. 2 introduces some achievements and related work. After discussing the DSFN model in Sect. 3, the specifications about the datasets with the implementation and experimentation details are discussed in Sect. 4. The obtained performance results are also analyzed. Finally, we make a summary and set further goals of future work in Sect. 5.

2   Related Work

2.1   Object Detection Accuracy Measures

Intersection over Union (IoU) is a standard for measuring the accuracy of corresponding objects in a specific data set. It can be used to measure any task that gets a prediction range in the output. In lots of task detection and 2D/3D bounding box projects [6,7], IoU is the most commonly used to determine true positives and false positives in a set of prediction, which will be given a decision threshold when be selected. Similarly, our experiment also uses IoU as a measure of the performance of our 3D bounding box, and the threshold is set to 0.5. Actually, due to not sensitive to the scales of the target object, IoU, as the core evaluation index, can be used as the direct representation of the regression loss of bounding box. However, there are few experiments that use IoU as loss function directly, so we have verified the effect of idea that IoU as loss function in our experiments.

2.2   Bounding Box Representations and Losses

In object detection, learning bounding box parameters is crucial. Various kinds of bounding box representations and losses are proposed in the literatures. Redmon et al. in YOLO v1 [8] propose a direct regression on the bounding box parameters with a small tweak to predict square root of the bounding box size to remedy scale sensitivity. Girshick et al. [9] in R-CNN parameterize the bounding box representation by predicting location and size offsets from a prior bounding box calculated using a selective search algorithm [10]. Most popular object detectors [11–13] utilize some combination of the bounding box representations and losses mentioned above. These considerable efforts have yielded significant improvement in object detection. As their loss of bounding box regression is not a direct representation of the core evaluation indicators, there may be opportunities to further improve localization.

In 2019, based on IoU, [5] introduced $GIoU$ as a new loss and new measure to make up for the lack of IoU as the loss function, and verified its advantages in 2D environment. However, as far as we know, the performance of this new concept in 3D environment has not been explored much yet, so we made a attempt in our experiment and has gotten the optimized effect just like it in 2D experiment.

3   Deep Sensor Fusion Network Model

DSFN is expected to use image features extracted by a standard CNN, ResNet 50, and corresponding point cloud features generated by changed PointNet subnetworks as inputs to combine these functions and output a 3D boundary frames of the target objects. The backbone network of DSFN is showed as follows:

- A changed PointNet network to extract point cloud features.
- Resnet50 to extract image appearance features.
- The fusion network takes the fusion information of the two mentioned above as input, and then outputs 3D boundary prediction.

3.1 Fusion Subcomponent

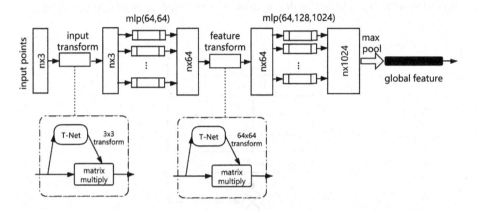

Fig. 2. Pointnet architecture: PointNet takes the lead in using the symmetric function (max pooling) to realize the permutation invariance in the processing of unordered 3D point sets. The modified model inherits this point, and then in order to improve the estimation performance of 3D bounding box, the original batch normalization layer is removed.

For the part of processing point cloud, we used a changed PointNet architecture, which was revised on the basis of Qi et al.'s model [14]. PointNet takes the lead in using the symmetric function (max pooling) to realize the permutation invariance in the processing of unordered 3D point sets. The modified model inherits this point, and then in order to improve the estimation performance of 3D bounding box, the original batch normalization layer is removed. But theme architecture has no fundamental change (see Fig. 2). The model takes the raw point cloud and learns a spatial coding of each point and the aggregated global point cloud features. Then these features are used for classification and semantic segmentation. PointNet has many ideal properties: it directly processes the original points without the lossy operation like voxelization or projection, and it is linearly proportional to the number of input points.

Resnet50 [15] (see Fig. 3) is used in the part of image feature extraction. The traditional convolution network or the all connected network have some problems, such as information loss, when information is transmitted. At the same time, it will cause the gradient to disappear or the gradient to explode, which makes the deep network unable to be trained. Resnet50 solves this problem to a certain extent by bypassing the input information directly to the output to protect the integrity of the information. The whole network only needs to learn the difference between the input and output to simplify the learning objectives and difficulties.

Our fusion model takes raw point cloud and image features preprocessed by the above two methods as input (we try the painless fusion function based on the previous experience, and find that the series of two feature vectors can obtain better performance), and then directly outputs the 3D positions of the eight corners of the target bounding box. The fusion network is implemented with three dense layers, where the dense network does well in regression problem.

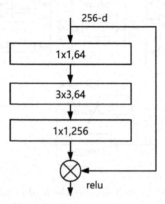

Fig. 3. ResNet50 architecture: Resnet50 solves some problems that will be caused by traditional convolution network to a certain extent by bypassing the input information directly to the output to protect the integrity of the information.

3.2   Loss Function

SmoothL1 Loss The loss function in the global fusion model we refer to is:

$$L = \sum_i smoothL1(X_i^*, X_i) + L_{stn} \tag{1}$$

where X_i^* are the ground-truth box corners, X_i are the predicted corner locations and L_{stn} is the spatial transformation regularization loss introduced in [14] to enforce the orthogonality of the learned spatial transform matrix. A major drawback of the global fusion network is that the variance of the regression target X_i is directly dependent on the particular scenario. For autonomous driving, the system may be expected to detect objects from 1 m to over 100 m. This variance places a burden on the network and results in suboptimal performance.

In order to solve this problem, we make two adjustments to the loss function in the model, namely, L_{IoU} and L_{GIoU}.

IoU Loss IoU is intersection over union, and it is the most commonly used index in object detection. In the anchor-based method, its function is not only to determine the positive and negative samples, but also to evaluate the deviation between the output box and the ground truth.

$$L_{IoU} = 1 - IoU = 1 - \frac{I(X)}{U(X)} \tag{2}$$

where $I(X)$ represents the intersection of the ground truth of target object and the 3D bounding box, and $U(X)$ is the union. As a loss function, *IoU* can directly reflect the detection effect of predicted box and ground truth. In addition, another good feature is scale invariance, that is, it is not sensitive to scale. In the region task, it can meet the nonnegative, identity, symmetry and triangle inequality of prediction results.

With the *IoU* loss is applied into the fusion model, it results a better regression effect than *SmoothL1* loss, but also shows some defects. One is that if two boxes with no overlap, according to the definition, $IoU = 0$, it cannot reflect the deviation between them (degree of coincidence). At the same time, when $IoU = 0$, the loss $= 1$, there is no gradient return, which stops the model from training.

GIoU (Generalized Intersection over Union) Loss In CVPR2019, [5] proposed the idea of *GIoU*. *IoU* is a concept of ratio, and it is not sensitive to the scale of the target. But it shows an obvious defect that doesn't take the situation that two boxes without overlap into consideration. *GIoU* can effective counters this situation through a more precise definition of the deviation between two boxes.

$$L_{GIoU} = 1 - GIoU = 1 - (IoU - \frac{A_c - U(X)}{A_c}) \tag{3}$$

Similar to *IoU*, *GIoU* is also a deviation measure. As a loss function, it meets the basic requirements of loss function: *GIoU* is not sensitive to scale. *GIoU* is the lower bound of *IoU*. For surrounding any group of ground truth and 3D bounding box, where A_c is the volume of the smallest box. In the case of infinite coincidence of two frames, $IoU = GIoU$. The value of *IoU* is within [0,1], but the value of *GIoU* has symmetric interval, and the value range is within $[-1,1]$. The maximum value is 1 when the two are coincident, and the minimum value is -1 when the two are not intersected and infinite, so *GIoU* is a very good deviation measure. Different from *IoU* only focusing on overlapping areas, *GIoU* not only focuses on overlapping areas, but also other non overlapping areas, which can better reflect the degree of coincidence between the two boxes.

4 Experiments

4.1 Dataset

KITTI The KITTI dataset [4] contains 2D and 3D labels of cars, pedestrians and cyclists in urban driving scenarios. The sensor configuration includes a wide-angle camera and velodyne $hdl-64e$ lidar. The official training collection contains

7481 images. In order to ensure the validity and credibility of the experimental results comparison, we follow the dataset processing in the comparison model, and divide the official training dataset into training set, development set and verification set. The size of each set is also consistent (see Table 1).

Table 1. Train-Dev-Test Split

	Train data	Dev.data	Test data
No.of examples	6750	365	366

4.2   Pre-processing

The velodyne setup on the station wagon is used to produce the points clouds. For details, it's a rotating 3D laser scanner that generates data points at a rate of 10 HZ, 64 beams, with 0.09° angular resolution, 2 cm distance accuracy, collecting 1.3 million point/second, with a horizontal and vertical field of view of 360° and 26.8 respectively. There are lots of points so that we need to trim down the input size for correspondence, feasibility and relevance. As a result, we filter the point clouds falling in the camera view angle and randomly sample 2048 points from them. Then, the points are fed through a Spatial Transformation Network in order to canonicalize the input space. Further, the ground truth labels are transformed to the velodyne coordinate for tractability in prediction.

4.3   Network Specifications

This subsection generally lists the network specifications resulted from extensive experimentations. The model we used has 1,808,207 trainable parameters, where most of them belong to the PointNet [14] architecture. Moreover, in the fusion network, we finally choose a simple and effective architecture which is consisted of 3 hidden layers having 512, 128 and 128 units, respectively, the fusion layer gives the box-corner locations as output.

4.4   Results

First of all, we completely restored Global Fusion network which is the component of Pointfusion [1]. SmoothL1 as the loss function of Global Fusion network has been introduced in 3.2. the model was trained to give the loss curves, presented in the figure below (see Fig. 4).

The output of Global Fusion has two aspects: classification and 3D bounding box regression. In the process of recurrence experiment, we found that the classification accuracy reached 96.17%, which was hardly to be improved qualitatively, leading to the truth that 96.17% shows a good enough performance indeed. However, there is a lot of room for improvement in the performance

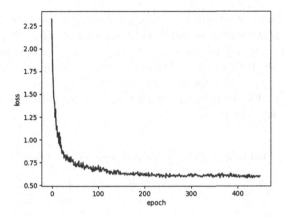

Fig. 4. Box output loss vs Epochs for training and development with SmoothL1

of 3D bounding box regression. So in our experiment, we focused on the 3D bounding box regression on the premise of ensuring classification accuracy.

IoU is the core index to measure 3D bounding box regression. The loss curves of applying IoU as loss function instead of SmoothL1 in DSFN is shown as follows (see Fig. 5):

$GIoU$ is a new concept based on IoU in 2019. An example has be given in [5] that the performance in 2D environment is indeed better than IoU. According to the research, $GIoU$ has not been widely used at present, especially in 3D bounding box regression. The performance of the DSFN using $GIoU$ as loss function is shown as follows. (see Fig. 5):

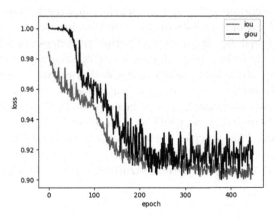

Fig. 5. Box output loss vs Epochs for training and development with L_{IoU} and L_{GIoU}

0 is the threshold of whether the target is framed in 3D bounding box. From many experimental results, we found that not all the values of IoU are bigger than 0. Thus, we decided to take the ratio of the number of samples with IOU bigger than 0 to the total number of test-set as the measurement of accuracy, and then calculated the average value of all the IOU bigger than 0 as the measurement of precision. Table 2 offers more information about different loss function's performance on the test set.

Table 2. Different Loss Function's Performance

	SmoothL1	L_{IoU}	L_{GIoU}
Test set	366	366	366
Number of simples (IoU>0)	201	85	343
Accuracy	54.9%	23.2%	**93.7%**
Precision (average IoU)	0.30	**0.49**	0.13

From the Fig. 4, we find that the model with SmoothL1 trends to converge roughly when the epoch exceeds 300. Moreover, there is a comparison between the model with L_{IoU} and L_{GIoU} as shown in Fig. 5. We can observe that the loss of L_{GIoU} with a faster convergence than L_{IoU}, while compared to the SmoothL1 show the same performance.

From Table 2 we find that model with loss of L_{IoU} shows a better precision than loss of SmoothL1, which has a precision improvement of 0.19. It can be explained by that L_{IoU} represents the expected aim at evaluation of model at the phase of training, which is more significant than SmoothL1. However, there is a accuracy descend of 31.7% of L_{IoU} when compared to SmoothL1, which is caused by that L_{IoU} only focuses on the case that intersection of two bounding boxes is bigger than 0, but ignores the case that the intersection is 0. For L_{GIoU}, we observe that there is a large improvement of accuracy to 93.7%, which is due to that L_{GIoU} take both cases of intersection of two bounding boxes into consideration. Nevertheless, an unsatisfactory effect occurs on the precision, we consider this maybe because L_{GIoU} is incomplete for the specific intersection position of 3D bounding box, which is a problem that does not exist in 2D verification before. This provides a research point for our follow-up work.

Further, figures (see Fig. 6 Fig. 7) demonstrate some correct result predicted by the model with L_{GIoU} (Red: ground truth; Ink Blue: 3D bounding box).

Image with 3D bounding box; IOU = 0.86 Class Probability = 0.97

Fig. 6. Correct prediction. (Color figure online)

Image with 3D bounding box; IOU = 0.87 Class Probability = 0.98

Fig. 7. Correct prediction. (Color figure online)

5   Conclution

In order to improve the performance of the bounding box regression in 3D environment, we optimized DSFN model based on Global Fusion model. Our model focuses on the optimization of loss function in training process. In experiments, we use IoU and $GIoU$ instead of SmoothL1 as the loss function of DSFN respectively. The results are obtained that IoU outperforms SmoothL1 on bounding box regression precision while $GIoU$ shows the best performance on bounding box regression accuracy.

References

1. Xu, D., Anguelov, D., Jain, A.: Pointfusion: deep sensor fusion for 3D bounding box estimation. In: 2018 IEEE Conference on Computer Vision and Pattern Recognition, CVPR 2018, Salt Lake City, UT, USA, 18–22 June 2018. IEEE Computer Society, pp. 244–253 (2018)
2. Lin, T., Goyal, P., Girshick, R.B., He, K., Dollár, P.: Focal loss for dense object detection. CoRR, vol. abs/1708.02002 (2017). http://arxiv.org/abs/1708.02002
3. He, K., Gkioxari, G., Dollár, P., Girshick, R.: Mask R-CNN (2017)
4. KITTI. http://www.cvlibs.net/datasets/kitti/raw_data.php

5. Rezatofighi, H., Tsoi, N., Gwak, J., Sadeghian, A., Reid, I.D., Savarese, S.: Generalized intersection over union: a metric and a loss for bounding box regression. In: IEEE Conference on Computer Vision and Pattern Recognition, CVPR 2019, Long Beach, CA, USA, 16–20 June 2019. Computer Vision Foundation/IEEE, pp. 658–666 (2019). https://arxiv.org/pdf/1902.09630.pdf

6. Everingham, M., Gool, L.V., Williams, C.K.I., Winn, J.M., Zisserman, A.: The pascal visual object classes (VOC) challenge. Int. J. Comput. Vis. **88**(2), 303–338 (2010). https://doi.org/10.1007/s11263-009-0275-4

7. Lin, Tsung-Yi., et al.: Microsoft COCO: common objects in context. In: Fleet, David, Pajdla, Tomas, Schiele, Bernt, Tuytelaars, Tinne (eds.) ECCV 2014, Part V. LNCS, vol. 8693, pp. 740–755. Springer, Cham (2014). https://doi.org/10.1007/978-3-319-10602-1_48

8. Redmon, J., Divvala, S.K., Girshick, R.B., Farhadi, A.: You only look once: unified, real-time object detection. In: 2016 IEEE Conference on Computer Vision and Pattern Recognition, CVPR 2016, Las Vegas, NV, USA, 27–30 June 2016. IEEE Computer Society, pp. 779–788 (2016). https://doi.org/10.1109/CVPR.2016.91

9. Girshick, R.B., Donahue, J., Darrell, T., Malik, J.: Rich feature hierarchies for accurate object detection and semantic segmentation. In: 2014 IEEE Conference on Computer Vision and Pattern Recognition, CVPR 2014, Columbus, OH, USA, 23–28 June 2014. IEEE Computer Society, pp. 580–587 (2014). https://doi.org/10.1109/CVPR.2014.81

10. Uijlings, J.R.R., van de Sande, K.E.A., Gevers, T.: Selective search for object recognition. Int. J. Comput. Vis. **104**(2), 154–171 (2013). https://doi.org/10.1007/s11263-013-0620-5

11. Redmon, J., Farhadi, A.: Yolov3: An incremental improvement. CoRR, vol. abs/1804.02767 (2018). http://arxiv.org/abs/1804.02767

12. Redmon, J.: Yolo9000: better, faster, stronger. In: 2017 IEEE Conference on Computer Vision and Pattern Recognition, CVPR 2017, Honolulu, HI, USA, 21–26 July 2017. IEEE Computer Society, pp. 6517–6525 (2017). https://doi.org/10.1109/CVPR.2017.690

13. Dai, J., Li, Y., He, K., Sun, J.: R-FCN: object detection via region-based fully convolutional networks. In: Lee, D.D., Sugiyama, M., von Luxburg, U., Guyon, I., Garnett, R. (eds.) Advances in Neural Information Processing Systems 29: Annual Conference on Neural Information Processing Systems 2016, 5–10 December 2016, Barcelona, Spain, pp. 379–387 (2016). http://papers.nips.cc/paper/6465-r-fcn-object-detection-via-region-based-fully-convolutional-networks

14. Qi, C.R., Su, H., Mo, K., Guibas, L.J.: Pointnet: deep learning on point sets for 3D classification and segmentation. In: 2017 IEEE Conference on Computer Vision and Pattern Recognition, CVPR 2017, Honolulu, HI, USA, 21–26 July 2017. IEEE Computer Society, pp. 77–85 (2017). https://doi.org/10.1109/CVPR.2017.16

15. He, K., Zhang, X., Ren, S., Sun, J.: Deep residual learning for image recognition. In: 2016 IEEE Conference on Computer Vision and Pattern Recognition, CVPR 2016, Las Vegas, NV, USA, 27–30 June 2016. IEEE Computer Society, pp. 770–778 (2016). https://doi.org/10.1109/CVPR.2016.90

A Hot/Cold Task Partition for Energy-Efficient Neural Network Deployment on Heterogeneous Edge Device

Jihe Wang$^{(\boxtimes)}$, Jiaxiang Zhao, and Danghui Wang

School of Computer Science, Northwestern PolyTechnical University,
Xi'an 710129, China
wangjihe@nwpu.edu.cn, zhaojiaxiang@mail.nwpu.edu.cn, wangdh@nwpu.edu.cn

Abstract. In recent years, neural network has made great achievements in the fields of image classification and object detection. At the same time, the rapid development of edge devices has also led many scholars to study the deployment of neural networks on resource-limited edge devices. However, the traditional neural network did not consider the specific application scenarios, which makes the data processing very inefficient. In this paper, we propose a method to deploy neural networks on heterogeneous edge devices. This method fully considers the specific applications scenarios of neural network, and proposes the idea of task partition. The whole dataset is divided into two classes: hot class and cold class, and then corresponding network models are trained respectively to handle the data. In order to speed up the execution efficiency, we propose a simple algorithm, which can balance the front-end and back-end load according to the processing capacity of edge devices. In addition, the time delay and energy consumption are considered in this paper. The experiments show that our method can save 36.4% energy consumption compared with the traditional method.

Keywords: Neural network · Task partition · Heterogeneous edge device · Model deployment · Load balance

1 Introduction

In recent years, the neural network has made great progress in many fields. At the same time, the rapid development of edge devices has accelerated the research of lightweight neural networks. However, the general neural network does not consider the specific application scenarios in training process. In fact, the data processed by the neural network has a great bias, that is, the frequency of some categories in the whole data set is relatively high in a specific application scenario. If it is processed by a general-purpose neural network, the effect on the high-frequency and low-frequency categories is the same, which is not what we

© Springer Nature Switzerland AG 2020
M. Qiu (Ed.): ICA3PP 2020, LNCS 12453, pp. 563–574, 2020.
https://doi.org/10.1007/978-3-030-60239-0_38

expected. In addition, the recognition effect of the network is related to the scale of the network model. The more categories can be distinguished, the higher the complexity of the network is. It is not suitable for resource-limited edge devices. In this paper, we aim to explore a method to accelerate network efficiency in specific scenario while achieve a faster response to high-frequency data with the lower energy consumption.

In order to solve the problem of deploying neural network on edge devices in a specific scenario, we propose the idea of task partition, in which the data will be divided into two categories according to the application scenario: hot class and cold class. The heterogeneous edge devices generally includes the front-end (CPU) and the back-end (GPU), the two network models with low power consumption and high precision are deployed to deal with the hot and cold class respectively. On the basis of task partition, we propose an algorithm to balance the front-end and back-end load. This algorithm allocate different load to front-end and back-end respectively to accelerate the execution efficiency according to the difference in computing power. Moreover, we use the concept of sample confidence, and take the front-end sample confidence as the basis for task partition of the front-end and back-end. (1) For the limited parallelism of the front-end, the simple network model is used for "early decision termination" to reduce the energy consumption of low sample confidence in the front-end; (2) For the limited resources of the back-end, the lightweight network model is deployed while satisfying the requirements of the storage and computing capacity.

To Summarize, our main contributions are as follows:

1. According to the data characteristics of specific application scenarios, we propose the concept of hot/cold class. The dataset in specific scenarios will be divided into hot and cold class according to actual frequency. The classes with higher frequency are regarded as hot class, and the remaining classes are regarded as cold class.
2. For heterogeneous edge devices, the front-end and back-end network models work together to process data based on the algorithm of load balance. Moreover, the concept of sample confidence is used as the basis of task partition.
3. This paper analyses the accuracy of different front-end lightweight network models, and considers the energy consumption and delay of edge devices. Finally, we determines a suitable network model. The experiments show that our method can complete specific tasks with lower energy consumption.

2   Background

2.1   Edge Devices and Neural Network

The application of a large number of mobile devices, such as smart phones, accelerates the development of edge devices. The edge devices generated massive data, it is inefficient to rely on the centralized server platform alone. In order to improve the data processing ability of edge devices, some hardware manufacturers add extra processing units. Heterogeneous edge devices usually have both general processing unit (CPU) and special processing unit (such as GPU), which are called front-end and back-end respectively. Although the resources of edge devices and storage are limited, the advantages of low power consumption and high performance provide the possibility to deploy neural network on edge devices [1].

CNN has made great progress in image classification, object detection and other fields. The success of AlexNet [2] in 2012 greatly promoted the development of deep learning. In order to pursue high precision of network model, the scale and complexity of network model become larger, such as AlexNet (8 layers), VGG [3] (19 layers) and ResNet [4] (152 layers) in ImageNet [5] classification competition. The training process needs to be carried out on the high-performance server platform while running the neural network also consumes a lot of memory and computing resources. Although the methods of group convolution, quantization [6] can reduce the number of parameters to a certain extent, it will also lose some accuracy. For resource-limited edge devices, there are some difficulties in deployment.

MobileNet [7] and ShuffleNet [8] are efficient convolutional neural networks for mobile and edge vision devices. The MobileNet uses depthwise separable convolutions to build lightweight network. The traditional convolution is divided into two parts: depthwise convolution and pointwise convolution, which can greatly reduce parameters and improve efficiency of the network model. However, a lot of 1×1 pointwise convolutions increased the computation complexity, so the ShuffleNet adds the group convolution to reduce this complexity. In addition, it adds a shuffle layer to help the information flowing among different feature groups. The ShuffleNet can be regarded as an extremely efficient network.

2.2 Task Partition

At present, there are few researches on task partition of neural network, and most of them are similar in principle. Taking BranchyNet [9] as an example, it is a branched neural network structure, which can realize the early exit of the network when it is running. The network set many exit points, and each exit point has different confidence levels. If the network output meets the confidence conditions, the result will be output directly and exit the network. DDNN [10] and Edgent [11] also refer to this method to optimize their design.

Sample confidence is the key concept to distinguish hot/cold class correctly. Generally speaking, each output result of image classification is a one-hot [12] vector. The value of the element in the vector represents the probability of a certain class. The larger the value is, the greater the probability for a certain class. The one-hot vector should be relatively clear, that is, the maximum value in the vector should be much larger than other values, the more clear one-hot vector is, the better classification effect is. On the contrary, the existence of multiple similar large values in a vector indicates that the model has not good classification effect for the test samples.

In a group of signals, the average uncertainty usually should consider all possible situations. If the signals have n values: $U_1, U_2, \ldots U_n$, the corresponding probability is: $P_1, P_2, ... P_n$, and the occurrence of various signals is independent of each other. The average uncertainty of all signals should be the statistical average value(E) of the each signal uncertainty $-logP_i$, which can be called information entropy [13]. Formula (1) shows the calculation method of information entropy.

$$H(P) = E[-logP_i] = -\sum_{i=0}^{n} P_i log P_i \tag{1}$$

Some researchers noticed that the formula (1) can exactly satisfy the one-hot vector of the network output. At the same time, the concept of information entropy is for measuring information. The more orderly a system is, the lower the value of information entropy is, and the higher the probability to a certain category is, which indicates the classification effect of the model is relatively good. The final formula is shown in formula (2).

$$T(x) = -\sum_{i=0}^{C} \frac{x_i log x_i}{log C} \tag{2}$$

The C represents all possible classes, the x_i represents the possibility of each class. The mapping of information entropy directly to one-hot vector will cause the uncertainty of entropy range, so some scholars limits the entropy value in the range $[0, 1]$. The entropy is normalized by dividing it by $logC$.

3   Method

In this section, we divide the dataset into hot class and cold class according to specific application scenarios, and deploy low-power and high-precision neural network replicas respectively on heterogeneous edge devices. We use the sample confidence of the front-end as the basis for task partition of the front-end and back-end. In the front-end, the parallelism is limited, and the simple network is used for "early decision termination" to reduce the energy consumption overhead of low confidence samples in the front-end. In the back-end, the resources are limited and the lightweight network model is deployed on the edge platform while satisfying the requirements of the storage and computing capacity. The basic flow chart is as Fig. 1.

3.1   Hot/Cold Class

Actually, the dataset used in training process is uniform, which makes the network have the ability to judge every class. However, in specific application scenario, the data to be processed by the network may not be uniform. There exist some certain bias, that is, the frequency of several classes is very high, and the frequency of the remaining is low. Taking traffic vehicles as an example, we suppose that there have a general network, which can complete the classification and identification tasks for various common vehicles. The network has the same

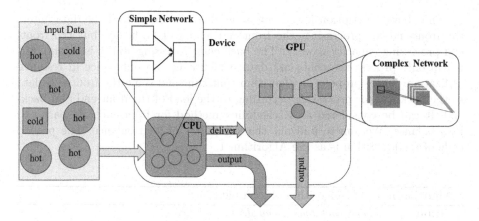

Fig. 1. The flow chart of basic design.

processing ability for each class. However, we expect that in the narrow road, the frequency of small vehicles is high rather than large vehicles. If we use the general network, it will become inefficient. We expect to use a simple network to process high frequency data efficiently instead of a general network. Therefore, this biased data feature brings a new idea to design neural network. The dataset will be divided according to specific application scenario and processed by different networks respectively, so as to improve the efficiency of data processing.

Therefore, This paper comes up with the concept of hot and cold class, which refers to the concept of cache [14] in memory. The hot class is one or several classes which often appear in a specific scenario, while the remaining classes are regarded as cold class. The purpose of task partition is to make use of the characteristics of heterogeneous edge devices. The heterogeneous edge devices contain general processor (CPU) and specific processor (such as GPU). Generally, the performance of general processor is low, and they can only support small-scale operations, which is not enough to run a complete network for all classes of data, while specific processor have higher performance, which can distinguish all classes. In order to make full use of the limited resource, the whole neural network is divided into two parts: the CPU is used as the front-end to deploy a very simple and efficient network which only handle the hot class, and the GPU is used as the back-end to deploy a more complex and lightweight network, which can recognize all classes. When a batch of data enters the network, hot class can be processed by the front-end efficient network, and then cold class will be processed precisely by the back-end complex network.

The method we proposed can use the uneven characteristics of the data in specific scenarios, the front-end efficient network can process more data, and only a small amount of cold data will be sent to the back-end complex network for processing. It can not only make full use of the characteristics of heterogeneous edge devices and reduce the amount of calculation, but also save the resource consumption and improve the energy efficiency.

The above description is the our basic idea. However, if all the data enters the front-end for processing, the front-end load will be heavy, but the front-end processing capacity is weak. The back-end processing ability is strong, but it can only process the cold-class data, which cause an imbalance in resource utilization. So we propose an algorithm that can send some the front-end data to the back-end for processing according to the load of the front-end and back-end. It can be considered as a scheduling method for front-end and back-end load balance. We aim to find out the key points of load balance. The pseudo code of the algorithm is as the Algorithm 1.

Algorithm 1: The algorithm of load balance

Input: $front\_load, back\_load, front\_delay, back\_delay$;
Output: $balance\_point$;
1   *Initialize all variables*;
2   **while** $(|front\_delay - back\_delay| > 5)$ **do**
3 | $front\_delay = time \ of \ processing \ front\_load$;
4 | $back\_delay = time \ of \ processing \ back\_load$;
5 | $balance\_point = front\_load$;
6 | $front\_load = front\_load - 5\%$;
7 | $back\_load = back\_load + 5\%$;
8   **end**
9   **return** $balance\_point$;

In the Algorithm 1, the $front\_load$ and $back\_load$ represent load of the front-end and back-end respectively. The load can be regarded as the amount of data to be processed. The $front\_delay$ and $back\_delay$ represent the delay of front-end and back-end respectively and their initial values are 0. In the initialization, the front-end load is 100% and the back-end load is 0. This algorithm move gradually some front-end load to the back-end according to the load proportion. We set the step size is 5% for reduction. When the front-end and the back-end processing delay are very close (less than 5 s), we think that the load distribution can meet the relevant requirements at this time. Through the above algorithm, we find the balance point of load and this load distribution can realize the front-end and back-end collaborative processing to improve the execution efficiency.

3.2 Network Design

The front-end network model we designed only has the convolution layer and the full connection layer since convolution operation will cost many memory and computing resources while reducing the number of convolution layers can effectively solve that CPU computing capacity is insufficient. In order to meet the characteristics of heterogeneous edge devices, the front-end network design must be simple and low power consumption. This paper trains and compares several lightweight networks, and finally determines a relatively simple and efficient network model with high accuracy. The design goal of the back-end network is to be able to process all classes. Although the GPU has relatively high performance, but its resources still have a huge gap with the traditional server, so the design of the network structure still needs to be lightweight. For this purpose, this paper uses the lighter network ShuffleNet as the back-end network. We improved the original ShuffleNet by cut out unnecessary parts based on the dataset we used while maintaining good accuracy.

3.3   Sample Confidence

From the definition of information entropy, it can be clear that the sample confidence is inversely proportional to the value of $T(x)$, which is contrary to general understanding. In order to make $T(x)$ reflect the sample confidence intuitively, this paper make a little adjustment to formula (2). The final sample confidence is determined by formula (3). The higher the sample confidence is, the more confident the network model is in the image classification.

$$T(x) = 1 + \sum_{i=0}^{C} \frac{x_i log x_i}{log C} \tag{3}$$

In this paper, we set the threshold T as a basis to distinguish hot/cold class. After the input data is processed by the front-end network, if the sample confidence is greater than or equal to T, it can output the results directly. Otherwise, it should be sent to the back-end for more precise processing. Through the sample confidence, the idea of task partition is more clear. Therefore, the collaborative processing of front-end and back-end can save a lot of resources because of the simplicity of the front-end network.

4   Experiments

4.1   Experiment Environment

In this paper, the server we used is equipped with 32 GB DDR4 memory, and the GPU is NVIDIA Titan XP with 12 GB DDR5 memory, Intel Core i7-8700k processor with 6 cores and 12 threads. The high performance hardware configuration of the server can greatly shorten the training time and improve the work efficiency. The dataset we used is Tiny-ImageNet. Tiny-Imagenet comes from ImageNet dataset, but compared with ImageNet, the former has smaller size. There are 200 classes of RGB images with the size of 64 × 64 for each image, while the size of each image in ImageNet is 224 × 224. The heterogeneous edge platform has limited resources, so the network can not be too complex, a small and medium data sets is more suitable for this scenario. In this paper, we choose 10 classes are randomly selected as hot classes, and the rest as cold classes for training respectively. The heterogeneous edge device used in this experiment is the board named Jetson TX1 [15] of NVIDIA. It is equipped with 4 GB LPDDR4 and ARM Cortex-A57 with 4 cores, and the GPU is Maxwell with 256 cores. In addition, we used the popular neural network framework Pytorch to design the neural network. The CPU and GPU are used for processing data in corresponding data respectively.

4.2   Evaluation

Sample Confidence and Load Balance: In this paper, we design some experiments to determine the threshold of sample confidence based on the formula (3). All the test samples are processed by the front-end network to obtain the network confidence. Table 1 show the sample confidence distribution which is obtained by processing 500 hot class samples and cold class samples respectively. In order to facilitate the processing, this paper merged the initial several intervals. The distribution of sample confidence is very different for hot and cold class. Obviously, the distribution of hot class samples presents a one-side state. On the contrary, the distribution of cold class samples tends to be uniform, and the number of samples in most intervals is similar.

Table 1. The confidence distribution of hot/cold class.

Class	Confidence level							
	$[0.0, 0.2]$	$[0.2, 0.4]$	$[0.4, 0.5]$	$[0.5, 0.6]$	$[0.6, 0.7]$	$[0.7, 0.8]$	$[0.8, 0.9]$	$[0.9, 1.0]$
Hot	0	3	8	25	30	64	109	261
Cold	19	46	71	95	89	31	41	108

In order to determine the confidence threshold, this paper compares and analyses several different sets of test data. Figure 2 shows the confidence distribution of three test samples, and the proportion of hot class in each test sample is 60%, 80% and 100%. The figure shows that setting the sample confidence threshold T to 0.7 is suitable, which is in line with the accuracy of the network model (the accuracy of the network model is about 98%). When the test samples contain 500 hot-class images, the number of samples whose confidence is greater than or equal to 0.7 is 452. It can roughly match the accuracy of the network model within acceptable deviation. Similarly, for the test samples contains 80% and 60% hot class, when the threshold value is 0.7, the result is consistent with the model accuracy. So this paper choose 0.7 as the threshold of sample confidence.

According to Algorithm 1, this paper verifies the load balance. For each front-end network, this paper get the corresponding balance point, and the specific data for each model as Table 2. In the follow-up experiments, we will allocate the appropriate amount of data to the front-end and back-end respectively according to the balance point. This method can maximize the advantages of the front and back-end, and improve the resource utilization.

Table 2. The balance point of different models

Model ID	1	2	3	4	5	6	7	8	9	10
Balance point	40%	40%	70%	30%	30%	60%	50%	30%	30%	15%

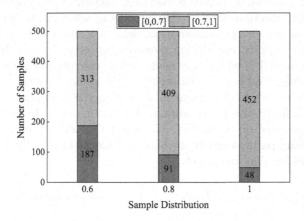

Fig. 2. The sample distribution under different hot/cold class proportion.

Table 3. Network structure and accuracy

Model ID	Network structure	Stride	Error rate
1	conv16/fc500/fc200/fc10	1	0.192
2	conv16/fc500/fc10	1	0.196
3	conv16/fc10	1	0.424
4	conv32/fc500/fc200/fc10	1	0.114
5	conv32/fc500/fc10	1	0.136
6	conv32/fc10	1	0.418
7	conv32/fc400/fc10	2	0.17
8	conv16/conv32/fc500/fc200/fc10	1	0.064
9	conv16/conv32/fc500/fc10	1	0.07
10	conv32/conv64/fc500/fc200/fc10	1	0.012

The design goal of the front-end network is to satisfy the limited CPU computing resources of the edge device as much as possible, which only needs to handle hot-class data. In order to achieve the above goal, this paper compares several small network models, and Table 3 shows the network structure and corresponding model accuracy, where the kernel size is 3×3 and the structure describes the specific operation and output channels. The model $1-7$ contains only one convolution layer, and the remaining models contain two convolution layers. At the same time, the number of full connection layers also has an impact on the network accuracy. It is better to add more than two full connection layers in the last stage of data processing. The network we designed is relative simple, and it satisfies the limited resource requirements and computing power. It can be seen from the Table 3 that the error rate of model 10 is the lowest, but the corresponding network structure is complex, including two convolution layers and three full connection layers. In the process of selecting the network model, we need to consider energy consumption, delay and other factors. Every choice is based on the specific application scenario and edge devices. In this paper, we choose the model 7, which only contains one convolution layer and two full connection layers. Compared with model 10, it is more lightweight. Although its accuracy is not as good as model 10, its energy consumption and delay are better than model 10, this paper will talk about it in next subsection. The back-end network is based on ShuffleNet. We adjust the network input and cut out some unnecessary parts to satisfy the back-end needs. Then we train the suitable network model for Tiny-ImageNet.

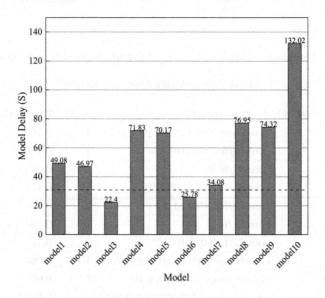

Fig. 3. The delay of different network models.

Delay and Energy: The delay is an important factor to evaluate the network model. This paper has tested 10, 000 images in which the number of hot-class samples accounts for 80% of the total samples. First, all the test data are processed using the complex network deployed in the back-end of the edge device and recorded delay. Then we use the front-end and back-end together to process the test data, and record the processing time, in this process, we refer the result in Table 2. The front-end network uses the ten network models as mentioned

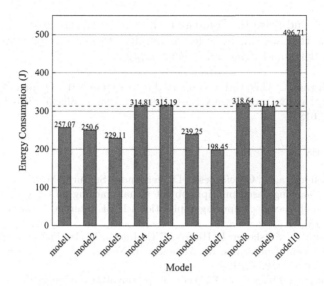

Fig. 4. The energy consumption of different network models.

above. Figure 3 shows the delay of each front-end model, and the delay of the traditional network is 30.98 which is shown by the dotted line. Different network models have different running time which varies from the minimum of 22 s to the maximum of 132 s, there has a huge gap between them.

The Fig. 4 shows the energy consumption of 10000 samples running in different network models. The dotted line in the figure represents the energy consumption within the traditional network, and its value is 312.9. We noticed that the energy consumption within some network models is even greater than the consumption by GPU alone. This is because the performance of the front-end CPU is too poor, and the CPU is not suitable for doing a large number of convolution operations, which resulting in more processing time and excessive energy consumption. However, the GPU has a natural advantage in convolution operation because of many cores. After considering the energy consumption and delay each network model, we choose its product as the final metrics and the model 7 has relatively low latency and energy consumption. Compared with the traditional network, the energy consumption is reduced by 36.4% and achieve high performance.

5   Conclusion

This paper proposes the concept of hot/cold class according to the data characteristics of specific application scenarios. We choose the sample confidence as the basis of task partition. We then propose algorithm to balance the load and analyses the energy consumption and delay of network models and determined a suitable front-end network. Compared with the traditional network, our experiments show that we can be able to complete tasks in specific scenario with relative low delay and lower power consumption.

Acknowledgment. This work is supported by National NSF of China (No. 61802312 and 61472322) and Natural Science Basic Research Plan in Shaanxi Province of China (No. 2019JQ-618).

References

1. Blanco-Filgueira, B., García-Lesta, D., Fernández-Sanjurjo, M., Brea, V.M., López, P.: Deep learning-based multiple object visual tracking on embedded system for IoT and mobile edge computing applications. IEEE Internet Things J. **6**(3), 5423–5431 (2019)
2. Krizhevsky, A., Sutskever, I., Hinton, G.E.: ImageNet classification with deep convolutional neural networks. In: Advances in Neural Information Processing Systems, pp. 1097–1105 (2012)
3. Simonyan, K., Zisserman, A.: Very deep convolutional networks for large-scale image recognition. arXiv preprint arXiv:1409.1556 (2014)
4. He, K., Zhang, X., Ren, S., Sun, J.: Deep residual learning for image recognition. In: Proceedings of the IEEE Conference on Computer Vision and Pattern Recognition, pp. 770–778 (2016)
5. Deng, J., Dong, W., Socher, R., Li, L.-J., Li, K., Fei-Fei, L.: ImageNet: a large-scale hierarchical image database. In: 2009 IEEE Conference on Computer Vision and Pattern Recognition, pp. 248–255. IEEE (2009)
6. Jacob, B., et al.: Quantization and training of neural networks for efficient integer-arithmetic-only inference. In: Proceedings of the IEEE Conference on Computer Vision and Pattern Recognition, pp. 2704–2713 (2018)
7. Howard, A.G., et al.: MobileNets: efficient convolutional neural networks for mobile vision applications. arXiv preprint arXiv:1704.04861 (2017)
8. Zhang, X., Zhou, X., Lin, M., Sun, J.: ShuffleNet: an extremely efficient convolutional neural network for mobile devices. In: Proceedings of the IEEE Conference on Computer Vision and Pattern Recognition, pp. 6848–6856 (2018)
9. Teerapittayanon, S., McDanel, B., Kung, H.-T.: BranchyNet: fast inference via early exiting from deep neural networks. In: 2016 23rd International Conference on Pattern Recognition (ICPR), pp. 2464–2469. IEEE (2016)
10. Lin, M., Hsieh, J., Du, D.H.-C., Thomas, J.P., MacDonald, J.A.: Distributed network computing over local ATM networks. IEEE J. Sel. Areas Commun. **13**(4), 733–748 (1995)
11. Li, E., Zhou, Z., Chen, X.: Edge intelligence: on-demand deep learning model co-inference with device-edge synergy. In: Proceedings of the 2018 Workshop on Mobile Edge Communications, pp. 31–36 (2018)
12. Finney, D.J.: Probit analysis, a statistical treatment of the sigmoid response curve. J. Roy. Stat. Soc. **110**(3), 263 (1947)
13. Shannon, C.E.: A mathematical theory of communication. Bell Syst. Tech. J. **27**(3), 379–423 (1948)
14. Smith, A.J.: Cache memories. ACM Comput. Surv. (CSUR) **14**(3), 473–530 (1982)
15. Otterness, N., et al.: An evaluation of the NVIDIA TX1 for supporting real-time computer-vision workloads. In: 2007 IEEE Real-Time and Embedded Technology and Applications Symposium (RTAS), pp. 353–364. IEEE (2017)

Towards Energy Efficient Architecture for Spaceborne Neural Networks Computation

Shiyu Wang[✉], Shengbing Zhang, Jihe Wang, and Xiaoping Huang

School of Computer Science, Northwestern Polytechnical University, Xi'an 710072, China
onion0709@mail.nwpu.edu.cn, {zhangsb,wangjihe,
huangxp}@nwpu.edu.cn

Abstract. Hybrid neural network (H-NN) has been applied in field of remote sensing. To satisfy demands of on-orbit processing that requires high throughput and restriction on power consumption, designing heterogeneous array processor becomes an effective way fulfilling various tasks. A heterogeneous array architecture is proposed to support the hybrid neural network based on characteristics of various computation types among neural network module types and of dynamic computation burden among layers. Firstly, a heterogeneous array structure consisting of different types of PEs is designed, enabling strong flexibility and high throughput. Secondly, multi-level on-chip memory structure and access strategy supporting different access modes are proposed. Thirdly, management strategy for heterogeneous computing array is designed, which combines pipelining and parallel processing to support efficient mapping of diverse hybrid neural networks. The processor has a peak throughput of up to 1.96 TOPS. The implementation on models of AlexNet, LRCN, VGG19-LSTM and CLDNN can achieve the throughput of 1.92 TOPS, 1.89 TOPS, 1.93 TOPS and 1.84 TOPS, respectively. Compared with similar neural network processor that is based on same technology, the throughput of AlexNet model is increased by 76.4%. The peak power consumption of single processor is 824 mW, to which the power restriction of on-orbit AI platform is satisfied.

Keywords: Remote sensing image · High energy efficiency · Neural networks · Specific architecture · Heterogeneous array

1 Introduction

Based on the all-weather airborne/satellite SAR high-precision remote sensing platform [1–4], intelligent remote sensing has exhibited its critical priorities in the field of real-time monitoring of large-scale dynamic scenes, tracking of moving targets as well as rapid classification. The emerging of deep neural network has offered a novel approach in solve these problems, which is suitable for space pattern recognition in the field of remote sensing. [5–7]. Hybrid neural network(H-NN) combined with different types of neural networks (NN) can far outperform a single neural network that exhibits certain disadvantageous shortcomings [8], making tremendous achievements in the field of

© Springer Nature Switzerland AG 2020
M. Qiu (Ed.): ICA3PP 2020, LNCS 12453, pp. 575–586, 2020.
https://doi.org/10.1007/978-3-030-60239-0_39

remote sensing. Therefore, it is necessary to develop a dedicated processor to efficiently accelerate the whole H-NN model, especially for remote sensing application platforms.

H-NN is mainly composed of CNN and RNN classes, to which it can be further subdivided into three types of computing tasks: convolution computing (CON), fully connected computing (FC) and RNN computing. The CNN module that is composed of four types of layers commonly known as convolutional, pooling activation and fully connected ones. The calculation model is shown in Eq. 1, where Conv represents convolution, Ac usually uses the ReLU function, Pooling represents the max (min) pooling process. Convolution computing task is computation intensive that requires a large amount of computing and parameter. Moreover, different CNN network models exhibit significant diversity in characteristics of computation load, differing from hundreds of MOPS to dozens of GOPS. In contrast to CON task, FC computing task is memory intensive, to which its required computation is greatly reduced to vector-matrix operations. RNN computing task is approximately similar to that of FC task, yet its computation process is more complicated. Owing to the fact that the RNN adopts a great deal of gate structures to achieve the sequence characteristics of information, such as the classic network model LSTM [9] which contains enormous nonlinear gate operations (Sigmoid and Tanh), therefore, it can be seen that the H-NN exhibits its characteristics in terms of large computation amount and of various computation loads, as well as dynamic characteristics of the computation load.

$$DCNN = \big[(Conv/Ac)i + (Pooling)\big]m + \big[(FC)j + (Ac)k\big]n \tag{1}$$

Note: Conv: convolution layer; Ac: activation function; FC: fully connected layer; i, m, j, k, n represents the cycle execution times of each functional layer respectively.

The second type is the recurrent neural networks (RNNs), to which the most widely used and most representative one is the long short-term memory (LSTM). This type of neural network primarily deals with applications involving sequence attributes, such as semantic analysis or time series event analysis. LSTM model with the calculation output of $O_{(t)}$ and $H_{(t)}$ is shown as follows,

$$O_{(t)} = Sigmoid\big(W_{io}X_{(t)} + W_{ho}H_{(t-1)} + b_o\big) \tag{2}$$

$$H_{(t)} = O_{(t)} \odot \tanh\big(C_{(t)}\big) \tag{3}$$

where,

$$C_{(t)} = I_{(t)} \odot \tanh\big(W_{ic}X_{(t)} + W_{hc}H_{(t-1)} + b_c\big) + f_{(t)} \odot C_{(t-1)} \tag{4}$$

$$f_{(t)} = Sigmoid\big(W_{if}X_{(t)} + W_{hf}H_{(t-1)} + b_f\big) \tag{5}$$

$$I_{(t)} = Sigmoid\big(W_{ii}X_{(t)} + W_{hi}H_{(t-1)} + b_i\big) \tag{6}$$

In the equations, $X_{(t)}$ is the input data at time t, $H_{(t)}$ is the output of the hidden layer at time t, $C_{(t)}$ is the state of the memory cell at time t, W_{ic} is the weight matrix from input layer to memory unit, W_{hc} is the weight matrix from hidden layer to memory unit, b_c is the bias vector at time t. $f_{(t)}$ is output value of forget gate, W_{if} is the weight matrix from input layer to forget gate, W_{hf} is the weight matrix from hidden layer to forget gate, b_f is the bias vector of this time. The $I_{(t)}$ is value of the input gate, W_{ii} is the weight matrix between the input layer and the input gate, W_{hi} is the weight matrix between the hidden layer and the input gate, b_i is the bias vector of this time.

Our study showed that significant differences occur between the computation model structure of CNN and the LSTM. First of all, convolution operations, activation operations and pooling operations are unique to CNN, whereas LSTM mainly includes vector operations and nonlinear function operations. Moreover, according to the logic models of the two networks, the output data of each layer of the CNN model flows to the end of the network in sequence, and the output data of each layer is only related to the input of the next layer. In addition, the computation in the same layer is relatively independent without any data dependence. The LSTM computation involving previous output data and current input data, etc., performs the iterative computation process in the hidden layer.

Despite researchers have proposed lots of acceleration schemes and hardware architectures for the convolution computation process in CNN [10–12], yet, few proposals were made for H-NN special processors [13]. In the study of accelerating neural networks, various neural network processing unit (NPU) solutions have emerged. However, most NPUs are mainly geared towards CNN acceleration. For example, Eyeriss [11] is an accelerator that is directly optimized for convolution computation. ENVISION [13] proposed a CNN processor with adjustable energy efficiency based on DVAFS. However, few solutions relating to H-NN acceleration processors have been made accordingly [14, 15]. DNPU [15] used a single chip to simultaneously support mixed operations of CNN and RNN, whereas in fact, its architecture was still composed of two sets of processor frameworks that are relatively independent, performing CNN and RNN computations respectively. Thinker [14] is a low-power NN computing chip based on a reconfigurable computing architecture that supports H-NN computing, to which the PE unit in the processor supports bit width adaptation, and the computing array can be reconfigured.

Great practical significance has proven to be prioritized in studying the dedicated architecture for H-NN operation in low-power domains. In this paper, we proposed a dedicated processor for hybrid neural networks in space field. The dedicated processor has the following features:

- The heterogeneous array structure consisting of different types of PE units, which exhibits excellence performance with respect to high parallelism, flexible resource scheduling, good scalability and tailoring. The power consumption of single processor meets the power restriction of on-orbit AI platform
- A multi-level on-chip storage structure and access strategy supporting different access modes is presented, which can reduce the bandwidth requirements of off-chip data access while improving the computing efficiency.
- A resource management strategy for heterogeneous computing arrays is presented, which combines pipeline processing with parallel processing, enabling efficient mapping of different types of H-NN.

The remainder of the paper is organized as follows. Section 2 highlights the design of the heterogeneous architecture implementation. Section 3 presents the evaluation of experimental results. Section 4 summarizes conclusions.

2 Proposed Processor Architecture

Figure 1. shows the top-level architecture of the proposed H-NN processor. The architecture consists of three main components: hybrid PE array, on-chip buffer module and Computation Flow & Resource Controller.

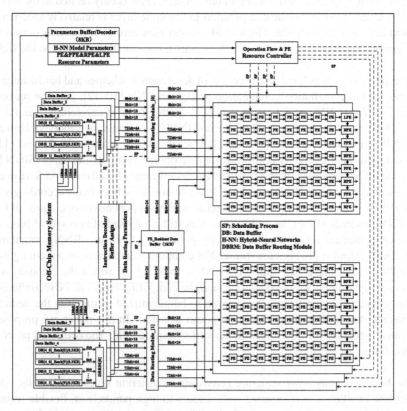

Fig. 1. Top-level architecture of processor.

As shown in Fig. 1, a miniaturized heterogeneous array scheme with an array size of 8 × 9 is proposed. A single heterogeneous array contains 8 × 8 PEs and 8 × 1 hybrid-PEs.

The processor contains four types of heterogeneous computing resources: PE, RPE, PPE and LPE, all of which support fixed-point operations. PE is used to perform all MAC operations in the H-NN, RPE is used to accumulate the intermediate results computed by PE and to perform ReLU activation operations. PPE adds pooling operations on the basis of RPE to perform pooling layer operations. The RPE and PPE collaborates to perform activation (ReLU) and pooling operations.

During the process of continuous convolution, the data accessing of adjacent PEs exhibits characteristics of adjacent address overlapping, as shown in Fig. 2. To support this data access model, a multibank distributed data placement strategy is therefore adopted. As shown in Fig. 2, each bank is used to store a row of pipeline data (corresponding to a row in the image data matrix).

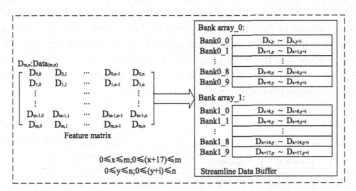

Fig. 2. Data placement polices (process of convolution operation).

For matrix-matrix or vector-matrix operations involved in the fully connected/RNN, vector (matrix) data is categorized into PE-resident data and pipeline data. Moreover, the long vector that exists in the computation process is segmented, for which the fixed bank batch flow placement strategy is therefore adopted. As shown in Fig. 3, the same column of vectors is accordingly segmented with respect to the segmentation results, of which are put into the same buffer bank in sequence, respectively.

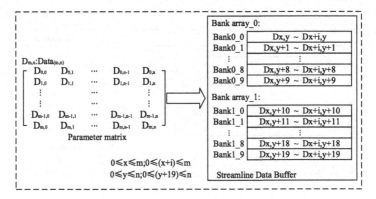

Fig. 3. Data placement polices (FC/RNN operation process).

3 Computation Process Mapping

In the convolutional neural network module, the operation process of convolutional network layer can be subdivided into three procedures: convolution, activation, pooling. According to the algorithm flow, each heterogeneous array can perform parallel pipeline processing on convolution, activation, and pooling operations. In order to organize the pipeline processing of the three operations engaged in the above process, a computation process is proposed on the basis of Deep Time-Nest Loop flow (Deep T-NL Flow), as shown in Fig. 4. In the multi-task nested computation process, multiple computing tasks can be operated in parallel, thereby ensuring that none of the PE units is idle. In regard to the time domain, the data is continuously sent to the processing unit, and the computation process will not be blocked due to waiting for data. With the help of T-NL Flow, computation procedures involving convolution, activation and pooling operations can therefore be completed within a continuous process.

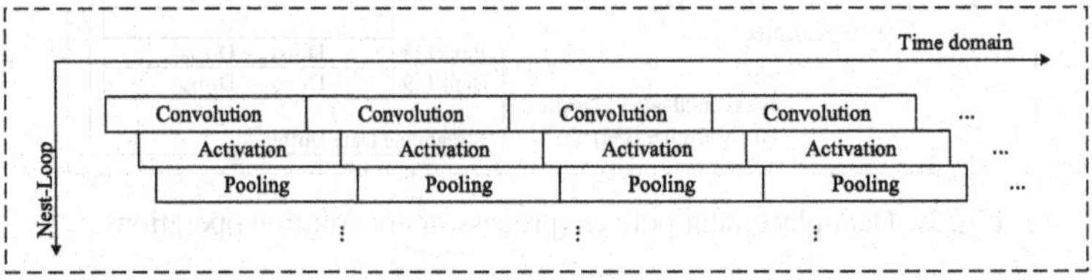

Fig. 4. Conv-Net computation flow (execution pipeline of single line PEs)

FC/RNN involves a large number of long vector-matrix or matrix-matrix operations. In addition, RNN includes the Look-up table operation of nonlinear functions. We normalize operations of vector-matrix or of matrix-matrix into multi vector-vector operations, to which the same level of vector-vector operation adopts the full parallel pipeline computation process, as shown in Fig. 5.

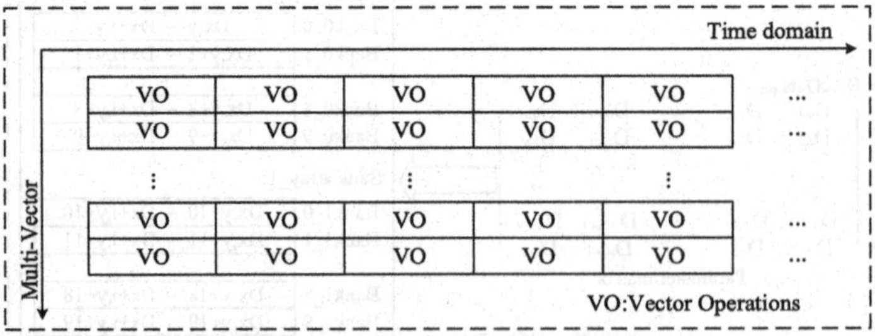

Fig. 5. FC-Net computation flow (execute parallel computing by all PE arrays).

As shown in Fig. 6, in order to organize computation task of the RNN module, the Shallow T-NL Flow computation process is adopted, of whose computation process is similar to that of the convolution module. During the nested loop process, both vector-vector computation and nonlinear function operation exhibit advantageous nestable parallelism, thereby validating that the computation process of shallow T-NL Flow can efficiently perform computation tasks of the RNN module.

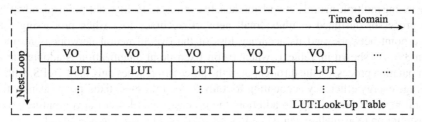

Fig. 6. RNN computation flow (shallow T-NL flow computing strategy).

4   Evaluation on Processor Performance

In this study, the hybrid-neural network processor is implemented by using Synopsys tools, which adopts 65-nm CMOS technology with 1.2 V supply voltage. The AlexNet, CLDNN, LRCN and VGG-LSTM were used as benchmarks in the evaluation.

4.1   Performance Analysis

Table 1. System performance assessment with different fixed-point length NN-m.

T-NL flow							
NN-m	AlexNet (B-S:)		LRCN (B-S:)		VGG19+LSTM (B-S:)		CLDNN (B-S:)	
A-u	97.95%		96.47%		98.13%		94.33%	
B-w	8-bit	16-bit	8-bit	16-bit	8-bit	16-bit	8-bit	16-bit
T-p	1.92	1.04	1.89	1.01	1.93	1.10	1.84	0.98
P-w	812	818	807	813	814	816	804	807
E-e	2.36	1.27	2.34	1.24	2.37	1.35	2.28	1.21
A-l	< 2%	–	< 2%	–	< 2.8%	–	2%	–

(Note: 1. NN-m: Neural network model; 2. C-t: Computing task 3. A-u: Array utilizations; 4. T-p: Throughput (TOPS); 5. P-w: Power (mW); 6. E-e: Energy efficiency (TOP/W); 7. C: Convolution layer; 8. FC/R: Full connection/RNN;9. B-w: Bit width; 10. A-l: Accuracy loss; 11. B-S: Batch size)

In this section, the processor is configured as a fixed-point heterogeneous PE array. The performance of the processor was evaluated at 200 MHz under the condition of different fixed-point lengths. The neural network models that are used for testing include AlexNet, CLDNN, LRCN and VGG-LSTM. Heterogeneous PE can execute multiple calculation operations of different neural network models in parallel, through which not only the computing power can be fully utilized, but also the throughput is therefore improved. The proposed heterogeneous processor can realize a throughput of 1.93 TOPS when processing multiple neural network models. In contrast to 16-bit fixed-point quantization, the processor designed exhibits small network accuracy loss when processing 8-bit fixed-point network, and its accuracy loss of the overall neural network is no more than 3%. As shown in Table 1, when 16-bit fixed-point quantization is adopted in all computation processes, the throughput of the processor reaches only 0.98 TOPS, with its average energy efficiency accounting for only 53% of the 8-bit fixed-point quantization neural network processing. In addition, the processor can lower the computing power consumption to a small extent when low bit fixed-point operation is selected.

4.2 Array Utilization Analysis

Nowadays, most of processors adopt neural network processing based on time flow (TI Flow). As can be seen from the comparison results of Table 1 and Table 2, during algorithm processing, the array utilization rate of the two-dimensional nested parallel pipeline based T-NL Flow outperforms that of one-dimensional T-I flow. In regard to processors, high utilization of the array can improve the throughput of the system. Therefore, the proposed processor architecture can flexibly allocate computing resources and data bandwidth according to different computing tasks, during which computation process the data can be flexibly intercepted and spliced. As shown in Table 1, in T-NL flow mode, a variety of neural network models was tested, suggesting that the resource utilization rate, throughput, and energy efficiency of the processor can reach up to 98.13%, 1.83TOPS, and 2.37TOP/W, respectively.

As shown in Table 2, in TI flow mode, input images enter the network in sequence, and various computing tasks in the neural network are executed successively. For example, in the VGG19-LSTM network model, three task modules of convolution? Full connection? RNN are executed successively. During computation process, convolution operation is a computation intensive task while FC/RNN is that of I/O intensive, to which the former one's high computation resource utilization rate can reach up to more than 98% whereas the latter one exhibits extremely low rate when it is executed alone. It can be seen that based on TI flow calculation mode, the FC/RNN operation will result in a serious plunge in the comprehensive utilization rate of computing resources throughout the whole computation process. As shown in Table 2, with respect to the LRCN model, the comprehensive utilization rate of computing resources is only 24.02%, and the comprehensive energy efficiency is merely 0.71 TOP/W. On the contrary, the task of convolution operation in VGG19-LSTM is relatively high, to which the comprehensive utilization rate of computing resources rises up to 77.4%, and the comprehensive energy efficiency reaches up to 2.02 TOP/W.

As shown in Table 1, the resource utilization and energy efficiency of T-NL flow mode are significantly improved compared with T-I flow mode. In contrast to the figure in Table 2, the throughput of computation process of AlexNet model, of LRCN model

Table 2. Array utilization with TI flow

TI flow						
NN-m	AlexNet			LRCN		
C-t	C	FC/R	C	FC/R	C	FC/R
A-u	98.87%	10.83%	98.87%	10.83%	98.87%	10.83%
T-p	1.94	0.21	1.94	0.21	1.94	0.21
P-w	–	–	–	–	–	–
E-e	–	–	–	–	–	–
NN-m	VGG19+LSTM			CLDNN		
C-t	C	FC/R	OA	C	FC/R	OA
A-u	99.12%	12.1%	77.4%	98.59%	11.9%	21.7%
T-p	1.95	0.24	1.51	1.93	0.23	0.42
P-w	–	–	747	–	–	631
E-e	–	–	2.02	–	–	0.67

(Note: 1. NN-m: Neural network model; 2. C-t: Computing task 3.
A-u: Array utilizations; 4. T-p: Throughput (TOPS); 5. P-w: Power
(mW); 6. E-e: Energy efficiency (TOP/W); 7. C: Convolution layer;
8. FC/R: Full connection/RNN)

and of VGG19-LSTM model is increased by 34.27%, 302.13% and 27.81%, respectively, of whose comprehensive power consumption is increased by 10.7%, 21.9% and 27.42%, respectively. It can be seen that the H-NN model will be featured with significant advantages by adopting the T-NL flow computation mode, thereby not only flexibly matching the computing tasks of various neural network models but also maximizing the energy efficiency of computing resources.

4.3 Comparison with State-of-the-Art Schemes

At the frequency of 200 MHz, the proposed architecture exhibits its peak throughput of 1.96 TOPS, power consumption of 824 mW, and peak energy efficiency of 2.38 TOPS/W. Table 3 lists out the processor's performance on throughput and on energy efficiency under different neural network models. It can be seen that the proposed scheme greatly improves the computational energy efficiency of the H-NN. The throughput and processor energy efficiency of other studies are also listed in Table 3. Under same frequency and benchmark (AlexNet), the proposed processor is compared with previous schemes. It can be seen from Table 3 that under 65 nm technology, the proposed architecture possesses the highest energy efficiency and throughput for the 8-bit quantitative computation process. With respect to the 28 nm technology, although Reference [20] and Reference [22] exhibit the lowest power consumption, the throughput of the proposed scheme is increased by 24.94 times compared with Reference [22]. Compared with Reference [25] that adopts a reconfigurable scheme, the throughput and energy efficiency of the proposed scheme is increased by 5.33 times and 2.21 times, respectively.

Table 3. Comparison with previous works.

Architectural model	Operating frequency	Power consumption	NN-m	Throughput (TOPS)	Energy efficiency (TOPS/W)
Proposed solution @(65 nm)	200 MHz	812 mW	AlexNet	1.92 @(8_bit)	2.36 @(8_bit)
		807 mW	LRCN	1.89 @(8_bit)	2.34 @(8_bit)
		814 mW	VGG19+LSTM	1.93 @(8_bit)	2.37 @(8_bit)
		804 mW	CLDNN	1.84 @(8_bit)	2.28 @(8_bit)
Eyeriss [10]	200 MHz	278 mW	AlexNet	0.0462	0.166
@(65 nm)	200 MHz	279 mW	AlexNet	@(16_bit)	@(16_bit)
DNPU [14]	200 MHz	290 mW	AlexNet	1.088	3.9
@(65 nm)	200 MHz	44 mW	AlexNet	@(4_bit)	@(4_bit)
Thinks [15]		41 mW	AlexNet	0.386	1.06
@(65 nm)				@(8_bit)	@(8_bit)
Envision [14]				0.076	3.8
@(28 nm)				@(4_bit)	@(4_bit)
SOC [16]				0.077	1.89
@(28 nm)				@(8_bit)	@(8_bit)

4.4 Analyzes of Array Scalability

Theoretically, a large-sized PE array can provide higher computational throughput. However, the H-NN involves both computationally intensive and I/O intensive two kinds of computational tasks, which proportion of computing load varies greatly. The on-chip cache capacity and off-chip storage data bandwidth limit the supply of data in the calculation process, resulting in the computing resources of large PE array cannot be fully utilized. The process of computing resource scheduling and data organization is more complex when a single large PE array is partitioned across task areas. Therefore, for the hybrid neural network model, without loss of energy efficiency, with the dedicated storage system, the high-density & compact PE array has the best flexibility and the highest resource utilization.

Through using the VGG19+LSTM hybrid neural network model, the performance of heterogeneous arrays with different scales is evaluated, to which eight arrays are designed in the processor and the T-NL computation strategy is adopted. As shown in Table 4, the throughput of the processor significantly increases with the increase in the size of a single array. However, as the size of array increases, the required storage bandwidth surges.

Taking into account the access bandwidth, the requirements of data supply engaged in computation process cannot be met if the array is abundantly over configured, thereby resulting in waste of computing resources. Similarly, lower throughput of the processor

Table 4. Analysis on scalability of array

Array scale	Throughput	Memory bandwidth requirement
2 × 2	0.11TOPS	7.11 GB/s
4 × 4	0.46TOPS	28.9 GB/s
6 × 6	1.03TOPS	64.9 GB/s
8 × 8	1.84TOPS	115.3 GB/s
12 × 12	4.14 TOPS	259.5 GB/s

Note: Analysis process based on VGG19+LSTM hybrid neural network model

can be attributed to the inadequately under configuration of the array, to which the demand in computation of the network model is hardly satisfied. Consequently, the size of a single heterogeneous processing array is designed to be 8 9 after seriously weighing among complexity of chip implementation, limitations in memory bandwidth and processing performance.

5 Conclusions

This paper proposed a heterogeneous hybrid neural network processor that adopts fixed-point heterogeneous PE array and parallel pipeline acceleration scheme, which can be applied to the aerospace field for target recognition and/or target classification. The processor consists of eight 8 × 9 heterogeneous arrays, which is capable of providing 1.96 TOPS peak throughput. Each array supports a variety of operations in H-NN, in which computing resources can be fully utilized, thereby greatly increasing the computing throughput of the processor. During the computation process of AlexNet model, LRCN model, VGG19-LSTM model and CLDNN model, the system throughput can reach up to 1.92TOPS, 1.89TOPS, 1.93TOPS and 1.84TOPS respectively. In order to improve energy efficiency, PE array can be subdivided into multi-task rows, and the algorithm is reasonably mapped to the hardware architecture, which improves not only the utilization rate of hardware resources but also computing energy efficiency. The power consumption of single processor meets the power restriction of on-orbit AI platform

References

1. Percivall, G.S., Alameh, N.S., Caumont, H., et al.: Improving disaster management using earth observations—GEOSS and CEOS activities. IEEE J. Sel. Top. Appl. Earth Obs. Remote Sens. **6**(3), 1368–1375 (2013)
2. Lou, Y., Clark, D., Marks, P., et al.: Onboard radar processor development for rapid response to natural hazards. IEEE J. Sel. Top. Appl. Earth Obs. Remote Sens. **9**(6), 2770–2776 (2016)
3. Tralli, D.M., Blom, R.G., Zlotnicki, V., et al.: Satellite remote sensing of earthquake, volcano, flood, landslide and coastal inundation hazards. ISPRS J. Photogramm. Remote Sens. **59**(4), 185–198 (2005)

4. Gierull, C.H., Vachon, P.W.: Foreword to the special issue on multichannel space-based SAR. IEEE J. Sel. Top. Appl. Earth Obs. Remote Sens. **8**(11), 4995–4997 (2015)
5. Xiaobing, H., Yanfei, Z., Liqin, C., et al.: Pre-trained AlexNet architecture with pyramid pooling and supervision for high spatial resolution remote sensing image scene classification. Remote Sens. **9**(8), 848 (2017)
6. Komiske, P.T., Metodiev, E.M., Schwartz, M.D.: Deep learning in color: towards automated quark/gluon jet discrimination. J. High Energy Phys. **1**, 110 (2017)
7. Shi, X., Chen, Z., Wang, H., et al.: Convolutional LSTM network: a machine learning approach for precipitation nowcasting (2015)
8. Kaiser, L., Gomez, A.N., Shazeer, N., et al.: One model to learn them all. arXiv: Learning (2017)
9. Hochreiter, S., Schmidhuber, J.: Long short-term memory. Neural Comput. **9**(8), 1735–1780 (1997)
10. Chen, Y.H., Krishna, T., Emer, J.S., et al.: Eyeriss: an energy-efficient reconfigurable accelerator for deep convolutional neural networks. In: IEEE International Solid State Circuits Conference. IEEE (2016)
11. Moons, B., Verhelst, M.: A 0.3–2.6 TOPS/W precision-scalable processor for real-time large-scale ConvNets. In: 2016 IEEE Symposium on VLSI Circuits (VLSI-Circuits), Honolulu, HI, pp. 1–2 (2016)
12. Song, L., Wang, Y., Han, Y., et al.: C-Brain: a deep learning accelerator that tames the diversity of CNNs through adaptive data-level parallelization. In: Design Automation Conference. IEEE (2016)
13. Moons, B., Uytterhoeven, R., Dehaene, W., et al.: 14.5 envision: a 0.26-to-10 TOPS/W subword-parallel dynamic-voltage-accuracy-frequency-scalable convolutional neural network processor in 28nm FDSOI. In: Solid-State Circuits Conference. IEEE (2017)
14. Yin, S., et al.: A high energy efficient reconfigurable hybrid neural network processor for deep learning applications. IEEE J. Solid-State Circuits **53**(4), 968–982 (2018)
15. Shin, D., Lee, J., Lee, J., et al.: 14.2 DNPU: an 8.1TOPS/W reconfigurable CNN-RNN processor for general-purpose deep neural networks. In: IEEE International Solid-State Circuits Conference. IEEE (2017)
16. Desoli, G., et al.: 14.1 a 2.9TOPS/W deep convolutional neural network SoC in FD-SOI 28nm for intelligent embedded systems. In: 2017 IEEE International Solid-State Circuits Conference (ISSCC), San Francisco, CA, pp. 238–239 (2017)

Roda: A Flexible Framework for Real-Time On-demand Data Aggregation

Jiawei Xu[1], Weidong Zhu[2], Shiyou Qian[1(✉)], Guangtao Xue[1], Jian Cao[1], Yanmin Zhu[1], Zongyao Zhu[3], and Junwei Zhu[3]

[1] Shanghai Jiao Tong University, Shanghai, China
{titan_xjw,qshiyou,gt_xue,cao-jian,yzhug}@sjtu.edu.cn
[2] Xuzhou University of Technology, Xuzhou, Jiangsu, China
zweidong@xzit.edu.cn
[3] Shanghai Big Data Center, Shanghai, China
{zyzhu,jwzhu}@shanghai.gov.cn

Abstract. It is critical to aggregate data from multiple sources to support real-time decision making in several fields, such as anti-telecommunications fraud detection. However, as data sources are distributed, heterogeneous and autonomous, it is challenging to ensure that data aggregation satisfies the requirements of real-time, on-demand and flexibility. In this paper, we propose a real-time on-demand data aggregation (Roda) framework, which is designed to be flexible enough to support the dynamic joining of new data sources, the immediate updating of aggregation rules and the quick adaptation to data velocity. We implement a prototype of Roda based on Kafka and Docker using the overlay network technique. To evaluate the effectiveness and performance of Roda, we conduct a series of experiments based on real trace data. The experiment results show that Roda can guarantee data aggregation latency at the millisecond scale, easily achieving our design goals.

Keywords: Data aggregation · Real-time · On-demand · Flexibility

1 Introduction

Increasing cross-organization cooperation motivates the need to aggregate data from multiple information systems, namely data sources, to support real-time decision making. For example, as shown in Fig. 1, the real-time detection of telecommunication fraud involves aggregating data from multiple organizations, including banks, telecommunication companies, transportation companies and the police department [8]. However, it is challenging to build an effective data aggregation system involving multiple spatially distributed heterogeneous information systems. In addition, the high velocity, large volume, and structural variety of big data makes data aggregation more difficult.

To support real-time decision making, an effective data aggregation system should satisfy two requirements in addition to data security [21]: (1) selectivity. Decision making may only need a part of the data based on prior experience which is expressed in the form of a set of aggregation rules. Therefore, a

© Springer Nature Switzerland AG 2020
M. Qiu (Ed.): ICA3PP 2020, LNCS 12453, pp. 587–602, 2020.
https://doi.org/10.1007/978-3-030-60239-0_40

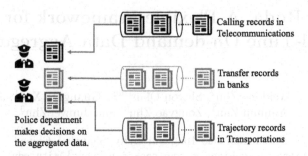

Fig. 1. Aggregating data from multiple departments to support the real-time detection of telecommunication fraud.

fine-grained data filter on sources is needed for selective data aggregation; (2) flexibility. Firstly, as new data sources may join the system, the aggregation system must be scalable to handle such a situation. Secondly, the aggregation rules maintained by the filter may need to be updated, which means the demand for rules takes effect immediately. Thirdly, the data velocity in sources may vary, so to guarantee real-time requirements, the filter should have the ability to quickly adjust its performance to adapt to the fluctuating workload. Traditional aggregation solutions are usually implemented by dedicated software or interfaces (usually HTTP or FTP) for the sink system to call. These solutions can not meet the requirements of real-time, on-demand and flexibility.

In this paper, we address these issues and propose an effective data aggregation framework named Roda. Roda is designed to be flexible enough to support the dynamic joining of new data sources, the immediate updating of aggregation rules and the quick adaptation to data velocity. We also implemented a prototype of Roda based on Kafka and Docker. To evaluate the effectiveness and performance of Roda, we build a testing environment which comprises three systems: one sink system and two source systems. We conduct a series of experiments based on real trace data. The experiment results show that Roda can guarantee the data aggregation latency at the millisecond scale, easily achieving our design goals.

The main contributions of this paper are as follows:

- We propose an effective framework called Roda for data aggregation to meet real-time on-demand requirements.
- We implement a prototype of Roda based on Kafka and Docker, which is easily deployable and scalable.
- We conduct experiments on real-world trace data to evaluate the effectiveness and performance of Roda.

The rest of the paper is organized as follows. Section 2 discusses the related work. Section 3 details the design of Roda. Section 4 describes the implementation of Roda. Section 5 analyzes the experiment results. We conclude the paper in Section 6.

2 Related Work

2.1 Aggregation Based on Service-Oriented Architecture

Traditional solutions to data aggregation usually adopt a service-oriented architecture (SOA), such as the network on terminal architecture. These solutions can diminish the problem of heterogeneity among distributed systems, but the participants need to have expert knowledge on the systems to be aggregated. They focus on forming a standard information exchange formulation so as to reduce the communication cost. The work in [11] defines data interoperability maturity level, expands capabilities, and develops a data portfolio which can help guide organizations as they move toward ubiquitous information sharing. Janssen et al. try to address interoperability issues in electronic governance [9]. Homann et al. [10] propose an interoperability framework for integrating banking systems and present a case study on two European banks using web services.

2.2 Data Fusion

Data fusion is a multilevel, multifaceted process dealing with the automatic detection, association, correlation, estimation and combination of data and information from single or multiple sources. Data fusion aims to achieve a refined position and identity estimates, and the complete and timely assessments of situations and threats and their significance. One of the most widely used frameworks is the Joint Directors of Laboratories (JDL) Data Fusion Framework [27]. This framework has three components: sources, human-computer interaction (HCI) and a database management system. Five levels of data processing are defined in JDL: source processing, object refinement, situation assessment, impact assessment and process refinement. Although the data fusion framework has many revisions and extensions [13,14], it focuses on the data process level instead of the data collection process.

2.3 Multi-domain Publish/Subscribe Systems

Publish/subscribe systems have been widely studied in recent years for data distribution. The multi-domain publish/subscribe system [6] tries to combine several clusters together to build a larger-scale system which has some similarity with our framework. In [24], the SDN framework is tightly coupled with the pub/sub system for high performance. The security issues of multi-domain pub/sub systems are discussed in [25,26], and several access control mechanisms have been proposed [15,16,23]. For content-based selective data distribution, the matching algorithm is critical to system performance. Many efficient matching algorithms have been proposed to promote the matching speed based on different data structures, such as TAMA [29], OpIndex [28], REIN [17,18], H-Tree [19], Be-Tree [22], Siena [5], Ada-REIN [20] and Gem [7].

In summary, there are no data aggregation frameworks that address the requirements of real-time, on-demand and flexibility to support real-time decision making.

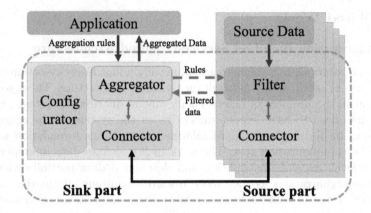

Fig. 2. The framework of Roda.

3   Design of Roda

3.1   Design Objectives

When designing Roda, we consider three objectives. Firstly, as Roda mainly focuses on aggregating stream data from multiple sources, aggregation latency should be minimized in order to support real-time decision making. Secondly, Roda should support fine-grained data filtering to obtain the desired data and to avoid being overwhelmed by large-scale data. Thirdly, Roda should be flexible enough to adapt to dynamic environments, which is manifested in three aspects: (1) new data sources may join the aggregation middleware; (2) the aggregation rules issued by applications may change over time; and (3) the data rate of streams in sources may vary.

The framework of Roda is shown in Fig. 2, which is composed of two parts: a sink part and source part. The sink part is deployed on a node in the sink information system which needs to aggregate data from other systems. The source part is deployed on a node in the source information system that provides the data to be aggregated. The sink part consists of three modules: a connector module for constructing an overlay to secure data transmission, an aggregator module for accepting aggregation rules from applications and providing interfaces for applications to access aggregated data, and a configurator module for the rapid joining of new source systems. The source part consists of two modules: a connector module and a filter module to realize selective data aggregation according to application requirements. Figure 2 presents the framework for the single-sink multi-source scenario, which can also be extended to contain multiple sinks as described in Sect. 4.5.

3.2   Connector

Organizations usually deploy their own information systems inside their own local area networks (LAN) with private IP addresses. This kind of deployment solves the problems of security and the insufficiency of IPv4 addresses, as the network traffic between inner hosts and outside hosts must be transmitted by a router or a powerful firewall which typically deploys an intrusion-detection system (IDS) or intrusion prevention system (IPS) to protect LANs. Although using network address translation (NAT) can alleviate the insufficiency of IPv4 addresses, this scheme causes difficulty for data aggregation because hosts with private IPv4 addresses in different LANs cannot connect to each other directly.

We adopt the overlay network in Roda to guarantee secure communication between hosts located in different LANs. An overlay network is a computer network that is layered on top of another network [4]. In this way, hosts in the same overlay network can communicate with each other directly. With such an abstraction of networks, the complex underlay network configuration, like LAN, firewall and IPv4 addresses allocated by internet service provider (ISP), can be ignored.

In Roda, connectors are the fundamental layer, which are deployed in both source systems and sink systems. Together, they construct an overlay network so as to abstract the underlay physic network. The connector is decoupled with other components of Roda, which means we can dynamically modify its configuration or choose a more powerful implementation to build a more efficient overlay network. The upper-layer components in Roda do not need to consider network communication problems, which greatly simplifies the design. When a new source system is going to join Roda, it first deploys a connector and then a filter. There is no need to modify the components in other current sources or sink systems, which means Roda is easily scalable. When the connector is successfully deployed in a sink system and source systems, the filters in the source systems can access the aggregator in the sink system directly.

3.3   Filter

To avoid being overwhelmed by large-scale data, selective aggregation is important to gather data on-demand. Therefore, for a real-time data aggregation system, a data filter is needed where some fine-grained data selection rules can be configured. We design a content-based filter in Roda to support on-demand data aggregation. A data filter maintains the data aggregation rules of applications in a data structure and performs a fast filtering operation on the stream data in the sources. As the volume of filtered data is reduced to some degree after filtering, the network traffic pressure is decreased.

Roda leverages the matching algorithm of content-based publish/subscribe systems [18,28,29] in our aggregation framework. The filter has two types of inputs: data records from sources and data aggregation rules from the sink system, which are defined as follows:

Definition 1. *Records: each record in the source data is a conjunction of multiple attribute-value pairs, e.g. $e = \{av_1, av_2, ..., av_M\}$, where av_i is an attribute-value pair in the form of $a_i = v_i$ and the number of attributes in e is denoted by M. We assume that each attribute appears only once in one data record.*

Definition 2. *Aggregation rules: a rule expresses the interests of applications to data which contains multiple interval predicates that specify constraints on attributes. An interval predicate is defined as a tuple $\{a, v_1, v_2, type\}$, where a is an attribute, v_1 and v_2 are two values defined by an interval on a, and v_1 is not larger than v_2. The types of attributes can be an integer, double or string.*

As shown in Fig. 2, the filter firstly receives aggregation rules from the aggregator in the sink system, and then retrieves data records through the interfaces/software of a target source system, which is denoted by the source data. The value of the attributes in each record are checked with the aggregation rules maintained by the filter, and the matched records are sent to the aggregator in the sink system.

Roda is designed to support real-time decision-making, which means the matched data should be transferred from the source to the sink with low latency. From Fig. 2, the transfer path of source data in Roda is: *Source Data →️ Filter →️ Aggregator*. The performance of the two network transfer procedures depends on the network state which is usually good nowadays. To ensure low latency of aggregation, the filter is the potential performance bottleneck when the data generation rate in the sources is high and the number of data aggregation rules is large. To enhance the filtering performance, we leverage a parallel method *PhSIH* which is able to adjust the parallel degree according to performance expectation [12], providing a balance between filtering performance and resource utilization.

3.4 Aggregator

The aggregator is an important component in Roda, which provides northbound APIs to external applications to collect data aggregation rules and distribute matched data to applications. As shown in Fig. 2, the application layer is beyond Roda in a sink system, which may consist of hundreds of user applications such as machine learning and database software. From the perspective of Roda's users, their applications submit data aggregation rules to the aggregator and obtain the desired data from it through northbound APIs. In the interior of Roda, the aggregator forwards data aggregation rules to all filters deployed in the source systems, and simultaneously receives the matched data from the filters. The logical connection between the aggregator and the filter is shown in Fig. 2 with the dotted line. This network traffic is delivered by the connector introduced in the previous section.

In real-time scenarios, applications may change their logic and in turn, change the aggregation rules in the data. In Roda, the creation, update or deletion of data aggregation rules are all supported through the northbound APIs of the aggregator. Filters in different sources will retrieve such modifications and update their inner data structures dynamically and rapidly. With such a scheme, the upper applications are decoupled from the aggregation framework. Thus, the logic modification of upper applications will not influence Roda.

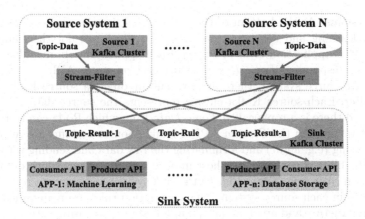

Fig. 3. Roda implementation based on Kafka.

3.5   Configurator

The configurator is a centrally visible web console to help operators manage Roda. It can configure the aggregator and connector in a sink system. The monitoring information from the filters is also sent to the configurator, therefore operators can monitor the performance of filters deployed in source systems visibly on web pages.

The design of Roda takes scalability into consideration. With expanding businesses or the increasing complexity of application logics, some new source systems may join the aggregation system. As Roda adopts the overlay network, new source systems only need to configure and deploy a connector and a filter. There is no need to modify the components in other existing source or sink systems. With the help of container techniques, the scalability of Roda is strengthened. In the configurator, the images of new filters can be automatically configured and built. Then, remote source systems can download images and run these rapidly and automatically. The deployment and update of filters in remote source systems can be operated in the configurator visibly and flexibly.

4   Implementation

4.1   Framework Implementation Based on Kafka

System Design. We design the implementation of Roda based on the architecture and API of Kafka, which is shown in Fig. 3. Overall, the system comprises two parts: source system and sink system.

The sink Kafka cluster in the middle of Fig. 3 plays the role of aggregator. There are two kinds of topics in the sink Kafka cluster: *Topic-Rule* and *Topic-Result-i*. *Topic-Rule* stores the aggregation rules issued by applications through the *Producer* API. For each application i, we create a topic called *Topic-Result-i* which stores the filtered data satisfying the application's aggregation rules from source systems. At the bottom of the sink system lies the applications such as machine learning and consistent data storage, which are not the components of Roda (the application layer in Fig. 2). Each application obtains filtered data from its corresponding *Topic-Result-i* through *Consumer* API.

In each source system, there are two sub-modules: a source Kafka cluster and a filter. Each source Kafka cluster represents the source data in a source system and there is a topic called *Topic-Data* from which Roda can retrieve the source data through the *Stream* API of Kafka. Source systems are responsible for processing (e.g. audit) and formulating the raw data and sending it to *Topic-Data*. The filter is deployed in a host in each source system, which obtains the aggregation rules from *Topic-Rule* in the sink system and the data records from *Topic-data* in each source system. The aggregation rules are maintained locally in a data structure and are dynamically updated by associating with *Topic-Rule* in the sink system. Data records in *Topic-Data* firstly enter the filter, then they are matched against the rules and finally the matched records are sent to each *Topic-Result-i* in the sink system.

With the help of Kafka, it is easy for source systems to integrate the data aggregation framework into their original information systems. They only need to deploy an *access point* on which a connector and a filter are deployed. Through the interface opened by the source systems to the filter, the data records in the source data can be efficiently retrieved from *Topic-data* and the matched results are sent to corresponding *Topic-Result-i* of applications.

4.2   Implementation of the Filter

We implement the matching algorithm OpIndex [28] as the filter. OpIndex is designed for scalability with respect to the volume, velocity and variety of data. It can handle high-dimensional and sparse datasets. Therefore, OpIndex can deal with different data types. In the big data scenario, the data to be aggregated is usually sparse, which cannot be efficiently processed by most matching algorithms. Furthermore, OpIndex has low memory requirements and maintenance costs, and can be easily extended to support more expressive subscriptions.

OpIndex builds a data structure to index aggregation rules, which consists of two indexing layers. The first layer is built on pivot attributes (the attribute with the smallest frequency in each rule) and the second layer is on operators. Each rule associates a counter that is used to judge whether a rule is matched by a data record in the filtering process. For each operator, a bucket list is constructed to store predicates and their links to counters. When filtering a data record, for each pivot attribute in the data record, OpIndex goes through each operator level in the second indexing layer to determine the buckets which store the predicates and evaluates them. When a predicate is satisfied by the attribute value, the counters associated with the rules that contain the predicate are decremented by one. When a counter is reduced to 0, the associated rule is judged as a match of the data record.

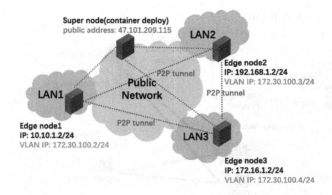

Fig. 4. N2N overlay network.

In the implementation of the filter, we make two optimizations. Firstly, we utilize the parallel method PhSIH [12] to guarantee the filtering performance. Given a performance expectation, PhSIH is able to adjust the parallel degree to adapt to data velocity, which is important to realize real-time data aggregation. Secondly, when a data record is matched with multiple applications, it will be sent multiple times to the aggregator in the sink system. To improve network transfer efficiency, we combine the matched applications into a string which is pinned with the data record. When such encoded data arrives at the sink system, a helper program decodes it and sends a copy of the data record to the topic corresponding to each application. This transfer is operated in a local area network which could reduce public traffic workloads.

4.3   Overlay Network Implemented by N2N

N2N [3] is a peer-to-peer virtual private network that can allow users to overcome some limitations of LANs. N2N users can create and manage their own secure and geographically distributed overlay network without the need for central administration which is typical in most virtual private network systems. There are two roles in an N2N network as shown in Fig. 4: (1) Edge nodes: each edge node is a virtual network interface and is allocated a VLAN IP address. A peer-to-peer tunnel is built between the edge nodes for data transmission. (2) A super node: a super node is in charge of announcing and discovering other nodes. It must have a port which is publicly accessible on the Internet.

A virtual network shared among multiple edge nodes in N2N is called a community. A single super node can relay multiple communities and a host running as a super node can be part of multiple communities at the same time. An encryption key can be used by the edge nodes to encrypt the packets within their community. Each edge node can be seen as a router. For example, as shown in Fig. 4, the other hosts in LAN1 can also communicate with hosts in LAN2 through edge node1. N2N tries to establish a direct P2P connection between the

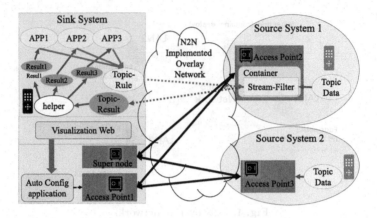

Fig. 5. A deployment example of Roda

edge nodes when possible. When this is not possible (usually due to special NAT devices), the super node is used to relay packets.

Figure 5 shows a deployment example of Roda in which we deploy N2N on an *access point* to generate a VLAN. Each *access point* is a host in a source system which deploys Roda components, e.g. the filter and N2N edge node. Therefore, an *access point* could directly access the source Kafka cluster in each source system to obtain the data records. Access points can communicate with the sink system in a logical VLAN to transfer the aggregation rules and filtered data (dotted lines in Fig. 5). In fact, the logic traffic is delivered through the peer-to-peer tunnels built by N2N (black lines).

In Roda, as the overlay network is essential, we deploy three super nodes in the containers to ensure its high reliability and availability. A primary super node is deployed on one master host and uses port mapping to provide the service. Another two backup super nodes are also deployed. The configurator will communicate with the primary super node and backup super nodes through heartbeats to check their health. When the primary super node crashes, one of the two backup super nodes takes over immediately, becoming the new primary super node.

4.4 Central Implementation of the Configurator

When adopting Kafka in the implementation, there are a lot of remote access points and filters running on source systems, as shown in Fig. 5. In order to monitor the system and to ensure it is easy to scale and update, we leverage the Docker technology in the implementation of Roda. The Docker is an open platform for developing, shipping, and running applications. Docker engine can pull application packages, called images, from the remote registry and run these in an isolated namespace of the operating system. Docker provides rich remote APIs to manage containers in the Docker engine. To cope with the thousands of containers deployed in hundreds of hosts, container scheduling platforms are

Table 1. Virtual machine list

System name	Configuration(n vCPUs + m GB RAM)
Sink system	7 VMs(4 + 16)
Source System1	4 VMs(4 + 16), 1 VM(32 + 64)
Source System2	4 VMs(4 + 16), 1 VM(32 + 64)

created. Swarm [1] is the official scheduling platform developed by Docker, but Kubernetes [2] is the most well-known and is maintained by Google.

However, both Swarm and Kubernetes are too heavy for Roda, as lots of their functions are not necessary. Based on Docker, we implement a central configurator in web form to show and manage images visibly. Furthermore, building and reconfiguring the images of filters also becomes easier with the help of Docker. We can automatically reconfigure and build new filter images through the visible web.

4.5   System Extension

Roda is originally designed for the single-sink multi-source situation, and it is easily scalable with more source systems. However, there may exist multi-sink multi-source situations. With some modifications, Roda can meet this demand. We can decouple the configurator component to form a new part which can be deployed on a new *access point* that joins the overlay network. Another modification is the filter. As there are multiple sink systems, each filter will receive the aggregation rules of different applications from different sink systems to construct a unified data structure. In order to track the applications belonging to different sink systems, an extra hash map is added. When a filter sends back the matched data, it first checks the hash map to send the data to the right sink system.

5   Experiments

Testbeds. We deploy Roda on public cloud and simulate organization LAN isolation through different private sub-networks. Specifically, three information systems are simulated: sink system, source system1 and source system2, as shown in Fig. 5. In each system, there are 3 virtual machines (VM) equipped with 4 vCPUs and 16 GB RAM where Kafka clusters are deployed individually. In the sink system, 3 VMs (4 vCPUs, 16 GB RAM) are running as a N2N high reliability super node cluster. A public IPv4 address with 200 Mb bandwidth is also bound with this cluster. Furthermore, one VM (4 vCPUs, 16 GB RAM) plays the role of access point where a N2N edge node is deployed. In the two source systems, one VM (32 vCPUs, 64 GB RAM) is also used as the access point and 1 VM (4 vCPUs, 16 GB RAM) plays the role of generating source data. The access points in the two source systems are different from those in the sink system as filters run on them, which requires more resources to guarantee the high performance of filtering. The configuration of these three systems is shown in Table 1. The number of predicates contained in the rules are randomly generated in the range [1, 10]. The parameters used in this paper are listed in Table 2.

Table 2. Experiment parameters

Name	Values
Sink system quantity	1
Source system quantity	2
Number of applications	1,000
Number of aggregation rules	250,000
Number of predicates in rules	[1, 10]
Number of attributes in stock data	57
Data records generated per second	10
Overlay network bandwidth	200 Mb

5.1 Environment Setup

Data Generation. Our experiments are conducted on real-world data. Source system1 generates financial stock data and source system2 produces telecommunication data. For each source system, we simulate the condition where there are 1,000 applications issuing 250,000 aggregation rules in total. As the two source systems are similar, in the following section, we introduce the result of source system1 (the stock data where each data record has 57 attributes) as an example. We generate data aggregation rules on the attributes of stock data uniformly, which means these rules are not centralized on several attributes.

Metrics. To evaluate the performance of Roda, we use the following metrics: (i) overall aggregation delay; (ii) time for new aggregation rules to take effect; (iii) time for the new source system to join the aggregation system, which comprises three parts: image building time, image pulling time and container startup time.

5.2   Results

Aggregation Delay. We conduct a 30-min experiment in which the source generates data records at the rate of 10 records/s. The results of aggregation delay are shown in Fig. 6.

From this figure, we find that the aggregation delay of Roda is at the millisecond level for cross-system LAN transfer. The average delay is 48.725 ms, which contributes to the employment of the efficient matching algorithm OpIndex and parallel method PhSIH. In addition, the standard deviation of the delay is 4.62 which means the delay is stable. From these results, we can validate that the filter and data transmission are efficient and stable. The performance partly depends on the network quality of the aggregation system. Using a high-speed network is beneficial to reduce aggregation delay.

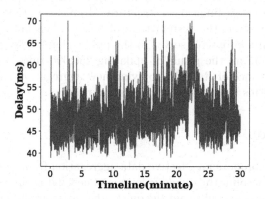

Fig. 6. Overall aggregation delay

Table 3. Maintenance cost

	Standard deviation	Average values
Image building time(s)	10.107	0.106
Image pulling time(s)	53.332	1.02
Image startup time(s)	0.455	0.024
Rules applying time(ms)	1.878	1.498

Maintenance Cost. The design of Roda takes the dynamics in consideration: joining new data sources and updating aggregation rules. This experiment tests the time consumed by new source systems to join and the time it takes to update the data aggregation rules. Table 3 shows these time costs of Roda. Each result is the average of 100 experiments.

When a new source system joins Roda or the image of filters needs to be updated, there are three steps to launch a filter in the source system: (a) building a new filter image for the new source system; (b) pulling the new image in the remote access point of the new source system; and (c) running the new image in a container in the access point. The first three lines in Table 3 list the execution time of these three steps. The pulling time is 53.334 s on average as the size of the filter images is about 960 MB and is transferred through the overlay network. The bandwidth of the super node in the experiment is only 200 Mb, which results in a relatively high pulling time. Due to the lightweight property of Docker, starting up an image is very fast, taking just 0.024 s on average. These results show that Roda is flexible for scaling up new source systems.

The last line in Table 3 measures the situation when applications have new data aggregation rules which are first sent to the aggregator in the sink system and then the aggregator distributes them to all the filters deployed in the source systems to update their inner data structures. To measure the time required for the rules to take effect, the system time of hosts is synchronized to the same NTP (network time protocol) server in the public cloud. The average time and standard deviation are the result of updating 250,000 rules, which verifies the real-time performance of Roda in applying new aggregation rules. On average, Roda takes 1.498 ms to update an aggregation rule.

6   Conclusion

In this paper, we propose an effective data aggregation framework named Roda, which takes into account the dynamic joining of new data sources, the immediate updating of aggregation rules and the quick adaptation to data velocity. We also implement Roda based on Kafka, Docker and the overlay network. To evaluate the effectiveness and performance of Roda, we conduct experiments based on real-world stock and telecommunication data. The experiment results show that Roda can guarantee the data aggregation delay at the millisecond scale, easily achieving our design goals.

Acknowledgments. This work was supported by National Key R&D Program of China (2017YFC0803700), the National Science Foundation of China (61772334, 61702151), and the Joint Key Project of the National Natural Science Foundation of China (U1736207).

References

1. Docker Swarm Overview. https://docs.docker.com/engine/swarm/
2. Kubernetes Github. https://github.com/kubernetes
3. n2n github. https://github.com/ntop/n2n
4. Overlay network. https://en.wikipedia.org/wiki/Overlay_network
5. Carzaniga, A., Wolf, A.L.: Forwarding in a content-based network. In: ACM SIG-COMM, pp. 163–174. ACM (2003)

6. Chandramouli, B., Yang, J., Agarwal, P., Yu, A., Zheng, Y.: ProSem: scalable wide-area publish/subscribe. In: ACM SIGMOD, pp. 1315–1318 (2008)
7. Fan, W., Liu, Y., Tang, B.: GEM: an analytic geometrical approach to fast event matching for multi-dimensional content-based publish/subscribe services. In: IEEE INFOCOM, pp. 1–9 (2016)
8. Farvaresh, H., Sepehri, M.M.: A data mining framework for detecting subscription fraud in telecommunication. Eng. Appl. Artif. Intell. **24**(1), 182–194 (2011)
9. Guijarro, L.: Semantic interoperability in eGovernment initiatives. Comput. Stand. Interfaces **31**, 174–180 (2009)
10. Homann, U., Rill, M., Wimmer, A.: Flexible value structures - in banking. Commun. ACM **47**, 34–36 (2004)
11. Janssen, M., Estevez, E., Janowski, T.: Interoperability in big, open, and linked data-organizational maturity, capabilities, and data portfolios. IEEE Comput. **47**(10), 44–49 (2014)
12. Liao, Z., et al.: PhSIH: a lightweight parallelization of event matching in content-based pub/sub systems. In: Proceedings of the 48th International Conference on Parallel Processing (ICPP), pp. 1–10 (2019)
13. Llinas, J., Bowman, C., Rogova, G., Steinberg, A., Waltz, E., White, F.: Revisions and extensions to the JDL data fusion model II. In: Proceedings of the 7th International Conference on Information Fusion, pp. 1218–1230 (2004)
14. Llinas, J., Hall, D.L.: An introduction to multi-sensor data fusion. In: IEEE International Symposium on Circuits & Systems (1998)
15. Pesonen, L.I.W., Eyers, D.M., Bacon, J.: A capability-based access control architecture for multi-domain publish/subscribe systems. In: International Symposium on Applications and the Internet (SAINT), pp. 222–228 (2006)
16. Pesonen, L.I.W., Eyers, D.M., Bacon, J.: Encryption-enforced access control in dynamic multi-domain publish/subscribe networks. In: ACM DEBS, pp. 104–115 (2007)
17. Qian, S., et al.: A fast and anti-matchability matching algorithm for content-based publish/subscribe systems. Comput. Netw. **149**, 213–225 (2019)
18. Qian, S., Cao, J., Zhu, Y., Li, M.: REIN: a fast event matching approach for content-based publish/subscribe systems. In: IEEE INFOCOM, pp. 2058–2066. IEEE (2014)
19. Qian, S., Cao, J., Zhu, Y., Li, M., Wang, J.: H-tree: an efficient index structure for event matching in content-based publish/subscribe systems. IEEE Trans. Parallel Distrib. Syst. **26**(6), 1622–1632 (2015)
20. Qian, S., Mao, W., Cao, J., Le Mouël, F., Li, M.: Adjusting matching algorithm to adapt to workload fluctuations in content-based publish/subscribe systems. In: IEEE INFOCOM, pp. 1936–1944 (2019)
21. Qiu, H., Noura, H., Qiu, M., Ming, Z., Memmi, G.: A user-centric data protection method for cloud storage based on invertible DWT. IEEE Trans. Cloud Comput. (2019, early access)
22. Sadoghi, M., Jacobsen, H.A.: BE-tree: an index structure to efficiently match Boolean expressions over high-dimensional discrete space. In: ACM SIGMOD, pp. 637–648 (2011)
23. Singh, J., Eyers, D., Bacon, J.: Disclosure control in multi-domain publish/subscribe systems. In: ACM DEBS, pp. 159–170 (2011)
24. Tariq, M.A., Koldehofe, B., Bhowmik, S., Rothermel, K.: PLEROMA: a SDN-based high performance publish/subscribe middleware. In: Proceedings of the 15th International Middleware Conference, pp. 217–228 (2014)

25. Visala, K., Lagutin, D., Tarkoma, S.: Security design for an inter-domain publish/subscribe architecture. In: The Future Internet - Future Internet Assembly 2011: Achievements and Technological Promises, pp. 167–176 (2011)
26. Wang, C., Carzaniga, A., Evans, D., Wolf, A.L.: Security issues and requirements for internet-scale publish-subscribe systems. In: Proceedings of the 35th Annual Hawaii International Conference on System Sciences, pp. 3940–3947. IEEE (2002)
27. White, F.E.: Data fusion lexicon. Technical report. Joint Directors of Labs, Washington DC (1991)
28. Zhang, D., Chan, C.Y., Tan, K.L.: An efficient publish/subscribe index for e-commerce databases. Proc. VLDB Endow. **7**(8), 613–624 (2014)
29. Zhao, Y., Wu, J.: Towards approximate event processing in a large-scale content-based network. In: IEEE ICDCS, pp. 790–799 (2011)

Structured Data Encoder for Neural Networks Based on Gradient Boosting Decision Tree

Wenhui Hu[1], Xueyang Liu[1(✉)], Yu Huang[1], Yu Wang[2], Minghui Zhang[3], and Hui Zhao[4]

[1] National Engineering Research Center for Software Engineering, Peking University, Beijing, China
liuxueyang@pku.edu.cn
[2] ChinaSoft International Co., Ltd., Hong Kong, China
[3] Handan Institute of Innovation, Peking University, Handan, Beijing, China
[4] School of Software, Henan University, Kaifeng, China

Abstract. Features are very important for machine learning tasks, therefore, feature engineering has been widely adopted to obtain effective handcrafted features, which is, however, labor-intensive and in need of expert knowledge. Therefore, feature learning using neural networks has been used to obviate the need of manual feature engineering and achieved great successes in the image and sequential data processing. However, its performance in processing structured data is usually unsatisfactory. In order to tackle this problem and learn good feature representations for structured data, in this work, we propose a structured data encoder (SDE) based on Gradient Boost Decision Tree (GBDT) to learn feature representations from structured data both effectively and efficiently. Then, PCA is further employed to extract the most useful information and to reduce the dimensionality for the following classification or regression tasks. Extensive experimental studies have been conducted to show the superior performances of the proposed SDE solution in learning representations of structured data.

Keywords: Feature learning · Structured data encoder · Gradient boosting decision tree · Neural network

1 Introduction

Features engineering are normally very important for machine learning tasks; they have great and sometimes even crucial influence on the performances of machine learning models [1]. Consequently, feature engineering has been widely adopted to obtain effective handcrafted features in machine learning. However, feature engineering is usually difficult, time-consuming, and requires expert knowledge, which makes "applied machine learning" becomes the tasks of feature engineering in most cases [2]. Consequently, in recent years, automated feature learning has been used to obviate the need of manual feature engineering, and has achieved great successes in many research fields.

M. Qiu (Ed.): ICA3PP 2020, LNCS 12453, pp. 603–618, 2020.
https://doi.org/10.1007/978-3-030-60239-0_41

Multilayer neural networks are typical methods for feature learning, where the latten feature representations of input data are automatically learned at the hidden layers without the need of human efforts. Those latten feature representations usually can better disentangle the underlying factors of variation [3], making classification or regression more accurate. For example, convolutional neural networks (CNNs) and recurrent neural networks (RNNs) have been proved to be able to learn good feature representations from image and sequential data [4–9], respectively. The successes of CNNs and RNNS mainly come from the following reason: the input data of CNNs and RNNs usually consist of the same type of features, e.g., images are composed of pixels and text data (a classic example of sequential data) are mainly composed of characters; consequently, it is reasonable to directly map the features into a same feature space and then apply convolutional or recurrent operations to capture the spatial or ordinal dependency information as higher-level features from a large amount of "homogenous" features.

However, using neural networks to learn good feature representations from structured data is usually difficult; one reason is that structured data usually includes many categorical attributes belonging to different feature spaces, which can be seen as "heterogeneous" features and cannot be easily handled by neural networks. Table 1 shows an example of structured data with six samples and five attributes, which are all categorical except Age. Though not accurate, we can intuitively assume that we could learn higher-level features of Salary information from Age, Sex, University and Employment, and Geographical Preference information from University and Birth Place. It is obvious that the feature University is entangled with multiple factors of variation, while the high-level features Salary and Geographical Preference are obviously better disentangled representations. Performing these feature transformations by neural networks, however, is not as straightforward as processing images and text data. For structured data, we usually feed neural networks with one hot encoding or carefully designed category embedding, increasing both the computational complexity and the chance of overfitting, and inevitably degrading the quality of resulted feature.

Table 1. An example of structure data

ID	Age	Sex	Birth Place	Employment	University
1	18	Female	Shanghai	No	Fudan Univ.
2	17	Male	Beijing	No	Peking Univ.
3	33	Male	Beijing	Yes	Tsinghua Univ.
4	28	Female	Sichuan	No	Sichuan Univ.
5	45	Male	Sichuan	Yes	Sichuan Univ.
6	50	Male	Xiamen	No	Xiamen Univ.

In addition, as above-mentioned, the successes of neural networks in feature learning mainly rely on their capabilities to discover the dependency information between (or within) input data, which can be utilized to enhance the accuracy of latten feature representations. However, the dependency between attributes of structured data are usually very sparse, which thus limits the feature learning performances of neural networks.

Although neural networks seems to have strong representation performance, tree-based model like GBDT can better handle categorical variables which doesn't including normalization and time-consuming hyper-parameter tuning. Gradient Boost Decision Tree (GBDT) is the most popular machine learning algorithm which has many successful applications in many fields. Unlike neural network, tree-based methods could split the shallow feature clearly by the given laws [10], so they can handle both numerical and categorical variables uniformly. Since tree-based methods utilize their hierarchical structure to partition the feature space into a set of hyper-rectangles, they can be seen as a neural network with one hidden layer [11], with better capabilities of modeling categorical data. More importantly, ensembles of trees add a third level to the architecture which allows the model to discriminate among a number of regions exponential in the number of parameters [3]. They implicitly form a distributed representation with the output of all the trees. Each tree in an ensemble can be associated with a discrete symbol identifying the leaf/region in which the input example falls for that tree. The tuple form by identities of leaf nodes can be seen as a transformation of the input into a new space, where the statistical structure of the data and the factors of variation in it could be disentangled [11], which is also what we would like a deep architecture to capture.

Consequently, in this work, we propose a Structured Data Encoder (abbreviated as SDE) based on Gradient Boost Decision Tree (GBDT) to learn feature representations from structured data both effectively and efficiently. We utilize the leaf node information of trees in GBDT to transform heterogeneous features into the same feature space. Then PCA is employed in order to reduce the dimension of transformed features, making subsequent processing of neural networks more convenient both mathematically and computationally. The experiments results showed that compared with other representation methods, SDE has great performance on feature representations. Meanwhile, SDE is easier to train than a pure neural network model, saving a lot of tuning time.

The rest of the paper is organized as follows. Section II gives the review of related works. Section III give the preliminaries support of our proposed method and the methodology of structured data encoder, while Section IV provides the comparison experiment results using various algorithms. In the end, Section V draws conclusions of this paper.

2 Related Work

Feature representation is a set of techniques that can automatically learn useful representations needed for machine learning tasks. This replaces manual feature engineering and allows a machine to both learn the features and use them to perform a specific task. The success of machine learning algorithms generally depends on data representation [12]. In [13], the author used supervised curvilinear segmentation to removes the requirement for hand-designed features, and it outperforms state-of-the-art other methods on both 2D images and 3D images stacks. In [14], Semi-supervised feature generation was employed, and had a better performance than supervised baselines in their experimental data set. In [15], the author proposed a novel approach that using k-means to do feature representation. In [16], the author explored an original strategy for building deep neural networks, based on stacking layers of denoising autoencoders which are trained locally to denoise corrupted versions of their inputs, denoising autoencoders performance pretty well in many problems. As for [17, 18], the author used CNN and RNN to do feature representation and get remarkable result.

To solve the feature space challenges and automatical extracting useful features, autoencoder [19] has proposed to use an encoder-decoder structure. The encoder maps the input to an intermediate representation, and the decoder takes this intermediate representation and reconstructs the original input. This feature learning approach has successfully conquer the traditional issue where required large quality of semantic features which advance the advancement of neural machine translation [8, 18, 20–22]. Traditionally, autoencoder is based on neural networks, while some of those were based on decision tree [23, 24] to extract hierarchical representations. Also, encoder-decoder structure has widely used in dimensionality reduction [25, 26]. Autoencoder is a non-linear extension of traditional subspace-based methods, such as Principal Component Analysis (PCA) and Independent Component Analysis (ICA), which minimizes the difference between raw data and its reconstruction. The linear-based approaches like PCA find the direction of the largest variance in the data set, and represent each data point by coordinating along with each of these directions, which excludes sparse features and leads to weakness in text-related tasks performance.

However, although autoencoders can use unsupervised methods to extract features, they are still time-consuming, as most of them are neural network based models that require extensive work on hyperparameter tuning and large-scale dataset.

To sum up, most methods are used in image data and sequential data, besides k-means feature representation is susceptible to abnormal samples, the parameters in autoencoders are difficult to fine tune and very time-comsuming, which have great influence on the performance of denoising autoencoders. There are very few feature representation methods that suitable for non-image and non-sequential data.

3 Preliminaries and Methodology

3.1 Problem Definition

We study the use of category and continuous numerical features to predict the label of sample. Most structured data were a tabular dataset (like what had proposed in Table 2). For a given dataset with n samples and m features $\mathcal{D} = (x_i, y_i)$, the intent would be using the existed $\mathcal{D}$ to learn the corresponding mapping information to solve the case where given the feature x to predict the state y. There are many applications for types of the problem including store sales prediction; click through rate prediction and customer behavior prediction, etc.

The most popular algorithm to deal with this problem would be Gradient Boost Decision Tree (GBDT) which has strong ability to handle category and numerical features. The main cost in GBDT lies in to find the best split points of each nodes. Usually, in order to achieve better performance, well-designed handcrafted feature engineering has been introduced.

3.2 Gradient Boost Decision Tree

Gradient Boost Decision Tree (GBDT) is an ensemble model of decision trees, which are trained in sequence. Gradient boost decision tree learns the decision trees by fitting the residual errors. The derivation follows from the same idea in existing literatures in gradient boosting.

We denote $\hat{y}_i^{(t)}$ be the prediction of the i-th instance at the t-th iteration, we will need to add f_t to minimize the Eq. 1.

$$\mathcal{L}^{(t)} = \sum_{i=1}^{n} l\left(y_i, \hat{y}_i^{(t-1)} + f_t(x_i)\right) + \Omega(f_t) \tag{1}$$

$$\mathcal{L}(\phi) = \sum_i l\left(\hat{y}_i, y_i\right) + \sum_k \Omega(f_k) \tag{2}$$

$$where\ \Omega(f) = \gamma T + \frac{1}{2}\lambda\|\omega\|^2$$

Here l is a differentiable convex loss function that measures the difference between the prediction $\hat{y}_i$ and the target y_i. The second term Ω penalizes the complexity of the model (i.e., the regression tree functions). The additional regularization term helps to smooth the final learnt weights to avoid over-fitting. Intuitively, the regularized objective will tend to select a model employing simple and predictive functions. A similar regularization technique has been used in Regularized Greedy Forest (RGF) model. Our objective and the corresponding learning algorithm is simpler than RGF and easier to parallelize. When the regularization parameter is set to zero, the objective falls back to the traditional gradient tree boosting. We thus utilize the spatial topology information of the GBDT.

3.3 Structured Data Encoder (SDE) Based on GBDT and PCA

In this paper, GBDT and PCA are used as the raw data invert unit, this process is called Structured Data Encoder (SDE). SDE is the most important part in our proposed method. Suppose we have a classification or regression task with structured data, like what had proposed in Table 2, there many different categorical and numbrical features, and also has labels that we what to predict. First, we employ GBDT to modeling the input data, that is, by adjusting the parameters, GBDT using the features to predict the labels. After that, instead of predicting the labels directly, for each instance (or sample), GBDT will get the position data $p \in \mathbb{R}^j$, also called the Leaf Node data from GBDT. Which denotes the position information that predicted by every decision tree in the model, where j is the number of trees in GBDT. Because P is the position information from every decision tree for every instance (or sample), the values in P are categorical features, it is not suitable to fed P into neural network for prediction task directly, so one-hot encoding is employed to handle this problem. Finally, PCA is used to deal with the curse of dimensionality.

The process of the SED includes the following three steps.

1. Transforming the input data into the Leaf Node data using GBDT.

If the data set has n rows, for our GBDT model, we set the number of leaves to i, and the number of boosting round to j, so we can get a GBDT model with j trees, and each tree has at most i leaf nodes.

For every instance, in prediction stage, using this model to predict the leaf node index information. All the decision trees in GBDT model will get its prediction, because the GBDT model has j trees, so we will get j predictions, the values range of the predictions are $[0, i-1]$, so we will get the position data $p \in \mathbb{R}^j$ for every instance.

For all input data, we will get a matrix $p \in \mathbb{R}^{n*j}$, with n rows and j columns, in each column, the range of values are from 0 to $i-1$, and the type of these values are all integer. As denoted as follows:

$$P = [p_1, p_2, p_3, \ldots, p_n] \tag{3}$$

2. Using One-Hot Encoding to process Leaf Node data.

The values in Leaf_Node data P are all categorical features, so it's not suitable to use them directly in many machine learning algorithms, such as neural network. So One-Hot Encoding is employed to solve this problem. For every instance, the position data p has j columns, and the values in p is from 0 to $i-1$, after one-hot Encoding, we get the new position data $\hat{p} \in \mathbb{R}^{(i*j)}$. For all input data, we will get a new matrix $\hat{P} \in \mathbb{R}^{n*(i*j)}$, with n rows and $i*j$ columns, that is:

$$\hat{p} = One\_HotEnding(p) \tag{4}$$

$$\hat{P} = One\_HotEnding(P) \tag{5}$$

3. Using PCA to get the most important information and reduce the dimension of transformed features.

It is also inappropriately to use $\hat{P}$ directly in many machine learning algorithms, because $\hat{P}$ is very spares, and the dimension is too large. So we using PCA to keep the most useful information and reduce the dimension. If the components number in PCA is m, then for each instance, we will get a new position data $\bar{p}$ which dimension is m, and for all input data, a new position data $\bar{P}$ is also generated, the dimension is $n*m$. $\bar{P}$ is the finally representation that we obtain from SDE. Notable, the components number in PCA is a hyper-parameter that we can adjust.

$$\bar{p} = PCA(\hat{p}) \tag{6}$$

$$\bar{P} = PCA(\hat{P}) \tag{7}$$

3.4 Case Analysis

This section would give a case analysis based on the data in the Table 2 to illustrate our proposed model.

The definition of Gini index would be:

Table 2. Dataset Information

ID	Age	Employment	Housing	Credit	Type
1	Teenager	No	No	Bad	No
2	Teenager	Yes	No	Medium	Yes
3	Middle Aged	No	No	Medium	No
4	Middle Ahed	No	Yes	Good	Yes
5	Elderly	Yes	No	Medium	Yes
6	Elderly	No	No	Bad	No

$$G(t) = \sum_{j=1}^{k} p_j (1 - p_j) = 1 - \sum_{j=1}^{k} p^2(j|t)$$

We denote t as the node and k as the number of categories of output variables; $p(j|t)$ is the normalized probability of the sample output variable in node t taking the j-th class:

$$p(j|t) = \frac{p(j, t)}{\sum_j p(j, t)}, p(j, t) = \frac{N_{j,t}}{N_j}$$

$p(j, t)$ is the probability that the node t contains j-th class; $N_{j,t}$ is the sample size of node t that containing the j-th class; N_j is the probability of the output of the entire sample variable where the value is j-th class. Normalization is performed to make the probabilities of each leaf node comparable.

It's obvious that the difference in output variable values is minimal when the output variables of the node samples are all grouped into the same category, in which the Gini Index equal to zero. On the contrary, when the probability of each type of output is the same, the value of the output variable is the most different.

The classification tree measures the degree of heterogeneity degradation by using the reduction of the Gini Index.

$$\Delta G(t) = G(t) - \frac{N_r}{N} G(t_r) - \frac{N_l}{N} G(t_l)$$

the $G(t)$ and N are denoted as the Gini index and sample size of the output variable before grouping. The optimal grouping variable and optimal split point would be the variables and split point where maximize the $\Delta G(t)$.

When the category is binary, the Gini Index would has this characteristic:

$$Gini(Type, Binary\ Category = True) = Gini(Type, Bininary\ Category = False)\quad (8)$$

1. The first split Step

    a. Calculate the Gini Index of the Age category.

$$Gini(Type, Age = Teenager) = \frac{|D_r|}{|D|}Gini(D_r) + \frac{|D_l|}{|D|}Gini(D_l)$$

$$= \frac{2}{6}\left(1 - \sum_{i=1}^{k} p_i^2\right) + \frac{4}{6}\left(1 - \sum_{j=1}^{k} q_j^2\right)$$

$$= \frac{1}{3}\left[1 - \left(\frac{1}{2}\right)^2 - \left(\frac{1}{2}\right)^2\right] + \frac{2}{3}\left[1 - \left(\frac{1}{2}\right)^2 - \left(\frac{1}{2}\right)^2\right] = \frac{1}{2}$$

$$Gini(Type, Age = MiddleAge) = \frac{1}{2}$$

$$Gini(Type, Age = Elderly) = \frac{1}{2}$$

    b. Calculate the Gini Index of Employment category.
       Because this is a binary category, based on Eq. 8

$$Gini(Type, Employment = Yes) = Gini(Type, Employment = False)$$

$$Gini(Type, Employmrnt = No) = \frac{|D_r|}{|D|}Gini(D_r) + \frac{|D_l|}{|D|}Gini(D_l)$$

$$= \frac{4}{6}\left(1 - \sum_{i=1}^{k} p_i^2\right) + \frac{2}{6}\left(1 - \sum_{j=1}^{k} q_j^2\right)$$

$$= \frac{2}{3}\left[1 - \left(\frac{3}{4}\right)^2 - \left(\frac{1}{4}\right)^2\right] + \frac{1}{3}(1 - 1 - 0) = \frac{1}{4}$$

    c. Calculate the Gini Index of the Housing category.
       Based on Eq. 8, we only need to calculate one value.

$$Gini(Type, Housing = No) = \frac{|D_r|}{|D|}Gini(D_r) + \frac{|D_l|}{|D|}Gini(D_l)$$

$$= \frac{5}{6}\left[1 - \left(\frac{2}{5}\right)^2 - \left(\frac{3}{5}\right)^2\right] + \frac{1}{6}(1 - 1 - 0) = \frac{2}{5}$$

    d. Calculate the Gini Index of the *Credit ψ* category.

$$ini(Type, Credit = Bad) = \frac{2}{6}(1 - 1 - 0) + \frac{4}{6}\left[1 - \left(\frac{3}{4}\right)^2 - \left(\frac{1}{4}\right)^2\right] = \frac{1}{4}$$

$$Gini(Type, Credit = Medium) = \frac{1}{2}\left[1 - \left(\frac{1}{3}\right)^2 - \left(\frac{2}{3}\right)^2\right] + \frac{1}{2}\left[1 - \left(\frac{1}{3}\right)^2 - \left(\frac{2}{3}\right)^2\right] = \frac{4}{9}$$

$$Gini(Type, Credit = Good) = \frac{1}{6}(1 - 1 - 0) + \frac{5}{6}\left[1 - \left(\frac{3}{5}\right)^2 - \left(\frac{2}{5}\right)^2\right] = \frac{2}{5}$$

Because the smallest Gini Index of all category Employment is 16, so the first feature that considered to be split is Employment.
After first split, as Fig. 1 shows, the data are separated into two parts.

2. Second Split Step 1

    a. After the first split, we only have three categories (Age, Housing, Credit)
       Calculate the Gini Index of the Age category.

$$Gini(Type, Age = Teenager) = \frac{1}{4}(1 - 1 - 0) + \frac{3}{4}\left[1 - \left(\frac{2}{3}\right)^2 - \left(\frac{1}{3}\right)^2\right] = \frac{1}{3}$$

$$Gini(Type, Age = MediumAge) = \frac{1}{2}\left[1 - \left(\frac{1}{2}\right)^2 - \left(\frac{1}{2}\right)^2\right] + \frac{1}{2}(1 - 1 - 0) = \frac{1}{4}$$

$$Gini(Type, Age = Elderly) = \frac{1}{4}(1 - 1 - 0) + \frac{3}{4}\left[1 - \left(\frac{2}{3}\right)^2 - \left(\frac{1}{3}\right)^2\right] = \frac{1}{3}$$

Example ID corresponding to the ID shown in Table 1, Classes(2,0) means there are two examples in this leaf, the type of them are two Yes and zero No.

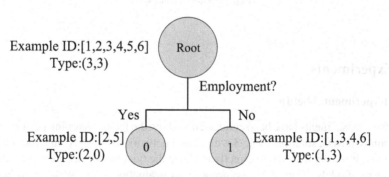

Fig. 1. After first split

    b. Calculate the Gini Index of the Housing category. Based on Eq. 8, we only need
       to calculate one value.

$$Gini(Type, Housing = No) = \frac{3}{4}(1 - 1 - 0) + \frac{1}{4}(1 - 1 - 0) = 0$$

c. Calculate the Gini Index of the Credit category.

$$Gini(Type, Credit = Bad) = \frac{1}{2}(1 - 1 - 0) + \frac{1}{2}\left[1 - \left(\frac{1}{2}\right)^2 - \left(\frac{1}{2}\right)^2\right] = \frac{1}{4}$$

$$Gini(Type, Credit = Meduum) = \frac{1}{4}(1 - 1 - 0) + \frac{3}{4}\left[1 - \left(\frac{2}{3}\right)^2 - \left(\frac{1}{3}\right)^2\right] = \frac{1}{2}$$

$$Gini(Type, Credit = Good) = \frac{1}{4}(1 - 1 - 0) + \frac{3}{4}(1 - 1 - 0) = 0$$

After second split, the data are separated as Fig. 2 shows.

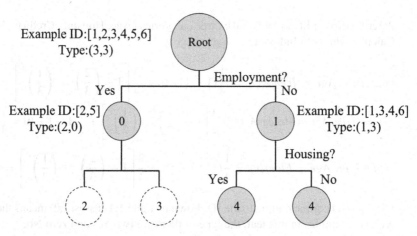

Fig. 2. After second split

4 Experiments

4.1 Experimental Setup

Extensive experiments have been conducted to show the feature learning performance of our proposed SDE solution Table 3 with neural network (SDE+NN) in structured data. Generally, it is very difficult to directly evaluate the qualities of feature representations learned by models. Therefore, we propose an evaluation framework to evaluate the quality of feature representations indirectly, where the classification accuracy of a model is used as indirect metric of the performances of its underlying feature learning module.

Consequently, we create a hybrid classification model, SDE+NN, which use the proposed SDE as the feature learning module and use multilayer neural networks as the classification module. SDE+NN is then comparing its classification performances with four baselines, GBDT, fully-connected neural networks (NN), fully-connected neural networks with batch normalization (NN+BN), autoencoder, on various structured datasets.

Some important training details are as follows and summarized in Table 3: (i) In GBDT, the the maximum number of leaves is 7, the learning rate to be *0.07*, the sample random sampling rate and feature random sampling rate is *0.8*. (ii) In Neural Networks, we have trained a *32 neurals* in the networks, *Dropout = 0.1*. The neural networks are learned using the Adam optimizer with learning rate $\alpha = 0.0003$. The activation function we use is *Relu*. (iii) The AutoEncoder use *300 hidden layers* and trained for *500 epochs*. The neural networks are learned using the SGD optimizer with learning rate $\alpha = 0.001$. The loss function that we used is MSE. Since our proposed method utilizes both neural networks and GBDT, we use the same hyperparameters when we choose the neural network model and GBDT for comparative experiments. Comparing the feature representation performance of AutoEncoder, the dimension reduction of AutoEncoder is identical to the dimension of PCA in SDE, and the configurations of the neural network in the model are also identical to other neural networks. In these experiments, we perform a five-fold cross validation and report the average of results. Please note that no feature engineering is conducted in this experimental study.

Table 3. Parameters of Experiment Model

Model	Category	Value
GBDT	Maximum Number of Leaves	7
	Learning Rate	0.07
	Sample Random Sampling rate	0.8
	Feature Random Sampling rate	0.8
	PCA Dimension	300
Neural networks	Num of Neurals	32
	Dropout	0.1
	Optimizer	Adam
	Learning Rate	0.0003
	Activiation Function	Relu
Auto-encoder	Num of Features in Hidden Layers	300
	Num of Epochs	500
	Activiation Function	Relu
	Optimizer	SGD
	Learning Rate	0.001
	Loss Function	MSE

4.2 Datasets

To simulate a variety of real-world machine learning scenarios as much as possible, three datasets, such as the size of the dataset, the richness of the features, and so on, were selected on the basis of careful consideration of a number of factors. The statistic information of the datasets is shown in Table 4, and the detailed descriptions are as follows:

Census Income Dataset[1]. The task is to predict whether income exceeds $50 K/yr based on census data. This dataset usually considered as unbiased data for machine learning tasks [27] for the reason that the data is clean by a strict laws.

Covertype Data Set[2]. The task is to predict the type of forest cover only from cartographic variables with total of 581,012 observations encompassing 52,291 hectares of the study area. Each observation contained information for every independent and response variable used in the study. This dataset is chosen because it has large amount of categorical features and the sample capacity is large enough to support the models.

Online Shoppers' Purchasing Intention Dataset Data Set[3]. In this dataset, 84.5% (10,422) were negative class samples that did not end with shopping, and the rest (1908) were positive class samples ending with shopping. This is also the only dataset with a label imbalance in all of the three datasets. Moreover, the dataset contains fewer categorical features than the previous two datasets, and the sample capacity is smaller than the previous two datasets.

Table 4. The Datasets Used Experiments

Dataset	Size	Feature	Feature types	Num. of label
Census income	32,561	14	Categorical(8), Numerical(6)	2
Covertype	581,012	54	Categorical(44), Numerical(10)	7
Shoppers	581,012	17	Categorical(7), Numerical(10)	2

4.3 Results

Since SDE+NN is a hybrid model based on GBDT and NN, we first compare the classification performances of SDE+NN with those of GBDT, NN, and NN+BN on three datasets in terms of classification accuracy and standard deviation. To keep fair comparison, the neural networks used in SDE+NN, NN, and NN+BN has the identical parameter settings. The experimental results are shown in Figs. 3, 4, 5, 6, 7 and 8.

In Figs. 3, 4, 5, 6, 7 and 8, it can be easily observed from the experimental results that the classification performance of the SDE-based hybrid classification model, SDE+NN, is much better than the baselines, GBDT, NN, and NN+BN, on all three datasets and in terms of both classification accuracy and standard deviation. Therefore, we have the following conclusions: (1) comparing to those NN and NN+BN, the superior performances of SDE+NN indicates that our proposed method, SDE, has very good feature

---

[1] http://archive.ics.uci.edu/ml/datasets/Adult.

[2] http://archive.ics.uci.edu/ml/datasets/Covertype.

[3] http://archive.ics.uci.edu/ml/datasets/Online+Shoppers%27+Purchasing+Intention+Dataset.

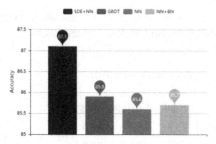

Fig. 3. Accuracy of census income

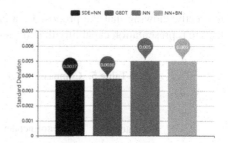

Fig. 4. Standard deviation of census income

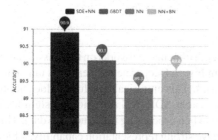

Fig. 5. Accuracy of shopper

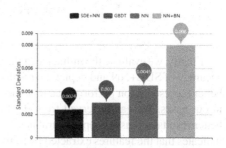

Fig. 6. Standard deviation of shopper

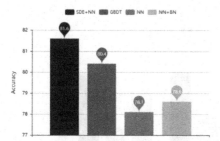

Fig. 7. Accuracy of covertype

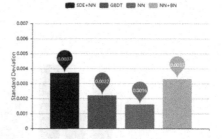

Fig. 8. Standard deviation of covertype

learning capability, resulting in great enhance of classification performances using neural networks. (2) Comparing to GBDT, the advantage of SDE comes from its flexible feature representations, making it easy to reuse its extracted features to achieve better performances.

In addition, we further evaluate the feature learning performance of SDE with the state-of-the-art deep-structure-based feature learning solution, autoencoders. To keep fair comparison, after obtaining the feature representations using autoencoders, a neural network with the same setting as that used in SDE+NN is utilized to finish the classification tasks, the resulting hybrid model is denoted as *AutoEncoder+NN*. The comparison results are shown in Table 5. As shown in Table 5, under the same setting, SDE+NN has much better feature classification performance than *AutoEncoder+NN*, indirectly showing that the proposed feature learning solution, SDE, has much better feature representation capability than the state-of-the-art baseline, autoencoders.

Table 5. Compassion between AutoEncoder and SDE

Model		Adult	Covtype	Shoppers
Accuracy	AutoEncoder+NN	85.3	78.7	88.3
	SDE+NN	87.1	86.1	90.9
Standard deviation	AutoEncoder+NN	5.2e-3	3.8e-3	4.1e-3
	SDE+NN	3.7e-3	3.7e-3	2.4e-3

Finally, we evaluate the influence of the depth of neural networks on the classification accuracy of SDE+NN and depict the results in Fig. 9. Deeper neural networks may have better representation performance in some cases. This empirical conclusion does not fit in our cases. We can observed from the Fig. 9 that SDE+NN does not get significant better results as it uses deeper neural networks when processing the structured data. This indicates that the features extracted by SDE have strong representation capability, which do not require multilayer neural networks. With SDE, single-layer neural networks can achieve satisfactory performance.

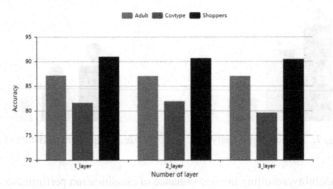

Fig. 9. The influence of different depth neural networks on accuracy in SDE

5   Conclusion

A novel method for feature learning on structured data, Structured Data Encoder (SDE), was proposed in this work, where Gradient Boost Decision Tree (GBDT) was used to transform heterogeneous features of structured data into the same feature space. This tree-based model solved the entanglement state of features in structured data effectively and efficiently by utilizing the spatial topology of tree-based model. In addition, PCA was further employed to reduce the dimension of transformed features, making subsequent processing of neural networks more convenient both mathematically and computationally. The experiments results showed that SDE has achieved much better performance than the baselines in feature learning accuracy on structured data. Moreover, by applying SDE to first transform and learn the features, the learning efficiency of the hybrid classification model, SDE+NN, is much better than conventional neural networks. In future works, we will integrate SDE on more real world structured datasets to investigate its performances and robustness on various practical situations. In addition, we will also investigate how to further improve the feature learning performances of SDE by importing more optimization methods.

Acknowledge. National Key R&D Program of China (No. 2017YFB1103003)

PKUSER-CRHMS Medical AI Co-Lab

National Science and Technology Major Project for IND (investigational new drug) 2018ZX09201-014

References

1. Domingos, P.: A few useful things to know about machine learning. Commun. ACM **55**(10), 78–87 (2012)
2. Ng, A.: Machine learning and ai via brain simulations (2013)
3. Bengio, Y., Delalleau, O., Simard, C.: Decision trees do not generaliz to new variations. Comput. Intell. **26**(4), 449–467 (2010)
4. Krizhevsky, A., Sutskever, I., Hinton, G.E.: Imagenet classification with deep convolutional neural networks. In: Advances in Neural Information Processing Systems, pp. 1097–1105 (2012)
5. Szegedy, C., Liu, W., Jia, Y., Sermanet, P., Reed, S., Anguelov, D., Erhan, D., Vanhoucke, V., Rabinovich, A.: Going deeper with convolutions. In: Proceedings of the IEEE Conference on Computer Vision and Pattern Recognition, pp. 1–9 (2015)
6. Simonyan, K., Zisserman, A.: Very deep convolutional networks for large-scale image recognition (2014). arXiv preprint arXiv:1409.1556
7. Mikolov, T., Karafiát, M., Burget, L., Cernocký,, J., Khudanpur, S.: Recurrent neural network based language model. In: Eleventh Annual Conference of the International Speech Communication Association (2010)
8. Sutskever, I., Vinyals, O., Le, Q.V.: Sequence to sequence learning with neural networks. In: Advances in Neural Information Processing Systems, pp. 3104–3112 (2014)
9. Chung, J., Gulcehre, C., Cho, K., Bengio, Y.: Empirical evaluation of gated recurrent neural networks on sequence modeling (2014). arXiv preprint arXiv:1412.3555
10. Ripley, B.D.: Pattern Recognition and Neural Networks. Cambridge University Press (2007)
11. Bengio, Y., et al.: Learning deep architectures for ai. Found. Trends® Mach. Learn. **2**(1), 1–127 (2009)
12. Bengio, Y., Courville, A., Vincent, P.: Representation learning: a review and new perspectives. IEEE Trans. Patt. Anal. Mach. Intell. **35**(8), 1798–1828 (2012)
13. Becker, C., Rigamonti, R., Lepetit, V., Fua, P.: Supervised Feature Learning for Curvilinear Structure Segmentation. Springer, Berlin Heidelberg (2013)
14. Li, Y., Hu, X., Lin, H., Yang, Z.: A framework for semisupervised feature generation and its applications in biomedical literature mining. IEEE/ACM Trans. Comput. Biol. Bioinform. **8**(2), 294–307 (2011)

15. Coates, A., Ng, Andrew Y.: Learning feature representations with K-means. In: Montavon, G., Orr, Geneviève B., Müller, K.-R. (eds.) Neural Networks: Tricks of the Trade. LNCS, vol. 7700, pp. 561–580. Springer, Heidelberg (2012). https://doi.org/10.1007/978-3-642-35289-8_30

16. Vincent, P., Larochelle, H., Lajoie, I., Bengio, Y., Manzagol, P.A.: Stacked denoising autoencoders: Learning useful representations in a deep network with a local denoising criterion. J. Mach. Learn. Res. **11**(12), 3371–3408 (2010)

17. Athiwaratkun, B., Kang, K.: Feature representation in convolutional neural networks. Comput. Sci. **32**(1), 41–45 (2015)

18. Cho, K., Van Merriënboer, B., Gulcehre, C., Bahdanau, D., Bougares, F., Schwenk, H., Bengio, Y.: Learning phrase representations using rnn encoder-decoder for statistical machine translation (2014). arXiv preprint arXiv:1406.1078

19. Bengio, Y., Yao, L., Alain, G., Vincent, P.: Generalized denoising auto-encoders as generative models. In: Advances in Neural Information Processing Systems, pp. 899–907 (2013)

20. Cho, K., Van Merriënboer, B., Bahdanau, D., Bengio, Y.: On the properties of neural machine translation: Encoder-decoder approaches (2014). arXiv preprint arXiv:1409.1259

21. Luong, M.-T., Pham, H., Manning, C.D.: Effective approaches to attention-based neural machine translation (2015) arXiv preprint arXiv:1508.04025

22. Wu, Y., Schuster, M., Chen, Z., Le, Q.V., Norouzi, M., Macherey, W., Krikun, M., Cao, Y., Gao, Q., Macherey, K., et al.: Google's neural machine translation system: Bridging the gap between human and machine translation (2016). arXiv preprint arXiv:1609.08144

23. Irsoy, O., Alpaydın, E.: Unsupervised feature extraction with autoencoder trees. Neurocomputing **258**, 63–73 (2017)

24. Feng, J, Zhou, Z.-H.: Autoencoder by forest (2017). arXiv preprint arXiv:1709.09018

25. Hinton, G.E., Salakhutdinov, R.R.: Reducing the dimensionality of data with neural networks. Science **313**(5786), 504–507 (2006)

26. Hinton, G.E., Osindero, S., Teh, Y.-W.: A fast learning algorithm for deep belief nets. Neural Comput. **18**(7), 1527–1554 (2006)

27. Zadrozny, B.: Learning and evaluating classifiers under sample selection bias. In: Proceedings of the Twenty-First International Conference on Machine Learning. ACM, p. 114 (2004)

Stochastic Model-Based Quantitative Analysis of Edge UPF Service Dependability

Haoran Zhu[1], Jing Bai[1], Xiaolin Chang[1(✉)], Jelena Mišić[2], Vojislav Mišić[2], and Yang Yang[1]

[1] Beijing Key Laboratory of Security and Privacy in Intelligent Transportation, Beijing Jiaotong University, Beijing 100044, China
{16281060,19112047,xlchang,16112082}@bjtu.edu.cn
[2] Computer Science Department, Ryerson University, Toronto, Canada
{jmisic,vmisic}@ryerson.ca

Abstract. Multi-Access Edge Computing (MEC), making computing resources extend to the edge continuously, plays a key role in 5G (Fifth Generation) networks. User Plane Function (UPF), as an important network function, can steer the traffic to the MEC applications and 5G core network in the integration of MEC and 5G architecture. Edge UPF (EUPF) runs in a virtual machine (VM) hosted by a virtual machine monitor (VMM) at the edge of the network. However, EUPF is likely to suffer more frequent network attacks than in the core network. The vulnerability of the virtualized environment itself also expands the attack surface for adversaries. Security protection of EUPF is urgently needed. Rejuvenation techniques play a key role in defense and service recovery. The existing researches focused on the deployment of EUPF and ignored the dependability analysis of EUPF service.

In this paper, we construct the continuous time Markov model to capture the behaviors of the EUPF service system deploying rejuvenation techniques for recovery. We also derive the formulas for computing transient availability and steady-state availability of EUPF service, EUPF service's first failure time and the mean time to failure (MTTF). Finally, we find that VMM recovery ability has an obvious impact on transient and steady-state availability and VM defense ability has a great influence on MTTF. Both transient analysis and steady-state analysis of the EUPF service are carried out to help EUPF service providers provide stable and robust EUPF services.

Keywords: Dependability · Markov model · Multi-Access Edge Computing · Quantitative analysis · User plane function

1 Introduction

In recent years, Multi-Access Edge Computing (MEC) has helped computing resources extend towards the edge continuously, occupying an important position in the process of intelligence and automation for Industry 4.0, 5G (Fifth Generation) commercialization, and intelligence of everything. It also provides new experiences for delay-sensitive services such as Internet of Vehicle (IoV) and auto-driving, augmented reality and virtual reality (AR/VR), manufacturing process automation and monitoring, and intelligent

© Springer Nature Switzerland AG 2020
M. Qiu (Ed.): ICA3PP 2020, LNCS 12453, pp. 619–632, 2020.
https://doi.org/10.1007/978-3-030-60239-0_42

transportation systems. MEC is developing rapidly recently. According to Gartner, the number of IoT (Internet of Things) devices will triple from 2020 to 2021 [1]. By 2025, 75% of enterprise production data will be created and processed in the edge cloud and the global edge computing market will reach $43.4 billion by 2027 [2]. In the 5G control-/user-plane separation architecture, the User Plane Function (UPF) is an important network function of user plane [3]. It can steer the traffic to the MEC applications and 5G core network in the integration of MEC and 5G architecture, so as to optimize traffic flows and reduce service delay and transmission cost [4].

Edge UPF (EUPF) runs in a virtual machine (VM) supported by a virtual machine monitor (VMM) [5]. These virtualized infrastructures are deployed at the edge of the network near the user side, reducing network latency while exposing themselves into an environment that lacks security protection compared to 5G core network. Thus, they are more likely to be attacked by external adversaries, resulting in EUPF service dependability (availability and reliability [6]) degradation and then bringing losses to users and service providers [7]. For example, adversaries can exploit the vulnerability of virtualization software to attack the VM where EUPF service is deployed, causing the data transmission between EUPF and MEC or 5G core network to be tampered, intercepted, replayed, and others, thereby further attacking the MEC system and 5G core network. In addition, MEC has less resources than the central cloud, so adversaries may execute programs that are not related to the current EUPF service on VMs to exhaust the network, computing and storage resources of host, and then affect the normal operation of EUPF service and the speed of data forwarding. Moreover, if adversaries invade the VMM, it will pose a threat to all the VMs running on the VMM, thereby causing irreparable losses.

With the large-scale integration of MEC and key applications, security protection of EUPF is urgently needed. Rejuvenation techniques can perform protective operations before EUPF service is interrupted to improve the availability and reliability of EUPF service. The existing researches focused on the deployment of EUPF and ignored the dependability analysis of EUPF service in terms of transient and steady-state. Analytical modeling is an effective approach for evaluating the rejuvenation techniques from transient analysis and steady-state analysis. When adversaries attack EUPF service system, their capabilities may change over time, and the defense capability of the system may also be changeable at the same time. Therefore, the transient QoS (Quality of Service) of EUPF service is worthy of attention. In the long term, after EUPF service system is stable for the long-term running, it is also important to evaluate its steady-state QoS.

All the above discussions form the motivation of the work in this paper. We consider an EUPF service system composed of a host for executing EUPF service. This system deploys VM reboot, VMM reboot and failover techniques for rejuvenation. For such an EUPF service system, we explore an analytical modeling approach to quantitatively analyze EUPF service dependability. The main contributions of this paper are summarized as follows:

1) We develop continuous-time Markov chain (CTMC) models for capturing the behaviors of EUPF service system deploying VM reboot, VMM reboot and failover techniques for rejuvenation.

2) We quantitatively analyze EUPF service dependability by deriving the formulas for computing transient availability of EUPF service, steady-state availability of EUPF service, EUPF service's first failure time (namely, the first failure time from available states to unavailable states) and the mean time to failure (MTTF). Among them, EUPF service's first failure time and MTTF are used to evaluate EUPF service reliability.

3) We conduct analytical experiments for investigating the impact of parameters on the availability and reliability of EUPF service. It can help EUPF service providers make decisions to maximize their interests.

The rest of paper is organized as follows. Section 2 introduces the related work. The Markov model for transient and steady-state analysis is constructed in Sect. 3. Section 4 presents the experimental results. In Sect. 5, we summarize the most important findings and discuss the future work.

2 Related Work

The existing researches focused on the deployment of EUPF. Fattore *et al.* [8] presented a method for deploying light UPF and this method did not affect autonomy and power consumption. In this paper, we analyze quantitatively the impact of the rejuvenation techniques on EUPF dependability.

The quantitative analysis methods of evaluating the impact of rejuvenation techniques on QoS include analytical modeling approach and measurement-based approach. The analytical modeling approach are divided into state-space model and non-state-space model. Attack graph and reliability block diagram (RBD) are typical non-state-space model. Ge *et al.* [9] analyzed the change of security and performance of IoT-system deploying the defense mechanisms based on attack graph. Kwan *et al.* [10] constructed RBD to analyze the system reliability. However, it was difficult to describe the dynamic interaction between the attack behaviors of adversary and the recovery behaviors of systems because non-state-space models did not allow time-dependence. In this paper, we use CTMC model to capture the behaviors of EUPF service system with time in order to evaluate the dependability of EUPF service quantitatively.

CTMC model is the typical state-space model. Chang *et al.* [11] analyzed the impact of rejuvenation techniques on VM availability based on CTMC model. In [12] they used stochastic reward net (SRN) model to capture typical behaviors of the adversary, the vulnerable system and a job, in order to MDP technique on job completion time in a system under attack. Bai *et al.* [13] captured the aging and rejuvenation behaviors of the virtualized system by using semi-Markov process and evaluated AS availability and job completion time. Jiang *et al.* [14] analyzed the impact of failover technique on steady-state QoS when 5G-AKA authentication services faced the service failure based on CTMC. These studies only evaluated one aspect of transient analysis and steady-state analysis. Differently, we analyze the dependability analysis of EUPF service in terms of transient and steady-state in this paper.

Researchers also explored the measurement-based approach. Bovenzi *et al.* [15] studied reboot techniques in Linux-based operating system and measured performance penalty, downtime overhead reduction and the recovery coverage. We focus on the quantitative analysis of EUPF service dependability and consider the system that apply VM reboot, VMM reboot and failover technique for recovery. Their experimental results can be complementary to our work for better analyzing the dependability of EUPF service.

3   System Description and Model

This section first presents the EUPF service system considered in this paper. Then the CTMC model is constructed. Finally, we present the methods for computing transient availability of EUPF service, steady-state availability of EUPF service, EUPF service's first failure time and MTTF in Sects. 3.3 and 3.4.

3.1   System Description

Figure 1 shows the system architecture considered in this paper. The EUPF service system consists Management Host and a host, which includes the active VM with EUPF service running and backup VM supported by VMM. Management host includes the monitoring tool for monitoring the behaviors of the host with EUPF service running. Initially, EUPF service is running normally. However, adversaries may attack the hosts and we divide all kinds of attacks into two categories, namely attacks for the active VM and attacks for the VMM. The former is divided into light attack and severe attack. The light attack can lead to the decrease of EUPF service processing capacity and is able to be detected by Management host. When light attack happens, failover technique is triggered and backup VM take charge of EUPF service. The active VM is rebooted immediately. If the failover technique is not triggered immediately, the active VM crashes due to attack. At this time, the active VM is fixed and rebooted. The severe attack can lead to interruption of EUPF service. EUPF service is rebooted after the VMM is fixed and rebooted. The latter will pose a threat to all the VMs running on the VMM, thereby causing irreparable losses. When this kind of attacks occur, VMM reboots at once and then EUPF service reboots.

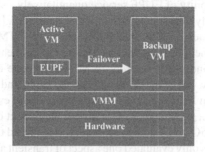

Fig. 1. System architecture

3.2 CTMC Model

The above description suggests that the system state space can be described by four states. We assume that attack time and recovery time follow exponential distributions. It is easy to know that this system model is CTMC model. Let $I = \{1, 2, 3, 4\}$ be the finite system state space of this random process. $X(t)$ represents the state of EUPF at time t. The meaning of each system state is given as follows:

1) **System State 1 (Security)**: There is no attack and the EUPF service is running normally.
2) **System State 2 (Danger)**: The EUPF can run, but the active VM with EUPF running is under attack.
3) **System State 3 (VM Failure)**: At this system state, the active VM with EUPF service running crashes due to attack. This system state is unavailable.
4) **System State 4 (VMM Failure)**: At this system state, the VMM crashes due to attack. This system state is unavailable.

Figure 2 shows the CTMC model for capturing the behaviors of the EUPF service system. Table 1 gives the definition of variables to be used in the rest of the paper. We can get the generator matrix Q and transition rate matrix R, which can be described as Eqs. (1) and (2).

$$Q = \begin{bmatrix} -(\xi_{ud} + \xi_{vb}) & \xi_{ud} & 0 & \xi_{vb} \\ \eta_{ud} & -(\eta_{ud} + \xi_{ub} + \xi_{vb}) & \xi_{ub} & \xi_{vb} \\ \eta_{ub} & 0 & -\eta_{ub} & 0 \\ \eta_{vb} & 0 & 0 & -\eta_{vb} \end{bmatrix} \tag{1}$$

$$R = \begin{bmatrix} 0 & \xi_{ud} & 0 & \xi_{vb} \\ \eta_{ud} & 0 & \xi_{ub} & \xi_{vb} \\ \eta_{ub} & 0 & 0 & 0 \\ \eta_{vb} & 0 & 0 & 0 \end{bmatrix} \tag{2}$$

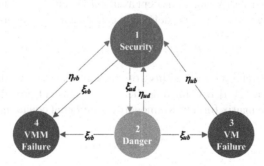

Fig. 2. CTMC model for the EUPF service system

Table 1. Definition of Variables

Notation	Definition	Value in default
$X(t)$	Stochastic Process	–
I	State Space	–
$1/\xi_{ud}$	Mean time for light attack	2 days
$1/\xi_{ub}$	Mean time for VM failure	7 days
$1/\xi_{vb}$	Mean time for VMM failure	30 days
$1/\eta_{ud}$	Mean time for failover	1 day
$1/\eta_{ub}$	Mean time for VM rejuvenation	6 h
$1/\eta_{vb}$	Mean time for VMM rejuvenation	1 day
Q	Infinitesimal generator matrix	–
R	Transition rate matrix	–
$\pi_i(t)$	Transient probability of system state i at time t	–
$\pi_a(t)$	Transient availability at time t	–
$\pi(t)$	Transient probability vector	–
$T_{i \to j}$	First passage time from state i to state j	–
π_i	Steady-state probability for state i	–
π_a	Steady-state availability	–
$L_i(t)$	Residence time in state i	–
$\widehat{Q}$	Generator matrix with available states	–
$\widehat{\pi}(t)$	Transient probability vector with available states	–

3.3 Transient Analysis

For EUPF service systems that may be attacked by adversary and can be recovered in real time, transient analysis is one of the most important evaluation dimensions. Before the system reaches a steady-state, transient availability could be an important reference for service providers. The Eq. (3) is defined to calculate the transient availability.

$$\pi_i(t) = P\{X(t) = i\}, i \in \{1, 2, 3, 4\}, t > 0 \tag{3}$$

$\pi_i(t)$ is the transient probability of system state i at time t. $\pi(t) = [\pi_1(t), \pi_2(t), \pi_3(t), \pi_4(t)]$ is defined as the system transient probability vector. $\pi(t)$ is calculated by the matrix form of Kolmogorov differential equations (Eq. (4)).

$$\frac{d\pi(t)}{dt} = \pi(t)Q \tag{4}$$

Then we can obtain the general equation of calculating $\pi(t)$ as follows:

$$\pi(t) = \pi(0)e^{Qt} \tag{5}$$

where $\pi(0)$ is the initial probability vector. The initial probability vector of the system is $\pi(0) = [1, 0, 0, 0]$ because there is no attack at the beginning. Then we can get the transient availability $\pi_a(t)$ of EUPF service at time t (Eq. (6)).

$$\pi_a(t) = \pi_1(t) + \pi_2(t) \tag{6}$$

It is difficult to calculate transient probability directly by Eq. (4). Generally speaking, the methods of calculating transient probability include matrix exponent method [16] and uniformization method [17]. The matrix exponent method directly multiplies the matrix, which has a large amount of calculation and low efficiency, and is prone to errors during the calculation process. Although the uniformization method simplifies the former to a certain extent, the different values of the uniform rate parameter q will finally cause deviations. In addition, when transient probability is calculated at different time, the selection of the censored number M is also difficult to unify. Choosing a small value of M may lead to an inaccurate result when t is large. Choosing a large value of M may cause a lot of useless calculations when t is small. In this paper, we use transient differential equations [18] to calculate the transient probability. This method can get accurate calculation results. According to the generator matrix Q, we can get the transient probability differential equation for each system state:

$$\begin{cases} \frac{d}{dx}\pi_1(t) = -(\xi_{ud} + \xi_{vb})\pi_1(t) + \eta_{ud}\pi_2(t) \\ \qquad\qquad + \eta_{ub}\pi_3(t) + \eta_{vb}\pi_4(t) \\ \frac{d}{dx}\pi_2(t) = \xi_{ud}\pi_1(t) - (\eta_{ud} + \xi_{ub} + \xi_{vb})\pi_2(t) \\ \frac{d}{dx}\pi_3(t) = \xi_{ub}\pi_2(t) - \eta_{ub}\pi_3(t) \\ \frac{d}{dx}\pi_4(t) = \xi_{vb}\pi_1(t) + \xi_{vb}\pi_2(t) - \eta_{vb}\pi_4(t) \end{cases} \tag{7}$$

The transient probability and transient availability of EUPF service at any time can be quickly calculated by Eq. (7).

3.4 Steady-State Analysis

Steady-state analysis focuses on the long-term stable behavior of EUPF service system. This section presents the methods for computing steady-state availability of EUPF service, EUPF service's first failure time and MTTF in Sects. 3.4.1, 3.4.2 and 3.4.3, respectively.

3.4.1 Steady-State Availability of EUPF Service and Steady-State Probability In order to calculate the steady-state availability, we can first calculate the steady-state probability of each state, and then add the probabilities of all available states. The methods of calculating steady-state probability include power method, Successive Overrelaxation (SOR) and so on. These methods are suitable for the case where the matrix is relatively sparse and the state space is large. However, in this paper, the transition rate matrix has only four dimensions and actually not sparse, so we can solve the linear equations directly to calculate steady-state availability of EUPF service and steady-state probability. The steady-state probability of state i is defined as follows:

$$\pi_i = \lim_{t \to \infty} \pi_i(t), i \in I \tag{8}$$

If the limit exists, then

$$\lim_{t \to \infty} \frac{d\pi_i}{dt} = 0 \tag{9}$$

The steady-state probability distribution vector is $\pi = [\pi_1, \pi_2, \pi_3, \pi_4]$. We can get the Eq. (10) by solving Eq. (4) and (9).

$$\begin{cases} \pi Q = 0 \\ \sum_{i=1}^{4} \pi_i = 1 \end{cases} \tag{10}$$

Then we can obtain the following linear equation:

$$\begin{bmatrix} \xi_{ud} + \xi_{vb} & -\eta_{ud} & -\eta_{ub} & -\eta_{vb} \\ -\xi_{ud} & \eta_{ud} + \xi_{ub} + \xi_{vb} & 0 & 0 \\ 0 & -\xi_{ub} & -\eta_{ub} & 0 \\ -\xi_{vb} & -\xi_{vb} & 0 & \eta_{vb} \\ 1 & 1 & 1 & 1 \end{bmatrix} \pi = \begin{bmatrix} 0 \\ 0 \\ 0 \\ 0 \\ 1 \end{bmatrix} \tag{11}$$

The steady-state probability of each system state is calculated as follows:

$$\begin{cases} \pi_1 = \frac{\eta_{ub}\eta_{vb}(\eta_{ud} + \xi_{ub} + \xi_{vb})}{(\eta_{vb} + \xi_{vb})(\xi_{ub} + \xi_{ud} + \xi_{vb} + \eta_{ud})\eta_{ub} + \eta_{vb}\xi_{ub}\xi_{ud}} \\ \pi_2 = \frac{\xi_{ud}\eta_{ub}\eta_{vb}}{(\eta_{vb} + \xi_{vb})(\xi_{ub} + \xi_{ud} + \xi_{vb} + \eta_{ud})\eta_{ub} + \eta_{vb}\xi_{ub}\xi_{ud}} \\ \pi_3 = \frac{\xi_{ub}\xi_{ud}\eta_{vb}}{(\eta_{vb} + \xi_{vb})(\xi_{ub} + \xi_{ud} + \xi_{vb} + \eta_{ud})\eta_{ub} + \eta_{vb}\xi_{ub}\xi_{ud}} \\ \pi_4 = \frac{\eta_{ub}\eta_{vb}(\xi_{ub} + \xi_{ud} + \xi_{vb} + \eta_{ud})}{(\eta_{vb} + \xi_{vb})(\xi_{ub} + \xi_{ud} + \xi_{vb} + \eta_{ud})\eta_{ub} + \eta_{vb}\xi_{ub}\xi_{ud}} \end{cases} \tag{12}$$

Finally, we can calculate the steady-state availability of EUPF service (Eq. (13)).

$$\pi_a = \pi_1 + \pi_2 \tag{13}$$

3.4.2 EUPF Service's First Failure Time We can calculate EUPF service's first failure time by first passage time. First passage time is the time when the system first enters a certain state, and the average first passage time (AFPT) is its expected value. The calculation of the AFPT can be used to predict the time when the system will first enter a certain state (such as state 2) in the future, so as to make relevant preparations in advance. The AFPT $T_{i \to j}$ from state i to state j is defined as follows:

$$T_{i \to j} = E(T | X(0) = i) \tag{14}$$

For the non-absorption Markov process, the AFPT from state i to state k satisfies the following equation [17] :

$$p_i T_{i \to k} = 1 + \sum_{i \neq j} p_{ij} T_{j \to k}, k, j \in I, i \neq k \tag{15}$$

where p_{ij} is the element of the transition rate matrix R, $p_i = \sum\limits_{j=1}^{N} p_{ij}$. When $i = k$, the AFPT from state k to itself is 0, so $T_{k \to k} = 0$. The initial probability of the system is $\pi(0) = [1, 0, 0, 0]$, and the AFPT from state 1 to itself is 0. The AFPT for state 2 can be obtained according to the transition rate matrix R:

$$\begin{cases} (\xi_{ud} + \xi_{vb})T_{1 \to 2} = 1 + \xi_{vb}T_{4 \to 2} \\ \eta_{ub}T_{3 \to 2} = 1 + \eta_{ub}T_{1 \to 2} \\ \eta_{vb}T_{4 \to 2} = 1 + \eta_{vb}T_{1 \to 2} \end{cases} \tag{16}$$

The AFPT for state 3 can be obtained as follows:

$$\begin{cases} (\xi_{ud} + \xi_{vb})T_{1 \to 3} = 1 + \xi_{ud}T_{2 \to 3} + \xi_{vb}T_{4 \to 3} \\ (\xi_{ub} + \xi_{vb} + \eta_{ud})T_{2 \to 3} = 1 + \eta_{ud}T_{1 \to 3} + \xi_{vb}T_{4 \to 3} \\ \eta_{vb}T_{4 \to 3} = 1 + \eta_{vb}T_{1 \to 3} \end{cases} \tag{17}$$

The AFPT for state 4 can be obtained as follows:

$$\begin{cases} (\xi_{ud} + \xi_{vb})T_{1 \to 4} = 1 + \xi_{ud}T_{2 \to 4} \\ (\xi_{ub} + \xi_{vb} + \eta_{ub})T_{2 \to 4} = 1 + \eta_{ub}T_{1 \to 4} + \xi_{ub}T_{3 \to 4} \\ \eta_{ub}T_{3 \to 4} = 1 + \eta_{ub}T_{1 \to 4} \end{cases} \tag{18}$$

By solving the above equations, we get the average first passage time between each state. In practice, service providers often pay more attention to when the system is unavailable, namely, the AFPT from the available state to unavailable state representing by $T_{1 \to 3}$, $T_{2 \to 3}$, $T_{1 \to 4}$, $T_{2 \to 4}$ in this paper. Their expressions are obtained as follows:

$$\begin{cases} T_{1 \to 3} = \frac{(\eta_{vb} + \xi_{vb})(\xi_{ub} + \xi_{ud} + \xi_{vb} + \eta_{ud})}{\xi_{ud}\xi_{ub}\eta_{vb}} \\ T_{2 \to 3} = \frac{(\eta_{vb} + \xi_{vb})(\xi_{ud} + \xi_{vb} + \eta_{ud})}{\xi_{ud}\xi_{ub}\eta_{vb}} \\ T_{1 \to 4} = \frac{\eta_{ub}^2 + \eta_{ub}(\xi_{ub} + \xi_{ud} + \xi_{vb}) + \xi_{ub}\xi_{ud}}{\eta_{ub}\xi_{vb}(\eta_{ub} + \xi_{ub} + \xi_{ud} + \xi_{vb})} \\ T_{2 \to 4} = \frac{(\eta_{ub} + \xi_{ub})(\xi_{ud} + \xi_{vb} + \eta_{ub})}{\eta_{ub}\xi_{vb}(\eta_{ub} + \xi_{ub} + \xi_{ud} + \xi_{vb})} \end{cases} \tag{19}$$

3.4.3 MTTF

MTTF is another indicator to evaluate EUPF service reliability. We can define all unavailable states as absorbing states for calculating MTTF. Figure 3 shows the state transition of EUPF service system with absorbing states. In Fig. 3, F represents the set of all unavailable states, namely State 3 and State 4. Let $\widehat{Q}$ be the submatrix of the generator matrix Q.

$$\widehat{Q} = \begin{bmatrix} -(\xi_{ud} + \xi_{vb}) & \xi_{ud} \\ \eta_{ud} & -(\eta_{ud} + \xi_{ub} + \xi_{vb}) \end{bmatrix} \tag{20}$$

$\widehat{\pi}(0)$ is the initial probability vector with available states and $\widehat{\pi}(0) = [1, 0]$. τ_i is defined as the expected time of the system in state i until the system enters the unavailable states, where $L_i(t) = \int\limits_{0}^{t} \pi_i(x)dx$.

$$\tau_i = \lim_{t \to \infty} L_i(t) \tag{21}$$

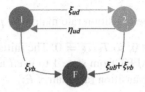

Fig. 3. System state transition of EUPF service system with absorbing states

Then we can get the Eq. (22):

$$\tau \widehat{Q} = -\widehat{\pi}(0) \tag{22}$$

Finally, the MTTF can be obtained as follows:

$$MTTF = \frac{2\eta_{ud} + \xi_{ub} + \xi_{vb}}{\xi_{vb}^2 + (\eta_{ud} + \xi_{ub} + \xi_{vd})\xi_{vb} + \xi_{ub}\xi_{ud}} \tag{23}$$

4   Analytical Experiments

This section aims to investigate the impact of parameters on EUPF service dependability in terms of transient availability, steady-state availability, EUPF service's first failure time and MTTF. Without changing the adversary's attack ability and strength, we change the recovery ability of EUPF service system, namely η_{ud}, η_{ub} and η_{vb}. Default values of the model parameters are summarized in Table 1 for analytical experiments. Analytical experiments are conducted on MAPLE [19]. The results and discussions of transient availability, steady-state availability, EUPF service's first failure time and MTTF are presented in Sects. 4.1, 4.2, 4.3 and 4.4, respectively.

4.1   Transient Availability of EUPF Service

This section presents the transient availability of EUPF service varying η_{ud}, η_{ub} and η_{vb}. Figure 4 shows the result of the transient availability over η_{ud}, η_{ub} and η_{vb}. The transient availability gradually increases with the increasing η_{ud} (that is, the ability of the VM to resist attacks enhanced) in Fig. 4 (a). At the same time, the transient availability rapidly decreases in the first two days. In Fig. 4 (b), when η_{ub} gradually increases, which means the ability of the VM with EUPF running to recover from the unavailable state enhanced, the transient availability increases gradually. This shows that the change of η_{ub} has little effect on EUPF service. When η_{vb} changes (namely, VMM recovery ability is enhanced), the transient availability is significantly increased in Fig. 4 (c). Specifically, the transient availability increases from 0.961 at $\eta_{vb} = 1$ to 0.984 at $\eta_{vb} = 3$ on day 8. This result shows that the ability of VMM to resist attacks has a profound impact on the transient availability of EUPF service.

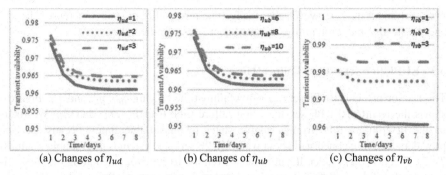

(a) Changes of η_{ud} (b) Changes of η_{ub} (c) Changes of η_{vb}

Fig. 4. Transient availability over η_{ud}, η_{ub} and η_{vb}

4.2 Steady-State Availability of EUPF Service and Steady-State Probability

This section presents the steady-state availability of EUPF service and probability varying η_{ud}, η_{ub} and η_{vb}. Figure 5 shows the result of the steady-state availability and probability of each state over η_{ud}, η_{ub} and η_{vb}. π_a increases from 0.961 to 0.965 with the increasing η_{ud}. There is a similar result for η_{ub} that π_a increases from 0.961 to 0.964. Steady-state availability has the greatest relationship with η_{vb} and it increases from 0.961 to 0.982 with the increasing η_{vb}.

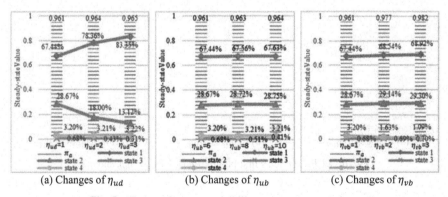

(a) Changes of η_{ud} (b) Changes of η_{ub} (c) Changes of η_{vb}

Fig. 5. The steady-state probability over η_{ud}, η_{ub} and η_{vb}

When parameters change, the steady-state probability change of state 3 and state 4 are not obvious and the steady-state probability change of state 1 and state 2 are more obvious. After adjusting η_{ud}, the steady-state probability of state 1 increases significantly, while the steady-state probability of state 2 decreases significantly in Fig. 5 (a). The steady-state probability of state 4 has not obvious change and is around 0.032. The steady-state probability of state 3 decreases from 0.0068 to 0.0031. The steady-state probability of each state does not obvious change with the increasing η_{ub} in Fig. 5 (b). The steady-state probabilities of state 1, state 2 and state 4 are about 0.675, 0.287 and 0.032, respectively. The steady-state probability of state 3 is always less than 0.007. Figure 5 (c) shows that the increase of η_{vb} has little effect on the steady-state probabilities of state 1 and state 2. However, the steady-state probability of state 3 increases slightly and the steady-state probability of state 4 decreases significantly.

4.3   EUPF Service's First Failure Time

This section presents EUPF service's first failure time varying η_{ud}, η_{ub} and η_{vb}. Figure 6 shows the result of EUPF service's first failure time over η_{ud}, η_{ub} and η_{vb}. We focus on the first failure time from available states to unavailable states, because service providers are more concerned about when EUPF service system is unavailable (namely, $T_{1\to3}$, $T_{2\to3}$, $T_{1\to4}$, $T_{2\to4}$). It can be seen from the formula of calculating the first failure time that $T_{1\to4}$ and $T_{2\to4}$ are only related to η_{ub}. Therefore, $T_{1\to4}$ and $T_{2\to4}$ are not changing with the increase of η_{ud} and η_{vb}. However, $T_{1\to4}$ and $T_{2\to4}$ decreases slowly with the increasing η_{ub} and their values are around 30 days. Similarly, $T_{1\to3}$ and $T_{2\to3}$ are not related to η_{ub}. The increase of η_{ud} has a significant effect on $T_{1\to3}$ and $T_{2\to3}$ and the increase of η_{vb} has little effect on them. $T_{1\to3}$ and $T_{2\to3}$ both increase exponentially with the increasing η_{ud} and decrease slowly with the increasing η_{vb}. Specifically, $T_{1\to3}$ increases from 24.25 to 53.18 days and $T_{2\to3}$ increases from 22.18 to 51.12 days with the increasing η_{ud}. When η_{vb} increases, $T_{1\to3}$ decreases slightly from 24.25 to 23.73 days and $T_{2\to3}$ decreases from 22.18 to 21.71 days.

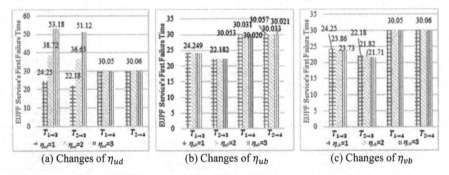

(a) Changes of η_{ud} (b) Changes of η_{ub} (c) Changes of η_{vb}

Fig. 6. EUPF service's first failure time over η_{ud}, η_{ub} and η_{vb}

4.4   MTTF

This section presents MTTF varying η_{ud}, η_{ub} and η_{vb}. Figure 7 shows the result of MTTF over η_{ud}, η_{ub} and η_{vb}. It can be seen from the formula of calculating MTTF that MTTF is related to η_{ud}, not to η_{ub} and η_{vb}. Therefore, changing the latter does not affect the results of MTTF. MTTF increases from 17.09 to 31.84 with the increasing η_{ud}. It can be explained that the change of η_{ud} has a huge impact on MTTF, which is also consistent with the analysis of EUPF service's first failure time and steady-state probability described above.

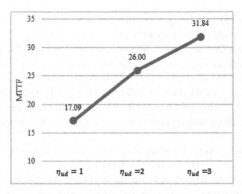

Fig. 7. MTTF over η_{ud}

5 Conclusion and Future Work

This paper explores analytical modeling approaches to quantitatively analyze the dependability of EUPF service. We first present the CTMC models capturing the behaviors of EUPF service system. Then we present the methods for computing transient availability of EUPF service, steady-state availability of EUPF service, EUPF service's first failure time and MTTF. Finally, analytical experiments are applied to illustrate the impact of various parameters on transient availability, steady-state availability, EUPF service's first failure time and MTTF.

Future work will relax the assumption that all time intervals are exponentially distributed and explore the semi-Markov process that the time intervals follow general distributions in order to make the models better capture the operational details of EUPF service when facing attack. Moreover, we plan to consider the recovery policy in our model and evaluate the impact of recovery policy on EUPF service dependability.

Acknowledge. This research was supported by the National Natural Science Foundation of China under Grant U1836105.

References

1. Weissberger, A.: Gartner: 5G IoT Endpoints to Triple between 2020 and 2021; Surveillance Cameras to be Largest Market over Next 3 Years, Ocotober 2019. https://techblog.comsoc.org/2019/10/17/gartner-5g-iot-endpoints-to-triple-between-2020-and-2021-surveillance-cameras-to-be-largest-market-over-next-3-years/
2. Stephanie Overby, Edge Computing by the Numbers: 9 Compelling Stats, April 2020. https://enterprisersproject.com/article/2020/4/edge-computing-9-compelling-stats
3. 3rd Generation Partnership Project(3GPP) TS 23.501 v15.7.0, System Architecture for the 5G System; stage 2(release 15), October 2019. http://www.3gpp.org/ftp/specs/archive/23series/23_501/
4. ETSI White Paper: MEC Deployments in 4G and Evolution Towards 5G. ETSI (2018)
5. Lee, D.,Park, J., Hiremath, C., Mangan, J., Lynch, M.: Accelerating the Virtualized User Plane for 5G Core Network Readiness. Intel solution brief (2018)

6. Trivedi, K.S., Bobbio, A.: Reliability and Availability Engineering: Modeling, Analysis, Applications. Cambridge University Press, Cambridge (2017)
7. Yang C.: Building a Comprehensive Security System for MEC, March 2020. https://www.zte.com.cn/global/about/magazine/zte-technologies/2020/2-en/Special-Topic/3.html
8. Fattore, U., Liebsch, M., Bernardos, C.J.: UPFlight: An Enabler for avionic MEC in a drone-extended 5G mobile network. In: VTC2020-Spring, pp. 1–7. IEEE (2020)
9. Ge, M., Hong, J.B., Yusuf, S.E., Kim, D.S.: Proactive defense mechanisms for the software-defined internet of things with non-patchable vulnerabilities. Future Gener. Comput. Syst. **78**, 568–582 (2018)
10. Kwan, W., Sanguk, C., Sang, C.: Software architecture of a reliability prediction system. Int. J. Appl. Eng. Res. **13**(23), 16199–16203 (2018)
11. Chang, X., Wang, T., Rodríguez, R.J., Zhang, Z.: Modeling and analysis of high availability techniques in a virtualized system. Comput. J. **61**(2), 180–198 (2018)
12. Chang, X., Shi, Y., Zhang Z., Xu Z., Trivedi, K.: Job completion time under migration-based dynamic platform technique. In: IEEE Transactions on Services Computing (2020) (Early Access). (https://doi.org/10.1109/tsc.2020.2989215)
13. Bai, J., Chang, X., Machida, F., Trivedi, K., Han, Z.: Analyzing software rejuvenation techniques in a virtualized system: service provider and user views. IEEE Access **8**, 6448–6459 (2020)
14. Jiang, L., Chang, X., Bai, J., Mišić, J., Mišić, V., Chen, Z.: Dependability analysis of 5G-AKA authentication service from server and user perspectives. IEEE Access **8**, 89562–89574 (2020)
15. Bovenzi, A., Alonso, J., Yamada, H., Russo, S., Trivedi, K.S.: Towards fast OS rejuvenation: An experimental evaluation of fast OS reboot techniques. In: ISSRE 2013, pp. 61–70. IEEE (2013)
16. Trivedi, K.: Probability and Statistics with Reliability, Queuing, and Computer Science Applications. Prentice-hall, Englewood Cliffs (1982)
17. Kulkarni, V.: Introduction to Modeling and Analysis of Stochastic Systems. Springer New York (2011)
18. Chang, X., Zhang, Z., Li, X., Trivedi, K.S.: Model-based survivability analysis of a virtualized system. In: LCN 2016, pp. 611–614. IEEE (2016)
19. Maplesoft: The Essential Tool for Mathematics. http://www.maplesoft.com/products/maple

QoE Estimation of DASH-Based Mobile Video Application Using Deep Reinforcement Learning

Biao Hou and Junxing Zhang$^{(\boxtimes)}$

Inner Mongolia University, Hohhot, China
houbiao@mail.imu.edu.cn, junxing@imu.edu.cn

Abstract. An increasing number of video content providers have adopted adaptive bitrate (ABR) streaming via the HTTP protocol. The client players usually run an ABR algorithm to determine the optimal quality of video playback in the next few seconds. Faced with unpredictable bandwidth variability, the latest ABR algorithm attempts to achieve the best balance between competing goals of high bitrate, less rebuffering and high smoothness. However, there is no guarantee that optimal resource utilization ensures a high quality of experience (QoE). QoE is also affected by users' preferences for video content. Even for the same movie clip, different users have varied preferences for characters, scenes, plots and other content. In this paper, we propose a Deep-Q Learning Network (DQN) based ABR algorithm to optimize the use of network and client resources in video playback and also improve QoE of users.

Keywords: MPEG-DASH protocol · ABR algorithm · QoE model · DQN algorithm

1 Introduction

In recent years, the traffic of video streaming applications based on the HTTP protocol has become an increasing share of global Internet traffic. The MPEG-DASH protocol (dynamic adaptive streaming based on HTTP) is widely employed by large content providers such as YouTube and Netflix [1]. It can play videos adaptively for users, can flexibly adjust video quality in real-time, and can be easily implemented on existing HTTP infrastructure (such as Content Delivery Network or CDN). These advantages make it widely adopted by the video streaming industry.

Adaptive bitrate algorithm generally applies the DASH protocol based on HTTP. Video content providers mainly use dynamic adaptive video streaming (DASH) based on the HTTP protocol to provide video content to their subscribers. In the DASH protocol, the server usually divides a video into chunks with fixed playing time, and each chunk is encoded into a different code rate and resolution, and stored in the server with different quality (such as 720P, 1080P). Each video chunk contains a few seconds of content. The client video player predicts the next optimal video chunk based on the current playback conditions (1) network conditions (throughput, bandwidth, etc.) (2) client player environment (buffer size), and uses the adaptive bitrate algorithm to select the

© Springer Nature Switzerland AG 2020
M. Qiu (Ed.): ICA3PP 2020, LNCS 12453, pp. 633–645, 2020.
https://doi.org/10.1007/978-3-030-60239-0_43

best for the next chunk quality. However, when the prediction is inaccurate, the player may suffer from problems such as freezes. If the client player requests to download high-quality video chunks when there are insufficient bandwidth resources, the playback will pause for a few seconds (until the chunks are completely downloaded). In addition to rebuffering events, continuous changes in the quality of video chunks will also impact the user's viewing experience. Therefore, the video streaming algorithm attempts to decrease the frequency of sudden changes, thereby improving the smoothness of playback. The core problem is to download high-quality video chunks as much as possible under the current playback conditions to ensure a smooth playback process and optimize the user's quality of experience. Therefore, the recent state-of-the-art ABR algorithms (such as MPC [2] and Pensieve [3]) optimize the objective function (or reward) of three influencing factors: (1) higher bitrate, (2) less rebuffering events, and (3) higher smoothness. As a result, during the video playback, the quality of the video chunks varies with the available bandwidth resources, and when the player buffer is abundant, the player downloads some certain video chunks at a higher quality.

The users' demand for video content with higher QoE is growing rapidly, and failure to provide users with satisfactory QoE will cause huge losses to service providers. Therefore, video service providers need to create personalized video viewing experiences for users. Because different users have different viewing experiences on the same video, it leads to the complexity of measuring user quality of experience. A large number of studies have indicated that the quality of users' video experience may depend on various factors, including gender, age, content, etc. [4, 5]. Therefore, the requirement for the video streaming strategy is that, in addition to the optimal use of bandwidth, etc., the video streaming strategy also needs to consider the user's content preferences. In this paper, we suggest that by optimizing the bitrate selection of network resources (bandwidth, etc.) and client resources (buffer size), use deep reinforcement learning Deep-Q learning network algorithm to optimize the ABR algorithm and improve the user quality of experience.

To our knowledge, our main contributions in this article are as follows:

1. Identify video chunk prefetch opportunities: We have implemented a cascaded neural network model that can empirically learn (that is, perform deep reinforcement learning) to solve the research challenge of intelligently detecting the download timing of video chunks. The DRL model further ensures that video chunks do not cause more rebuffering events, nor do they cause excessive degradation of the quality and playback smoothness of conventional chunks.
2. Achieve the balance of QoE indicators: In this article, we focus on introducing QoE estimation within the network, including real-time QoE estimation technology within the network. We choose network indicators and client player resources to build a QoE evaluation model, which can accurately reflect the user quality of experience.

The remainder of the paper is organized as follows. Section 2 presents relevant background on MPEG-DASH protocol, ABR algorithm and QoE model. Section 3 presents an overview of our system architecture. Section 4 presents details on the system implementation, while Sect. 5 is dedicated to evaluation results. Section 6 describes the paper concludes and future plan.

2 Background Review

In this section, we provide the background review of adaptive bitrate video streaming using MPEG-DASH protocol and the research of the ABR algorithm. We introduce the research on the QoE model finally.

2.1 MPEG-DASH Protocol

When watching a video, we often encounter the problem that the video is stuck or the bitrate is not high-definition due to various network reasons. At this time, everyone generally chooses the video clarity in the lower right corner to solve the problems. Many video content providers have launched adaptive bitrate streaming technology to improve user experience. The adaptive bitrate algorithm generally uses the DASH protocol based on HTTP. The basic operation process is to first store video chunks, which have been encoded according to different bitrates in the CDN server. Then the client continuously requests the video chunks at a certain bitrate from the server. Each video chunk contains a few seconds of video content. The server transmits the video to the client that stores the video chunk in the local buffer. In Fig. 1, The purple chunk is regarded as a video chunk, the duration is two seconds, and the height is its bitrate. During video playback, the client player will consume buffer in real-time. If your download rate does not match the bitrate at present, there will appear the empty buffer or the accumulated buffer. When the bandwidth is far below the bitrate, there will be a pause, so we should cope with the use of an adaptive bitrate algorithm to select the appropriate bitrate according to the current network conditions and buffer size to optimize the user's quality of experience.

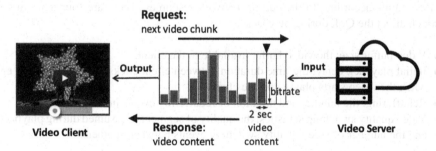

Fig. 1. An overview of HTTP adaptive video streaming

2.2 ABR Algorithm

The research of the ABR algorithm has been going on for many years. The rate-based method [6] is a classic traditional ABR algorithm. It estimates throughput and then selects a bitrate close to or slightly lower than throughput. The BBA (buffer-based approach [7]) algorithm proposed in 2014 is based on buffers to make bitrate selection decisions. The reason for choosing buffers is because the throughput jitter is very large and it is difficult to predict it. There is also a method MPC based on throughput and buffer for bitrate decision, which adopts throughput and buffer size to determine the future QoE that can be obtained at different bitrates so that this model can be used to plan future predictions. By directly optimizing the required QoE goals, the performance of MPC is better than the method using a fixed heuristic algorithm, but the performance of MPC depends on accurate system dynamics models, especially the prediction of future network throughput, which makes MPC is sensitive to throughput prediction errors and the length of the optimized range. These solutions require a lot of adjustments and cannot be extended to different network conditions and QoE goals.

2.3   QoE Model

QoE is a user experience indicator. The design concept of the QoE indicator is to be closer to the user's true feelings. The behavior data collected by the user is mainly used to measure the user's subjective experience. To cope with evaluate the QoE of the DASH client player, the solutions in the existing research are mainly divided into 3 categories: 1) direct feedback from users; 2) QoE evaluation using network indicators and 3) QoS parameters (such as bandwidth, delay, etc.) are mapped to the QoE indicator through the mean opinion score (MOS) [8]. When the video is spread using the MPEG-DASH standard [9], it is encoded and compressed by the server into different representations with different bitrates (i.e., video quality). Metadata (such as URLs) about video representations and their fragments are stored in manifest files called media presentation files(MPD files). When the client uses a streaming video, it will first request the MPD list from the server so that the client can understand the available representation and resource structure of the video. Then the client player decides to transmit the optimal video chunk according to the current network resources. There are four main metric which affect the QoE during playback.

- Video buffer size: the video buffer available to the client.
- Initial playback delay: the time duration between client clicks the "Play" button and when the video starts playing on the client.
- Rebuffering: the number and duration of rebuffering events in the playback.
- Video quality switching statistics: the quality of segments consumed during playback, and the number of video quality switching events between segments.

3   System Review

In this section, we mainly design the DASH streaming media decision engine and use deep reinforcement learning (DRL) model to achieve the optimal bitrate prefetching and decision-making. During the video chunk download, the decision engine has the following permissions: (1) decide whether to prefetch the next video chunk, (2) decide the bitrate of the next video chunk. The basic logic of reinforcement learning is to learn by constantly interacting between agents and the environment. The agent continuously observes the state from the environment, make some decision actions based on the state and take action to the environment, the environment transfer to the next state based on the current state and decision, and give the agent reward feedback, the agent learns from this maximize future cumulative reward accumulation. We can optimize the bitrate selection decision through the expected reward.

The DASH player mainly consists of two components: (1) dash.js player, which is based on the open-source DASH player dash.js[1] implementation [10], and (2) DASH decision engine, the current video chunk after the download is complete, dash.js player provides a set of parameters (for example, the bitrate of the last chunk, throughput, available buffer size, the next chunk size, etc.), and these parameters are combined to playback status information. Send this information to an external video server via an HTTP request. The DASH decision engine responds whether to prefetch the next video chunk and the best bitrate for the next chunk. The client dash.js player downloads the corresponding video chunk according to the decision information. In the following sections, we will delve into the design of the DASH decision engine (Fig. 2).

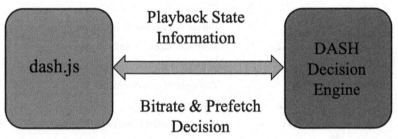

Fig. 2. Our system consists of 2 components, the prefetch-enabled player dash.js, and the DASH decision engine, which computes optimal bitrate and prefetch decisions.

3.1 Perfetch and Bitrate Decision

It is not easy for the server to make the optimal-expected decision, because the improper prefetching decision may cause rebuffering to occur and increase the stall time. De facto, the following factors need to be referenced to make prefetch and bitrate decisions at a certain point in time: (1) current network conditions, such as available bandwidth, and time delay, etc. (2) current player operating status, such as local buffer size, etc. (3) detailed information of the next video chunk (chunk size, chunk bitrate).

3.2 Cascaded Design

We cascade designed the decision model of DASH. During the download of video chunks, the ABR controller in dash.js player generates the following set of state parameters: (S1) player environment, (S2) video chunk information, (S3) network information. Combine state S1 (player environment), state S2 (video chunk information), and state S3

---

[1] dash.js, https://github.com/Dash-Industry-Forum/dash.js.

(network information) to form the input of the bitrate selection model, determine whether to prefetch the next video chunk and choose the optimal bitrate for it. According to the two execution combinations the bitrate decision model form S4 (bitrate decision).

Based on the required four sets of states generated, their combination is used as input to our prefetch decision model (the exact parameters used have been summarized in Table 1). The output is whether to prefetch the next video chunk and the bitrate of the video chunk. The prefetch decision and the bitrate decision together constitute the output of the DASH decision engine. dash.js sent it to ABRController in the server, which starts the corresponding download and completes the download of the video chunk.

Table 1. DASH prefetch and decision engine state parameters

Category	Param	Significance
Player environment (S1)	D_t	Download time for the last chunk
	B_t	Player buffer occupancy
	P_t	Bitrate of last download chunk
Video chunk state (S2)	N_t	Available sizes for next chunk
	C_t	No. of chunks remaining in video
Network state (S3)	T_t	Throughput measurement
	L_t	Latency of network
Bitrate decision (S4)	A_t^y	Bitrate returned if next chunk is prefetch
	A_t^n	Bitrate returned if next chunk isn't prefetch

3.3 Deep Reinforcement Learning

The above cascade design of DASH requires us to design a suitable prefetch and bitrate-select decision model. The ABR algorithm has long-term planning characteristics, which is very suitable for deep reinforcement learning itself. Deep neural networks can also process multi-dimensional input states to assist in making decisions. Because there are a large number of highly variable input parameters that affect prefetching and bitrate select decisions, it is particularly difficult to design a model with a simple optimization function. At the same time, the network situation is also complicated and uncontrollable. Therefore, we use reinforcement learning algorithms to deal with various external environmental conditions and learn from past decisions, optimize decisions based on cumulative rewards, and use deep neural networks to process high-dimensional input signals to assist in optimization decisions.

In this paper, we use a deep reinforcement learning algorithm, called deep-Q learning network, which originated from Q-learning and combined with the deep neural network. The DRL algorithm runs in environment E, takes state S as input, takes action A, and gets a reward R from its action. When the algorithm iterates over a large set of states and continuously receives rewards for all operations it performs, it will learn the most optimized operations for a particular state, thereby maximizing the rewards. For the bitrate selection model, the video player is environment E, S1 + S2 + S3 + S4 form the input state, the QoE metric used to optimize the model is reward R, and the corresponding bitrate decision is operation A. In the case of the DASH decision engine, the environment E remains unchanged, and the combination of S1 to S3 forms the input state S. The QoE indicator specifically designed to train it is reward R, and the binary prefetch decision is action A.

4 System Implementation

In this section, we will delve into the implementation details of each component of DASH, the implementation of DQN algorithm and the establishment of QoE evaluation model.

4.1 Dash.js Player Implementation

In our system, we implement the client video player on top of the implementation of dash.js based on the open-source DASH player. Normally, dash.js performs video playback in a sequential stop-and-wait manner: once video playback starts, the scheduler starts downloading video chunks, and then adds the chunk completely download to the player's local buffer, and then it starts downloading the next chunk that is the next in-order chunk in the session. Besides, the prefetch chunk data requires two specific additional functions: (1) video chunk data according to the playback request, and (2) management of the local buffer so that the out-of-order prefetch chunk data can be accommodated and appropriately rendered for playback. We describe in detail how to achieve each of the above requirements in the DASH player.

(1) Player dash.js with prefetch function: We have implemented a 'specific clip request rule', which can download video chunks by time, this time is the playtime of the chunk when there is not adequate buffer. We have further rewritten ABRController and ScheduleController to implement the functions of prefetch decision and bitrate selection decision.

(2) Player buffer management: The dash.js player is equipped with a local buffer. In addition to the regular chunks arranged in sequence, it can also store unsynchronized chunks. We implement the buffer management function in the dash.js player. We mainly consider two parameters: (1) Play buffer size and (2) Total buffer size. The playback buffer refers to a part of the internal buffer. This part sequentially stores the video chunk data for playback. These video chunks have been downloaded but not yet rendered (played on the video player). Note that to play seamlessly, the buffer must play all video chunks in order. However, dash.js player finally downloads completely out-of-sequence video chunks when it executes prefetch decision. These prefetched chunks are also stored in the buffer, and these buffer segments are combined with the playback buffer to form the total buffer. Therefore, the total buffer is the actual measure occupied by the buffer, and the playback buffer is the measure of time that the video player can continue to play without pausing (rebuffering). Whenever a video chunk is downloaded, the total buffer increases. When the video chunks are played continuously, the playback buffer and total buffer will be consumed.

4.2 Perfetch and Bitrate Decision Implementation

According to the above discussion, the DASH decision engine uses an RL-based model, that is, prefetching and bitrate selection decision model. We use the most advanced deep reinforcement learning algorithm deep Q learning network to make end-to-end rate decision. We use deep neural networks to process multi-dimensional input signals, reinforcement learning to adapt sequence decisions, interactive learning to remove redundant model assumptions, and reward functions to support optimized QoE goals. The DQN algorithm can be summarized as the pseudo-code shown in Algorithm 1. In the early days of DQN model training, various inappropriate decision-making behaviors will appear. As the number of pieces of training increases, the model's decision-making effect becomes better and better.

Algorithm 1 DASH Deep Q-learning Network(DASH-DQN)

Initialize replay memory D to capacity N

Initialize action-value function Q with random weight θ

Initialize target action-value function Q' with random weight $\theta' = \theta$

For episode1 ,M do Initialize $s_t = \{D_t, B_t, P_t, N_t, C_t, T_t, L_t, A_t^y, A_t^n,\}$

    For t=1,T do

        With probability ε select a random action a_t

        Otherwise select $a_t = \arg max_a\ Q(\emptyset(s_t), a; \theta)$

        Execute action a_t and observe reward r_t and last chunk bitrate P_t

        Set $s_{t+1} = \{s_t, a_t, P_t\}$ and preprocess $\emptyset_{t+1} = \emptyset(S_{t+1})$

        Store experience $(s_t, a_t, r_t, \emptyset_{t+1})$ in D

        Sample random minibatch of experiences $(\emptyset_j, a_j, r_j, \emptyset_{j+1})$ from D

        Set $y_i = \begin{cases} r_j & \text{if episode terminates at step } j+1 \\ r_j + ymax_{a'}Q'(\emptyset_{j+1}, a'; \theta') & \text{otherwise} \end{cases}$

        Perform a gradient descent step on $(y_j - Q(\emptyset_j, a_j; \theta))$

        Every C steps reset $Q' = Q$

    End for

End

On the other hand, bypassing 8 vector input values to a 1D convolutional layer with 128 filters (each with a size of 4 and a stride of 1), prefetching and bitrate decision models are implemented in TensorFlow. The remaining scalar parameters passed to another 1D-CNN have the same shape. The results obtained from these layers are then aggregated into a hidden layer, which uses 128 neurons and applies the softmax function, whose final output is a linear neuron. In the training parameters, the discount rate y = 0.95, the score collection interval is every 10,000 rounds, the planned training is 1,000,000 rounds, the initial and final values of μ are 1e−1 and 1e−5, respectively, and the random sample set size is mini-batch times = 32.

4.3 QoE Model

QoE is defined for videos with n chunks; R_i is the bitrate of video chunk value, $q(R_i)$ maps this bitrate to utility value, which represents the quality perceived by the user. Rebuffering times T_i gives the time it takes to download a chunk at the bitrate R_i; μ is a weight that determines the severity of the rebuffering event. The last item describes the fluency of playback, which affects user QoE when the continuous bitrate changes suddenly. In short, QoE increases with high bitrate but decreases with rebuffering and lack of smoothness [18].

$$\text{QoE} = \sum_{i=1}^{n} q(R_i) - \mu \sum_{i=1}^{n} T_i - \sum_{i=1}^{n} |q(R_{i+1}) - q(R_i)| \qquad (1)$$

We consider three different variations of QoE for our performance evaluation, each of which defines a different $q(R_i)$, and a different value of μ.

1) QoE_{lin}: In the linear case,

$$q(R_i) = q_{lin}(R_i) = R_i; \; \mu = \mu_{lin} = 4.3 \qquad (2)$$

2) QoE_{log}: In the logarithmic case,

$$q(R_i) = q_{log}(R_i) = log(R/R_{min}); \; \mu = \mu_{log} = 2.66 \qquad (3)$$

3) QoE_{hd}: In the high definition (HD) case, $q(R_i)$ assigns static scores for different bitrate values (Table 2):

Table 2. Bitrate utility of QoE_{hd}

Name	Bitrate utility ($q(R_i) \rightarrow q_{hd}(R_i)$)
QoE_{hd}	0.30 → 1; 0.75 → 2; 1.20 → 3 1.85 → 12; 2.85 → 15; 4.3 → 20

$$q(R_i) = q_{hd}(R_i) = \mu = \mu_{hd} = 8 \qquad (4)$$

QoE Metric used for Training DASH: We train our DRL model in the DASH decision engine, using a variant of QoE_{lin} and QoE_{hd} as the DRL reward metric. Let a target video consisting of n chunks be represented by the ordered set V = $\{v_1, v_2, ..., v_n\}$. We define an ordered set H ∈ V, which consists of all m high quality chunks present in the video: H = $\{h_1, h_2, ..., h_m; m \leq n\}$, We recall that the objective of our model is to determine prefetch opportunities, such that the high chunk bitrates are maximized, but at the same time, other objectives (maximum regular chunk bitrates, minimum rebuffering, and maximum smoothness) are not compromised. We incorporate this notion in the QoE metric in the QoE metric by defining QoE_{DQN}, a variant of QoE_{lin}, such that:

$$\text{QoE} = \sum_{h \in H} q_{hd}(R_h) + \sum_{v \in V-H} q_{lin}(R_v) - \mu \sum_{i=1}^{n} T_i - \sum_{i=1}^{n-1} |q_{lin}(R_{i+1}) - q_{lin}(R_i)| \quad (5)$$

The DRL model is encouraged to download prefetch chunks at higher bitrates, by using the HD reward function q_{hd} (.) to determine bitrate utility of chunks. In contrast, the bitrate utility for regular chunks is determined using the linear function q_{lin} (.). The rest of the terms remain the same as in QoE_{lin}, while keeping other QoE indicators intact.

5 Evaluation Results

To complement the proposed DASH framework, we produce a dataset using our framework covering a large variety of scenarios. As stated, we gather the output of the generated testbed log files and create a DASH dataset, While this dataset can be used for further analysis and comparison with other adaptive algorithms. In our setup, the ABR server and the video player were run on different client machines in the same local network. The player dash.js was configured with a (total) buffer capacity of 60 s. We used the Envivio-dash3 [12] video for all our experiments; this video has been encoded using the H.264/MPEG-4 codec, and is available at 6 qualities – 240p (bitrate: 300 Kbps), 360p (bitrate: 750 Kbps), 480p (bitrate: 1200 Kbps), 720p (bitrate: 1850 Kbps), 1080p (bitrate: 2850 Kbps), and 1440p (bitrate: 4300 Kbps). The video consists of 48 chunks of approx. 4 s each, and has a total playback time of 193 s. The video chunk server was implemented in the same workstation (Ubuntu 16.04 LTS) where the video player was run. The chunks were served using Apache, while the client player used a Google Chrome browser. The network emulation tool Mahimahi [13] was used to emulate network conditions represented by the throughput traces generated.

It depicts all scenarios we have run with DASH. For bandwidth traces, we randomly select thirty 4G traces [14], and twenty five 3G traces [14]. Additionally, we collect five traces over WiFi. Selected traces are included with the framework. All experiments are five minutes long (we only stream first five minutes of clips and not for whole duration).

We compare DASH-DQN algorithm to the following algorithms which collectively represent the state-of-the-art in bitrate adaptation: (1) Buffer-based: choose the bitrate to make the buffer take up more than 5 s, and select the highest selectable bitrate when the buffer takes more than 15 s. (2) Rate-based: When downloading the last 5 chunks, use the harmonic mean value of the measured throughput to predict the next throughput, and then select the highest selectable bitrate lower than the predicted throughput. (3) MPC: Use buffer occupancy and throughput for simultaneous prediction (the prediction method is the same as RB) to select the bitrate that can maximize the QoE metric for the next 5 chunks.

We started with a series of standard experiments, that is, running a single algorithm for wireless trajectories. This experiment provides a comparison between the performance of each algorithm. In addition, all algorithms are evaluated for different segment durations, so that the impact of different segment durations on the overall algorithm performance can be quantified. Next, repeat the experiment using the same client (each client uses a different algorithm) under the same conditions. As can be seen from Fig. 3, we use the DASH-DQN algorithm to optimize the video chunk bitrate selection. During playback, the bitrate is higher than the other three algorithms, and the average video bitrate is 9.7% higher than the best MPC algorithm. In other words, using the DASH-DQN algorithm can get a higher bitrate (video resolution) during playing a video

Next, under the same conditions, we use a single client to download the video stream to the video server and use four different ABR optimized algorithms to select the video chunk bitrate. From Fig. 4, we can obtain better QoE through the DASH-DQN algorithm. DASH-DQN algorithm achieves an improvement of 9.6% over the best performing baseline in terms of average QoE. From Fig. 5, comparing with existing ABR algorithm on the QoE model, DASH-DQN has a better performance.

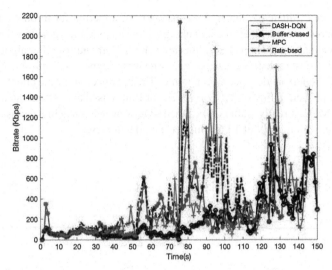

Fig. 3. Comparing DASH-DQN algorithm with existing ABR algorithms by analyzing their performance on the bitrate during video chunk playback

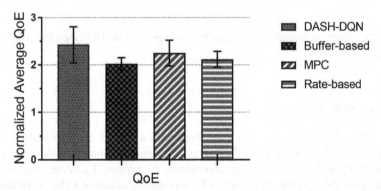

Fig. 4. Comparing DASH-DQN algorithm with existing ABR algorithms by analyzing their performance on the individual components in the general QoE definition.

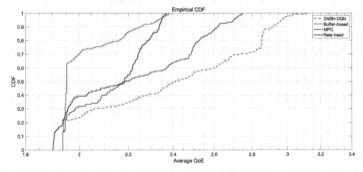

Fig. 5. Comparing DASH-DQN algorithm with existing ABR algorithms on the QoE metrics listed in Eq. 5. Average QoE values are listed for each ABR algorithm.

Finally, we evaluate two heterogeneous clients competing for throughput resources and network resource, with each client using a same adaptation algorithm. These experiments permit analysis on how algorithms are actually fair against other algorithms. This scenario is the most realistic one. In practice, one would expect multiple users are streaming from different providers at the same time. The results on multiple client competition scenario are plotted in Fig. 6, the QoE results of Rate-based and buffer-rated algorithms are similar, MPC is really stable QoE, DASH-DQN algorithm achieve close-to-optimal performance, which is about 13.6% higher than the average.

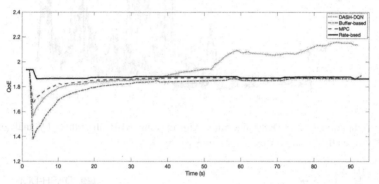

Fig. 6. The results on multiple client competition scenario, DASH-DQN algorithm achieve close-to-optimal performance.

6   Conclusion

In this paper, we supplement the framework with a dataset of generated content for the implemented algorithms across a range of scenarios, thus providing a means of benchmarking and validating future algorithms. We proposed a system that takes into consideration the content preferences of users during adaptive video streaming over HTTP. The prefetch and bitrate decisions are taken in an opportune manner, such that higher bitrates are selected for video chunks, but other QoE considerations, such as bitrate, rebuffering, and smoothness, are not heavily compromised. This is made possible by the application of deep Q-learning network to dynamically learn optimal prefetch decisions and bitrate decisions through experience, over a large corpus of network traces. DASH-DQN algorithm achieves an improvement of 13.6% over the best performing baseline in terms of average QoE.

Finally, we will explore the resource costs of using deep reinforcement learning (such as CPU and battery consumption), and at the same time, consider combining edge computing to preprocess the video at the edge in the future work.

Acknowledgement. This work was partially supported by the National Natural Science Foundation of China (Grant No. 61261019), the Inner Mongolia Science & Technology Plan (Grant No. 201802027), and the Inner Mongolia Autonomous Region Natural Science Foundation (Grant No. 2018MS06023).

References

1. Cisco Visual Networking Index: Cisco Visual Networking Index: Forecast and Trends, 2017–2022 white paper, Document ID, vol. 1551296909190103 (2019)
2. Yin, X., Jindal, A., Sekar, V., Sinopoli, B.: A control-theoretic approach for dynamic adaptive video streaming over http. ACM SIGCOMM Comput. Commun. Rev. **45**(4), 325–338 (2015)
3. Mao, H., Netravali, R., Alizadeh, M.: Neural adaptive video streaming with pensieve. In: Proceedings of the Conference of the ACM Special Interest Group on Data Communication, pp. 197–210. ACM (2017)
4. Dobrian, F., et al.: Understanding the impact of video quality on user engagement. ACM SIGCOMM Comput. Commun. Rev. **41**(4), 362–373 (2011)
5. Duanmu, Z., Ma, K., Wang, Z.: Quality-of-experience of adaptive video streaming: exploring the space of adaptations. In: Proceedings of the 2017 ACM Multimedia Conference (2017)
6. Ketykó, I.: et al. QoE measurement of mobile youtube video streaming. In: Proceedings of the 3rd Workshop on Mobile Video Delivery, pp. 27–32. ACM (2010)
7. Huang, T.Y., et al.: A buffer-based approach to rate adaptation: evidence from a large video streaming service. In: SIGCOMM. ACM (2014)
8. Juluri, P., Tamarapalli, V., Medhi, D.: Measurement of quality of experience of video-on-demand services: a survey. IEEE Commun. Surv. Tutor. **18**(1), 401–418 (2016)
9. ISO. ISO/IEC 23009-1:2014 dynamic adaptive streaming over HTTP (DASH) – Part 1: Media presentation description and segment formats
10. Dash-Industry-Forum/dash.js (2017). https://github.com/Dash-Industry-Forum/dash.js
11. Spiteri, K., Urgaonkar, R., Sitaraman, R.K.: Bola: near-optimal bitrate adaptation for online videos. In: Proceedings of the 35th Annual IEEE International Conference on Computer Communications, pp. 1–9. IEEE (2016)
12. Envivio-dash3. http://dash.edgesuite.net/envivio/Envivio-dash2
13. Netravali, R., et al.: Mahimahi: Accurate record-and-replay for HTTP. In: USENIX Annual Technical Conference, pp. 417–429 (2015)
14. Riiser, H., Vigmostad, P., Griwodz, C., Halvorsen, P.: Commute path bandwidth traces from 3G networks: analysis and applications. In: Proceedings of the 4th ACM Multimedia Systems Conference, MMSys 2013, New York, NY, USA, pp. 114–118. ACM (2013)
15. Raca, D., Quinlan, J.J., Zahran, A.H., Sreenan, C.J.: Beyond throughput: a 4G LTE dataset with channel and context metrics. In: Proceedings of the 9th ACM Multimedia Systems Conference, MMSys 2018, New York, NY, USA, pp. 460–465 (2018)
16. Chinchali, S., et al.: Cellular network traffic scheduling with deep reinforcement learning. In: Proceedings of the 32nd AAAI Conference on Artificial Intelligence, pp. 1–9 (2018)
17. Szegedy, C., Toshev, A., Erhan, D.: Deep neural networks for object detection. In: Proceedings of the 2013 Advances in Neural Information Processing Systems, NIPS, Lake Tahoe, Nevada (2013)
18. Sengupta, S., Ganguly, N., Chakraborty, S., De, P.: HotDASH: hotspot aware adaptive video streaming using deep reinforcement learning. IEEE (2018). ISBN 978-1-5386-6043-0/18. https://doi.org/10.1109/icnp.2018.00026

Modeling and Analyzing for Data Durability Towards Cloud Storage Services

Feng Jiang[1], Yongyang Cheng[1(✉)], Zhao Hui[1], and Ruibo Yan[2]

[1] China Telecom Cloud Computing Corporation, Beijing, China
chengyy2@chinatelecom.cn
[2] State Key Laboratory of Networking and Switching Technology, Beijing University of Posts and Telecommunications, Beijing, China

Abstract. Based on the consideration of economic cost and system performance, the distributed storage technology using multiple data replicas has been widely applied in cloud storage applications. This redundant storage mode could ensure that the data loss event occurs only when all data replicas deployed on disks are damaged. In this case, the data durability is determined by the failure recovery model and replica organization strategy. However, the traditional approaches have poor performance in resisting data loss when associated failures occur. In this paper, we propose a novel modeling and analyzing approach for data durability towards cloud storage services. Initially, we model the processes of data failure and recovery using a no-retrogressive Markov chain. Furthermore, we present a routing table-based replica organization strategy to reduce the data loss caused by associated failures. Finally, our approach presented in this paper has been validated to be effective through a series of quantitative evaluations in the simulation environment.

Keywords: Cloud services · Distributed system · Data durability modeling · No-retrogressive markov · Cluster capacity.

1 Introduction

With the development of information and communication technology, the storage requirements of accumulated data are increasing, which might be structured documents or unstructured pictures and videos [1,2]. Thus, the storage and access of massive data has become a research hotspot in the field of cloud services. In the traditional way, people use a larger capacity of hard disk to store a lot of data, which not only produces additional hardware resource cost, but also limits the mobility of data access [3]. However, with the development of cloud storage technology, users could upload their data to the cloud and deliver the management of data to cloud service providers. In this way, users could

---

F. Jiang and Y. Cheng—Co-first authors of the article.

© Springer Nature Switzerland AG 2020
M. Qiu (Ed.): ICA3PP 2020, LNCS 12453, pp. 646–661, 2020.
https://doi.org/10.1007/978-3-030-60239-0_44

access their data anytime and anywhere without carrying any physical devices. Although cloud storage technology makes it more convenient for users to control data, data durability is always the focus of users' concern, which is also an important basis for the service level evaluation of cloud service providers [4–6]. In general, the concept of data durability is vague, which usually contains the meaning of durability and availability. We call the feature that data could be accessed availability. When the data is temporarily inaccessible and could be found after a period of time, such a state would not be called available, but it still belongs to durability. Only the data is completely lost and could never be found, it's state would be called unavailable. The distributed storage technology using multiple replicas has been widely applied in cloud storage services to improve data durability. If one replica is damaged, the data could be recovered from other replicas. This redundant storage mode could ensure that the data loss event only occurs when all data replicas are damaged deployed on disks.

Although some traditional approaches could support modeling and analyzing for data durability [7–12], challenges still remain to be solved. First of all, Markov chains are usually used to model the failure and recovery processes of a redundant cloud storage architecture. However, most of these approaches are retrogressive, which is not applicable when all replicas are lost (at this time, the data has been lost and could not be recovered). Furthermore, the traditionally random replica organization strategy has poor performance in resisting data loss when associated failures occur (e.g. power outage, network congestion, earthquake, etc). Finally, in a specific scenario, the data durability of a redundant cloud storage architecture has certain extremum. However, most of existing approaches are based on qualitative analysis, lacking of quantitative calculation and giving the optimal cluster capacity to ensure a maximum data durability.

The main contributions of our work contain the following.

- Based on the processes of data failure and recovery, we propose a novel approach to model the data durability using a no-retrogressive Markov chain.
- A routing table-based data replica organization strategy is designed to reduce the probability of data loss caused by associated failures.
- We quantitatively evaluate our proposed approach by a series of experiments and discuss the optimal cluster capacity to ensure a maximum durability.

The rest of this paper is organized as follows: We discuss related works and clearly point out the similarities and differences in Sect. 2. In Sect. 3, we model the processes of data failure and recovery using a no-retrogressive Markov chain. Section 4 presents a routing table-based replica organization strategy to reduce the probability of data loss caused by associated failures. In Sect. 5, we evaluate our proposed approach by quantitative experiments and discuss the optimal cluster capacity to ensure a maximum data durability. Concluding remarks are made in Sect. 6.

2   Related Works

In this section, we mainly compare our proposed approach with other existing
approaches. We will clearly point out the similarities and differences with their
works. Tang et al. [7] survey different cloud storage services and related data
redundancy schemes that are introduced to overcome the impact of host churn
on data reliability. Furthermore, simulations based on real traces are performed
to evaluate their proposed approach. However, this demonstration work is based
on the simulated experiment environment and lacks theoretical analysis. Nguyen
et al. [8] propose a hierarchical modeling framework for the data durability evaluation, which includes durability graphs to model the master system, fault-trees to
model the subsystems and stochastic reward nets to capture the behaviors of the
components. Although graph is a common method to formalize business process,
it could not directly reflect the processes of data failure and recovery. Machida
et al. [9] present a model for RAID storage services using Markov regenerative
process, which allows them to model the generally distributed storage services
by assigning reward rates to each state based on the real storage benchmark
results. However, in a real case, data would not be recovered if all replicas have
been damaged. Thus, we need to extend the traditional Markov model to fit the
data failure and recovery processes. Li et al. [10] explore the problem of placing object replicas on nodes in a distributed system to maximize the number of
objects that remain available when node failures occur. However, this random
placement-based replication is almost guaranteed to lose data in the common
scenario of simultaneous node failures. Differ to them, we propose a routing
table-based replica organization strategy to reduce the possibility of data loss.
Kang et al. [11] present an efficient replication scheme that allows a Cloud Service Provider (CSP) to determine the optimal number of replicas for each user
depending on the user's budgetary constraint and the CSP's resource capacity while maximizing the financial benefit of the CSP. However, this approach
only supports a single data redundant mode and is not suitable for unstructured
data storage applications, in which the data size could range from a few bytes
to hundreds of megabytes. Dieye et al. [12] present a heterogeneity-aware data
management scheme to satisfy Service Level Agreement (SLA) requirements in
terms of data durability. In a specific scenario, the value of data durability of
a redundant storage architecture has certain extremum. However, this work is
based on qualitative analysis, lacking of quantitative calculation and giving the
optimal cluster capacity to ensure a maximum data durability.

3   Modeling for Data Failure and Recovery

Data durability refers to an ability that data could be accessed within the specified time. Data durability is usually expressed by probability or Mean Time To
Failure ($MTTF$) according to actual demands. For a disk, the available time T
is a random variable. We could define its durability as $R(t)$ and express it as
Eq. (1), where t defines the specified time. Thus, the durability at time t refers

to the probability that data could be accessed between time $[0, t]$. Similarly, we could define the nondurability of a disk as $F(t)$ and express it as Eq. (2), which refers to the probability that data could not be accessed between $[0, t]$.

$$R(t) = P \quad (T > t) \tag{1}$$

$$F(t) = P \quad (T \le t) \tag{2}$$

For a finite object set, we assume that the total number of disks is N_0 and the number of fault disks is $r(t)$. Then, the durability and nondurability of a disk between time $[0, t]$ could be expressed as Eq. (3) and (4).

$$R(t) = \frac{N_0 - r(t)}{N_0} \tag{3}$$

$$F(t) = \frac{r(t)}{N_0} \tag{4}$$

Obviously, the relationship between $R(t)$ and $F(t)$ is complementary set expressed as Eq. (5). In addition, with the increase of time t, the number of fault disks is increasing. Thus, the value of $R(t)$ gradually reduces from $R(0) = 1$ to $R(\infty) = 0$ and the value of $F(t)$ gradually increases from $F(0) = 0$ to $F(\infty) = 1$.

$$R(t) + F(t) = 1 \tag{5}$$

Based on the definition of $F(t)$, we could calculate the failure probability density function $f(t)$. For a finite object set, we assume that the total number of disks is N_0. The number of fault disks after time t is $r(t)$ and after time $t + \delta t$ is $r(t + \delta t)$. Then, the value of $f(t)$ could be expressed as Eq. (6).

$$f(t) = \frac{dF(t)}{dt} = \frac{r(t + \delta t) - r(t)}{N_0 \delta t} \tag{6}$$

Failure probability refers to the probability that a disk normally runs at any time t but fails at the next time $t + \delta t$. Similarly, for a finite object set, we assume that the total number of disks is N_0. The number of fault disks after time t is $r(t)$ and after time $t + \delta t$ is $r(t + \delta t)$. Then, we could define the failure probability a physical disk as $\lambda(t)$ and express it as Eq. (7).

$$\lambda(t) = \frac{r(t + \delta t) - r(t)}{[N_0 - r(t)]\delta t} \tag{7}$$

If the number of disks is large enough $(N_0 \to \infty)$ and specified time is short enough $(\delta t \to 0)$, the failure probability $\lambda(t)$ could be expressed as Eq. (8).

$$\lambda(t) = \lim_{\substack{\delta t \to 0 \\ N_0 \to \infty}} \frac{\dfrac{r(t + \delta t) - r(t)}{N_0 \delta t}}{\dfrac{N_0 - r(t)}{N_0}} = \frac{f(t)}{1 - F(t)} = \frac{f(t)}{R(t)} \tag{8}$$

As mentioned above, $MTTF$ is a key index that is usually used to evaluate data durability. Since the available time of a disk is a continuous random variable, we could calculate the value of $MTTF$ by Eq. (9).

$$MTTF = \int_0^\infty tf(t)\, dt = -[tR(t)]\,|_0^\infty + \int_0^\infty R(t)\, dt = \int_0^\infty R(t)\, dt \qquad (9)$$

Through the statistics of a large number of physical disks, we find that their available time T is exponential distribution. Thus, the corresponding failure probability density function of a disk $f(t)$ is expressed as Eq. (10).

$$f(t) = \lambda e^{-\lambda t} \quad (t \geq 0, \lambda > 0) \qquad (10)$$

Based on this and Eq. (5), we could calculate the value of $R(t)$, which is expressed as Eq. (11). Then, according to Eq. (8), we could get the value of failure probability $\lambda(t)$, which is a constant with a value of λ.

$$R(t) = 1 - F(t) = 1 - \int_0^t f(t)\, dt = \int_t^\infty f(t)\, dt = e^{-\lambda t} \qquad (11)$$

Thus, the value of $MTTF$ could also be expressed as Eq. (12). In this way, we build the relationship between $MTTF$ and the failure probability λ.

$$MTTF = \int_0^\infty R(t)\, dt = \int_0^\infty e^{-\lambda t}\, dt = -\frac{1}{\lambda} e^{-\lambda t}\,|_0^\infty = \frac{1}{\lambda} \qquad (12)$$

We have built the relationship between $MTTF$ and the failure probability λ, then we would model the processes of failure and recovery of a disk using the no-retrogressive Markov chain. For a Markov chain, if its state at time t_i is known, then the possibility of all states at another time t_j ($j > i$) is completely determined, which is not affected by the state at any time before t_i. Differ to the traditional approach, if all replicas on disks have been damaged, the data would not be recovered again. Thus, the final state in our proposed Markov chain is no-retrogressive. In order to build a failure and recovery model using the no-retrogressive Markov chain, we have the following assumptions.

1. The probability of a disk fails in time $[t, t + \delta t]$ is λ, which is a constant;
2. The probability of a disk is recovered in time $[t, t + \delta t]$ is μ, which is a constant;
3. In time $[t, t + \delta t]$, the probability of a disk fails or is recovered two times is 0.
4. The final state in this Markov chain is no-retrogressive.

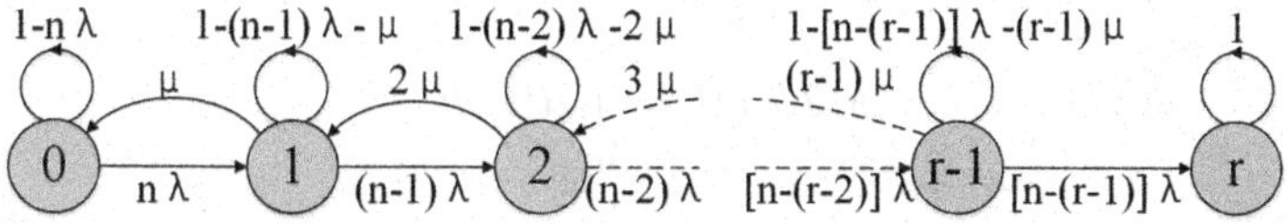

Fig. 1. Processes of data failure and recovery using no-retrogressive Markov chain

We assume that the total number of physical disks in a distributed storage system is n and the data redundancy is r, that is to say, every data object has r replicas. For a data object, its failure probability is λ and recovery probability is μ. The concept of data recovery means that the data on a damaged physical disk would be migrated to other available disks to provide services. Thus, data loss occurs only if all of its replicas are damaged. Figure 1 illustrates the processes of data failure and recovery using no-retrogressive Markov chain. Based on this, we could get the system transition probability matrix $P[(r+1) \times (r+1)]$.

$$P = \begin{bmatrix} 1 - n\lambda & n\lambda & 0 & \cdots & 0 \\ \mu & 1 - (n-1)\lambda - \mu & (n-1)\lambda & \cdots & 0 \\ 0 & 2\mu & 1 - (n-2)\lambda - 2\mu & \cdots & 0 \\ \vdots & \vdots & \vdots & \vdots & \vdots \\ 0 & 0 & 0 & \cdots & n - (r-1)\lambda \\ 0 & 0 & 0 & \cdots & 1 \end{bmatrix}$$

Table 1. The dependency relationships among indicators

State	Failure Probability	Recovery Probability	Durability
0	$n\lambda$	——	√
1	$(n-1)\lambda$	μ	√
2	$(n-2)\lambda$	2μ	√
⋮	⋮	⋮	⋮
$r-1$	$n - (r-1)\lambda$	$(r-1)\mu$	√
r	——	0	×

Table 1 shows the dependency relationships among system state, failure probability, recovery probability and data durability. We take the state 1 as an instance. There is already one damaged physical disk waiting to be recovered. Thus, the recovery probability is μ and the failure probability of remaining disks is $(n-1)\lambda$. Since there is still one available replica, the system is data durability.

As mentioned above, we use $MTTF$ to describe the data durability of one single physical disk. Similarly, we introduce the concept of Mean Time To Data Loss ($MTTDL$) to describe the data durability of the whole storage system, which could be calculated by Eq. (13).

$$MTTDL = \int_0^\infty R(t)\, dt \tag{13}$$

For a redundant distributed storage system with r data replicas, there are $r+1$ states in our proposed no-retrogressive Markov chain, of which the data durability states are: state 0, 1, 2, ... ,$r-1$ and the data nondurability state is: state r. Thus, $R(t)$ refers to the sum of the probabilities of the data durability states, which could be expressed as Eq. (14).

$$R(t) = P_0(t) + P_1(t) + \cdots + P_{r-2}(t) + P_{r-1}(t) \tag{14}$$

In addition, Laplace transform of R(t) is given by:

$$R^*(s) = \int_0^\infty R(t)e^{-st}\, dt \tag{15}$$

By assigning the value of s to 0 in Eq. (15), we could find $MTTDL$ is an aggregate measure of data durability, which is expressed as Eq. (16).

$$R^*(0) = \int_0^\infty R(t)\, dt = MTTDL \tag{16}$$

Thus, according to Eqs. (13) to (16), the expression of $MTTDL$ could be calculated by Eq. (17).

$$MTTDL = P_0^*(0) + P_1^*(0) + \cdots + P_{r-1}^*(0) \tag{17}$$

In most real could storage services, we would use three replicas to achieve redundant storage of data. This pattern could not only ensure the data durability, but also reduce the storage cost as much as possible [13]. Thus, we would take three replicas as an instance to introduce our proposed approach in the following paper. Based on our proposed no-retrogressive Markov chain and the transition probability matrix P, we could get the conventional Kolomogorov equations in time domain for this model as following:

$$\begin{cases} P_0'(t) = \mu P_1(t) - n\lambda P_0(t) \\ P_1'(t) = 2\mu P_2(t) - [(n-1) + \mu]P_1(t) + n\lambda P_0(t) \\ P_2'(t) = -[(n-2)\lambda + 2\mu]P_2(t) + (n-1)\lambda P_1(t) \end{cases} \tag{18}$$

According to the differential properties of Laplace transform in Eq. (19), we could convert the function $P_i'(t)$ with parameter real number t into a function with parameter complex s. Eq. (20) shows this Laplace transform.

$$\mathcal{L}[F'(t)] = sF(s) - F(0) \tag{19}$$

$$\mathcal{L}_i = \mathcal{L}[P_i'(t)] = sP_i^*(s) - P_i(0) \quad (i = 0,1,2) \tag{20}$$

Thus, Eq. (18) could be converted to Eq. (21).

$$\begin{cases} \mathcal{L}_0 = \mu P_1^*(s) - n\lambda P_0^*(s) \\ \mathcal{L}_1 = 2\mu P_2^*(s) - [(n-1)\lambda + \mu]P_1^*(s) + n\lambda P_0^*(s) \\ \mathcal{L}_2 = -[(n-2)\lambda + 2\mu]P_2^*(s) + (n-1)\lambda P_1^*(s) \end{cases} \tag{21}$$

When a three replicas-based storage system is in its initial state (the value of s is assigned 0), we could get the value of $P_j(0)$ expressed as Eq. (22).

$$P_j(0) = \begin{cases} 1 & (j = 0) \\ 0 & (j > 0) \end{cases} \tag{22}$$

Therefore, we could get the following Eq. (23).

$$\begin{cases} 1 = n\lambda P_0^*(0) - \mu P_1^*(0) \\ 0 = 2\mu P_2^*(0) - [(n-1)\lambda + \mu]P_1^*(0) + n\lambda P_0^*(0) \\ 0 = -[(n-2)\lambda + 2\mu]P_2^*(0) + (n-1)\lambda P_1^*(0) \end{cases} \tag{23}$$

Then, the values of $P_r^*(0)$ for $r = 0, 1$ and 2 could be calculated explicitly and they are given by Eq. (24).

$$\begin{cases} P_0^*(0) = \dfrac{\lambda^2(n-2)(n-1) + \lambda\mu(n-2) + 2\mu^2}{\lambda^3 n(n-1)(n-2)} \\[3mm] P_1^*(0) = \dfrac{\lambda(n-2) + 2\mu}{\lambda^2(n-1)(n-2)} \\[3mm] P_2^*(0) = \dfrac{1}{\lambda(n-2)} \end{cases} \tag{24}$$

According to Eq. (17), the value of $MTTDL$ is the sum of $P_0^*(0)$, $P_1^*(0)$ and $P_2^*(0)$. Thus, the basic data durability P_b of our proposed no-retrogressive Markov chain is calculated explicitly and it is shown by Eq. (25).

$$P_b = \frac{\lambda^3 n(n-1)(n-2)}{\lambda^2(3n^2 - 6n + 2) + \lambda(3\mu n - 2\mu) + 2\mu^2} \tag{25}$$

4   Table-Based Replica Organization Strategy

We have introduced the modeling approach based on no-retrogressive Markov chain in previous section. From this model, we could know that the number of replicas is an important factor to determine the basic data durability. A cloud storage system could improve its data durability by increasing the number of replicas. However, this means that the type of data stored in a unit storage space would be greatly reduced [14]. Thus, with the same number of replicas, their organization strategy plays a core role.

Random replica placement strategy is widely used in cloud storage services [15–17]. This is a simplest replication organization strategy to prevent data loss. They process the data in the form of r replicas and randomly select r physical disks on different servers in the cluster for storage. In this way, if one replica is damaged, users could still get data from other replicas, which improve the durability of the whole storage system. However, there might be some associated failures (e.g. power outage, network congestion, earthquake, etc). According to the data statistics in work [18], about 0.5% to 1% nodes could not come back to life after the power has been restored. Thus, the randomness of storage disk selection would greatly increase the probability that all r replicas of at least one data object are not available. Taking Hadoop Distributed File System ($HDFS$) as an instance, we assume that the number of nodes in $HDFS$ is n, the number of replication is r and the percentage of nodes that could not survive a power outage is 1%. Differ to their work, we consider that the data size might range from a few bytes to hundreds of megabytes. To make the simulation reflect the real system performance as much as possible, we assume that all files stored in the system are some ones. Thus, once the size of cluster nodes beyond $n_{max}(n_{max}=100r)$, the data loss event is nearly guaranteed to occur.

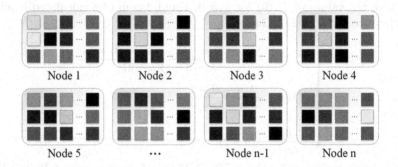

Fig. 2. The possible distribution of data based on random placement strategy

To further explain the results, we assume that there are n nodes in the system, each of which stores enough small data objects. If a randomness-based replica placement strategy is adopted, the finally possible distribution of data objects is illustrated in Fig. 2, in which a block represents a data replica. Those with the same color belong to the same data object and those with different colors belong to different data objects. In this case, the total number of combinations based on r damaged nodes could be calculated as Eq. (26).

$$C_n^r = \binom{n}{r} = \frac{n!}{r!(n-r)!} \tag{26}$$

Thus, taking the example mentioned above, if the number of nodes exceeds 300 and the number of replications is 3 in $HDFS$, the minimum number of combinations based on 3 damaged nodes is $C_{300}^3 = 4455100$, which means that the data loss event is nearly guaranteed to occur. It is a tradeoff between the event to have fewer incidents with more data and the event to have more incidents with less data. As mentioned in Sect. 3, we could recover the damaged data replica based on our proposed no-retrogressive Markov chain. However, the main cost of this process is due to the time to check the available disks, seek time and rotational delays [19]. Thus, it is much better to have fewer incidents of data loss. In order to achieve this goal, we should distribute the data replicas according to certain rules as much as possible. The most constrained case is that every 3 nodes belong to a logical group. A logical group is used to store the three replicas of a data object and the intersection among different groups is an empty set, which is illustrated in Fig. 3. In this case, the total number of logical groups is 100. If 3 nodes are damaged, the data would loss only when 1 of the above 100 groups happens. The probability of data loss is 1/44551.

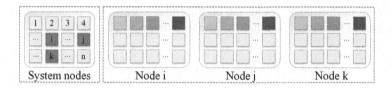

Fig. 3. The distribution of data objects based on group

Although this logical group-based data replica placement strategy minimizes the probability of data loss if associated failures occur, it still brings some challenges. As we know, the most significant advantage of a redundant storage system is that a damaged data replica could be recovered by other available ones. We define that $Capacity_{disk}$ represents the capacity of a disk node, $Usage_{disk}$ represents the usage of a disk node, $WriteRate_{disk}$ represents the write-rate of a disk node and $AvailableDisk_{disk}$ represents the available disk nodes in a cluster. Then, the recovery time $T_{recovery}$ could be calculated by Eq. (27).

$$T_{recovery} = \frac{Capacity_{disk} \cdot Usage_{disk}}{WriteRate_{disk} \cdot AvailableDisk_{disk}} \qquad (27)$$

For a specific cluster, the capacity of a disk node, the usage of a disk node and the write-rate of a disk node are fixed. Then, the value of $T_{recovery}$ would be determined by the value of available disk nodes in this cluster. Due to the number of available disk nodes (only two) is relatively reduced using the logical group-based data replica placement strategy, the failure recovery time would be naturally longer. To address this issue, we propose a routing table-based replica organization strategy to increase the number of available disk nodes. Initially, we design a routing table with 65536 rows and r columns, where r represents the number of data replicas. The reason to select 65536 as the rows of routing table is related to the resource deploying algorithm. Resources include hosts and data stored on them. However, the number of hosts might change (reducing the number of hosts due to failures or increasing the number of hosts according to business requirements). In order to reduce frequent data migration, we take advantage of Dynamic Consistent Hashing-based (DCH) resource deploying algorithm [20]. As we know, it is a world of 0 and 1 in the computer, therefore, it is easier to distribute and balance resources with 2^m (m is a constant) when partitioning. Furthermore, this DCH-based resource deploying algorithm uses IP address of a host as the Key to conduct operation $hash(IP\,address)\%(2^m)$, which has 32 bits. Thus, the value of m should be a divisor of 32:

$$m \in \{1,\ 2,\ 4,\ 8,\ 16,\ 32\} \tag{28}$$

Table 2 shows the performances of capacity and memory when a routing-table takes different values of its rows. As we know, the $HDFS$-based storage system could support 4000 to 10000 nodes in a cluster. If the value of $m \leq 8$, the number of hosted hosts is far less than the recommended capacity, meaning that the system performance could not be fully utilized and the number of available disk nodes is relatively insufficient (the maximum is $2^8 = 256$).

Table 2. The performances of capacity and memory

Value (m)	1	2	4	8	16	32
$Capacity$	2	4	16	256	64 K	4 GB
$Memory$	8 B	16 B	64 B	1 KB	256 KB	16 GB

Besides the capacity, the memory is another factor that restricts the value of m in this routing table. We consider that the sorting time is composed of comparison time and memory access time. In the process of comparison, about 1/2 of the branch prediction errors would occur. According to Table 3, if the value of $m = 32$, the total amount of data in the memory is $16\,GB$. Thus, the number of comparisons is $2^{32} \times \log(2^{32}) \approx 2^{37}$ and the comparison time is $\frac{1}{2} \times 2^{37} \times 5ns$ = 336 s. In addition, every split operation in the sorting needs to scan memory once. We assume that the sequential access rate of memory is $4\,GB/s$. Then, the memory access time is $32 \times 16\,GB/(4\,GB/s) = 128\,s$. Therefore, the total sorting time in the memory is $336\,s + 128\,s = 464\,s$. However, if the value of $m = 32$, the corresponding data capacity and soring time in the memory are unacceptable. On the contrary, if the value of $m=16$, the total amount of data in the memory is $256\,KB$. The number of comparisons is $2^{16} \times \log(2^{16}) \approx 2^{21}$, the comparison time is $\frac{1}{2} \times 2^{21} \times 5ns = 5.13\,ms$ and the memory access time is $16 \times 256\,KB/(4GB/s)$ $= 0.98\,ms$. Thus, if the value of $m=16$, the total sorting time in the memory is $5.13\,ms + 0.98\,ms = 6.11\,ms$, which is suitable for high performance storage services. Based on the above discussion, we select $2^{16} = 65536$ as the rows in our proposed routing table-based replica organization strategy.

Table 3. The data structure of our designed routing table.

Virtual Disk	Disk 0	Disk 1	Disk 2	$\cdots$	Disk r-1
Vd_0	Pd_1	Pd_2	Pd_3	$\cdots$	Pd_5
Vd_1	Pd_2	Pd_3	Pd_4	$\cdots$	Pd_6
Vd_2	Pd_3	Pd_4	Pd_5	$\cdots$	Pd_7
$\vdots$	$\vdots$	$\vdots$	$\vdots$	$\vdots$	$\vdots$
Vd_{65535}	Pd_i	Pd_j	Pd_k	$\cdots$	Pd_m

Table 3 shows the data structure of our designed routing table. During the system initialization, we build a virtual disk group from r physical disks belonging to different hosts. We further subdivide the total number of disks (n) in a cluster into the number of hosts $(hosts)$ and the number of disks attached to a host $(disks)$, where $n = hosts \times disks$. In this case, to recover damaged data replicas, the available disks $AvailableDisk_{disk}$ could be calculated as following:

$$\begin{cases} f_1 = \dfrac{65536r(r-1)}{hosts \cdot disks} \\ f_2 = (hosts - r) \cdot disks + (disks - 1) \\ AvailableDisk_{disk} = min\{f_1, f_2\} \end{cases} \tag{29}$$

If there are r damaged disks, the data loss event would occur only all damaged disks appear in a same virtual disk group of the routing table. Thus, the maximum data loss probability P_r cloud be calculated as Eq. (30).

$$P_r = \frac{2^{16}}{C_n^r} = \frac{65536 \cdot n!}{r!(n-r)!} = \frac{65536 \cdot (hosts \cdot disks)!}{r!(hosts \cdot disks - r)!} \tag{30}$$

Table 4. Data loss probability with different data replicas

Nodes	Replication 3	Replication 4	Replication 5
0	0	0	0
100	0	0	0
200	0	0	0
300	$14.7e-3$	0	0
400	$6.19e-3$	$6.24e-5$	0
500	$3.16e-3$	$2.55e-5$	$2.57e-7$
600	$1.83e-3$	$1.23e-5$	$1.03e-7$
700	$1.12e-3$	$6.62e-6$	$4.75e-8$
800	$7.71e-4$	$3.87e-6$	$2.43e-8$
900	$5.41e-4$	$2.41e-6$	$1.35e-8$
1000	$3.94e-4$	$1.58e-6$	$7.94e-9$

Taking the example mentioned above, if the number of disks is 300 and replicas are 3, the maximum data loss probability $P_{r=3}$ is $65536/4455100 \approx$ 0.0147, meaning the 85.3% reduction in the probability of data loss. Table 4 shows the data loss probability with different data replicas ($r = 3$, 4 and 5) and Fig. 5 illustrates the corresponding trend of data loss probability using our proposed routing table-based replica organization strategy.

5   Further Experimental Evaluation

5.1   Final System Data Durability Description

We have discussed the basic data durability P_b in Sect. 3 and the data loss probability P_r in Sect. 4. For a cloud storage system with data recovery mechanism, the data loss event would occur only when r damaged disks hit a same virtual disk group of the routing table and corresponding data replicas could not be recovered in time. Thus, according to Eqs. (25) and (30), we take the most widely applied three replicas as an instance, the final system data durability P_s could be quantitatively calculated as following:

$$P_s = 1 - P_r(1 - P_b) \quad (r = 3) \tag{31}$$

5.2   Simulation Experiments and Result Evaluations

To further evaluate our proposed approach, we conducted a series of simulation experiments on the basis of larger data sets. The simulation experiments were deployed on a PC, which had $16\,GB$ of RAM, $i7$-$8550U$ of Core, 3.80 GHz of CPU and 1 TB of disk space. We randomly generated data packets to simulate different data objects that ranged from $1\,KB$ to $10\,KB$. These data objects would be stored in 3 replicas on different disk nodes. To get closer to the real scene, we specified that $Hosts = 36$, $Disks = 28$, $Usage = 0.75$ and $WriteRate = 300\,MB/s$ in the topology. In this case, the total amount of simulated disk nodes was $36 \times 28 = 1008$, which could fully simulate the real situations. Furthermore, according to the relevant statistics of disk lifecycle, the common value of $MTTF$ was $1200000\,h$. To simulate the processes of data failure and recovery, we randomly suspend 3 nodes every $10\,ms$ and the sampling time was $10000\,s$. Due to the diversity of disk capacities, we repeatedly conducted the experiments in the condition that the values of $Capacities$ were set to $2\,TB$, $4\,TB$ and $6\,TB$.

The left part of Fig. 4 illustrated the total amount of lost data object based on different replica organization strategies. According to the experimental results, the average amount of lost data objects of the randomness-based placement strategy, the logical group-based placement strategy and the routing table-based placement strategy were $10.12\,KB$, $8.84\,KB$ and $7.76\,KB$. The total amount of lost data object based on our proposed approach was 23.32% and 12.22% less than the other two strategies. In the randomness-based strategy, the minimum

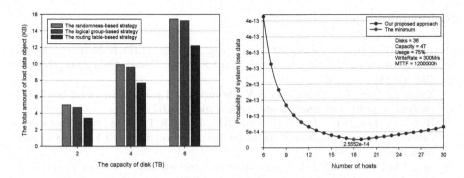

Fig. 4. Left: the trend of data loss probability (r = 3, 4 and 5); Right: the optimal capacity of hosts in a cluster to ensure a maximum data durability

combinations based on 3 damaged nodes were C_{1008}^3, meaning that the probability of data loss was about 100%. In the randomness-based strategy, if the data loss event occurred, all data on the disk node ($2T$, $4T$ or $6T$) would be lost. Furthermore, due to available disk nodes were relatively reduced, the failure recovery time would be longer. In our proposed approach, we distributed data replicas on different nodes and improved available disk nodes based on routing table, which not only reduced the probability of data loss, but also ensured that the amount of lost data each time was not too large. Thus, the average total amount of lost data object in our proposed approach was the smallest. In a specific scenario, the system data durability of a cloud storage had certain extremum. From the perspective of disk storage cost and access efficiency, the disk with $4T$ capacity was widely applied. The experimental results showed that the system data durability had maximum: P_s = 1-2.5552e-14 = 99.999999999997% when *Hosts* = 18, which is illustrated in the right part of Fig. 4.

6 Conclusions

In this paper, we have proposed a novel modeling and analyzing approach for data durability towards cloud storage services. To reach this goal, we initially model the processes of data failure and recovery using a no-retrogressive Markov chain. Then, we present a routing table-based replica organization strategy to reduce the probability of data loss caused by associated failures. Finally, we validate our proposed approach by a series of quantitative experiments and give the optimal capacity of hosts in a cluster to ensure a maximum system data durability. However, the updating of routing table directly determines the efficiency of disks' offline or online. Thus, we plan to leverage our proposed approach based on a more efficient disk replacement algorithm.

Acknowledgment. The authors would like to thank Ke Yan who works in Baidu for helpful discussions during the initial phase of this work.

References

1. Zhang, H., Chen, G., Ooi, B.C., Tan, K.L., Zhang, M.H.: In-Memory big data management and processing: a survey. IEEE Trans. Knowl. Data Eng. **27**, 1920–1948 (2015)
2. Zhou, J., Cao, Z.F., Vasilakos, A.V.: Security and privacy for cloud-based IoT: challenges. IEEE Commun. Magazine **55**, 26–33 (2017)
3. Ryu, J.H., Lee, D.E., Han, C.H., Shin, H.S., Kang, K.T.: File-system-level storage tiering for faster application launches on logical hybrid disks. IEEE Access **4**, 3688–3696 (2016)
4. Liu, J.W., Shen, H.Y.: A popularity-aware cost-effective replication scheme for high data durability in cloud storage. In: IEEE International Conference on Big Data, DC, USA, Washington (2016)
5. Galinanes, V.E., Felber, P.: Ensuring data durability with increasingly interdependent content. In: IEEE International Conference on Cluster Computing. IL, USA, Chicago (2015)
6. Simon, V., Monnet, S., Feuillet, M., Robert, P., Sens, P.: Scattering and placing data replicas to enhance long-term durability. In: IEEE International Symposium on Network Computing and Applications, MA, USA, Cambridge (2015)
7. Tang, B., Fedak, G.: Analysis of data reliability tradeoffs in hybrid distributed storage systems. In: IEEE 26th International Parallel and Distributed Processing Symposium Workshops & PhD Forum, Shanghai, China (2012)
8. Nguyen, T.A., Min, D., Choi, E., Tran, T.D.: Reliability and availability evaluation for cloud data center networks using hierarchical models. IEEE Access **4**, 9273–9313 (2019)
9. Machida, F., Xia, R.F., Trivedi, K.S.: Performability modeling for RAID storage systems by markov regenerative process. IEEE Trans. Dependable and Secure Comput. **15**, 138–150 (2018)
10. Li, P., Gao, D.B., Reiter, M.K.: Replica placement for availability in the worst case. In: IEEE 35th International Conference on Distributed Computing Systems. Columbus, OH, USA (2015)
11. Kang, S., Veeravalli, B., Aung, K.M.M., Jin, C.: An efficient scheme to ensure data availability for a cloud service provider. In: IEEE International Conference on Big Data. DC, USA, Washington (2014)
12. Dieye, M., Zhani, M.F., Elbiaze, H.: On achieving high data availability in heterogeneous cloud storage systems. In: IFIP/IEEE Symposium on Integrated Network and Service Management, Lisbon, Portugal (2017)
13. Ibrahim, I.A., Dai, W., Bassiouni, M.: Intelligent data placement mechanism for replicas distribution in cloud storage systems. In: IEEE International Conference on Smart Cloud. NY, USA, New York (2016)
14. Kermarrec, A.M., Merrer, E.L., Straub, G., Kempen, A.V.: Availability-based methods for distributed storage systems. In: IEEE 31st Symposium on Reliable Distributed Systems. Irvine, CA, USA (2012)
15. Singal, S.M., Rakesh, N., Matam, R.: Coding strategies to avoid data loss in cloud storage systems. In: Fourth International Conference on Parallel. Distributed and Grid Computing, Waknaghat, India (2016)
16. Lin, B., Li, S.S., Liao, X.K., Wu, Q.B., Yang, S.Z.: eStor: energy efficient and resilient data center storage. In: International Conference on Cloud and Service Computing. China, Hong Kong (2011)

17. He, Z.B., Cai, Z.P., Yu, J.G.: Latent-data privacy preserving with customized data utility for social network data. IEEE Trans. Vehicular Technol. **67**, 665–673 (2018)
18. Cidon, A., Rumble, S.M., Stutsman, R., Katti, S., Ousterhout, J., Rosenblum, M.: Copysets: reducing the frequency of data loss in cloud storage. In: USENIX Annual Technical Conference. SAN JOSE, CA, USA (2013)
19. Kaneko, H.: Failure recovery cost reduction of disk arrays using adaptive erasure correction coding and data compression. In: IEEE 21st Pacific Rim International Symposium on Dependable Computing. Zhangjiajie, China (2015)
20. Cheng, Y.Y., Zhao, S., Cheng, B., Chen, X.W., Chen, J.L.: Modeling and deploying IoT-aware business process applications in sensor networks. Sensors, **19**(1), 125–152 (2018)

CC-MOEA: A Parallel Multi-objective Evolutionary Algorithm for Recommendation Systems

Guoshuai Wei and Quanwang Wu[✉]

College of Computer Science, Chongqing University, Chongqing, China
wgs0208@foxmail.com, wqw@cqu.edu.cn

Abstract. Recommender systems have nowadays been widely used in a variety of applications such as Amazon and Ebay. Traditional recommendation techniques mainly focus on recommendation accuracy only. In reality, other metrics such as diversity and novelty also play a key role for modern recommendation systems. Although some works based on multi-objective evolutionary algorithm have been proposed for multi-objective recommendation, they are usually very time-consuming because of the large data size of the RSs and the long-term evolution iterations and hence it greatly limits their application in practice. To address these shortcomings, this paper first designs a multi-objective recommendation system, taking into account diversity and novelty as well as accuracy. Then, a novel parallel multi-objective evolutionary algorithm called CC-MOEA is proposed to optimize these conflicting metrics. CC-MOEA is devised grounded on NSGA-II and a cooperative coevolutionary island model, and a parallel global non-dominated selection method is introduced to reduce the runtime of finding the global optimal individuals. Furthermore, a new initialization method and a crossover operator are specifically designed. The experimental results reveal that CC-MOEA outperforms some state-of-the-art algorithms in terms of hypervolume and runtime.

Keywords: Accuracy · Cooperative coevolutionary · Multi-objective evolutionary algorithm · Parallel · Recommendation system

1 Introduction

Recommendation systems (RSs) is an inevitable product of the development of e-commerce and Internet technology. It is widely used in news websites, decision support, learning recommendation, digital library, e-commerce and other aspects. At present, many recommendation algorithms [1] have been proposed and they can be mainly classified as follows: content-based recommendation [2, 3], collaborative filtering recommendation [4, 5], network-based recommendation [6, 7], hybrid recommendation [8, 9] and so

This work is supported in part by the National Natural Science Foundation of China under Grant 61702060 and 61672117, and the Fundamental Research Funds for the Central Universities of China under Grant 2019CDXYJSJ0021.

© Springer Nature Switzerland AG 2020
M. Qiu (Ed.): ICA3PP 2020, LNCS 12453, pp. 662–676, 2020.
https://doi.org/10.1007/978-3-030-60239-0_45

on. The above traditional recommendation techniques mainly concentrate on recommendation accuracy as the only objective. However, some other metrics such as novelty and diversity also matter for RS. Up to present, only quite a few multi-objective recommendation systems have been proposed [10–12]. These works differ in considered objectives and used optimizing algorithms to achieve a good balance between the considered multiple objectives. However, due to the large data size of the RSs and the long-term evolution iterations, these algorithms are very time-consuming, which greatly limits their application in reality. Although parallelization is a potential way to reduce runtime, the challenge is how to reduce the runtime greatly while guaranteeing the quality of the solutions.

To address the above problems, this paper first sets up a recommendation system model, involving three metrics of accuracy, diversity and novelty. A novel parallel multi-objective evolutionary algorithm called CC-MOEA is proposed to optimize these conflicting metrics. CC-MOEA is devised grounded on NSGA-II and a cooperative coevolutionary island model, and a parallel global non-dominated selection method is introduced to reduce the runtime of finding the global optimal individuals. Furthermore, a new initialization method and a crossover operator are specifically designed to improve the performance. The experimental results reveal that CC-MOEA outperforms some state-of-the-art multi-objective recommendation algorithms as it always achieves the highest hypervolume with the least runtime.

The remainder of this paper is organized as follows. In Sect. 2, some back-grounds are given including recommendation algorithms, multi-objective optimization, and parallel multi-objective evolutionary algorithm. Section 3 describes the multi-objective recommendation model and the CC-MOEA algorithm in detail. In Sect. 4 experimental studies are presented and then conclusions are given in Sect. 5.

2 Background

2.1 Classical Recommendation Algorithms

The traditional recommendation system aims to find items that users are most interested in and recommend them to users. Generally, the recommendation systems can get users' evaluations of some items. RSs must predict users' preferences for unrated items based on users' previous behavior and recommend a number of items which have the highest predicted preference to the target user.

At present, the most widely used recommendation method is the user-based collaborative filtering algorithm (User-Based-CF) [13]. It calculates the target user's predicted item rating based on the item ratings of k most similar users to the target user. The predicted rating $pr_{u,i}$ is calculated as follows:

$$pr_{u,i} = \frac{\sum_{v \in s(u,k)} sim_{u,v} \times r_{v,i}}{\sum_{v \in s(u,k)} ||sim_{u,v}||} \tag{1}$$

where, $s(u, k)$ is the set of k users who are the most similar to user u, $sim_{u,v}$ is the similarity between users u and v, and $r_{v,i}$ is the rating of item i given by user v. Let r_u be the rating vector of user u. The similarity between users can be calculated with the cosine similarity, which is shown as follows:

$$sim_{u,v} = \frac{r_u \cdot r_v}{|r_u| \cdot |r_v|} \tag{2}$$

2.2 Multi-objective Optimization

Multi-objective optimization problems often consist of several conflicting objectives, which can be stated as follows:

$$\begin{aligned} \text{maximize} \quad & F(x) = (f_1(x), \dots, f_m(x))^T \\ \text{subject to} \quad & x \in \Omega \end{aligned} \tag{3}$$

where, Ω is the decision (variable) space, $F: \Omega \to R^m$ consists of m real-valued objective functions and R^m is called the objective space.

Usually the objectives contradict with each other, and it is generally impossible to find a solution with all the objective values maximal. Hence, we have to balance them and the best tradeoffs among the objectives can be defined in terms of Pareto dominance. Formally, a solution $F(x_A)$ dominates $F(x_B)$, denoted as $F(x_A) \prec F(x_B)$, if:

$$\forall i \in \{1, \dots, m\}, \; f_i(x_A) \geq f_i(x_B), \quad \exists j \in \{1, \dots, m\} \; f_j(x_A) > f_j(x_B) \tag{4}$$

A solution is called Pareto optimal, if no solution in the objective space can dominate it. A Pareto optimal solution set, also known as the Pareto front (PF), consists of all the Pareto optimal solutions.

Computing the set of the Pareto optimal solutions is usually very difficult. It is more realistic to find an approximate set for the Pareto optimal solution set. Multi-objective evolution algorithm (MOEA) is an effective method to find this approximate set, and they can be divided into three categories according to the selection mode: Pareto dominance-based (e.g., NSGA-II [14], SPEA2 [15], MOELS [16]), decomposition-based (e.g., MOEA/D [17]) and indicator-based (e.g., SMS-EMOA [18]). NSGA-II [14] takes dominance relation and crowding distance as the selection criterion, and SPEA2 [15] uses the combination of the domination number and the number of being dominated. MOEA/D [17] decomposes a multi-objective optimization problem into a group of single-objective problems. By contrast, SMS-EMOA [18] adopts hypervolume contribution as the secondary selection criterion.

2.3 Parallel Multi-objective Evolutionary Algorithm

A traditional sequential MOEA is unable to provide satisfactory results within a reasonable time when the population size is very large, or the computational cost of fitness evaluation and evolution is extremely high. Parallelizing the multi-objective evolutionary algorithm can be an effective way in these cases [19]. Generally, parallel MOEAs can be divided into four models according to the parallel topology: master-slave model [20, 21], island model [22], cellular model [23], hierarchical model [24].

In the master-slave model [20, 21], a commonly used implementation is a coarse-grained master-slave model in which each slave processor contains a subpopulation and performs a complete multi-objective evolutionary algorithm. After a certain interval the master receives the best individuals from each slave and distributes the global best information to all slaves. In e-DNSGA-II [20], the master shares all the extreme individuals to guide the evolution in each slave. However, it may also make the slaves form the same search habit so that the search is repeated between slaves and the global efficiency may be decreased.

In the island model (e.g. SLA-MOEA [22]), all island processors have their own subpopulation to perform a complete MOEA and individuals migrate between is-lands after a certain interval. Each individual is split into n (the number of islands) chunks. Island p performs crossover and mutation on the $p^{th} - c$ to $p^{th} + c$ parts of individuals, where c is the number of overlapping chunks. This method greatly reduces the time cost of crossover and mutation. However, the migration topology usually cannot guarantee the global search capability.

The cellular model is fine-grained and spatially structured [23]: it has only one population and individuals are arranged on the grid, ideally one individual per processor. Each individual on the grid can only compete and mate within its neighborhood.

The hierarchical model [24], also known as hybrid model, combines two or more distributed models hierarchically to take advantages of both models for improving scalability and problem-solving capability.

3 Multi-objective Recommendation Model and Optimization

3.1 Multi-objective Recommendation Systems

In this section, the multi-objective recommendation system is modelled which takes into account three optimization objectives, i.e., accuracy, diversity and novelty. These objective functions are described in detail below.

Objective Function of Accuracy. The accuracy function is used to measure the user's acceptance probability (preference) of the items. We first adopt collaborative filtering to calculate the predicted rating $pr_{u,i}$ via Eq. (1), and then use $pr_{u,i}$ to measure the user's acceptance probability. Specifically, the accuracy function of a recommendation list can be defined as follows:

$$PR = \frac{\sum_{u \in U} \sum_{i=1}^{L} pr_{u,i}}{|U| \cdot L} \tag{5}$$

where $|U|$ is the user number in the user set U, and L is the length of the recommendation list.

Objective Function of Novelty. The novelty function measures how many unknown items are included in a recommendation list. Here, the metric of self-information [10] is adopted to evaluate the novelty of recommended items. The self-information of item i can be defined as:

$$N_i = \log_2\left(\frac{|U|}{k_i}\right) \tag{6}$$

where k_i represents the number of users who rate item i. Let S_u be the recommendation list of user u. A user-relative novelty function is defined by calculating the average self-information of items in the target user's recommendation list, which can be calculated as follows:

$$N(L) = \frac{1}{|U| \cdot L} \sum_{u=1}^{|U|} \sum_{i \in S_u} N_i \tag{7}$$

Objective Function of Diversity. The diversity function is used to measure the difference between items in a recommendation result. Classical diversity metrics for a recommendation system include inter-user diversity, intra-user diversity and coverage [10]. Here, the coverage is chosen to evaluate the diversity of a recommendation result, which can be calculated as:

$$D = \frac{R_{dif}}{|I|} \tag{8}$$

where, R_{dif} is the number of non-duplicated items in the recommendation lists and $|I|$ is the total number of items.

3.2 Proposed Parallel Multi-objective Evolutionary Algorithm (CC-MOEA)

Individual Encoding and Initialization. In CC-MOEA, the individual is encoded as a matrix, where each entry represents an item *ID*, and the ith row represents a recommendation list for the ith user. Given that there are $|U|$ users and L items will be recommended to each user, the scale of the matrix is $|U|*L$. Generally, a recommendation system cannot recommend an item to a user twice, and hence each row of an individual cannot contain duplicate numbers. In addition, rated items by user u cannot be recommended to u again. We give an instance of an individual in Fig. 1. The individual consists of $|U|$ recommendation lists, and each list is recommended to a user. For example, items with IDs of 15, 44, 32, 56, …, and 03 will be recommended to user 1.

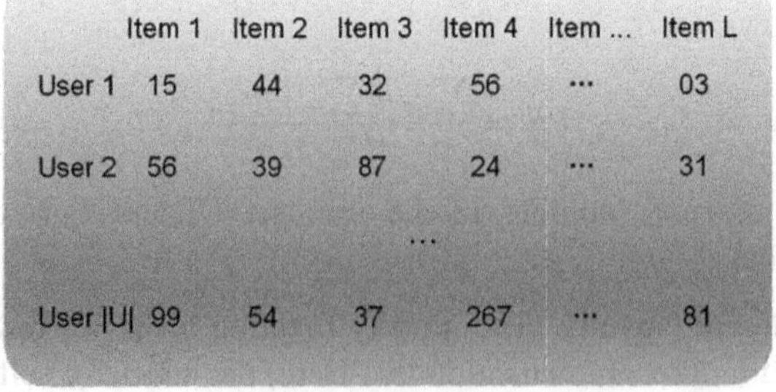

	Item 1	Item 2	Item 3	Item 4	Item ...	Item L
User 1	15	44	32	56	···	03
User 2	56	39	87	24	···	31
			···			
User $|U|$	99	54	37	267	···	81

Fig. 1. An instance of an individual

To initialize individuals, it is usually effective for improving evolution performance by making use of some prior knowledge. Hence, we initialize some extreme individuals with the optimal weighted sum of accuracy and novelty here. Given a weight vector $\omega = (\omega_0, \omega_1)$, we select the top L items with the optimal weighted sum of $pr_{u,i}$ and novelty N_i as the recommended items set $S_{u,\omega}$ for user u, and the optimal weighted sum PN_ω can be calculated as,

$$PN_\omega = \frac{1}{|U| \cdot L} \sum_{u=1}^{|U|} \sum_{i \in S_{u,\omega}} (\omega_0 \times pr_{u,i} + \omega_1 \times N_i) \qquad (9)$$

Each weight value in ω ranges in [0, 1] and their sum is equal to 1. When ω_0 is 1, an extreme individual with the optimal value of accuracy is initialized. On the contrary, when ω_0 is 0, an extreme individual with the optimal value of novelty is initialized. Meanwhile, in order to get an extreme individual with optimal diversity, the items that have been recommended to previous users are removed and no longer pushed to the other users.

Individuals are then initialized in two manners. In the first one, we generate w uniformly distributed two-dimensional weight vectors and obtain w weighted extreme individuals according to these weight vectors. Besides, an extreme individual with optimal diversity is generated. In the second manner, individuals are initialized by randomly selecting different items. It should be noted that in the parallel framework, $w + 1$ extreme individuals generated in the first manner are evenly distributed among the islands.

Genetic Operators. The genetic operators in CC-MOEA include crossover and mutation, and they are carried out to generate new individuals. In this paper, a new uniform crossover operator is designed which gives a stronger parent individual a higher genetic probability and a weaker one a lower genetic probability. Firstly, the child individual inherits the genes which exist in both two parents. Then, we calculate the genetic probability $\lambda(A, B)$ of parent individuals A and B, and based on $\lambda(A, B)$ we determine which genes of the parents are passed to the child. An illustration of the crossover operator is shown in Fig. 2, where only one row is used for focus and simplicity. The child first inherits item 02 from their parents as it exists in both of them. Then the genetic probability of parents A and B is calculated, which is 0.7 here. We generate a random number within the range of [0, 1] for each remaining gene in the child individual. The random number at the 2nd and 4th genes of the child are less than $\lambda(A, B)$, and therefore the child inherits items 44, 89 from parent A. The random number at the 3rd gene is larger than $\lambda(A, B)$, and hence the child inherits item 07 from B.

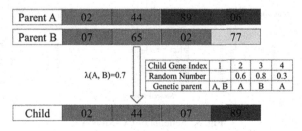

Fig. 2. An illustration of the crossover operator

CC-MOEA is devised grounded on NSGA-II, which takes non-dominated rank and crowding distance as the selection criterion. Hence, we also use the non-dominated rank and crowding distance to compare two individuals. Specifically, for two individuals with different non-dominated ranks, we prefer the individual with the lower rank. Otherwise, if both individuals belong to the same rank, then we prefer the individual that has a larger crowding distance. The genetic probability of individuals A and B can be calculated as follows,

$$\lambda(A, B) = \begin{cases} \frac{rank(A)}{rank(A)+rank(B)} & \text{if} \quad rank(A) \neq rank(B) \\ \frac{dist(A)}{dist(A)+dist(B)} & \text{else} \end{cases} \tag{10}$$

where, $rank(A)$ denotes the non-dominated rank of individual A, and $dist(A)$ denotes its crowding distance.

As for mutation, the standard single-point mutation operator is adopted.

Cooperative Coevolutionary Island Model. In the traditional master-slave model, the master processor must make global selection at regular intervals to take advantage of global information, which results in a performance bottleneck at the main processor. Although traditional island model does not suffer from this drawback, it is usually difficult for its migration strategy to guarantee a full use of global information. By making all islands cooperate to finish the task which should be done by the master processor, the proposed cooperative coevolutionary island model greatly alleviates the drawback of the master-slave model and guarantee a full utilization of global information.

Figure 3 shows the schema of the cooperative coevolutionary island model. Each island has its own subpopulation and performs a complete NSGA-II. In order to take advantage of the global non-dominated individuals, we need to select them from all islands. Hence, after a certain time of iterations, each island processor sends all their local non-dominated individuals to a local pool. Then, a parallel global non-dominated selection method is proposed to make all islands cooperate for this end.

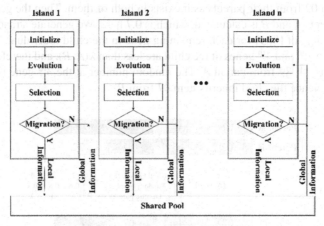

Fig. 3. Schema of CC-MOEA

Algorithm 1 Parallel Global Non-dominated Selection Method

Input: island ID k, local non-dominated individual set *local-pool*
Output: global non-dominated individual set *global-pool*
1: **for** each individual I in *local-pool*[k] **do**
2:    *mark* = true;
3:    **for** each individual J in the *local-pool* of other islands **do**
4:       **if** I is dominated by J **do**
5:          *mark=false*;
6:          break;
7:       **end if**
8:    **end for**
9:    **if** *mark=true* **do**
10:      Set the rank of I to 0;
11:      Add I to *global-pool*[k];
12:   **end if**
13:**end for**

Algorithm 1 shows the detail of the parallel global non-domination selection method. The local pool is implemented as a two-dimensional array, and local non-dominated individuals of the kth island are stored in the kth entry of the local pool. Each island is only responsible for detecting whether its local non-dominated individuals are dominated by local non-dominated individuals of other islands. When a global non-dominated individual is found, its rank is set to 0, which indicates a highest priority in mating and selection as the rank of local non-dominated individual is 1. The global non-dominated individuals are then added into *global-pool*, which is also a two-dimensional array. Afterwards, global non-dominated individuals from *global-pool* are sent to each island. In order to avoid repetitive search, all individuals from *global-pool* are divided equally into n (the number of islands) subsets, and the ith island receives individuals of the ith subset.

Furthermore, to reduce runtime of generating a new population, the individual is split into n (the number of islands) fragments and each fragment has $|U|/n$ users. Island p only performs crossover and mutation on the pth fragment of individuals. By contrast, the traditional all-crossing method can often achieve a more comprehensive exploration. To balance exploitation and exploration for the evolution algorithm, we use the fragmentation method to generate $1/n$ individuals in each island, other individuals are generated by performing crossover and mutation on all users.

Whole Procedure of CC-MOEA. Algorithm 2 shows the whole procedure of the proposed algorithm CC-MOEA. The initialization step in Line 1 not only initializes population P and Q, but also determines the location of the gene fragments used to cross and mutate in the kth island. Firstly, a combined population $R = P \cup Q$ is formed. The objective values of individuals in R are calculated and normalized via being divided by their maximal values in Line 5. Then, a fast-non-dominated sort is then performed and returns F, which contains all the individuals following their non-dominated rank.

Algorithm 2 CC-MOEA

Input: island ID k
Output: population P
1: Initialize populations P, Q, and determine $location_k$;
2: **while** *termination condition not met* **do**
3:   **while** *migration condition not met* **do**
4:     $R = P \cup Q$;
5:     Evaluate individuals in R and F = Fast-non-dominated-sort (R);
6:     **if** *migration condition not met in next iteration* **do**
7:       Select individuals from F to construct a new P;
8:       Generate a new population Q from P via proposed genetic operators;
9:   **end if**
10: **end while**
11: Send local non-dominated individuals to *local-pool*;
12: Synchronize all threads;
13: Parallel global non-dominated selection $(k, local\text{-}pool)$;
14: Synchronize all threads;
15: Receive $(k, R, global\text{-}pool)$ and F = Fast-non-dominated-sort (R);
16: Select individuals from F to construct a new P;
17: Generate a new population Q from P via proposed genetic operators;
18:**end while**

Lines 7 and 8 perform the selection process and produce a new population Q from P. These two processes do not need to be executed in the last iteration before migration because they will be executed in Lines 16 and 17. When the migration condition is met, each island sends its local non-dominated individuals to *local-pool* in Line 11. Then, all the islands cooperate to perform parallel global non-dominated selection and the global non-dominated individuals are stored in *global-pool*. All individuals from *global-pool* are divided equally into n (the number of islands) subsets, the ith island receives individuals of the ith subset and store them in set R.

4 Experiments

4.1 Experiment Settings

In our experiments, two classical benchmark data sets (i.e., Movielens and Netflix) are used. The Movielens data set, publicly available at the website of GroupLens Research (http://www.grouplens.org/), contains 943 users, 1682 movies and 100000 users' rating. The Netflix data set (http://www.netflixprize.com) is much sparser and unevenly distributed, and a subset from the data set is used for evaluations, which contains 44370 ratings of 10000 users on 3952 movies. Each rating in the two data sets is an integer ranging from 1 to 5. Here, a binary rating system is adopted by using "like" or "dislike": if the rating is not less than 3 then we mark it as "like", otherwise we mark it as "dislike". Each data set is sampled at the ratio of 8/2, forming training sets and test sets.

Evaluation experiments are conducted by implementing the proposed algorithms on a PC with Core i7 4.00 GHz, 8 GB RAM, Windows 10, Visual Studio 2019 C++.

4.2 Performance Metrics

Hypervolume. It is non-trivial to compare the performance of different multi-objective evolutionary algorithms. The most widely-used metric is hypervolume [25], also called as the S metric or the Lebesgue measure. The hypervolume of a set of solutions measures the size of the portion of objective space that is dominated by those solutions collectively. It is the only unary measure that can indicate that one non-dominated solution set is not worse than the other. Formally, it is computed as follows:

$$HV(S) = VOL(\bigcup_{x \in s} [f_1(x), z_1^r] \times \cdots \times [f_m(x), z_m^r]) \tag{11}$$

where, VOL($\cdot$) denotes the Lebesgue measure, which is a standard method for assigning a length, area, or volume to a subset of Euclidean space. In hypervolume computation, it is the union of all spaces dominated by non-dominated solutions. $z^r = (z_1^r, \ldots, z_m^r)$ denotes the reference point in the objective space which is dominated by all solutions in the true Pareto front. It is set to (0, 0) in this paper.

Real Accuracy, Diversity and Novelty. Real Accuracy P_{real} is the most important metric for evaluating RSs, and it is defined as the ratio of relevant items which are correctly recommended, as calculated by:

$$P_{real} = \frac{1}{|U| \cdot L} \sum_{u \in U} |A_u| \tag{12}$$

where, A_u is the set of items which exist in both u's recommendation list and the test set. Objective function of novelty in (7) is used to measure novelty and objective function of coverage in (8) is used to measure the diversity.

4.3 Compared Algorithms and Parameter Settings

CC-MOEA is compared with three multi-objective recommendation algorithms from the literature: MOEA-Probs [10], PMOEA [11] and MOEA-EPG [12] and two state-of-the-art parallel multi-objective evolutionary models: e-DNSGA-II [20] and SLA-MOEA [22]. Their parameter settings are listed in Table 1. Among them, L is the length of recommendation list, N is the population size, T is the number of iterations, p_m is the mutation probability, k is the number of neighbors, p_c is the crossover probability, and v is the migration interval. The parameter K is only used in MOEA-EPG and it represents the ratio of the most similar users to the target users within a cluster. For a parallel algorithm, N is the sum of subpopulation sizes from all islands. Since a too small subpopulation will have a negative impact towards evolution, N is set to 400, which can guarantee a population size of 50 per island even when the thread number is 8. w is the number of weight vectors used in the proposed initialization method, and it is set to 10 after basic tuning tests.

Table 1. Related parameters of each algorithm

Algorithm	Parameters setting
MOEA-Probs	$L = 10, N = 400, T = 3000, p_m = 1/L, k = 20, p_c = 0.8$
PMOEA	$L = 10, N = 400, T = 3000, p_m = 1/L, k = 20, p_n = 5$
MOEA-EPG	$L = 10, N = 400, T = 3000, p_m = 1/L, K = 0.5, p_c = 0.8$
e-DNSGA-II	$L = 10, N = 400, T = 3000, p_m = 1/L, k = 20, w = 10, v = 5$
SLA-MOEA	$L = 10, N = 400, T = 3000, p_m = 1/L, k = 20, w = 10, v = 5$
CC-MOEA	$L = 10, N = 400, T = 3000, p_m = 1/L, k = 20, w = 10, v = 5$

4.4 Experiment Results

Performance Comparison. In the first stage, serial CC-MOEA is compared with three multi-objective recommendation algorithm algorithms: MOEA-Probs, PMOEA and MOEA-EPG on two data sets. To test the validity of the proposed crossover operator, two CC-MOEA variants are also introduced: CC-MOEA-1 is a variant using the crossover operator from MOEA-probs and MOEA-EPG, and CC-MOEA-2 is a variant using the crossover operator from PMOEA.

The evaluations are run for 30 times. By setting the HV value of CC-MOEA as the baseline, Table 2 lists the average HV ratio of all algorithms. The HV value of CCMOEA is larger than that of CC-MOEA-1 and CC-MOEA-2, which verifies the superiority of the proposed crossover operator. Moreover, CC-MOEA-1 achieves a higher HV value than MOEA-probs and MOEA-EPG, and CC-MOEA-2 outperforms PMOEA, showing the effectiveness of the proposed initialization method.

Table 2. Mean HV ratio of serial algorithms

	MOEA-Probs	PMOEA	MOEA-EPG	CC-MOEA-1	CC-MOEA-2	CC-MOEA
Movielens	0.582	0.571	0.954	0.965	0.963	1
Netflix	0.701	0.732	0.922	0.932	0.934	1

Then, the cooperative coevolutionary island model in CC-MOEA is compared with two state-of-the-art parallel models: e-DNSGA-II and SLA-MOEA. For the sake of fairness, all parallel models adopt the proposed initialization method and crossover operator. The evaluations are run for 30 times, and Table 3 lists the average HV ratio of each model, with the HV value of CC-MOEA set as the baseline. The results in Table 3 indicate that the parallel model of CC-MOEA performs better for multi-objective recommendation system.

Through the above comparison experiments, we respectively verify the effectiveness of the proposed initialization method, crossover operator and collaborative island model. Finally, Fig. 4 gives the compared algorithms' final recommendation results with the best

Table 3. Mean HV ratio of parallel algorithms

# of threads	e-DNSGA-II		SLA-MOEA		CC-MOEA	
	Movielens	Netflix	Movielens	Netflix	Movielens	Netflix
2	0.946	0.956	0.999	0.986	1	1
4	0.942	0.944	0.914	0.923	1	1
6	0.951	0.922	0.962	0.870	1	1
8	0.993	0.904	0.966	0.939	1	1

HV values regarding the predicted rating-diversity-novelty and real accuracy-diversity-novelty spaces, respectively. We can easily see that in both spaces, the non-dominated frontiers of the other three algorithms are almost surrounded by the non-dominated frontiers of CC-MOEA. Therefore, we can conclude that CC-MOEA can always find a better recommendation than others in most cases.

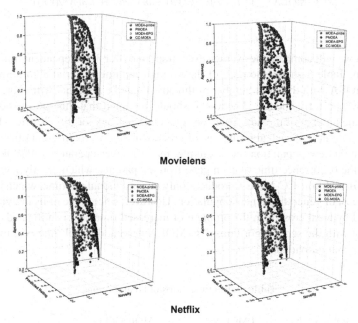

Movielens

Netflix

Fig. 4. Pareto fronts of compared algorithms

Scalability. In this section, we compare the computational complexity and runtime of each multi-objective recommendation algorithm. CC-MOEA can be divided into two parts: evolution and migration. In the former, the complexity of fast-non-dominated sort is $O(m \cdot (N/n)^2)$. Since the dimension (number of users) of individual is much larger, it also has a great impact on the complexity of generating a new population. In CC-MOEA, the complexity of generating a new individual is $O(|U| \cdot L^2)$, and n islands produce N individuals in parallel, and thus the complexity of generating next generation is $O((N \cdot |U| \cdot L^2)/n)$. In the migration period, CC-MOEA performs a parallel global non-dominated selection, whose complexity is $O(m \cdot N^2/n)$. It is important to note that the migration process is performed only T/v times. The complexity of other compared algorithms is determined by non-dominated sorting and the process of generating new populations, whose computational complexity is $O(m \cdot N^2)$ and $O(N \cdot |U| \cdot L^2)$, respectively. Table 4 shows the computational complexity of all algorithms.

Table 4. Computational complexities of all algorithms

Algorithm	Computational complexity
MOEA-probs	$O(T \cdot (N \cdot |U| \cdot L^2 + m \cdot N^2))$
PMOEA	$O(T \cdot (N \cdot |U| \cdot L^2 + m \cdot N^2))$
MOEA-EPG	$O(T \cdot (N \cdot |U| \cdot L^2 + m \cdot N^2))$
CC-MOEA	$O(T \cdot ((N \cdot |U| \cdot L^2)/n) + m \cdot (N/n)^2) + T \cdot m \cdot N^2/(v \cdot n))$

Next, we will compare the average execution time over 30 independent trials of each algorithm. Table 5 shows the ratio of the average runtime of serial algorithms to that of CC-MOEA with different number of threads. It can be seen that with more threads, CC-MOEA performs better. However, it should be noted that this ratio is not simply equal to the number of threads. There are two main reasons for this situation. First, the crossover operator of each algorithm is different, which leads to different time overhead to generate a new population. For example, the crossover operator of PMOEA needs to count the occurrence time of each gene from p_n parents, which greatly increases its runtime. In addition, CC-MOEA produces only one individual at a time, which increases the time cost of generating a new population. However, since some of its individuals are generated by local crossover, the problem of increased time cost is mitigated. Besides, compared with the serial algorithms, CC-MOEA need additional time cost to find the global optimal individuals.

Table 5. Runtime ratios of all algorithms

# of threads	MOEA-probs		PMOEA		MOEA-EPG		CC-MOEA	
	Movielens	Netflix	Movielens	Netflix	Movielens	Netflix	Movielens	Netflix
2	1.33	1.43	5.33	5.43	1.54	1.64	1	1
4	2.84	2.84	7.61	7.63	3.02	3.09	1	1
6	4.6	4.7	8.83	8.84	4.9	5.1	1	1
8	5.6	5.6	9.15	9.25	5.8	5.7	1	1

5   Conclusion

This paper first sets up a parallel multi-objective recommendation system, taking into account diversity and novelty as well as accuracy. A novel parallel multi-objective evolutionary algorithm called CC-MOEA is proposed to optimize these conflicting metrics. Then, a parallel global selection method is introduced to reduce the runtime of finding the global optimal individuals. Furthermore, a new initialization method and a crossover operator are proposed to improve the performance. The experimental results reveal that CC-MOEA outperforms some state-of-the-art multi-objective recommendation algorithms as it always achieves the highest hypervolume with the least running time.

In the future, we are interested in adapting the parallel multi-objective evolutionary algorithm to a distributed platform such as clusters and clouds.

References

1. Bobadilla, J., Ortega, F., Hernando, A., Gutiérrez, A.: Recommender systems survey. Knowledge-based systems **46**, 109–132 (2013)
2. Son, J., Kim, S.B.: Content-based filtering for recommendation systems using multiattribute networks. Expert Syst. Appl. **89**, 404–412 (2017)
3. Shu, J., Shen, X., Liu, H., Yi, B., Zhang, Z.: A content-based recommendation algorithm for learning resources. Multimedia Syst. **24**(2), 163–173 (2017). https://doi.org/10.1007/s00530-017-0539-8
4. Juan, W., Yue-xin, L., Chun-ying, W.: Survey of recommendation based on collaborative filtering. In: Journal of Physics: Conference Series, pp. 012078. IOP Publishing (2019)
5. Najafabadi, M.K., Mahrin, M.N., Chuprat, S., Sarkan, H.M.: Improving the accuracy of collaborative filtering recommendations using clustering and association rules mining on implicit data. Comput. Hum. Behav. **67**, 113–128 (2017)
6. Wang, X., Liu, Y., Xiong, F.: Improved personalized recommendation based on a similarity network. Physica A **456**, 271–280 (2016)
7. Dai, X., Cui, Y., Chen, Z., Yang, Y.: A network-based recommendation algorithm. In: 2018 3rd International Conference on Computational Intelligence and Applications (ICCIA), pp. 52–58. IEEE (2018)
8. Wan, S., Niu, Z.: A hybrid E-learning recommendation approach based on learners' influence propagation. IEEE Trans. Knowl. Data Eng. **32**(5), 827–840 (2019)
9. Chu, W.-T., Tsai, Y.-L.: A hybrid recommendation system considering visual information for predicting favorite restaurants. World Wide Web **20**(6), 1313–1331 (2017). https://doi.org/10.1007/s11280-017-0437-1
10. Zuo, Y., Gong, M., Zeng, J., Ma, L., Jiao, L.: Personalized recommendation based on evolutionary multi-objective optimization. IEEE Comput. Intell. Mag. **10**(1), 52–62 (2015)
11. Cui, L., Ou, P., Fu, X., Wen, Z., Lu, N.: A novel multi-objective evolutionary algorithm for recommendation systems. J. Parallel Distrib. Comput. **103**, 53–63 (2017)
12. Lin, Q., Wang, X., Hu, B., Ma, L., Chen, F., Li, J.: Multiobjective personalized recommendation algorithm using extreme point guided evolutionary computation. Complexity, **2018**, 18 (2018)
13. Sneha, C., Varma, G.: User-based collaborative-filtering recommendation (2015)
14. Deb, K., Pratap, A., Agarwal, S., Meyarivan, T.: A fast and elitist multiobjective genetic algorithm: NSGA-II. IEEE Trans. Evol. Comput. **6**(2), 182–197 (2002)

15. Zitzler, E., Laumanns, M., Thiele, L.: SPEA2: Improving the strength Pareto evolutionary algorithm. TIK-report, 103 (2001)
16. Wu, Q., Zhou, M., Zhu, Q., Xia, Y., Wen, J.: Moels: multiobjective evolutionary list scheduling for cloud workflows. IEEE Trans. Autom. Sci. Eng. 17(1), 166–176 (2019)
17. Zhang, Q., Li, H.: MOEA/D: a multiobjective evolutionary algorithm based on decomposition. IEEE Trans. Evol. Comput. 11(6), 712–731 (2007)
18. Beume, N., Naujoks, B., Emmerich, M.: SMS-EMOA: multiobjective selection based on dominated hypervolume. Eur. J. Oper. Res. 181(3), 1653–1669 (2007)
19. Talbi, E.-G.: A unified view of parallel multi-objective evolutionary algorithms. J. Parallel Distrib. Comput. 133, 349–358 (2019)
20. Sato, Y., Sato, M., Miyakawa, M.: Distributed NSGA-II sharing extreme non-dominated solutions for improving accuracy and achieving speed-up. In: 2019 IEEE Congress on Evolutionary Computation (CEC), pp. 3086–3093. IEEE (2019)
21. Santander-Jiménez, S., Vega-Rodriguez, M.A.: Comparative analysis of intra-algorithm parallel multiobjective evolutionary algorithms: taxonomy implications on bioinformatics scenarios. IEEE Trans. Parallel Distrib. Syst. 30(1), 63–78 (2018)
22. García-Sánchez, P., Ortega, J., González, J., Castillo, P., Merelo, J.: Distributed multi-objective evolutionary optimization using island-based selective operator application. Appl. Soft Comput. 85, 105757 (2019)
23. Alba, E., Dorronsoro, B., Giacobini, M., Tomassini, M.: Decentralized cellular evolutionary algorithms. Handbook Bioinspired Algorithms Applications 7, 103–120 (2005)
24. Burczynski, T., Kus, W.: Optimization of structures using distributed and parallel evolutionary algorithms. In: International Conference on Parallel Processing and Applied Mathematics, pp. 572–579. Springer, Berlin (2003). https://doi.org/10.1007/978-3-540-24669-5_75
25. While, L., Hingston, P., Barone, L., Huband, S.: A faster algorithm for calculating hypervolume. IEEE Trans. Evol. Comput. 10(1), 29–38 (2006)

CS-Dict: Accurate Indoor Localization with CSI Selective Amplitude and Phase Based Regularized Dictionary Learning

Jian-guo Jiang[1,2](✉), Shang Jiang[1,2](✉), Bo-bai Zhao[1,2], Si-ye Wang[1,2], Meng-nan Cai[1,2], and Yan-fang Zhang[1,2]

[1] School of Cyber Security, University of Chinese Academy of Sciences, Beijing, China
[2] Institute of Information Engineering, Chinese Academy of Sciences, Beijing, China
{jiangjianguo,jiangshang,zhaobobai,wangsiye,caimengnan, zhangyanfang}@iie.ac.cn

Abstract. With the increasingly growing demand for indoor location-based services(LBS) in the field of wireless sensing, Wi-Fi have been the mainstream method in indoor localization for the reasons of easy deployment and the popularity of signal. Channel State Information (CSI) is extracted from the physical layer of WiFi network interface cards and includes more fine-grained signal characteristics than received signal strength index (RSSI) which is commonly used in the literature. In this paper, we propose CSI selective dictionary (CS-Dict), an accurate model-free indoor localization algorithm using only one access point simultaneously. CS-Dict mainly contains two parts: CSI feature enhancement and over-complete dictionary learning. In the feature enhancement, CSI features with high reliability are selected as the input for dictionary learning. In the over-complete dictionary learning, we utilize the regularized K-SVD to perform a dictionary representation of selective CSI features in each reference point. Finally, a similarity measurement between the real-time measured CSI and the learned dictionary is performed to find the best match for position estimation. An extensive experiment is deployed in two typical indoor environments, the results show that the mean error are 0.12 m and 0.23 m respectively.

Keywords: Indoor localization · WiFi channel state information · Regularized dictionary learning

1 Introduction

Internet of Things (IoT) has developed rapidly in the past decade and its applications have been widely used in several fields such as human healthcare monitoring, industrial automation and supply chain [1]. In the Internet of Things, based on these application fields mentioned above, indoor localization plays an important role. However, unlike outdoor localization where GPS system can obtain

© Springer Nature Switzerland AG 2020
M. Qiu (Ed.): ICA3PP 2020, LNCS 12453, pp. 677–689, 2020.
https://doi.org/10.1007/978-3-030-60239-0_46

the accurate position, indoor localization confronts numerous challenges due to the presence of obstacles, multi-path effect and signal fluctuation [2]. Recently, with the development of wireless communication technology, cutting-edge indoor localization performance becomes possible. A wide range of emerging technologies which are related to wireless communications, radio frequency identification devices (RFID) [3], ultra-wideband (UWB) [4], wireless sensor networks (WSN) [5], Wi-Fi [6] have been generally utilized for accurate indoor localization.

Compared with other indoor localization technologies, Wi-Fi can achieve good performance in location-based services with convenient availability and low deployment cost. Recently, Wi-Fi based localization systems have achieved good localization accuracy, which can generally achieve the localization accuracy of decimeter-level in the literature [7]. This technology acquires the signal characteristics of received signal strength indicator (RSSI) or the channel state information (CSI) from a communication link and identify the location of a target. RSSI, which comes from the MAC layer, is a representation of the energy information in the average state of links of communication and has been widely used in Wi-Fi based localization schemes because of its low hardware requirements, such as GreenLoc [8] and D-Log [9]. However, RSSI is not a better choice than CSI which is used in this paper because RSSI is a so coarse-grained feature in indoor environment that it is unable to portray multipath fading, space-time dynamic and other signal degradation factors of the received Wi-Fi signal. Halperin et al. [10], [11] proposed a toolkit which can extract fine-grained CSI information in PHY layer from Intel 5300 Wi-Fi chips. Since CSI is sensitive to the presence of objects in indoor environment, the identifiable and quantifiable characteristics of CSI can be used to extract reliable feature for indoor localization and provide the possibility for improving localization accuracy [12].

Wi-Fi CSI based indoor localization can be divided into two categories: the range-based methods which acquiring geometric model parameters such as distance or angle from the location of the area of interest by trilateration or triangulation and the range-free methods which using fingerprints that relies on a pattern-matching approach. Comparing to the range-based methods, the rang-free methods have promising effects in mitigating multipath and the condition of none line-of-sight. At the same time, the rang-free methods have no demands for large and complex computation with respect to calculating the accurate geometric model parameters. With the limited devices in the indoor environments, how to reduce the interference from adjacent reference points and enhance the fingerprint discrimination is the key to improve indoor localization accuracy based on the range-free methods.

In this paper, we propose an over-complete dictionary learning based indoor localization algorithm, CS-Dict. In our proposed method, the collected high dimension CSI measurements are described by sparse linear combinations of the dictionary atoms (columns). We choose regularized K-SVD [13] as the dictionary learning method, and it can reduce the representation error from other reference points, which is one of the improved version based on the well-known algorithm K-SVD [14]. The processing of our proposed CS-Dict mainly includes two essential parts: Firstly, the noise of amplitude and phase that comes from the environment and hardware are alleviated respectively, and the sanitized features are selected from the best discriminative sub-carriers by the principle of information theory entropy; Secondly, we train the amplitude and phase dictionaries respectively in each reference point and combine two dictionaries together with different weights as the fingerprint of each reference point. Finally, we get the estimated position which is the optimum similarity between the online data and one of the trained fingerprints.

The contributions of this paper are summarized as follows:

1. An effective pre-processing mechanism for CSI amplitude and phase is proposed to alleviate the noise of environment and hardware and is able to improve the resolution of features.
2. The improved regularized K-SVD algorithm is utilized to represent amplitude and phase fingerprint sparsely through over-complete dictionary learning. Extensive experiments in two typical indoor environments shows that the mean error are 0.12 m and 0.23 m respectively in complex indoor environments.
3. Our prototype design is implemented on the COTS 802.11 network interface card(NIC) and has good compatibility. Using this algorithm can effectively reduce the density and cost of hardware deployment which only needs an access point and a laptop.

The remainder of the paper is organized as follows: we review related work in Sect. 2. Then the system structure and details of CS-Dict are presented in Sect. 3. Experimental deployment, results and analysis will be discussed in Sects. 4 and 5. Section 6 summarizes this paper.

2   Related Work

A number of approaches have been introduced to address the indoor localization-related fundamental issues and to improve the localization performance by combining the WiFi technology with CSI based localization algorithms. As mentioned above, WiFi CSI based localization algorithms as other common indoor localization technologies can be divided into two categories one is based on fingerprint matching and the other is based on distance or angle measuring.

In [15], Kotaru et al. proposed SpotFi, an angle of arrival (AoA) based localization system. To improve the performance of localization, SpotFi incorporated super-resolution algorithms that can accurately compute the angle of arrival (AoA) and time-of-flight of multi-path components and identified the direct path. It achieves a median accuracy of 40 cm but needs at least three access points. In [16], Vasisht et al. proposed Chronos, a novel algorithm that can compute subnanosecond time-of-flight using commodity Wi-Fi card, and obtain decimetre-level localization accuracy. This method will bring additional computation and the equipment deployment because of bandwidth stitching of multichannel.

Compared with the geometric parameter measurement method, the finger-print based localization methods which relatively need less equipment deployment, maps the extracted data matching process into a classification problem and its localization accuracy can be improved by optimizing the discernibility of features and classifiers. In [17], Wang *et al.* designed a deep network with three hidden layers to train the calibrated phase values, and employed the weights of the deep network to represent fingerprints. In [18], the processed phase features were fed into an ensemble of extreme learning machine to output their respective predictions. Two above methods which based on the CSI fingerprint only utilize CSI phase or amplitude. As a result, WiFi CSI based fingerprint localization can still undergo significant improvement to enhance localization accuracy.

3   System Design

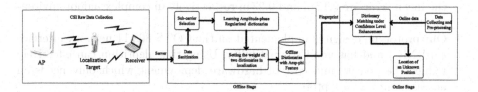

Fig. 1. The block diagram of CS-Dict system.

As shown in Fig. 1, we plot the block diagram of CS-Dict system. The system is composed of an access point (AP), a PC with modified Intel 5300 NIC as a receiver and a server for processing data in a WLAN infrastructure. The position estimation is performed through the following two procedures:

3.1   Offline Stage

In the offline stage, the area of interest is evenly divided into I reference points and we collect CSI data of each reference point using the CSI tool. However, due to the wireless channel instability and environmental noise, the feature enhancement of collected data are implemented in the data sanitization and sub-carrier selection.

Data Sanitization. The collected CSI data is the CFR of sub-carriers, which contains amplitude and phase information. First of all, we pre-process the amplitude and phase, respectively.

For the collected amplitude, we use Hampel filter to remove outliers. In the collected data set, an observation window is generated for each amplitude, which contains the data itself such as A_s and six surrounding samples that three on each side. We compute the local median m_s, and standard deviation σ_s over the current window of data, and then A_s is compared with $n_\sigma \times \sigma_s$, where n_σ is the predefined threshold value. If $\|A_s - m_s\| > n_\sigma \times \sigma_s$, the filter identifies the current sample A_s as an outlier and replaces it with the median value m_s. In addition, in order to construct a complete window at the beginning and the end of the processing, the filter algorithm prepends the data set with 3 zeros at the beginning and end, respectively. According to the Fig. 2 (a)–(b), the outliers have been removed and the curve becomes smoother after the Hampel filtering.

For the phase, through our key observation, it jumps regularly between $-\pi$ and π. Therefore, we unwrap the phase to restore the processed phase. However, duo to the phase offset [19], such as Symbol Time Offset (STO), Sampling Frequency Offset (SFO) and noise interference in the indoor environment, the unwrapped phase can not be used for indoor localization, and we need to further eliminate these offsets. The measured phase ϕ_k for the k-th sub-carrier can be expressed as:

$$\phi_k = \tilde{\phi}_k + \phi_{STO} + \phi_{k_{SFO}} + \phi_n$$
$$\phi_{k_{SFO}} = -2\pi \frac{t_k}{Q} \delta \tag{1}$$

where ϕ_k is the tangled phase, $\tilde{\phi}_k$ is the true phase of the k-th sub-carrier in theory. Q is the FFT size and is 64 in IEEE 802.11.n. t_k is the sub-carrier index of the k-th sub-carrier, and ϕ_n is the added phase shift caused by noise interference.

The main idea is to eliminate $\phi_{k_{SFO}}$ according to SFO is the linear growth relationship with the sub-carrier index. We utilize unitary linear regression $at + b$ to remove the phase offset. Assume that n sub-carrier serial numbers are increasing, the a and b are defined as follows:

$$a = \frac{\tilde{\phi}_n - \tilde{\phi}_1}{t_n - t_1} = \frac{\phi_n - \phi_1}{t_n - t_1} - \frac{2\pi}{Q} \delta \tag{2}$$

$$b = \frac{1}{n} \sum_{j=1}^{n} \phi_j - \frac{2\pi\delta}{nQ} \sum_{j=1}^{n} t_j + \phi_{STO} \tag{3}$$

According to the IEEE 802.11n specification, the sub-carrier frequency is symmetric at $40\,MHz$ bandwidth, which indicates $\sum_{j=1}^{n} t_j = 0$. b is further expressed as $b = \frac{1}{n} \sum_{j=1}^{n} \phi_j + \phi_{STO}$. We obtain a linear combination of true phases, which is denoted by :

$$\tilde{\phi}_i = \phi_i - at - b = \phi_i - \frac{\phi_n - \phi_1}{t_n - t_1} t_i - \frac{1}{n} \sum_{j=1}^{n} \phi_j \tag{4}$$

Figure 2 (c)–(d) respectively represent the raw phase and results of sanitization. From the figure, we can see that the phase information is more realistic so that improving the localization accuracy to some extent.

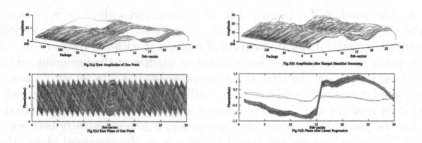

Fig. 2. The Data sanitization of amplitude and phase.

Sub-carrier Selection. After data sanitization, we get relatively true data. The size of the amplitude or the phase is $m_t \times n_r \times 30$ correspondingly, where m_t is the number of transmitting antennas, n_r is the number of receiving antennas and 30 is the number of sub-carriers. In our system, m_t is 2 and n_r is 3. However, high-dimensional CSI not only incurs over-fitting problem [20] but also reduces the position estimation accuracy. In order to solve the problem, we adopt the method based on the entropy evaluation and histogram equalization(HEQ). Specifically, we assume that complex sanitized CSI measurements in a reference point are given as Eq. (5):

$$\widetilde{H} = [\widetilde{H}_{11}(f_1), \cdots , \widetilde{H}_{mn}(f_{30})] \tag{5}$$

To select discriminative sub-carriers, we first normalize the CSI using the HEQ technique (*i.e.*, converting the coefficients into a reference). HEQ provides a transformation that converts the probability density function of an original variable into a reference probability density function which we use a typical Gaussian as a reference. Then we calculate the information gain of the k-th sub-carrier in position estimation beforehand in the Eq. (5) and rank the sub-carriers in the descending order:

$$Gain(\widetilde{H}_{mn}(f_k)) = I(D) - I(\widetilde{H}_{mn}(f_k)) \tag{6}$$

where $I(D)$ is entropy of the reference location without $\widetilde{H}_{mn}(f_k)$, and $I(\widetilde{H}_{mn}(f_k))$ is the conditional entropy of the the reference location given $\widetilde{H}_{mn}(f_k)$.

Finally, we select the most discriminative 50 sub-carrier amplitude and phase as the input of dictionary learning.

Learning Amplitude and Phase Regularized Dictionaries. In this section, we take advantage of enhanced amplitude and phase under credible sub-carriers to learn the dictionary respectively. Then, we consider the amplitude and phase dictionaries as the fingerprint of a reference point. The phase/amplitude dictionary learning problem can be expressed as:

$$f_\mu\left(D_i, X_i\right) = \|Y_i - D_iX_i\|_F^2 + \mu\|X_i\|_F^2 \tag{7}$$
$$\text{s.t. } \|x_\ell\|_0 \leq s, \ell = 1 : N$$

Given the sequence of amplitude or phase measurements $Y_i \in \mathbb{R}^{m \times N}$ up to the time instant m in each point $i \in I$ and N is the size of the measurements, $D_i \in \mathbb{R}^{m \times n}$ is the dictionary to be learned, which n denotes the number of atoms. We denote by x_l the l-th column of the representation matrix $X_i \in \mathbb{R}^{n \times N}$. μ is a given parameter for regularized coefficients and the aim of adding $\mu\|X\|_F^2$ is to constrain the large sparse coefficients, which avoids the possible bottlenecks due to nearly dependent atoms.

There are some existing scalable algorithms to solve the problem in Eq. (7). Here we make use of the regularized K-SVD algorithm to solve the sparse coefficients and the columns of the dictionary in an iterative fashion. The proposed algorithm for solving Eq. (7) is summarized in Algorithm 1. To be specific, starting with a random initial dictionary, we first fix the dictionary and solve for the sparse coefficients through the orthogonal matching pursuit (OMP). Then we update the dictionary, which is carried out in a column-wise fashion. Here the j-th column of D_i is found through the minimization of the representation error as follows:

$$E_j = min(Y_i - \sum_{\alpha \neq j}^{n} d_\alpha x_\alpha) \tag{8}$$

We make the assumption that all atoms are fixed except d_j, then we erase the contribution of d_j and compute the representation error. Regularized K-SVD aims to optimize the following error function:

$$\varphi_j = \|E_j - d_j x_j\|_F^2 + \mu\|x_j\|^2 \tag{9}$$

As a result, we find the optimal atom d_j and the corresponding optimal representation coefficients x_j. Here, the singular value decomposition (SVD) is performed for Eq. (8):

$$E_j = \sum_{\alpha=1}^{r} \sigma_\alpha u_\alpha v_\alpha^T \tag{10}$$

where the rank $r \geq 1$ is irrelevant, the solution of Eq. (9) is:

$$d_j = u_1, x_j = \frac{1}{1+\mu}\sigma_1 v_1 \tag{11}$$

The Algorithm 1 iterates between the sparse coding step and the dictionary update step until it achieves the convergence or the maximum number of iterations is reached. Finally, we obtain both amplitude dictionary and phase dictionary in each reference point.

Algorithm 1. The regularized K-SVD based fingerprints learning in reference point i.

Input:

    amplitude or phase measurements with m packages $Y_i \in \mathbb{R}^{m \times N}$

    initial dictionary $D_i \in \mathbb{R}^{m \times n}$

    number of iterations K

    parameter μ

Output:

    fingerprint dictionary D_i

    sparse coefficients X_i

1: **for** $i = 1$ to I **do**

2:    $f_\mu(D_i, X_i) = \|Y_i - D_i X_i\|_F^2 + \mu\|X_i\|_F^2$

3:    **for** $k = 1$ to K **do**

4:        **sparse coding**:fixing D_i,compute coefficients X_i with orthogonal matching pursuit

5:        **dictionary update**:

6:        **for** $j = 1$ to n **do**

7:            $d_j, x_j \overset{SVD}{\longleftarrow} argmin\|E_j - d_j x_j\|_F^2 + \mu\|x_j\|^2$

8:        **end for**

9:    **end for**

10: **end for**

3.2 Online Stage

For the localization problem at hand, D_i serves as a fingerprint for the physical location of reference point i. The overall regularized dictionary learning entails the construction of nodal fingerprints $\{D_i\}_{i=1}^I$ during the offline stage. In the online localization stage, the measurements at an arbitrary reference point are denoted by $Y_l = \{y_\tau\}_{\tau=1}^l$, where $y_\tau \in \mathbb{R}^{m \times N}$ is the element of measurements and l is the number of measurements. Given Y_l and a 2-D Cartesian coordinate system, the location estimate $(\hat{x}_\tau, \hat{y}_\tau)$ for the measurement y_τ is obtained by using all available dictionaries from the amplitude and phase database:

$$\hat{X}_{i\tau} = \arg\min_{X \geq 0} \|y_\tau - D_i X\|_F^2 + \mu\|X\|_F \tag{12}$$

where i is the index of amplitude/phase dictionaries. Subsequently, the reference point minimizes the average reconstruction error is selected as the position estimation through the single amplitude or phase dictionary:

$$\hat{i} = \arg\min_{i \in I} E(Y_l, D_i)$$

$$= \frac{1}{l} \sum_{\tau=1}^l \|y_\tau - D_i \hat{X}_{i\tau}\|_F^2 \tag{13}$$

Finally, we integrate the discriminant results of the two dictionaries and get the final localization results.

$$(\hat{x}, \hat{y}) = max(\omega_{i_{amp}}(\hat{x}, \hat{y})_{amp} + \omega_{i_{phi}}(\hat{x}, \hat{y})_{phi}) \tag{14}$$

where ω is the weight of the confidence level of the corresponding dictionary and we set it in advance. The full algorithm used for the online localization stage is summarized in Algorithm 2.

Algorithm 2. Online localization stage.

Input: Y_l ,$\{D_{i-amp}, D_{i-phi}\}_{i=1}^{I}$
Output: coordinates $(\hat{x}, \hat{y})$
 1: set the weights of the amplitude and phase dictionary fingerprints according to the actual environment
 2: calculate the sparse coefficients $\hat{X}_{i\tau}$ separately
 3: calculate results at the highest confidence level from the combination of amplitude and phase dictionaries
 4: obtain the position estimation from the weighted results

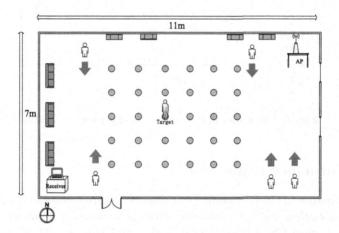

Fig. 3. The layout of yoga room experiment scenario.

4   Experimental Setup

In our experiment, the reference points are evenly distributed in the entire room and the test points are randomly chosen. We utilize a TP-link wireless AP(Model:TL-WDR5620) as a transmitter, which operates in the 5 GHz frequency band. The transmission rate was set 100 packages per second. The receiver is a Lenovo laptop integrated with Intel Wi-Fi Wireless Link 5300 Card with three antennas and it is also used as a server to estimate the position of our target when the target stands on a random test point in real time. Then, our proposed algorithm is evaluated in two typical indoor environments: a yoga room and a meeting room.

The floor plan of the yoga room is shown in Fig. 3. One access point is deployed in the upper right corner of the figure and the receiver is in the lower left corner. We divide this room into 30 reference points, and the distance between adjacent points is 0.8 m. At each reference point, we collect the raw CSI values of 3 min for our fingerprint training while there are few people walking about the room at random in order to make our scheme more persuasive.

The floor plan of the meeting room is shown in Fig. 4 and the AP and receiver are deployed in the corner diagonally. We divide this room into 28 reference points, and the distance between adjacent points is 0.8 m. In the same way, we collect the raw CSI values of 3 min in each reference point while there are few people walking about the room at random.

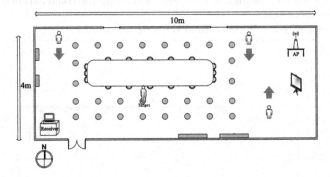

Fig. 4. The layout of meeting room experiment scenario.

5   Performance Evaluation

The performance of CS-Dict system is evaluated in terms of mean distance error, standard deviation and are compared with the CSI-based approach Pilot [21] which is a typical CSI localization algorithm and FuseLoc [22] which is a CSI fingerprint method combine phase and amplitude features with Canonical Correspondence Analysis.

Table 1 shows the mean error of CS-Dict is 0.12 m and the STD error is 0.17 m for all test points in the yoga room. In addition, for the meeting scenario where there exists abundant multipath and shadowing effect, the mean error of CS-Dict is 0.23 m and the STD error is 0.26 m. As shown in Tables 1 and 2, CS-Dict both outperforms two other methods with the smallest mean error, as well as with the smallest standard deviation error in both scenarios.

Figure 5 (a) presents the cumulative distribution function (CDF) of distance error for different methods in the yoga room. CS-Dict has 70% of the test locations having an error less than or equal to 0.2 m, while other is 33% or less. We also find that approximately 90% of the test locations for CS-Dict have an error under 0.31 m, while the percentage of test locations having a smaller error than 0.31 m are 42%, 17% for FuseLoc and Pilot respectively. In addition, our algorithm also performs better than other algorithms in the meeting room as shown

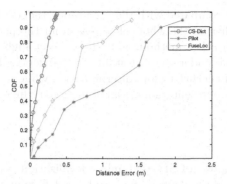

(a) CDF of distance error for yoga room scenario.

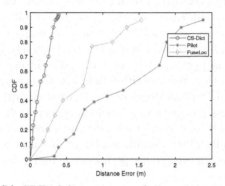

(b) CDF of distance error for meeting room scenario.

Fig. 5. CDF in different scenarios.

Table 1. Comparison of mean distance error and standard deviation for different schemes in the yoga room environment.

Algorithms	Mean error(m)	Std. dev.(m)
CS-Dict	0.12	0.17
Pilot	1.02	0.92
FuseLoc	0.63	0.70

Table 2. Comparison of mean distance error and standard deviation for different schemes in the meeting room environment.

Algorithms	Mean error(m)	Std. dev.(m)
CS-Dict	0.23	0.26
Pilot	1.30	0.76
FuseLoc	0.73	0.76

in the Fig. 5 (b). But compared with the yoga room, the localization error of the meeting room is higher on the whole. The reason for the performance improvement compared with other methods is that we select high credible amplitude and phase information under selective stable sub-carriers and the regularized K-SVD make sure the fingerprint more discrimination while reducing the adjacent interference and the representation errors of dictionary fingerprints.

6   Conclusion

In this paper, we propose a CSI based indoor localization system CS-Dict which utilizes both amplitude and phase for regularized dictionary learning. In CS-Dict, we firstly use a Hampel filter to remove amplitude outliers and a unitary linear transformation to remove noise in the phase. Then, based on the principle of information entropy, sub-carriers with strong discernibility are selected. Next, with the credible features, regularized dictionary training is performed at each reference point of division to obtain fingerprints respectively. Furthermore, online dictionary matching is performed to perform position estimation. Finally, we experimentally prove that the algorithm achieves higher localization accuracy than two other typical CSI-based indoor localization algorithms under the same experimental condition. Evaluations achieve a mean error distance of 0.12 m and a probability of 90% within 0.31 m in the yoga room and a mean error distance of 0.23 m and a probability of 90% within 0.34 m in the meeting room. The results show that our scheme achieves better performance under the different scenarios.

Acknowledgement. This work is supported by National Key Research and Development Project under grant No.2018YFF0301202, which is 'High precision, low delay electromagnetic spectrum monitoring and comprehensive situation analysis system'.

References

1. Li, S., Xu, L.D., Zhao, S.: The internet of things: a survey. Inf. Syst. Front. **17**(2), 243–259 (2014). https://doi.org/10.1007/s10796-014-9492-7
2. Basri, C., El Khadimi, A.: Survey on indoor localization system and recent advances of WIFI fingerprinting technique. In: 2016 5th International Conference on Multimedia Computing and Systems (ICMCS). IEEE (2016)
3. Huang, W., Ding, C., Wang, S., Lin, J., Zhu, S., Cui, Y.: Design and realization of an indoor positioning algorithm based on differential positioning method. In: Ma, L., Khreishah, A., Zhang, Y., Yan, M. (eds.) WASA 2017. LNCS, vol. 10251, pp. 546–558. Springer, Cham (2017). https://doi.org/10.1007/978-3-319-60033-8_47
4. Yu, K., et al.: A novel NLOS mitigation algorithm for UWB localization in harsh indoor environments. IEEE Trans. Veh. Technol. **68**(1), 686–699 (2018)
5. Boukerche, A., et al.: Localization systems for wireless sensor networks. IEEE Wirel. Commun. **14**(6), 6–12 (2007)
6. Zhao, B., et al.: Convolutional neural network and dual-factor enhanced variational Bayes adaptive Kalman filter based indoor localization with Wi-Fi. Comput. Netw. **162**, 106864 (2019)

7. Ma, Y., Zhou, G., Wang, S.: WiFi sensing with channel state information: a survey. ACM Comput. Surv. (CSUR) **52**(3), 1–36 (2019)
8. Abdellatif, M., et al.: GreenLoc: an energy efficient architecture for WiFi-based indoor localization on mobile phones. In: 2013 IEEE International Conference on Communications (ICC). IEEE (2013)
9. Ren, Y., et al.: D-Log: a WiFi Log-based differential scheme for enhanced indoor localization with single RSSI source and infrequent sampling rate. Pervasive Mob. Comput. **37**, 94–114 (2017)
10. Halperin, D., et al.: Predictable 802.11 packet delivery from wireless channel measurements. ACM SIGCOMM Comput. Commun. Rev. **40**(4), 159–170 (2010)
11. Halperin, D., et al.: Tool release: gathering 802.11 n traces with channel state information. ACM SIGCOMM Comput. Commun. Rev. **41**(1), 53–53 (2011)
12. Yang, Z., Zhou, Z., Liu, Y.: From RSSI to CSI: indoor localization via channel response. ACM Comput. Surv. (CSUR) **46**(2), 1–32 (2013)
13. Dumitrescu, B., Irofti, P.: Regularized K-SVD. IEEE Sig. Process. Lett. **24**(3), 309–313 (2017)
14. Aharon, M., Elad, M., Bruckstein, A.: K-SVD: an algorithm for designing overcomplete dictionaries for sparse representation. IEEE Trans. Sig. Process. **54**(11), 4311–4322 (2006)
15. Kotaru, M., et al.: Spotfi: decimeter level localization using wifi. In: Proceedings of the 2015 ACM Conference on Special Interest Group on Data Communication (2015)
16. Vasisht, D., Kumar, S., Katabi, D.: Decimeter-level localization with a single WiFi access point. In: 13th USENIX Symposium on Networked Systems Design and Implementation (NSDI 16) (2016)
17. Wang, X., Gao, L., Mao, S.: CSI phase fingerprinting for indoor localization with a deep learning approach. IEEE Internet Things J. **3**(6), 1113–1123 (2016)
18. Gao, R., et al.: Extreme learning machine ensemble for CSI based device-free indoor localization. In: 2019 28th Wireless and Optical Communications Conference (WOCC). IEEE (2019)
19. Sen, S., et al.: You are facing the Mona Lisa: spot localization using PHY layer information. In: Proceedings of the 10th International Conference on Mobile systems, Applications, and services (2012)
20. Taso, Y., et al.: Subcarrier selection for efficient CSI-based indoor localization. In: IOP Conference Series: Materials Science and Engineering, vol. 383. no. 1. IOP Publishing (2018)
21. Xiao, J., et al.: Pilot: passive device-free indoor localization using channel state information. In: 2013 IEEE 33rd International Conference on Distributed Computing Systems, pp. 236–245. IEEE (2013)
22. Sanam, T.F., Godrich, H.: FuseLoc: a CCA based information fusion for indoor localization using CSI phase and amplitude of WiFi Signals. In: ICASSP 2019–2019 IEEE International Conference on Acoustics, Speech and Signal Processing (ICASSP). IEEE (2019)

Recommendation with Temporal Dynamics Based on Sequence Similarity Search

Guang Yang$^{(\boxtimes)}$ (iD), Xiaoguang Hong$^{(\boxtimes)}$, and Zhaohui Peng$^{(\boxtimes)}$

School of Computer Science and Technology, Shandong University, Jinan, China
loggyt@yeah.net, {hxg,pzh}@sdu.edu.cn

Abstract. Recommender system is playing an indispensable role in our daily lives as well as in the Internet industry for the problem of information overload. Similarity search is a crucial operation in networks, database and Web search, which is usually used in recommendation. With the advent of the era of mass information that consists of great time span data, it is important to study similarity search combining temporal information. Intuitively, the older the data is, the less time weight will be in similarity calculation, so the conventional research always use the forgetting curve to model the time factor. However, these tasks only take time as a common attribute rather than a dimension. Each interaction actually is not independent in the time dimension and their chronological order contains a lot of information, utilizing the symbolic sequence relationship among the interaction and the overall structure of the data network will use these contexts efficiently and benefit the measure precision of similarity search. In this paper, a recommendation framework called SeqSim is proposed, which can synthetically utilizes the interaction information and its chronological order information to measure sequence similarity, then measures the item similarity based on sequence similarity and finally makes recommendations by preference curve. Empirical studies on real-world have shown that our new algorithm substantially improves the precision of measuring the similarity and meet the special requirements of the similarity calculation in more applications.

Keywords: Editing distance · Preference curve · Sequence similarity · Similarity search · Temporal recommend

1 Introduction

Currently the overloaded online information overwhelms users. In order to tackle the information overload problem, Recommender Systems are widely employed to guide users.

Similarity search plays a key role in recommender systems. However, data collection is regarded as static in traditional approaches [2, 4, 5, and 6]. Time information of interaction log (a user's interaction with the system) has been ignored and interactive records at different times are weighted equally. That is to say, changes in attributes or behavior of these objects are not taken into consideration. Conventional similarity measure algorithms suppose these data have the same weights in the search of similar crowd.

© Springer Nature Switzerland AG 2020
M. Qiu (Ed.): ICA3PP 2020, LNCS 12453, pp. 690–704, 2020.
https://doi.org/10.1007/978-3-030-60239-0_47

Apparently, people with similar taste who use these methods to find are not really similar. From the perspective of intuition, an interactive record recently formed should have a bigger impact on similarity search than an interactive record formed a long time ago. To address this problem, sliding window is proposed as a simple and popular approach. The approach just uses new data in current sliding window and discards old data. However, this approach can loss accuracy and aggravate sparsity. As a result of people's interest and buying habits change significantly, recommender system pays special attention to the problem. The forgetting curve is used to describe the different efficiency of the interactive record associated with the passage of time in existing studies [14, 15, and 16]. Existing forgetting curve models usually use monotonically decreasing function, e.g., linear function, exponential function, to calculate timeliness of information. Above tasks independent view of the time of each interactive record. But the chronological order of interactive records also contain huge information.

In contrast, sequential recommender systems treat the user-item interactions as a dynamic sequence and take the sequential dependencies into account to capture the current and recent preference of a user for more accurate recommendation [17]. We can simply regard sequence recommendation as sequential dependencies or sequential patterns mining. However, two problems also perplex the sequential recommender system. First, in the sequential recommender systems, the short-term sequential pattern may cover the long-term sequential pattern. Taking art purchase as an example, the sequential recommender system tends to recommend other works of the same author. But in reality, with the improvement of aesthetic taste, consumers may prefer other advanced works. Secondly, both the users' preference and items' popularity are dynamic rather than static over time. In fact, a user's preference and taste may change over time. The popularity of items is deeply influenced by seasonal, holiday, weather and other factors. For example, as people grow older their interest in action movies will become less and less. On the contrary, drama films have become the favorite of many people. Another example, many people want to watch horror movies when Halloween comes and love movies are always popular on Valentine's Day. Such dynamics are of great significance for precisely profiling a user or an item for more accurate recommendations. Most of the existing sequential recommender systems can only capture simply the dynamic change in the short term.

In this paper, combining the advantages and disadvantages of sequential recommender system and similarity search, a recommendation framework called SeqSim is proposed. As shown in Fig. 1, the specific operational steps of our new recommendation framework are as follows:

- We generate sequences for data modeling. The simplest sequence is to sort interaction records directly in chronological order. But the dimension of sequence data is higher than that of common data, so the data sparsity is increased. In order to solve this problem, we design a new clustering algorithm, which transforms data sequence into cluster sequence.
- Calculate sequence similarity. Sequence similarity calculation is a new calculation mode, and different from the traditional sequence, the sequence item in our sequence model is not a simple data or symbol. So a new algorithm is proposed to calculate

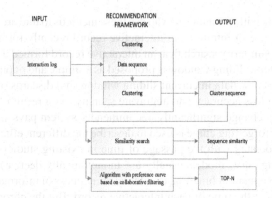

Fig. 1. SeqSim recommendation framework

sequence similarity. We use the interaction sequence of users or items to describe users or items, so sequence similarity means user or item similarity.
- Complete the recommendation. The recommendation result of collaborative filtering or other algorithms based on sequence similarity is cluster. However, the recommended result of our requirements is the item. At the same time, the trend of preference evolution also affects the accuracy of recommendation. Therefore, we propose a new algorithm with preference curve based on collaborative filtering to make recommendations.

To summarize, the major contributions of this paper are as follows:

- We propose a recommendation framework called SeqSim, It can solve the temporal problem and preference evolution problem of recommendation system.
- In SeqSim, in order to improve the efficiency of similarity search, we design a new temporal clustering algorithm to transform item sequence into cluster sequence. In order to measure sequence similarity, a new similarity algorithm is proposed. And we propose a new algorithm with preference curve based on collaborative filtering to make recommendations.
- We systematically compare the proposed SeqSim approach with other algorithms on the dataset of Movielens and Amazon. The results confirm that our new method substantially improves the precision of recommendation.

2   Temporal Clustering Algorithm and Similarity Measure

In our recommendation framework, we first need to model the data and build the sequence. Take movie data of Movielens as an example, interaction record expressed as a quadruple $q_{ui} = (user, item, rating, timestamp)$. $S_u = \{q_{u1}, q_{u2}, \ldots, q_{ul}\}$ is the rating sequence of user u for the movie and each interaction record is arranged in chronological order. Now we will initially convert the log data into a sequence and change the size of data to m in a $m \times n$-size data space, and m represents the number of users and n represents the number of items. In practical application, we observed that the above operations make the data sparse and the accuracy of similarity measurement reduced. Since the number of data cannot be increased, we choose to reduce the size of the data space. Reducing the number of users will reduce the number of data at the same time, so we can only try to reduce the number of items. To sum up, we have chosen to cluster the items to solve this problem. According to the characteristics of data, Hierarchical Agglomerative Clustering (HAC) is a good choice. Choosing a metric to measure the distance between two clusters is the key issue of hierarchical clustering. We choose to use similarity instead of distance. The more similar the two items are, the closer their distance will be.

Similarity measure is divided into three parts: rating similarity measure, feature similarity measure and temporal similarity measure.

2.1 Rating Similarity Measure

We decide to use the classic cosine similarity algorithm to calculate rating similarity. We can compute the first similarity measure between item c and d as follows:

$$PR_{cd} = \frac{\sum_{e \in I(c, d)} (r_{ec} - \bar{r}_c)(r_{ed} - \bar{r}_d)}{\sqrt{\sum_{e \in I(c, d)} (r_{ec} - \bar{r}_c)^2}\sqrt{\sum_{e \in I(c, d)} (r_{ed} - \bar{r}_d)^2}} \tag{1}$$

$$Sim_{cd}^R = e^{-\omega PR_{cd}} (\omega > 0) \tag{2}$$

Where ω is a predefined parameter. Given two items c and d, if they have commonly rating users, $I(c, d)$ represents the set of common users. $|I(c, d)|$ represents the size of the set. r_{ec} represents the user's rating of the item c and $\bar{r}_c$ represents the average score of item c. Here, we adopt an exponential function to transform user's rating difference into a similarity value. The greater the value of Sim, the greater the similarity.

2.2 Feature Similarity Measure

In Movielens websites, users and movies have many characteristics besides the rating records. For example, the movie has many characteristics, such as type, director, actor and screenwriter, etc. These features are just as important as the rating matrix but often ignored by similarity measure. We decided to use similarity search on graph to deal with these similarity measure [8]. First of all, we use movie features to create a graph. Movies and their features are the nodes on the graph. An edge indicates that the movie owns the feature. If two nodes have similar neighbors in the network, we think they are similar. We can compute the second similarity measure between node c and d as follows:

$$Sim_{cd}^F = \frac{\sum_{i=1}^{|I(c)|} \sum_{j=1}^{|I(d)|} Sim^F (I_i(c), I_j(d))}{|I(c)||I(d)|} \tag{3}$$

$$Sim_{cd}^F = 0, \quad if \ I(c) = \emptyset \ or \ I(d) = \emptyset \tag{4}$$

$$Sim_{cc}^F = 1 \tag{5}$$

$$Sim_{cd}^F = Sim_{dc}^F, \quad \text{symmetric} \tag{6}$$

Where $I(c)$ represents the entry neighborhood of node c, $|I(c)|$ represents the size of the neighborhood.

2.3 Temporal Similarity Measure

With the advent of the era of mass information that consists of great time span data, the importance of temporal information is highlighted. Intuitively, the older the data is, the less time weight will be in similarity calculation, so the conventional research always use the forgetting curve to model the time factor. However, the time difference between different events or behaviors also contains important information. Sequence can represent relative time difference, but absolute time difference is often ignored. In the case of movies, we can measure the similarity of two movies by the number of people watching them together. But there are two difficult problems in the actual calculation. First of all, for many similar films, we can no longer distinguish the similarity differences in detail by traditional algorithms. Secondly, most of the time, what we are looking for is the next movie that we will see immediately after this movie, rather than the movie that will be seen eventually. So we created a model using time difference to compute similarity. The basic idea is that the closer the two movies are viewed, the more relevant they are. Take another example of a computer journal, the time interval between the author's papers published on TOC and TPDS is very short. Therefore, we can think that TOC and TPDS themes are very similar, and the level is very close. In order to solve the above problems, we created a time similarity measure model as follows:

$$\Delta T_{c,d|u_i} = \frac{\sum_{j=1}^{k_{cd}} g\left(t_{c|u_i} - t_{d|u_i}\right)}{k_{cd} t_m} \tag{7}$$

$$\Delta T_{c,d} = \frac{|I(c)||I(d)|}{I(c,d)} \cdot \frac{\sum_{i=1}^{|I(c,d)|} \Delta T_{c,d|u_i}}{|I(c,d)|} \tag{8}$$

$$Sim_{cd}^T = e^{-\mu \Delta T_{c,d}} (\mu > 0) \tag{9}$$

Where μ is a predefined parameter. Take shopping behavior as an example, $\Delta T_{c,d|u_i}$ represents the average time interval between the purchase of items c and d by user u_i. $t_{c|u_i}$ represents the time when user u_i purchases commodity c at one time. t_m represents the average time interval of user u_i' s shopping behavior. Because users may buy the same product multiple times, k_{cd} represents the number of times user u_i purchases commodity c. $g\left(t_{c|u_i} - t_{d|u_i}\right)$ indicates that for each $t_{c|u_i}$, we select the closest $t_{d|u_i}$ to calculate the time difference. When there are two closest times, select the one that can get a positive value. The pseudocode of the proposed $\Delta T_{c,d|u_i}$ algorithm is shown in Algorithm 1. $I(c,d)$ represents a set of users who jointly purchase two items. $|I(c,d)|$ represents the size of the set. $I(c)$ represents a set of users who purchased item c. $\Delta T_{c,d}$ represents the average time interval between the purchase of items c and d by all common users. The first half of the formula ensures that hot users/items do not affect the accuracy of the formula in Eq. 8. Here, we adopt an exponential function to transform user's rating difference into a similarity value.

Algorithm 1 $\Delta T_{c,d|u_i}$ distance

Input : $log = (r_{ul})$: user and item interaction logs.
t_m: the average time interval of user u_i's interaction behavior.
Output: $\Delta T_{c,d|u_i}$: distance between items c and d based on user u_i.
1:     $j = 1$, $\Delta t = 0$, sum $= 0$;
2:     **while** $j \leq k_{cd}$ **do**
3:        **for** each time stamp $t_{c|u_i}$ in log **do**
4:           **for** each time stamp $t_{d|u_i}$ in log **do**
5:              $\Delta t' = t_{c|u_i} - t_{d|u_i}$;
6:              **if** $|\Delta t| < |\Delta t'|$ **then**
7:              $\Delta t = \Delta t'$;
8:              **if** $|\Delta t| == |\Delta t'|$ and $\Delta t < \Delta t'$ **then**
9:              $\Delta t = \Delta t'$;
10:       $g\left(t_{c|u_i} - t_{d|u_i}\right) = \Delta t$;
11:       $sum = sum + g\left(t_{c|u_i} - t_{d|u_i}\right)$;
12:       $j++$;
13:    $\Delta T_{c,d|u_i} = sum/k_{cd}t_m$;
14:    **return** $\Delta T_{c,d|u_i}$;

In order to verify the effect of the new similarity algorithm, we use collaborative filtering with different similarity algorithms to predict movie ratings in the Amazon movie dataset and utilize RMSE to evaluate the performance of all similarity algorithms. The similar algorithms we compare are Euclidean Distance (ED), Chebyshev Distance (CD), Cosine Similarity (CS), Personalized PageRank (PPR), SimRank (SR), Jaccard Coefficient (JC) and our Time Similarity Measure (TS). The experimental results are shown in Fig. 2 and our algorithm shows great superiority.

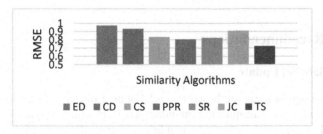

Fig. 2. Similarity algorithm performance

2.4 Joint Similarity

$$Sim = \alpha Sim^R + \beta Sim^F + \gamma Sim^T \tag{10}$$

Where α, β, and γ are the weights to control the importance of the three parts and Sim is the similarity we finally get.

3   Sequence Similarity

After completing the clustering using the HAC, we update the quadratic group to a triad $q'_{uc} = (user, cluster, timestamp)$. $S_u = \left\{ q'_{u1}, q'_{u2}, \ldots, q'_{uk} \right\}$ is the new cluster sequence. To facilitate the similarity calculation, the like items in the sequence are combined to form a new sequence $S_u = \{Q_{u1}, Q_{u2}, \ldots, Q_{uk}\}$, where $Q_{uc} = (user, cluster, size, time)$, $size$ is the number of q'_{uc} contained in each Q_{uc} in the sequence and $time$ is the median of all q'_{uc}'s timestamps. In order to calculate the similarity of the above sequences, we propose an algorithm called SeqSim Editing Distance (SSED). The dynamic programming formula of SSED is as follows:

$$
\text{SSED}[i][j] = \begin{cases} 0; & \textit{if } i = 0 \textit{ and } j = 0 \\ |b|'; & \textit{if } i = 0 \textit{ and } j > 0 \\ |a|'; & \textit{if } i > 0 \textit{ and } j = 0 \\ min\left\{ \text{SSED}[i-1][j] + \frac{size[i]}{|a|'}, \text{SSED}[i][j-1] + \frac{size[j]}{|b|'}, \text{SSED}[i-1][j-1] + f'(i,j) \right\}; & \textit{if } i \geq 1 \textit{ and } j \geq 1 \end{cases}
$$

(11)

$$
f'(i,j) = \begin{cases} \left(\frac{size[i]}{|a|'} + \frac{size[j]}{|b|'} \right)/2; & \textit{if } a[i] \neq b[j] \\ \left| size[i] - size[j] \right| / \left| |a|' - |b|' \right|; & \textit{if } a[i] = b[j] \textit{ and } size[i] \neq size[j] \\ 0; & \textit{if } a[i] = b[j] \textit{ and } size[i] = size[j] \end{cases}
$$

(12)

Where $\text{SSED}[i][j]$ represents the editing distance between sequence S_a of user a with length i and sequence S_b of user b with length j. $|a|' = size_{a1} + size_{a2} + \ldots + size_{ak}$ indicates the original length of the sequence S_a. $size[i]$ denotes the value of $size$ of the i-th term of the sequence S_a. $a[i]$ denotes the value of $cluster$ of the i-th term of the sequence S_a.

4   Make Recommendations

4.1   Candidate Set Update

Our recommendation ideas are based on collaborative filtering, therefore, in order to improve efficiency, we determine the candidate sets V_u of similar sequences and clusters in advance. For user u, S_u is its corresponding sequence. $V_u = \left(c_j, S_{u_i}(c_j) \right)$, where c_j c indicates a cluster and $S_{u_i}(c_j)$ indicates the sequence of user u_i whose last sequence term is c_j. c^{-k} represents the element of the last k in the sequence and $S_{u_i}^{-k}$ represents a new sequence with the last k elements removed. $\text{SSED}(S_u, S_{u_i})$ represents the similarity between S_u and S_{u_i}. Algorithm 2 is the set update algorithm.

Algorithm 2 Candidate set update
Input : S: all sequences for all users.
Output: V_u: the candidate set after the update.
1:      $k=1$, V_u=null;
2:      **for** each sequence S_{u_i} in S **do**
3:          **while** $SSED\left(S_u, S_{u_i}^{-k}\right) \geq SSED\left(S_u, S_{u_i}\right)$ **do**
4:              add $\left(c^{-k+1}, S_{u_i}^{-k+1}\right)$ in V_u;
5:              $k++$;
6:      return V_u;

4.2 Preference Curve

The change of preference has a great influence on the recommendation system, so we use curve fitting to model preference. In our model, a cluster represents a preference. Figure 3 shows an example of a preference change. Through the observation of the data set, we find that the change of user's preference generally follows the pattern in the legend. Some of the preferences are on the rise, some are in a stable stage, and some are cyclical. Therefore, logistic curve is used for curve fitting and its derivative is the rate of change of preference. In the example shown in Fig. 2, $a = 1.73698545968e + 001$, $b = 1.98965958209e + 001$, and $c = 2.79296646597e - 001$. We set up a unique interest curve for each user on each cluster.

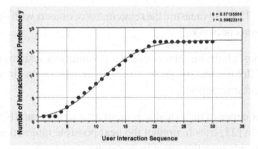

Fig. 3. An example of a preference change

Logistic curve $f_{log}^c(x)$ and its derivation $f_{log}^c{}'(x)$ of preference c are as follows:

$$f_{log}^c(x) = \frac{a}{1 + be^{-cx}} \tag{13}$$

$$f_{log}^c{}'(x) = \left(\frac{a}{1 + be^{-cx}}\right)' = acg(x)\left[1 - g(x)\right], \quad g(x) = \frac{1}{1 + b \cdot e^{-cx}} \tag{14}$$

4.3 Recommendation

SeqSim is composed of three parts. First, the similarity between the target user u and the cluster in the candidate set is obtained. Secondly, the change rate of the target user's preference in S_u is combined with the similarity of the previous step. Finally, we calculate the similarity between the items in the cluster in the candidate set and the target item.

$$P_{u,i} = \sum_{(c,S_{u_j}(c)) \in V_u} SSED(S_u, S_{u_j}) e^{f_{log}^{c}{}'(x)} \sum_{y \in c} Sim(i, y) \tag{15}$$

Where u is the target user, i is target item, and x is the current point in time that we have chosen. If there is no corresponding preference in user's historical data, $f_{log}'(x) = 0$. Finally, we sort $P_{u,i}$ and then return top-N items to user u.

5   Experiments

5.1   Experiments Settings and Metric

We use the Amazon dataset [21] and Movielens dataset to evaluate the performance of our model and baselines. Among them, Amazon dataset contains repeated shopping data, while Movielens dataset cannot contain repeated scoring records because of its particularity. There are ratings, simple attribute and reviews in total spanning from May 1996 to July 2014 in Amazon dataset. The Amazon dataset contains 7032 users who have issued 906132 explicit ratings, referring to 8033 items. The Movielens dataset contains 6040 users who have issued 999209 explicit ratings on a 5-point scale, referring to 3883 movies. We compare our model with the following baselines:

HeteSim: HeteSim [13] can measure the relatedness of objects with the same or different types in heterogeneous networks.

Markov-RS: Markov chain-based sequential recommender systems [22] adopt Markov chain models to model the transitions over user-item interactions in a sequence, for the prediction of the next interaction.

LSTM: Given a sequence of historical user-item interactions, an RNN-based sequential recommender system [23] tries to predict the next possible interaction by modelling the sequential dependencies over the given interactions.

TItemKNN: TItemKNN [14] is a time weighted item-based collaborative filtering method by reducing the influence of old data when predicting users' further behavior.

We adopt Root Mean Square Error (RMSE) and Hit Ratio defined as follows as the evaluation metrics.

$$\text{Hit Ratio} = \frac{\sum_u G(T_u \in R(u, t))}{|U|} \tag{16}$$

Where $G(\cdot)$ is an indicator function, $R(u, t)$ is a set of items recommended to user u at a specified time period t, T_u is the test item that user u accessed at a specified time period t and $|U|$ is size of all test sets. If the test item appears in the recommendation set, we call it a hit.

$$\text{RMSE} = \sqrt{\sum_{r_{ac} \in I_{test}} \frac{(r_{ac} - \hat{r}_{ac})^2}{|I_{test}|}} \tag{17}$$

Where I_{test} is the set of test ratings. r_{ac} denotes an observed rating in I_{test}. $\hat{r}_{ac}$ represents the predictive value of r_{ac}. $|I_{test}|$ is the number of test ratings.

5.2 Experimental Results

Parameter Sensitivity Analysis

There are a total of 5 parameters that need to be analyzed in the item similarity metric and they are α, β, γ, ω, and μ. $\alpha + \beta + \gamma = 1$ and $\omega = \mu = 1$. We use the classic collaborative filtering algorithm to predict the score using the Movielens dataset, and use RMSE to evaluate the effect of the score prediction in Table 1. From the table, we can see that when $\alpha = 0.1$, $\beta = 0.1$, and $\gamma = 0.8$, the score prediction is the best. This also shows that three similarity measures are essential.

Table 1. The influence of parameters α, β, and γ on scoring accuracy

RMSE											
α	γ										
	0	0.1	0.2	0.3	0.4	0.5	0.6	0.7	0.8	0.9	1
0	1.1136	1.1003	1.0977	1.052	1.0037	0.9952	0.9845	0.9617	0.9432	0.9218	0.9307
0.1	1.0998	1.0901	1.0824	1.0464	1.0001	0.9832	0.9619	0.9307	**0.9143**	0.9215	–
0.2	1.0724	1.0643	1.0528	1.0332	0.9911	0.9763	0.9527	0.9212	0.9189	–	–
0.3	1.0515	1.0502	1.0368	1.0179	0.9877	0.9632	0.9417	0.9195	–	–	–
0.4	0.9989	0.9986	0.9931	0.9893	0.9766	0.9543	0.9355	–	–	–	–
0.5	0.9907	0.9898	0.9841	0.9765	0.9635	0.9482	–	–	–	–	–
0.6	0.9811	0.9763	0.9721	0.9703	0.9531	–	–	–	–	–	–
0.7	0.9799	0.9694	0.9657	0.9581	–	–	–	–	–	–	–
0.8	0.9782	0.9725	0.9506	–	–	–	–	–	–	–	–
0.9	0.9796	0.9729	–	–	–	–	–	–	–	–	–
1	0.9845	–	–	–	–	–	–	–	–	–	–

When clustering, we need to consider the influence of clustering granularity on the experimental results. We set a similarity threshold K, when the similarity is less than Kmax(Sim), we stop merging the similar items. The value range of Sim is 0 to 1, and the value range of K is also 0 to 1. Figures 4 and 5 show the effect of K value on Hit Ratio, when N = 20 and 40. It can be seen from the figure that whether N = 20 or N = 40, the recommended effect is the best on Movielens when k = 0.7, and on Amazon, the optimal value of K is 0.6.

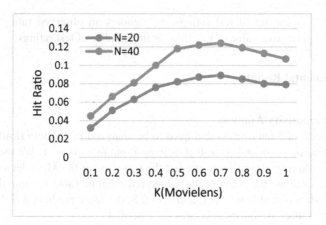

Fig. 4. The effect of K value on Hit Ratio in Movielens

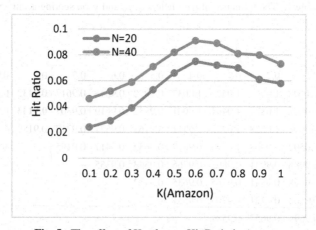

Fig. 5. The effect of K value on Hit Ratio in Amazon

Overall Accuracy Performance

Through the above experiments, we choose the optimal parameters. Then, we compare SeqSim with other excellent recommendation algorithms. Among them, HeteSim is good at recommendation based on similarity calculation, Markov chain depicts the change of preference, LSTM is the representative of sequence recommendation, and TItemKNN is the classic algorithm of sequence recommendation. The experimental results are shown in Figs. 6, 7, 8, and 9, which shows the superiority of our algorithm on all datasets. Figures 6 and 7 compare the effect of each algorithm on Hit Ratio. As common sense, in all algorithms, with the increase of N, the hit rate increases. No matter how N changes, SeqSim is the best in all algorithms. Figures 8 and 9 compare the effect of each algorithm on RMSE. As before, our algorithm SeqSim still keeps the best performance on RMSE. By comparing the two evaluation criteria, we find that our algorithm SeqSim outperforms other algorithms on Hit Ratio.

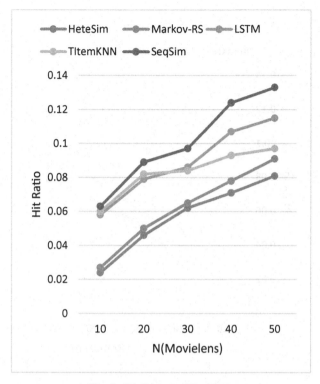

Fig. 6. Hit Ratio on Movielens

6   Related Work

Many applications require a measure of "similarity" between objects, such as document query [3], web search [1] and recommender system [2, 4, 5, and 6]. Some studies model the data into networks and calculate the similarity between nodes. Conventional study focuses on homogeneous networks, such as personalized PageRank [7], SimRank [8] and SCAN [9]. Subsequently, many studies extend to heterogeneous networks, such as ObjectRank [10], PopRank [11], Pathsim [12] and HeteSim [13]. Timing information has received increasing attention in a number of subsequent studies [14, 15, and 16]. The chronological order of interactive records also contain huge information, but the above studies and recommendation systems based on them ignore this.

In contrast, the sequence recommendation system [18, 19, and 20] satisfies the concern for chronological order, but ignores absolute time information. At the same time, they are not sensitive to preference evolution.

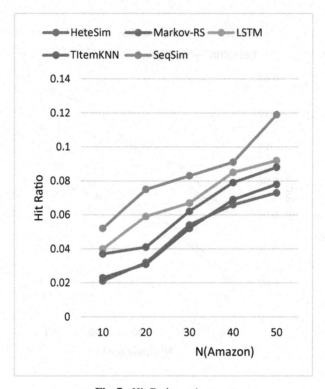

Fig. 7. Hit Ratio on Amazon

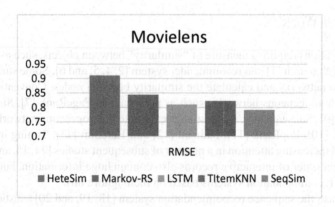

Fig. 8. RMSE on Movielens

Combining the above two categories of recommendations, we have launched the recommendation system of SeqSim in this paper and it greatly improves the accuracy of recommendation.

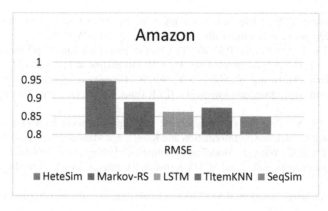

Fig. 9. RMSE on Amazon

7 Conclusion and Future Work

We propose a recommendation framework called SeqSim, It can solve the temporal problem and preference evolution problem of recommendation system. The experimental results show a significant improvement in the accuracy of our top-N recommendation method compared with the baselines. In the future work, we will try to find new unknown preferences to expand the user's recommendation horizon.

References

1. Page, L., Brin, S., Motwani, R., Winograd, T.: The PageRank Citation Ranking: Bringing Order to the Web. Stanford InfoLab (1999)
2. Konstan, J.A., et al.: GroupLens: applying collaborative filtering to Usenet news. Commun. ACM **40**(3), 77–87 (1997)
3. Baeza-Yates, R., Ribeiro-Neto, B.: Modern Information Retrieval, vol. 463. ACM press, New York (1999)
4. Goldberg, D., Nichols, D., Oki, B.M., Terry, D.: Using collaborative filtering to weave an information tapestry. Commun. ACM **35**(12), 61–70 (1992)
5. Konstan, J.A., et al.: GroupLens: applying collaborative filtering to Usenet news. Commun. ACM **40**(3), 77–87 (1997)
6. Shardanand, U., Maes, P.: Social information filtering: algorithms for automating "word of mouth". In: Proceedings of the SIGCHI Conference on Human Factors in Computing Systems, pp. 210–217 (1995)
7. Jeh, G., Widom, J.: Scaling personalized web search. In: Proceedings of the 12th International Conference on World Wide Web, pp. 271–279 (2003)
8. Jeh, G., Widom, J.: SimRank: a measure of structural-context similarity. In: Proceedings of the Eighth ACM SIGKDD International Conference on Knowledge Discovery and Data Mining, pp. 538–543 (2002)
9. Xu, X., Yuruk, N., Feng, Z., Schweiger, T.A.: Scan: a structural clustering algorithm for networks. In: Proceedings of the 13th ACM SIGKDD International Conference on Knowledge Discovery and Data Mining, pp. 824–833 (2007)
10. Balmin, A., Hristidis, V., Papakonstantinou, Y.: Objectrank: Authority-based keyword search in databases. In: VLDB, vol. 4, pp. 564–575 (2004)

11. Nie, Z., Zhang, Y., Wen, J.R., Ma, W.Y.: Object-level ranking: bringing order to web objects. In: Proceedings of the 14th International Conference on World Wide Web, pp. 567–574 (2005)

12. Sun, Y., Han, J., Yan, X., Yu, P.S., Wu, T.: PathSim: meta path-based top-k similarity search in heterogeneous information networks. Proc. VLDB Endow. **4**(11), 992–1003 (2011)

13. Shi, C., Kong, X., Huang, Y., Philip, S.Y., Wu, B.: Hetesim: a general framework for relevance measure in heterogeneous networks. IEEE Trans. Knowl. Data Eng. **26**(10), 2479–2492 (2014)

14. Ding, Y., Li, X.: Time weight collaborative filtering. In: Proceedings of the 14th ACM International Conference on Information and Knowledge Management, pp. 485–492 (2005)

15. Chen, J., Wang, C., Wang, J.: Modeling the interest-forgetting curve for music recommendation. In: Proceedings of the 22nd ACM International Conference on Multimedia, pp. 921–924 (2014)

16. Töscher, A., Jahrer, M., Bell, R.M.: The bigchaos solution to the netflix grand prize. Netflix Prize Doc. 1–52 (2009)

17. Chen, X., et al.: Sequential recommendation with user memory networks. In: Proceedings of the Eleventh ACM International Conference on Web Search and Data Mining, pp. 108–116 (2018)

18. Yap, G.E., Li, X.L., Philip, S.Y.: Effective next-items recommendation via personalized sequential pattern mining. In: Lee, S., et al. (eds.) International Conference on Database Systems for Advanced Applications. Lecture Notes in Computer Science, vol. 7239, pp. 48–64. Springer, Berlin, Heidelberg (2012). https://doi.org/10.1007/978-3-642-29035-0_4

19. Rendle, S., Freudenthaler, C., Schmidt-Thieme, L.: Factorizing personalized markov chains for next-basket recommendation. In: Proceedings of the 19th International Conference on World Wide Web, pp. 811–820 (2010)

20. Wu, C.Y., Ahmed, A., Beutel, A., Smola, A.J., Jing, H.: Recurrent recommender networks. In: Proceedings of the Tenth ACM International Conference on Web Search and Data Mining, pp. 495–503 (2017)

21. McAuley, J., Pandey, R., Leskovec, J.: Inferring networks of substitutable and complementary products. In: Proceedings of the 21th ACM SIGKDD International Conference on Knowledge Discovery and Data Mining, pp. 785–794 (2015)

22. Feng, S., et al.: Personalized ranking metric embedding for next new poi recommendation. In: Twenty-Fourth International Joint Conference on Artificial Intelligence (2015)

23. Hochreiter, S., Schmidhuber, J.: Long short-term memory. Neural Comput. **9**(8), 1735–1780 (1997)

A Software Stack for Composable Cloud Robotics System

Yuan Xu[1,2]([✉]), Tianwei Zhang[3], Sa Wang[1,2], and Yungang Bao[1,2]

[1] State Key Laboratory of Computer Architecture, ICT, CAS, Beijing, China
xuyuan@ict.ac.cn
[2] University of Chinese Academy of Sciences, Beijing, China
[3] Nanyang Technological University, Singapore, Singapore

Abstract. Modern cloud robotic applications face new challenges in managing today's highly distributed and heterogeneous environment. For example, the application programmers must make numerous systematical decisions between the local robot and the cloud server, such as computation deployment, data sharing and function integration.

In this paper, we propose ROBOTCENTER, a composable cloud robotics operating system for developing and deploying robotics applications. ROBOTCENTER provides three key functionalities: runtime management, data management and programming abstraction. With these functionalities, ROBOTCENTER enables application programmers to easily develop powerful and diverse robotics applications. Meanwhile, it can efficiently execute these applications with high performance and low energy consumption. We implement a prototype of the design above and use an example of AGV/UAV cooperative transport application to illustrate the feasibility of ROBOTCENTER. In the experiment, we reduce the total energy consumption and mission completion time up to 41.2% and 51.5%, respectively.

Keywords: Cloud robotics · Runtime management · Data management · Programming abstraction · Composable system

1 Introduction

Over the past decade, we have witnessed two revolutionary hardware trends: the massive-scale cloud computing platforms and plentiful, cheap wireless networks. These technologies have triggered the killer applications of edge-cloud revolution – cloud robotics [6,27,30,31]. On one hand, robotic applications, such as Microsoft FarmBeats [46] and Dji FlightGub [12], enable robots to publish and share information with other edge robots or sensors. On the other hand, Rapyuta [37] and Amazon IoT Greengrass [8] provide a platform for robots to offload computation-intensive tasks on cloud servers. The rise in demands for those applications has led more researchers and practitioners to develop cloud robotics systems for computation offloading and data sharing.

© Springer Nature Switzerland AG 2020
M. Qiu (Ed.): ICA3PP 2020, LNCS 12453, pp. 705–720, 2020.
https://doi.org/10.1007/978-3-030-60239-0_48

The state-of-the-art cloud robotics systems mainly fall into two ends of the spectrum in the robotic community. One end is the monolithic systems, which tightly integrate edge robots with cloud servers for specific tasks, including automated logistics (e.g. Amazon Kiva [22]) and precision agriculture (e.g. Microsoft FarmBeats [7]). However, these systems are usually ad-hoc and hard to be extended with other kinds of robots or customized tasks [44].

The other end is the robotic middlewares (e.g. ROS [2], Player [20]). They provide an abstraction layer to facilitate coordination and message passing between different modules of robots [23]. All the modules are loose-coupled and easy to be integrated with third-party libraries, such as OpenCV [10], PCL [11] and so forth. Those flexible robotic middlewares have drawn much attention and became the de facto operating systems for developing robotic applications. Those middlewares work well for managing one or a few robots. However, there are still a lot of unresolved challenges when managing a group of heterogeneous robots. Cloud robotics application programmers have to take into consideration plenty of complicated resource management details by their own, e.g. availability, adaptability, fault-tolerance and reactivity.

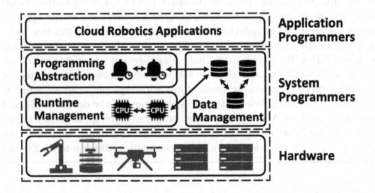

Fig. 1. The robot application environments and system architectures.

In this paper, we argue that the cloud robotics applications need a new software stack to manage and share multi-robots resources. Figure 1 shows the robot application environments and system architectures. The bottom layer is the hardware, containing computing processor, memory, sensors, etc. The middle layer is the software stack, developed by system programmers. The top layer is the robotics applications developed by application programmers. The software stack should provide at least three key functionalities: 1) **Runtime Management**: the stack should make decisions when (dynamic migration) and where (robots or cloud) to deploy user-defined jobs. 2) **Data Management**: the stack should maintain a bunch of state information about each robot, such as ID, heartbeats, maps and so forth, which can be shared among multiple robots. 3) **Programming Abstraction**: the stack should provide a high-level and easy-to-use interface for application programmers to manage the underlying robot resources.

We further propose ROBOTCENTER, a software stack for *cloud robotics composable systems*. Specifically, (1) for *runtime management*, we implement `Avalon`, an adaptive resource management system, to provide the functionalities of computation offloading for better performance and energy usage. (2) For *data management*, we introduce Shared Knowledge Base (`SKBase`), a shared memory database to maintain the data for application programmers and provide data abstraction for availability. (3) For *programming abstraction*, we propose `ActionFlow`, a cloud robotics programming framework to provide modularity through a collection of pluggable planner modules that extend the software stack to support application-specific scheduling needs.

The combination of `Avalon`, `SKBase` and `ActionFlow` provide a composable cloud robotics system that manages distributed computation and data for cloud robotics applications. Programmers can develop their applications with modularized interfaces without considering availability, adaptability, fault-tolerance and reactivity. Our prototype show that the processing time of path panning process in cloud based multi-robot cooperative transport application can be reduced by 52%.

The rest of this paper is organized as follows: In Sect. 2, we discuss the main challenges and requirements for designing a cloud robotics OS. The mechanisms and functionalities of ROBOTCENTER framework is presented in Sect. 3. In Sect. 4, we use an example of AGV/UAV cooperative transport application to illustrate the feasibility and experimental results of ROBOTCENTER. Section 5 discusses some open problems and our future work. Section 6 concludes this paper.

2 Background

2.1 Cloud Robotics Application

This section first describes the characteristics of developing a cloud robotics application, and then discusses the new challenges and requirements for designing a cloud robotics OS.

Development Characteristics. Today, robotic applications have become inherently *distributed*, with data and computation spreading across edge robots and cloud servers. As a result, modern robotic application development has acquired two new characteristics:

1. **A distributed computation deployment.** Different from past robotic applications that run on a single robot, modern cloud robotics applications distribute computation across edge robots and cloud servers with wireless network. However, due to uncontrollable environmental factors, such as instability of wireless connections, fluctuation of communication bandwidth, and robot mobility, programmers must consider an adaptive computation scheduling policy under the highly unreliable and unpredictable data transmission.

2. **New reliable and persistence guarantees.** Different from past robotic applications that store robot's states in local file systems, modern cloud robotics applications shift the states storage from edge to cloud back-ends. The definition of the term "states" is the collection of states of a robot and its environment. Such a distributed data storage can incur more frequent faults, and make it harder to recover them. It is challenging to satisfy users' requirements of executing cloud robotics applications reliably without process crash or data loss.

Table 1. Cloud robotics OS requirements and challenges

Characteristic	Requirement	
Distributed computation deployment	**Availability**:	Remain usable with unreachable nodes
	Adaptability:	Respond with low latency under variable performance
Reliable and persistence guarantees	**Fault-tolerance**:	Ensure data is not lost on crashes
	Reactivity:	Propagate state updates automatically
Inherent application design	**Modularity**:	Simple to reuse at low cost

Application Development. As shown in Table 1, the two characteristics lead to a set of new challenges for programmers while developing cloud robotic applications. These challenges dictate the requirements for a cloud robotics OS which guide our stack design.

The first two challenges and requirements stem from the distributed computation deployment. Since mobile robots execute tasks in a highly unpredictable environment, the interaction between edge robots and cloud is influenced by many incontrollable factors. For example, As a robot keeps moving, its uplink and downlink bandwidth can fluctuates frequently, or even get disconnection, due to the absorption of electromagnetic waves or wireless interference with other devices. As a result, programmers must ensure that their applications are *available* so the robot can continuously execute tasks with unreachable nodes in the cloud. Meanwhile, the benefit from computation offloading can be decreased in the poor wireless environment, thus the cloud robotics applications should be *adaptive* to migrate between edge robots and cloud servers.

The next two challenges and requirements arise from new reliable and persistence guarantees. Robotic applications are used to be executed in a closed-loop streaming processing model, e.g. sensing data from the environment, interpreting and making decision, navigating or modifying the environment. Each component generates large amounts of states that need to be stored in either edge robots or cloud. Consequently, cloud robotics applications must periodically checkpoint or log to the storage system for *fault-tolerance*, and keep *reactive* to automatically propagate the state updates from each component to persistent storage.

The final challenge and requirement is a consequence of inherent cloud robotics application design. *Modularity* plays an important role in the rapid development of applications, as it allows programmers to develop a new application from existing components. So, programmers must make their code easier to be integrated in other robotic applications to meet different demands of dynamic environment.

2.2   Cloud Robotics OS Abstractions

It is necessary to have a cloud robotics OS to meet the five requirements for developing applications. In this section, we describe the related work from three OS functionalities: runtime management, data management and programming abstraction. We further discuss the issues of current technologies to address these challenges.

Runtime Management. Cloud robotics OSes manage runtime processes of applications and decides when and where to deploy these computations dynamically. Recently, the RoboEarth project [3] introduces a centralized task controller to manage a group of robots [29]. It also uses Rapyuta [37] to offload computation to clouds and speed up computation-intensive robotic applications such as navigation [29] and SLAM [39,40]. Amazon recently released a cloud robotics platform, RoboMaker [9], which provides a cloud-based platform for robotic application development and simulation. Programmers can deploy their applications in the edge robot and cloud through IoT Greengrass [8].

These cloud runtime systems support lightweight containers that allow application programmers to easily execute customized processes in the cloud (*modularity*). They provide basic runtime management that monitors and updates states of each robot periodically (*reactivity*), detects and restarts crashed containers (*fault-tolerance*).

Issue #1: Both Rapyuta and RoboMaker systems are in short of *adaptivity* and *availability*. Specifically, they require programmers to set up and deploy containers without automatically computation migration for better performance or energy usage, and neglect the network disconnection due to unpredictable wireless network (Table 2).

Data Management. Cloud robotics OSes also manage the storage distribution of robots' states. For example, the KnowRob [15,45] and the RoboBrain [5] intend to build a robotic wikipedia to share information that guide robots to perform actions and interact with objects.

Table 2. Comparison of our stack to prior work for cloud robotics application development

OS	Availability	Adaptability	Fault-tolerance	Reactivity	Modularity
Rapyuta [37]			✓	✓	✓
RoboMaker [9]			✓	✓	✓
Cloud Database [5,45]			✓	✓	✓
ROS [2]					✓
SOA [32,35,36,40,42]					✓
Monolithic OS [7,22]	✓	✓	✓	✓	
Our Framework: RobotCenter	SKBase ($3.2)	Avalon ($3.1)	SKBase ($3.2)	SKBase ($3.2)	ActionFlow ($3.3)

These cloud database systems store the logic data in the relation database (e.g. Sesame [17]) and the binary data in the distributed storage systems (e.g. Apache HDFS [1]). The robot is regarded as a completely stateless mobile client that can query and analyze knowledge representation languages (e.g. Web Ontology Language [4]) that are related to objects and maps of applications. Application programmers can continuously checkpoint to these storage systems for *fault-tolerance* and periodically poll to update *reactive* state updates.

Issue #2: These cloud database systems do not meet the requirements of *adaptivity* and *availability*. While applications are implemented with stateless mobile clients, users cannot access the data in cloud storage under poor wireless bandwidth. Moreover, replicating data to the cloud increases the availability of the applications at the cost of responsiveness. It is ineffective for some tasks with strong real time constraints for fault recovery.

Programming Abstraction. The key role of an Cloud Robotics OS is to provide abstractions and APIs, hiding the underlying complex implementations of robotic functions and hardware features for the programmers. ROS (Robot Operating System [2]) makes a great progress towards programming abstraction through providing a series of common libraries for low-level device control and inter-process communication. Application programmers achieve functions via launching a set of concurrent long-running processes, called "nodes". The communication between two nodes can be either a two-way request/reply pattern called *rosservice* or a many-to-many publish/subscribe pattern called *rostopic*. Application programmers can easily achieve these functions by sending/subscribing to each function with ROS-type message. Besides, researchers in the robotic community have proposed many cloud robotics services to meet the needs of different applications, such as 3D Mapping [40,42] and grasp planning [32,35,36].

Issue #3: Unfortunately, ROS just provides the basic communication and hardware control abstraction. Application programmers need to implement all cloud robotics application requirements for themselves. Besides, current Service-Oriented Architectures (SOA) frameworks focus on the efficiency of algorithm implementations and cannot meet other OS requirements.

3   RobotCenter Framework

This section describes the mechanisms and functionalities of ROBOTCENTER framework. We present how ROBOTCENTER meets five requirements discussed in Sect. 2.

3.1   Runtime Management: Avalon

We designed a resource management framework, Avalon, to achieve *adaptability* in ROBOTCENTER. Avalon implements an adaptive energy-efficient scheduler to migrate the computation across edge robots and cloud servers automatically under unpredictable wireless transmission latency. Although adaptive computation offloading problem is widely studied in mobile cloud computing domain (e.g. CloneCloud [19], MAUI [21] and ThinkAir [33]), they are not well suited for robotics applications because of several reasons. First, robotic applications are multi-process while mobile applications are multi-thread without exposing sockets. Second, the offloading decision model of robotic application depends on not only data transmission latency, task processing time and CPU frequency, but also some robotic factors, e.g. velocity and decision accuracy.

To ensure the responsiveness of various robotic applications, Avalon delegates control over scheduling to the framework with a customized energy-efficient utility model. Specifically, the energy-efficient utility function is characterized along two dimensions: *total completion time* and *total energy consumption*. The total completion time (TT) denotes the time a mobile robot completes the mission, which depends on the traveling distance and robot's velocity. In contrast, the total energy consumption (TE) means the energy consumed during the mission execution. Thus, we propose the energy-efficient cost (EEC) by the following definition.

Definition 1 (Energy-Efficiency Cost (EEC)): The energy-efficiency cost (EEC) is defined as the weight sum of total completion time and total energy consumption. Thus, the EEC of a robotic workload is given by

$$Z = \gamma^T TT + \gamma^E TE \tag{1}$$

where $0 \leq \gamma^T, \gamma^E \leq 1$ denotes the weights of total completion time and totally energy consumption for the workload execution of the mobile robot. In this equation, we normalize the measurement of TT and TE so that the value of Z will fall in $[0, 1]$. To provide a scalable and resilient core for enabling various mobile robots to efficiently perform the workload, we allow that different framework programmers can customize different weighting parameters in the decision making. For example, when a mobile robot has a long journey at a low battery state, the robot prefers to set a larger weight on TE to save more energy. When a mobile robot executes some applications that are sensitive to the delay, such as search and rescue, choosing a larger weight on TT would navigate the robot to the destination as soon as possible.

The utility function of local mode Z_{local} (robot only) and remote mode $Z_{remotel}$ (robot + cloud) is different due to the resource characteristics. Thus, the intersection of two utility curves determines the migration time to maximize performance. As discussed in recent works of mobile cloud offloading [18,26], the utility function Z_{local} and $Z_{remotel}$ depend on many factors, such as cpu frequency, transmission time, processing time of computations in local and remote server. Here, we introduce a novel factor that only exists in robot workloads, i.e. decision accuracy. Unlike mobile phones, robots can sacrifice the decision accuracy to reduce computation utilization under constraints [16,44]. For instance, the increase in planning resolution makes the voxels of map larger, so space is represented more coarsely and less accurately, then performance improves due to the less needed computation. However, the mission may fail in a complex environment under a low accuracy. Hence, Avalon attaches the accuracy constraints to utility function and helps programmers to determine on redirecting a computation replication with proper accuracy.

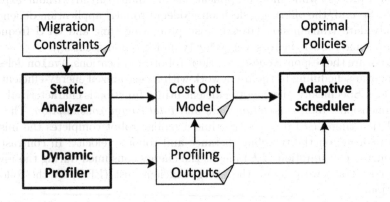

Fig. 2. Avalon architecture overview.

As Fig. 2 shows, the Avalon implements the utility model with three parts: *Static Analyzer*, *Dynamic Profiler* and *Adaptive Scheduler*. The *Static Analyzer* identifies the legal range of outputs in the utility model, according to a set of customized migration constraints. For example, maximum CPU frequency and maximum velocity limit the upper bound of frequency and speed scaling policy. The *Dynamic Profiler* collects the robot's states (e.g. CPU load, network latency) as the input of the model and sends to the cloud servers periodically. Finally, the *Adaptive Scheduler* finds the optimal decision by minimizing the utility function, including computation migration, edge robot's CPU frequency and planning accuracy control.

3.2 Data Management: SKBase

We propose SKBase, a shared knowledge database, to achieve *availability, fault-tolerance* and *reactivity* for data management. SKBase keeps multiple copies of states on edge robots and cloud servers for fault recovery, and automatically propagate updates across these copies.

To ensure the *availability* of application effectively, SKBase provides shared memory among processes in both edge and cloud. Further, SKBase introduces a new data abstraction, called *shared data object* (SDO) as the sharing unit. SDOs support various data type, including simple primitives (e.g. string), collections (e.g. list) and ROS messages (e.g. geometry_msgs). Thus, application programmers can encapsulate robot's states without modifying original data type.

Due to various real time requirements for fault recovery of processes, SDOs should be deposited in different places. For example, the localization process should be auto-restarted with initial position on crash. The application might place the current position state on edge robot for low latency because the robot's position is still changing while moving. On the other hand, states like generated map could be placed on a cloud server for sharing with other robots. Thus, *SKBase* proposes a new data mapping interface to place copies in a cloud-based distributed storage system or edge robot's disk. Application programmers use this interface to link in-memory states to keys in different storage systems and all SDOs with the same key will be automatically synchronized.

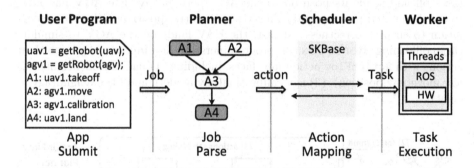

Fig. 3. ActionFlow execution process.

3.3   Programming Abstraction: ActionFlow

We propose ActionFlow, a ROS based *modularized* cloud robotics programming framework, which provides a RPC-like action abstraction for job scheduling. ActionFlow delegates data management and resource management to SKBase and Avalon respectively. Thus, ROBOTCENTER meets all five cloud robotics application requirements.

ActionFlow encapsulates the SOA frameworks as a collection of pluggable modules that support application-specific scheduling needs. Figure 3 presents the basic components and execution workflow in ActionFlow. Specifically, application programmers submit their work to ActionFlow in the form of *jobs*, each of which consists of one or more *actions* that all run the same program. System programmers design different planners to parse jobs into different actionDAGs. The scheduler then maps each action into a corresponding ROS task through SKBase and sends it to available robots. After the worker in each robot completes the task, it returns the results to the planner and updates the data. Each robot works in a stateless process, which means they are agnostic about the job collaboration, while the action scheduling is decided by the planner.

Another contribution of ActionFlow is the RPC-like action abstraction based on action mapping, which hides ROS relevant interfaces for interacting with robots. ActionFlow provides various APIs for different types of actions to simplify development of cloud robotics frameworks through storing mapping from its action interface to the a specific ROS task.

4   Use Case and Evaluation

We implemented a prototype[1] of the design above to illustrate the feasibility of ROBOTCENTER. We use the example of AGV/UAV cooperative transport application. In this workload, both AGV (automatic guided vehicle) and UAV (unmanned aerial vehicle) are put in an obstacle-filled environment. The UAV takes off and sends its position to the AGV periodically. The AGV navigates to the designated position and calibrates the relative position with vision recognition to ensure correctness. At last, the UAV lands on the AGV to simulate cargo offloading. ROBOTCENTER programmers can decompose the job into four actions with ActionFlow action interfaces, including (1) uav.takeoff(height), (2) agv.move(uav.position), (3) agv.calibration(uav.model), and (4) uav.land(agv.height).

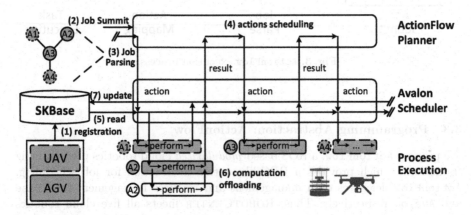

Fig. 4. A time-line sketch of cooperative transport. All white blocks construct Robotcenter base built in gateway, and grey blocks are launched in robots.

---

[1] The experiment video can be found in https://youtu.be/KeYyS6lZxo0.

In our experiment, We deployed the Dji Spark and turtlebot in our lab. The turtlebot is equipped with an Intel i5-2450M CPU @ 2.5 GHz, 1.8 GB of RAM netbook and a Dji guidance over the top board. The two robots were connected to a powerful gateway server (Intel i7-7700K CPU @ 4.2 GHz with 16 GB of RAM and a GeForce GTX 1080) with a passive 5 GHz band wireless network. In this application, we controlled the turtlebot to create a 2D occupancy grid map of our lab environment and upload it to the SKBase as a shared data object.

Figure 4 shows the runtime of the collaboration between the turtlebot and UAV. We describe each step executed (the numbers in Fig. 4 correspond to the numbers in the following list):

1. The AGV and UAV are registered to Avalon and the static knowledge is stored in SKBase.
2. A cooperative transport job is submitted through the application with some parameters. The application passes the job to the Planner in ActionFlow.
3. The Planner parses the job into an actionDAG, including six action primitives shown in Fig. 3.
4. The Planner forwards each action to the Scheduler in order and waits for the feedback of these actions.
5. The Scheduler queries SKBase through ActionFlow scheduler API to allocate the two robots and map each action to ROS task.
6. The Scheduler sends ROS tasks to the turtlebot and UAV with Avalon API. Notice that the worker in turtlebot detects that the processing time of the path planning node is too high and migrates it to the gateway server.
7. The Avalon worker executes each task in an ROS environment and returns the result and updates the related information in SKBase. The Planner determines the next action until the job is finished.

In the whole life cycle of above job, the 'move' and 'calibration' actions are important because they use Avalon to offload computation nodes and SKbase to share knowledge respectively. Thus, we discuss the detail implementation of these two actions in the next subsections.

4.1 Move Action

The function of action "move" aims to navigate the robot through an obstacle-filled environment to reach some arbitrary destination. Specifically, it plans an efficient collision-free path in the map and follows. For each movement, it simulates multiple trajectories based on some mechanical characteristics (velocity, acceleration limit), and identifies each trajectory whether conflicts with obstacles. Obviously, this action is a computation-intensive task that consumes much energy of the robot. Moreover, the robot has to sacrifice velocity to meet some constraints of each functional node, such as the accuracy loss in localization and the conflict possibility in obstacle avoidance. The reduction of velocity further prolongs the mission completion time in navigation task. In summary, the limited resource constrains both total energy and mission completion time of this action.

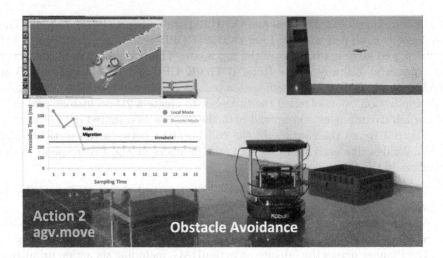

Fig. 5. The computation offloading example of "move" action.

Figure 5 shows the computation offloading example of "move" action in our experiment. We implement our "move" action with ROS navigation stack [13]. In our initial prototype, we simply consider processing time as the only factor to affect our utility function. We preset offloading threshold of processing time of local robot to 250 ms through *Static Analyzer*, which means while the processing time of any functional node in local robot is bigger than 250 ms, it would be migrated to the gateway server for energy and performance optimization through *Adaptive Scheduler*. The *Dynamic Profiler* monitors the processing time of each node in runtime.

While the turtlebot received the position of the UAV from SKBase, it launches a group of ROS functional nodes to setup action runtime environment, such as amcl, move_base node. From the figure, we can observe that the processing time of move_base node exceeds our threshold, the Avalon automatically migrates the move_base node to the remote gateway server and synchronizes related data (e.g. UAV position, node status) to the SKBase. Based on these operations, the processing time of move_base node reduces almost 64%. Besides, the total energy and mission completion time reduce up to 41.2% and 51.5% during the life cycle of "move" action execution.

4.2 Calibration Action

The function of action "calibration" is the process of identifying relative position between the turtlebot and the UAV based on the sensor data. Specifically, the turtlebot moves to the bottom of the UAV and receives the UAV's image through Dji guidance in five directions. To reduce unnecessary computation, once recognizing UAV in one direction, the turtlebot will only process the images from this direction. All the received images are used to predict the UAV position relative to itself through an on-line image recognition model.

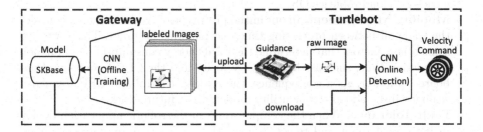

Fig. 6. The implementation of UAV-AGV "calibration" action.

Figure 6 shows the implementation of UAV-AGV "calibration" action. The edge gateway is responsible for receiving images from the turtlebot and we label the UAV position in each image. These preprocessing images will be used for off-line training in a convolutional neural network. The trained model is later saved in the SKBase and downloaded by the turtlebot for the on-line detection when the turtlebot makes "calibration" action. For each receiving raw image, the turtlebot identifies the related position of the UAV and then computes the velocity command for calibration.

We use TensorFlow framework [14] to implement an SSD-MobileNet [34, 41] as our image recognition model. The reason we choose the SSD-MobileNet is because the process of on-line detection is sensitive to latency. This model decomposes the standard convolutional layer into depth-wise convolutional layer and point-wise convolutional layer, which decreases the numbers of parameter, reduces the computation complexity, and further shortens the detection time. In our experiment, we trained 500 labeled images with the UAV through the TensorFlow Objection Detection interface [28] and reduced the loss up to 1.213 after 2700 iterative operations. In the turtlebot side, the detection time reduced up to almost 280 ms. Note that although the processing time of on-line detection is bigger than 250 ms, we did not migration this node because of the real-time constraint.

5   Discussion and Future Work

ROBOTCENTER is our initial attempt to build a general-purpose and composable cloud robotics framework for robotic applications. While this system addresses most of drawbacks of current robotic systems, there are still some open problems listed below, which are our future work.

- **Multi-robot Challenge.** In our use case, we only consider the robotic application with two heterogeneous robots. While the number of robots scales up to

718   Y. Xu et al.

hundreds and access concurrently, ROBOTCENTER must coordinate synchronized accesses to shared data in real time and ensures data sharing without
conflicts. Thus, *scalability* and *consistency* will be new requirements and challenges for cloud robotics OS.

- **Mobility Management.** In our initial prototype of Avalon, we only consider
the trade-off between processing time and planning accuracy. There are also
some others factors that determine the total energy consumption and mission
completion time, such as the robot's velocity and the CPU frequency of onboard computer. As a consequence, the velocity control and CPU frequency
scaling will be added to our utility model in the future work.
- **Fault Tolerance.** When the computations in the robotic application deploy
in both local robot and remote server, the mission failure will occur more
frequently due to the unstable wireless communication and robot's mobility.
For example, the network will be disconnected while the robot moves far
away from the wireless access point. So, how to make ROBOTCENTER robust
to adapt for various real-world conditions is a big challenge.
- **Security and Privacy.** The computation offloading and data sharing in the
cloud also raises a range of privacy and security concerns [24,25,38,43]. For
example, robots may be hacked to send images or video data from private
homes or corporate trade secrets to the cloud. Besides, an adversary can
also take over a robot to disrupt functionality or cause damage. As a result,
ROBOTCENTER should also ensure a secure and private environment for end
users.

6   Conclusion

In this paper, we proposed ROBOTCENTER, rethink and reconstruct a novel
cloud robotics system in a composable perspective. We described the challenges
of designing a cloud robotics system in three levels: *runtime management, data
management* and *programming abstraction*. We then provided the solutions to
address each of these challenges.

We implemented a prototype of the design above three components and use
an example of AGV/UAV cooperative transport application to illustrate the
feasibility of ROBOTCENTER. In the experiment, we reduced the total energy
consumption and mission completion time up to 41.2% and 51.5%, respectively.
We believe that the goals we outlined in this paper have a wide range of technical value for software systems and robotic communities. The ROBOTCENTER
software stack represents a promising direction for future research of robotics
systems.

References

1. Apache Hadoop project. http://hadoop.apache.org/ (2009)
2. Open source robot operating system. http://www.ros.org/ (2009)
3. Roboearth project. http://roboearth.ethz.ch/ (2009)

4. Web ontology language. https://www.w3.org/OWL/ (2013)
5. The robo briain project. http://robobrain.me/ (2015)
6. Mit technology review: Robots that teach each other. https://www.technologyreview.com/s/600768/10-breakthrough-technologies-2016-robots-that-teach-each-other/ (2016)
7. Microsoft farmbeats. https://www.microsoft.com/en-us/research/project/farmbeats-iot-agriculture/ (2017)
8. Amazon aws iot greengrass. https://aws.amazon.com/greengrass/ (2018)
9. Amazon aws robomaker. https://aws.amazon.com/robomaker/ (2018)
10. Open source computer vision library. https://opencv.org/ (2018)
11. The point cloud library. http://pointclouds.org/ (2018)
12. Dji flightgub. https://www.dji.com/flighthub/ (2019)
13. Ros navigation stack. http://wiki.ros.org/navigation/ (2019)
14. Abadi, M., et al.: Tensorflow: a system for large-scale machine learning. In: Operating Systems Design and Implementation (OSDI) (2016)
15. Beetz, M., Beßler, D., Haidu, A., Pomarlan, M., Bozcuoğlu, A.K., Bartels, G.: Know Rob 2.0–a 2nd generation knowledge processing framework for cognition-enabled robotic agents. In: International Conference on Robotics and Automation (ICRA) (2018)
16. Boroujerdian, B., Genc, H., Krishnan, S., Cui, W., Faust, A., Reddi, V.J.: Mavbench: micro aerial vehicle benchmarking. In: MICRO (2018)
17. Broekstra, J., Kampman, A., van Harmelen, F.: Sesame: a generic architecture for storing and querying RDF and RDF schema. In: Horrocks, I., Hendler, J. (eds.) ISWC 2002. LNCS, vol. 2342, pp. 54–68. Springer, Heidelberg (2002). https://doi.org/10.1007/3-540-48005-6_7
18. Chen, X.: Decentralized computation offloading game for mobile cloud computing. IEEE Trans. Parallel Distrib. Syst. 26(4), 974–983 (2015)
19. Chun, B.G., Ihm, S., Maniatis, P., Naik, M., Patti, A.: Clonecloud: elastic execution between mobile device and cloud. In: European Conference on Computer Systems (EuroSys) (2011)
20. Collett, T.H.J., MacDonald, B.A., Gerkey, B.: Player 2.0: toward a practical robot programming framework. In: Proceedings of the Australasian Conference on Robotics and Automation (ACRA) (2005)
21. Cuervo, E., et al.: Maui: making smartphones last longer with code offload. In: International Conference on Mobile Systems, Applications, and Services (MobiSys) (2010)
22. D'Andrea, R.: Guest editorial: a revolution in the warehouse: a retrospective on kiva systems and the grand challenges ahead. IEEE Trans. Autom. Sci. Eng. (T-ASE) 9(4), 638–639 (2012)
23. Elkady, A.Y., Sobh, T.M.: Robotics middleware: a comprehensive literature survey and attribute-based bibliography. J. Robot. (2012)
24. Elkady, A.Y., Sobh, T.M.: A user-centric data protection method for cloud storage based on invertible DWT. IEEE Trans. Cloud Comput. pp. 1–1 (2017)
25. Elkady, A.Y., Sobh, T.M.: A user-centric data protection method for cloud storage based on invertible DWT. IEEE Trans. Cloud Comput. pp. 1–1 (2019)
26. Guo, S., Xiao, B., Yang, Y., Yang, Y.: Energy-efficient dynamic offloading and resource scheduling in mobile cloud computing. In: IEEE International Conference on Computer Communications (INFOCOM) (2016)
27. Hu, G., Tay, W.P., Wen, Y.: Cloud robotics: architecture, challenges and applications. IEEE Network 26(3), 21–28 (2012)

28. Huang, J., et al.: Speed/accuracy trade-offs for modern convolutional object detectors. In: Computer Vision and Pattern Recognition (CVPR) (2017)
29. Janssen, R., van de Molengraft, R., Bruyninckx, H., Steinbuch, M.: Cloud based centralized task control for human domain multi-robot operations. Intel. Serv. Robot. 9(1), 63–77 (2015). https://doi.org/10.1007/s11370-015-0185-y
30. Kamei, K., Nishio, S., Hagita, N., Sato, M.: Cloud networked robotics. IEEE Network 12(2), 432–443 (2012)
31. Kehoe, B., Abbeel, P., Goldberg, K.: A survey of research on cloud robotics and automation. IEEE Trans. Autom. Sci. Eng. (T-ASE) 12(2), 398–409 (2015)
32. Kehoe, B., Matsukawa, A., Candido, S., Kuffner, J., Goldberg, K.: Cloud-based robot grasping with the google object recognition engine. In: International Conference on Robotics and Automation (ICRA) (2013)
33. Kosta, S., Aucinas, A., Hui, P., Mortier, R., Zhang, X.: Thinkair: dynamic resource allocation and parallel execution in the cloud for mobile code offloading. In: IEEE International Conference on Computer Communications (INFOCOM) (2012)
34. Liu, W., et al.: SSD: single shot multiBox detector. In: Leibe, B., Matas, J., Sebe, N., Welling, M. (eds.) ECCV 2016. LNCS, vol. 9905, pp. 21–37. Springer, Cham (2016). https://doi.org/10.1007/978-3-319-46448-0_2
35. Mahler, J., et al.: Dex-Net 2.0: deep learning to plan robust grasps with synthetic point clouds and analytic grasp metrics. In: Robotics: Science and Systems (RSS) (2017)
36. Mahler, J., Matl, M., Liu, X., Li, A., Gealy, D.V., Goldberg, K.: Dex-Net 3.0: computing robust vacuum suction grasp targets in point clouds using a new analytic model and deep learning. In: International Conference on Robotics and Automation (ICRA) (2018)
37. Mohanarajah, G., Hunziker, D., D'Andrea, R., Waibel, M.: Rapyuta: a cloud robotics platform. IEEE Trans. Autom. Sci. Eng. (T-ASE) 12(2), 481–493 (2015)
38. Ren, K., Wang, C., Wang, Q.: Security challenges for the public cloud. IEEE Internet Comput. 16(1), 69–73 (2012)
39. Riazuelo, L., Civera, J., Montiel, J.M.M.: C2tam: a cloud framework for cooperative tracking and mapping. In: Robotics and Autonomous Systems (RSS) (2014)
40. Riazuelo, L., et al.: Roboearth semantic mapping: a cloud enabled knowledge-based approach. IEEE Trans. Autom. Sci. Eng. (T-ASE) 10(3), 643–651 (2015)
41. Sandler, M., Howard, A.G., Zhu, M., Zhmoginov, A., Chen, L.C.: Mobilenets: efficient convolutional neural networks for mobile vision applications. In: CoRR abs/1704.04861 (2017)
42. Schneider, T., et al.: Maplab: an open framework for research in visual-inertial mapping and localization. IEEE Rob. Autom. Lett. 3(3), 1425–1428 (2018)
43. Shao, Z., et al.: Security protection and checking for embedded system integration against buffer overflow attacks via hardware/software. IEEE Trans. Comput. 55(4), 443–453 (2006)
44. Stoica, I., et al.: A Berkeley view of systems challenges for AI. In: CoRR abs/1712.05855 (2017)
45. Tenorth, M., Perzylo, A.C., Lafrenz, R., Beetz, M.: Representation and exchange of knowledge about actions, objects, and environments in the RoboEarth framework. IEEE Trans. Autom. Sci. Eng. (T-ASE) 10(3), 643–651 (2013)
46. Vasisht, D., et al.: A cloud robot system using the dexterity network and Berkeley robotics and automation as a service (Brass). In: USENIX Symposium on Networked Systems Design and Implementation (NSDI) (2017)

Poster Paper

Poster Paper

An OpenMP-Based Parallel Execution of Neural Networks Specified in NNEF

Nakhoon Baek[1]([⊠])[iD] and Seung-Jong Park[2]

[1] School of Computer Science and Engineering, Kyungpook National University,
Daegu 41566, Republic of Korea
`nbaek@knu.ac.kr`
[2] Computer Science and Engineering, Center for Computation and Technology,
Louisiana State University, Baton Rouge, LA 70803, USA
`sjpark@lsu.edu`

Abstract. Recently, we have many research works on the neural networks and their related issues. For exchangeability of neural network frameworks, the Neural Network Exchange Format (NNEF) specification is now widely used. Due to very large size of these neural networks, their accelerations are actively explored, and can be achieved through parallel processing techniques. In this work, we present a prototype implementation of C++ code generator with parallel-processing accelerations based on OpenMP, for the NNEF specification files. Our implementation shows remarkable accelerations, in comparison to the original C++ template-based execution. We will tune the prototype acceleration to achieve more remarkable speed ups.

Keywords: OpenMP · Parallel processing · Neural network · NNEF · Code generation · Prototype · Acceleration

1 Introduction

Recently, the artificial intelligence techniques, especially the neural network and its related methods are widespread. As the result, we now have many neural network frameworks, including *TensorFlow* [1], *Caffe* [2], *Keras* [3], and others. To increase the portability and also the exchangeability of these neural network frameworks, *Khronos Group*, which is a de facto industrial standard organization, announced the *Neural Network Exchange Format* (NNEF) [7].

NNEF originally works as a file format for information exchange among neural network frameworks. Recently, there are also research trends to execute the neural network described in NNEF. The Khronos NNEF official *GitHub* site recently provides an NNEF interpreter, implemented in C++ templates [4]. Another implementation [8] executes NNEF convolution operations with

This work has supported by Basic Science Research Program through the National Research Foundation of Korea (NRF) funded by the Ministry of Education (Grand No. NRF-2019R1I1A3A01061310).

M. Qiu (Ed.): ICA3PP 2020, LNCS 12453, pp. 723–726, 2020.
https://doi.org/10.1007/978-3-030-60239-0

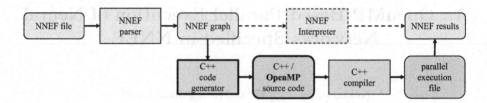

Fig. 1. Overall flows: our source code generator converts original NNEF files to C++ source codes with OpenMP directives, instead of previous NNEF interpreters.

OpenCL (open compute language) [5] parallel processing features. In this work, we present another approach: NNEF-to-C++ source code translator, as shown in Fig. 1. Additionally we apply *OpenMP* [6] parallel processing features, to achieve more accelerations.

2   Implementation and Results

For rapid prototyping, we use the NNEF parser of the Khronos NNEF interpreter [4]. After successful parsing, it provides a graph structure with *NNEF operations* and *NNEF data* (or tensors, in NNEF terminology). We implement the C++ source code generation features based on a graph traversal algorithm, with the NNEF parsed graph.

Actually, the NNEF specification provides more than 80 operations. At this time, we have implemented many of them, and we are adding more NNEF operations to build up a complete NNEF framework.

Table 1. Experimental results from our prototype implementation.

Implementation	a Windows10 PC		a Linux system	
	case A	case B	case A	case B
(a) original C++ template	426	622	268	401
(b) memory coalesced C++ template	281	418	184	276
(c) C++ code generation	450	331	185	278
(d) C++ code, with OpenMP accel.	207	346	162	246
acceleration ratio ((d)/(a))	2.058	1.798	1.654	1.630

(unit: msec)

One more distinguished point of our implementation is the OpenMP-based accelerations. Table 1 shows the current results of our prototype implementation. We compared 4 implementations:

(a) the original Khronos NNEF interpreter, using C++ templates

(b) a modified NNEF interpreter with improved memory allocation: most memory allocations are coalesced into a single one, to enhance the memory access patterns.

(c) C++ code generator, using plain C++ templates: some operations are enhanced with our rewritten algorithms and modified implementations.

(d) C++ code generator, with OpenMP acceleration directives.

In the case of (c) and (d), the C++ source codes are compiled and executed alone. We show two sample cases in Table 1, even though we have more demonstration cases.

case A: a sequence of basic arithmetic operations and a single convolution operation for $1 \times 3 \times 224 \times 244$ input tensors, with $64 \times 3 \times 7 \times 7$ filter tensors.

case B: another sequence of arithmetic and windowing operations, followed by another convolution operation for $1 \times 3 \times 224 \times 224$ input tensors, with $96 \times 3 \times 7 \times 7$ filter tensor and $(2, 2)$ stride values.

Table 1 shows the elapsed times for some samples, on both of a Windows-10 PC with dual-core 3GHz CPU, and a Linux system with quad-core 3.4GHz CPU. As shown in Table 1, our final OpenMP-accelerated C++ codes can execute the same NNEF files more than 1.63 to 2.06 times faster.

3   Conclusion

In this work, we present a C++ code generator from NNEF specification files. More precisely, our system focused on the C++ code generation from the NNEF parse graph, and we also applied OpenMP parallel processing features to the generated C++ codes. Experimental results from our prototype implementation shows that we achieved at most 2.06 times faster, with basic-level OpenMP directive-based accelerations. In the future, we will tune the OpenMP and C++ codes, and finally plan to make a parallel processing framework for NNEF and its related neural network processing.

References

1. Abadi, M., et al.: TensorFlow: Large-Scale Machine Learning on Heterogeneous Systems. white paper available from tensorflow.org (2015)
2. Jia, Y., et al.: Caffe: Convolutional architecture for fast feature embedding. In: Proceedings of the 22nd ACM International Conference on Multimedia (MM 2014), pp. 675-678 (2014)
3. Keras Homepage: Keras: the Python deep learning API. http://www.keras.io (2020). Accessed May 2020
4. Khronos Group: NNEF-Tools. https://github.com/KhronosGroup/NNEF-Tools (2020). Accessed May 2020
5. Khronos OpenCL Working Group: The OpenCL Specification, Version 3.0.1. Khronos Group (2020)

6. OpenMP Team: OpenMP Homepage. http://www.openmp.org/ (2020). Accessed May 2020
7. The Khronos NNEF Working Group: Neural Network Exchange Format, Version 1.0.2. Khronos Group (2019)
8. Yu, M., Chen, T., Lee, J.: Accelerating NNEF framework on OpenCL devices using clDNN. In: Proceedings of the International Workshop on OpenCL (IWOCL 2020), pp. 1–2 (2020)

Author Index

Printed in the United States
By Bookmasters